C000026156

Air Pollution Modeling and Its Application XII

NATO • Challenges of Modern Society

A series of volumes comprising multifaceted studies of contemporary problems facing our society, assembled in cooperation with NATO Committee on the Challenges of Modern Society.

Recent volumes in this series:

Volume 6 REMOTE SENSING FOR THE CONTROL OF MARINE POLLUTION
Edited by Jean-Marie Massin

Volume 7 AIR POLLUTION MODELING AND ITS APPLICATION IV
Edited by C. De Wispelaere

Volume 8 CONTAMINATED LAND: Reclamation and Treatment
Edited by Michael A. Smith

Volume 9 INTERREGIONAL AIR POLLUTION MODELING: The State of the Art
Edited by S. Zwerver and J. van Ham

Volume 10 AIR POLLUTION MODELING AND ITS APPLICATION V
Edited by C. De Wispelaere, Francis A. Schiermeier, and Noor V. Gillani

Volume 11 AIR POLLUTION MODELING AND ITS APPLICATION VI
Edited by Han van Dop

Volume 12 RISK MANAGEMENT OF CHEMICALS IN THE ENVIRONMENT
Edited by Hans M. Seip and Anders B. Heiberg

Volume 13 AIR POLLUTION MODELING AND ITS APPLICATION VII
Edited by Han van Dop

Volume 14 HEALTH AND MEDICAL ASPECTS OF DISASTER PREPAREDNESS
Edited by John C. Duffy

Volume 15 AIR POLLUTION MODELING AND ITS APPLICATION VIII
Edited by Han van Dop and Douw G. Steyn

Volume 16 DIOXIN PERSPECTIVES: A Pilot Study on International Information Exchange on Dioxins and Related Compounds
Edited by Erich W. Bretthauer, Heinrich W. Kraus, and Alessandro di Domenico

Volume 17 AIR POLLUTION MODELING AND ITS APPLICATION IX
Edited by Han van Dop and George Kallos

Volume 18 AIR POLLUTION MODELING AND ITS APPLICATION X
Edited by Sven-Erik Gryning and Millán M. Millán

Volume 19 METHODS OF PESTICIDE EXPOSURE ASSESSMENT
Edited by Patricia B. Curry, Sesh Iyengar, Pamela A. Maloney, and Marco Maroni

Volume 20 PLANNING ESTUARIES
Cees-Jan van Westen and Reinier Jan Scheele

Volume 21 AIR POLLUTION MODELING AND ITS APPLICATION XI
Edited by Sven-Erik Gryning and Francis A. Schiermeier

Volume 22 AIR POLLUTION MODELING AND ITS APPLICATION XII
Edited by Sven-Erik Gryning and Nadine Chaumerliac

Air Pollution Modeling and Its Application XII

Edited by

Sven-Erik Gryning
Risø National Laboratory
Roskilde, Denmark

and

Nadine Chaumerliac
Université Blaise Pascal–CNRS
Clermont-Ferrand, France

Published in cooperation with
NATO Committee on the Challenges of Modern Society

PLENUM PRESS • NEW YORK AND LONDON

Library of Congress Cataloging in Publication Data

Air pollution modeling and its application XII / edited by Sven-Erik Gryning and Nadine Chaumerliac.
p. cm.—(NATO challenges of modern society; v. 22.)
"Published in cooperation with NATO Committee on the Challenges of Modern Society."
Proceedings of the Twenty-Second NATO/CCMS International Technical Meeting on Air Pollution Modeling and its Application, held June 2–6, 1997, in Clermont-Ferrand, France.
Includes bibliographical references and index.
ISBN 0-306-45821-7
1. Air—Pollution—Mathematical models—Congresses. 2. Atmospheric diffusion—Mathematical models—Congresses. I. Gryning, Sven-Erik. II. Chaumerliac, Nadine. III. North Atlantic Treaty Organization. Committee on the Challenges of Modern Society. IV. NATO/CCMS International Technical Meeting on Air Pollution Modeling and its Application (22nd: 1997: Clermont-Ferrand, France) V. Series.
TD881.A47523 1998
628.5′ 3′ 015118—dc21 98-7493
CIP

Proceedings of the Twenty-Second NATO/CCMS International Technical Meeting on Air Pollution Modeling and Its Application, held June 2–6, 1997, in Clermont-Ferrand, France

ISBN 0-306-45821-7

A Division of Plenum Publishing Corporation
233 Spring Street, New York, N.Y. 10013

http://www.plenum.com

10 9 8 7 6 5 4 3 2 1

Printed in the United States of America

PREFACE

In 1969 the North Atlantic Treaty Organization (NATO) established the Committee on Challenges of Modern Society (CCMS). The subject of air pollution was from the the very beginning one of the priority problems under deliberation within the framework of various pilot studies undertaken by this committee. The organization of a yearly conference dealing with air pollution modelling and its application has become one of the main activities within the pilot study relating to air pollution. Please see the listing on the next page for completed NATO/CCMS Pilot Studies and the International Technical Meetings (ITM) on Air Pollution Modelling and Its Application.

This volume contains the papers at the 22ND ITM, being held in Clermont-Ferrand, France during June 2–6, 1997. It was attended by 152 participants representing 33 countries. This ITM was jointly organized by the Risø National Laboratory of Denmark (pilot country); the Laboratoire Associé de Météorologie Physique, associated with the Centre National de la Recherche Scientifique, the Université Blaise Pascal, and the Observatoire de Physique du Globe de Clermont-Ferrand, France (host country).

We wish to express our gratitude to the sponsors that made this conference possible. In addition to financial support from NATO/CCMS, the conference received contributions from Centre National de Recherche Scientifique, Observatoire de Physique du Globe de Clermont-Ferrand, Universite Blaise Pascal, Electricite de France, Institut Francais du Petrole, Conseil Regional d'Auvergne, Conseil Général du Puy de Dôme, Mairie de Clermont-Ferrand, SATCAR Semaine des arts techniques et culture de l'automobile et de la route, European Association for the Sciencies of Air Pollution, Risø National Laboratory, and the Danish Environmental Research Programme.

On behalf of the ITM Scientific Committee and the organizers, we express our gratitude to those who worked to make this conference a reality and to the active participants representing numerous countries.

Sven-Erik Gryning and Nadine Chaumerliac

THE SCIENTIFIC BOARD OF THE 22nd NATO/CCMS INTERNATIONAL TECHNICAL MEETING ON AIR POLLUTION MODELING AND ITS APPLICATION

HISTORY OF NATO/CCMS AIR POLLUTION PILOT STUDIES

Pilot Study on Air Pollution: International Technical Meetings (ITM) on Air Pollution Modeling and Its Application

Dates of Completed Pilot Studies

- 1969–1974 Air Pollution Pilot Study (United States Pilot Country)
- 1975–1979 Air Pollution Assessment Methodologies and Modeling (Germany)
- 1980–1984 Air Pollution Control Strategies and Impact Modeling (Germany)

Dates and Locations of Pilot Study Follow-Up Meetings

Pilot Country—United States (R.A. McCormick, L.E. Niemeyer)

Feb 1971, Eindhoven, The Netherlands—First Conference on Low Pollution Power Systems Development
Jul 1971, Paris, France—Second Meeting of the Expert Panel on Air Pollution Modeling

---

All following meetings were entitled NATO/CCMS International Technical Meetings (ITM) on Air Pollution Modeling and Its Application

---

Oct 1972, Paris, France—Third ITM
May 1973, Oberursel, Federal Republic of Germany—Fourth ITM
Jun 1974, Roskilde, Denmark—Fifth ITM

Pilot Country—Germany (Erich Weber)

Sep 1975, Frankfurt, Federal Republic of Germany—Sixth ITM
Sep 1976, Airlie House, Virginia, USA—Seventh ITM
Sep 1977, Louvain-La-Meuve, Belgium—Eighth ITM
Aug 1978, Toronto, Ontario, Canada—Ninth ITM
Oct 1979, Rome, Italy—Tenth ITM

Pilot Country—Belgium (C. De Wispelaere)

Nov 1980, Amsterdam, The Netherlands—Eleventh ITM
Aug 1981, Menlo Park, California, USA—Twelfth ITM
Sep 1982, Ile des Embiez, France—Thirteenth ITM
Sep 1983, Copenhagen, Denmark—Fourteenth ITM
Apr 1985, St. Louis, Missouri, USA—Fifteenth ITM

Pilot Country—The Netherlands (Han van Dop)

Apr 1987, Lindau, Federal Republic of Germany—Sixteenth ITM
Sep 1988, Cambridge, United Kingdom—Seventeenth ITM
May 1990, Vancouver, British Columbia, Canada—Eighteenth ITM
Sep 1991, Ierapetra, Crete, Greece—Nineteenth ITM

Pilot Country—Denmark (Sven-Erik Gryning)

Nov 1993, Valencia, Spain—Twentieth ITM
Nov 1995, Baltimore, Maryland, USA—Twenty-first ITM
Jun 1997, Clermont-Ferrand, France—Twenty-second ITM

CONTENTS

INTEGRATED REGIONAL MODELLING

GLOBAL AND LONG-RANGE TRANSPORT

ETEX SESSION

NEW DEVELOPMENTS

ACCIDENTAL RELEASES

MODEL ASSESSMENT AND VERIFICATION

POSTER SESSION

Air Pollution Modeling and Its Application XII

INTEGRATED REGIONAL MODELLING

chairmen:	S. E. Gryning N. Chaumerliac
rapporteur:	A. G. Straume J. Langner

MODELING OF A SAHARAN DUST EVENT

G. Cautenet [1], F. Guillard[3], B. Marticorena[2], G. Bergametti[2]
F. Dulac[3] and J. Edy[1]
[1]LaMP - Université Blaise-Pascal - CNRS
24, Avenue des Landais - 63177 Aubière - France
[2]LISA - Universités Paris 7 &12
61, Avenue du Général de Gaulle 94010 Créteil - France
[3]CFR/LMCE

ABSTRACT

We have coupled a nonhydrostatic mesoscale model with a simple but comprehensive mineral aerosol source scheme, along with a spectral sedimentation scheme. We present a simulation of a Saharan dust transport event (4 days), including mass uptake estimates, 3D transport and dry deposition. The model is initialized with ECMWF data. Meteosat imagery is used to check the dust cloud uptake and trajectory.

1 - INTRODUCTION

The value of the natural aerosol flux is not accurately known, but it is about some gigatons/year. According to Andreae (1994), its primary source may be divided almost equally into mineral aerosol: 1.5 Gigatons/year, and sea salt aerosol: 1.3 Gigatons/year (however, recent estimates give much larger values for sea aerosol flux: see e.g. Gong et al., 1997). The North African desert areas (Sahara and Sahel) are probably the most important mineral aerosol source, with 0.7 gigatons/year (d'Almeida, 1987).

The environmental impact of desert dust is far from negligible. We could mention its geochemical role, in particular in ocean basins fertilization, whereas its radiative effects must be considered in climate change studies. This latter problem is quite complex: besides the direct effect (scattering of the incoming shortwave radiation), and indirect effect is possible. The mixing of desert dust with non-desert material (carbon particles from bush or forest fires, sea salt, industrial aerosol, for instance) may modify strongly its properties, including condensation ability. As mineral dust strongly impacts on the environment, some of the tasks pointed by international programs (WCRP, IGBP) refer explicitly to the mineral dust budget and effects. It is quite important to model dust transport as accurately as possible from the source area to the most remote deposition zones. For global transport assessments and climate effects, a General Circulation Model is really a good tool (Jousseaume, 1990), but this is more questionable for small or mesoscale ranges. Prior of all, some complex dynamical processes may be involved in dust deflation processes. A classical model assumes that the main dust raise motor is momentum transfer from the AEJ (African Eastern Jet) to surface layer. This transfer is governed by the atmospheric instability in early afternoon. Westphal et al. (1988) report evidence that this simple mechanism may not explain some observations, so that the atmospheric layer from surface to 600 hPa should be accurately modeled, in order to represent the low-level jets. A mesoscale model is quite appropriate to describe the boundary-layer processes. Another question is how to deal accurately with complex local circulations, e.g. near the ITCZ front. Obviously, this may be more easy in the framework of a non-hydrostatic mesoscale model. Finally, modeling the deflation itself requires an accurate value of wind

Air Pollution Modeling and Its Application XII
Edited by Sven-Erik Gryning and Nadine Chaumerliac, Plenum Press, New York, 1998

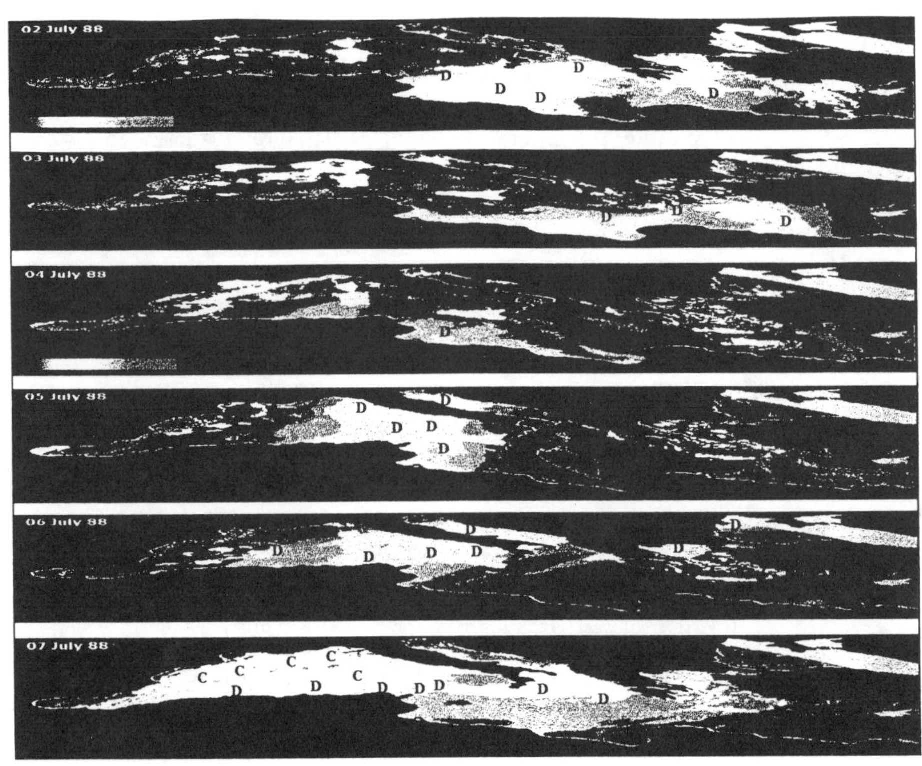

Figure 1. Visible Meteosat imagery over Mediterranean (from Dulac et al., 1992).
D: desert dust cloud; C: (water) cloud cover.

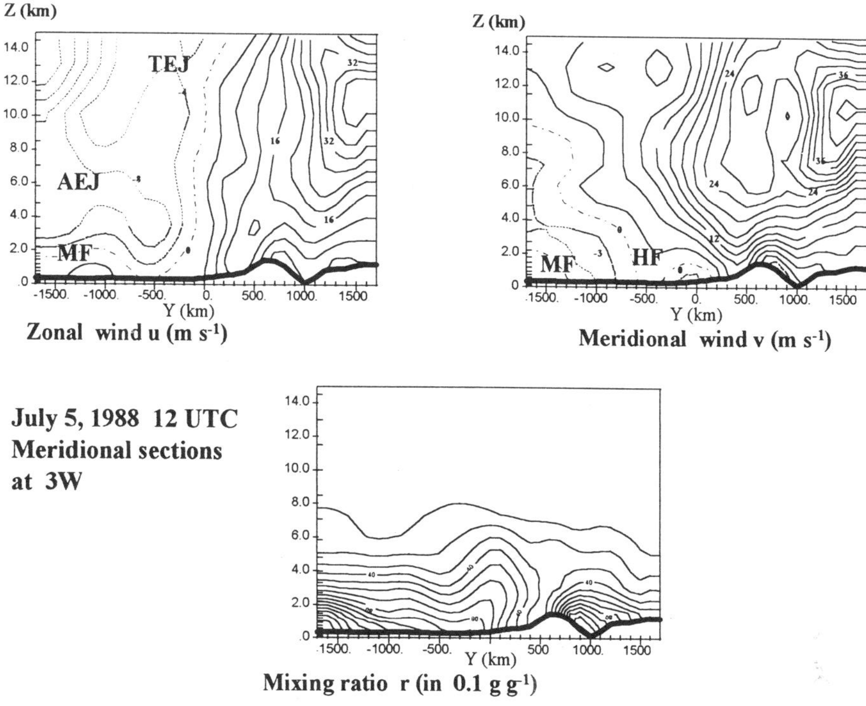

Figure 2. some meteorological issues from RAMS model (MF: Monsoon Flux; HF: Harmattan Flux; AEJ: African Eastern Jet; TEJ: Tropical Eastern Jet). The solid heavy line represents the surface.

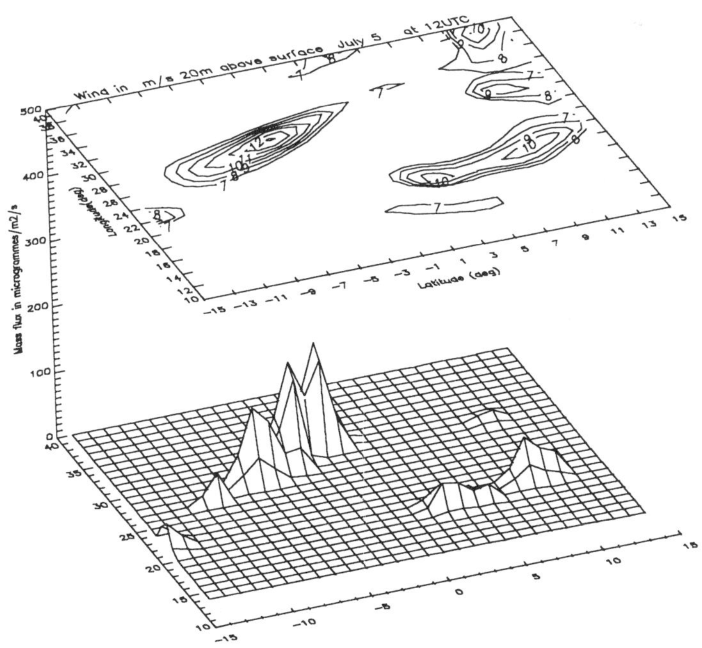

Figure 3. relationship between dust source intensity and wind near surface

speed near surface. This paper deals with the results of a nonhydrostatic mesoscale model coupled with a simple but comprehensive mineral aerosol source scheme designed by Marticorena and Bergametti (1995), along with a spectral sedimentation scheme. We present a simulation of a Saharan dust event (4 days during July 1988). The model is initialized with ECMWF analysis data. Meteosat imagery is used to check the dust cloud trajectory.

2 - A SHORT DESCRIPTION OF THE MODELS

The mesoscale model is the Colorado State University RAMS (Regional Atmospheric Modeling System; Pielke et al., 1992), an eulerian, nonhydrostatic mesoscale model. The dust source scheme (Marticorena and Bergametti, 1995), hereafter referred to as MB scheme, considers the size distribution of the erodible part of soil surface and the surface roughness, which controls the part of wind energy available for dust uptake. It involves a variable threshold in friction velocity in horizontal mobilization of soil grains (the minimal drag required for material mobilization is a function of the grain size). This scheme has been qualified over Sahara, with a quite satisfactory efficiency score (Marticorena et al., 1997). The MB scheme has been coupled with the RAMS model, which provides the wind speed 10 m above surface and allows the aerosol 3D transport. We use a fully spectral scheme with 20 bins, from 0.01 to 30 micrometers, and a truncation for upper sizes. We fit the aerosol mass estimated by the MB scheme along an assumed initial distribution derived from d'Almeida's data (1987). It is a 3-mode, lognormal distribution. The three mass mode radii are: 0.105 μm (.1% in mass); 5 μm (57.9% in mass); 35 μm (42% in mass). In fact, due to the experimental cut-off of the cascade impactors, the MB scheme allows to represent the fraction of aerosol uptake less than about 10 micrometers in radius, so that our initial fitting underestimates probably the relative importance of the small-size aerosol fraction and, therefore, the amount of long-range aerosol. In this paper however, we do not intend to perform any accurate quantitative qualification of the model results, but only check the coherence of the complete model (RAMS plus MB scheme), including a qualitative comparison to satellite imagery. Moreover, such an initial spectrum allows us to test how quickly the largest particles are removed by dry deposition (the only removal process considered here). The settling velocity of any particle in a particular bin is calculated using an iterative procedure described in Seinfeld (1986).

The model domain ranges from 10°N to 40°N, and from 15°W to 15°W. Aklthough the RAMS model provides a grid-nesting capability (zooming), we used here only a single grid: the spatial step is 100km, in agreement with the MB surface properties mapping, which covers, in the present state, the area with a space step of 1°x1°. The timestep is 60s. ECMWF database allows initialization and nudging every 6 hours.

3 - THE DUST EVENT

We model a dust event over Northern Africa and Mediterranean Basin from July 4 to July 7, 1988. Figures 1 from visible METEOSAT imagery at 12000 UTC (Dulac et al., 1992) present the dust outbreak over Mediterranean. On July 2 and 3, a previous event is quickly decaying over the sea. On July 4, a new outbreak is detected, which reaches the Adriatic Sea on July 5. On July 6 and 7, a reversal seems to occur; on July 8, the dust cloud over Mediterranean is weak again.

4 - MODEL RESULTS

4-1. Meteorological features

A preliminary result concerns the ability of RAMS to represent the complex African meteorological features. In figures 2-a to 2-c, we note that the model issues retrieve the main large-scale characteristic patterns of this region: the jets (AEJ, TEJ), the Monsoon and Harmattan fluxes, the ITCZ are clearly observed. At smaller scales, the surface model included in RAMS allows realistic scenarii (not presented here) of the atmospheric surface layer diurnal cycle (stability/unstability). All these scales are involved in the 3D desert dust transport, so that it is quite important to ensure the quality of the modeled atmospheric fields.

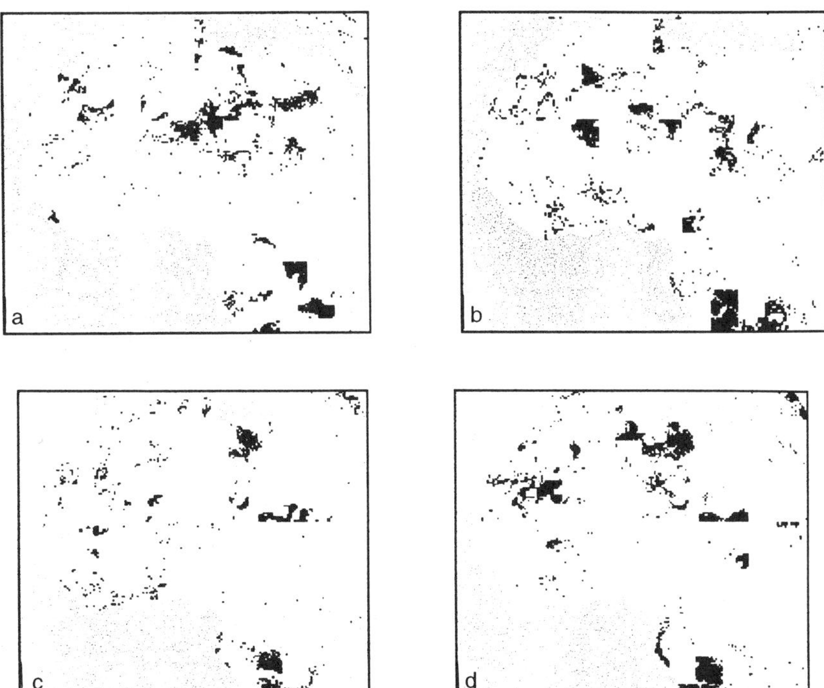

Figure 4: IDDI from Meteosat imagery (Legrand, 1990)
Grey: radiative count from 26 to 35; dark: radiative count from 36 to 90

A qualitative comparison is performed against IR Meteosat data at 1200 UTC every day. We use the IDDI (Infrared Difference Dust Index) imagery processed by Legrand (1990) analysis.

We examine now the main results from our simulation from July 2 at 00 UTC to July 7 at 2400 UTC. Recall that the timestep is 60s. At 15-minute intervals, the dust source scheme is called in view to refresh the dust uptake estimate over all the model surface gridpoints. The dry deposition scheme is activated every time step and we derive continuously a new spectrum at any 3D gridpoint. We start from a clear state of the atmosphere (no dust).

4-2. Mapping the dust sources

Figure 3 shows an example of the relationship between surface wind and dust source intensity. Obviously, the wind enhances the source flux but surface properties (locally modifying the erodibility) act to spatially modulate this flux.

High values of the IDDI, associated to grey and dark colours, suggest a dust raise event, or very high dust concentrations (the lower values of radiative count are assumed not relevant to dust raise). Although a persistent cloud deck over a large part of impedes a complete survey during this period, the source dust intensity (figure 4-a to 4-d) looks satisfactorily related to the IDDI pattern (figure 5-a to 5-d) in most cases.

4-3. Mapping the columnar dust content

The columnar dust content is an important parameter because (i) it informs on the total aerosol mater above surface and (ii) it is related to the scene as viewed from a satellite. Here we display only the mass content, i.e. the vertically integrated mass concentration over the whole column, but our spectral scheme would allow us to estimate any spectral content, relevant to radiative interactions problems. The columnar dust content at 12UTC every day is

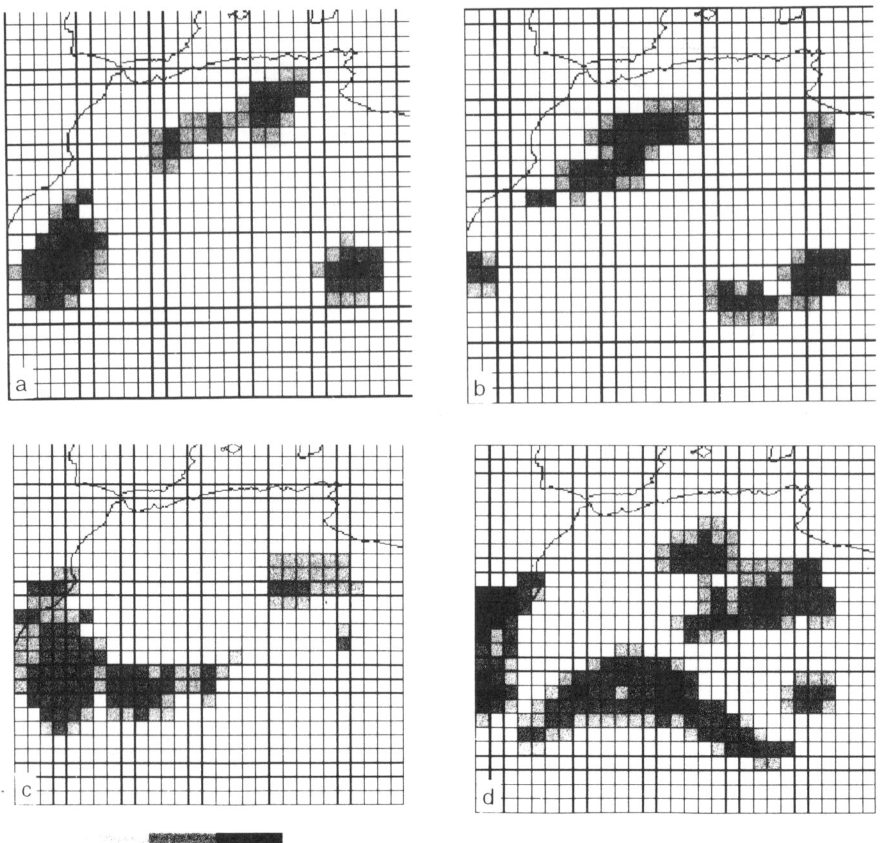

Figure 5. Modelled dust source intensity from RAMS
light grey: 0-0.1 medium grey: 0.1-10 dark grey: 10-100 black: 100-1000 μg/m2/s

displayed on figures 6-a to 6-d (we limit to 12 UTC the displayed figures). Recall that we assume an initial clear state of the atmosphere (no dust) in our simulation. This fact explains that on July 4, the modelled dust concentration is quite low. On the eveneing however, the dust cloud reaches the Mediterranean coast (isolines 0.2 g/m2). On July 5, the dust cloud has crossed the Mediterranean, following a SW-NE trajectory. During the next day, the dust cloud seems to stretch in the West-East direction and weakens in the Northern part of the zone. This may be explained by a decrease in the dust sources activities. During the evenig however, in agreement with a revival of the dust uptake over Algeria around 1500 UTC (not displayed here), an increase is observed over the Sea. On the last day (July 7), after some revival, the dust concentration seems to decay on the evening. These observations agree roughly with the VIS imagery (figures 1).

4-4. Dust content and streamlines

The modeled streamlines pattern at 2 km above surface at 12 UTC (figures 7-a to 7-d) roughly explains the satellite observations and model calculations: this pattern is globally oriented NW on July 4 and 5; on July 6 and 7, only the dust sources lying in the northern part of Algeria is involved in the NW transport, which causes the decay of dust concentration on the northern part of the domain, and the increase in the southern part.

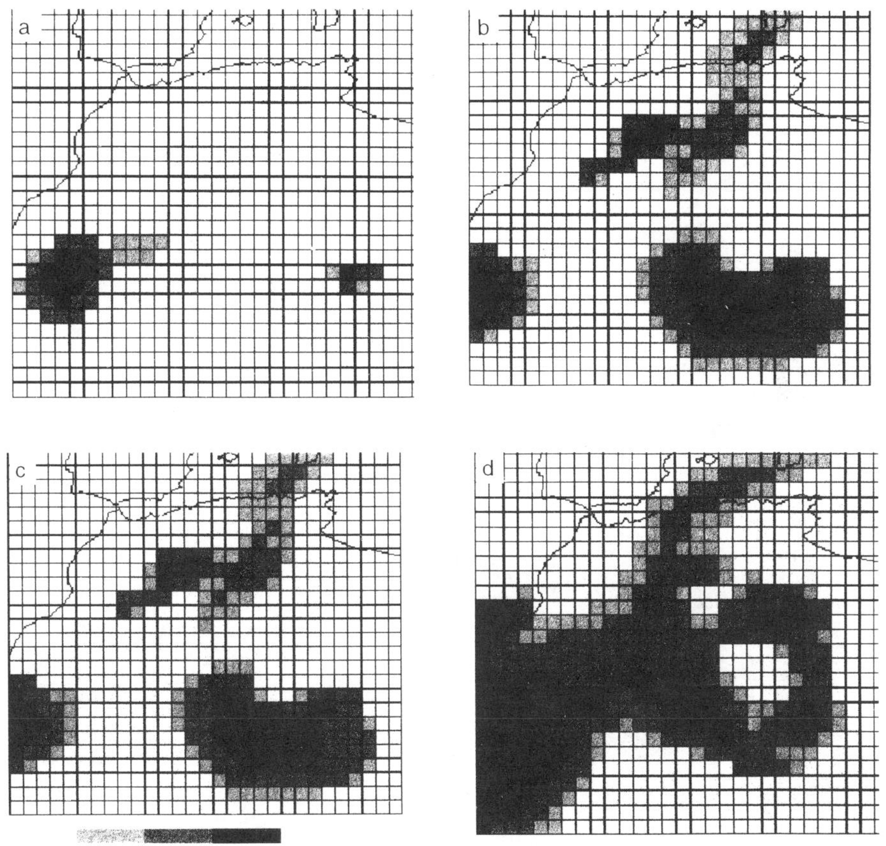

Figure 6. Columnar dust content at 12UTC in g/m2
light grey: 0.02-0.15 medium grey: 0.15-0.20 dark grey: 0.20-0.4 black:: >0.40 g/m2

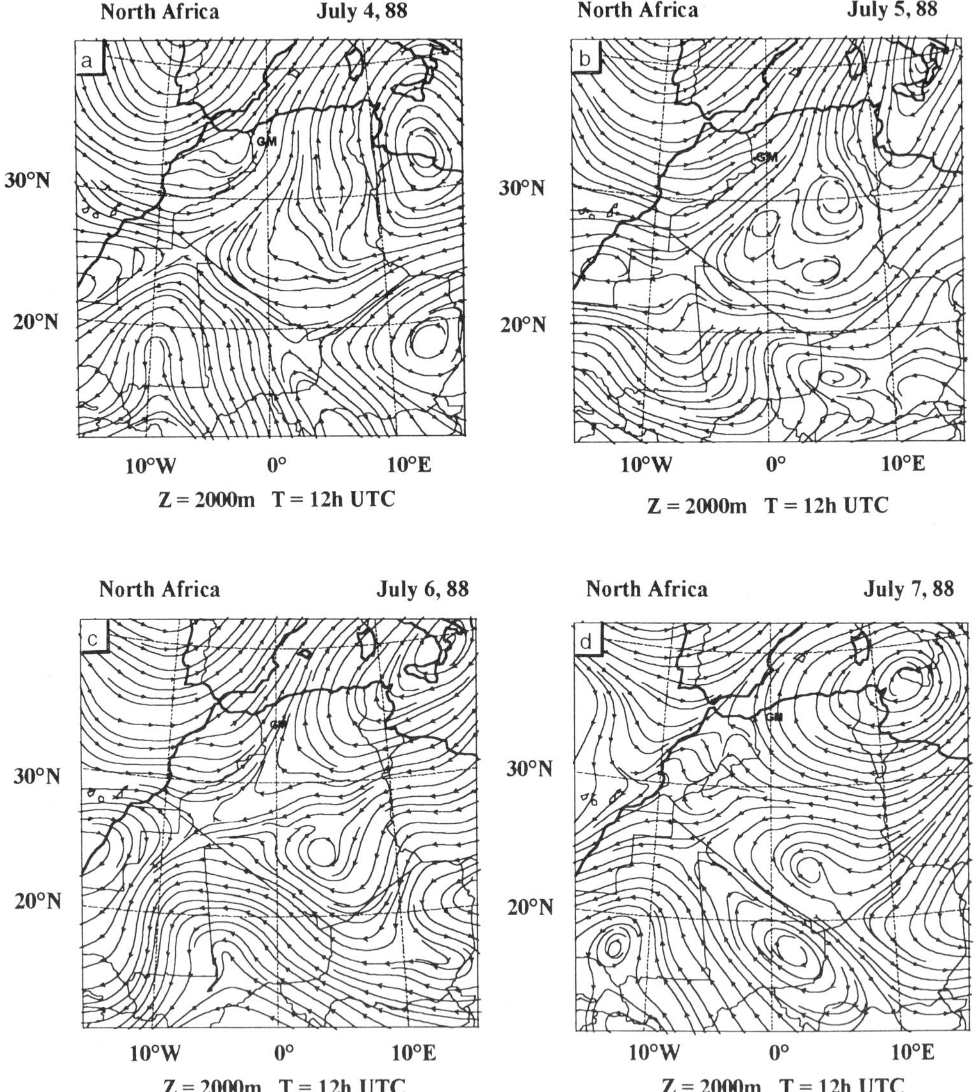

Figure 7. Streamlines at 2km above surface level from RAMS at 12 UTC

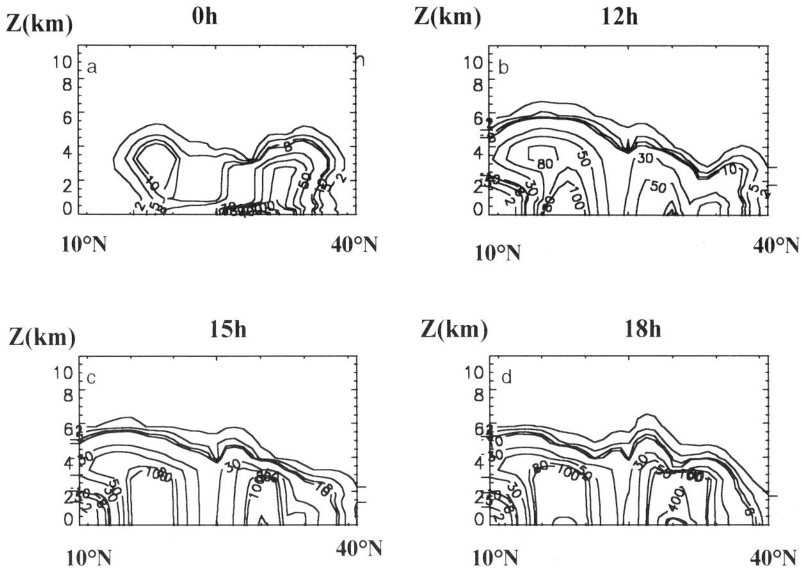

Figure 8. meridional cross section at 2°E of dust mass concentrations from RAMS (μg/m3) on July 5

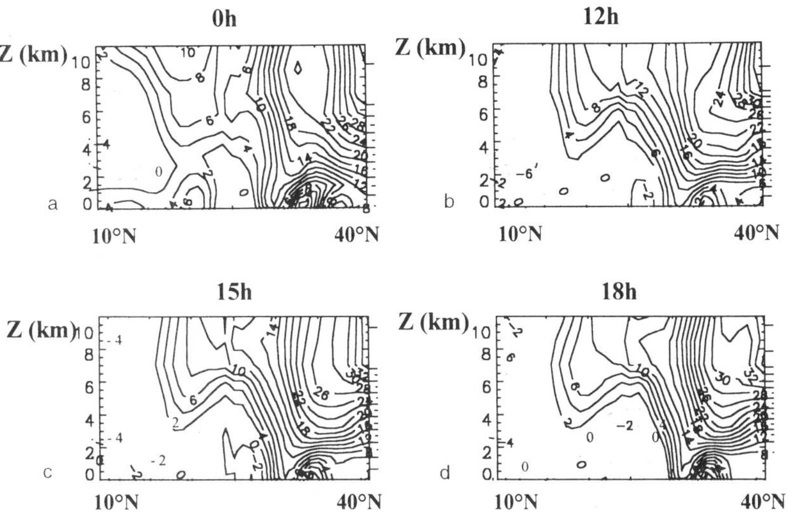

Figure 9. meridional cross section at 2°E of « v » wind component from RAMS in m/s on July 5

4-5. A brief analysis of dust outbreaks

In the example displayed on figures 8 and 9, we analyze how the atmospheric stability and the zonal "v" wind component reversal explain the zonal dust outbreaks in the model. We focus on a meridional section at 2°E on July 5. Along this transect, there are two emission regions which are not always active: the first near 20N, the second near 32N. During nighttime, a stable stratification exists near surface, whereas well-mixed layers may be observed above: they may be explained as a residual of the dry convection during the previous day (**figure 8**).

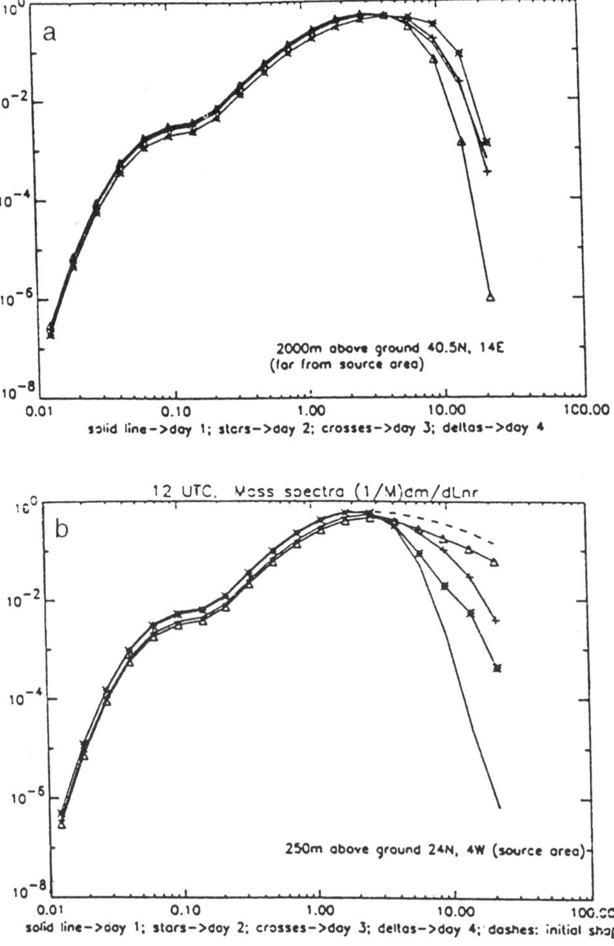

Figure 10. time and space evolution of the modelled mass spectrum. Dashed line in figure 10-a represents the shape of the initial spectrum (near source area)

In the southern part of the transect, dust cloud seems trapped during nighttime, and the vertical mass concentrations undergo strong gradients (two orders of magnitude) due to atmospheric stability (figure 8-a). Moreover, the meridional « v » wind component is not likely to provide a strong transport towards south during nighttime (figure 9-a), because the area which exhibits negative (southwards) « v » values is limited to a zone roughly included in a rectangle between 2 and 10 km in the vertical and 10 to 15°N in latitude, with weaker dust concentrations. On the other hands, at the same time, we note that dust transport towards north is possible (positive « v » values). During daytime, the vertical gradients in dust concentration strongly weaken (figures 8-b to 8-d) due to atmopsheric instability. In the same time, the negative values of v (figures 9-b to 9-d) are found in a much larger zone, in the southern part of the area. The lower left part of the figures (the "monsoon corner"), where « v » remains weakly positive (figures 9-b to 9-d) approximately tracks the ITCZ near surface and represents a well-known dynamical barrier in the meridional exchanges near surface. In the case under study, this corner, « v » remains negative during daytime and the dust isolines (figures 8-b to 8-d) are quite horizontal: the large dust flux towards south represents a jump above the "monsoon corner. This outbreak (a southwards jump over the monsoon barrier) seems governed by vertical dust transport associated to the diurnal vertical instability, and the southwards reversal of "v".

4-6. A short insight in spectral aspects

We finally show (figures 10) that the initial modelled spectrum quickly loses its largest particles and evolves towards a more realistic shape, since various investigators (see e.g. Gomes et al, 1990) found generally a main mode near 1 micrometer in radius far from sources (background aerosol). We display two examples of modelled spectra at two different locations. The former (figure 10-a) is far from sources (40N, 14E); the latter (figure 10-b) is close to a dust source area (24N, 4W). Obviously, the spectrum is strongly variable (at least for the largest particles) near the sources and near the surface (the data in the source area are taken near ground surface). Far from sources and from surface (the spectrum is taken 2 km above surface for the remote point), the variability is less obvious.

4-7. Global budget

The average raised dust amount is 3Mt/day. The deposited mass is about 60% of the raised mass. Note that this ratio must strongly depend on the initial spectrum shape.

5 - CONCLUSION

In the introduction, we have tried to present some of the various problems associated with the phenomenon of desert dust raise and transport, and what could be some of its possible climatic effects. The model presented here is the first state of a tool we are developing to investigate these problems from a mesoscale viewpoint. We show that RAMS is able to retrieve the complex behaviour of the african low-latitude circulation using ECMWF data as input. The association of this mesoscale model with the dust source parameterization looks satisfactory. A spectral scheme is used and the spectrum evolution is realistic. On the other hand, the constraints on model results look somewhat loose, as it is not easy to use satellite imagery when clouds are present, so that more tests are planned. At smaller scale, we plan also to compare the modeled size spectrum with spectral measurements in order to enhance the constraint. The nexts steps of this work deal with basic problems. Some of them are listed below:

- (i) the original radiative schemes are not suitable for high dust concentrations, so that we are replacing them with new ones which take the dust radiative properties into account. These will allow us to model the complex radiation-dynamics feedbacks in source areas;
-(ii) another problem associated with small-scale dynamics is the role of the low-level jets in dust uptake: it may be investigated with this model, taking advantage of the RAMS "zoom" facility (use of imbedded grids);
- (iii) obviously, it is necessary to investigate the evolution of the dust flux when grids smaller than 1°x1° are considered. Due to nonlinearities, it seems possible that dividing such a square area into smaller areas could induce significant differences in the dust mass budget. Such sensitivity tests, associated with perturbation tests on surface parameters (roughness, pedology,...) will help us to provide dust source parameterizations for GCM (with grids reaching several square degrees);
- (iv) the indirect radiative effect is also to be studied soon. It would be useful to study the cycles of capture/evaporation of dust by cloud or rain drops, associated with the "coating" phenomena: when water evaporates, a chemical complex deposits on the dried insoluble phase of the dust particle. New properties probably appear: radiative (absorptivity, ...) and hygroscopicity. The latter remark suggests that the particle is likely to become a CCN to some extent.

REFERENCES

Andreae, M. O., Climate effects of changing atmospheric aerosol levels, *World Survey of Climatology*, **XX**, **Future Climate of the World**, Henderson-Sellers Ed., 1994

d'Almeida, G. A., A model for Saharan dust transport *J. Clim. Appl. Meteor.*, **24**, 903-916, 1986.

d'Almeida, G. A., On the variability of desert aerosol radiative characteristics, *J. Geophys. Res.*, **92**, 3017-3026, 1987.

Dulac, F., D. Tanré, G. Bergametti, P. Buat-Ménard, M. Desbois and D. Sutton, Assessment of the African airborne dust mass over Western Mediterranean sea using Meteosat data, *J. Geophys. Res.*, **97**, 2489-2506, 1992.

Gomes L., G. Bergametti, G. Coudé-Gaussen and P. Rognon, Submicron desert dusts: a sandblasting Process, *J. Geophys. Res.*, **95**, 13927-13935, 1990.

Gong, S. L., L. A. Barrie, J.-P. Blanchet and L. Spacek, Modeling size-distributed sea-salt aerosols in the atmosphere: an application using canadian climate models, *22nd NATO/CMS International Technical Meeting on Air Pollution Modelling and its Applications*, 244-251, June 2-6, 1997, Clermont-Ferrand, France.

Jousseaume, S., Three-dimensional simulations of the atmospheric cycle of desert dust particles using a general circulation model, *J. Geophys. Res.*, **95**, 1909-1941, 1990.

Legrand, M., Etude des aérosols sahariens au-dessus de l'Afrique à l'aide du canal à 10 microns de Météosat: validation, interprétation et modélisation, *Thèse de Doctorat d'Etat*, Université des Sciences et Technologies de Lille, 1990.

Marticorena, B. and G. Bergametti, Modeling the atmospheric dust cycle: 1- Design of a soil-derived dust emission scheme. *J. Geophys. Res.*, **100**, 16415-16430, 1995.

Marticorena, B., G. Bergametti, B. Aumont, Y. Callot, C. N'Doumé and M. Legrand, Modeling the atmospheric dust cycle: 2- Simulation of Saharan dust sources. *J. Geophys. Res.*, **102**, 4387-4404, 1997.

Pielke, R. A., W. R. Cotton, R.L. Walko, C. J. Tremback, W. A. Lyons, L. D. Grasso, M. E. Nicholls, M. D. Moran, D. A. Wesley, T. J. Lee and J. H. Copeland, A comprehensive meteorological modeling system - RAMS. *Meteorol. Atmos. Phys.*, **49**, 69-91, 1992.

Seinfeld, J., Atmospheric Chemistry and Physics of Air Pollution, J. Wiley & s, 1986, 738 pp.

TEMPORAL AND SPATIAL SCALES FOR TRANSPORT AND TRANSFORMATION PROCESSES IN THE EASTERN MEDITERRANEAN

G. Kallos[1] V. Kotroni[1] K. Lagouvardos[1] A. Papadopoulos[1], M. Varinou[1], O. Kakaliagou[1], M. Luria[2], M. Peleg[2], A. Wanger[2], and M. Uliasz[3]

[1] University of Athens, Department of Applied Physics, Meteorology Laboratory, Panepistimioupolis, PHYS-5, 15784 Athens, Greece
[2] The Hebrew University of Jerusalem, School of Applied Science, Jerusalem 91904, Israel
[3] Colorado State University, Department of Atmospheric Science, Fort Collins, Co 80523, USA

INTRODUCTION

In several studies during the past, the urban plumes have been extensively considered. In these studies, the spatial and temporal scales of episodic conditions have been described and emphasis was given to the formation and evolution of air pollution episodes within city limits (or in an area covering a few tens of kilometers around the city) and for a time period of one to two days. Moreover, the weather phenomena exhibiting strong diurnal variations (e.g. sea/land-breezes, upslope/downslope and drainage flows, orographic effects, heat islands etc.) were emphasized. The influence of the regional scale phenomena in such cases was not considered on a systematic manner. Actually, the role of phenomena with wavelengths larger than a few tens of kilometers was considered as not important for the formation of a specific air quality over the city of consideration. During the last few years, the influence of regional scale forcing on the formation of specific air quality conditions was found to be important. Kallos et al. (1993) reported that the regional scale phenomena should contribute significantly in the formation of specific air quality conditions in the Greater Athens Area (GAA). Luria et al. (1996) showed that significant degradation of the air quality in some areas should be attributed to regional scale transport phenomena. While the physico-chemical properties of various urban plumes have been described at the urban scale with the aid of organized experimental campaigns and/or mesoscale and photochemical modeling (e.g. Ziomas, 1996), not enough attention was paid to the properties of the urban plume as it is passing to areas relatively far from its origin. Consequently, the urban plume impact on remote locations has not been extensively studied. Such phenomena should be considered as very important in some cases, especially in areas with specific characteristics like the Mediterranean Region.

As it was found in the pioneering work of Millan et al., (1993), urban plumes are transported inland over the Iberian Peninsula for several tens (or even a few hundreds) of kilometers without being significantly diluted. In addition, significant concentrations of primary and secondary pollutants were found in multiple-layer structure up to a few kilometers above the ground. These processes were found to have a repetition time scale of 2-4 days. While such phenomena have been documented in the Western part of the Mediterranean at around 1990, there was no information at all for the Eastern Mediterranean. During the last five years, some evidence was found for the occurrence of large scale transport in the Eastern Mediterranean. More specifically, Luria et al., (1996) documented that the amount of sulphates measured in some areas in Israel during summer are very high and they cannot be considered as produced locally. In addition to this, the climatic conditions in the area were known to have significant regional scale characteristics capable of long-range transport (Kallos et al., 1993).

Based on this evidence and the experience gained from the extensive work over the Iberian Peninsula by Millan et al. (1993), two new proposals were funded by the DG-XII of the EU namely South European Cycles of Air Pollutants (SECAP, Kallos et al., 1995) and Transport and Transformation of Air Pollutants over the East Mediterranean (T-TRAPEM, Kallos et al., 1996a). The first of them was focused on the near-urban scale while the second at the regional scale. More specifically, at the T-TRAPEM project, the transport and transformation mechanisms associated with urban plumes from selected areas were extensively studied with surface and airborne measurements and extensive meteorological and dispersion simulations. Emphasis was given to the description of the mechanisms associated with the injection of the urban plume of Athens in to the free troposphere and the marine boundary layer. The plume was tracked for hundreds of kilometers, for several days over the East Mediterranean waters. In addition, urban or industrial plumes having their origin in various areas in Southern and Eastern Europe were tracked and the influenced areas were identified too. Some of these results are presented in the next sections.

GENERAL CHARACTERISTICS

The Mediterranean Sea is closed from all sides and it is surrounded by high peninsulas and important mountain barriers. The gaps between these major mountainous regions act as channels for the air mass transport towards the Mediterranean. These topographic features along with the significant variation of the physiographic characteristics are partially responsible for the development of various-scale atmospheric circulations (Weather in the Mediterranean, 1962).

The climatic conditions in the Eastern Mediterranean should be roughly divided in cold and warm periods. During the cold period, cyclogenesis is a common characteristic which takes place in preferable locations. The anticyclonic type of circulation during this period is associated with a cold core anticyclone laying over the Central Europe or the Balkan. During the warm period of the year, the Mediterranean region and South Europe are covered by an anticyclonic system which is relatively shallow (large-scale subsidence). Trade winds persist over areas like the Aegean Sea (Etesians). The Etesians is a regional-scale phenomenon with significant diurnal variation (Kallos et al. 1993, among others).

During both the cold and the warm period of the year, the general trend of the flow fields is from North to South across the Mediterranean with significant variations in each area. This is mainly due to the differential heating between the two sides (the Northern and Southern) of the Mediterranean Basin. Because of these complicated flow patterns, the air pollutants released from various sources located in the surrounding areas can be transferred in long distances, in a complicated way (Kallos et al., 1996b; Kallos et al. 1995; Luria et al., 1996). Since these phenomena are considered as very important because they affect the air quality of areas like the North African coast and the Middle East on a significant way, and because of the lack of systematic information concerning the air quality in several Mediterranean areas, it is considered as very important to define characteristic spatial and temporal scales of such transport and transformation phenomena.

In this presentation, an attempt will be made in order to provide some information about these characteristic scales in the Eastern Mediterranean during the warm period of the year. These are results of the projects T-TRAPEM and SECAP. More specifically, some of the two-year surface and airborne air quality observations are shown. The regional-scale transport mechanisms are investigated by using the Regional Atmospheric Modeling System (RAMS) and the Hybrid Particle Concentration and Transport (HYPACT) models.

EXPERIMENTAL CAMPAIGNS

During the summer-months of the period 1993-1995 approximately 180 flight-hours were completed and the data were analyzed. Most of the flights were performed at an altitude of approximately 300 m above sea level (within the boundary layer). For preselected areas, measurements of up to 2000 m were performed in order to gain information on the concentrations above the boundary layer. The aircraft used is a twin-engine Cessna 310 equipped with an instrument package which includes SO_2, NO_x and O_3 monitors, tandem aerosol sulphate/nitrate filter sampler, temperature, relative humidity and altitude sensors. The aircraft research flight measurements were performed over the areas shown in Fig. 1.

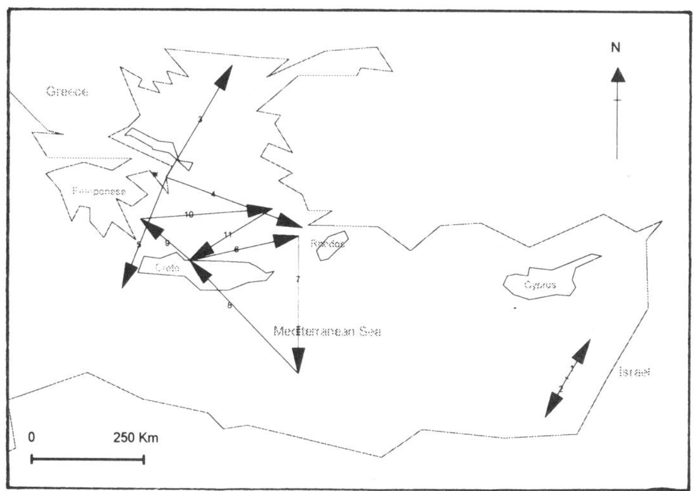

Figure 1. Flight paths for the three-year campaign, 1993-1995.

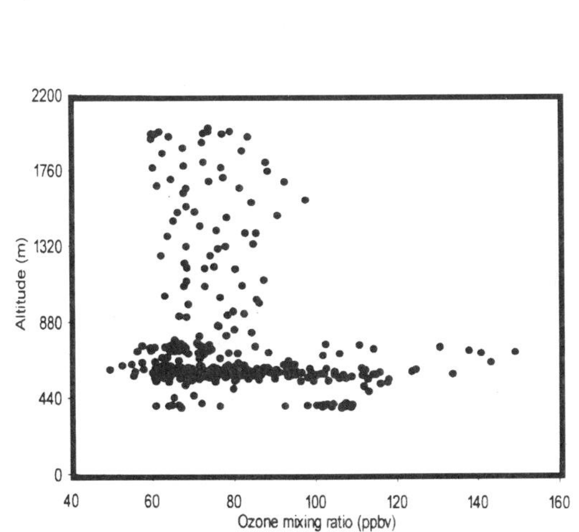

Figure 2. Ozone level variations versus altitude over GAA measured during the eight flights over the area from 6 to 8 July 1994.

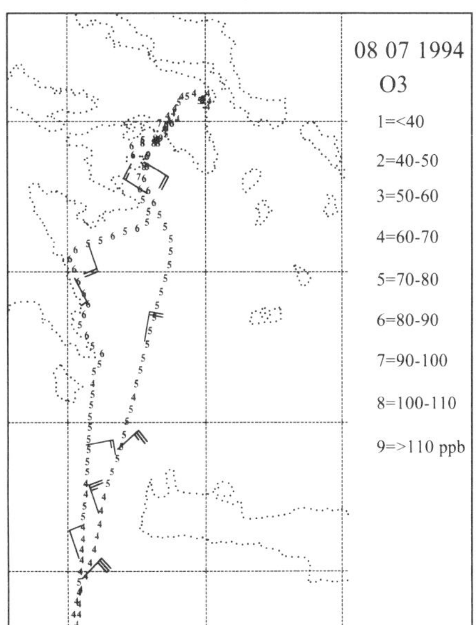

Figure 3. Ozone concentrations measured by the research aircraft on 8 July 1994 for the time period 1005 to 1355 LT (UTC+3h). One full wind-barb equals 5 ms^{-1} and one half is 2.5 ms^{-1}.

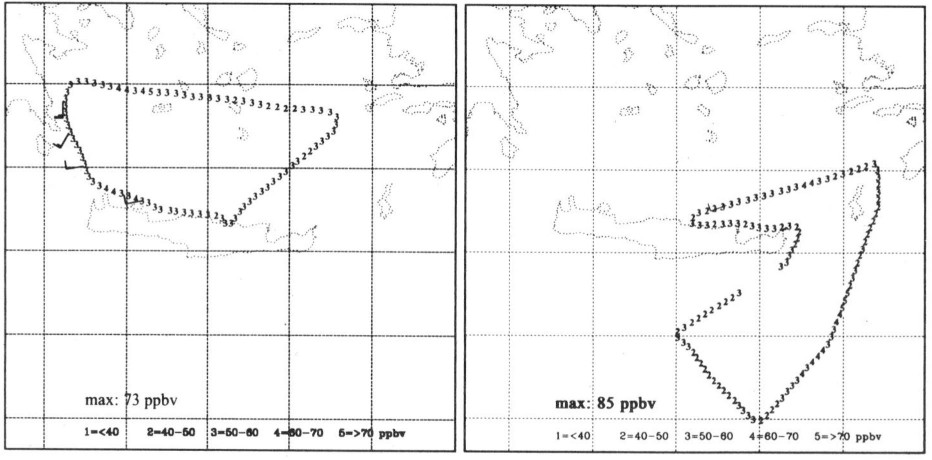

Figure 4. Ozone concentrations measured by the research aircraft on (a-left) 4 July 1995 for the time period 1630 to 1930 LT (UTC+3h) and (b-right) on 5 July 1995 for the time period 1110 to 1440 LT (UTC+3h). One full wind-barb equals 5 ms^{-1} and one half is 2.5 ms^{-1}.

During the period 6-8 July 1994 an air pollution episode started to build-up over the Greater Athens Area (GAA). For these three days, intensive flights were performed (approximately 20 hours) at various altitudes (from approximately 200 m up to 2000 m AGL), over Athens, the Saronic Gulf and all the way downwind following the urban plume. As it was found, the urban plume of Athens is injected either within the marine boundary layer or into the free troposphere, mainly with the aid of the upslope flows during the day-hours (Fig. 2).

In Fig. 3, the O3 concentrations monitored during the second day of the air pollution episode in Athens (8 July 1994) is shown. As it is seen, the plume travels within the marine boundary layer for several hundreds of kilometers keeping most of its characteristics. For this case, the model results have been presented in Kallos et al., 1996a,b. More characteristic is the event of 3-5 July 1995 where the plume of Athens was tracked for three days. As it is seen in Figs 4, consequent O_3 maxima were detected all the way from the Saronic Gulf to the area approximately 250 km SE of Crete within a period of approximately three days. The data collected through these experimental campaigns are tabulated and discussed in Kallos et al., 1996). A brief presentation of some of these results is given below.

As it concerns SO_2 it appears that over the southern part of the Aegean Sea the highest concentrations were recorded, especially between Peloponese and Crete (path 5) while as regards to NO_x it is not possible from the available results to designate any particular area over Greece as being more polluted. The results from the flights over Greece are much more definitive for O_3, with higher ozone levels over the southern part of the Aegean Sea in the vicinity of Crete. Analysis of the sulphates showed that over Greece all of the paths gave similar values with path 5 between the Peloponese and Crete being slightly higher. Finally as it concerns HC the amount of samples collected did not allow to derive definite conclusions about it. As it was also found, the O_3 over the Israeli is always higher than in other areas. In the same area, the sulfates were found to exeed even the 200 nmoles/m^3 which is considered as very high.

MODEL DESCRIPTION - CONFIGURATION

The model used for this study is the Regional Atmospheric Modelling System (RAMS), developed at Colorado State University. A general description of the model and its capacities is given in Pielke et al (1992).

For dispersion calculations, the Hybrid Particle and Concentration Transport Package (HYPACT) was used. It is a combination of a Lagrangian particle model and an Eulerian concentration transport model. It was developed at the Colorado State University and ASTeR, Inc (Tremback et al., 1993).

In order to simulate the summer-type atmospheric circulations on an accurate way, RAMS and HYPACT were configured on different ways. For most of the simulated cases, an outer grid covering an area 4000 km x 3000 km, with 60 km horizontal grid and thirty vertical levels following the topography were defined. For selected cases, finer resolution outer grids (30 or 20 km) were used over the Eastern Mediterranean and Middle East. One or two nested grids were usually configured over the areas of interest and mainly over the Greek Peninsula and the Aegean Sea. For specific cases, a very fine nested grid with 2 km horizontal grid resolution was used in an area approximately 200 km x300 km around Athens. The horizontal grid increments selected were found to be adequate for describing all scales of thermal circulations developed in this region. The model was initialized using 1x1 degree ECMWF analysis files available every 6 hours at the standard pressure levels. For SST, climatological fields of 1x1 degree were used. Simulations were performed for several cases.

Because the purpose of this study is to define some characteristic scales of dispersion over the Mediterranean, qualitative-type dispersion simulations were performed. For the performed simulations, the sources under consideration are mainly large urban areas or areas known for their industrial activities. The areas from where releases were made are mainly Athens, Thessaloniki, Istanbul, Constanza, Burgas, Milano, Messina, Barcelona and Catania.

MODEL RESULTS

In Kallos et al. (1996b) model simulations describing the major paths of transport of air pollutants from Europe to the East Mediterranean Region were presented. Two paths for such a transport were identified: one across the Aegean Sea and the other from the western to the eastern part of the Mediterranean over the sraights between Sicily and North African Coast. In addition to these, secondary roots are also possible but again, the Middle East and the Northeastern African coastal area is the final receptor. The difference in such cases is in the temporal scales of such transport.

As it was discussed in Kallos et al., (1996) during the cases where the etesians are in full development, the areas affected at the most are these of Libya, Egypt and the Middle East. The main affect is from sources located in southern Italy, Greece, Turkey and on a secondary way from the countries surrounding the Black Sea (Fig. 5). The characteristic time scales for such a transport is approximately 3 days. During these days, because of the relatively strong horizontal component of advection, the plumes (urban or industrial) from sources located near the coast are injected almost entirely within the marine boundary layer and stay within it until they reach the southern or southeastern coast of the Mediterranean. Air pollutants having their origin in the area around the Black Sea are aged by one or two days more because they are usually trapped within the stable layers over the sea before they start advected towards South.

The transport from the western to the eastern part of the Mediterranean usually occurs during summer with the existence of a relatively strong westerly component of the synoptic flow. The origin of polluted air masses in this case is either the northwestern part of the Mediterranean coast or the western coast of Italy. Usually the transported air masses are aged by 2-4 days over the Mediterranean waters before they are started to move eastward. In such cases, pollutants from sources located in the southern part of Italy (e.g. industrial areas of Messina, Catania, urban plumes from Italian cities) are injected directly within the marine boundary layer and are transported eastward. Therefore, during the transport from the western towards the eastern Mediterranean the plumes should be both “aged” (2-4 days) and “new” (fresh releases). The origin of the anchor system which circumnavigates such kind of flow defines also the path which will be followed by the air pollutants. If it is a low passing over northern Italy or central Europe, it usually transports polluted air masses towards the Greek Peninsula (over or around it) but usually, they are ending again in the southeastern corner of the Mediterranean (Kallos et al., 1996). In general, the time scale for the transport from the southern Italy and Sicily (and from the area south of Sicily) towards the Greek Peninsula is usually less than two days. In order the plumes to reach the Middle East, it usually takes an extra 2-3 days.

In the case where the anchor system is located in the Atlantic or western Europe, the transport occurs southeasterly but it deflects towards the African coast quickly. The affected

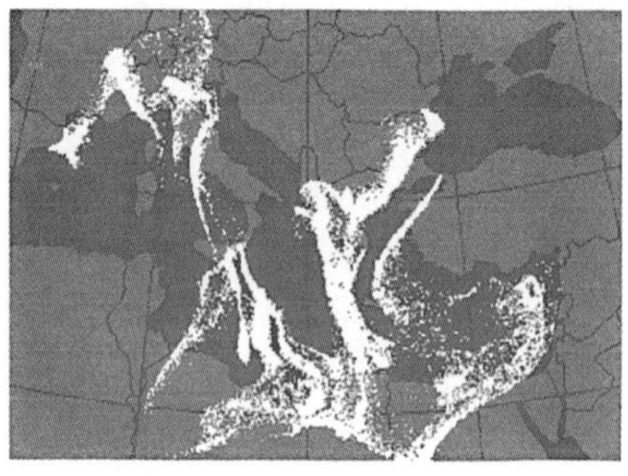

Figure 5. Particle projection at the first model level from the HYPACT dispersion model at 1500 UTC, 4 July 1994, (after 63 hours of particle release).

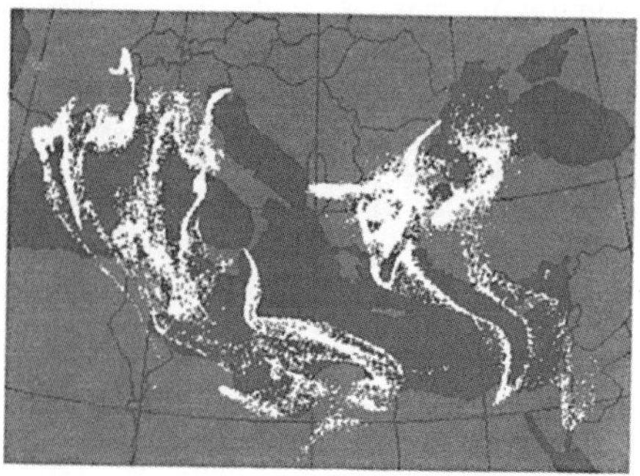

Figure 6. Particle projection at 1200 UTC, 12 July 1994, after 60 hours of particle release.

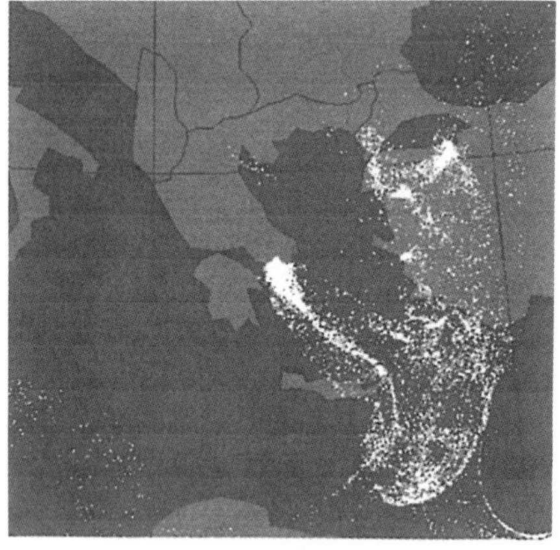

Figure 7. Particle projection at 1000 UTC, 5 July 1995, after 58 hours of particle release (from Athens), from HYPACT model.

areas are these of Tunis and Libya. The Middle East is rarely affected by such kind of transport (Fig. 6).

As it was mentioned in the previous chapter, during the days with relatively weak pressure gradients over the Eastern Mediterranean, the flow is still from North to South but at the same time, the mesoscale components (e.g. the sea/land breezes, upslopes/downslopes) are becoming significant and they play significant role in the dispersion from urban and industrial sources. In such cases, while the followed paths are approximately the same as previously, the temporal scales are significantly longer. As a good example is the case discussed previously (July 3-5, 1995). In this case, as the airborne measurements showed and the model simulations verified, the time scale for the transport of the urban plume of Athens towards the area southeast of Crete where relatively high concentrations of pollutants were measured, is approximately two and a half days (Fig. 7) and an extra day and a half to reach the Middle East. This time scale is longer than it is regularly found in cases with etesians by at least one day.

As it was described above, in most of the examined summer cases, the final receptor area is the southeastern Mediterranean and mainly the Egypt and the Middle East. One question raised after these findings: how deep inland in Africa and the Middle East these plumes should

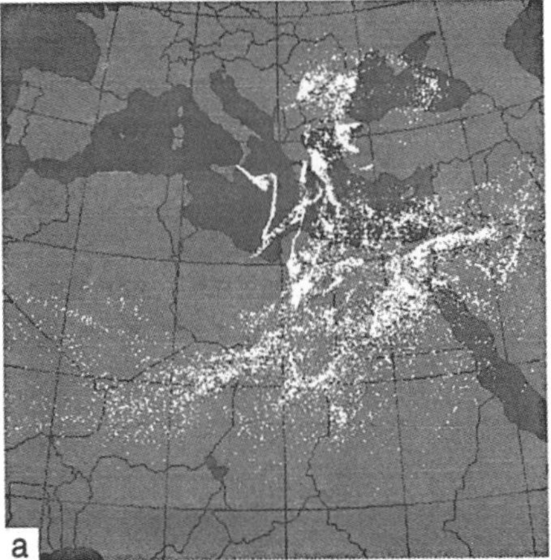

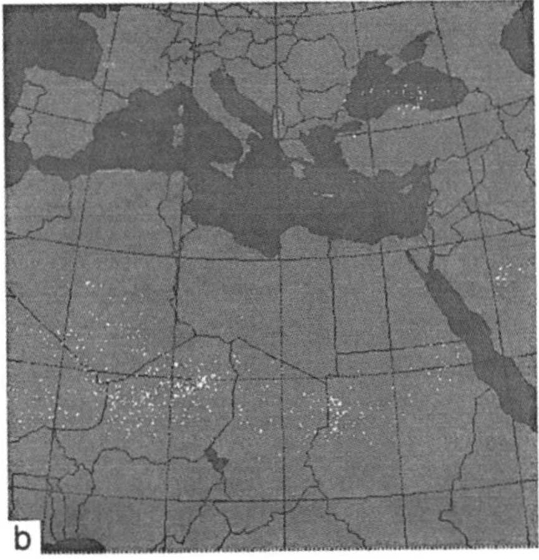

Figure 8. (a) Particle projection from HYPACT dispersion model, at 1800 UTC, 7 July 1994 (after 138 hours of particle release). (b) As in (a) but only particles located at altitudes higher than 5 km.

penetrate? In order to answer to the question raised, additional model simulations were performed. The RAMS and HYPACT models were configured in order to cover a very large area from North Europe to the Equatorial Zone and from the Atlantic to the Persian Gulf. The grid increment used was 38 km which is sufficient to resolve the important regional scale circulations which exhibit a diurnal cycle (etesians). The simulation was performed for 7 days. As it is seen in Fig. 8, particles released in locations such as Athens, Istanbul and Messina are reaching areas located more than 1000 km inland in Africa and a few hundreds of kilometers inland over the Middle East. The locations where these plumes are dispersed significantly coincide with the position of the Intertropical Convergence Zone (ITCZ). The ITCZ is reached in a time increment of approximately 5 days which is still within the life cycle of some air pollutants. As it is shown in Fig. 8b, the particles are reaching high altitudes over Africa and simultaneously are moving towards the West due to the easterlies usually observed in these locations. In such time scales, most of the primary pollutants (e.g. SO_2, NO_x, HC) have been already transformed, mainly in particles. As it was mentioned previously, most of the transport occurs within the marine boundary layer. This is the main reason for the existence of relatively high concentrations of various air pollutants in long distances from their origin. As long as the pollutants, in their form after their long trip, are reaching the North African coast, they are entering in to a deep boundary layer during the day-hours (which usually is extended vertically up to 4-5 km).

SOME CONCLUDING REMARKS

In this presentation, the observational evidence (surface and airborne measurements) and model simulations performed at the framework of two EU projects (SECAP and T-TRAPEM) are discussed. The model simulations include 3-D atmospheric and dispersion modeling at various scales simultaneously. Based on the experimental evidence and the model simulations some useful conclusions should be drawn:

- The transport of air masses from Europe towards the Eastern Mediterranean occurs during all seasons with the summer to be the most efficient. Most of the transport occurs within the marine PBL. Horizontal advection was found to be more important in the Eastern part of the Mediterranean. This is in contrast to what was found in the Western Mediterranean and the Iberian Peninsula. Because of the stable conditions occurring over the Mediterranean waters, plumes from various sources, especially those located near the coast, should travel for several hundreds of kilometres preserving some of their characteristics.

- The combination of some weather systems should form regional-scale flow patterns appropriate for long-range transport. The interaction between local thermal circulations with regional and synoptic ones is considered as very important for such a long-range transport.
- Two main paths of transport have been identified: the main path is from the Eastern part of Europe towards the Middle East and North Africa, through the Black Sea and the Aegean. The second path is from the Central and Western part of the Mediterranean towards the East. The first one was found to be more efficient. Some locations are acting as "temporal reservoirs" where air pollutants are "concentrated", "aged enough" before they are re-advected. The model simulations for the cases analyzed showed that the direct transport from the Western Mediterranean to the Middle East is almost not possible to occur.
- The transformation of air pollutants occurs on a very complicated matter. Mixing between plumes from different origins and ages makes the physico-chemical processes even more complicated.
- The time scales for transport of air masses from Southeastern Europe and the Black Sea towards the Middle East is approximately 2-3 days. This time period should be even longer for cases with relatively weak the synoptic and regional scale component of the atmospheric circulation. The transport from the Western part of the Mediterranean towards the Middle East is in general longer. If the flow is directed towards the northeastern part of the Mediterranean, the time scales for transport from Central Mediterranean to the East is in general less than 2 days while for reaching the Middle East is approximately double. For the direct transport from western Mediterranean to the North African coasts the time scales are approximately 2-3 days. These time scales are comparable to these for transformation of some primary pollutants.
- Air masses from Europe should reach the mid-tropospheric layers of the Equatorial Zone within a time period of a few days (4-6). During summer, the ITCZ is located in Northern Africa South of the Mediterranean Coast, over Southern Libya and Egypt where there are some strong convergence lines. This results in a massive upward transport of various aged pollutants. This transport within the ITCZ needs farther examination.
- As it was found, the air quality in urban and rural areas around the Mediterranean are influenced significantly by this kind of transport and transformation processes. There are indications that these multi-scale transport and transformation processes might have significant climatic impacts. More specifically, they might be effects on rain and therefore the water balance. This is possible through the increase of the number of CCN and through the direct warming of the lower tropospheric layers (up to about 3 km) without an increase in the specific humidity. Of course, these processes are further more complicated because of the appearance of desert dust particles in the atmosphere which, in a wet environment, should be coated by sulphates and on that way they become very effective CCN.

Acknowledgments

This research was supported by the research projects SECAP, contract number EV5VCT910050 and the AVICENNE Initiative (T-TRAPEM), contract number EV*-CT92-0005 of the DG-XII of EU. Acknowledgement is also made to the National Centre for Atmospheric Research, which is sponsored by the National Science Foundation, for the data computing time used in this research.

REFERENCES

Kallos, G., P. Kassomenos, and R.A. Pielke, 1993: Synoptic and mesoscale weather conditions during air pollution episodes in Athens, Greece. Boundary-Layer Meteorol. 62, 163-184.

Kallos, G., Kotroni V., Lagouvardos K., and Varinou M., A. Papadopoulos, 1995: South European Cycles of Air Pollution(SECAP). Final Report prepared for the DGXII, EU.

Kallos, G., Kotroni V., Lagouvardos K., and Varinou M., M. Luria, M. Peleg, G. Sharf, V. Matveev, D. Alper-SimanTov, A. Vanger, G Tuncel, S. Tuncel, N. Aras, G. Gullu, M. Idrees, F. Al-Momani, 1996a: Transport and Transformation of Air Pollutants from Europe to the East Mediterranean. Environmental Research Program AVICENNE. Final Report for the DGXII of EU. Pp 352.

Kallos G., V. Kotroni, K. Lagouvardos, M. Varinou, and A. Papadopoulos, 1996b: "Possible mechanisms for long range transport in the eastern Mediterranean" Proc. of the 21st

NATO/CCMS Int. Techn. Meeting on Air Pollution Modelling and Its Application, 6-10 November, Baltimore, USA, pp.99-107.

Luria M., M. Peleg, G. Sharf, D. Siman Tov-Alper, N. Schpitz, Y. Ben Ami, Z. Gawi, B. Lifschitz, A. Yitzchaki, and I. Seter, 1996: Atmospheric Sulphur over the East Mediterranean region. J. Geophys. Res., 101, 25917-25930.

Meteorological Office, 1962: Weather in the Mediterranean. Vol. I, General Meteorology H.M. Stat. Office, London. Second Edition.

Millan, M., B. Artinano, L. Alonso, M. Castro, R. Fernandez-Patier and J. Goberna, 1992: Mesometeorological Cycles of Air Pollution in the Iberian Peninsula. Final Report for the DGXII of EU. Pp 219.

Pielke, R.A., W.R. Cotton, R.L. Walko, C.J. Tremback, W.A. Lyons, L.D. Grasso, M.E. Nicholls, M.D. Moran, D.A.Wesley,T.J. Lee, and J.H. Copeland, 1992: A comprehensive meteorological modelling system - RAMS. Meteorol. Atmos. Phys. 49, 69-91.

Tremback, C.J., W.A. Lyons, W.P. Thorson, and R.L. Walko, 1993: An emergency response and local weather forecasting software system. Preproceedings of the 20th ITM on Air Pollution and its Application, November 29 - December 3, 1993. Valencia, Spain. pp 8.

Ziomas, I., 1996: Mediterranean Campaign of Photochemical Tracers-Transport and Chemical Evolution (MEDCAPHOT-TRACE): An outline. Atmos. Environ. (accepted for publication).

DISCUSSION

S.T. RAO: Could you please define the influence function and describe how it is computed? Can this procedure identify the potential source-receptor relationship?

G. KALLOS: You can find a good description in publications from M. Uliasz. One of his recent publications is in one of the previous ITMs. The influence function is a good tool to be used for source-receptor relationships.

M. TAYANC: You showed us the results of your simulations for summer days. The model includes emissions from large urban areas in Eastern Mediterranean region, included Istanbul. We know that the air pollution concentrations, in terms of sulphur dioxide and particulate matter in Istanbul are low for summer days. What kind of emission inventory or air pollution concentrations for Istanbul did you use in your simulations?

G. KALLOS: As I said during my presentation, we performed meteorological simulations and based on the outputs we performed additional simulations with the aid of a Lagrangian Particle Dispersion Model. Since we are looking for spatial and temporal scales of long range transport from large urban conglomerates we do not need emission inventories. We want to show the paths followed by the urban plumes and the time scales. This is why we made just continuous particle releases from selected areas, Istanbul included.

ASSIMILATION OF SATELLITE DATA IN REGIONAL AIR QUALITY MODELS

Richard T. McNider, William B. Norris, and Daniel M. Casey

Earth System Science Laboratory
University of Alabama in Huntsville
Huntsville, Alabama

Jonathan E. Pleim and Shawn J. Roselle

Atmospheric Science Modeling Division
Air Resources Laboratory
National Oceanic and Atmospheric Administration
Research Triangle Park, North Carolina
(on assignment to the National Exposure Research Laboratory, U.S. EPA)

William M. Lapenta

NASA Marshall Space Flight Center
Global Hydrology and Climate Center
Huntsville, Alabama

INTRODUCTION

In regional-scale air-pollution models probably no other source of uncertainty ranks higher than the current ability to specify clouds and soil moisture. Because modeled clouds are highly parameterized, the ability of models to predict the magnitude and spatial distribution of radiative characteristics is highly suspect and subject to large error. While considerable advances have been made in the assimilation of winds and temperatures into regional models (Stauffer and Seaman, 1990), the poor representation of cloud fields from point measurements at National Weather Service stations and the almost total absence of observations of surface moisture availability has made assimilation of these variables difficult if not impossible. Yet, the correct inclusion of clouds and surface moisture are of first-order importance in regional-scale photochemistry. Consider the following points relative to these variables.

Clouds dominate the availability of actinic flux, which directly affects the photolysis rates in photochemical models (Seinfeld, 1988; Dunker, 1980). Inaccurate specification of clouds can lead to the inaccurate specification of photolysis rates and can, in a first-order fashion, affect model performance. Clouds are also the prime controller of surface temperature through their effect on solar insolation. Biogenic isoprene emissions and some anthropogenic emissions are nonlinearly dependent upon surface temperature (Tingey *et al.*, 1979; Zimmerman *et al.*, 1988). Errors in cloud cover and distribution can drastically change the modeled chemistry of the environment.

Air Pollution Modeling and Its Application XII
Edited by Sven-Erik Gryning and Nadine Chaumerliac, Plenum Press, New York, 1998

Surface moisture availability is probably second only to clouds in controlling surface temperature (Wetzel *et al.*, 1984; Carlson *et al.*, 1981; Pleim and Xiu, 1995). Moist surfaces or actively transpiring vegetation can sharply reduce temperatures over that of dry surfaces. Mixing heights in regional-scale models are also highly dependent on surface temperature through the surface sensible heat flux (Deardorff, 1974), which is controlled by insolation and surface moisture availability. Errors in specification of clouds and surface moisture can substantially alter air-pollution concentrations. Also, cumulus cloud convection can alter boundary-layer concentrations by effectively deepening the mixing height to include a large portion of the troposphere and can inject precursors into an environment where the chemistry can be efficient and the chemical chain lengths long in the absence of surface losses.

Considerable emphasis has been placed on wind direction in regional-scale models. Winds affect the distribution of pollutants, but clouds and moisture availability greatly affect photochemical production. (The clouds and moisture also affect winds--see below). The domino effect of clouds and moisture through photolysis rates, emissions, mixing heights, etc. makes them a pivotal element in regional models.

The advent of earth-observing satellites has opened the door to new approaches for reducing or overcoming the shortcomings in regional modeling that have just been described. It is the purpose of this paper to describe methods of satellite remote sensing that can be used to specify clouds and surface moisture in reconstructive photochemical modeling on the regional scale with improved fidelity over traditional methods. Specifically, geostationary satellite data are used because of the extent of their temporal and spatial coverage. The following summarizes techniques for estimating insolation, photolysis rates, and surface moisture.

INSOLATION

The net solar radiation at the surface provides the prime source of energy controlling the diurnal variation in temperature in the surface energy budget in regional-scale models. In addition to astronomical factors, the net radiation at the surface is determined by the surface albedo and by reflection and absorption of solar radiation by both the clear and cloudy atmosphere (including aerosols).

Surface Albedo

The surface albedo in regional-scale models such as MM5 (Grell *et al.*, 1994) or RAMS (Pielke, 1992) is usually estimated from land-use types. However, the relationship between gross land-use characteristics and the radiative properties of the surface is not always well defined. Also, the albedo can change due to meteorological conditions and anthropogenic activities (such a harvesting), which are not included in routine land-use data bases. Satellites, on the other hand, provide a direct measurement of reflected radiation, although interpretation due to view angle and bi-directional effects must be considered (Gautier *et al.*, 1980).

The GOES series of geostationary satellites operated by NOAA over the last two decades (the most recent being GOES-8, launched in April 1994) provide observations of the U.S. that can be used in regional models. These satellites return the magnitude of upwelling radiance in the visible portion of the spectrum as brightness counts. The lowest counts arise when little or no cloud is present and the reflection is primarily, if not entirely, from the earth's surface. Using the physical retrieval method of Gautier *et al.* (1980) and Diak and Gautier (1983), the surface albedos can be recovered from the brightness counts. The technique requires hourly surface albedos obtained from clear-sky brightness counts. If a single, cloud-free image were available for each hour of daylight, the brightness counts could be obtained directly from them. However, because cloudy skies are so common, a single, cloud-free image is usually not available. Experience has shown that for a given daylight hour images over a period of 20-30 days are needed to obtain a stable minimum brightness count for that hour, especially in the Southeast during summer when cumulus clouds are ubiquitous. Brightness counts are converted into reflectances using a calibration curve unique to each satellite. This approach has the inherent ability to account for both spatial and temporal differences in albedo due to soil type, vegetation, and time of day and year. Figure 1 shows a clear-sky albedo derived using this technique and converted to an 80-km grid used in MM5.

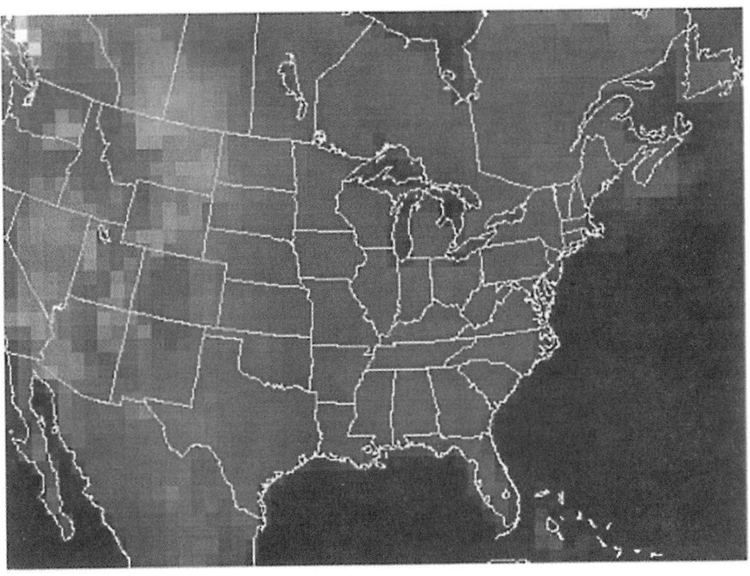

Figure 1. Clear-sky albedo using minimum brightness from GOES-7 images during July 1988 gridded to an 80-km MM5 domain.

Cloud Albedo

Once the clear-sky albedo and the brightness count are known for a given satellite-image pixel, surface insolation at the pixel scale (usually 1, 4, or 8 km) can be calculated from the simplified radiative-transfer model of Gautier *et al.* (1980) as described in McNider *et al.* (1995). The model assumes a single cloud layer. Above the cloud layer, radiation is Rayleigh scattered and absorbed by water vapor. In the cloud layer, radiation is scattered and absorbed. Below the cloud layer, radiation is absorbed by water vapor. For the scattering coefficients, we use the parameterization originally presented by Kondratyev (1969) and modified by Atwater and Brown (1974). For the water-vapor absorption coefficients, we use an empirical formulation of MacDonald (1960). For in-cloud absorption we use a step function that depends on brightness count (McNider *et al.*, 1995). This results in a quadratic equation in cloud albedo. Once known, the cloud albedo can be used to calculate downwelling solar radiation and insolation at the surface.

A flow chart of the computational procedure is shown in Figure 2. The procedure yields surface insolation at each pixel in a satellite image. Such images can be gridded and values for all pixels within a grid cell averaged to produce hourly input fields for assimilation into photochemical or meteorological models. Figure 3 shows the surface insolations computed by this technique for Julian day 216 on an 80-km MM5 grid covering most of the U.S.

While the GOES visible sensors do not have on-board calibration, pre-launch calibration curves (Raphael and Hay, 1984) and ground-truth data (Tarpley, 1979) can approximate the needed calibration information. To provide some understanding of the fidelity of the technique, a comparison was made of observed versus satellite-derived insolation values for the period 31 July - 8 August, 1988, using GOES-7 measurements. The observed data were taken from available surface pyranometer archives (National Renewable Energy Laboratory, 1992). Because the GOES sensor measures near-instantaneous values and the pyranometer data are hourly averages, exact agreement cannot be expected. Also, the absolute navigation of the GOES-7 image is of order 10 km. Thus, in making the comparison the best-fit satellite pixel value within 20 km of a pyranometer site was selected. Figure 4 shows this comparison. While scatter exists, the key fact is that almost no bias is present. Because the insolation calculations are based on straightforward radiative principles and insolation is fairly insensitive to uncertain specifications such as cloud absorption (McNider *et al.*, 1995), it is felt that absolute and especially the relative spatial accuracy is quite high.

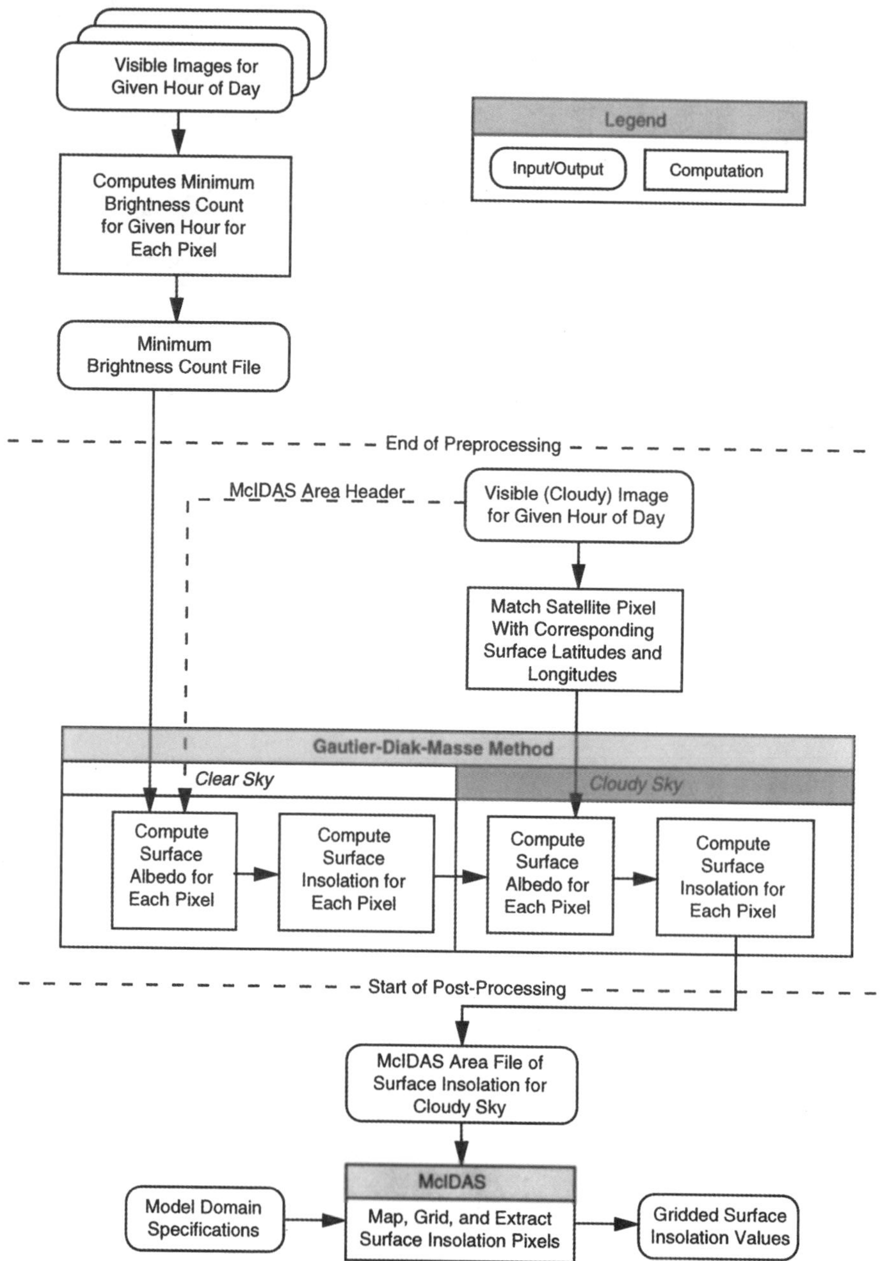

Figure 2. Flow chart of computational procedures for determining surface insolation.

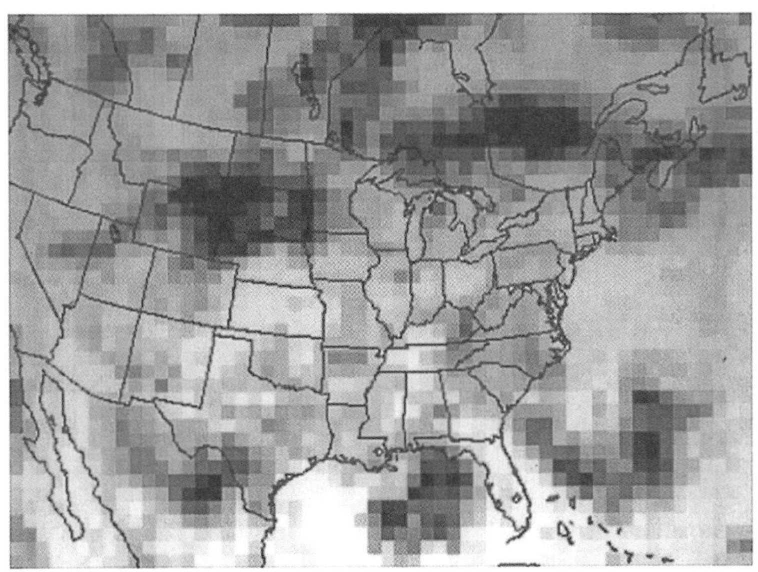

Figure 3. Surface insolation computed from a satellite image for an 80-km grid for 1800 GMT Julian Day 216, 1988. Lighter areas indicate higher surface insolation.

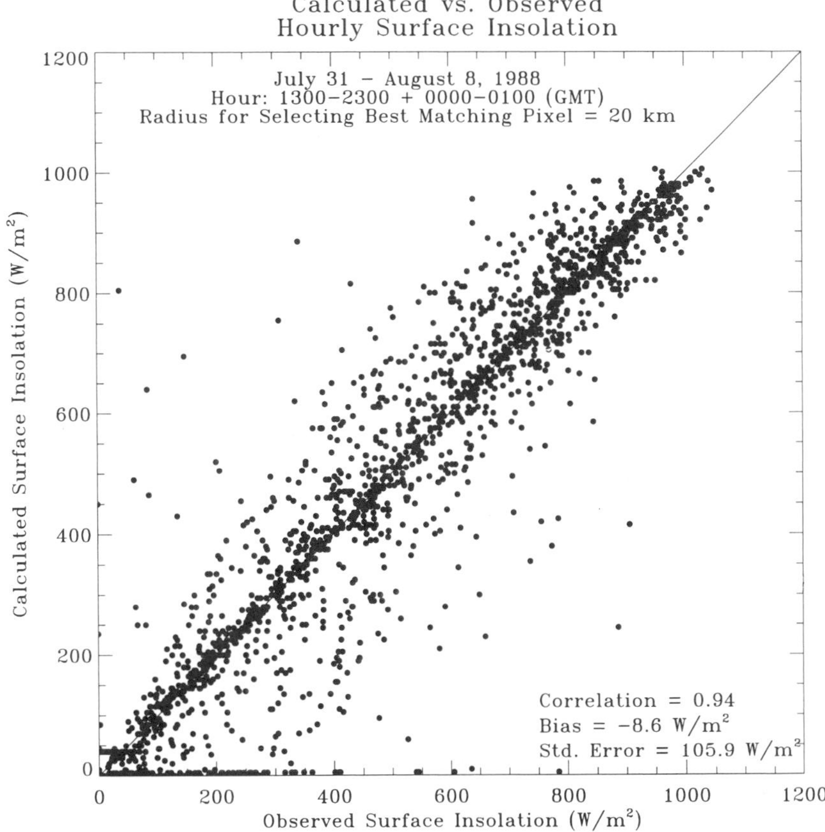

Figure 4. Comparison between satellite-derived and surface pyranometer-measured insolation values for the period 31 July - 8 August, 1988. The best-fit pixel within a 20-km radius of the observation site was used.

PHOTOLYSIS RATES

Photochemical modeling systems differ in their sources of photodissociation constants. In some cases these are obtained from radiative transfer models that include the scattering, absorbing, or reflecting properties of atmospheric aerosols and clouds (*e.g.*, Ruggaber *et al.*, 1994; Roselle *et al.*, 1995). In other cases radiative transfer models assume clear skies, and correction factors are later applied to the results to obtain the desired cloudy-sky values (*e.g.*, Chang *et al.*, 1987). In either approach cloud information is essential.

To compute cloud effects on photolysis rates three fundamental parameters are needed: (1) cloud-layer transmittance, (2) cloud-top elevation, and (3) cloud-bottom elevation. In the past in photochemical models these parameters were estimated from hydrological information from the meteorological processor. Often the information used to recover these parameters is fairly indirect. For example, the regional photochemical model RADM approximates the cloud optical depth τ with the parameterization

$$\tau = 3L_{con}\Delta z_{cld}/2\rho_{H_2O}\, r$$

where L_{con} is the mean condensed water content, Δz_{cld} is the mean depth of the cloud layer, r is the mean cloud drop radius, and ρ_{H_2O} is the density of water (Chang *et al.*, 1987). In RADM, constant values are assumed for each factor except Δz_{cld}, which is obtained from the meteorological model. This cloud optical depth is then used in a Beer's Law formulation to estimate cloud transmittance. In practice the cloud top is often determined from relative humidity thresholds and cloud base from computed lifting condensation levels. We now propose techniques that use satellite data to estimate two of the needed parameters--cloud transmittance and cloud-top elevations and discuss options for determining cloud base.

Satellite-Derived Transmittance

Cloud transmittance is fundamentally a radiative property and can be recovered from the requirement that cloud transmittance, reflection and absorption sum to unity. We make the assumption that the broad-band transmittance determined from the satellite sensor (0.52 - 0.72 μm) is the approximate transmittance required in the ultimate photolysis calculations. Since cloud absorptivity is parameterized as a function of brightness count, and cloud reflectance is determined during the calculation of insolation, then cloud transmittance can be

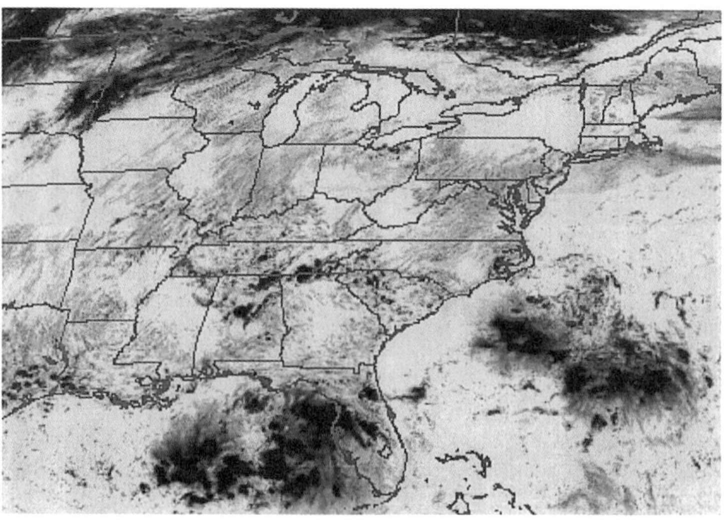

Figure 5. Satellite-derived transmittance field for 1800 GMT Julian Day 216, 1988. Lighter areas indicate higher transmittance.

computed. Absorptivity is the least well-known parameter in the present scheme. McNider *et al.* (1995) showed in sensitivity tests that even making extreme changes in the absorptivity function changes the transmittance only slightly. Figure 5 shows the transmittance field determined by satellite using the techniques described above.

Satellite-Derived Cloud-Top Elevations

Cloud-top elevations can be estimated using infrared satellite images from atmospheric clear-window channels. Just as with visible images, the GOES satellites return infrared measurements in discrete counts. These values correspond to blackbody temperatures and in clear-window channels represent the temperature of the Earth's surface in clear regions and cloud-top temperatures in cloudy regions. If a temperature sounding is available near the

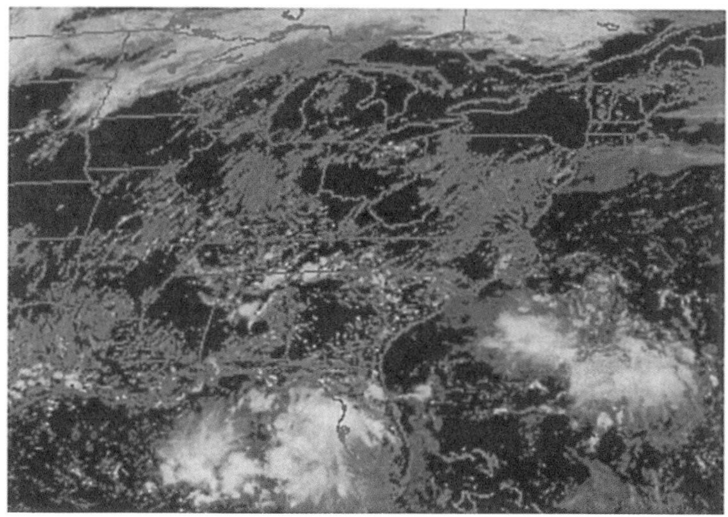

Figure 6. Satellite-derived cloud-top elevations for 1800 GMT Julian Day 216, 1988. Black areas are the surface. Lighter areas are higher cloud tops.

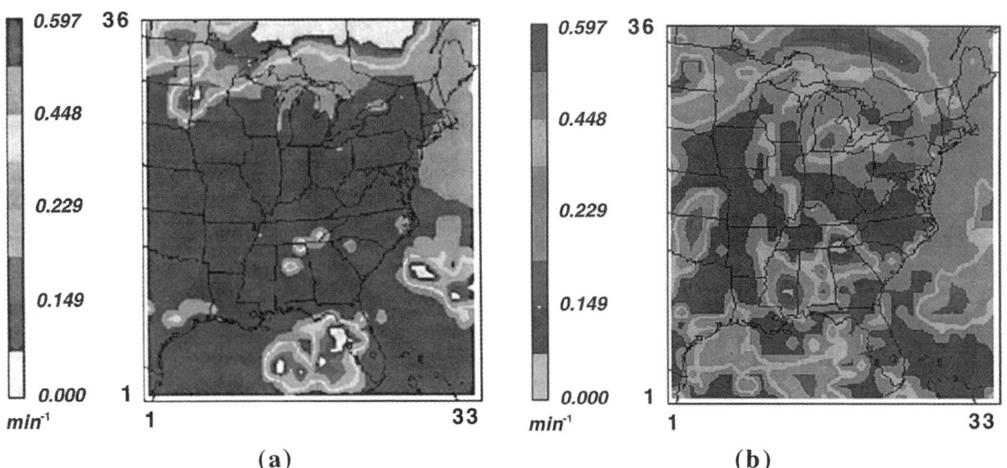

Figure 7. Photolysis fields (J_{NO2}) derived from (a) satellite images and (b) prognostic model output for 1800 GMT Julian Day 216, 1988.

location of a pixel at the time of the satellite measurement, the pressure level, and hence the elevation, at cloud top can be determined. At this point in the development of the technique, for a given pixel we are using the grid-point sounding from the meteorological model. Figure 6 shows satellite-derived cloud top elevations.

Cloud-Base Elevations

GOES satellites cannot provide any guidance on the cloud base since infrared techniques cannot determine the depth of clouds. However, auxiliary thermodynamic structure information can be used to compute the lifting condensation level, which gives a lower limit on the cloud-base elevation. This auxiliary information can come from prognostic mesoscale models or rawinsonde observations.

Satellite-Derived Photolysis Fields

The cloud characteristics defined above were used in a photolysis model incorporated in a new version of RADM. Figure 7 shows a comparison between the photolysis fields generated from the satellite-derived cloud characteristics and the diagnostic determination of cloud characteristics from the prognostic meteorological model fields. Relative to the satellite image, the degree to which the diagnostic model overestimates the coverage of clouds that significantly attenuate actinic flux for this hour is dramatic. No direct measurements of the photolysis fields are available for comparison with Figure 7. However, the similarity in the appearance of the surface insolation (Figure 3), the transmittance (Figure 6), and the satellite-derived photolysis (Figure 7a), coupled with the near absence of bias in the comparison of satellite-derived surface insolation and pyranometer observations (Figure 4), indicates that Figure 7a represents reality better than Figure 7b. Analysis of this case is continuing, but it appears that the highly parameterized approach to cloud development in meteorological models does not capture the actual cloud fields with fidelity.

SURFACE MOISTURE AVAILABILITY

Biogenic and soil NO_x emission rates are strongly dependent on surface temperatures, as are mixing heights and turbulent dispersion. In meteorological models surface temperatures

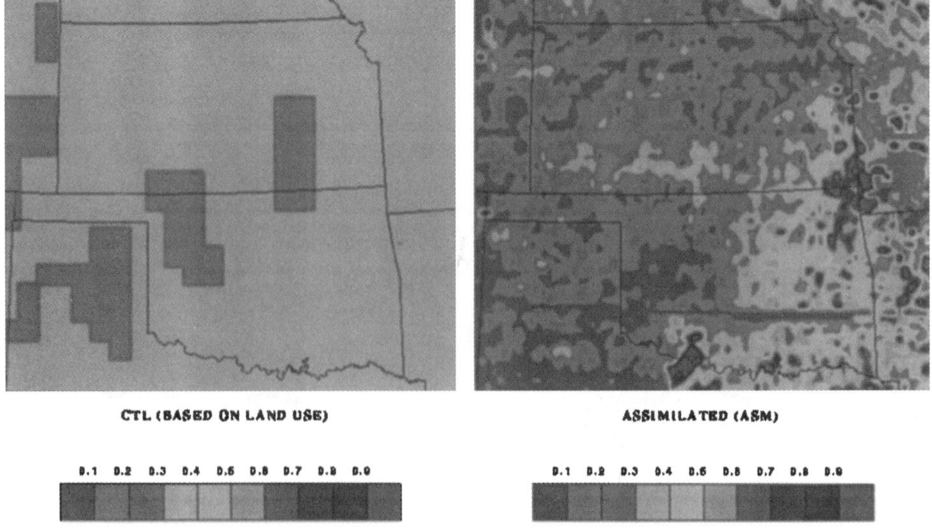

Figure 8. Model moisture availability, *M*, representing the fraction of possible evaporation for a saturated surface. Left panel: *M* based upon land-use categories. Right panel: *M* after assimilating GOES data.

are highly sensitive to the specification of surface moisture (Pleim and Xiu, 1995), yet observed moisture values are not readily available and must be estimated from vegetation and soil-moisture parameterizations that contain difficult-to-specify parameters such as deep root-zone moisture and stomatal conductance.

To avoid these problems we have developed a method using infrared satellite images to recover soil moisture (McNider *et al.*, 1994). The method accomplishes this by assimilating satellite-observed surface temperature rates of change into boundary-layer models in a thermodynamically consistent manner. It requires the use of GOES infrared images, land surface temperatures (LSTs), surface albedo, and insolation over the time period of interest. It adjusts the model's bulk moisture availability so that the model's rate of change of LST agrees more closely with that derived from satellite observations. A critical assumption is that the availability of moisture (either from the soil or vegetation) during mid-morning hours is the least known term in the model's surface energy budget (Wetzel *et al.*, 1984; Carlson 1986). Therefore, the simulated latent heat flux, which is a function of surface moisture availability, is adjusted based upon differences between the modeled and satellite-derived LST tendencies.

Here we present results from the assimilation technique as applied within MM5 using a grid resolution of 12 km. The region of interest was Oklahoma and Kansas, where a west-east vegetation gradient existed (wheat stubble to the west, deciduous forest to the east). Given this variation in vegetation, one would expect the satellite-observed surface thermal response to differ across the region. The LST is derived via a physical split window technique (Guillory *et al.*, 1993) from GOES 8 data between 1200 and 2300 UTC on 7 July 1995. The LST tendencies are fairly uniform across the region from 1200 to 1400 UTC. However, between 1400 and 1700 UTC the LST tendency in eastern Oklahoma (5°C per 3 h) is considerably lower than that farther west (10°C per 3 h). This spatial variation is consistent with the west-east vegetation gradient that existed across the region.

The assimilation scheme in MM5 adjusts the evaporative flux by altering the moisture availability parameter, M, which represents the fraction of possible evaporation for a saturated surface. When the standard procedure of specifying M as a function of land-use category was used to initialize the model, the field exhibits an unrealistic, blocky structure (Figure 8, left panel).

After the model assimilated GOES LST tendencies from 1400 UTC to 1800 UTC (900 to 1300 local standard time), the adjusted M exhibits a more realistic structure (Figure 8, right panel) which matched well with Normalized Difference Vegetation Index (NDVI) (not shown). A more stringent check of the technique was done by comparing modeled surface evaporative flux data from observations obtained via Energy Balance Bowen Ratio (EBBR) systems deployed at the Southern Great Plains (SGP) Cloud and Radiation Testbed (CART) sites for the U.S. Department of Energy's Atmospheric Radiation (ARM) Program (Splitt *et al.*, 1995). Without the assimilation, MM5 overestimates the daytime evaporative flux by 300 Wm^{-2} (150%) on both 7 and 8 July. However, use of the satellite data produces a much more realistic time series with an average difference for the 36-h period of less than 40 Wm^{-2}.

DISCLAIMER

This paper has been reviewed in accordance with the U. S. Environmental Protection Agency's peer and administrative review policies and approved for presentation and publication. Mention of trade names or commercial products does not constitute endorsements or recommendations.

ACKNOWLEDGMENT

This work was partially supported by the U.S. Environmental Protection Agency under Cooperative Agreement CR822765-01-0.

REFERENCES

Atwater, M. A. and P. S. Brown, Jr., 1974, Numerical calculation of the latitudinal variation of solar radiation for an atmosphere of varying opacity, *J. Appl. Meteor.*, **13**:289-297.

Carlson, T. N., 1986, Regional scale estimates of surface moisture availability and thermal inertia using remote thermal measurements. *Remote Sensing Rev.*, **1:**197-246.

Carlson, T. N., J. K. Dodd, S. G. Benjamin, and J. N. Cooper, 1981, Satellite estimation of the surface energy balance, moisture availability and thermal inertia, *J. Appl. Meteor.,* **20:**67-87.

Chang, J. S., R. A. Brost, I. S. A. Isaksen, S. Madronich, P. Middleton, W. R. Stockwell, and C. J. Walcek, 1987, A three-dimensional eulerian acid deposition model: physical concepts and formulation, *J. Geophys. Res.*, **92**:D12, 14681-14700.

Deardorff, J.,1974, Three-dimensional numerical study of the height and mean structure of the planetary boundary layer, *Bound-layer Meteor.* **15:**1241-1251.

Diak, G. R. and C. Gautier, 1983, Improvements to a simple physical model for estimating insolation from GOES data, *J. Appl. Meteor.*, **22:**505-508.

Dunker, A. M., 1980, The response of an atmospheric re/action-transport model to changes in input function, *Atmos. Environ.*, **14:**671-679.

Gautier, C., G. Diak, and S. Masse, 1980, A simple physical model to estimate incident solar radiation at the surface from GOES satellite data, *J. Appl. Meteor.*, **19**:1005-1012.

Grell, G. A., J. Dudhia, and D. R. Stauffer, 1994, *A Description of the Fifth-Generation Penn State/NCAR Mesoscale Model (MM5).* NCAR Technical Note NCAR/TN-398+STR, Nationial Center for Atmospheric Research, Boulder, Colorado.

Guillory, A. R., G. J. Jedlovec, and H. E. Fuelberg, 1993, A technique for deriving column-integrated water content using VAS split-window data, *J. Appl. Meteor.*, **32:**1226-1241.

Kondratyev, K. Y., 1969, *Radiation in the Atmosphere.*, Academic Press, New York.

McDonald, J. E., 1960, Direct absorption of solar radiation by atmospheric water vapor, *J. of Meteor.*, **17**:319-328.

McNider, R. T., A. J. Song, D. M. Casey, P. J. Wetzel, W. L. Crosson, and R. M. Rabin, 1994, Toward a dynamic-thermodynamic assimilation of satellite surface temperature in numerical atmospheric models, *Mon. Wea. Rev.*, **12**:2784-2803.

McNider, R. T., J. A. Song, and S. Q. Kidder, 1995, Assimilation of GOES-derived solar insolation into a mesoscale model for studies of cloud shading effects, *Int. J. Remote Sensing*, **16**:2207-2231.

National Renewable Energy Laboratory, 1992, *National Solar Radiation Data Base User's Manual (1961-1990),* Golden, Colorado.

Pielke R. A., W. R. Cotton, R. L. Walko, C. J. Tremback, W. A. Lyons, L. D. Grasso, M. E. Nicholls, M. D. Moran, D. A. Wesley, T. J. Lee, and J. H. Copeland, 1992, A comprehenseive meteorological modeling system--RAMS, *Meteor. Atmos. Phys.*, **49**:69-91.

Pleim, J. E. and A. Xiu, 1995, Development and testing of a surface flux and planetary boundary layer model for application in mesoscale models. *J. Appl. Meteor.*, **34**:16-32.

Raphael, C., and Hay, J. E., 1984, An assessment of models which use satellite dta to estimate solar irradiance at the earth's surface, *J. Climate and Appl. Meteor.*, **23:**832-844.

Roselle, S. J., A. F. Hanna, Y. Lu, J. C. Jang, K. L. Schere, J. E. Pleim, 1995, Refined photolysis rates for advanced air quality modeling systems, in *Proceedings of the A&WMA Conference on the Applications of Air Pollution Meteorology.*

Ruggaber, A., R. Dlugi, and T. Nakajima, 1994, Modelling radiation quantities and photolysis frequencies in the troposphere, *J. Atmos. Chem.*, **18**:171-210.

Seinfeld, J. H., 1988, Ozone Air Quality Models: A critical review, *J. Air Poll. Control Assoc.*, **38:**616-645.

Splitt, M. E. and D. L. Sisterson, 1995, Site Scientific Mission Plan for the Southern Great Plains CART Site: July-December 1995, ARM-95-002, Argonne National Laboratory, Argonne, Illinois.

Stauffer, D. R. and N. L. Seaman, 1990, Use of four-dimensional data assimilation in a limited-area mesoscale model. Part I: Experiments with synoptic-scale data, *Mon. Wea. Rev.*, **118:**1250-1277.

Tarpley, J. D., 1979, Estimating incident solar radiation at the surface from geostationary satellite data, *J. of Appl. Meteor.*, **18:**1172-1181.

Tingey, D. T., M. Manning, L. C. Grothaus, and W. F. Burns, 1979, The influence of light and temperature on isoprene emission rates from live oak, *Physiol. Plant*, **47**:112-118.

Wetzel, P. J., 1984, Determining soil moisture from geosynchronous satellite infrared data: A feasibility study, *J. Climate Appl. Meteor.*, **23:**375-391.

Zimmerman, P. R., J. P. Greenberg, and C. E. Westberg, 1988, Measurements of atmospheric hydrocarbons and biogenic emission fluxes in the Amazon boundary layer, *J. Geophys. Res.*, **93**:1407-1416.

DISCUSSION

P. QUANDALLE: How complex and time consuming are the image processing techniques you are using? Also, what is the availability and cost of the required images?

R. T. McNIDER: The major computational obstacle for the images is usually storage space since each of the images are quite large. It takes about a half a day on a workstation to process the one-month clear sky images hourly and compute the hourly solar insolation and photolysis input.

S.T. RAO: You touched upon a very important and subject, namely, the role of clouds in photochemical model simulations. Do you have any idea how the ozone concentrations produced by regional models might be affected by the uncertainties in the spatial/temporal specification of clouds?

R. T. McNIDER: We don't know the integrated effect at this time. We are currently making simulations with and without the satellite data and hope to have these available by late fall of this year.

EVALUATION OF AN AEROSOL MODEL: RESPONSES TO METEOROLOGY AND EMISSION SCENARIOS

S.C. Pryor† and R.J. Barthelmie†‡

†Climate and Meteorology Program, Department of Geography,
Indiana University, Bloomington, IN47405.
‡Wind Energy and Atmospheric Physics, Risoe National Laboratory, Denmark.

INTRODUCTION

The Fraser Valley is a complex topographic coastal environment which episodically experiences visibility degradation (and elevated aerosol concentrations) (Pryor et al., 1997). To examine concentrations and speciation of secondary inorganic aerosols and ozone in the transition between an oxidant event and period of elevated aerosol concentrations, numerical simulations were performed using a modified version of the ACDEP (Atmospheric Chemistry and Deposition) model. ACDEP is a lagrangian model which contains detailed and fully coupled gas-aerosol phase chemistry (Hertel et al., 1995). The modeling period is August 5-8, 1993, and the modeling domain is shown in Figure 1. Results calculated for two receptor sites are shown herein (locations specified in Figure 1). PIME is located in the western reaches of the valley directly east of the Vancouver metropolitan area, and CHIL is located in a region of mixed agricultural land, approximately 80 km from the Vancouver urban core. The modified ACDEP model was applied at a horizontal resolution of 5 km and with 10 layers in the vertical, increasing logarithmically from 2 m to 2 km. The meteorological parameterizations within ACDEP have been extended such that stability parameters and mixed layer depth are calculated using routinely available meteorological data.

METEOROLOGICAL PARAMETERIZATIONS

To remove the necessity for input of variables such as the mixed layer depth and heat flux, the KNMI codes (Beljaars et al., 1989) have been incorporated within ACDEP to calculate stability parameters. The Monin-Obukhov length, L, is now calculated from roughness length, temperature, wind speed and cloud cover. Over sea, the sea surface temperature is also required and the roughness length is calculated iteratively using the Charnock formula (Charnock, 1955). Over land an effective roughness length was calculated for each grid square based on the fraction of different land use categories (Figure 1).

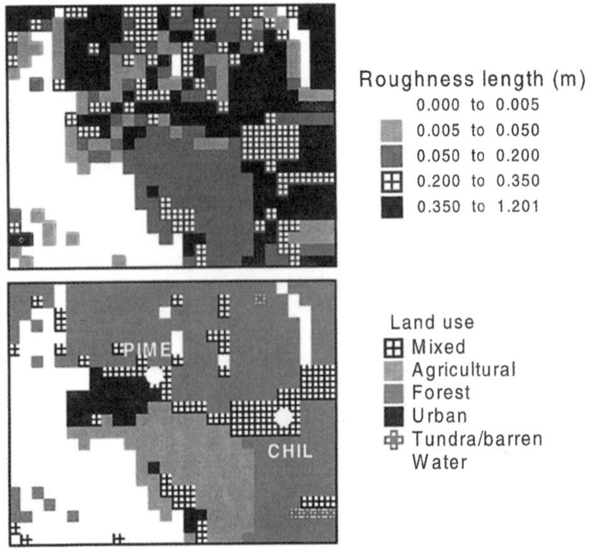

Figure 1. Roughness length and land-use (category > 50% of grid cell) in the modeling domain. The locations of the receptor sites (PIME and CHIL) are shown as the white dots. The grid indicates 5 km in the horizontal.

Mixed layer depth

Three new parameterizations for mixed layer depth have been introduced for convective, stable and near-neutral conditions. For convective boundary layers the parameterization (Table 1) incorporated is from (Batchvarova and Gryning, 1991) where the growth of the mixed layer (h) with time (t) is dh/dt. Since the kinematic heat flux ($\theta'\omega'$) is calculated within the KNMI routines, only the potential temperature gradient above the mixed layer, γ is specified based on observations collected using a radiosonde released in the center of the valley (approx. 8 K/km). The subsidence velocity is disregarded as in (Batchvarova and Gryning, 1991) giving equation 1 in Table 1. The depth of the stable or nocturnal boundary layer (NBL) can be defined either in terms of the depth of turbulence or the inversion height which is usually higher (Garratt, 1992). In the Fraser Valley the NBL has been shown to be strongly influenced by topographically induced flow (Banta et al., 1997). As an approximation, equation 2 is used to define the depth of the equilibrium NBL (Garratt, 1992). A formulation for a near-neutral boundary layer is required for high wind speeds or in the transition between stable/unstable boundary layers. Equation 3 (see (Garratt, 1992)) is adopted if: L becomes very large indicating that conditions are close to neutral, or the calculation of L or friction velocity (u_*) fails to converge, or L is negative with a negative sensible heat flux .

Figure 2 shows the calculated mixed layer depth in meters at 1200, 1800, 2400 PST on August 5/6 1993 and the method selected by ACDEP to determine mixed layer depth (criteria shown in Table 1). Note that the meteorological parameters within each grid square dictate the method used and hence the value of the mixed layer depth can vary from hour to hour and from grid square to grid square. Although this can introduce discontinuities, evaluation of the current scheme suggests that the methodology produces values which are reasonable and vary according to surface type. Overnight the stable or nocturnal boundary layer method is selected over land and during the day the convective scheme is used. As shown, over sea the contrast between varying air temperature and nearly constant sea temperatures produces stable boundary layers during the day and unstable boundary layers overnight. This is in accord with modeling and observations reported in Barthelmie et al. (1996).

Table 1. Mixed layer depth equations and selection criteria

Method	Monin-Obukhov length (m)	Sensible heat flux	Equation
1. Convective	$L < 0$, $L > -1500$	Positive	$\left\{\frac{h^2}{(1+2A)h-2BkL}+\frac{Cu_*^2 T}{\gamma\, g[(1+A)h-BkL]}\right\}\left(\frac{dh}{dt}\right)=\frac{\left(\overline{w'\theta'}\right)_s}{\gamma}$ A, B and C = 0.2, 2.5 and 8. k = von karman constant (0.4). g/T = buoyancy parameter.
2. Stable /nocturnal	$L > 0$, $L < 1500$	Any	$h=\alpha\left(\frac{u_* L}{f}\right)^{1/2}$ $\alpha = 0.4$. f = Coriolis parameter.
3. Near-neutral	$L > \pm 1500$	Any	$h = 0.3\, u_*/f$

MODEL RESULTS

The base case model simulations are briefly summarized in Table 2. As has been previously described the modeled aerosol concentrations are in reasonable accord with the measurements with the exception that sulphate concentrations are consistently under-predicted (Pryor et al., 1996). It has been suggested that this is due to under-estimation of sulphur dioxide oxidation (either in the gas phase or within cloud). Figure 3 shows the results of simulations where cloud cover varies from 0-100% (also shown is the base case simulation performed using observed cloud cover). Whilst these simulations show individual hours during which sulphate concentrations are sensitive to cloud amount (via modification of the actinic flux or in-cloud oxidation), generally the results indicate that sulphate concentrations are comparatively insensitive to cloud cover during this particular episode. Although this may be partially due to the simplified parameterization of cloud processes, other contributory factors

Table 2. a. 24 hour average aerosol concentrations (μgm^{-3}) and maximum O_3 (ppb) at PIME.

	Modeled results				Measured values			
	NH_4^+	NO_3^-	SO_4^{2-}	Max. O_3	NH_4^+	NO_3^-	SO_4^{2-}	Max. O_3
Aug 5-6	1.0	3.2	2.6	102	0.9	1.1	3.0	69
Aug 6-7	1.2	1.1	3.0	47	1.5	1.5	4.1	46
Aug 7-8	1.8	2.0	4.1	54	1.7	1.2	5.0	43

Table 2. b. 24 hour average aerosol concentrations (μgm^{-3}) and maximum O_3 (ppb) at CHIL.

	Modeled results			Measured values		
	NH_4^+	NO_3^-	SO_4^{2-}	NH_4^+	NO_3^-	SO_4^{2-}
Aug 5-6	1.0	3.8	1.6	1.2	2.1	2.5
Aug 6-7	2.1	5.4	2.4	3.9	7.5	6.2
Aug 7-8	1.3	5.4	1.5	3.2	5.4	5.9

Measured O_3 values are not available for the CHIL site.

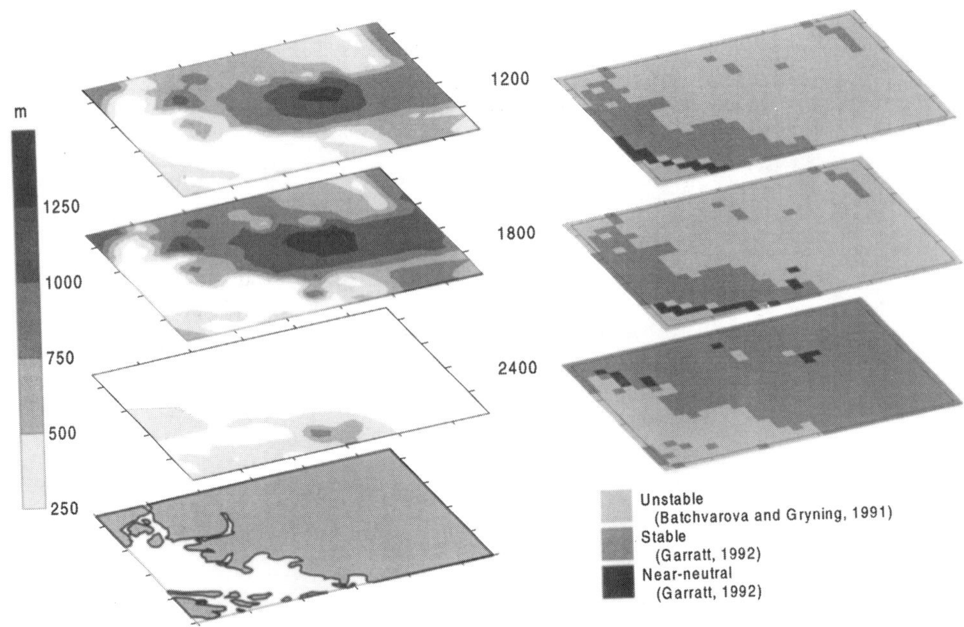

Figure 2. Mixed layer depth and the method used to calculate MLD for each grid square in the domain for 1200, 1800, 2400 PST on August 5/6 1993.

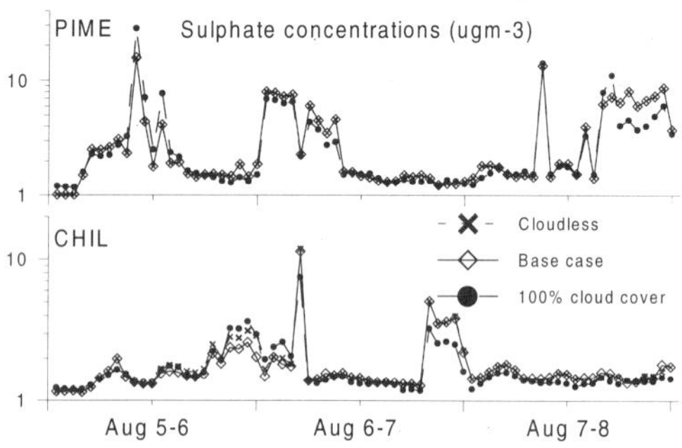

Figure 3. The results of sensitivity analyses performed to examine the impact of cloud cover on hourly modeled sulphate concentrations (shown in μgm^{-3}).

include the uniformly low modeled sulphur dioxide concentrations resulting from either the omission of biogenic sulphur species from the emission inventory or the influence of sulphur dioxide/sulphate sources outside the modeling domain. Work is underway to modify the cloud parameterization and also to extend the chemical mechanism to include dimethyl-sulphide (DMS) emissions and chemistry and apply ACDEP in a nested-grid.

MODEL SENSITIVITY TO PRECURSOR EMISSIONS

Figure 4 shows model simulations to examine the sensitivity of secondary inorganic aerosols and ozone concentrations to changes in precursor emission amounts. The figures indicate change in the modeled species (sulphate SO_4^{2-}, nitrate NO_3^-, ozone O_3) as a percent increase or decrease from base case during each simulation hour. The following emission scenarios are shown; -50%, -25%, as inventoried, +25%, +50% of sulphur dioxide (SO_2),

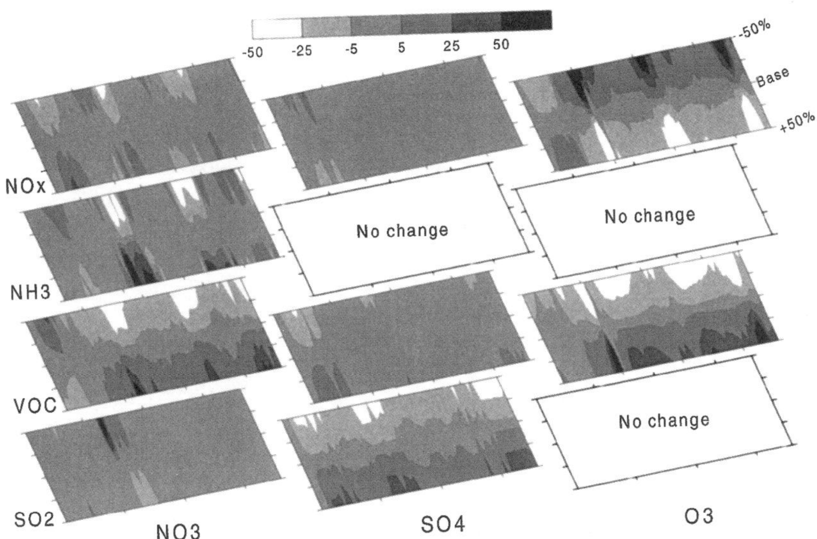

Figure 4a. The results of sensitivity analyses for PIME showing aerosol and ozone response to changing precursor emissions. Each frame shows time series for modeled species (identified at the bottom of the page) for changing emissions of the precursor species (identified on the left). From top to bottom in each frame the emission changes are -50%, -25%, +25%, +50%. Concentration changes are shown as percent change from base case. No change indicates all hours <5% change relative to base case.

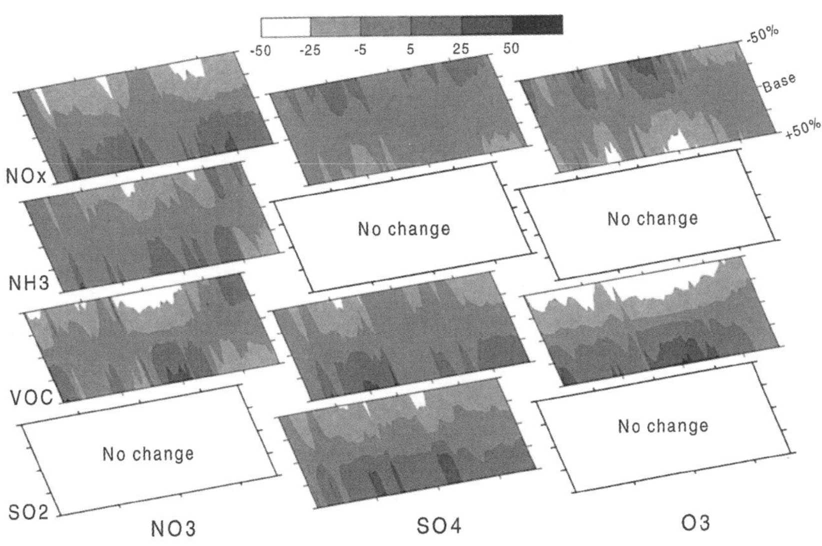

Figure 4b. The results of sensitivity analyses for CHIL showing aerosol and ozone response to changing precursor emissions. Each frame shows time series for modeled species (identified at the bottom of the page) for changing emissions of the precursor species (identified on the left). From top to bottom in each frame the emission changes are -50%, -25%, base case, +25%, +50%. Concentration changes are shown as percent change from base case. No change indicates all hours <5% change relative to base case.

volatile organic compounds (VOC), ammonia (NH_3) and oxides of nitrogen (NO_x). The results indicate that at PIME modeled nitrate concentrations are increased by increasing VOC emissions. This is indicative of a system which is oxidant sensitive. Nitrate concentrations are insensitive to SO_2 emissions, are enhanced during daylight hours by increased NH_3 availability, and are only weakly affected by changing NO_x emissions. Modeled sulphate concentrations are most sensitive to SO_2 availability. Increasing VOC emissions increases modeled O_3 but the reverse is true for increased NO_x. This is in accord with findings in Pottier (1997, this volume) that O_3 concentrations are VOC sensitive in the west of the valley and hypotheses that scavenging by NO plays a dominant role in determining O_3 concentrations near to the Vancouver urban core (Pryor and Steyn, 1995). The results from CHIL indicate nitrate concentrations in the east of the valley are less VOC sensitive and more NO_x sensitive, and respond more consistently to changes in NH_3. This is in accord with suggestions that under mesoscale flows, air from the west of the valley is advected into the central valley during which time NO_x is converted into HNO_3 and subsequently reacts with NH_3 to form (NH_4NO_3) (Barthelmie and Pryor, 1997). Sulphate shows high sensitivity to SO_2 availability. Results for O_3 indicate a weaker response to VOC and NO_x emissions.

FUTURE DEVELOPMENTS

Modification of the model chemical code is currently focusing on incorporating secondary organic aerosol (SOA) formation and biogenic sulphur chemistry. Preliminary simulations performed to predict SOA concentrations from monoterpenes indicate that during the episode described above biogenic sources contribute upto 1/3 of aerosol organic carbon. Future modifications will include replacing the sectional approach to particle dynamics with parameterizations using the modal method (based on Binkowski and Shankar (1996)).

ACKNOWLEDGMENTS

The research presented herein was funded by Environment Canada. The authors wish to express their gratitude to Frank Binkowski, Bruce Thomson, Ekaterina Batchvarova, Sven-Erik Gryning, Ole Hertel, and Mark Hedley for their assistance and support.

REFERENCES

Banta, R. et al., 1997. Nocturnal cleansing flows in a tributary valley. *Atmos. Environ.*, 31: 2147-2162.

Barthelmie, R.J., Grisogono, B. and Pryor, S.C., 1996. Observations and simulations of diurnal cycles of near-surface wind speeds over land and sea. *J. Geophys. Res.*, 101(D16): 21,327-21,337.

Barthelmie, R.J. and Pryor, S.C., 1997. Ammonia emissions: implications for fine aerosol formation and visibility degradation. *Atmos. Environ.*, Conditionally accepted.

Batchvarova, E. and Gryning, S.E., 1991. Applied model for the growth of the daytime mixed layer. *Boundary Layer Meteorology*, 56(3): 261-274.

Beljaars, A.C.M., Holtslag, A.A.M. and van Westrhenen, R.M., 1989. *Description of a software library for the calculation of surface fluxes. Technical report TR-112, KNMI, De Bilt, Netherlands.*

Binkowski, F.S. and Shankar, U., 1996. The regional particulate matter model: Part I, Model description and preliminary results. *J. Geophys. Res.*, 100(12): 26191-26205.

Charnock, H., 1955. Wind stress on a water surface. *Quart. J. Royal Meteor. Soc.*, 81: 639.

Garratt, J.R., 1992. *The atmospheric boundary layer. Cambridge atmospheric and space science series. Cambridge University Press, Cambridge, 316 pp.*

Hertel, O. et al., 1995. Development and testing of a new variable scale air pollution model - ACDEP. *Atmos. Environ.*, 29: 1267-1290.

Pryor, S.C., Barthelmie, R.J. and Hertel, O., 1996. Modelling fine aerosol concentrations in the Lower Fraser Valley, British Columbia., *12th AMS Biometeorology and Aerobiology Conference. AMS, Atlanta.*

Pryor, S.C., Simpson, R., Guise-Bagley, L., Hoff, R. and Sakiyama, S., 1997. Visibility and aerosol composition in the Fraser Valley during REVEAL. *J. Air Waste Manage. Assoc.*, 41(2): 147-156.

Pryor, S.C. and Steyn, D.G., 1995. Hebdomadal and diurnal cycles in ozone time series from the Lower Fraser Valley, B.C. *Atmos. Environ*, 29: 1007-1019.

MESOSCALE FLOW AND POLLUTANTS TRANSPORT MODELLING IN NORTH-EAST SICILY

P. Grossi, G. Graziani, C. Cerutti

Environment Institute, JRC Ispra, I-21020

INTRODUCTION

The area around Milazzo, a city located on a narrow peninsula, north-east Sicily, is heavily polluted by the presence of power plants and refinery. In this work, the atmospheric pollution levels resulting from two circulation regimes are simulated. First, westerly synoptic winds are considered, as the most likely meteorological conditions (Graziani et al., 1997). Then, a less frequent but most dangerous situation with southerly winds is examined. In this meteorological condition, weak winds can develop that may favour the accumulation of pollutants in the most populated area of the city of Milazzo. The RAMS model (version 3b) is used to calculate the flow, in its non-hydrostatic configuration, due to the highly complex terrain. RAMS is initialised and nudged with data from ECMWF model and a multiple nested grid configuration is selected to solve both local flow and large scale circulation over the Mediterranean. The time-dependent RAMS output is used to drive the Lagrangian particle model MONTECARLO, written in terrain following co-ordinates. In the model, buoyancy of hot emissions in convective conditions is taken into account. Non-reactive SO_2 sources are only considered. Both meteorological and concentration model outputs are compared with observations performed during the 1996 Spring campaign (Cerutti et al., 1996).

THE METEOROLOGICAL MODEL

The meteorological model used is non-hydrostatic-anelastic configuration of RAMS (Pielke et al., 1992), a primitive equation atmospheric mesoscale model in terrain influenced co-ordinates. Model is applied with a first order turbulence closure (Smagorinsky). Three nested grids are used to capture the main features of the large scale flows over a part of the Mediterranean Sea, as well as the local circulation driven by the presence of the Sicilian mountains and of the Italian peninsula. The grid configuration for this study is shown in Fig 1a. Horizontal spacing decreases from 16 to 1 km, passing from

Air Pollution Modeling and Its Application XII
Edited by Sven-Erik Gryning and Nadine Chaumerliac, Plenum Press, New York, 1998

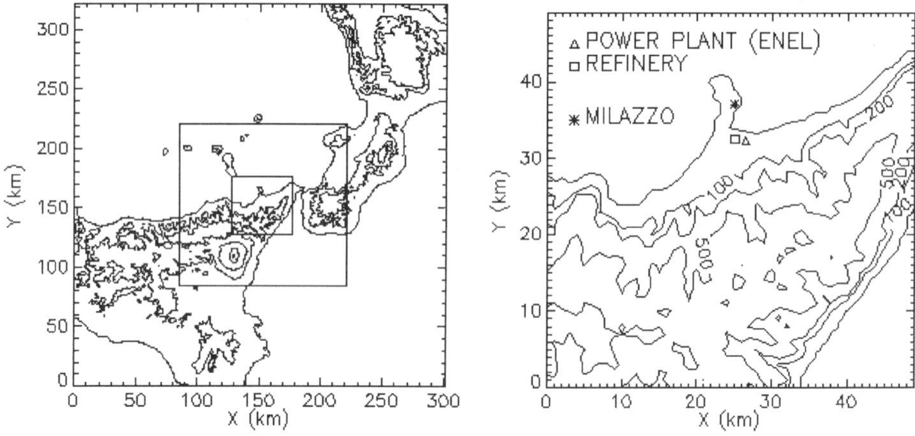

Figure 1a. The region considered and the grids used

Figure 1b. Milazzo and the main SO_2 sources of the area

the outermost to the innermost grid. The latter is centred on the Milazzo city, where SO_2 sources are located. Vertical nesting is employed for the internal grid to solve in greater detail the circulation near the ground: the first level is fixed at 100 m (horizontal wind, pressure, temperature and humidity are then calculated at 50 m over the ground). The top of the three grids is set at 15,000 m. Detailed information on vegetation and land use are considered in the soil sub-model, which then allow to correctly calculate the sensible and latent heat fluxes between atmosphere and ground. The sea temperature is set at 17 °C, referring to some measurements taken during the campaign. RAMS is initialised at 0000 UTC (corresponding to 0200 LST) and nudged every six hours with the 4DDA technique, using output of synoptic model from ECMWF. Each calculation lasts 24 hours.

Results of the calculation are compared with observations taken at S. Filippo del Mela (S.FdM) power plant, on the coast, and Pace del Mela (PdM), more inland, at an elevation of 100 m a.s.l., roughly corresponding to the power plant stack height (Fig. 2a-b).

Meteorological modelling results: westerly synoptic winds (May 1, 1996)

Westerly synoptic winds are one of the most likely meteorological conditions in Sicily during the whole year and also in the period under study, that is April-May 1996. During the night, the RAMS model generates drainage flows along the mountains slopes, both in the inner grid (Mt. Peloritani) and in the intermediate grid, where the Etna volcano is present. In the region where the plants are located, synoptic forcing merges nocturnal winds, resulting in a low level circulation directed towards N-N-E. The transition between nocturnal and diurnal circulation occurs at 1000 LST, when the flow over the hills changes direction and some up-valley breezes start to develop. At 1400 LST, one can observe the valley breeze climbing the northern slopes of Peloritani, whilst almost calm conditions are calculated at the south-eastern corner of the domain. In fact, around the north-eastern corner of the Sicily island, the proximity of Italian peninsula and the presence of the Messina channel and of Etna volcano considerably influence the flow. Due to the small temperature gradient between land and sea, no sea-breeze could be observed in the calculations at that time of the year (Fig. 2a). The potential temperature profile shows the almost neutral conditions over the sea and inland convective conditions. Comparison with experimental data is shown in Fig. 3a-b and 4a-b. They indicate the good agreement between

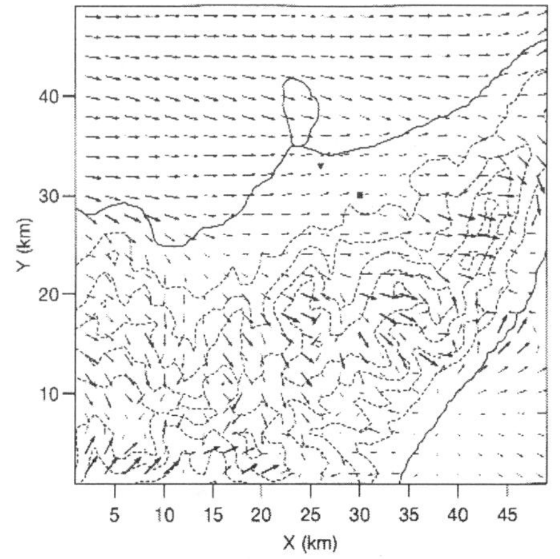

Figure 2a. May 1, 1400 LST; wind field at the 1st level. (S.FdM is the point on the coast, PdM the one more inland)

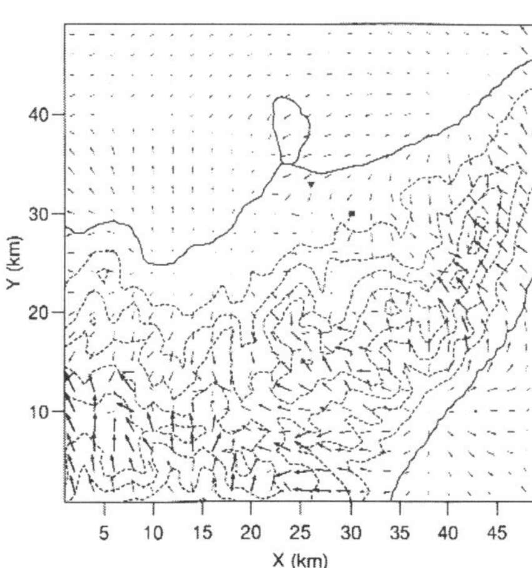

Figure 2b. May 4, 1400 LST; wind field at the 1st level (S.FdM is the point on the coast, PdM the one more inland)

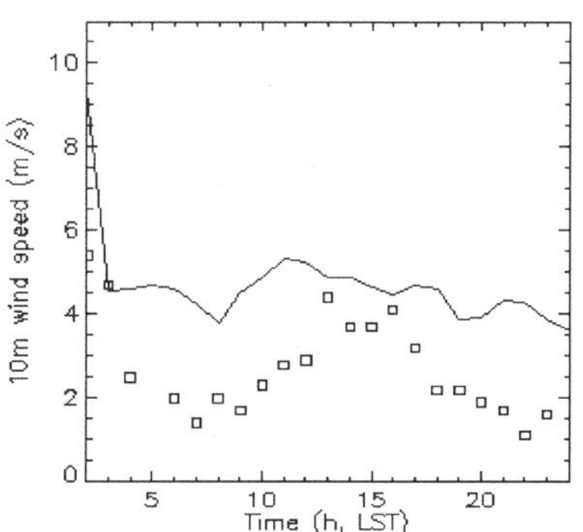

Figure 3a. Modelled (line) and observed (dot) wind intensity at PdM on May 1

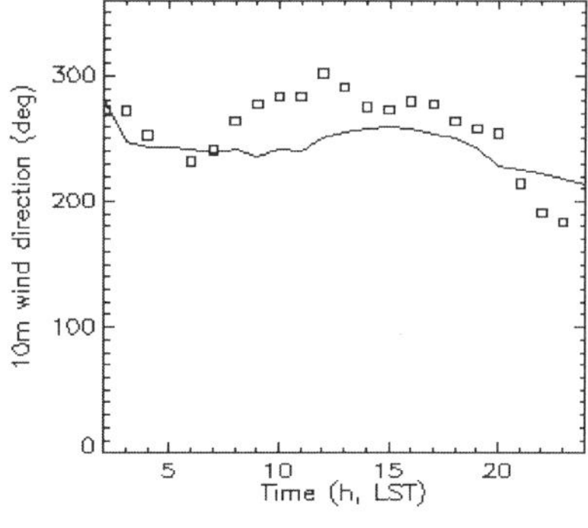

Figure 3b. Modelled (line) and observed (dot) wind direction at PdM on May 1

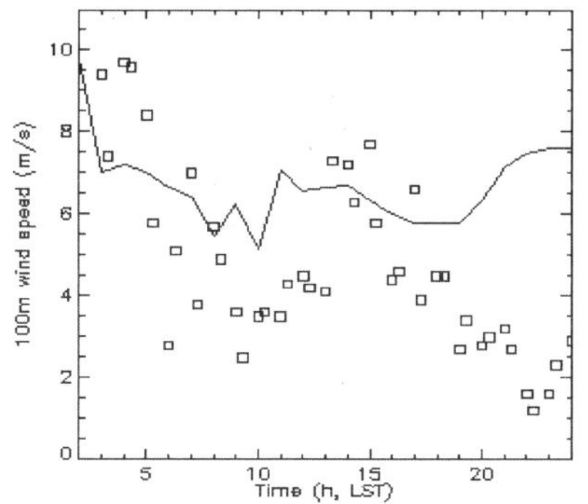

Figure 4a. Modelled (line) and observed (dot) wind intensity at S.FdM on May 1

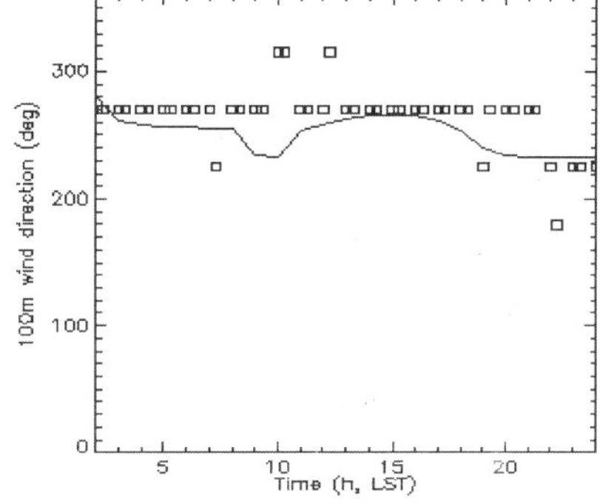

Figure 4b. Modelled (line) and observed (dot) wind direction at S.FdM on May 1

observations and calculations, particularly during the daily hours. The rather strong wind observed at power plant stack (S.FdM) is well reproduced, both in intensity (up to 8 m/s) and direction (270 degrees), as well as the air temperature daily evolution. At Pace del Mela, a stronger wind is calculated than observed: this is probably due to the fact that measurements are taken at 10 m, while the first atmospheric model level is at 50 m. The difference in direction from measurements to model results (about 20 degrees, from 270 to 250 degrees) can be attributed to the prevailing synoptic forcing on the local thermal flow in the model.

Meteorological modelling results: southerly synoptic winds (May 4, 1996)

Southerly synoptic winds are a less frequent meteorological scenario for the region. This condition is however one of the most dangerous for pollutants dispersion. During the night, drainage flows are again calculated along the slopes of Etna volcano and of Mt. Peloritani. They tend to reinforce the synoptic winds at the sea-border line and off-shore. As a result, one finds that the plant emissions do not pass over the inhabited area. The transition between nocturnal and diurnal circulation still occurs at 1000 LST. A weak sea breeze develops nearby Milazzo peninsula, directed towards SW. Also in this case, the thermal forcing produces up-slopes and up valleys winds which develop in the mountainous inland region. At 1400 LST they are well established (Fig. 2b). On the southern slope of the Peloritani, synoptic winds are reinforced by the slope breeze. On their northern slope, the two tend to cancel out (to the West), or the breeze is even prevailing at ground (to the East). This latter effect is due to the turning of the wind caused by the presence of the Italian peninsula. On the south-eastern edge of the domain, the weak wind is due to the proximity of Etna volcano that acts as an obstacle to the synoptic winds. The inland turning to the South only of the ground wind is evidenced by the vertical cross-section. The potential temperature values are higher than before of about 3 °C, as confirmed also by observations. The comparison of model results with observations is shown in Fig. 5a-b and 6a-b. The wind velocity is somewhat under-estimated at 100 m height and the well developed north-easterly sea-breeze, observed during day-hours, is not reproduced. At Pace del Mela, the wind strength is well reproduced, but the direction is still not correctly calculated. Model results indicate a valley breeze starting at 1200 LST and lasting until 1800 LST; measurements also show a wind rotation of about 180 degrees, but starting 2 hours later and lasting until 1500 LST. This may be due to some local effects that model is not able to capture.

THE DISPERSION MODEL

The Lagrangian particle model MONTECARLO uses the mean component of the wind evaluated by RAMS and semi-random velocity following the Langevin stochastic differential equation. Mean flow data are supplied at 30' interval. In convective conditions, fluctuations of vertical velocity are calculated using a generalisation of the scheme proposed by Hurley and Physics (1993), to take into account for different probability of up-drafts and down-drafts. The rising of buoyant emissions is evaluated using a generalisation of the Briggs scheme derived by Anfossi (1985). Turbulent parameters, as friction velocity and Monin-Obukhov length, are also derived by RAMS output and used to calculate wind fluctuations and lagrangian time-scales following the Hanna schemes for homogeneous terrain (1982). The height of mixing layer is determined by the Richardson number Ri, where this increases beyond the critical value Ri_c=1.3. The SO_2 emissions from the power plants (ENEL) and a refinery group are only considered. The positions of the sources in the

Table 1. Power and refinery plants stacks and emission characteristics

	ENEL power plant			Milazzo refinery			
	stack 1	stack 2	stack 3	Vacuum	Cracking		CTE
Stack height (m)	100	100	125	57.5	54.35	47.35	106
Stack inner diameter (m)	5.2	5.2	7.1	3.5	3.12	2.15	4.08
SO_2 flow (g/s) *	152	152	304	4.7	17	1.3	6.
Emission temp.(C) *	140	140	70	380	275	470	155.

* time dependent

map are indicated in Fig. 1b, as the two points along the shoreline. The characteristics of the emitting stacks are summarised in Table 1. Simulation of MONTECARLO starts at 7:00 LST and lasts for 10 hours.

Dispersion modelling results: westerly synoptic winds (May 1, 1996)

During the transition from night to day, the pollutant is transported in West-East direction parallel to the coast. At 1000 LST, when valley breezes start to develop, plume turns a little to South, but a cloud of pollutant can still be observed along the shoreline. The plume centre is located among 50 and 250m agl. At 1200 LST, particles start to climb over Mt. Peloritani and then to be dispersed to higher levels by the flow convergence at the mountain top. Vertically integrated particle positions at 1400 LST as resulting from MONTECARLO simulation are shown in Fig. 7a. It is possible to notice the little horizontal dispersion. One might therefore expect that a small error in wind direction can influence considerably the position of the peak. Ground concentration peaks around 300 $\mu g/m^3$ are observed somewhat inland during different displacements of the mobile measuring unit both in the late morning and in the afternoon. This happens after the development of some valley breeze that bends the wind a little to the south. Results of calculations show about the same peak on the shoreline, as shown by Fig. 8a. Comparison with the observed values for two of the routes taken by the mobile unit is presented in Fig. 9a and 9b (10 $\mu g/m^3$ is threshold of measurements). Along the first route (route C), parallel to cloud displacement and approaching the sources, peak is well reproduced, in position and intensity. Along the other route (route F), orthogonal to the plume, a larger cross wind dispersion is calculated.

Dispersion modelling results: southerly synoptic winds (May 4, 1996)

Results of the dispersion model indicate that during the transition from night to day, the pollutant is transported to the north. The majority of it is dispersed to the sea and only part goes into the small Milazzo peninsula. During the day, the upper fraction of the plume is still transported toward north by synoptic winds; the lower part of the cloud is displaced by the sea breeze at ground and turns to S or S-E in the direction of Pace del Mela and other villages. This situation tends to last for the entire afternoon. Vertically integrated particle positions at 1400 LST as resulting from MONTECARLO simulation are shown in Fig. 7b. Measured ground SO_2 concentrations are generally higher than in the previous case. Peak values of about 600 $\mu g/m^3$ are found only later in the day, both westerly and south-easterly of the source. Results of calculations are somewhat lower (Fig. 8b). An example of the comparison between observed and calculated concentration is shown in Fig. 10a-b for two routes. The first is orthogonal to the coast, leaving the site and going to SW. In this case, calculations produce a larger and less pronounced peak, not so displaced to south as the measured. The second route is also orthogonal to the coast, more towards East, and

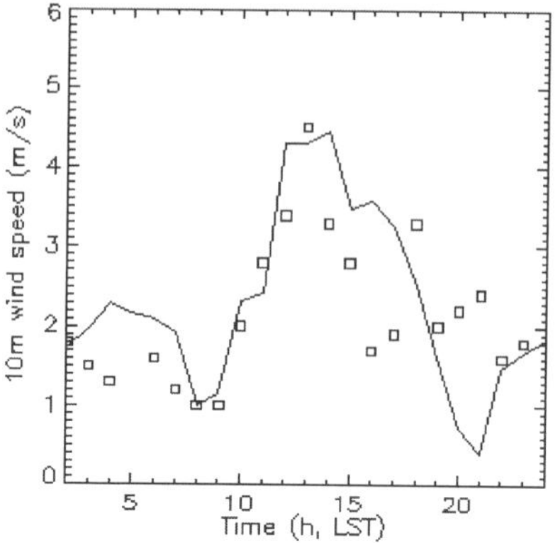

Figure 5a. Modelled (line) and observed (dot) wind intensity at PdM on May 4

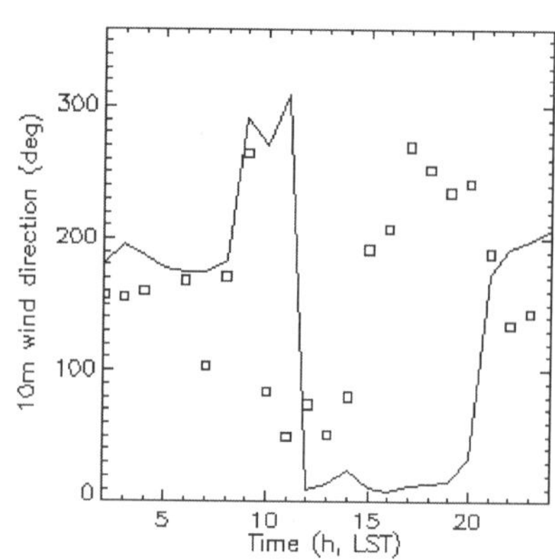

Figure 5b. Modelled (line) and observed (dot) wind direction at PdM on May 4

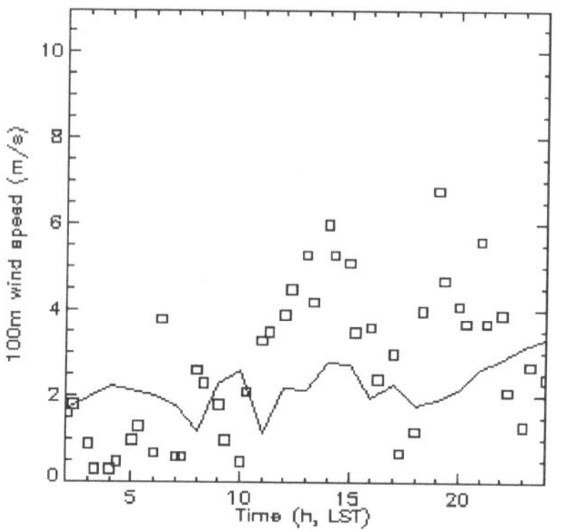

Figure 6a. Modelled (line) and observed (dot) wind intensity at S.FdM on May 4

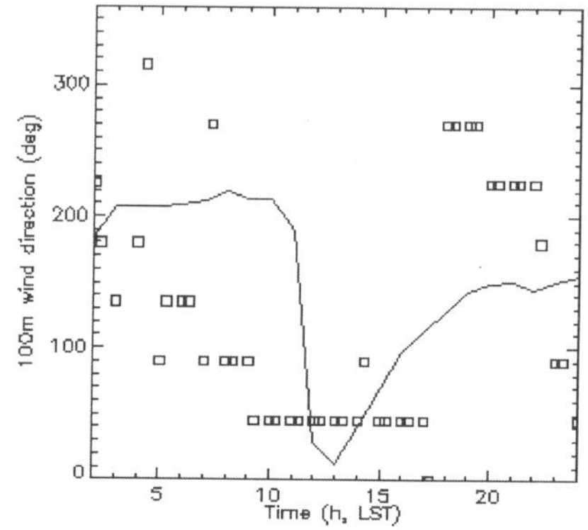

Figure 6b. Modelled (line) and observed (dot) wind direction at S.FdM on May 4

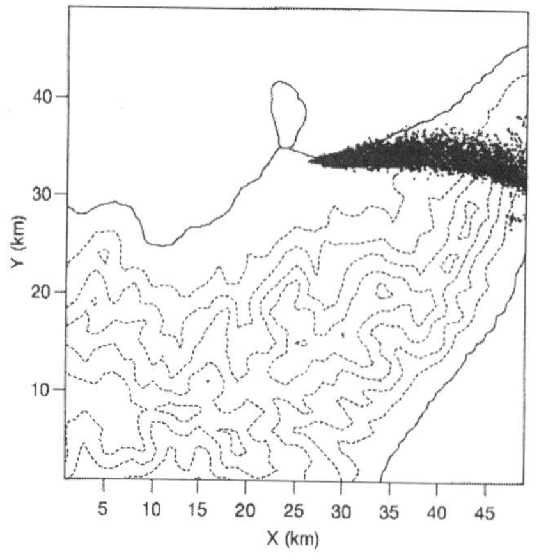

Figure 7a. Positions of particles on May 1, 1400 LST

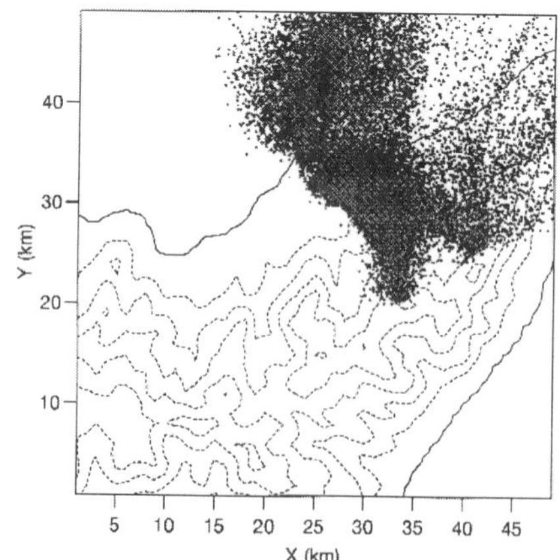

Figure 7b. Positions of particles on May 4, 1400 LST

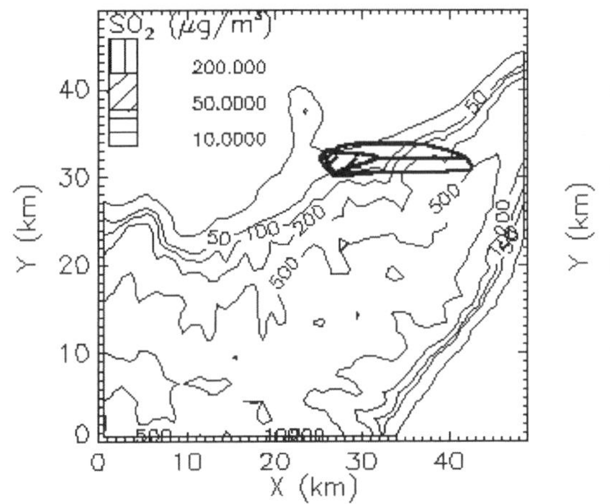

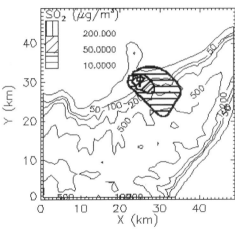

Figure 8a. Ground concentration, May 1, 1400 LST

Figure 8b. Ground concentration, May 4, 1400 LST

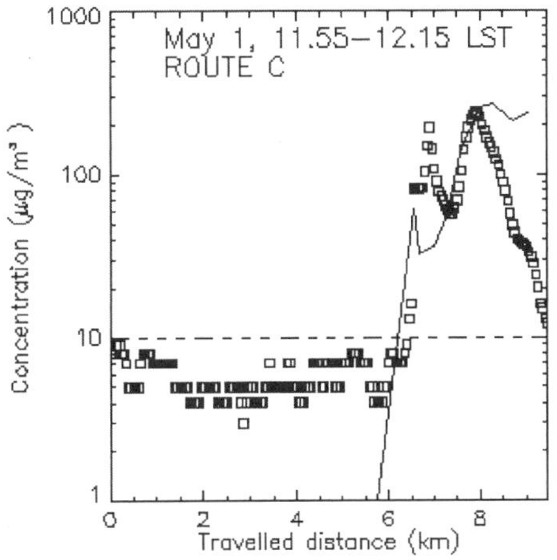

Figure 9a. Model (line) and observations (dots).

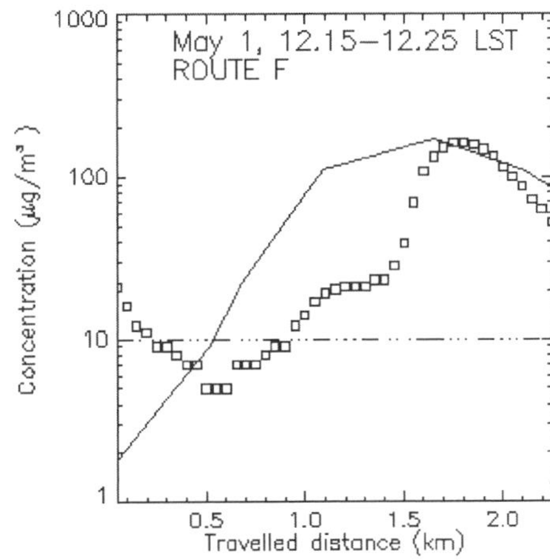

Figure 9b. Model (line) and observations (dots).

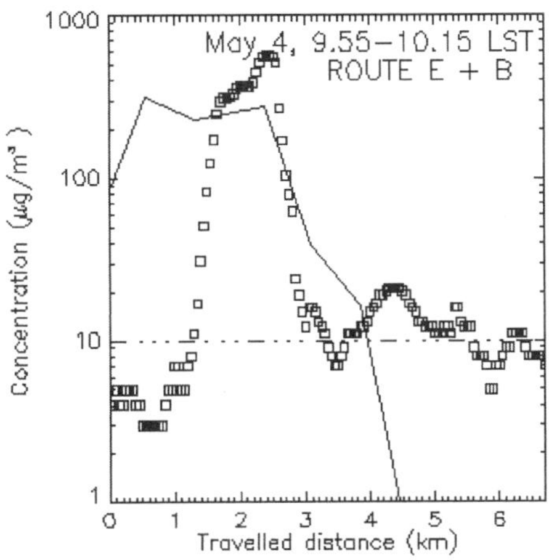

Figure 10a. Model (line) and observations (dots)

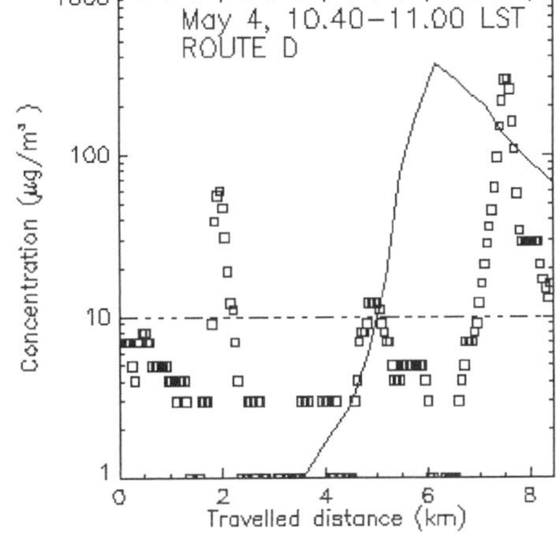

Figure 10b. Model (line) and observations (dots).

approaching the site from inland. Only the largest concentration peak, close to the sources, is reproduced by calculations; in fact, the other one is due to spurious sources in the zone.

CONCLUSIONS

The results of this study demonstrated that meteorological and dispersion models allow to describe and better understand pollutant's transport in the area of Milazzo. In fact, the main features of the dispersion in two completely different meteorological situations are satisfactorily reproduced. However, small differences in calculated wind direction, due to the rather complex terrain, could result in large differences in the population dose. This is what is found for circulation of May 1. During May 4, particles are dispersed both NW and SE to the release point. However, in the former direction the majority of particles are remaining above the first level. Only on SE they are brought to ground, following the breeze development, which is then correctly described in the simulation. Elevated concentration values are therefore lasting only for the duration of the breeze and should then still be considered within the regulation limit of 250 $\mu g/m^3$ (Official J. of E.C., L 229/34 of 30.8.80), when averaged over 24 hours. The good performances of the two models are obtained when the 4-dimensional data assimilation technique is used to determine the flow in the region and when boundary layer depth is evaluated by means of the critical Richardson number. It should be noted that these results refer to two particular meteorological conditions, which were the most frequent during the period of the campaign (April-May 1996). Other dangerous situations may occur which are not considered in the present report. However, the established procedure coupled with the information on local climatology can be eventually used in other scenarios to evaluate population hazard map for the area.

ACKNOWLEDGMENTS

This work was partly supported in the frame of POP Sicily, contract 101.22.94.03 TIPC.JSP.I, commissioned by Sicily Region. Thanks are due to Ing. A. Cappadonia and Dr. C. Musumeci (Province of Messina) for information supplied.

REFERENCES

Anfossi D. (1985) : Analysis of plume rise data from five TVA steam plants, *Journal of Climate and Applied Meteorology,* **24**, 1225-1236.

Cerutti C., Sandroni S., Payrissat M. and Sieja B. (1996) : Indagine sulle ricadute al suolo di inquinanti atmosferici nella zona industriale di Milazzo (Sicilia), EUR 17267IT.

Graziani G., Pareschi M.T., Ranci M.: Synoptic and local circulation over Sicily with a zoom on Etna area, submitted to "Il Nuovo Cimento".

Hanna S. R. (1982) in *Atmospheric Turbulence and Air Pollution Meteorology*, eds. Nieuwstadt F.M.T. and Van Dop H. Reidel, Dordrecht, The Netherlands.

Hurley P. and Physics W. (1993) : A skewed homogeneous Lagrangian particle model for convective conditions, *Atmospheric Environment*, **27(A)**, 619-624.

Pielke W. R., W. R. Cotton, R. L. Walko, C. J. Tremback, W. A. Lyons, L. D. Grasso, M. E. Nicholls, M. D. Moran, D. A. Wesley, T. J. Lee and J. H. Coperland (1992) : A Comprehensive Meteorological Modelling System - RAMS, *Meteorol. Atmos. Phys.*, **49**, 69-91.

DISCUSSION

R. ROMANOWICH:

1. What were the measurement errors ?
2. Is it planned to use geostatistical interpolation methods to infer the spatial distribution of concentrations from observations (to compare with model predictions)?

G. GRAZIANI:

1. The measurement errors were in the order of background, i.e. 10 mg/m^3.
2. Not for the moment, since the measurements are sparse in time and space.

D. STEVENSON:

Do the SO_2 emissions from Mount Etna influence the observation and model results?

G. GRAZIANI:

Mount Etna influence only the circulation in the region; its SO_2 emissions are at the main cone, at an altitude of 3.000 m asl and do not influence the SO_2 concentration levels at sea level. The only volcanic emissions from the slopes are of CO_2 (5% of he world total natural CO_2 emissions).

P. SEIBERT:

Considering the aims of the project and the needs of the authorities financing it, how do you rate the relative contributions of measurements and modelling to reach these aims?

G. GRAZIANI:

Costing of the project can be quantified in 75% cost of the campaign and 25% of modelling.

MESOSCALE MODELING OF TRANSPORT AND DEPOSITION OF HEAVY METALS IN SOUTHERN POLAND

Marek Uliasz[1], Krzysztof Olendrzyński[2] and Jerzy Bartnicki[2]

[1]Mission Research Corporation, *ASTER Division
P.O.Box 466, Fort Collins, CO 80526-0466, USA

[2]Norwegian Meteorological Institute (DNMI)
P.O. Box 43 Blindern, N-0313 Oslo, NORWAY

INTRODUCTION

The Katowice province in southern Poland experiences serious air pollution problems including deposition of heavy metals which is among the highest in Europe (Bartnicki et al., 1996; Olendrzyński et al., 1996). Heavy metals deposit onto surfaces at relatively low rates, however, due to their toxicity and accumulation in soils, long-term deposition needs to be evaluated. Long-range transport models applied to the whole of Europe cannot simulate high values of local deposition fluxes because of their low spatial resolution. A typical grid cell of the long-range models can cover the whole Katowice province. Therefore, it is necessary to apply a high resolution transport and deposition model linked to a three-dimensional mesoscale/regional meteorological model. Recent advances in computer technology, especially availability of modern workstations, allows one to integrate 3-dimensional mesoscale meteorological models over extended time periods, months or even years, to provide necessary input fields for long term dispersion modeling (Uliasz et al., 1996; Pielke and Uliasz, 1997). This paper presents selected results from the METKAT (heavy METals in the KATowice province) project launched by the International Institute for Applied Systems Analysis in cooperation with two Polish research institutions, Institute for Ecology of Industrial Areas (IEIA), Katowice, and Institute of Meteorology and Water Management (IMWM) in Warsaw. The goal of the project was to investigate high local deposition fluxes of arsenic (As), cadmium (Cd), lead (Pb) and zinc (Zn) in the Katowice province with the aid of mesoscale modeling. The extended description of the METKAT study is provided by Uliasz and Olendrzyński (1996).

NUMERICAL MODELS

For the purpose of the METKAT project, a new dispersion model called the Lagrangian Particle Dispersion and Deposition (LPDD) model was derived from a family of Lagrangian Particle Dispersion models being a part of a Mesoscale Dispersion Modeling System (MDMS) (Uliasz, 1994; Uliasz et al., 1996). A simplified advection algorithm for particles based on a fully random walk scheme instead of a Markov chain scheme adopted in the LPDD model allows one to perform long-term particle simulations even for multiple emission sources. The LPDD model was run over the deposition modeling domain which corresponds to one 150×150 km cell of the EMEP grid covering the Katowice province (Fig.1). A grid spacing, $\Delta x = 5$ km, for deposition calculations was chosen in relation to a spatial resolution of available emission and terrain data. The LPDD model is driven by meteorological fields simulated by

Air Pollution Modeling and Its Application XII
Edited by Sven-Erik Gryning and Nadine Chaumerliac, Plenum Press, New York, 1998

the MESO model (Uliasz, 1990; Uliasz, 1993). Condensation processes in the atmosphere are not included in this relatively simple, hydrostatic model. However, precipitation estimated from observation is treated as input to its soil-vegetation submodel. The 3-dimensional MESO model can be applied to mesoscale domains with horizontal sizes of a few hundred kilometers using a regular or telescoping grid. Within the time limits of the project, a series of long-term meteorological simulations was performed with the aid of a one-dimensional version of the MESO model. This simplification allowed us to run several sensitivity experiments to investigate the role of emission data accuracy and aggregation, land use representation and interannual variability of meteorological conditions.

Some idealized short-term 3-dimensional meteorological simulations were also performed with the CSU RAMS (Colorado State University Regional Atmospheric Modeling System) (Pielke et al., 1992). This modeling system can be configured with different options for physical parameterizations and with any number of nested grids where the finest grid is usually located in the area of interest and the coarse grid is covering much larger regional domain. This approach allows for interaction between mesoscale phenomena and synoptic scale circulation. The CSU RAMS can be used for long term simulations, however, it is much more computer expensive than the MESO model. The modeling domain used in the meteorological simulations with the CSU RAMS is larger than the deposition domain and covers the mountains south of the Katowice province (Fig.1).

The deposition background of heavy metals from all European emission sources located outside the mesoscale deposition domain was calculated with the aid of the Heavy Metals Eulerian Transport (HMET) model (Bartnicki et al., 1993; Bartnicki, 1994; Bartnicki, 1996). The HMET model simulates transport and deposition of heavy metals on the EMEP grid system with horizontal grid steps of approximately 150 km and of 50 km in its newer version (Bartnicki and Olendrzyński, 1996). Parameterizations of dry and wet deposition developed for the HMET model were adapted with some modifications to the LPDD model.

INPUT DATA

Routine meteorological data from six Polish synoptic stations and radiosonding data from the aerological station located about 200 km northeast of the area were used in the project. The radiosounding data were utilized to initialize each meteorological simulation and then were continuously assimilated into the MESO model with the aid of the nudging

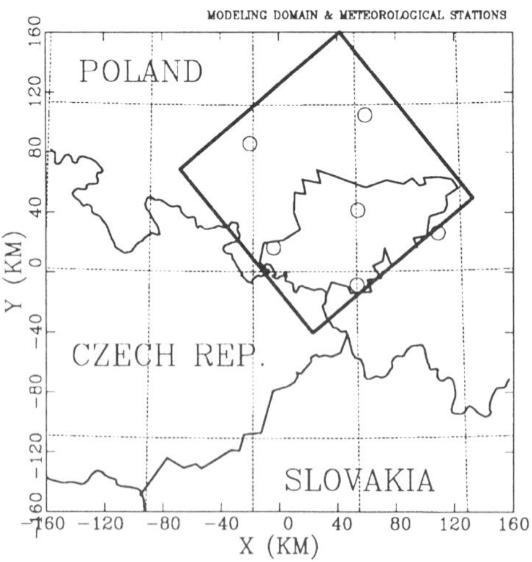

Figure 1. Modeling domain used in 3-dimensional meteorological simulations. The 150×150 km square, (25,19) cell of the EMEP grid, covering the Katowice province in a southern Poland is the deposition modeling domain. Meteorological stations used to derive precipitation and cloud cover fields are marked by circles.

Table 1. Summary of Cd total depositions calculated in sensitivity experiments (D_{min}, D_{mean}, D_{max} - minimum, mean, and maximum total deposition in mg/m^2)

EFFECT	EXPERIMENT	D_{min}	D_{mean}	D_{max}	D_{max}/D_{mean}
	reference simulation	0.009	0.126	0.758	6.0
emission aggregation	150×150 km source	0.020	0.105	0.162	1.5
	50×50 km grid	0.011	0.103	0.350	3.4
	10×10 km grid	0.007	0.103	0.503	4.9
plume rise for major point sources	plume rise	0.002	0.055	0.352	6.4
	no plume rise	0.001	0.059	0.407	6.9
land use	bare soil	0.007	0.113	0.582	5.2
	agriculture land	0.007	0.124	0.695	5.6
interannual variability (Apr-Dec)	1992	0.007	0.099	0.634	6.4
	1993	0.010	0.111	0.651	5.9

technique. Precipitation and cloud cover data from the synoptic stations were interpolated on a 5 km grid covering the deposition modeling domain using optimal interpolation methods. This fields were used as input for the both MESO (soil-vegetation parameterization) and LPDD (wet deposition parameterization) models. The terrain height data were extracted from the GLOBE Project (Global Land One-Kilometer Base Elevation). In the preliminary meteorological and deposition simulations reported here, the land use data from the Baltic Sea Drainage Basin GIS Database were used.

The available emission data for As, Cd, Pb, and Zn included a detailed emission inventory for the Katowice province prepared by the IEIA (point sources and 5×5 km grid data), information on major point emission sources in the region and estimation of emission outside the Katowice province as proportional to the population density. The LPDD model can handle releases from multiple emission sources with arbitrary geometry and time characteristics. There are no numerical or oversmoothing problems which appear in numerical grid transport models in the case of narrow plumes released from point or line sources. For buoyant sources plume rise is calculated and particles are released from the effective stack height. Computer time required by the LPDD model is dependent on the number of particles involved in the calculation, and in turn, on the number of emission sources. Therefore, some aggregation of emission sources is necessary for practical reasons, especially, for the long term simulations. In the reference deposition simulations only the major point sources with emission rate above arbitrary assumed threshold were treated separately and smaller point sources were included into 5×5×0.05 km volume sources. In the next aggregation step the major 5×5×0.05 km sources were selected and the rest of these sources with smaller emission rates contributed to 10×10×0.05 km volume sources. Additionally, some other source emission aggregations were used in the sensitivity experiments.

TRANSPORT AND DEPOSITION SIMULATIONS

The primary goal of transport and deposition simulations was to perform the reference deposition calculations for all four metals in the entire year of 1992 using a detailed emission information (Uliasz and Olendrzyński, 1996). The additional goal was to investigate the role of different factors in uncertainty of the modeling results by running a series of sensitivity experiments for deposition of cadmium with special focus on resolution of emission data (Table 1). A series of deposition simulations were performed with different aggregation level of Cd emission sources (Fig.2). First, the total emission of Cd was distributed uniformly within the 150x150 km modeling domain and in the layer from 0 to 500 m. This setup corresponds to the emission treatment in the HMET-150 model. However, in contrast to the HMET model, the precipitation and surface roughness are variable within the modeling domain. This experiment resulted in the nearly uniform deposition field. Next, the Cd emission was aggregated into a grid of 50x50 km volume sources which is similar to the emission treatment in the new HMET-50 model. The pattern of deposition in this simulation starts to correspond to the location of real emission sources in the region and maximum values are much higher

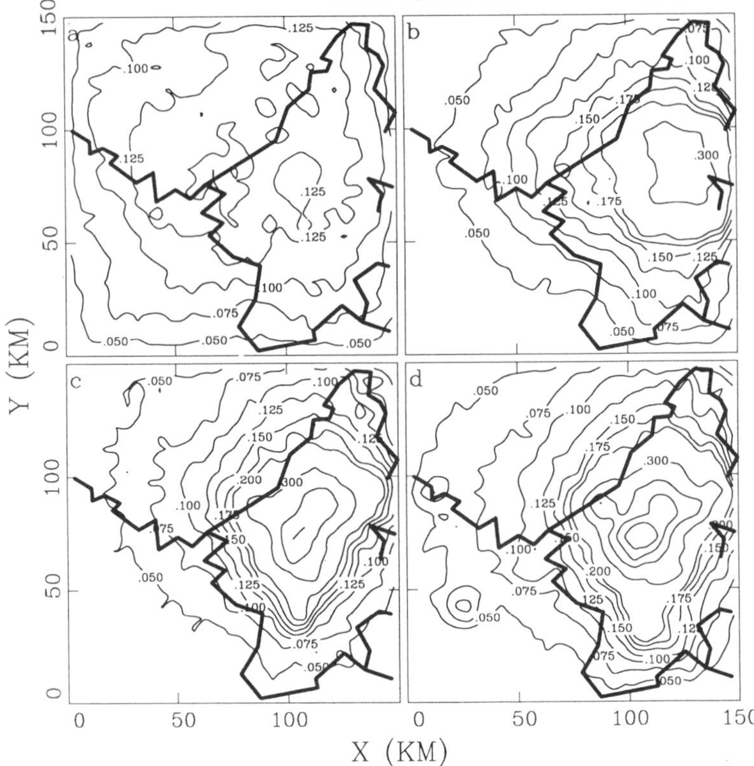

Figure 2. Total deposition of Cd $[mg/m^2]$ in 1992 calculated with different aggregation level of emission sources: a) a single 150×150×0.5 km volume source, b) a grid of 50×150×0.5 km volume sources, c) a grid of 0×10×0.5 km volume sources, d) a grid of 5×5×0.05 km volume sources and major point sources (reference simulation)

than in the previous experiment. Finally, the total emission of Cd was aggregated into a grid of 10x10 km volume sources. The increased resolution of the emission field results in a further increase of the local maximum value of deposition fluxes. The mean deposition flux over the 150×150 km modeling domain is nearly the same in the above experiments and the reference simulation, while its maximum values increases with the increased resolution of the emission field as illustrated by the ratio of the maximum to mean deposition flux (1.5, 3.4, 4.9, and 6.0).

Additional numerical experiments were performed in order to demonstrate the importance of the proper treatment of major point sources. Taking into account the plume rise lowers significantly values of the total deposition. The results from other sensitivity experiments in reference to interannual variability of meteorology and land use representation as well as the results from the deposition simulation with the HMET model with 150 and 50 km grid spacing are reported in (Uliasz and Olendrzyński, 1996).

The maximum value of the Cd deposition flux obtained in the reference simulation is 0.76 mg/m^2. This value should be increased by the background values calculated by the HMET model (0.13 mg/m^2). The deposition measurements in the Katowice province from the SANEPID (Provincial Sanitary Board) network in 1992 show values which are higher by up to two orders of magnitude. According to these measurements there are two separate zones with very high Cd deposition: a small zone around the Szopienice zinc smelters with the

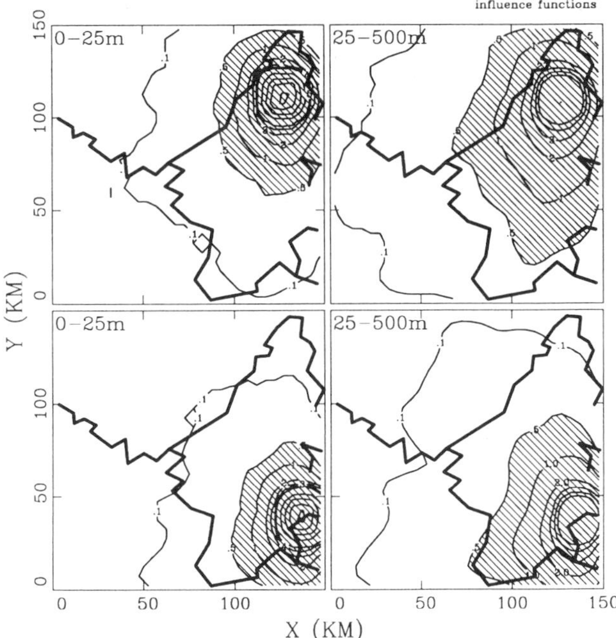

Figure 3. Influence function calculated for the average air concentration of Cd at the receptors located in the Olkusz mining region (top) and the Goczałkowice Reservoir (bottom). These influence function shows the potential impact from the low emission sources located within the layer 0-25 m (left) and within the layer 25-500 m (right).

maximum above 70 mg/m^2, and a much larger zone near the zinc smelters in Bukowno and an adjusted mining area with maximum values above 20 mg/m^2. The Cd emissions reported for each of these smelters are about 100 $kg/year$. In an attempt to explain this discrepancy between the model and observations an additional deposition simulation was performed for only these two point sources. We took emission levels reported for these two zinc smelters in the early 1980's namely 25 $tons/year$ for each of them. The calculated maximum values of the Cd deposition fluxes are about 5 mg/m^2, which still cannot fully explain the observed values. We conclude that emission data for these specific sources are not reliable. On the other hand, the deposition measurements in the areas with the high deposition of heavy metals may be contaminated by local conditions, i.e., mineral dust.

The LPDD model allows one to perform backward in time simulations in order to calculate influence functions as an alternative approach to atmospheric transport modeling (Uliasz, 1994). The influence function provides information on the potential impact of any emission source to pollution at a selected receptor. Pollution characteristics at the receptor may be defined in various ways depending on the application, e.g., as the average concentration or long-term deposition flux. The influence function characterizes atmospheric transport and deposition processes from the point of view of the considered receptor. A value of the influence function at the location of the emission source, multiplied by its emission rate, gives a contribution of this source to the pollution at the receptor. The influence functions seem to be a very useful tool for application in the METKAT or similar projects. In particular, they may be used for the identification of emission sources from measurements. They can also help us to eliminate the sources which, due to atmospheric transport and deposition, can not have a significant impact on the pollution observed at the considered receptor. This approach would allow us to learn more about transport and deposition in the Katowice region, where emission data seem to be uncertain, and it is not clear how deposition measurements are affected by local mineral dust in the areas of current and old zinc mining.

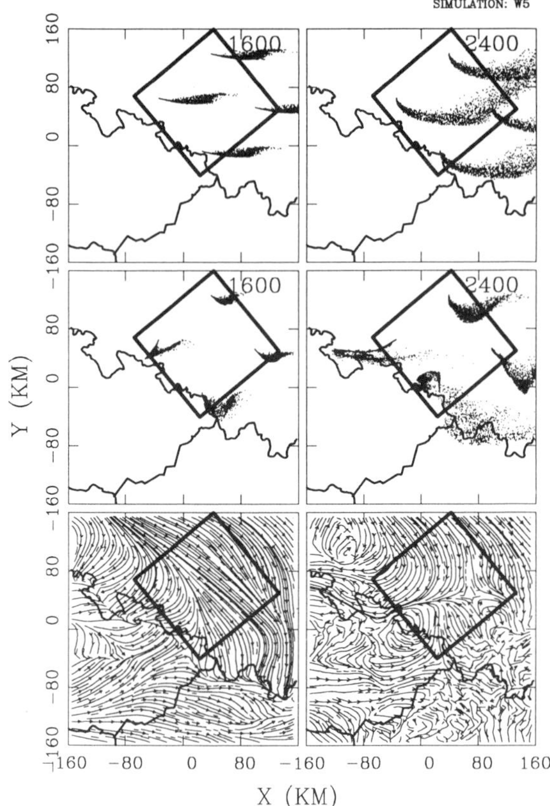

Figure 4. Plumes of passive tracer simulated using 1-dimensional (top) and 3-dimensional meteorological fields (Δx = 5 km), and streamlines (bottom) at 200 m model level derived from the same 3-dimensional meteorological simulation at 16:00 (left) and 24:00 (right) GMT for $U_g = 5\ m/s$ and $V_g = 0\ m/s$.

Simple examples of the influence functions were calculated for the Cd air concentration in 1992 at two receptors: (1) the 5×5×0.25 km receptor located in the Olkusz/Bukowno mining region with high values of measured Cd deposition, and (2) the 5×10×0.25 km receptor covering the Goczałkowice Reservoir which is the major reservoir of drinking water in the region. The influence functions are presented in Fig.3 for the specific layer of the atmosphere, i.e., the influence function for the 0-25 m layer shows the potential impact from low emission sources, and the influence function for the 25-500 m layer determines the potential impact from the elevated sources located within this layer.

SIMULATIONS WITH THE CSU RAMS

Terrain topography within the Katowice province and the 150×150 km deposition domain is not very complex. However, this region is border on south by the Sudeten and Carpathian Mountains, with a wide Moravian Pass just southwest from the Katowice province. The potential effect of terrain topography on atmospheric transport and deposition in this region was investigated with the aid of 3-dimensional meteorological simulations performed for idealized synoptic conditions with the CSU RAMS. In order to isolate the topography effect, a uniform ground surface (agriculture land) was assumed. The CSU RAMS was configured with a single grid to reproduce conditions typical for the middle of July. Simulations were run for 24 hours starting at 00.00 GMT for different geostrophic wind speeds (5 and 20 m/s) and directions (N,E,S,W) and with different horizontal resolution: Δx = 5 km, (65×65×30 grid points) and Δx = 10 km, (33×33×30 grid points) A series of 1-dimensional simulations (no topography effect) was also performed for the reference. The LPDD model was used as a flow visualization tool to illustrate the complexity of transport conditions in the performed simulations. Plumes of a passive tracer were released from four point sources located at the corners of the deposition modeling domain (Fig.4).

The simulations indicate that the mountains may strongly affect flow and dispersion conditions in the Katowice region, especially, in the case of low winds. This is true for all wind directions, though, the topography effect is less significant for stronger winds. The obtained results have important implications for the selection of modeling domain for 3-dimensional meteorological simulations in this region. In order to correctly include the effects of topography, the meteorological modeling domain must cover area much larger than the Katowice province or the 150×150 km deposition modeling domain and should include at least a part of the Sudeten and Carpathian Mountains. Most of the terrain effects on plume dispersion, were reproduced by meteorological simulations on a coarser grid (Δx = 10 km). Therefore, for long-term simulations it seems possible to use a coarser and more computationally efficient grid. A practical compromise can be achieved by using nested grids (in RAMS) or a telescoping grid (in MESO).

CONCLUSIONS

The modeling approach proposed for the Katowice province, allows one to simulate atmospheric transport and deposition of heavy metals in mesoscale taking into account landscape variability, local atmospheric circulation, precipitation fields and detailed information about emission sources. The performed simulations demonstrated that maximum deposition fluxes of heavy metals are very sensitive to aggregation of emission sources and treatment of major point sources. The influence function method together with the improved transport and deposition models may be very useful in investigating the uncertainty of the emission inventory and reliability of deposition observations in the Katowice province. Further research should include implementation of a three-dimensional mesoscale or regional meteorological model linked to the LPDD model for long term transport and deposition simulations. A simple model limited to a small mesoscale domain, like the MESO model, provides an economic option for this task. However, it is also feasible to run a nested grid regional meteorological model, like the CSU RAMS, with a coarse grid over the whole of Europe and a finer grid covering the Katowice region.

Acknowledgments. Several persons were involved in preparing meteorological, land use and emission input data for the METKAT project: Marek Korcz, Stanisław Hławiczka (IEIA), Andrzej Mazur, Jarosław Hrehoruk (IMWM), Sylvia Prieler (IIASA), Józef Pacyna (NILU). Special thanks for continuing support and assistance are due to Prof. Ewa Marchwińska, the director of the IEIA, and William Stigliani and Stefan Anderberg, co-leaders of IIASA's Regional Material Balance Approaches to Long-Term Environmental Policy Planning (IND) project, of which the METKAT project was a part.

References

Bartnicki, J., 1994: An Eulerian model for atmospheric transport of heavy metals over Europe: Model description and preliminary results. *Water, Air and Soil Pollut.*, **75**, 227–263.

Bartnicki, J., 1996: Computing atmospheric transport and deposition of heavy metals over Europe: Country budgets for 1985. *Water, Air and Soil Pollut.*, **92**, 343–373.

Bartnicki, J., H. Modzelewski, H. Szewczyk-Bartnicka, J. Saltbones, E. Berge, and A. Bott, 1993: An Eulerian model for atmospheric transport of heavy metals over Europe: Model development and testing. Technical Report 17, Det Norske Meteorologiske Institut, Oslo, Norway.

Bartnicki, J. and K. Olendrzyński, 1996: Modeling atmospheric transport of heavy metals in the 50km grid system. IIASA Working Paper, WP-96-141, International Institute for Applied Systems Analysis, A-2361 Laxenburg, Austria.

Bartnicki, J., K. Olendrzyński, J. Pacyna, S. Anderberg, and W. Stigliani, 1996: An operational model for long-term, long-range atmospheric transport of heavy metals over Europe. In *5th Intern. Atmospheric Sciences and Application to Air Quality Conference*, June 18-20, Seattle, WA, USA.

Olendrzyński, K., S. Anderberg, J. Bartnicki, J. Pacyna, and W. Stigliani, 1996: Atmospheric emissions and depositions of cadmium, lead and zinc in Europe during the period 1955-1987. *Env. Rev.*, 4, 300–320.

Pielke, R. A., W. R. Cotton, R. L. Walko, C. J. Tremback, M. E. Nicholls, M. D. Moran, D. A. Wesley, T. J. Lee, and J. H. Copeland, 1992: A comprehensive meteorological modeling system - RAMS. *Meteor. Atmos. Phys.*, **49**, 69–91.

Pielke, R. A. and M. Uliasz, 1997: Use of meteorological models as input to regional and mesoscale air quality models - limitations and strengths. *Atmos. Environ.* (accepted).

Uliasz, M., 1990: Development of the mesoscale dispersion modeling system using personal computers. Part I: Models and computer implementation. *Zeitschrift für Meteorologie*, **40**, 104–114.

Uliasz, M., 1993: The atmospheric mesoscale dispersion modeling system. *J. Appl. Meteor*, **32**, 139–149.

Uliasz, M., 1994: Lagrangian particle dispersion modeling in mesoscale applications. In *Environmental Modeling II*, Zannetti, P., Editor, Computational Mechanics Publications, 71-102.

Uliasz, M. and K. Olendrzyński, 1996: Modeling of atmospheric transport and deposition of heavy metals in the Katowice province. IIASA Working Paper, WP-96-123, International Institute for Applied Systems Analysis, A-2361 Laxenburg, Austria.

Uliasz, M., R. A. Stocker, and R. A. Pielke, 1996: Regional modeling of air pollution transport in the southwestern United States. In *Environmental Modeling III*, Zannetti, P., Editor, Computational Mechanics Publications, 145-182.

DISCUSSION

D. STEYN: If the results of the study are to be used in guiding emission reduction strategies, model output must be shown to agree with observations. Does the model output match observation?

K. OLENDRZYNSKI: The model results were compared with available measurements of heavy metals deposition in the region. In general, the model underestimated observed depositions, especially, the local maxima at two locations. There can be two reasons for this. First, the anthropogenic emissions used for the study were underestimated. Secondly, we did not account for the possible high contribution of reemission from the surface. We propose to use the influence function approach to further study the uncertainty of heavy metals emission inventory and reliability of deposition observation in the Katowice region.

S. GONG: Particle size distribution affects particle transport and deposition. Have you considered heavy metal particle in a size-distributed way?

K: OLENDRZYNSKI: The size distribution is accounted for by means of mass median diameter (MMD). For cadmium, for example, we applied MMD=0.86 μm. This and the values for other metals were taken from literature. Unfortunately, at the time of the study, there were no reliable measurements of particle size distribution in the region in question.

BIOMASS BURNING : LOCAL AND REGIONAL REDISTRIBUTION

J. Edy and S. Cautenet

LaMP/OPGC, CNRS and Université Blaise Pascal,
24 Av. des Landais
63177 Aubière, France

INTRODUCTION

Tropical biomass biogeochemistry is one of the most poorly understood on the earth. The tropics account for about 60% of the global annual net primary productivity and this enormous productivity is characterized by many chemical species which are emitted as gas or aerosols and can modify the global radiative balance. Biomass burning is associated with agricultural activity in the savannah, the destruction of tropical forests and the use of wood as "fuel". They release into the atmosphere large quantities of CO2 and a variety of chemically active species such as CO, NOx, N_2O, CH_4, and others expressed in ([1], [2]). The biomass annually burned in the world represents about 1.8 to 4.7 GT. of carbon ([3]), savannah fires being the dominant component with about 1-1.6 GT. of carbon burned. Savannah fires alone contribute approximately 10% of the global CO emissions. This phenomenon is especially important in Africa. The contribution of African savannah fires to global emission of trace gas and aerosols has been estimated by ([4]). During dry season, pollution events are similar in magnitude to those observed in industrialised regions as observed in high levels of acid precipitation that were reported in these region ([5]).

In tropical regions, over West Africa, these trace gas are emitted by biomass burning, in savana zone (Harmattan flow). There, they are mixing and can suffer chemical transformations. They are ultimely removed from the atmosphere by one of these two mechanisms, dry deposition (the most important process) or wet deposition (the clouds are developped close to ITCZ). The crossing of chemical species between Harmattan and Monsoon fluxes is weak. The clouds located near ITCZ allow the venting and the transfer of trace species from boundary layer to the free troposphere

During dry season linked to biomass burning occurrence, aircraft measurements were achieved over West Africa (TROPOZ II experiment). on January 1991 The observations have shown on the other hand, the composition of chemical species like NOx, O3, CO, H2O2 were not the sàme in monsoon flow or harmattan flow and on the other hand, the occurrence of strong concentrations of CO at high altitude (10000m), in presence of clouds. The redistribution of chemical species is simulated using a three dimensional meso-scale model, RAMS (Regional Atmospheric Modeling System) in its nonhydrostatic version coupled with chemical model. It has been initialised with ECMWF data, two nested grids are considered in order to take into account the dynamic due to cloud convection which is associated with ITCZ. We retrieve the complex dynamic over West Africa, the chemical

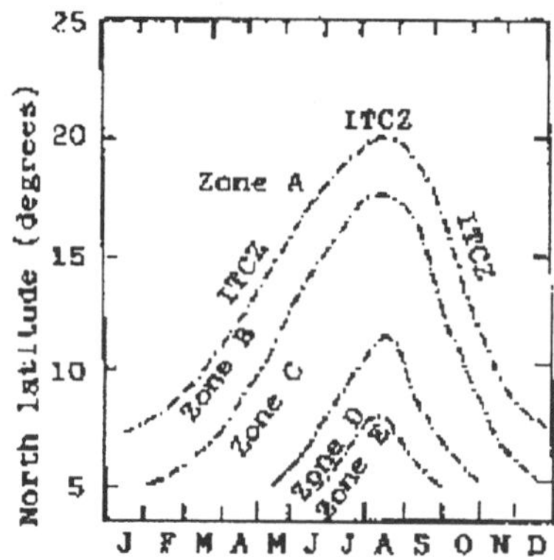

Figure 1 Walker's diagram. Zone A: Saharan cloudiness air, zone B: W/SW winds, shallow maritime air, thunderstorms; zone C: intense rainfalls associated with SW monsoon flow; zone D: stable, cold and very cloudy atmosphere with continuous weak rainfalls; zone E above the coastal lands accompanied by low temperatures and only a few rains.

fields associated with Harmattan and Monsoon fluxes and the strong convective cloud developpement which is linked to the concentrations of CO observed in high troposphere.

ATMOSPHERIC DYNAMIC OVER WEST AFRICA

The synoptic meteorology of West Africa is greatly influenced by the seasonnal migration of the Inter Tropical Convergence Zone (ITCZ) and the associated African easterly jet (AEJ) located at an altitude of approximately 3.5 km. The dry season, from December to March, is also marked by the presence of the South Tropical Jet (STJ) located at an altitude of 15km. Around 0° longitude, in the north hemisphere, airflow near the surface over West Africa is characterized by the simultaneous occurrence of the Northeastern trade wind "Harmattan" and the monsoon flow from the southwest. The ITCZ associated with those two flow regimes give rise to a strong gradient of moisture between the dry Saharan air advected by the Harmattan and the oceanic air of the monsoon flow. The front is located near 8°N latitude in January as shown in Walker's diagram (Figure 1).During the dry season, the Guinea Golf region lies in the B or C zones in this diagram.

TROPOZ II CAMPAIGN

TROPOZ II campaign, has been acomplished in January 1991. This campaign is performed with aircraft measurements in order to study the composition of the high troposphere. Thirty flights have been done all over the World and among these flights, five take place over west Africa. During every flight, meteorogical and chemical data were

Table 1 . Chemical compounds distribution in the differents air masses over west Africa

chemical species	monsoon flux	far from fires	close to fires	free troposphere
CO	160 ppb	200 ppb	275 ppb	100 ppb
NO	50 ppt	250 ppt	400 ppt	50 ppt
O3	25 ppb	55 ppb	75 ppb	45 ppb
H2O2	2 ppb	4 ppb	7 ppb	2 ppb
HCHO	1 ppb	1.5 ppb	2.5 ppb	1 ppb

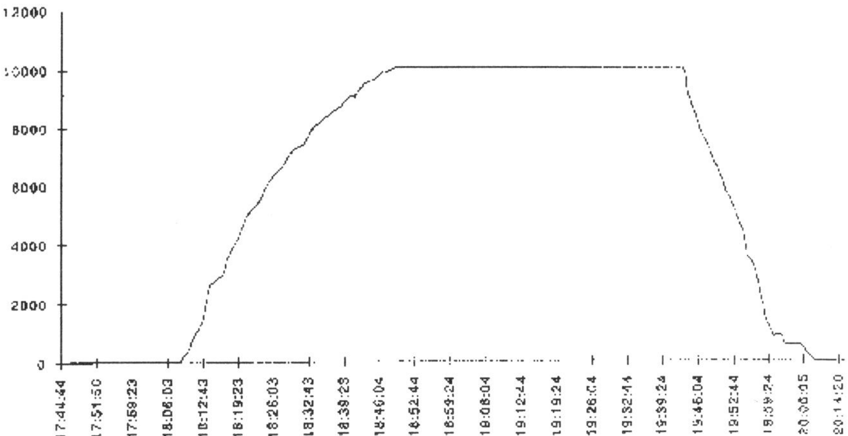

Figure 2a. Altitude of flight for January 27, 1991

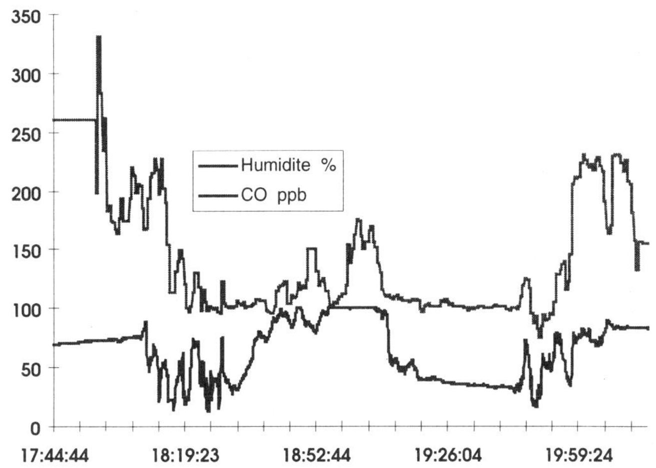

Figure 2b. Aircrafts measurements for January 27, 1991 (relative humidity, CO concentration)

measured: the altitude of the aircraft, the temperature, , the relative humidity, the concentrations of trace gases as O3, CH4, CO, NOx, H202.... More details about this campaign can be found in (6).

The analysis of the five differents fligths achieved over West Africa shown that the chemical compounds were distributed according to many layers lying in horizontal surface (monsoon flow, far from fires or close to fires in Harmattan flow) or in vertical level (above free troposphere, from 2500 m to 10000 m) (Tableau 1). We can see that close to fires the level of concentration had a typical value for a polluted area and in monsoon flux, that of an unpolluted area. This study has shown that the vertical venting of trace gases emitted from biomass burning is limited to 2500 m altitude

During the flight of January 27, 1991, we can see on the figure 2a and 2b, the CO concentrations, the relative humidity and the altitude of the plane. The plane flied between Freetown and Abidjan over Ivory Coast. When the plane flied at 10000 m, measurements have shown strong variations of CO concentration values.

The high concentrations of CO were associated with a relative humidity equal to 100%. It means that high concentrations of CO were linked with the presence of cloud at 10000 m altitude.

NUMERICAL MODEL

West Africa modeling is realized using RAMS. A detailled description of the model, including the basic equations, is given in (7). The model is used in nonhydrostatic mode. The initialization is performed from ECWMF data, for January the 27, 1991 and the 28, 1991. A nudging is prescribed every 6 hours. At ground surface, topography and land-sea mask were obtained from ECMWF data. Local vegetation is introduced following to the main vegetation zones and using the vegetation classes which are defined in RAMS, the same in BATS (Biosphere-Atmosphere Transfert Scheme) (8)
Three runs are performed.

Run 1 is with one grid, the RAMS model is coupled with chemical and dry deposition modules. The chemical model monitors twelve chemical species in gaseous phases as introduced by (9) and (10). It describes the main oxidation chains of CH_4 and CO in presence of NO_x in a remote troposphere. The dry deposition module (11) calculates the deposition velocity for 15 chemical species and 11 vegetation types. This run is initialized with a homogenous chemical field with the value of compounds equal to these observed in monsoon zone (Tableau 1). The emissions on the ground in savanna zone are taken into account by : the CO flux equal to 0.5 10^{13} molecules $cm^{-2}s^{-1}$ and the NO flux equal to 0.5 10^{11} molecules cm^{-2} s^{-1} .

Run 2 only considering the redistribution of CO which is assumed as a passive tracer due to is a very long residence time in the troposphere. It is a 48 hours simulation. The initialisation is realised using the results obtened from the run 1 (CO is capted in the harmattan flux below 2 km).

Run 3 is with two nested grids. The cloud convection model of Tripoli and Cotton (1982). CO is still considered as a passive tracer to study the redistribution of CO by clouds. The initialization of its concentration is performed from the analysis of the values observed during TROPOZ II experiment (Tableau 1). Two layers are taken : the first, from the ground to 2500 m with CO concentration equal to 250 ppb, this second layer from 2500 m to 10000 m with homogeneneous concentrations equal to 120 ppb.
The coarse grid domain is bounded by 10°E to 20°W longitude and 20°N to 10°S latitude.

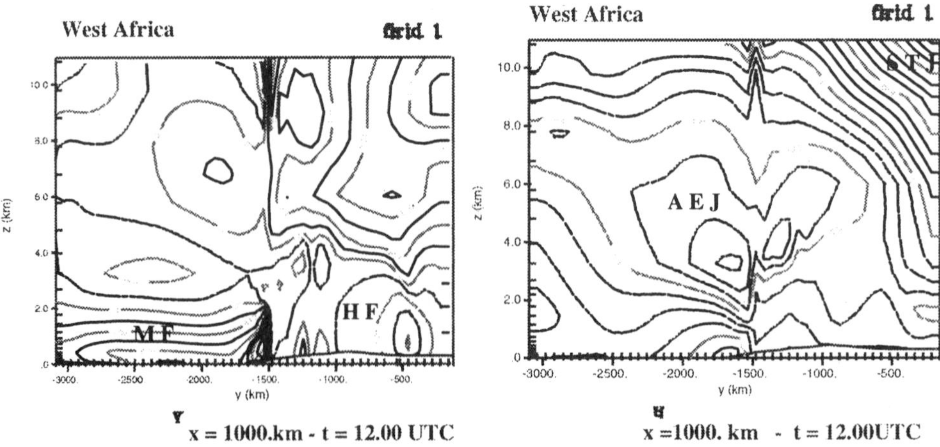

Figure 3 Vertical cross sections of the meridional and zonal wind for the large grid over west Africa.

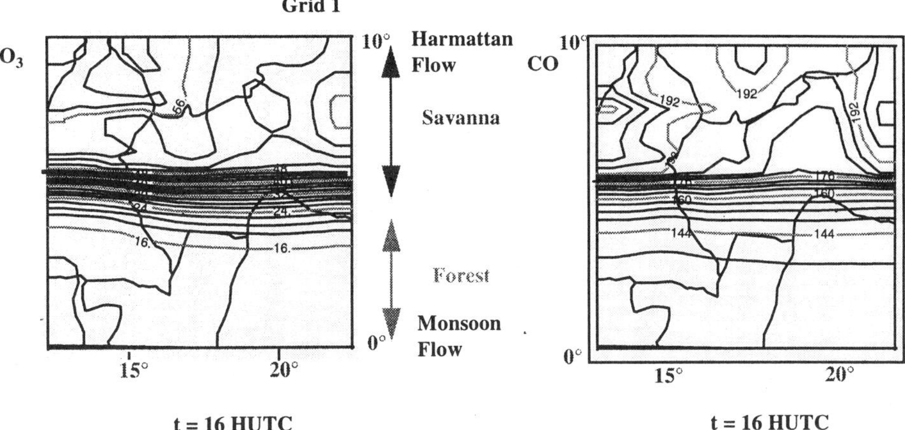

Figure 4. Horizontal cross section of O3 and CO concentrations (ppb) at 16 H. The values have been integrated betwenn 0 an 500m.

The coarse grid or grid 1 spacing was 60km in both east-west and north-south directions. There are 50 grid points in the x-direction and 50 grid points in the y-direction with 30 sigma-z levels. The fine grid or grid 2 has for center the point 8°N-8°O. This second grid has the same number of grids points in the x, y and z directions.The grid spacing is 20 km in both directions. The two grids communicate with each other (12).The nesting grid allows a wider range of motions scales to be modeled simultaneously and interactively. The time step for the grid 2 is 30 seconds.

RESULTS AND DISCUSSION

Large scale dynamical fields: The first objective of this study is to examine if the model initialized with ECMWF data can restitute the complex dynamical fields of the West Africa during the dry season. The vertical cross sections of the meridional and zonal wind are presented on figures 3. We note the location of the main African jets (African Easterly Jet AEJ, South Tropical Jet STJ and the Monsoon flux MF, the Harmathan flux HF, in the low levels. These results show that RAMS is able to restitute with a good agreement the complex circulation observed over the west Africa.

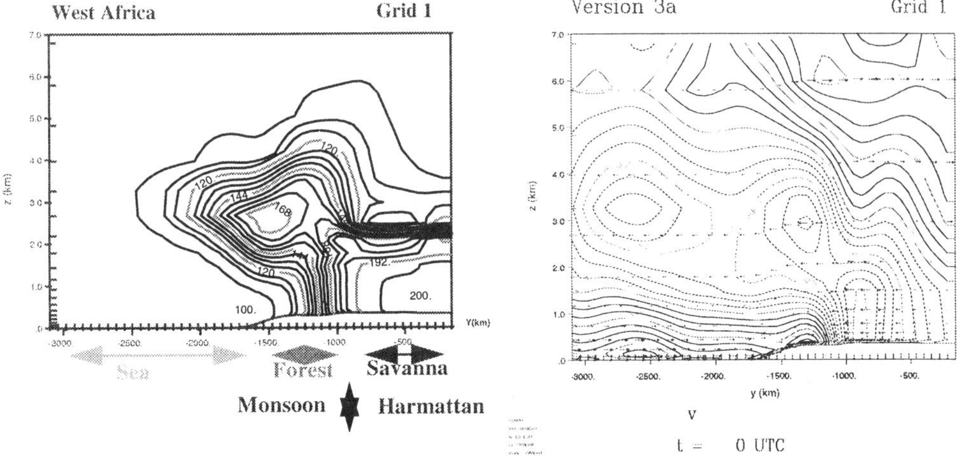

Figure 5. Vertical meridional cross section of CO concentrations and meridional wind after 44 hours of simulation

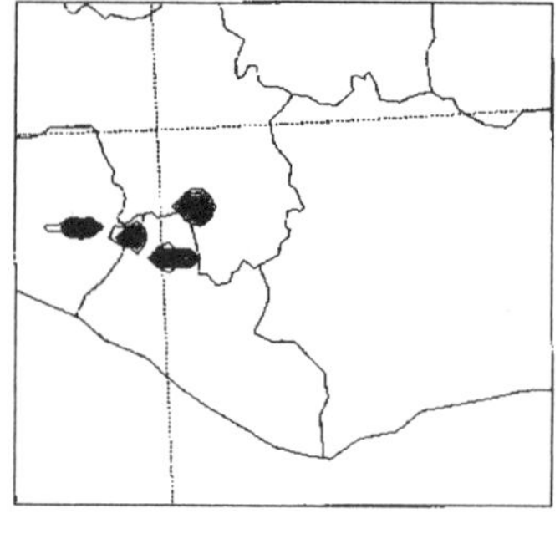

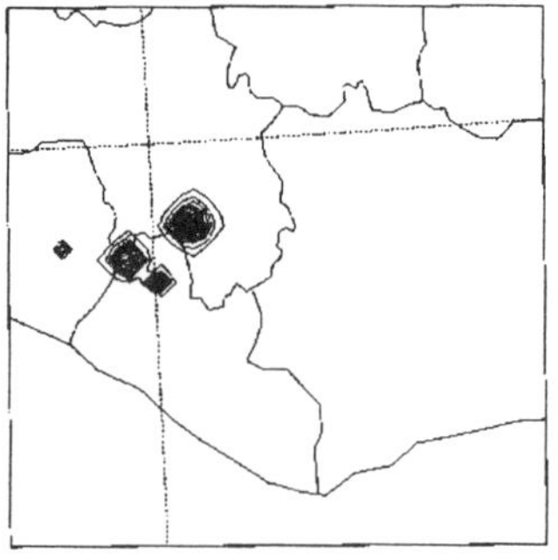

Figure 6. Grid 2. at 10000m the horizontal cross sectin of condensed total water and of CO concentrations at 20h

Run 1 : The figure 4 shows the distribution of CO and O3 concentration at 16 H in a horizontal level. The values are integrated between ground and 500m. We can see a strong gradient which is located at the same place as ITCZ. The low values are in south in monsoon flux (16 ppb for O3 and 144 ppb for CO) and the high values are in North in Harmattan values (56 ppb for O3 and 192 ppb for CO). The isocontours follow the topography and the vegetation zones because the dry deposition is taken into account. This distribution is due to ITCZ occurrence which blocks of the crossing chemical species

Run 2 : The figure 5 shows the vertical redistribution of CO after 44 hours of simulation.We notice that the strong concentrations of CO which are initialy localised in the Harmattan flux below 2km altitude now can be observed over the monsoon flux. These concentrations have been advected by the North-South wind of the Harmattan flux which is located over the monsoon flux.

Run 3 : The redistribution of CO is examined with grid 2. A strong convective development occurres from 2000m to 14000 m. It is associated with strong vertical winds.The high values of CO concentration due to biomass burning in lower layers are brought up to the high troposphere by upwards linked to the cloud development. On figure 6 are shown at an altitude of 10000m (altitude of flight) the condensed total water and the CO concentrations simulated at 20h. Areas of strong condensed water are associated with high levels of CO concentrations. This is due to the updraft of CO from low polluted layer. The aircraft's measurements on January 27, 1991 have observed the increasing of CO at the same level (10000m) (figure 2). These clouds developments are located to the zone B in Walker's diagram. They can efficiently act for the venting of the pollutants emitted by the savanna burning.

REFERENCES

(1) Crutzen, P. J., A. C. Delany, l. Greenberg, P. Haagenson, L. Heidt, R. Lueb, W. Pollock, W. Seiler. A. Wartburg, and P. Zimmerman,Tropospheric chemical composition measurements in Brazil during the dry season, *J. Atmos. Chem.*, *2*, 233-256 (1985).

(2) Greenberg et al.: Hydrocarbon and carbon monoxide emissions from biomass burning in Brazil. J. Geophy. Res.,89,1350-1354 (1984).

(3) Crutzen P.J. and Andreae M.O., Biomass burning in the tropics: Impact on atmospheric chemistry and biological cycles, Science, 1669-1677 (1990).

(4) Lacaux J.P.,Cachier H. and R. Delmas, Biomass Burning in Africa: an overview of its impact on atmospheric chemistry, Fire in the Environment: The ecological, atmospheric and climatic importanceof vegetation fires, edited by P.J.Crutzen and J.G.Goldammer (1993).

(5) Lacaux J.P., R. Delmas, G. Kouadio, B. Cros and M.O. Andreae : Precipitation chemistry in the Mayombe forest of equatorial Africa. J. Geophys. Res.,97, 6195-6206..

(6) Marenco A.,E. Perros, J. Sanak, J.C. Le Roulley and B. Bonsang, Campagne méridienne aéroportée TROPOZ II, final report ,75 p(1993).

(7) Pielke R.A., W.R. Cotton, R.L. Walko, C.J. Tremback, M.E. Nicholls, M.D. Moran, D.A. Wesley, T.J. Lee, and J.H. Copeland, A comprehensive meteorogical modeling system - RAMS, *Meteor. Atmos. Phys.*, 49, 69-91 (1992).

(8) Dickinson R.E., A. Henderson-Seller, et P.J. Kennedy., Biosphere-Atmosphere Transfer Scheme (BATS) Version 1e as Coupled to the NCAR Community Climate Model, NCAR Technical Note-387 emissions from biomass burning in Brazil.J. Geophys. Res., 89, 1350-1354 (1993).

(9) Lelieveld J and P.J. Crutzen, Influences of cloud photochemical processes on tropospheric ozone. Nature,343,227-233 (1990).

(10) Lelieveld J and P.J. Crutzen,The role of clouds in tropospheric photochemistry. J. Atm. Chem.,12, 229-267(1991).

(11) Wesely M.L.,Parameterization of surface resistance to gaseous dry deposition in regional scale numerical models. Atmos. Environ.,23,1293-1304 (1989).

(12) Clark, T.L. et R.D. Farley, Severe downslope windstorm calculations in two and three spatial dimensions using anelastic interactive grid nesting: A possible mecanism for gustiness., *J. Atmos. Sci.*, 41, 329-350 (1984).

DISCUSSION

A. HANSEN: Do you include oxidation of CO in your chemical module? I did not see it included in the scheme illustrated in your viewgraph. Measurements and modeling in the 1995 Nashville study indicate that in large areas of rural eastern U.S. the CO is the largest contributor to O_3 production. Therefore, you might consider including CO oxidation in your scheme.

J. EDY: The chemical module we use include the oxidation of CO. This chemical module is adapted from Lelieveld and Crutzen (1990). It considers 12 species and describes the oxydation chains of carbon monoxides and methane in the presence of nitrogens oxides in the troposphere. It has been applied in few studies Grégoire and al. (1994) and Edy and al. (1996). In our study, where high concentrations of CO are associated with biomass occurrence the production and the redistribution of O_3 are highly influenced by this CO.

Edy J., S. Cautenet and P. Brémaud, 1996: Modeling ozone and carbon monoxide redistribution by shallow convection over the amazonian rain forest., *J. Geophys. Res,* 101, 28,671-28,681.

Grégoire, P. J., N. Chaumerliac, and E. C. Nickerson, 1994: Impact of cloud dynamics on tropospheric chemistry: advances in modeling the interactions between microphysical and chemical processes. *J. Atm. Chem., 18,* 247-266.

Lelieveld, J., and P. J. Crutzen, 1990: Influences of cloud photochemical processes on tropospheric ozone. *Nature, 343*, 227-233.

R. ROMANOWICZ: Were there any quantitative comparisons made of simulated model results and observations?

J. EDY: In our study about African biomass burning, we used TROPOZ II data which consisted of profiles and flats measurements of chemical species (O_3, CO, CH_4, H_2O_2, NO_2) and meteorological data. These observations present a distribution of chemical species at a given time, we have no temporal evolution. This numerical study is a preliminary work which has allowed to restitute the observed heterogeneous distribution of chemical species between monsoon and harmattan fluxes. An EXPRESSO campaign might show the diurnal evolution in different regions: savanna or forest. This experience took place in december 97 and the results are in progress.

HIGH RESOLUTION, LONG-PERIOD MODELLING OF PHOTOCHEMICAL OXIDANTS OVER EUROPE

Joakim Langner, Christer Persson and Lennart Robertson

Swedish Meteorological and Hydrological Institute
S-601 76 Norrköping
Sweden

INTRODUCTION

Concentrations of surface ozone over Europe currently exceeds the critical levels over which damage to vegetation and health may occur in many locations. This is true also in northern Europe and over Sweden regarding critical levels for vegetation and also occasionally, during summer in southern Sweden, regarding critical levels for human health. Optimising measures to reduce these exeedances, both nationally and internationally, requires a better understanding of the importance of various types of precursor emissions and processes influencing the distribution of photochemical oxidants.

Work on developing models suitable to study control strategies and the contribution from individual countries or activities to the production of photochemical oxidants has been underway for several years in Europe. Both Lagrangian and Eulerian type models have been used for longer simulations (months) and scenario calculations over Europe (e.g. Simpson, 1992; Simpson, 1995; Builtjes, 1988; Zlatev et al., 1993). These studies have employed a rather coarse horizontal resolution (100-150 km) and limited vertical resolution. Studies over northern Europe using higher resolution over longer periods are lacking. The chemical system describing the production of photchemical oxidants is strongly nonlinear and the range of simulated concentration levels depend on model resolution. It is therefore of great interest to carry out calculations with higher horizontal and vertical resolution.

Over the last five years SMHI has developed an Eulerian atmospheric transport and chemistry modelling system called MATCH (Mesoscale Atmospheric Transport and Chemistry model). The MATCH system is used in a range of applications from high resolution assessment studies for sulfur and nitrogen compounds in regions of Sweden to continental scale studies in developing parts of the world as well as emergency response applications over Europe (Langner et al., 1995; Robertson, et al., 1995)

Currently a photochemical submodel is developed and the intention is to use MATCH as a tool for assessing the importance of different sources of pollutants to the levels of

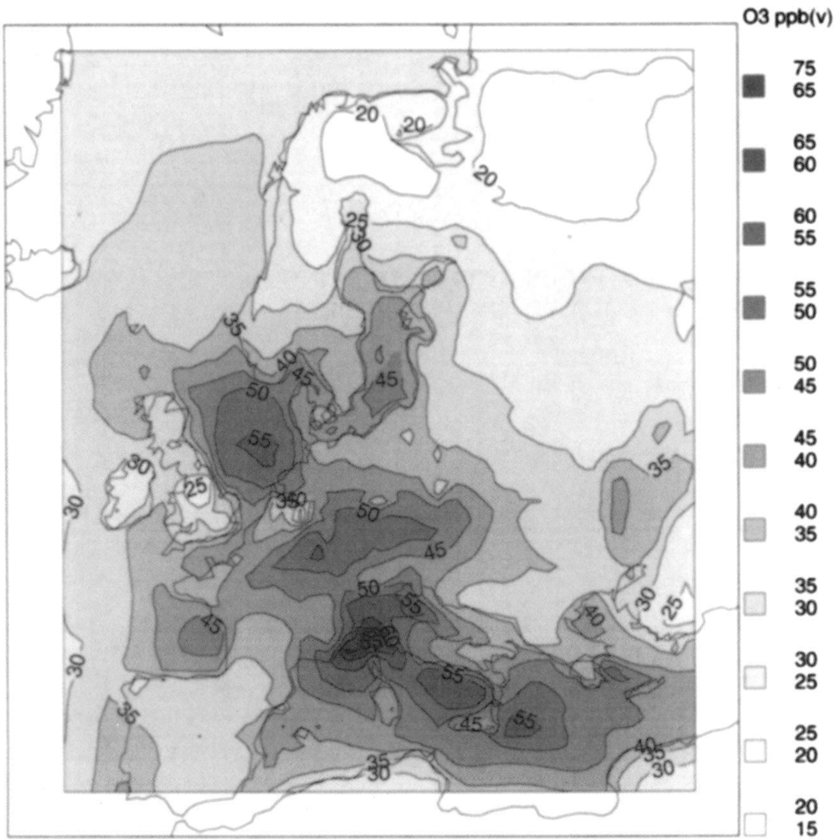

Figure 1. Calculated monthly average, surface concentration of ozone for July 1994. Units: ppb(v)

photochemical oxidants over Sweden and to study control strategies. This will involve nesting of a European scale model with a higher resolution model over Sweden. Once the model becomes operational the intention is to focus on seasonal modeling although detailed study of episodes will of course be required to establish the quality of the modelling system. Here we present some results from test runs with the European scale model.

THE MATCH MODELLING SYSTEM

Transport model

MATCH is a three-dimensional Eulerian atmospheric transport model. The model is a so called "off-line" model, meaning that meteorological data from an external archive at regular time intervals (usually three or six hours) is required in order to calculate transport, chemistry and deposition. The model is designed to be flexible with regard to horizontal and vertical resolution and caters for the most commonly used horizontal and vertical grid systems. It is used for several different applications in atmospheric transport modelling at SMHI, with horizontal grid resolutions ranging from 5 to 100 km.

Horizontal advection is calculated using a modified fourth order flux correction scheme (Bott 1989a, 1989b). The scheme utilises polynomial fitting between neighbouring grid points of the concentration field in order to calculate the advective fluxes through the

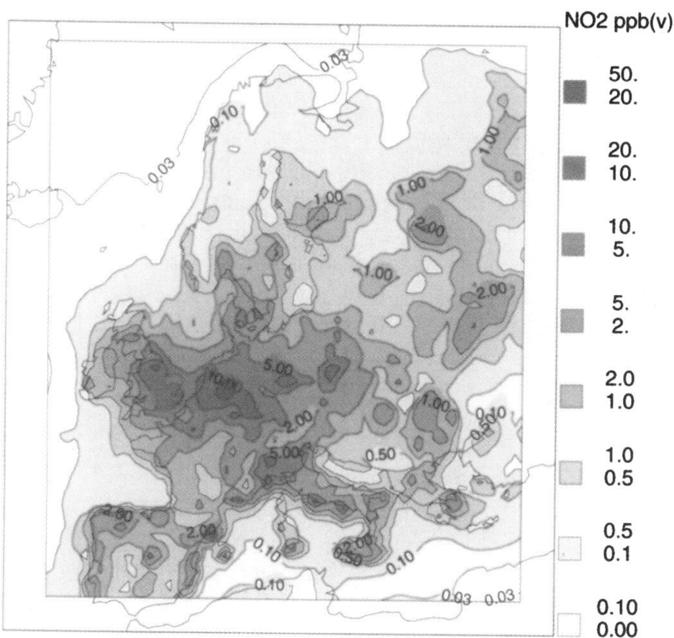

Figure 2. Calculated monthly average, surface concentration of NO_2 for July 1994. Units: ppb(v)

boundaries of adjacent grid boxes. It is a positive definite mass conserving scheme with low numerical diffusion. Vertical advection is calculated using an upstream scheme. Transport by vertical diffusion is also included. For the numerical solution of the combined horizontal and vertical transport, chemistry and deposition an operator split time integration scheme is used. A detailed description of the transport model is given in Robertson et al. (1996).

Emissions

Emissions can be specified both as area and point sources. Surface area sources are introduced into the lowest layer of the model. The initial dispersion from point sources is described with a Gaussian puff model including plume rise calculations. A new puff is introduced every hour for each source. The puffs are then advected until they have reached the size of the horizontal grid when they are merged into the large scale concentration field. In the calculations presented below all emissions were treated as area sources.

Chemistry

The gas-phase chemical mechanism used is identical to the one used in the EMEP-model (Simpson et al., 1993; Simpson et al., 1995). The mechanism includes ca. 130 thermal and photochemical reactions between 69 chemical components and is designed to provide a good descripition of the chemistry for both high and low NO_x conditions. Standard numerical integration techniques following the work by Verver et al. (1996) are used to integrate the chemical mechanism. This results in stable integrations where the accuracy of the calculations can be controlled. The use of a standard solver makes it easy to change the chemical mechnism whithout having to recode the solver.

Deposition processes

Pollutants are removed from the atmosphere by wet and dry deposition processes. Wet scavenging of the different species is taken as proportional to the precipitation rate and a species specific scavenging coefficient. Dry deposition is proportional to the concentration and a species specific dry deposition velocity at 1 m height. Since the lowest model layer has a thickness of ca. 75 m, the dry deposition flux calculation is transformed to the middle of that layer using standard similarity theory for the atmospheric surface layer. Dry deposition velocities are specified as a function of the surface characteristics (land or water.). Scavenging coefficients and deposition velocities have in most cases values close to those used in the EMEP calculations (Simpson et al., 1993).

EMISSION DATA

Anthropogenic emissions for the simulations presented below was derived from the 50x50 km emission data provided by EMEP MSC-W at the Norwegian Meteorological Institute. The emissions for 1993 for NO_x, SO_2, NMHC, and CO was used in the calculations. The EMEP emission data is divided into emissions below and above 100m. The emissions were assumed to be constant in time. No natural emissions were included.

METEOROLOGICAL DATA

Meteorological data for the period of July 1994 was taken from archived output from the operational numerical weather prediction model HIRLAM at SMHI. The data is available at ca. 55x55 km horizontal resolution with 16 levels in the vertical. Initialised analysis, 3 and 6 hour forcasts were utilized to get new input data every three hours. Precipitation was taken from the 6 hour forcasts. All meteorological data was interpolated to one hour time resolution inside the MATCH model.

RESULTS

Here we show some results from trial runs with the new gas-phase chemistry module. These results should only be taken as an indication about the present status of the development. There are still parts of the modelling system that will be changed and added. A thorough evaluation of the system will be reported in due course. Figure 1 shows the calculated monthly average, surface concentration of ozone for July 1994. Note the low average ozone concentrations simulated over major population centers in north-western Europe caused by the strong NO_x emissions in these regions. High ozone concentrations are simulated over water surfaces. This is a result of the low dry deposition velocity assumed for ozone over water surfaces in combination with the gradual production of ozone downwind from the source regions. Figure 2 shows the corresponding results for NO_2. In this case the highest concentrations are found over land areas reflecting the distribution of the emissions. Figure 3 shows a comparison between observed and model calculated hourly surface ozone concentrations at four locations; Harwell in southern England, Rörvik on the Swedish west coast, Logrono in northern Spain and Ispra in northern Italy for the period 94-06-10 to 94-07-21. Much of the episodic variations as well as the amplitude of the diurnal variations and the magnitude of the concentrations are captured, but there is a tendency towards underestimation at Logrono. The strong diurnal variation at Ispra is not captured in the beginnig of the period. These four stations cover a wide range of different

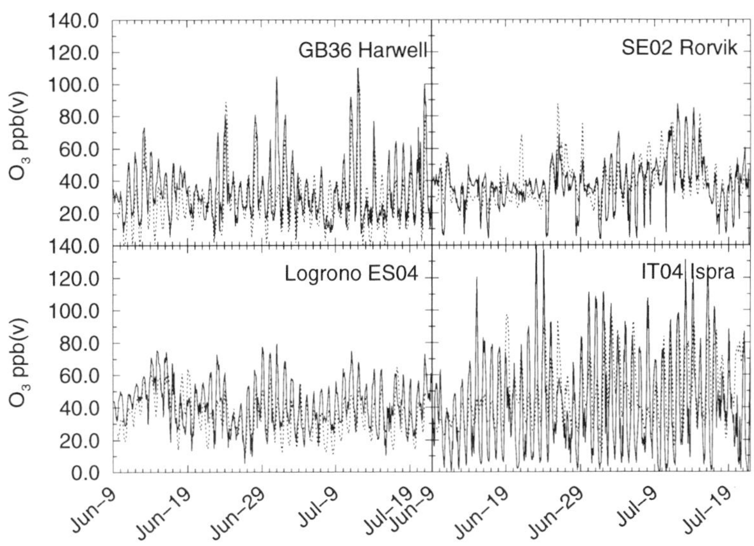

Figure 3. Observed (full line) and model calculated (dashed line) hourly surface ozone concentrations at four European stations for the period 94-06-10 to 94-07-21. Units: ppbv.

chemical and meteorological conditions. Given that the modelled values are averages for the lowest (ca. 75 m thick) model layer, that no natural emissions were included, and other uncertainties in emission data, we think that the results looks promising.

CONCLUSIONS

A gas-phase chemistry module has been implemented into the MATCH modelling system. The aim is to use MATCH as a tool for assessing the importance of different sources of pollutants to the levels of photochemical oxidants over Sweden and to study control strategies. The first preliminary results presented here indicate that the model is capable of providing a reasonable simulation of surface ozone for a forty day summer period over Europe.

ACKNOWLEDGMENT

The development of the gas-phase chemistry module is supported by the Swedish Environmental Protection Agency. The choice of chemical mechanism and verification of the integration procedure has been carried out in coopoeration with the Swedish Environmental Research Institute. Emission data was provided by the EMEP-Meteorological Synthesizing Centre West at the Norwegian Meteorological Institute. Observation data for ozone was provided by the Chemical Co-ordinating Centre of EMEP (EMEP/CCC) at The Norwegian Institute for Air Research.

REFERENCES

Bott, A., 1989a, A positive definite advection scheme obtained by nonlinear renormalization of the advective fluxes, *Mon. Wea. Rew.* 117:1006.

Bott, A., 1989b, Reply. Notes and correspondance, *Mon. Wea. Rew.* 117:2626.

Builtjes, P.J., Stern, R.M., and Pankrath, J., PHOXA: The use of a photochemical dispersion model for several episodes in North-Western Europe, in *Air Pollution Modelling and its Application, Vol VI*, H. van Dop, ed., Plenum Press, New York (1988).

Langner, J., Persson, C., and Robertson, L., 1995, Concentration and deposition of acidyfying air pollutants over Sweden: Estimates for 1991 based on the MATCH model and observations, *Water, Air, and Soil Pollut.* 85:2021.

Robertson, L., Langner J., and Enghardt M., 1996, MATCH-Meso-scale atmospheric transport and chemistry modelling system. Basic transport model description and control experiments with Rn^{222}. Swedish Meteorological and Hydrological Institute RMK-70.

Robertson, L., Rodhe, H., and Granat, L., 1995, Modelling of sulfur deposition in the southern Asian region, *Water, Air, and Soil Pollut.* 85:2337.

Simpson, D., 1992, Long-period modelling of photochemical oxidants in Europe. Model calculations for July 1985. *Atmos. Environ.* 26A:1609.

Simpson, D., 1995, Biogenic emissions in Europe. 2. Implications for ozone control strategies. *J. Geophys. Res.* 100:22,891.

Simpson, D., Andersson-Sköld Y. and Jenkin M. E., 1993, Updating the chemical scheme for the EMEP MSC-W oxidant model: current status. EMEP MSC-W Note 2/93.

Zlatev, Z, Christensen, J., and Eliassen, A., 1993, Studying high ozone concentrations by using the Danish eulerian model, *Atmos. Environ,* 27A:845.

Verver, J.G., Blom, J.G., van Loon, M., and Spee, E.J., A comparison of stiff ODE solvers fo atmospheric chemistry problems, *Atmos. Environ.* 30:49.

DISCUSSION

R. D. BORNSTEIN: Is your model really "terrain influenced" and not "terrain following"?

J. LANGNER: The way we use the hybrid coordinates the model is terrain following near the surface and terrain influenced or pressure level based higher up.

S. PRYOR: When you say emissions are invariant - does that mean a single diurnal cycle is repeated or midnight an mid day emissions are identical?

J. LANGNER: Midnight and mid day emissions are the same, but clearly this is a point we will improve.

R. ROMANOWICZ: Were the presented results from calibration or validation stages?

J. LANGNER: The results should be viewed as test simulations since all parts that will be available for the evaluation phase were not included.

R. ROMANOWICZ: What was the scale of measurements versus model grid?

J. LANGNER: All measurements were EMEP stations at background locations, i.e. away from large population centers and local sources.

APPLICATION OF THE URBAN AIRSHED MODEL IN THE FRASER VALLEY OF BRITISH COLUMIA, CANADA AND IMPLICATIONS TO LOCAL OZONE CONTROL STRATEGIES

Joanne L. Pottier

Aquatics and Atmospheric Sciences Division,
Environment Canada, 7th Floor, 1200 West 73rd Avenue,
Vancouver, British Columbia, Canada V6P 6H9
joanne.pottier@ec.gc.ca

INTRODUCTION

The Fraser Valley in the southwestern province of British Columbia, Canada is one of Canada's highest ozone concentration areas where the National Ambient Air Quality Objective of 82 ppb per hour is exceeded several times a year under high pressure ridge conditions (Taylor, 1991; McKendry, 1994). Modeling efforts are aimed at supporting local regulatory agencies in their attempt to apply appropriate control strategies to the problem. The complex nature of the Fraser Valley offers a modeling challenge. The irregular terrain (Figure 1) generates complex mesoscale flows and modeling is further complicated by changing land use from the urban centre of Vancouver in the west to agricultural farmland in the east. Located on the east coast of the Pacific Ocean, the area is not influenced by upstream transport and thus provides a unique closed system in which to assess and verify model behavior.

Model testing and performance evaluation were conducted using a limited 1985 dataset, weak in biogenic content in the emissions inventory and in meteorological data coverage. While a major field measurement program conducted in 1993 (Thomson et al., 1993) provides a comprehensive dataset for final model verification and in modeling future scenarios, results of the 1985 model runs are examined herein. In spite of limited data, model simulation of observed ozone patterns and wind fields in the valley was successful.

As a first attempt to model for air quality control scenarios, several percentage reductions were applied to input nitrogen oxides (NOx) and volatile organic compounds (VOC). The results agreed with similar studies conducted in the Los Angeles area (Milford et al., 1989) and showed non-linear response to across-the-board reductions in NOx and VOC precursors.

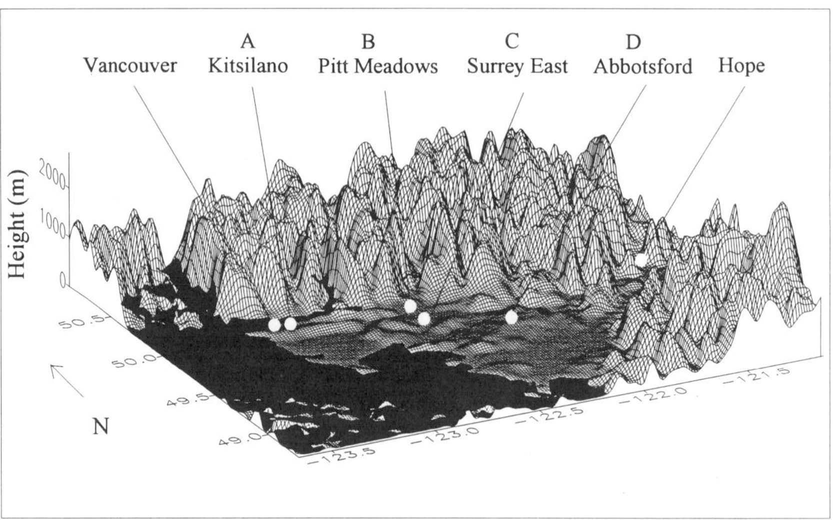

Figure 1. The Fraser Valley is a complex coastal delta which narrows from approximately 60 km across in the west to 2 km near Hope. Mountains rise sharply to 2000 m along the north and south sides of the valley, broken by several transverse tributary valleys.

UAM-V AND 1985 BASE CASE INPUT FEATURES

The widely applied Urban Airshed Model - Variable Grid (UAM-V) (Morris et al., 1992) made it an appropriate candidate for simulating the complex nature of land and sea breezes and mountain and valley flows. Other desirable features include two-way nested gridding, plume-in-grid capabilities, enhancements in vertical exchange coefficients, user defined vertical grid structure, and three dimensional input capability (Users Guide, 1995). Meteorological input fields to UAM-V were generated from the Systems Applications International Meteorological Model (SAIMM) and chemistry input fields were generated from the Emissions Preprocessor (EPS2), using the CB-IV mechanism.

The meteorology and emissions data used were for a 4 day ozone episode which occurred between July 17-20, 1985. The 1985 Greater Vancouver Regional District emissions inventory was used for the Canadian portion of the airshed while U.S. emissions were backcast from a 1990 inventory. Model emission preparation was performed by the National Research Council of Canada.

A two nested grid structure was employed, using a 5 km inner grid within an outer 10 km grid which covered an area of approximately 200 km x 230 km in southern British Columbia (Figure 2). The vertical structure of the SAIMM model consisted of 18 levels, and the model top was at 8000 meters, while the UAM-V model consisted of 8 layers, with a model top at 3200 meters. Both models contained more levels in the boundary layer, and used terrain-following coordinates, providing a vertical structure which minimized the need for vertical interpolation.

Four dimensional data assimilation was used in the SAIMM meteorological preprocessor to incorporate observed data into the model runs. This generated prognostic wind fields which were in close agreement with observations and were used confidently with the UAM-V simulations.

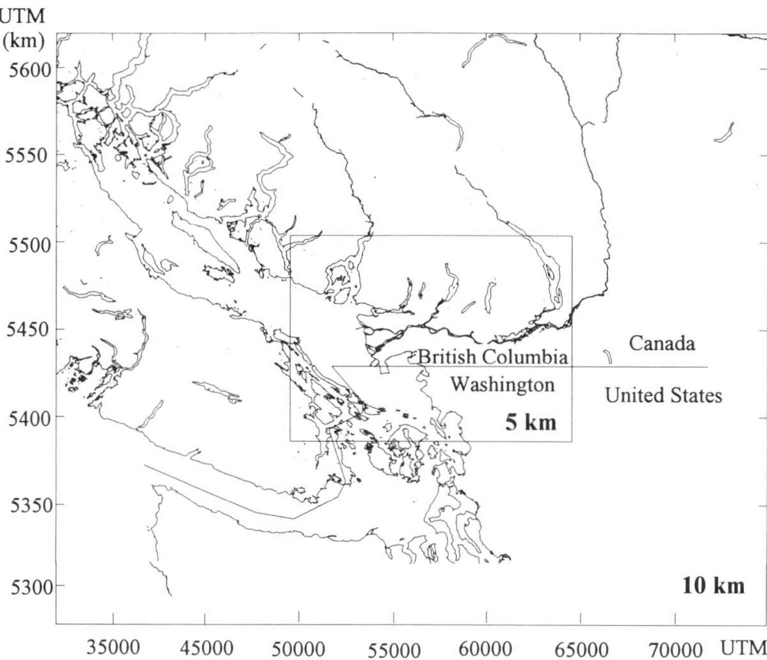

Figure 2. A 5 km inner grid was nested within an outer 10 km grid for the Fraser Valley modeling study.

Model output results are examined for the inner 5 km resolution domain, the area of highest activity and of most concern to air quality control agencies.

SIMULATED AND OBSERVED OZONE PATTERNS

Initial model results showed generally lower than observed ozone concentrations, attributed to low biogenic emission sources, a weakness not singular to the Fraser Valley emissions inventory (Magliano et al., 1991). Isoprene, a significant component of biogenics, was adjusted to more realistically depict the abundant sources of agricultural and tree covered areas in the eastern end of the valley. This resulted in an increase in simulated ozone concentrations across the airshed, though somewhat overpredicted in the west urban center and underpredicted in the east agricultural areas. The simulated peak location on July 19, the day of maximum ozone concentrations, remained in the central part of the valley while the observed peak was located farther east (Figures 3 and 4). While this could be due to a weakly simulated sea breeze flow failing to advect ozone far into the valley, poorly represented precursor emissions in the eastern part of the airshed are more highly suspected.

Overall, comparison of observed and simulated patterns showed several similarities. Figures 3 (observed) and 4 (simulated) both show areas of minimum concentration in the western urban area with increasing concentrations eastward. The simulated pattern shows a north-south axis of higher concentration across the central part of the valley. The observed pattern shows a similar southward extension of high concentrations from the central valley and no northward extension due to lack of data. It is likely that daytime up-valley flows in the transverse valleys tend to advect ozone northward, suggesting the modeled pattern may be more realistic than the surface plots due to missing data.

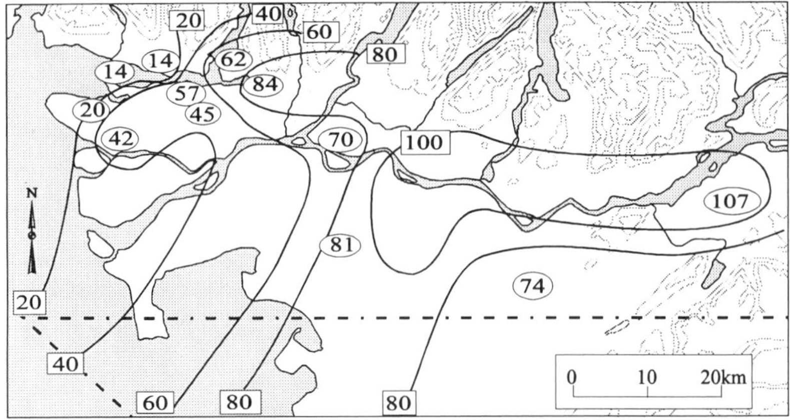

Figure 3. Observed ozone concentrations in the Fraser Valley on 19 July, 1985 at 1600 PST. Contours are incomplete in the north due to missing data.

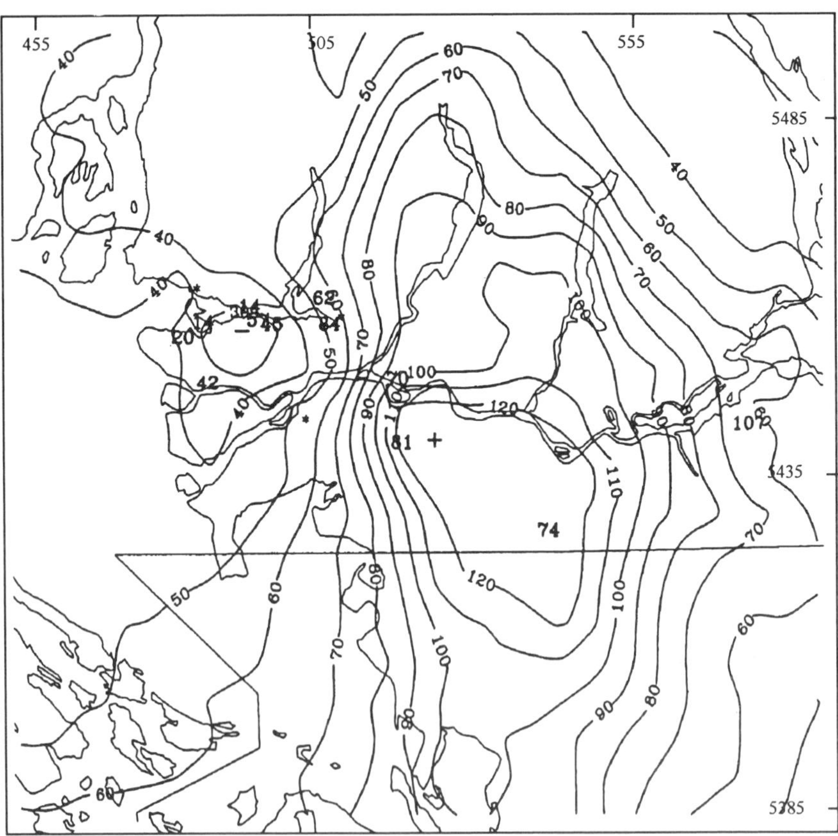

Figure 4. Maximum simulated ozone concentrations in the inner 5 km domain on 19 July 1985, averaged over the period 1500-1600 PST. The northward extension of high ozone concentration contours is reasonable since up-valley slope flows dominate during the daytime.

NOx AND VOC REDUCTION SCENARIOS

Model runs were performed with 25%, 50% and 75% reductions from base case input NOx and VOC emissions, as shown in Table 1. The nomenclature reads 25nox as a model run with 25% of base case NOx (NO and NO2) emissions (i.e. a 75% **reduction** in NOx). VOC reductions were applied to 11 species of non-methane hydrocarbons. Percentage changes refer to 24 hour maximum ozone concentrations resulting from changes in NOx and VOC input emissions. Notwithstanding the limitations of interpreting results for the entire airshed, some observations are made from Table 1.

In all reduction runs, little change occurred in the west side of the airshed, where concentrations remained near 40 ppb, likely representative of early unreacted chemistry since ozone precursors are released in the downtown urban area of Vancouver. The overall ozone contour pattern remained consistent over the four episode days. Highest concentrations generally occurred in the central part of the valley, with northward and southward extensions from the peak, as seen in Figure 4. Model runs with higher ozone peaks in the central part of the valley thus produced stronger east-west gradients across the western half of the airshed.

The non-linear relationship between NOx input emissions and resulting ozone concentrations is clearly seen in the three NOx variations of Table 1. While little change occurred in ozone peaks with 50% and 75% NOx **content** (157 and 154 ppb, respectively), a **reduction** to 25% NOx (25nox) produced a lower ozone peak of 120 ppb. Thus, a 25% to 50% reduction in NOx produced a 1.9% increase in peak ozone concentration while a 50% to 75% NOx reduction resulted in a 23.6% decrease in ozone. With reference to the 'base case', 25%, 50% and 75% reductions in NOx led to a 9.1% increase, a 10.8% increase, and a 16.7% decrease in ozone concentrations, respectively.

Table 1. UAM-V NOx and VOC Reductions from base case for 19 July, 1985.

NOx/VOC Mixture:	**BaseCase 100%NOx,100%VOC**		
Peak Ozone Conc. (ppb)	140		
NOx Content:	**25nox (75% Reduction)**	**50nox (50% Reduction)**	**75nox (25% Reduction)**
Peak Ozone Conc. (ppb)	120	157	154
Percentage Change in Conc. (Right to Left)	23.6% decrease	1.9% increase	
Percentage Change from Base Case (140 ppb)	16.7% decrease	10.8% increase	9.1% increase
VOC Content:	**25voc (75% Reduction)**	**50voc (50% Reduction)**	**75voc (25% Reduction)**
Peak Ozone Conc. (ppb)	78	94	117
Percentage Change in Conc. (Right to Left)	17.0% decrease	19.7% decrease	
Percentage Change from Base Case (140 ppb)	44.3% decrease	32.9% decrease	16.4% decrease

A slight eastward shift in the grid cell location of the ozone peaks occurred with NOx increases. Peak times lagged one hour with each 25% increase in NOx, peaking at 13 PST in the 25nox case and at 15 PST in the 75nox case. The time delay and the eastward shift in ozone peaks are as expected and related to eastward transport of precursors in the westerly sea breeze while being chemically reacted.

The variations in VOC showed a more linear drop in ozone with each decrease in VOC input. A 25% to 50% drop in input VOC resulted in a 19.7% decrease in output ozone concentration (117 ppb to 94 ppb). Further reduction of VOC from 50% to 75% led to a 17.0% decrease in ozone (94 ppb to 78 ppb). With reference to the 'base case', 25%, 50% and 75% reductions in VOC led to a 16.4% decrease, a 32.9% decrease, and a 44.3% decrease in ozone concentrations, respectively.

As Table 1 demonstrates, ozone peaks for most runs remained in the central part of the valley. All peaks occurred downwind of the urban area where ozone precursors are released, and generally showed south and eastward shifts in peak locations with increasing NOx, as expected. Reductions in VOC lead to more linear reductions in ozone.

OZONE ISOPLETH SURFACES

Overall, the model runs point to a more dramatic reduction in ozone concentration from VOC reductions, i.e. a VOC limited condition. Figure 5, modeled after Finlayson-Pitts and Pitts (1993) shows some similarities with model results from the Los Angeles case, with varying VOC/NOx ratios across the airshed. In the 3-dimensional ozone isopleths depicted for different locations in the airshed, there are some NOx and some VOC limited areas.

Figure 5 shows relatively low VOC/NOx ratios near the urban (Kitsilano) area, similar to that found near urban Los Angeles. Kitsilano is close to the downtown core where large motor vehicle precursor emissions are released. In this case VOC reductions at constant NOx values would appear be more effective than NOx reductions at constant VOC values, typical of a VOC limited situation. In following the technique of Finlayson-Pitts and Pitts, starting at point A, rolling down the ozone hill appears faster by VOC reduction than by NOx reduction.

The surfaces over Surrey East and Pitt Meadows also exhibit VOC limited features. Pitt Meadows is representative of an area downwind of several industrial sites. A departure from B or C would achieve a faster roll down the ozone hill by controlling VOCs without controlling NOx. Higher VOC/NOx ratios were expected in these rural areas, 30-50 km downwind of Kitsilano. Both Pitt Meadows and Surrey East, however, appear to remain influenced by upstream industrial sources, hence the lower than expected VOC/NOx ratio.

Farther inland, Abbotsford showed VOC limited conditions, though approaching a NOx limited case. Highest ozone concentrations were simulated and observed between Surrey East and Abbotsford, indicating reacted chemistry and leading one to suspect a higher VOC/NOx ratio and NOx limited conditions eastward. The isopleth surface indicates, however, that a departure from D would achieve lower ozone concentrations only slightly faster from VOC reductions than from NOx reductions, a slightly more VOC limited condition. Further investigation is in order before concluding that the entire airshed is VOC limited, as this is in contrast to the findings of the Los Angeles study. Inaccurate VOC emissions in the southeast corner of the domain is suspected of leading to higher VOC/NOx ratios than would be expected, and an overall exaggeration of the effect of decreasing VOC.

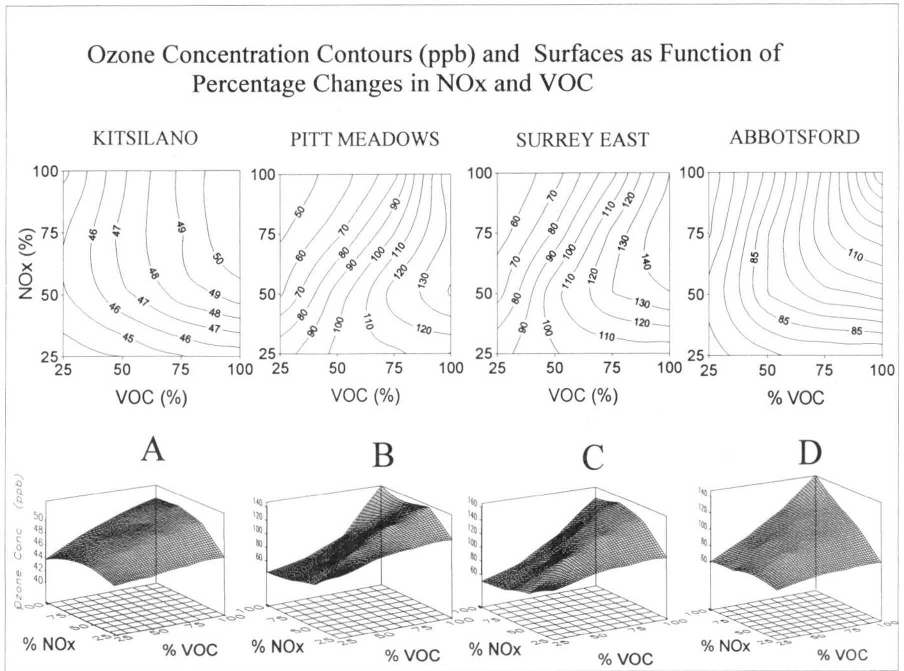

Figure 5. Ozone contours and isopleth surfaces at sites across the Fraser Valley. (Vertical exaggeration of 2x). Sites A, B, C, and D correspond to those in Figure 1, and lie west to east across the valley. The west urban area of Kitsilano and the downwind sites of Pitt Meadows and Surrey East all exhibit VOC limited conditions.

DISCUSSION

UAM-V model simulation of an ozone episode in the Fraser Valley has been successfully demonstrated. The 1985 simulated base case ozone contour patterns were in close agreement with observed patterns. Peak values, however, tended to be overpredicted in the west urban core and underpredicted in the eastern suburban agricultural areas, possibly due to an inaccurate emissions inventory. The overall pattern suggests the meteorology is being suitably represented with a west to east transport of precursors in the sea breeze flow. Enhanced meteorology and emissions datasets in the 1993 episode are expected to produce even closer agreement.

NO_x reductions applied across the airshed showed a non-linear relationship with output ozone concentrations. 25%, 50%, and 75% reductions in NOx resulted in a 9.1% increase, a 10.8 % increase, and a 16.7% decrease in ozone concentration from the base case, respectively. Ozone concentrations showed a more linear response to VOC reductions. 25%, 50%, and 75% reductions in VOC resulted in a 16.4% decrease, a 32.9% decrease, and a 44.3% decrease in ozone concentration from the base case, respectively.

Response to precursor emission reductions also varied with site location across the airshed (Figure 5). The western urban area near Kitsilano, and the sites downstream influenced by industrial sources showed VOC limited features. Pitt Meadows and Surrey East showed ozone **increases** from 25% and 50% reductions in NOx before showing a significant drop at 75%. At the eastern end of the airshed, Abbotsford showed more linear decreases in ozone with decreases in NOx. VOC reductions have the highest impact at all sites, but seem most effective in the central part of the valley.

The results of this study partially agree with the Finlayson-Pitts and Pitts study, which demonstrates that varying VOC and NOx controls would be appropriate in different parts of an airshed. In their study, however, urban areas showed relatively low VOC/NOx ratios and were VOC limited while downwind rural areas showed higher VOC/NOx ratios and were NOx limited. The Fraser Valley does not appear to exhibit a strong NOx limited situation at the east agricultural end of the valley, but remains slightly VOC limited, perhaps due to dissimilarities in emissions inventories. The eastern end of the Fraser Valley may also not be as rural as that in the Los Angeles area study.

The 1985 base case simulation provides a comfortable bases for continuing the Fraser Valley modeling efforts. Future modeling will continue to provide policy makers with scientific information on which to base sound air quality planning strategies for the area.

Acknowledgments

The author would like to thank the following who assisted in this project: J. Haney, S. Douglas, N. Lolk and G. Mansell, from S.A.I., for their assistance with implementation of the UAM-V model; R. McLaren and S. Bohme, National Research Council, for preparation of the emissions files; B. Thomson of Environment Canada for his guidance on the project.

REFERENCES

Finlayson-Pitts, B.J., and Pitts, J.N. Jr., 1993, Atmospheric chemistry of tropospheric ozone formation: scientific and regulatory implications, *Air & Waste.* 43:1091.

Magliano, K.L., and Chinkin, L.R., 1991, San Joaquin valley air quality study technical suport study number 5 - emission inventory assessment, Proceedings of the Air & Waste Management Association 84th Annual Meeting & Exhibition, Vancouver.

McKendry, I.G., 1994, Synoptic circulation and summertime ground-level ozone concentrations at Vancouver, British Columbia, *J. Appl. Meteor.* 33: 627.

Milford, J.B., Russell, A.G., and McRae, G.J., 1989, A new approach to photochemical pollution control: implications of spatial patterns in pollutant responses to reductions in nitrogen oxides and reactive organic gas emissions, *Environ. Sci. Technol.* 23:1290.

Taylor, E., 1991, Forecasting ground-level ozone in vancouver and the lower fraser valley of british columbia, Report PAES-91-3, Scientific Services Division, Pacific Region, Environment Canada, Vancouver.

Thomson, R.B., Bottenheim, J.W., and Steyn, D.G., 1993, The lower fraser valley oxidant study pacific '93, Proceedings of an A&WMA International Specialty Conference 'Regional Photochemical Measurement and Modeling Studies', San Diego.

User's guide to the variable-grid urban airshed model, (UAM-V), 1995, ICF Kaiser, Systems Applications International, SYSAPP-95/027, San Rafael.

DISCUSSION

J. LANGNER: Please explain the rationale for the suggested modeling of the 1993 Pacific episode using the 1985 meteorology. It seems to me that by using meteorology not related to the 1993 measurement date, you will have little basis for model evaluation.

J. L. POTTIER: Model evaluation and verification are being performed with two cases: 1)1985 meteorology and chemistry, and 2) 1993 meteorology and chemistry. The 1985 meteorology will be used with 1993 chemistry to model future scenarios. While the 1985 case consisted of fewer data points, the meteorology was more representative of a typical ozone episode than in 1993. The emission inventory which provides the chemical species, however, was much improved in the 1993 case. The aim is to model a typical ozone event, with the meteorology most conducive to ozone formation, and with the most accurate emissions, which can be accomplished by using the 1985 meteorology with the 1993 emissions.

P. QUANDALLE: What chemical scheme did you use with UAM-V?

J. L. POTTIER: The original chemistry was the Carbon Bond IV chemical mechanism, which has been updated to add the XO2-RO2 reaction along with new temperature effects from PAN reactions, and a new isoprene mechanism. Aqueous - phase chemistry can also be incorporated as an option in the model. A dry deposition algorithm has been incorporated, similar to that used by the Regional Acid Deposition Model (RADM).

D. W. BYUN: In your presentation, you told us that one of the reasons for using UAM-V is the two-way nesting technique. Coming from our own study, we encountered very disturbing problems of mass consistency of species which involved non-linear photochemical reactions. Applying current models for control strategy studies, it is better to locate high resolution nest domain appropriately avoiding complicated topographic features (such as large waterbodies), and apply 1-way nesting technique rather than 2-way nesting.

J. L. POTTIER: Only one-way nesting was used in this case. Two-way nesting would require more detailed and finer resolution input data, which were not available.

ESTIMATES OF SENSITIVITIES OF PHOTOCHEMICAL GRID MODELS TO UNCERTAINTIES IN INPUT PARAMETERS, AS APPLIED TO UAM-IV ON THE NEW YORK DOMAIN

Steven R. Hanna[1], Joseph C. Chang[1], Mark E. Fernau[1], and D. Alan Hansen[2]

[1]Earth Tech, Inc., 196 Baker Ave., Concord, MA 01742 USA
[2]EPRI, P.O. Box 10412, Palo Alto, CA 94303 USA

INTRODUCTION

Because photochemical grid models such as UAM-IV are being used to make policy decisions concerning emissions controls, it is important to know what confidence bounds we can place on the model predictions of, for example, how ozone will respond to changes in emissions. These are presently unknown. The factors influencing prediction error can be classified as input errors, as model formulation errors, or as random stochastic processes. In the present study we include among inputs such things as initial and boundary conditions, emissions, meteorological variables and chemical rate constants. Bias, imprecision and variability can contribute to the error. Formulation errors would include such things as inaccuracies in advection schemes, numerical solvers, process representations, and temporal and spatial resolution. Sometimes the distinction between input and formulation errors is not well drawn. The study described here has been limited by time and resource constraints to an examination of the prediction error associated only with input, not with model formulation, errors. We therefore implicitly assume, without justification, that the model physics and chemistry are correctly formulated. Since model formulation can influence not only simulation fidelity but how input errors are propagated through the model to output errors, we view the results of this study more as a methodological demonstration than a definitive uncertainty analysis.

As mentioned above, the third factor that influences prediction errors is the random stochastic component. The model predictions represent ensemble averages while the observations represent individual realizations. The stochastic component is impossible to predict with deterministic models and is the cause of much of the "scatter" seen when predictions are compared with observations. Since the current study does not deal with comparisons of predictions with observations, the stochastic component is ignored for the moment.

Thus, our objectives are to explore the effects of input error on model output error and to identify the input uncertainties that contribute most to the output uncertainty. To accomplish this we need to (1) characterize the functional form of the input uncertainties, (2) estimate the magnitudes of the uncertainties in the model predictions, (3) characterize output sensitivities to uncertainties in input parameters, and (4) determine whether these uncertainties and sensitivities change as VOC and NO_x emissions controls are applied (Hansen et al., 1994).

An uncertainty/sensitivity analysis system has been developed and has been tested using an application of UAM-IV to a 230 km by 290 km New York City domain for the 6-8 July 1988 ozone episode. In this application the horizontal grid cell size is 5 x 5 km, with 5 vertical layers: 2 below the mixing height and 3 above it. The analysis system is based on multiple repetitive model runs using Monte Carlo resampling of the input parameters (Hanna et al., 1997). 50 base runs are made using best estimates of 1988 NO_x and VOC emissions, then 50 runs are made for 50% anthropogenic NO_x emissions reductions, and 50 additional runs are made for 50% anthropogenic VOC emissions reductions. The results are analyzed by a combination of methods, including plotting the spatial location of the domain-wide peak hourly ozone concentration, determining the variance (uncertainty) in the ozone predictions at specific geographic positions, and carrying out correlation and regression analyses between the variations in predicted ozone and the variations in input parameters. The specific emphasis of this paper is on how the results change when NO_x or VOC emissions reductions are assumed.

METHODOLOGY

The first step in the process was to identify the key inputs to the UAM-IV photochemical grid model and to solicit the advice of experts concerning the magnitude of the uncertainty for each variable. Table 1 lists these 109 input variables, their uncertainty ranges, and whether the assumed distribution is normal or log-normal. It is seen that there are six input variables in the "Emissions Group", six in the "Initial and Boundary Chemical Concentrations Group", 11 in the "Meteorological Conditions Group", and 86 in the "Carbon-Bond IV Chemical Reaction Group". Note that the most common value of the uncertainty range (i.e., including 95% of data) in Table 1 is ±30%. Model control parameters such as grid size and time step are not included in the set of input parameters being varied, since it is a major effort to modify the UAM-IV input files following a change in grid size.

Many methods exist for the analysis of uncertainty in large environmental models. For example, Carmichael et al. (1997) discuss a sensitivity coefficient approach. The Monte Carlo uncertainty analysis method has been chosen for use here (see Ang and Tang, 1984, for an overview). With this method, each new Monte Carlo run with UAM-IV involves a random and independent selection of each of the 109 input variables. For simplicity, no correlations are assumed among pairs of input variables. It is desirable to make as many runs as possible within time and budget constraints. Since each three-day UAM-IV run took about 10 hours on our workstations, about 50 Monte Carlo runs could be made within our project constraints. This number is sufficient to estimate the variances of the output variables and to assess correlations and regression coefficients. Of course, even if infinite time and computer resources were available and the model ran very quickly, it would be of questionable value to make more than a few hundred Monte Carlo runs.

The analysis of the 50 Monte Carlo runs that used baseline emissions has been thoroughly described by Hanna et al. (1997). The main emphasis of the current paper is the NO_x and VOC emissions reduction runs. Two separate sets of 50 Monte Carlo runs were made - one with across-the-board 50% reductions in anthropogenic NO_x emissions, and another with across-the-board 50% reductions in anthropogenic VOC emissions. In order to help remove the effects of limited selections of input variables (due to only 50 runs), the same 50 sets of input variables (e.g., wind speeds, boundary conditions, etc.) for the baseline runs were used in the emissions reduction runs, and the same relative variations in anthropogenic NO_x and VOC emissions were assumed.

The sets of model predictions from the Monte Carlo runs contain millions of individual data points (i.e., many variables over a three-dimensional grid that varies in time over three days, for 150 model runs). It is imperative that the analysis be limited to a small subset of these data, so that the key technical questions can be answered by a few figures, tables, and calculations. This paper focuses on the predicted peak hourly-averaged ozone concentration during 7-8 July 1988

Table 1. Listing of 109 UAM-IV Input Variables and Assumed 95% Uncertainty Ranges (with Respect to the Medians). Log-Normal and Normal Distributions are denoted by L-N or N, Respectively.

Number	Input Variable	Uncertainty Range (includes 95% of data)	PDF
	Emissions Group		
1	Anthropogenic NO_x area source emissions	±40%	L-N
2	Anthropogenic VOC area source emissions	±80%	L-N
3	Anthropogenic NO_x point source emissions	±30%	L-N
4	Anthropogenic VOC point source emissions	±50%	L-N
5	Biogenic NO_x emissions	±factor of two	L-N
6	Biogenic VOC emissions	±factor of two	L-N
	Initial and Boundary Chemical Concentrations Group		
7	Initial ozone concentration	±30%	L-N
8	Initial NO_x **and** VOC concentration	±50%	L-N
9	Boundary ozone concentration	±30%	L-N
10	Boundary NO_x **and** VOC concentration	±80%	L-N
11	Top ozone concentration	±70%	L-N
12	Top NO_x **and** VOC concentration	±factor of two	L-N
	Meteorological Conditions Group		
13	Wind speed	±30%	L-N
14	Wind direction ±0.52 radians	(±30 degrees)	N
15	Mixing depth	±30%	L-N
16	Vertical diffusivity K_z	±50%	L-N
17	Relative humidity	±30%	N
18	Surface temperature	±3 K	N
19	Stability class	± one class	N
20	Temperature gradient in daytime mixed layer	±0.002 K/m	N
21	Temperature gradient at night and aloft	±0.005 K/m	N
22	Region top	±30%	L-N
23	Deposition velocity vd	±50%	L-N
	Chemical Reaction Group		
24-109	Chemical rate constants for Carbon Bond - IV reactions 1-86 (all assumed to have an uncertainty of ±30%, a log-normal distribution, and to vary independently)		

anywhere on the domain and at ten key receptor locations shown in Figure 1. The variations of the locations of the domain-wide maxima as emissions are reduced are presented in the first part of the analysis. Then the variations of the scatter of the ozone distributions at the ten receptors are shown as the emissions are reduced. Finally, the input variables that are most strongly correlated with variations in predicted ozone are described and compared for the emissions-reduction runs.

LOCATIONS OF PREDICTED DOMAIN-WIDE PEAK OZONE

Each new Monte Carlo UAM-IV run produces a single domain-wide value for the peak hourly-averaged ozone concentration for the 7-8 July 1988 period. This concentration represents an average over the 5 km x 5 km grid volume closest to the ground (its vertical depth is usually about 500 to 1500 m in the afternoon when the daily ozone peak occurs). Figure 1 contains the magnitudes (ppb) and locations of these 50 predicted peak ozone concentrations from the 50% VOC emission reduction runs. The predicted peaks range from 184 ppb to 237 ppb, and the locations are mostly clustered near the southwest (upwind) boundary. It can be concluded that, when 50% VOC anthropogenic emissions reductions are imposed, the domain-wide peaks are highly dependent on the boundary ozone concentrations. Note that the mean boundary concentrations were not modified in these emissions reduction runs. This is unrealistic, since if emissions were reduced inside the domain, they would likely be reduced to some extent outside the domain, too.

Similar maps were plotted for the base emission runs and for the 50% NO_x anthropogenic emissions reduction runs. They are not given here because of page limitations on this manuscript. The map showing the locations and magnitudes of the maximum predicted ozone concentrations for the base emissions runs, presented by Hanna et al. (1997), is different from Figure 1 in that it shows a major cluster of predicted maxima about 50 to 100 km to the north and northeast of New York City. A few maxima were located near the boundary as in Figure 1. The map for the 50% NO_x emissions reduction runs (not plotted here) is also different in that it shows that the maxima are clustered closer to New York City. That is, once NO_x emissions are reduced, the "ozone titration effect" in the source area around New York City is reduced, and consequently ozone concentrations increase in that area. The influence of the boundary ozone concentrations is diminished in the NO_x emissions reduction runs.

PREDICTED PEAK OZONE DISTRIBUTIONS AT RECEPTOR LOCATIONS AND CHANGES IN DISTRIBUTIONS FOR EMISSION REDUCTION RUNS

The distributions of the 50 predicted peak hourly-averaged ozone concentrations at a given receptor or across the domain can be simply expressed by "box plots", as shown in Figure 2. The significant points on the 33 little box plots represent the minimum, the 2.5 percentile, the 16 percentile, the median, the 85 percentile, the 97.5 percentile, and the maximum. Within each group of three box plots, the individual box plots represent (from left to right) the base emissions, the 50% VOC emission reduction runs, and the 50% NO_x emission reduction runs.

The base run box plot for the domain-maximum shows a variation in predicted peak ozone from about 180 to 330 ppb due to the variations in input parameters. At individual receptors, the base run variations were less (about ± 20 ppb) in rural areas and near boundaries, and were greater (about ± 60 or 80 ppb) in the urban plume and in the New York City source area.

Away from the upwind boundaries, the medians of the box plots for the 50% emissions reduction runs are significantly different than the medians of the box plots for the base emissions runs. In rural downwind areas in the urban plume (i.e., 2-Oxford and 3-East Hartford) the

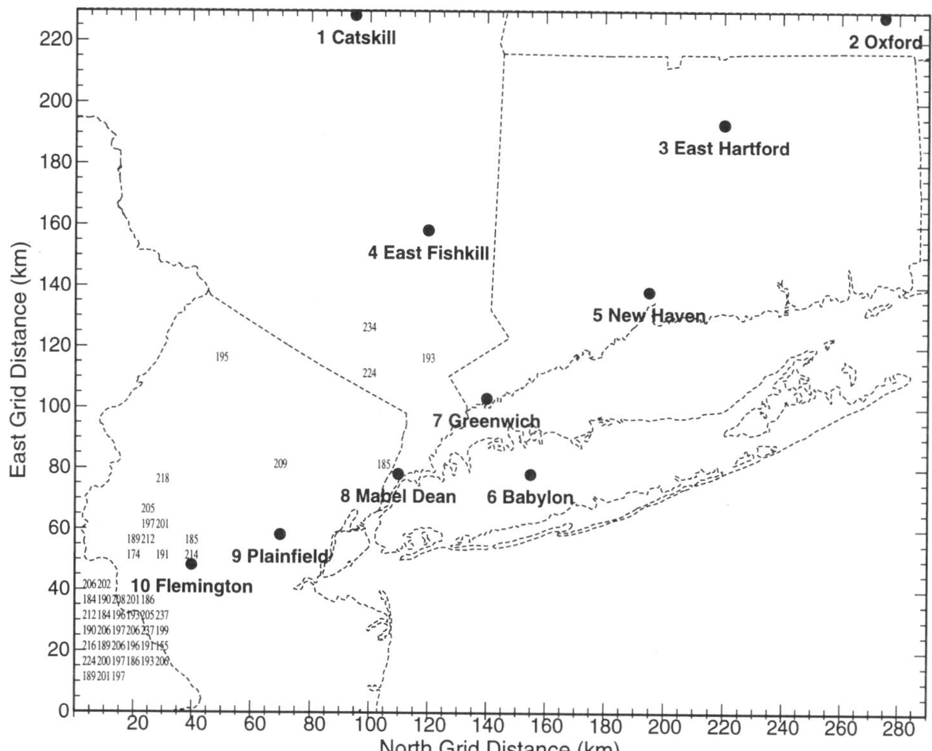

Figure 1. Map of New York UAM-IV domain, showing locations of ten key receptors, as well as locations and magnitudes (in ppb) of the peak domain-wide hourly-averaged ozone concentration for 7-8 July 1988 for the 50 Monte Carlo UAM-IV runs with 50% reductions in anthropogenic VOC emissions. Note that some of the numbers in the lower left corner were moved slightly for clarity, since they originally lay on top of each other.

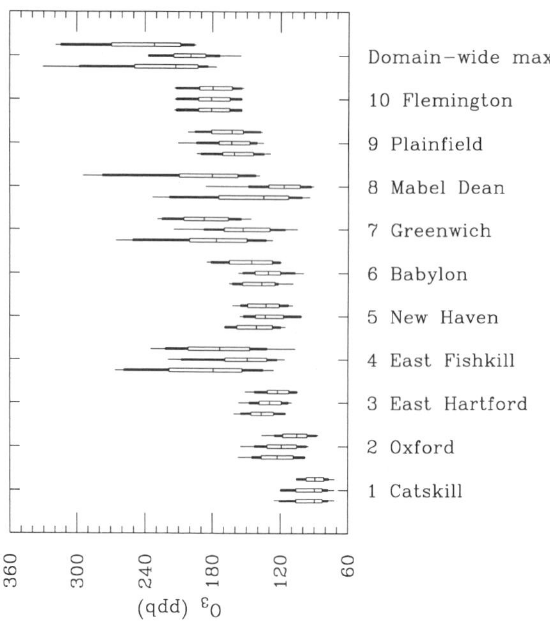

Figure 2. Box plots of distributions from the Monte Carlo runs of peak hourly-averaged ozone concentration for 7-8 July 1988 for the ten receptors shown in Figure 1 and for the domain-wide maximum. Each receptor has a group of three box plots: left - base emissions; middle - 50% VOC emissions reductions; right - 50% NO_x emissions reductions. Significant points on the box plots, from bottom to top, are: minimum, 2.5%, 16%, 50% (median), 84%, 97.5%, maximum.

median concentration drops by 5 to 10 ppb for 50% VOC reductions and by 10 to 20 ppb for 50% NO_x reductions. The variances or scatter remain the same. In the urban source area (i.e., 7-Greenwich and 8-Mabel Dean), the median concentration drops by 10 to 20 ppb for 50% VOC reductions but increases by 20 to 50 ppb for 50% NO_x reductions. This latter phenomenon has been seen in many other photochemical grid model sensitivity studies by other researchers, and is explained by the effect of the chemical reactions between NO and ozone near the NO_x sources (Finlayson-Pitts and Pitts, 1993).

The three box plots in the group for the domain-wide maximum (right side of Figure 2) show the same behavior as the three box plots in the groups for receptors in the source area (i.e., the predicted median ozone decreases by about 10 ppb with 50% VOC emissions reductions and increases by about 20 ppb with 50% NO_x emissions reductions). However, the variance or scatter for the 50% VOC emissions reduction runs is decreased by about a factor of two. This reduction is most likely due to the dominance of the upwind ozone boundary condition for the 50% VOC emissions reduction runs, and the fact that the assumed variability in the input ozone boundary condition is less than that for the input anthropogenic VOC emissions (± 30% vs. ± 80%, see Table 1).

It is also important to analyze Figure 2 to compare the shapes of the distributions shown by the box plots between the base runs and the VOC and NO_x emissions reduction scenarios. For example, did the distribution change from log-normal to normal? Hanna et al. (1997) demonstrated that the distributions for the base runs follow a three-parameter log-normal distribution. The distribution shapes can be qualitatively assessed by a visual inspection of Figure 2, since a box plot for a log-normal distribution has a long tail (i.e., it is skewed towards higher concentrations), while a box plot for a normal distribution has an even distribution towards high and low concentrations. In this exercise, log-normal distributions in predicted ozone are expected at most receptors since most of the input variables in Table 1 are assumed to have log-normal distributions. For example, Table 1 shows that the distribution of uncertainty in VOC anthropogenic area source emissions is log-normal with uncertainty of ± 80%. This expected pattern is seen to be verified for most of the receptors and box-plots in Figure 2. However, the distributions for the NO_x emissions reduction runs have a normal shape at some receptors, such as Catskill, East Fishkill, and Greenwich. At the Oxford, East Hartford, and Plainfield receptors, the distributions are consistently nearly-normal, since the dominant input variable is wind direction, which is also normally-distributed.

SENSITIVITY OF OZONE PREDICTIONS TO UNCERTAINTIES IN INDIVIDUAL INPUT VARIABLES

Correlations and regression coefficients were calculated between the set of 50 predictions of peak hourly-averaged ozone and the set of 50 individual values of each input variable selected in the 50 Monte Carlo runs. This was done for the domain-wide maximum and for the ten receptor locations in Figure 1. Separate sets of calculations were made for the base emissions, the 50% VOC emissions reductions, and the 50% NO_x emissions reductions. The results summarized below emphasize differences between the conclusions for the base emissions runs and the emissions reduction runs.

A major difference between the three sets of 50 runs concerns the input variable that gives the largest correlation with the variations in ozone predictions for the domain-wide maximum. Variations in VOC anthropogenic area source emissions have the strongest correlation (about 0.7) with variations in predicted domain-wide peak ozone for the base emissions runs and the 50% NO_x emissions reduction runs, where the location of the peak is usually near New York City or in the downwind plume. However, this result changes for the 50% VOC emissions reduction runs, for which the variations in ozone boundary conditions show the strongest correlations with peak ozone on the domain. This correlation could have been anticipated from Figure 1, where it can be seen that most of the locations of peak ozone are clustered near the upwind boundary.

Hanna et al. (1997) discuss the significant correlations and regression results for the base emissions Monte Carlo runs for the domain-wide maximum and for the ten receptors. The dominant input variable at a few of the receptors about 100 km to 200 km downwind of New York City was wind direction, since a 10¡ change in wind direction could shift the big urban plume on or off a receptor. At some receptors, variations in chemical reaction rates were found to be important (e.g., reaction 3: $O_3 + NO \rightarrow NO_2 + O_2$ and reaction 46: $C_2O_3 + NO \rightarrow NO_2 + XO2 + FORM + HO_2$). Changes in the sensitivities of predicted peak ozone to uncertainties in important input variables between the base emissions runs and the emissions reduction runs are briefly summarized below, by input variable group. Possible reasons for the changes (if known) are suggested in parentheses.

Emissions Group:
- No major changes, except slight drop-off in influence of emissions inputs for 50% VOC reductions.

BC/IC Group:
- Ozone B.C. more important for domain-wide maximum for 50% VOC reductions. (This result is caused by the fact that the maximum occurs near the boundary.)
- Ozone I.C. more important for receptors in NYC urban plume for 50% VOC reductions. (In the source region, the predicted ozone is dominated by the I.C. when VOC emissions are reduced.)
- NO_x and VOC B.C. more important in many locations for 50% VOC reductions; but less important for 50% NO_x reductions. (The location of the domain maximum is near the boundary for the 50% VOC runs and is near NYC for the 50% NO_x runs.)

Meteorology Group:

- Wind speed is important only for 50% NO_x reduction runs for receptors in and near New York City. (Advection and dilution are important near the source.)
- Temperature gradient at night and aloft is also important only for 50% NO_x reduction runs near New York City. (Near the source, vertical stability affects the plumes.)

Reaction Rate Group:

- R1 (NO_2 + hν -> NO + O), which is correlated with increases in peak ozone, is important only for the 50% VOC reduction runs and near New York City.
- R3 (O_3 + NO -> NO_2 + O_2), which is correlated with decreases in peak ozone, is more important for the 50% VOC reduction runs but is not important for the 50% NO_x reduction runs.
- R46 (C_2O_3 + NO -> NO_2 + XO2 + FORM + HO_2), which is correlated with increases in peak ozone, is less important for the 50% VOC reduction runs.
- R50 (C_2O_3 + HO_2 ->0.79 FORM + 0.79 XO2 + 0.79 HO_2 + 0.79 OH), which is correlated with decreases in peak ozone becomes important in the 50% NO_x reduction runs, in the urban plume close to New York City.
- R57 (OH + OLE -> FORM + ALD2 + XO2 + HO_2 - PAR), which is correlated with decreases in peak ozone, drops out entirely for the 50% VOC and 50% NO_x reduction runs.

The results outlined above suggest that future research should be directed towards the input variables that seem to have the largest effect on variations in predicted peak ozone. For example, the dominant meteorological input variables are wind direction, wind speed and night-time vertical temperature gradient. The reaction rates that should be carefully specified are R1, R3, R46, R50, R57 in the Carbon Bond-IV system. Of course these results may change for other geographic domains, other time periods, other models, and, perhaps if model formulation error had been accounted for.

Acknowledgments: The project is supported by the Electric Power Research Institute. Our consultants, H. Christopher Frey (North Carolina State University), Eduard Hofer (GRS, Munich, Germany), and Owen Hoffman (SENES Oak Ridge, Inc.) made major contributions to the design of the Monte Carlo project and the interpretation of the results. Dr. Peter Mayes of Earth Tech assisted with the analyses.

REFERENCES

Ang, A.H.-S. and Tang, W.H., 1984, Probability Concepts in Engineering Planning and Design, Volume 2: Decision, Risk, and Reliability. John Wiley and Sons, New York.

Carmichael, G.R., Sandu A. and Potra, F.A., 1997, Sensitivity analysis for atmospheric chemistry models via automatic differentiation. Atmos. Environ. 31, 475-489.

Finlayson-Pitts, B.J. and Pitts, J.N., Jr., 1993, Atmospheric Chemistry of tropospheric ozone formation: scientific and regulatory implications. J. Air and Waste Management Assoc., 43, 1091-1100.

Hanna, S.R., Chang, J.C., and M.E. Fernau, 1997, Monte Carlo estimates of uncertainties in predictions by a photochemical grid model (UAM-IV) due to uncertainties in input variables. Submitted to Atmospheric Environment.

Hansen, D.A., Dennis, R.L., Ebel, A., Hanna, S., Kaye, J., and Thuillier, R., 1994, The quest for an advanced regional air quality model. Environ. Sci. and Tech. 28, 560A-569A.

EFFECTS OF INITIAL AND BOUNDARY VALUES OF REACTIVE NITROGEN COMPOUNDS AND HYDROCARBONS ON THE OZONE CONCENTRATION IN THE FREE TROPOSPHERE

B. Schell[a], H. Feldmann[b], M. Memmesheimer[b], A. Ebel[b]

[a]Ford Forschungszentrum Aachen, Dennewartstr. 25, 52068 Aachen, Germany.
[b]University of Cologne, EURAD Project, Aachener Str. 201-209, 50931 Cologne, Germany.

INTRODUCTION

The initialisation and the treatment of the boundary conditions of a mesoscale chemistry-transport-model, covering a limited area, are of great importance. The choice of the initial and boundary values can significantly influence the results of a simulation, so that they should be determined as well as possible (NAPAP, 1991). For this reason it is important to provide realistic conditions, if possible derived from current measurements. Unfortunately trace species in the troposphere, especially in the middle and upper free troposphere, are not observed continuously so that relatively little is known about background concentrations. Usually there are no current observations available which can be used as input data for episodic simulations. The available measurements show a high variability in the concentrations of the trace species. To analyse and to quantify the effects of a variation of the initial and boundary values for model results sensitivity studies were carried out with the European Air Pollution Dispersion modeling system (EURAD) using different initial and boundary scenarios. Therefore the literature has been reviewed and a set of initial and boundary values were derived based on available observation data. With regard to the formation of ozone the focus was set on reactive nitrogen species and hydrocarbons, which are important photooxidant precursor species. First a set of simulations with different scenarios representing free tropospheric conditions is calculated with a boxmodel version of the EURAD model in order to determine the non-linear dependencies of the gas phase chemistry. Furthermore a sensitivity study with the full three dimensional model is performed for a summersmog episode.

THE EURAD MODELING SYSTEM

The EURAD modeling system contains three major modules: the mesoscale meteorological model MM5 (Grell et al., 1994), the EURAD emission model (Memmesheimer et al., 1991), and the chemistry-transport model CTM2 (Hass, 1991; Hass et al., 1993).

Air Pollution Modeling and Its Application XII
Edited by Sven-Erik Gryning and Nadine Chaumerliac, Plenum Press, New York, 1998

The CTM2 is shortly described in the following. Descriptions of the other model components and more details about the CTM2 can be found in the cited references.

The Chemistry Transport Model CTM2

The simulation of the chemistry and transport is performed with the CTM2, which is derived from the RADM model (Chang et al., 1987). A set of prognostic equations has to be solved for the concentrations C_n of each chemical constituent:

$$\frac{\partial C_n}{\partial t} = -\nabla(\mathbf{u}C_n) + \nabla(\mathbf{K}_e\nabla C_n) + \frac{\partial C_n}{\partial t}|_{dep} + \frac{\partial C_n}{\partial t}|_{cld} + \frac{\partial C_n}{\partial t}|_{aer} + P_n - L_n + E_n, \quad (1)$$

- $\nabla(\mathbf{u}C_n)$: advection
- $\nabla(\mathbf{K}_e\nabla C_n)$: turbulent diffusion
- $\frac{\partial C_n}{\partial t}|_{dep}$: deposition
- $\frac{\partial C_n}{\partial t}|_{cld}$: changes due to effects of clouds chemistry, transport, wash- and rainout
- $\frac{\partial C_n}{\partial t}|_{aer}$: changes due to aerosols
- P_n: gas-phase chemistry (production)
- L_n: gas-phase chemistry (loss)
- E_n: emissions.

The RADM2 chemical mechanism (Stockwell et al., 1990) is used for the gas phase chemistry. 39 of the 63 model species are treated as prognostic variables with 13 anorganic and 26 organic substances. 15 of the organic model species are representing classes of chemical compounds with similar chemical properties, where the others represent individual chemical compounds. This treatment of the organic compounds can cause some problems assigning measurements to the model classes. Usually the observations do not describe the model classes completely and often only the concentrations of the most abundant species are given.

The initialisation follows the approach of Chang et al. (1987). Average climatological vertical profiles of the described substances are generally used as initial and boundary conditions. Additionally two days prior to the actual simulation period are used as 'initialisation period' to calculated self consistent initial values.

Boxmodel

The boxmodel used for the simulations corresponds to an isolated gridbox of the three-dimensional CTM2. It is limited to the gas phase chemistry so that it is an useful tool to study the answer of the RADM2 mechanism to variations of the initial concentrations. The development of the concentrations is descibed by

$$\frac{\partial C_n}{\partial t} = P_n - L_n \quad (2)$$

no other effects are implemented. This allows a detailed study of the chemical effects without interference from transport, emissions or deposition. In a simple way it can be

Table 1: Description and names of the boxmodel simulations. k08 and k12 refer to 778 hPa, 281 K and 416 hPa, 252 K, respectively.

model run			description
run	level	note	
REF	k08	–	reference, NO_x = 0.054ppbv, VOC = 0.272ppbv
REF	k12	–	reference, NO_x = 0.051ppbv, VOC = 0.115ppbv
NOX	k08	50%	NO_x initial values 50% of reference
NOX	k12	100%	NO_x initial values 100% of reference
NOX	k08	500%	NO_x initial values 500% of reference
VOC	k12	50%	VOC initial values 50% of reference
VOC	k08	100%	VOC initial values 100% of reference
VOC	k12	500%	VOC initial values 500% of reference

considered as an air parcel advected from the boundary, so that one can get hints to possible reasons of effects in three dimensional simulations.

RESULTS AND DISCUSSION

Boxmodel Studies

The boxmodel was applied to simulate three days under summerly conditions. Photolysis rates were calculated for the 03.09.1991 at 50°N 0°W. Height dependend temperature and humidity profiles were averaged from three-dimensional simulation results. Model runs were conducted with different initial values and at different pressure levels, representing conditions found in the free troposphere. The initial values of reactive nitrogen compounds and VOC were varied to study the behaviour of the ozone concentration under different conditions (see tabel 1).

Figure 1 shows the temporal development of the ozone mixing ratio at 416 hPa and 252 K depending on the NO_x and VOC initial values. More NO_x leads due to

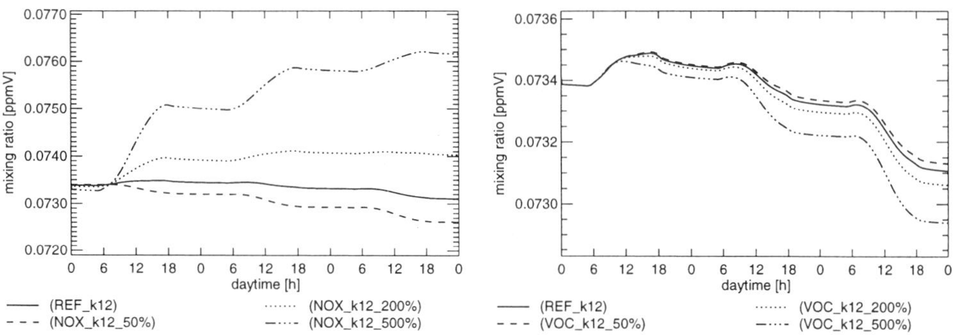

Figure 1. Boxmodel results: ozone mixing ratio in dependence on the variation of the initial values (T = 252K, p = 416hPa, see also table 1).

photochemical production to a higher ozone mixing ratio, whereas higher VOC initial values cause a decomposition of ozone in a low NO_x regime. The ozone dependence upon the VOC mixing ratio is weak (see figure 1, right frame), but after three days there is a stronger destruction of ozone if higher VOC initial conditions are used. In figure 2 the ozone isopleths are shown at p = 778hPa, T = 281K and 18:00 local time on the first model day. One can see that at this time the ozone chemistry is NO_x-limited over the whole range of the used initial values. This is supported by the results shown in figure 1. The ozone decomposition is more developed for all simulations at level k08.

A possible explanation is the reaction path of the peroxiradicals which are produced as a result of the decomposition of hydrocarbons. If one focus on the HO_2 radical there are generally two possible reaction paths in dependence on the NO_x mixing ratio which are of importance for the ozone mixing ratio:

$$\text{high } NO_x: \quad NO + HO_2 \rightarrow NO_2 + OH \tag{R1}$$

$$\text{low } NO_x: \quad O_3 + HO_2 \rightarrow OH + O_2 \tag{R2}$$

High NO_x mixing ratios support ozone formation whereas low NO_x mixing ratios lead to ozone decomposiotion through the reaction of the HO_2 radical with ozone. The ratio of the reaction rates R is

$$\frac{R_{R2}}{R_{R1}} = \frac{k_{R2}[O_3][HO_2]}{k_{R1}[NO][HO_2]} = \frac{k_{R2}[O_3]}{k_{R1}[NO]}, \tag{3}$$

where k_{R1} and k_{R2} are the rate constants of the reactions R1 and R2, respectively. If the ratio $[NO]/[O_3]$ is greater than k_{R2}/k_{R1} there is a net production of ozone, if it is lower an ozone destruction occurs. The ratios are roughly of the same order for the used initial values so that both, destruction and production of ozone, can occure. The trends shown in figure 1 can partially be explained by these ratios.

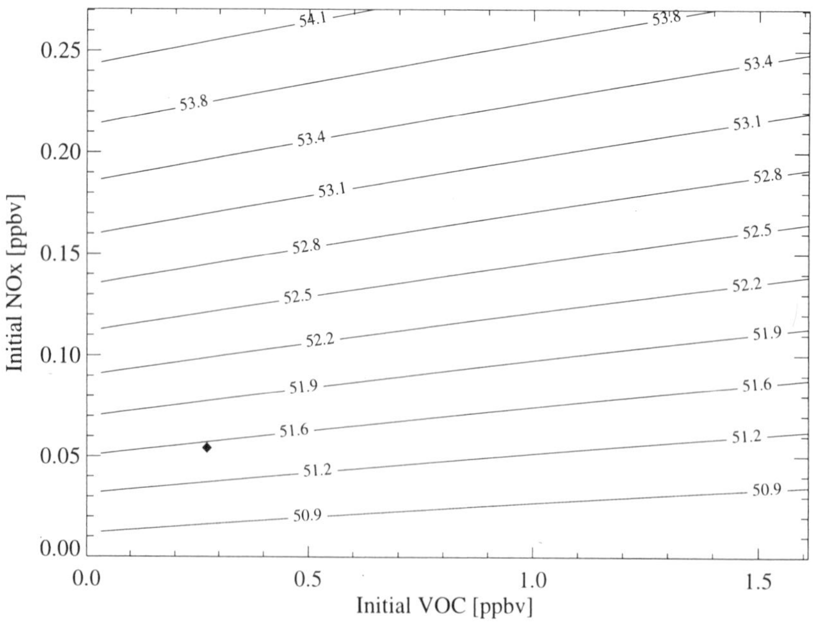

Figure 2. Ozone isopleth diagram for 18:00 local time (p = 778hPa, T = 281K), concentrations in ppbv. The diamond (◇) represents the reference initial values used for REF_k08.

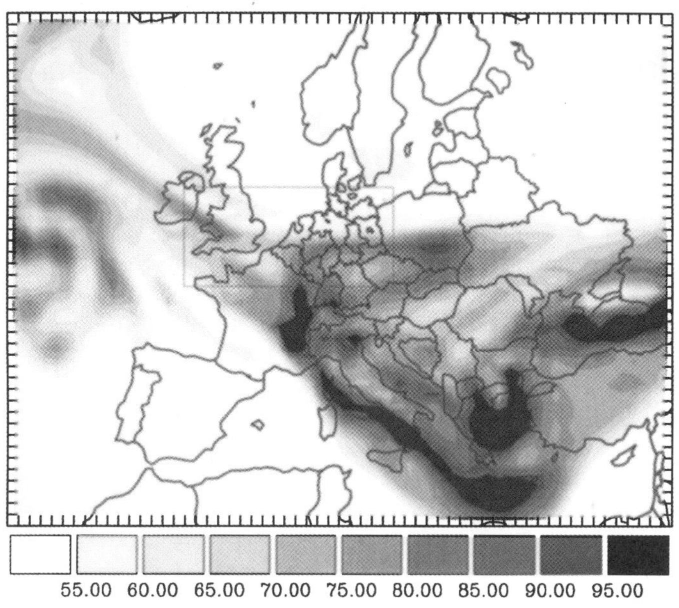

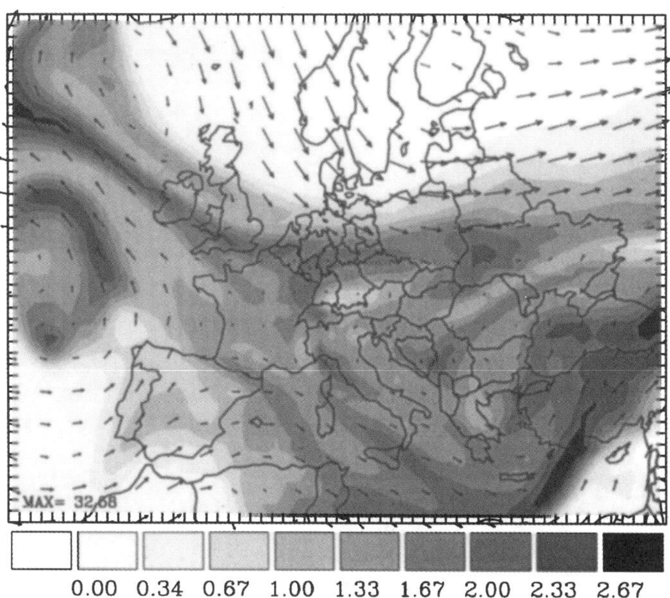

Figure 3. O_3 mixing ratio (ppbv) at 1200 GMT 05 September 1991 at aproximately 4150 m above ground for the BASE case simulation (top) and the difference ΔO_3 (ppbv) between the NOY case and the BASE case simulation (bottom). The corresponding wind vectors are shown for the simulation and the maximum wind velocity (m/s) of the domain is printed in the lower left corner.

CTM2 Model Studies

The EURAD modeling system has been applied to a summerly high pressure period in August/September 1991, the so called SANA3 episode (for details see Feldmann et al. (1995, 1996) and references therein), which lasts from 28. Aug. to 6. Sep. 1991. The first two days are used as initialisation periode for the CTM2. This summersmog episode is characterized by a high pressure system over eastern central Europe and it is terminated by a northerly flow of cold moist air after the 4th September. The horizontal grid spacing is 80 km and the vertical direction is resolved by 15 levels between the earth's surface and 100 hPa ($\sim$16 km) using a σ-coordinate system. The modelling domain covers most of Europe (see figure 3).

Previous studies (Feldmann et al., 1995) have shown, that the subsidience from the free troposphere contributes to the ozone budget for the boundary layer and the enhanced near surface concentrations during this episode. To test the influence of the initial and boundary values in the free troposphere sensitivity studies were carried out with different initial and boundary conditions derived from the literature based on available observation data. In this paper some results of the base case simulation and the so called NOY case simulation are presented (further refered to as BASE and NOY, respectively). For the NOY case PAN values are updated based on the recent literature (see e.g. Perros (1994) and references therein) and are implemented in the model. Besides the PAN values the initial and boundary values of the model species ONIT, representing organic nitrates, are changed. Observations indicate a shortfall of measured NO_y compared with the sum of the individual nitrogen oxides (e.g. Singh et al. (1996) and references therein). Fine nitrogen aerosols and/or complex organic nitrates are discussed as possible reasons. Hence, for this sensitivity study the NO_y shortfall has been added to ONIT. Figure 4 shows the used vertical profiles (left frame) of PAN and ONIT for the simulations BASE and NOY.

The simulated ozone concentration for the BASE case and the differences (NOY-BASE) are shown in figure 3 at aproximately 4150 m, 05. Sep. 1991, 1200 GMT. There are two distinct air masses in the model domain: over the continent relatively

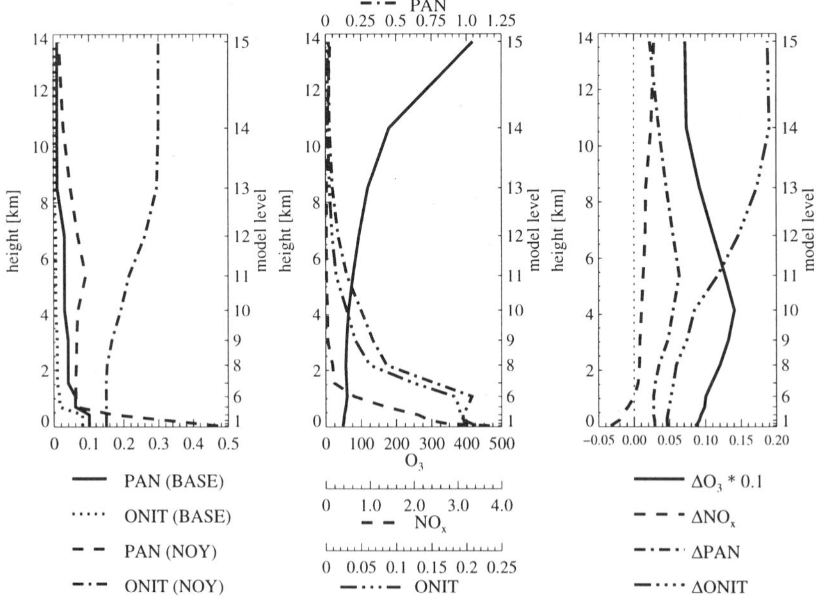

Figure 4. *Left*: vertical initial and boundary profiles of PAN and ONIT in ppbv for BASE and NOY simulation; *middle*: spatial averaged vertical profiles of O_3, NO_x and PAN (all in ppbv) at 1200 GMT 05 September 1991 for the BASE case simulation; *right*: differences NOY-BASE in ppbv of the averaged vertical profiles of O_3, NO_x and PAN. The averages were calculated for the domain shown in figure 3 (rectangle in the upper frame).

polluted aged air can be found whereas in the northern part much stronger winds advect unpolluted air into Europe. On 5th September the frontal zone is placed over northern Germany. The variation of the NOY initial and boundary values has a significant effect on the ozone concentration within the aged polluted air mass (see figure 3 bottom) whereas the effect is low in the unpolluted fresh air mass. In this air mass the ozone boundary value dominates. Few points excepted, the increase of the PAN and ONIT initial and boundary values results in an increase of the ozone mixing ratio in the aged air mass.

Spatial averaged vertical profiles and the averaged differences (NOY–BASE) of O_3, NO_x, PAN and ONIT are shown in figure 4. There is an increase of the ozone mixing ratio over the whole altitude range of the free troposphere. The maximum value of the averaged difference (NOY-BASE) is 1.406 ppbv at a height of about 4 km. The NO_x concentration is also increased in the free troposphere which is a consequence of the elevated PAN and ONIT values. This is an indication that PAN and ONIT act as a NO_x reservoir in the free troposphere and that they are important for the long-range-transport. Boxmodel studies showed that especially the increase of the ONIT values leads to an increase of the NO_x mixing ratio. If only the PAN initial and boundary values are changed the effect is less clear pronounced. The PAN mixing ratio is also elevated over the whole altitude range, showing the maximum difference of 0.064 ppbv at 5.4 km. This reflects the initial and boundary profile. The shape of the ONIT initial and boundary profile is reflected by the averaged profile, too. However, the concentrations are somewhat shifted to lower values over the whole altitude range. The experiment indicates that to a large extent the observed shortfall of NO_y could be explained by organic nitrates.

The direct influence of the boundary values caused by advective transport is clearly seen behind the frontal passage. Because of much stronger winds the effects are more pronounced in the free troposphere, whereas they are lower in the planetary boundary layer.

CONCLUSIONS

Initial and boundary value sensitivity studies have been performed with a boxmodel and the three dimensional EURAD model. The results have been analysed with respect to the effects on the ozone concentration in the free troposphere. The boxmodel results show that the ozone chemistry is NO_x-limited in the free troposphere and the sensitivity against variation of hydrocarbons' concentrations is low. The NO_x dependence can also be seen in the three dimensional simulation results of the considered episode. A set of initial and boundary values has been derived from available observation data in the literature for the model species PAN and ONIT. These species act as a NO_x reservoir in the free troposphere. The elevated PAN and ONIT values result in increased NO_x and O_3 concentrations.

The increased mixing ratios might be important features with regard to budget studies of photooxidants and their precursors (Feldmann et al., 1995; Memmesheimer et al., 1997). It is shown here that the influence of initial and boundary values can be significant, especially for precursor species such as PAN. Depending on the amount of reactive NO_y a tendency for chemical loss or production of ozone can be found within the free troposphere. Because of the high background concencentration the ozone mixing ratio is raised only about 3% in the NOY case. Further analysis is necessary to clarify the role of changes in boundary and initial values for the results of budget studies. More measurements are needed especially in the free troposphere and not only of species as ozone and NO_x to evaluate the presented results. Based on available observations it is not possible to reduce the uncertainties caused by initial and boundary values for episode simulations considerably. However, the range of possible uncertainties could be quantified. A denser network with specific emphasis on the improvement of the obsevational data base for NO_y in the free troposphere would help considerably to understand the contribution of an anthropogenically polluted area as Europe to the global atmosphere. Such a data base would also improve the evaluation of chemical and dynamical processes as simulated by mesoscale air pollution models.

REFERENCES

Chang, J.S., Brost, R.A., Isaksen, I.S.A., Madronich, S., Middelton, P., Stockwell, W.R., and Walcek, C.J., 1987, A three-dimensional eularian acid depositon model: physical concepts and formulation, *J. Geophys. Res.* 92(D12):14681–14700.

Feldmann, H., Ebel, A., Mass, H., Memmesheimer, M., and Jakobs, H.J., 1995, Analysis of polluted air masses effecting the area of Eastern Germany during a SANA episode, In: A. Ebel and N. Moussiopoulos, eds., *Air Polution III*, volume 4: Observation and simulation of air pollution: results from SANA and EUMAC (EUROTRAC), pp. 95–102, Southampton, CMP.

Feldmann, H., Hass, H., Memmesheimer, M., and Jakobs, H.J., 1996, Budgets of atmospheric sulfer for East Germany based on meso-α-scale simulations, *Meteorol. Zeitschrift* 5:193–204.

Grell, G.A., Dudhia, J., and Stauffer, D.R., 1994, A description of the fifth-generation Penn State/NCAR mesoscale model (MM5), Technical note TN-398+STR, NCAR, Boulder, Colorado.

Hass, H., 1991, Description of the eurad chemistry-transport-model version2 (CTM2), In: A. Ebel, F. Neubauer, and P. Speth, eds., *Mitteilungen aus dem Institut für Geophysik und Meteorologie der Universität zu Köln*, volume 83, Universität Köln, Köln.

Hass, H., Ebel, A., Feldmann, H., and Memmesheimer, M., 1993, Evaluation studies with a regional chemical transport model (EURAD) using air quality data from the EMEP monitoring network, *Atmos. Environ.* 27A(6):867–887.

Memmesheimer, M., Ebel, A., and Roemer, M., 1997, Budget calculations for ozone and its precursors: seasonal and episodic features based on model simulations, *J. Atmos. Chem.* in press.

Memmesheimer, M., Tippke, J., Ebel, A., Hass, H., Jacobs, H.J., and Laube, M., 1991, On the use of EMEP emission inventories for European scale air pollution modeling with the EURAD model, In: J. Pankrath, ed., *EMEP workshop on photooxidant modelling for long-range transport in relation to abatement strategies*, pp. 307–324, Umweltbundesamt, Berlin.

NAPAP, 1991, *Acid deposition: state of science and technology*, volume 1, emissions, atmospheric processes, and deposition, U.S. National Acid Precipitation Assesment Program, Washington, DC.

Perros, P.E., 1994, Large-scale distribution of peroxyacetylnitrate from aircraft measurements during the TROPOZ II experiment, *J. Geophys. Res.* 99(D4):8269–8279.

Singh, H.B., Herlth, D., Kolyer, R., Salas, L., Bradshaw, J.D., Sandholm, S.T., Davis, D.D., Crawford, J., Kondo, Y., Koike, M., Talbot, R., Gregory, G.L., Sachse, G.W., Browell, E., Blake, D.R., Rowland, F.S., Newell, R., Merrill, J., Heikes, B., Liu, S.C., Crutzen, P.J., and Kanakidou, M., 1996, Reactive nitrogen and ozone over the western Pacific: distribution, partitioning, and sources, *J. Geophys. Res.* 101(D1):1793–1808.

Stockwell, W.R., Middelton, P., and Chang, J.S., 1990, The second generation regional acid deposition model chemical mechanism for regional air quality modeling, *J. Geophys. Res.* 95(D10):16343–16367.

DISCUSSION

P. BUILTJES: Why don't you use for additional information of your measured initial and boundary conditions results from global models?

B. SCHELL: I thought about this when I started my research but I do not know a global model which is able to provide this information in the required spatial and temporal resolution for all the species included in the RADM2, especially for the VOCs, and not only for species like NO_x, ozone and SO_2. The aim of my studies was a set of initial and boundary conditions derived from actual measurements, but a coupling to global models will be an additional important issue for future work. Another point was that I tried to be independent of other models in order to compute fast several simulations.

P. QUANDALLE: What are the emission inventory data that you have used?

B. SCHELL: The emission input data are derived from the EMEP European Emission inventory for 1991 and additional biogenic emission data (Lübkert & Schöpp, 1989) using the EURAD Emission Model to generate the appropriate spatial and temporal resolution for the chemical compounds needed in the model. But the emission data did not change from one run to the other one. The only changes are the initial and boundary values of the model species ONIT and PAN.

A. HANSEN: Why include ONIT with PAN when changing boundary conditions, since it is not a reservoir species for NO_x?

B. SCHELL: The model species ONIT is built by the reactions of peroxiradicals of hydrocarbons with NO. The decomposition of ONIT undergoes photolysis and the reaction with the OH radical. Both, photolysis and the OH reaction, have NO_2 as a product. In comparison to PAN, there is no thermal decomposition in the RADM2 mechanism. The used ONIT values are in the range of the measurements and the studies with the boxmodel have shown that there is a higher level of NO_x in the model over the whole simulation period if higher ONIT values are used. This additional NO_x is partially released during and by the decomposition of ONIT. Therefore, ONIT can act as a NO_x source in the troposphere.

Fast Sensitivity Analysis of Three-Dimensional Photochemical Models

Y. J. Yang, J. G. Wilkinson, A. G. Russell

School of Civil and Environmental Engineering
Georgia Institute of Technology
Atlanta, GA 30332-0512
USA

INTRODUCTION

Photochemical air quality models increasingly are being used to understand the atmospheric dynamics of air pollutants and as the basis of emission control regulations. The response of the these model predictions to system parameters or emission controls provides valuable information for the strategy design to improve air quality. Such information can be pursued via sensitivity analysis, the systematic calculation of sensitivity coefficients, to quantitatively measure these dependencies. However, sensitivity analysis has not seen as wide of use as desired, in part because of the implementation complexity as well as computational limitations. For these reasons, sensitivity analysis has been applied primarily to subsystems of air quality models (e.g. Koda et al., 1974; Rabitz et al., 1983; Milford et al., 1992), or to limited aspects in air quality models (Cho et al., 1987). The "brute-force" method has been the most frequently used to determine model sensitivities, but it rapidly becomes less viable and prohibitively inefficient for a model when a large number of sensitivity coefficients needs to be computed.

A number of approaches have been developed to calculate sensitivity coefficients. One class of techniques is built on the coupled direct method. In this method, the sensitivity equations are derived from the model equations and solved simultaneously with the model equations. This method was found to be unstable and inefficient when applied to stiff equations found in many problems (Dunker, 1984). A second set of techniques relies on the Green's function (Rabitz et al., 1983; Cho et al., 1987; Dougherty et al., 1979) or adjoint method, in which the sensitivity coefficients are computed from integrals of the Green's function of sensitivity equations derived from the model equations. The implementation of this method is unwieldy for most current, comprehensive air quality models, especially for those having a modular algorithm structure.

Air Pollution Modeling and Its Application XII
Edited by Sven-Erik Gryning and Nadine Chaumerliac, Plenum Press, New York, 1998

Another approach for computing sensitivity coefficients is the decoupled direct method (DDM) (Dunker, 1981, 1984), in which the sensitivity equations are derived from the model equations, but solved separately from the model equations. DDM does not share the instability problem found with the direct methods or adjoint methods (Dunker, 1984). Further, the implementation of this method is more straightforward than the coupled direct or adjoint methods since the sensitivity equations are similar to the model equations. Therefore, only minimal modifications in the model algorithms are required to simultaneously calculate the sensitivity coefficients. DDM has been applied to zero-dimensional models (Milford et al., 1992; Dunker, 1984; Yang et al., 1995; Gao et al., 1995), and a somewhat different approach in a three-dimensional model (Dunker, 1981).

Another technique is ADIFOR (Automatic DIfferential in FORtran) (Bishof et al., 1992), which automatically translates large FORTRAN codes to a subprogram that includes the original functions as well as those for the desired sensitivity coefficients. This method has been used in past studies, notably for sensitivity analysis of the advection equation as used for atmospheric modeling (Hwang and Byun, 1996). Because ADIFOR is designed for general purpose sensitivity analysis, the expanded codes do not take advantage of the program structure and re-use of calculations.

In this paper, a fast, direct, multidimensional sensitivity analysis similar to the DDM was developed and applied to the CIT (California/Carnegie Institute of Technology) airshed model. The method is referred to the DDM-3D. The technique, as applied here, computes sensitivity coefficients, including those with respect to initial conditions, dry deposition velocities, rate constants, and ground-level NO_x and VOC emissions, in the SoCAB during the 27-29 August 1987 ozone episode. Furthermore, the approach can also be applied to other multidimensional air quality models.

MULTIDIMENSIONAL DIRECT DIFFERENTIATION METHOD

The formation and transport of chemically reacting atmospheric contaminants is described by the atmospheric diffusion equation (ADE) (McRae et al., 1982):

$$\frac{\partial c_i}{\partial t} = -\nabla(\mathbf{u}c_i) + \nabla(\mathbf{K}\nabla c_i) + R_i[c_1,\ c_2,\ \ldots,\ c_n;\ T,\ t] + S_i \quad i = 1,\ldots, N \tag{1}$$

where $c_i(\mathbf{x}, t)$ is the ensemble average concentration of species i, $\mathbf{u}(\mathbf{x}, t)$ is the wind field, $\mathbf{K}(\mathbf{x}, t)$ is a second order turbulent diffusivity tensor, $R_i = R_i[c_1, c_2, \ldots, c_n, t]$ is the net rate of chemical production, $S_i(\mathbf{x}, t)$ is the elevated source rate of emissions of specie i, and N is the number of total species. This equation is solved numerically subject to initial conditions and boundary conditions:

IC: $c_i(t_o) = c_i^o$

BC's:

(1) $\mathbf{u}c_i - \mathbf{K}\nabla c_i = \mathbf{u}c_i^b$ Horizontal inflow

(2) $-\nabla c_i = 0$ Horizontal outflow

(3) $v_g^i c_i - K_{zz}\dfrac{\partial c_i}{\partial z} = E_i$ $z = 0$

(4) $-\dfrac{\partial c_i}{\partial z} = 0$ $z = H$ (top of model domain)

where c_i^b is the concentration of compound i at the boundary, v_g^i is the dry deposition velocity, E_i is the ground level emission rate, and H is the height of model domain. The local sensitivity of a model output to a parameter can be calculated as the partial derivative of the output with respect to the parameter or input (p_j):

$$s_{ij}(t) = \frac{\partial c_i(t)}{\partial p_j}$$

It is useful to define and use semi-normalized sensitivity coefficients because sensitivity coefficients to different parameters may differ by several orders of magnitude. Given a model parameter, p_j, its variation is defined here as $p_j(\mathbf{x},t) = \varepsilon_j P_j(\mathbf{x},t)$, where $P_j(\mathbf{x},t)$ is the unperturbed field, which can vary in time and space, and ε_j is a scaling variable with a nominal value of one. Thus the semi-normalized sensitivity coefficient, s^*_{ij}, is calculated by the partial derivative of a species concentration, c_i, to the scaling variable of parameter j, ε_j:

$$s^*_{ij}(t) = P_j \frac{\partial c_i(t)}{\partial p_j} = P_j \frac{\partial c_i(t)}{\partial(\varepsilon_j P_j)} = \frac{\partial c_i(t)}{\partial \varepsilon_j} \tag{2}$$

By substituting Eq. (2) into Eq. (1), the auxiliary equations of sensitivity coefficients and initial and boundary conditions are obtained as follows:

$$\frac{\partial s^*_{ij}}{\partial t} = -\nabla(\mathbf{u}s^*_{ij}) + \nabla(\mathbf{K}\nabla s^*_{ij}) + J_{ik}s^*_{kj} + \frac{\partial R_i}{\partial \varepsilon_j} + \frac{\partial S_i}{\partial \varepsilon_j} - \nabla(\mathbf{u}c_i)\delta_{ij} + \nabla(\mathbf{K}\nabla c_i)\delta_{ij} \tag{3}$$

I.C.: $s^*_{ij}(t_o) = c^o_j \delta_{ij}$

B.C.'s: (1) $\mathbf{u}s^*_{ij} - \mathbf{K}\nabla s^*_{ij} = \mathbf{u}c^b_i \delta_{ij} - \mathbf{u}c_i \delta_{ij} + \mathbf{K}\nabla c_i \delta_{ij}$ Horizontal inflow

(2) $-\nabla s^*_{ij} = 0$ Horizontal outflow

(3) $v^i_g s^*_{ij} - K_{zz}\frac{\partial s^*_{ij}}{\partial z} = -v^i_g c_i \delta_{ij} + K_{zz}\frac{\partial c_i}{\partial z}\delta_{ij} + E_i \delta_{ij}$ $z = 0$

(4) $-\frac{\partial s^*_{ij}}{\partial z} = 0$ $z = H$

where **J** is the Jacobian matrix defined by $J_{ik} = \partial R_i/\partial c_k$, and depending on whether parameter j is a constituent of the term, all δ_{ij}'s on the right-hand-side will be either one or zero. If parameter j exists in the term, then δ_{ij} is one; otherwise δ_{ij} will be zero. As seen in Eq. (3), the sensitivity coefficients are functions of the concentrations and are linear in the sensitivity coefficients, though coupled. Also, while the sensitivity coefficients do depend on the concentrations, the converse is not the case; so the sensitivities can be integrated separately from the concentrations over the time step. Further, the temporal evolution of the concentrations, which are found before the sensitivity equations are integrated, can be used to stabilize the calculation of sensitivity coefficients and to use large time steps, when integrating Eq. (3).

NUMERICAL IMPLEMENTATION

Like most three-dimensional air quality models, CIT model uses an operator-splitting technique to reduce the multidimensional problem to a sequence of one-dimensional equations. The sequence of operator splitting for sensitivity coefficients at t=t+Δt is similar to one used for the concentrations:

$$s^*_{ij}(t+\Delta t) = L_H\left(\frac{\Delta t}{2}\right)L_V(\Delta t)L^P_{RS}(\Delta t)L_H\left(\frac{\Delta t}{2}\right)s^*_{ij}(t) \tag{4}$$

where

$$L_H(s^*_{ij}) = -\nabla_H(\mathbf{u}s^*_{ij}) + \nabla_H(\mathbf{K}\nabla_H s^*_{ij})$$

$$L_V(s^*_{ij}) = -\nabla_V(\mathbf{u}_z s^*_{ij}) + \nabla_V(\mathbf{K}\nabla_V s^*_{ij})$$

$$L^p_{RS}(s^*_{ij}) = \left(J_{ik}s^*_{kj} + \frac{\partial R_i}{\partial \varepsilon_j}\right) + \frac{\partial S_i}{\partial \varepsilon_j} - \nabla(\mathbf{u}c_i)\delta_{ij} + \nabla(\mathbf{K}\nabla c_i)\delta_{ij}$$

∇_H and ∇_V denote the del operator applied to horizontal and vertical transport, respectively; L^p_{RS} is the production term arising from the sensitivity to the chemistry, emission sources, and transport. The only difference in solution procedure between the concentrations and sensitivity coefficients will be the calculation of the sensitivity production terms in the operator L^p_{RS}. Thus, the same algorithms can be used for the transport process of the sensitivities, and the same general model structure is maintained. The last three terms in the L^p_{RS} operator are independent of the sensitivity coefficients, and can be treated as source/sink terms.

The most intensive computational aspect of atmospheric transport-chemistry problems is the task of solving the differential equations of chemical transformation. Unlike the equations governing the evolution of species concentrations, in which the integration is dominated by the nonlinear chemistry, the linear system of sensitivity equations with respect to chemistry can be solved very efficiently by LU decomposition using larger time steps than used during chemical integration. In particular, the time step used for integration of the L^p_{RS} step is twice the advection time step (which is determined by the Courant-Friedrichs-Lewy limit (McRae et al., 1982). The linear algebraic system of sensitivity equations associated with the chemistry is represented by:

$$\frac{\mathbf{s}^{*n+1} - \mathbf{s}^{*n}}{\Delta t} = \widetilde{\mathbf{J}}^{n+1}\left(\frac{\mathbf{s}^{*n+1} + \mathbf{s}^{*n}}{2}\right) + \frac{\partial R(\mathbf{c}^{n+1})}{\partial \varepsilon} \tag{5}$$

where $\widetilde{\mathbf{J}}^{n+1} = \mathbf{J}(\widetilde{\mathbf{c}})$ and $\widetilde{\mathbf{c}} = \dfrac{\mathbf{c}^{n+1} + \mathbf{c}^n}{2}$.

The solution of sensitivity coefficients in Eq. (5) at t=t+Δt is:

$$\mathbf{s}^{*n+1} = \left(\mathbf{I} - \frac{\Delta t}{2}\widetilde{\mathbf{J}}^{n+1}\right)^{-1}\left\{\left(\mathbf{I} + \frac{\Delta t}{2}\widetilde{\mathbf{J}}^{n+1}\right)\mathbf{s}^{*n} + \frac{\partial R(\mathbf{c}^{n+1})}{\partial \varepsilon}\right\} \tag{6}$$

Of note, the LU factorization need only be carried out once for all sensitivity coefficients calculated in each computational cell because both the Jacobian and the local parametric derivation matrix (i.e. $\partial\mathbf{R}/\partial\varepsilon$) are independent of sensitivity coefficients, only being functions of the concentrations and rate constants (see Eq. (3)).

APPLICATION TO SOUTHERN CALIFORNIA

Ozone formation in an airshed usually is a multiday event associated with transport and chemical transformation. In this study, the sensitivity analysis technique developed in the previous section is applied to the 27-29 August, 1987 episode, an intensive monitoring period of the Southern California Air Quality Study (SCAQS).

The parameters for which sensitivity coefficients were computed are summarized in Table 1. The spatial distributions of NO_x and VOC emissions over the model region are shown in Figure 1 and 2, respectively. The SAPRC90 chemical mechanism (Carter, 1990) was used in this study.

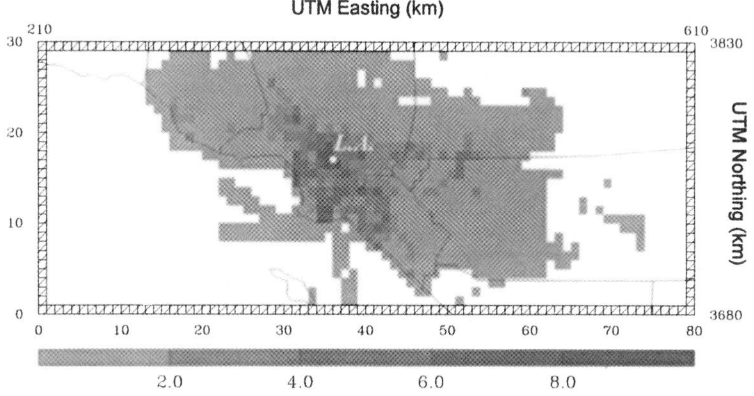

Figure 1. Distributed NO_x emissions (metric tons/day).

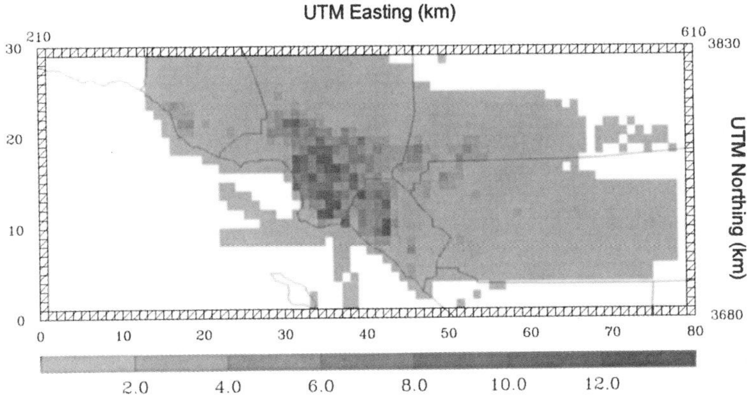

Figure 2. Distributed emissions of volatile organic compounds (metric tons/day).

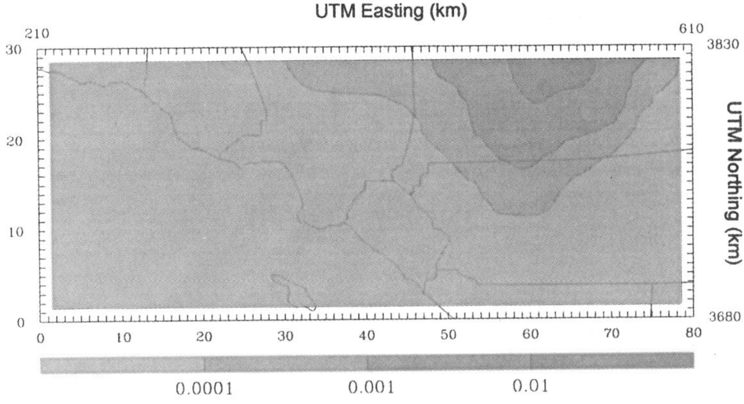

Figure 3. Ozone sensitivity (ppb-O_3) at 1500 hours, August 29, 1987, to the O_3 initial condition.

Table 1. Model parameters used for ozone sensitivity analysis

Parameters	Nominal values
O_3 initial concentration	60 ppb
Rate constant of $HO + NO_2 \rightarrow HNO_3$	1.7×10^{-4} ppm^{-1}min^{-1} [a]
Ground-level Emissions	
NO_x	1035 tons/day
VOC	1745 tons/day

[a] Evaluated at 300°K but varies at locations in the model region.

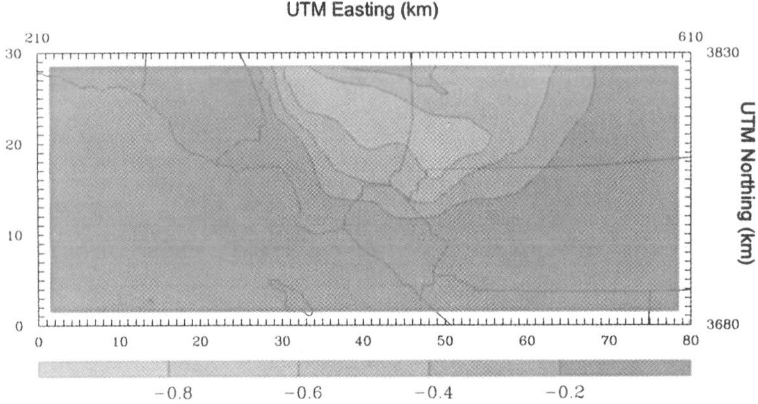

Figure 4. Ozone sensitivity (ppb-O_3) at 1500 hours, August 29, 1987, to the $HO+NO_2 \rightarrow HNO_3$ rate constant.

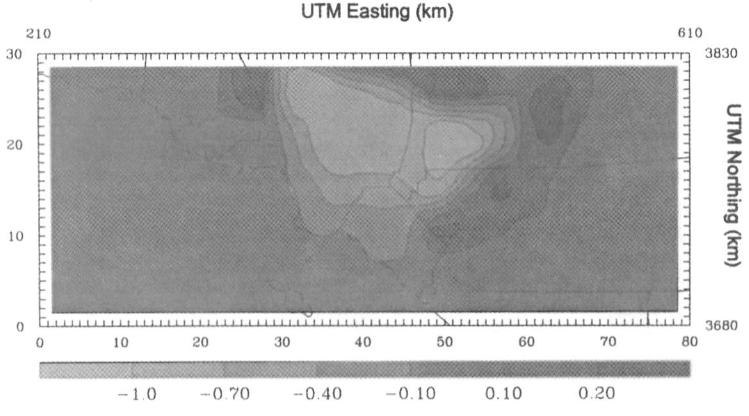

Figure 5. Ozone sensitivity (ppb-O_3) at 1500 hours, August 29, 1987, to the NO_x emissions.

RESULTS AND DISCUSSION

Sensitivity analysis results are given as the change in ozone levels (ppb) per 1% increase in the nominal value of the parameters, and results are shown at the time of predicted peak ozone occurrence on August 29. Figure 3 shows the ozone sensitivities to the ozone initial condition. As expected, an increase in initial ozone concentrations over the domain increases predicted ozone throughout the modeling region with higher sensitivities in the downwind area where the maximum ozone sensitivity is about 0.01 ppb-O_3. By the third day of the simulation, only slight effects over a small portion of the domain are predicted.

Figure 4 presents the spatial distributions of ozone sensitivities due to the reaction rate constant of $HO+NO_2 \rightarrow HNO_3$. This reaction is one of the most ozone-sensitive parameters in the chemical mechanism. Increasing this rate parameter decreases ozone concentrations in the downwind area because more hydroxyl radical and NO_2 in system are removed through the reaction. This agrees with that found by Yang et al. (1995) using a box model.

Figure 5 shows ozone sensitivities to ground-level NO_x emissions. In this case, increasing NO_x emissions has the effect of decreasing predicted ozone concentrations immediately downwind of the urban core (0-100 km). However, ozone concentrations tend to increase much further downwind of the urban core ($\geq$ 100 km). This is attributed to the local abundance of VOC and NO_x in the two regions (immediately downwind and much further downwind). In urban Los Angeles, the NO_x is relatively abundant and ozone production is VOC-limited. On the other hand, downwind of the urban core, the NO_x is less abundant. The maximum ozone sensitivities found for a 1% increase in NO_x emissions is about -1.3 ppb-O_3

Figure 6 shows the ozone sensitivities to increasing VOCs emissions. An increase in VOC emissions results in more peroxy radical generation through the VOC oxidation to convert NO to NO_2, and, consequently, increases ozone formation downwind. The maximum ozone sensitivities is about 1.6 ppb-O_3 occuring downwind of the urban core.

One of the attractive features of this method is the computational efficiency. All results were conducted on a SUN Ultra 1-170 workstation with 64 megabytes of RAM. A one parameter sensitivity analysis increases execution time 30% relative to concentration calculations alone. For the cases where ten and twenty parameters are calculated, the execution time increases about 50% and 80%, respectively. Recall, sensitivity coefficients are independent of the nonlinear coupling in the reactions and the LU factorization needs to

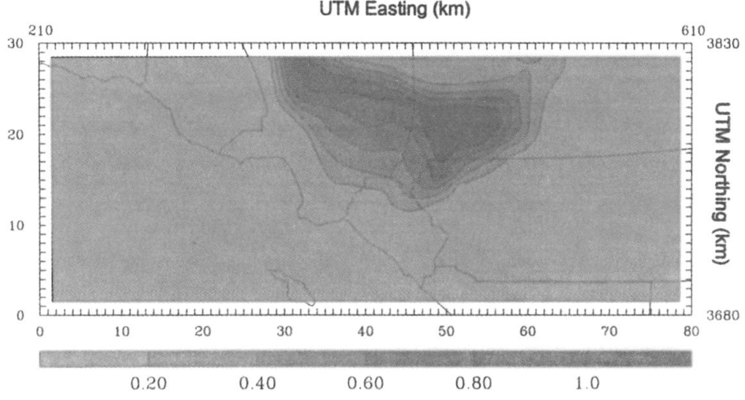

Figure 6. Ozone sensitivity (ppb-O_3) at 1500 hours, August 29, 1987, to the VOC emissions.

be done once when all the sensitivity coefficients are advanced from t_n to t_{n+1}. Therefore, solving additional sensitivity coefficients requires relatively less CPU time as compared to solving additional species concentrations.

CONCLUSIONS

A technique (DDM-3D) for fast, formal sensitivity analysis in multidimensional air quality models has been described and implemented in a three-dimensional airshed model. The results of an application to the 27-29 August 1987 SCAQS episode indicate that the new method is viable and can efficiently compute the sensitivity coefficients of all species simultaneously. The sensitivity equations directly derived from the model equations have a similar structure to that of the concentrations, and the general structure of the model is maintained while calculating the sensitivity coefficients. Therefore, the effort to solve the sensitivity equations separately only requires limited programming modifications from the original model code. As described, the implementation of the DDM-3D presented here can also be conceptually applied to other multidimensional air quality models.

REFERENCES

Bischof, C.A., Carle, A., Corliss, G., Griewank, A., and Hovland, P., 1992, ADIFOR: Generating Derivative Codes from Fortran Programs, *Scientific Programming*, 1:11.

Carter, W.P.L., 1990, A Detailed Mechanism for the Gas-Phase Atmospheric Reactions of Organic Compounds, *Atmos. Environ.* 24A:481.

Cho, S.-Y., Carmichael, G.R., and Rabitz, H., 1987, Sensitivity Analysis of the Atmospheric Reaction-Diffusion Equation, *Atmos. Environ.* 12:2589.

Dougherty, E.P., Hwang, J.T., and Rabitz, H., 1979, Further Developments and Applications of the Green's Functions, *J. Chem. Phys.*, 71:1794.

Dunker, A.M., 1981, Efficient Calculation of Sensitivity Coefficients for Complex Atmospheric Models, *Atmos. Environ.* 1981, 15:1155.

Dunker, A.M., 1984, The Decoupled Direct Method for Calculating Sensitivity Coefficients in Chemical Kinetics, *J. Chem. Phys.* 81:2385.

Gao, D., Stockwell, W.R., and Milford, J.B., 1995, First Order Sensitivity and Uncertainty Analysis for a Regional-Scale Gas-Phase Chemical Mechanism, *J. Geophys. Res.*, 100:23153.

Hwang, D., and Byun, D.W., 1996, On the Use of Automatic Differentiation for Sensitivity Analysis in Emission Control Process, Ninth Joint Conference on Applications of Air Pollution Meteorology with AWMA, January 28.

Koda, M., Dogru, A.H., and Seinfeld, J.H., 1974, Sensitivity Analysis of Partial Differential Equations with Application to Reaction and Diffusion Processes, *J. Computat. Phys.* 30:259.

McRae, G.J., Goodin, W.R., and Seinfeld, J.H., 1982, Development of a Second-Generation Mathematical Model for Urban Air Pollution - 1. Model Formulation, *Atmos. Environ.* 16:679.

Milford, J.B., Gao, D., Russell, A.G., and McRae, G.J., 1992, Use of Sensitivity Analysis To Compare Chemical Mechanisms for Air Quality Modeling, *Environ. Sci. Technol.* 26:1179.

Rabitz, H., Kraman, M.A., and Dacol, D., 1983, Sensitivity Analysis in Chemical Kinetics, *Ann. Rev. Phys. Chem.* 34:419.

Yang, Y.J., Stockwell, W.R., and Milford, J.B., 1995, Uncertainty Analysis in Incremental Reactivities of Volatile Organic Compounds, *Environ. Sci. Technol.*, 29:1336.

INTEGRATING OBSERVATIONS AND MODELING IN OZONE MANAGEMENT EFFORTS

S. Trivikrama Rao[1], Eric Zalewsky[1], Igor G. Zurbenko[1],
P. Steven Porter[2], Gopal Sistla[3], Winston Hao[3], Nianjun Zhou[3],
Jia-Yeong Ku[3], George Kallos[4], and D. Alan Hansen[5]

[1] State University of New York - Albany, Albany, New York 12222
[2] University of Idaho, Idaho Falls, Idaho 83405
[3] New York State Department of Environmental Conservation, Albany, NY 12233
[4] University of Athens, Athens 10680, Greece
[5] Electric Power Research Institute, Palo Alto, California

INTRODUCTION

Many urban areas in the Eastern United States have been classified to be in non-attainment for ozone, placing a high priority on finding cost-effective emission control measures for improving ambient ozone air quality. Recognizing the complexities associated with the nation's ozone non-attainment problem, the 1990 Clean Air Amendments mandated the use of grid-based photochemical models for evaluating emission control strategies in urban areas having a *serious* or higher designation. Given the influx of elevated concentrations of ozone and its precursors into the urban-scale modeling domains (regional-scale transport), many states in the Eastern U.S. were unable to demonstrate ozone attainment for urban areas in their 1994 State Implementation Plans (SIPs) submittal using the urban-scale models. The 1994 SIPs were based on the UAM-IV photochemical model (Morris et al., 1990), which is an urban-scale model that reflects the state-of-science of the late 1980's. Systems Applications International (SAI) recently developed the UAM-V, a regional-scale ozone air quality model, which contains some new features over the UAM-IV such as multi-scale modeling capability, grid nesting, plume-in-grid (PiG) treatment for point sources, etc. (SAI, 1995). Of particular interest is this model's treatment of subgrid-scale processes relating to the transport, transformation, and interaction of elevated plumes with the ground-level plume.

Given the complexities associated with the ozone problem and the fact that air quality models are not perfect simulators, innovative methods that can integrate the spatial information in the observations with photochemical modeling are needed for designing effective ozone management plans. Currently, the models are being applied to past ozone episodic events to examine the future ozone non-attainment problem. Once the model's ability to reproduce the observed ozone air quality has been evaluated, the model is applied to assess the control measures needed to

Air Pollution Modeling and Its Application XII
Edited by Sven-Erik Gryning and Nadine Chaumerliac, Plenum Press, New York, 1998

meet and maintain the ozone standards in the future using projected emissions inventories with historical episodic meteorological conditions. Since the historical episodic meteorological conditions under which the model has performed best may never occur in the future, there is an inherent uncertainty in the controls identified as required for meeting and maintaining the ozone standards. Assuming the atmospheric processes that affect ambient ozone levels are properly treated in the model, one would expect that the modeling results for average conditions to be more reliable than those based on extreme ozone events. When the current 1-hr ozone standard is revised, it might become necessary to apply models on a seasonal basis for examining longer-term (8-hr, Sum 06, seasonal average, etc.) ozone concentrations.

In this paper, as an initial step in the comparison of model outputs with observations, we extract the spatial scale for short-term ozone derived from observations and compare it with the spatial scale of the changes in ozone concentrations as predicted by UAM-V. Short-term (weather-related) ozone is separated from raw ozone data using an iterative moving average filter (Zurbenko, 1986). Model ozone concentration reductions result from a hypothetical elimination of anthropogenic emissions over selected subregions of the modeling domain. Spatial scales for a given process are described in terms of the decay of the process correlation in space (correlations between different stations as a function of distance) to a value of 1/e (the e-folding distance). The 1/e index of scale has also been used by the National Climatic Data Center to design monitoring networks (Wallis, 1996). The techniques for extracting the spatial scale from both ambient data and photochemical models are described and illustrated with an example based on the July 1995 ozone episode that occurred in the Eastern United States.

The results reveal that both short-term (weather-related) ozone in the Eastern United States and the impact of emissions reduction on ozone levels decays exponentially with distance and have an e-folding distance of about 300 miles (500 km). The large spatial extent of the ozone footprint in observations and model-predicted emission reduction impacts are further evidence that the ozone problem in the Eastern U.S. needs to addressed from a regional perspective.

ANALYSIS OF OZONE OBSERVATIONS

Data

Hourly ozone concentrations at various monitoring stations in the Eastern U.S. are obtained from the EPA's AIRS data base and time series of daily maximum 1-hr ozone concentrations for the 1984 to 1995 period are analyzed in this study.

Extraction of the Short-term (weather-related) Component of Ozone Observations

Meteorological and air quality variables can be decomposed into synoptic, seasonal, and long-term components each attributable to different physical phenomena. In ozone, the synoptic scale component is attributable to weather and short-term fluctuations in precursor emissions, seasonal scale to variation in the solar angle, and long-term scale to changes in climate, policy, and/or economics. Therefore, time series of ozone data will be represented by:

$$O(t) = e(t) + S(t) + W(t) \tag{1}$$

where O(t) is the natural logarithm of the original ozone time series, e(t) is the long-term (trend) component, S(t) is seasonal change, W(t) is short-term variation, and t is time. We used the Kolmogorov-Zurbenko (KZ) filter (Zurbenko, 1986) to separate the components described by

(1). The KZ filtration is accomplished by several iterations of a simple moving average filter. Properties of the filter as well as its applications to ozone and meteorological variables are described in Rao and Zurbenko (1994), Rao et al. (1996), Porter et al. (1996), Rao et al. (1997), and Eskridge et al. (1997).

Following Rao and Zurbenko (1994), separation of O(t,x,y) in time by the KZ filtration provides baseline (O^B) (deterministic) and short-term or synoptic (O^S) (stochastic) components in time t and space (x,y) defined by:

$$\begin{aligned} O(t,x,y) &= O^{B}(t,x,y) + O^{S}(t,x,y) \\ O^{B}(t,x,y) &= e(t) + S(t) \\ O^{S}(t,x,y) &= W(t) \end{aligned} \tag{2}$$

Expression (2) is practically realizable only when the O^B and O^S components are cleanly separated. Sample estimates of individual components can be obtained as follows.

$$\text{estimate of } W(t) = O(t) - KZ_{15,5} \tag{3}$$

$$\text{estimate of } e(t) = KZ_{365,3} \tag{4}$$

$$\text{estimate of } S(t) = KZ_{15,5} - KZ_{365,5} \tag{5}$$

where $KZ_{m,k}$ refers to k iterations of a simple moving average filter of width m days. The time series of separated components of ozone data during 1987 at Charlotte, NC are depicted in Figure 1.

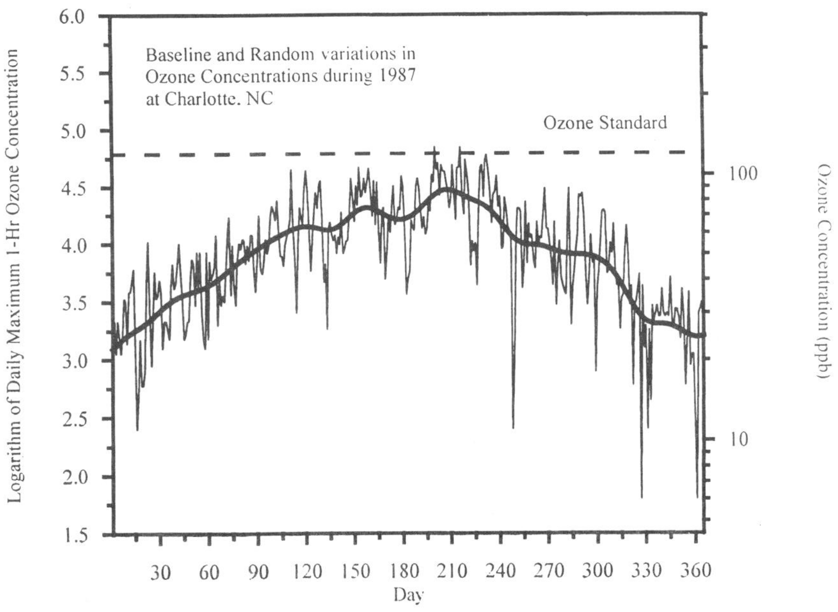

Figure 1. Ozone time series data at Charlotte, NC for 1987 decomposed into short-term and baseline.

The total variance of O(t) can be written as a sum of the variances and covariances of the ozone components separated by the KZ filter:

$$\sigma^2(O) = \sigma^2(e) + \sigma^2(S) + \sigma^2(W) + 2\cdot\text{cov}(e,S) + 2\cdot\text{cov}(e,W) + 2\cdot\text{cov}(S,W) \qquad (6)$$

The sum of covariance terms was less than 2% of the total variance, indicating good separation of components (Porter, et al., 1996; Rao, et al., 1997).

Spatial Scale in Ozone Observations

Synoptic ozone ($O^S(t,x,y)$) contains a very precise spatial law related to the directional decay of their spatial correlations (Rao, et al., 1995), namely, there is a strong relationship between exponential decay in the correlation between two monitors m_o, m' and their distance of separation, d, in direction ϕ:

$$\log \text{Cor}\left(O^S(t,m_o),\ O^S(t,m')\right) = -a(m_o,\phi)d + b \qquad (7)$$

where 'a' depends on reference monitor location (x_o ,y_o) and spatial direction ϕ. In other words, there is a Markov relationship between short-term (synoptic) components in space. The Markov property for a Gaussian process yields:

$$O^S(t,m') = \exp\{-ad + b\}\ O^S(t,m_o) + O^{loc}(t,m') \qquad (8)$$

where $O^{loc}(t,m')$ is a local contribution portion at the location m'. Therefore, expression (8) separates the transport portion of $O^S(t,m')$ from m_o to m' and local production. *The exponential term in (8) is related to the synoptic transport of the pollution, weather conditions, and the emissions which created ozone*. The spatial extent of the correlation structure among short-term ozone components for the Charlotte, NC monitor is extracted from analyzing the daily maximum 1-hr ozone concentrations over the 1984 to 1995 period (Figure 2). The spatial structure in the ozone "footprint" in Figure 2 is dictated by both weather and emission effects. The exponential decay of the correlations with distance along the direction of the prevailing wind at Charlotte, NC, presented in Figure 3, reveal an e-folding distance (scale) of 300 miles (500 km) for ozone. Similar results have been found at all other monitoring stations in urban areas of the Eastern U.S.

PHOTOCHEMICAL MODELING

Model Set Up

The UAM-V was designed for application to the Eastern United States with a fine (12 km) grid extending from 69.5^0 to 92^0 longitude and from 32^0 to 44^0 latitude and a coarse grid (36 km) extending from 67^0 to 99^0 longitude and from 26^0 to 47^0 latitude. Further details on the model structure and its applications can be found in OTAG (1996). A performance evaluation of this modeling system can be found in Lurmann and Kumar (1996).

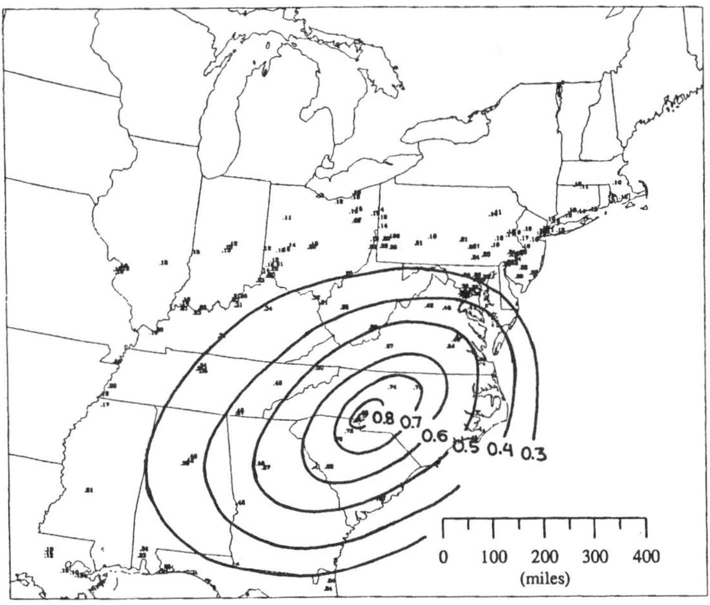

Figure 2. Isolines of correlation among ozone short-term components between Charlotte, NC and other monitors, depicting the spatial extent for the ozone footprint.

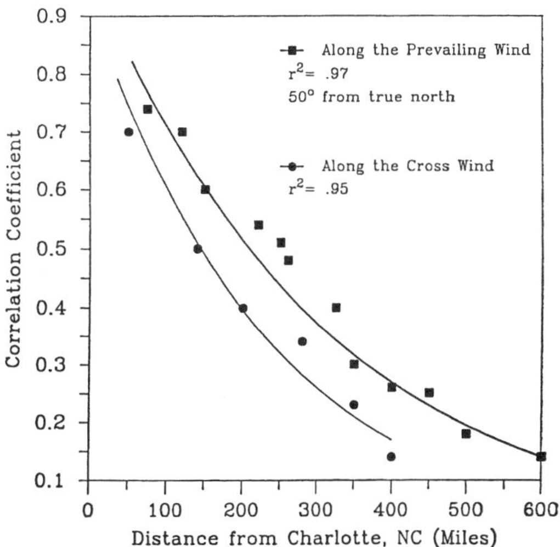

Figure 3. The exponential decay of correlations as a function of distance from Charlotte, NC.

Modeling Simulations

As part of the Ozone Transport Assessment Group's (OTAG) activities, several UAM-V simulations have been performed for the July 7-18, 1995 ozone episodic event. Since high concentrations of ozone were recorded throughout the Eastern United States between July 13-16 only, the initial modeling days are considered as ramp-up days. The meteorological data for the July 1995 episode were derived by applying the RAMS-version 3a with nested grids ranging in size from 12 to 36 to 108 km, where the outer grid covered most of North America. Anthropogenic emissions from the state-generated 1990 data and natural emissions using BEIS2 were assembled and adjusted for the episodic conditions simulated here. The modeling results for the base case and various control cases can be found on the OTAG Website (OTAG, 1996).

Analysis of Modeling Results

Several modeling simulations have been carried out to examine the sensitivity of the modeled ozone to changes in emissions. For this paper, we discuss a scenario in which all anthropogenic emissions over the States of North Carolina and South Carolina are eliminated. The impact of these selective emission reductions on ozone concentrations in the entire modeling domain are examined.

The spatial extent of ozone benefits is derived by subtracting the peak ozone concentrations predicted for the base case (no emission reductions) from the emission reduction scenario (Figure 4). Since the greatest benefit from emission reductions is expected in the near-field, and because we are interested in relative rather than absolute changes, the change in the peak ozone at each grid cell resulting from the emissions reduction is normalized by the maximum change (benefit) predicted; in this case, the maximum ozone improvement of 111 ppb from the base case (base case value was 156 ppb) was predicted near Charlotte, N.C. The decay of ozone benefits from Charlotte, NC along the direction of the prevailing wind was determined as a function of distance for this episode (Figure 5). Similar results have been found at all other locations where emissions were perturbed.

DISCUSSION

Observational analysis based on long-term data and modeling for a single episode provide similar spatial information. Along the direction of the prevailing wind, the spatial information extracted from observational analyses shown in Figure 2 is comparable to that derived from the model (Figure 4). In addition, the decay in ozone benefits resulting from the emissions reduction (Figure 5) is consistent with the decay of information in ambient ozone data (Figure 3): both observations (Figure 3) and modeling results for a single episode (Figure 5) reveal e-folding distances of about 300 miles (500 km), indicating that the ozone scale in the Eastern U.S., dictated by weather and emissions effects, is on the order of 300 miles. A superposition of the various ozone footprints extracted from ambient data at different urban monitoring locations as well as the superposition of the ozone footprints derived from the model by perturbing different emission source regions, suggest the need to address the ozone problem in the Eastern United States from a regional perspective. Further, since the spatial scale for ozone in the Eastern U.S. is about 300 miles, ozone concentrations predicted by the urban-scale models such as the UAM-IV can be expected to be strongly influenced by the boundary conditions. Application of the techniques described here to the seasonal modeling results would enable us to assess the model's ability to simulate the spatial and temporal features in ambient ozone data.

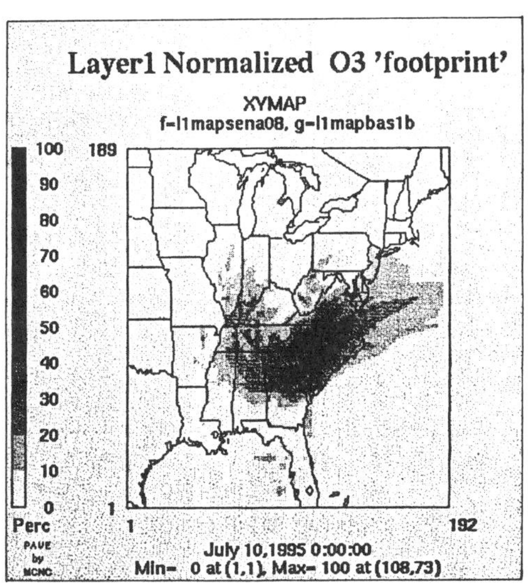

Figure 4. The footprint of ozone benefit resulting from the elimination of anthropogenic emissions in North Carolina and South Carolina. The maximum improvement of 111 ppb was predicted near Charlotte, NC.

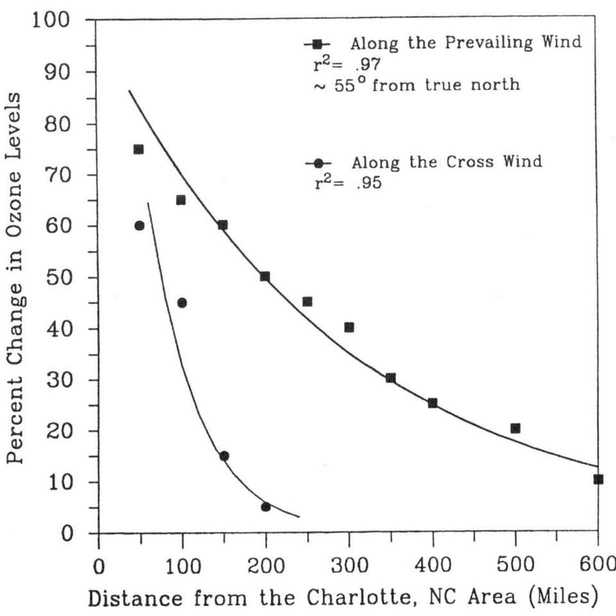

Figure 5. The exponential decay of ozone benefits as a function of distance from Charlotte, NC.

SUMMARY

Among the useful results of these analyses, unavailable in a meaningful form from raw ozone data, is a description of the spatial information in ozone data, needed for ozone management efforts. The short-term ozone component, attributable to weather fluctuations and emissions, is highly correlated in space; the correlation structure of short-term ozone permits highly accurate predictions of ozone concentrations up to distances of about 300 miles (500 km) from a monitor.

The UAM-V simulations for the July 7-18, 1995 ozone episode reveal that the improvement in ozone resulting from emission reductions decrease exponentially with distance downwind from the controlled region along the direction of the maximum transport. The e-folding distance for ozone benefits is also about 300 miles (500 km). Modeling results for a single ozone episode and ozone observations both point toward the need to address the ozone problem in the Eastern United States from the regional perspective. The techniques discussed here would be useful in evaluating the model's ability to simulate the spatial and temporal features in ozone observations.

ACKNOWLEDGMENTS

This research is supported by the Environmental Protection Agency under grant #R819328-01 and by the Electric Power Research Institute under contract #WO4447-01.

REFERENCES

Eskridge, R.E., Ku, J.Y., Rao, S.T., Porter, P.S. and I.G. Zurbenko. Separating different scales of motion in time series of meteorological variables. *Bull. Amer. Meteor. Soc.*, In Press.

Lurmann, F.W. and Kumar, N. Evaluation of the UAM-V model performance in OTAG simulations Phase I: Summary of performance against surface observations. Draft Final Report #STI-996120-1605-DFR, Sonoma Technology, Inc., September 30, 1996.

Morris R. E., Myers T.C., Carr E. L. Causley M. C., Douglas S. G., and Haney J. L. Users Guide for the Urban Airshed Model; Vol II: Users Manual for the UAM (CB-IV) Modeling Systems (Preprocessors), EPA-450/4-90-007B, 1990.

Systems Applications International. Users Guide to the Variable Grid Urban Airshed Model (UAM-V), San Rafael, CA, 1995.

Ozone Transport Assessment Group (OTAG) Modeling Report, Volume 1.1, February 12, 1997. OTAG Website (http://www.iceis.mcnc.org.80/OTAGDC/index.html)

Porter, P.S., Rao, S.T., Zurbenko, I.G., Zalewsky, E., Henry, R.F. and J.Y. Ku. Statistical characteristics of spectrally-decomposed ambient ozone time series data. 1996, OTAG Web Site (http://capita.wustl.edu/otag/reports).

Rao, S.T. and I.G. Zurbenko. Detecting and tracking changes in ozone air quality, *J. Air & Waste Manage. Assoc.*, 1994, p 1089.

Rao, S.T., Zalewsky, E. and I.G. Zurbenko. Determining spatial and temporal variations in ozone air quality, *J. Air & Waste Manage. Assoc.*, 1995, p 57.

Rao, S.T., Zurbenko, I.G., Porter, P.S., Ku, J.Y and R.F. Henry. Dealing with the ozone non-attainment problem in the Eastern United States, *Environmental manager*, January, 1996, pp 17-31.

Rao, S.T., Zurbenko, I.G., Neagu, R., Porter, P.S., Ku, J.Y and R.F. Henry. Space and time scales in ambient ozone data. Submitted to the *Bull. Amer. Meteor. Soc.*, March 1997

Wallis, T.W.R. Report on the search for core stations and suggested products from the Comprehensive Aerological Reference Data Set (CARDS), 1996.

Zurbenko, I.G. *The Spectral Analysis of Time Series*, North Holland, 1986.

DISCUSSION

R. MCNIDER: In terms of the drop off in ozone versus distance, can you tell how much is due to actual loss of ozone, e.g. deposition versus dispersion?

S. T. RAO: First, it should be noted that the long-term and seasonal variations embedded in the raw ozone data must be separated from the short-term variations since only the short-term component is related to synoptic-scale weather patterns. We accomplished this using a simple moving average filter. Had we not done this, the decay of spatial correlations among raw ozone data would be reflective more of climatic conditions than weather effects. In this paper, I have shown how the information in short-term (weather-related) ozone decays as a function of distance from a given location along the prevailing downwind and crosswind directions using ten years of observations in the Eastern United States. Since all physical and chemical processes are imbedded in the short-term component, it is difficult to identify the specific contribution from each physical and chemical process which affects ozone on this time scale with the type of analysis performed here. However, I think that the decay of ozone benefits resulting from the elimination of all anthropogenic emissions over a given area, extracted from the photochemical model, can be used to identify the relative contributions of deposition and dispersion. Although I have not done it in this paper, it is possible to re-run the model without deposition and compare the ozone footprint with that derived including deposition. Similarly, other physical processes can be turned on or off to gain insight into how these processes affect the spatial structure in the ozone footprint. Note, this is only a model-derived information. What is happening in real-world can only be extracted from analyzing the observations and not from model data. Models give you whatever that is put into the models. For example, one can use ambient data for two pollutants monitored simultaneously at each site that behave very differently in terms of deposition but very similarly in terms of dispersion to address the question you raised. Note, the analysis techniques identified in this paper can help us in extracting the information imbedded in ambient data.

D. STEYN To what extent is the spatial scale of ozone variability determined by the spatial scale of precursor emissions patterns as contrasted with scale determination by photochemical transform and advection processes?

S. T. RAO: The spatial structure in the ozone footprint, extracted from the observations, is related to the synoptic transport of the pollution, weather conditions, and the emissions which created ozone. If a dense monitoring network of ozone precursors, similar to that in ozone, were available, we would be able to identify the spatial scales of emissions and atmospheric chemical and physical processes. In our paper (October 1997 issue of the Bulletin of the American Meteorological Society), we have addressed the space and time scales in ozone data. Note, the spatial correlations in short-term ozone might be influenced by correlations in short-term weather variables (see our article in the July 1997 issue of Environmental Manager Journal). We have constructed the ozone footprint using temperature/dew point depression-independent short-term ozone data and compared the decay of information in this data with that in raw short-term ozone data. The spatial extent of the ozone footprint in the temperature/dew point depression-independent short-term ozone is slightly smaller that in the raw short-term ozone data. I should note that our preliminary analysis on available precursor data shows that the seasonal components in precursors, ozone and meteorological variables (e.g., radiation) are highly correlated, pointing the need to find a relation between weather-independent ozone and weather-independent precursors. We really need a large precursors database to help answer the questions you raise.

STUDY OF THE ROLE OF A STRATIFORM CLOUD LAYER ON THE REDISTRIBUTION OF HYDROGEN PEROXIDE

Nicole Audiffren, Emmanuel Buisson and Nadine Chaumerliac

Laboratoire de Météorologie Physique
Université Blaise Pascal/CNRS/OPGC
24, Av. des Landais
F-63177 Aubière cedex

INTRODUCTION

Due to its high density of urban and industrial sources, the eastern United States often experiences widespread pollution episodes during the summer (Logan, 1985; Vukovich and Fishman, 1986). The effects of such continental emmissions on the oxiding capacity of atmosphere over the North Atlantic have been studied for many years. Most of the earlier works (Zeller et al., 1977; Kelleher and Feder, 1978; Spicer, 1982) presented evidence for the transport of plumes from the eastern seaboard of the United States out over 100 km or more of the North Atlantic. Measurements at Kejimbuijk National Park in Canada (Brice et al., 1988; Beattie and Wepdale, 1989), begun in 1979, demonstrate transport of these plumes to central Nova Scotia, located at a distance of than 500 km more. The 1993 North Atlantic Regional Experiment (N.A.R.E.) intensive provided further evidences for the transport of anthropogenic pollutants and ozone precussors (CH_4, CO, ...) from the continent sources out over the Atlantic ocean (Fehsenfeld et al., 1996). We expect that many tropospheric photo-oxidants are generated by chemical reactions, in particular the hydrogen peroxide. This latter presents a real interest for acidification of the clouds and in gas-phase as a efficient source of OH.

This study deals with the effect of a stratiform cloud layer on the H_2O_2 concentration in gas phase offshore Nova Scotia. First, we use a box model to assess the chemical impact of the aqueous phase. Next, we have coupled a 2D transport model with a chemical module in order to account for the interactions between dynamical, microphysical, radiative and chemical processes. We consider that the photolysis rates vary with the zenithal angle, including the attenuation by standard ozone and aerosols columns and by the evolutive cloud layer.

A- UNIDIMENSIONAL STUDY OF THE BEHAVIOR OF THE HYDROGEN PEROXIDE

1- Objective and initialization

We use a "box" model for simulating the chemical processes which govern the tropospheric chemistry. This latter take into account gas-phase and aqueous-phase reactions as well as the transfert between the both phases. The numerical solver is based on Gear's method (1971).
The chemical mecanism describes a simplified photochemistry in both phases, of the ozone and its precussors NO_x, CO and CH_4 (Grégoire and al., 1994), and the chemistry of sulfur. 13 species and their counterparts in aqueous phase are followed. The chemical module consists of 29 gas-phase reactions, 14 aqueous-phase reactions. The mass transfer of the soluble species between both phases is explicitly written (Schwartz, 1986). The transfer term accounts for acid-base equilibrium.
The conditions of the simulation are those of the N.A.R.E. campaign. Liquid water content and temperatures are issued from vertical profiles given by airborne measurements during September 7th 1993. The initial spatial gaseous mixing ratio of species, typically of rural levels of NO_x (0.6 ppbv), is taken to be constant with height from the chemical measurements.

Photolysis rates are allowed to vary during the entire simulation. These rates are calculated every the 15 minutes, using Madronich's photolysis model (1995). This latter considers the vertical profile of temperature, ozone column and aerosol for a standard atmosphere (1976). In order to model the effect of a stratiform cloud layer on the radiative transfer, we take into account the vertical profile of the temperature and the liquid water content measured during the flights 45 and 46 (07/09/93) of the N.A.R.E. campaign. We note that below the cloud layer, the photolysis rates are weaker of a factor of 0.4 than in clear sky conditions. On the contrary, above the cloud there is an increase of 2.2. In summary, the stratiform cloud layer has a significant impact on the photolysis rates and thus, we expect an effect on the tropopheric chemistry.

The simulation lasts for 6 hours, between 14h and 20h UT, the first hour allowing the onset of gas-phase equilibrium. Box model runs at each altitude between the ground and 1250 meters, i.e. 11 several simulations. We consider different cases :

- run 1 : clear sky ; gas-phase chemistry and clear sky photolysis.
- run 2 : cloudy sky ; gas-phase chemistry and photolysis disturbed by the stratiform cloud layer.
- run 3 : cloudy sky ; gas-phase and aqueous-phase chemistries and photolysis disturbed by the stratiform cloud layer.

2- Model results

In clear sky (dotted line), the hydrogen peroxide concentration is destroyed by photodissociation and oxidation by OH reactions and produced via reaction with HO_2 radicals. The budget of the production and destruction rates shows a net production of H_2O_2 (figure 1). Moreover, the production rate decreases during the simulation because there are less HO_2 radicals present in the closed system, but, in the same time, the destruction rate also decreases owing to a fall of the photolysis and the consumption of OH radicals. Finally, the hydrogen peroxide concentration steadily increases during the simulation from 1.1 ppbv at 15h UT to 1.4 ppbv at 20h UT(figure 2).

In run 2 (dashed line), a cloud layer is now considered, evolving in time and with the altitude. The chemistry has been modified via the photolysis disturbed by the cloud.

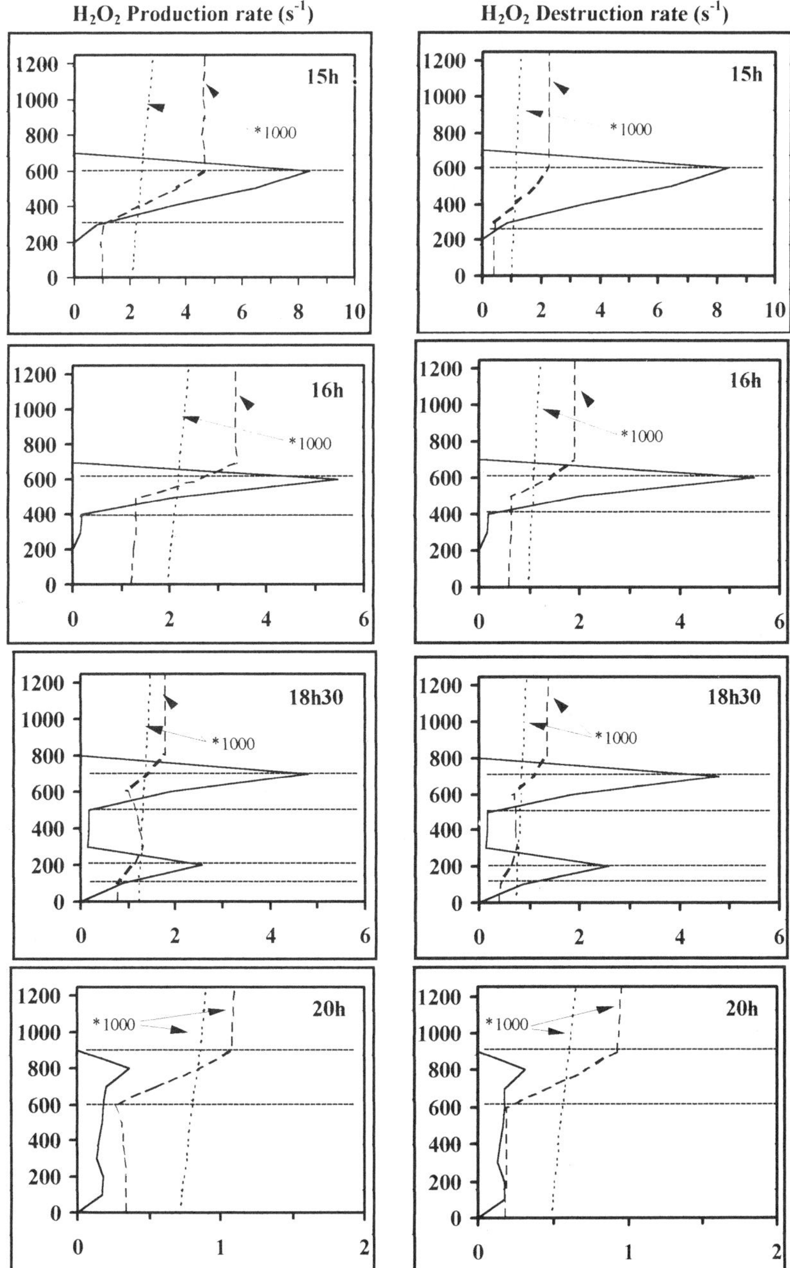

Figure 1. Temporal evolution of H_2O_2 production and destruction rates ; run 1 (dotted line), run 2 (dashed line) and run 3 (rigid line) with the box model.

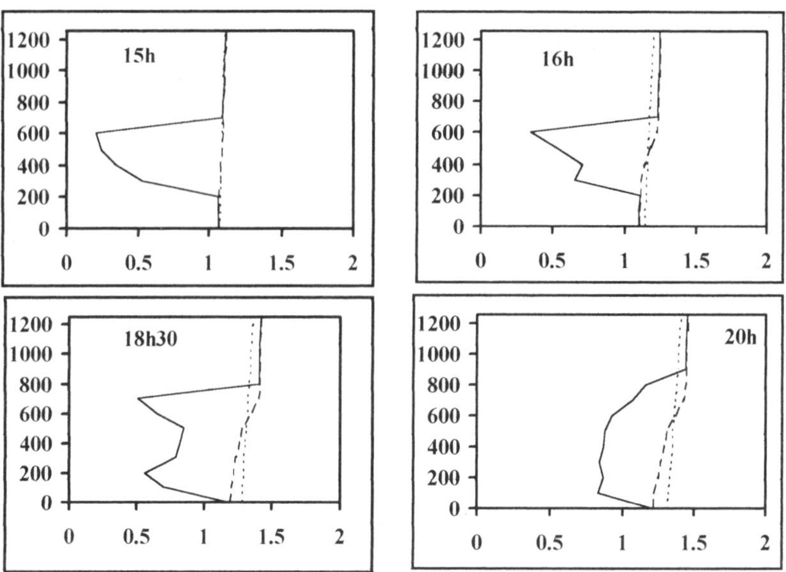

Figure 2. Temporal evolution of H_2O_2 gaseous concentration (in ppbv) ; run 1 (dotted line), run 2 (dashed line) and run 3 (rigid line) with the box model.

The vertical and temporal variations of H_2O_2 depend on its photodissociation in the atmosphere. The variations of the photolysis rates are important in presence of cloud and affect the chemistry. After six hours of simulation, below the cloud layer, H_2O_2 concentrations are lowered by about 6% and are enhanced above by less than 7% (figure 2). When photolysis rates are disturbed by the cloud layer a net production of H_2O_2 is observed at each vertical level, during the entire simulation.

For the run 3 (continuous line), the consideration of the aqueous phase and the mass transfer between both phases hightlights the impact of the aqueous chemistry on the gaseous concentrations of the hydrogen peroxide. H_2O_2 is a strong soluble compound. These production and destruction rates include transfer between both phases. Comparisons with the run 2 show that mass transfer is driving process (figure 1). Presumably, Henry's law equilibrium is not reached and water acts continuously as efficient sink for H_2O_2. Moreover, the hydrogen peroxide reacts, in aqueous phase, with OH radicals and mainly with HSO_3^-. This latter reaction is efficient for the consumption of H_2O_2. The loss of H_2O_2 in the cloud is about 60% at the beginning of the run and decrease to 30% at 20h UT (figure 2). In addition, the cloud layer continuously changes during the entire simulation. Hence, the model results show that the impact of the aqueous phase is not limited at cloud levels. The gaseous concentrations decrease below the cloud layer. For instance, at 20h UT the cloud is located between 600 and 850 meters, whereas the gaseous concentrations are altered since 100 m AMSL. In fact, the consideration of the aqueous phase in the "box" model involves irreversible loss of the concentrations. Effects of aqueous-phase reactions in the evolutive cloud layer are still felt after the passage of the cloud for H_2O_2, O_3 and SO_2 whereas RO_2 radicals and OH rapidly recover the values obtained in run 2.

In summary, the consideration of a stratiform cloud layer leads to modifications on the gaseous concentrations of H_2O_2 due, first to the variations of the photolysis rates modified by the cloud, and secondly to the effects of the aqueous phase. The impact is not limited at cloud levels but remains below the cloud layer after its passage due to the depletion of the chemical compound.

B- LONG-RANGE FLUX OF POLLUTANTS OVER A STRATIFORM CLOUD LAYER; 2D TRANSPORT/CHEMISTRY SIMULATIONS

The coupling between the meteorological code R.A.M.S. and the chemical module allows to take into account the interactions between the dynamical, microphysical, radiative and chemical processes.

1- Model description

The meteorological model. We use R.A.M.S. (Regional Atmospheric Modeling System) whose a full description is available from the article of Pielke et al. (1992). In this study, the mesoscale model runs in non-hydrostatic two-dimensional mode. The domain covers 300 km over sea with a grid spacing of 5 km. The grid is located in such a way that it intersects with the trajectory of the flights of September, 7th 1993. A stretched terrain-following coordinate is used with a grid spacing of 25 m between the surface and 1000 m above mean sea level (AMSL), above which the spacing increases to 200 m as far as the model top at an elevation of 5000 m AMSL. The dynamical, microphysical and radiative processes are fully interactive in the model. The microphysical parametrization considers only the cloud water and the evaporation and condensation processes. The time step is 10 s. The simulation lasts for 9 hours and is initialized by a rawinsounding released near Yarmouth at 11h UT.

The chemical model. The chemical solver used for the coupling with the transport model is of type QSSA (Hesstvedt, 1978). Indeed, for reasons of memory place and computing time we can not use a Gear's solver (Audiffren et al., 1997, submitted to Atm. Research). The coupling of the chemical module with the meteorological model has been realized of simply manner. At each time step (10 s) the different meteorological features (temperature, liquid water content, air density) are used for initializing the chemistry. The chemical time step is 5 s in clear sky and decreases to 0.5 s when the aqueous phase is accounted for. Two differential equations are simultaneously solved for gas-phase and aqueous-phase concentrations respectively. We use the same chemistry than in the box model and the photolysis rates vary during the entire simulation. The initial gaseous concentrations vary with altitude following the different air masses (Kleinman et al., 1996). Ozone and sulfur fluxes have been introduced for modeling the air mass arriving on the domain. These fluxes released at the west edge of the grid vary with altitude and are initialized from airborne measurements (Banic et al., 1996). Their maxima are located at about 1000 m AMSL. The emission works only during the first hour in order to characterize the air mass arriving at 19h UT over the measurements area. Different cases are considered, as we did for the box model. The first simulation treats chemical reactions in both phases with photolysis rates disturbed by the cloud layer. The another simulations consist of two tests of sensibility : a run in cloudy sky conditions without the aqueous-phase chemistry and a run in clear sky conditions.

2- Results

From the meteorological point of view, the cloud layer evolves during the simulation due to the radiative heating, from 650 m AMSL at 15h UT to 800 m AMSL at

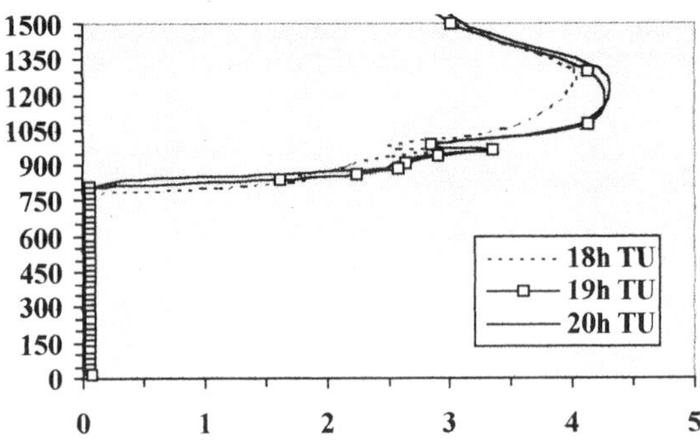

Figure 3. Temporal evolution of H_2O_2 gaseous concentration (in ppbv) when the gas-phase and aqueous-phase chemistries and the photolysis disturbed by the cloud layer are accounted for.

20h UT. This latter is limited by a inversion of temperature which starts at 700 m AMSL at 15h UT (850 m AMSL at 20h UT) occuring up to 950 m AMSL. Between 700 m and 950 m (850 m and 950 m at 20h UT), turbulent exchanges appear and generate an air mixing. We will afterwards call this turbulence layer the 'inversion layer'.

The introduction of ozone and sulfur fluxes modifies the hydrogen peroxide concentrations due to the production of HO_2 radicals (figure 3).
Above the 'inversion layer', the numerical results hightlight the H_2O_2 peak agree with the ozone and sulfur fluxes. The temporal evolution between 18h and 20h UT at about 1000 m AMSL shows a regular production of H_2O_2 from 4.1 ppbv at 18h to 4.5 ppbv at 20h UT. In fact, after the passage of the front of the fluxes, we expect a decrease of the concentration like for O_3 and SO_2 due to the dynamical process. In the contrary, the response of the hydrogen peroxide is longer because it depends on a chain of chemical reactions involving HO_2 radicals.

In the lower layers (<800 m AMSL), H_2O_2 being very soluble, its concentration is very weak owing to its absorption by aqueous phase and its destruction in aqueous-phase by reaction with HSO_3^-. The simulations with the box model have shown that the cloud has an impact on the gaseous concentration after its passage. The HO_2 radicals are destroyed by the passage of the cloud layer and the hydogen peroxide is not produced any more. Moreover, the strong inversion of temperature of the September 7th prevents exchanges of species below and above the 'inversion layer' : the values in the mixing layer stay lower than these measured. These weak concentrations could be explain by the bidimensional nature of the simulation. On an another hand, local observations report some drizzle below the cloud layer. In our simulation, only cloud water has been considered. The consideration of the rain water in the meteorological model may release soluble chemical compounds in the gas phase during the evaporation of the drops in the lower layers.

Another simulation was driven in order to assess the aqueous-phase effect of the stratiform cloud layer. The conditions of the simulation are similar to run 2 of the box model, i.e. only gas-phase reactions with photolysis rates in cloudy sky. The cloud layer being located to alitudess lower than 900 m AMSL, we will discuss numerical results only in the lower layers. The strong inversion of temperature limits exchanges of compounds between the mixing layer and the free troposphere. Results point out the impact of the aqueous phase on the gaseous concentration (figure 4).
The H_2O_2 concentration decreases from about 0.9 ppbv to less than 0.1 ppbv when the aqueous phase is activated. The passage of the cloud layer leads to destroy the soluble compound as we saw in box model. Two chemical phenomena occur : the depletion of HO_2 radicals by cloud weakens H_2O_2 gas-phase production; this aspect joined to the absorption

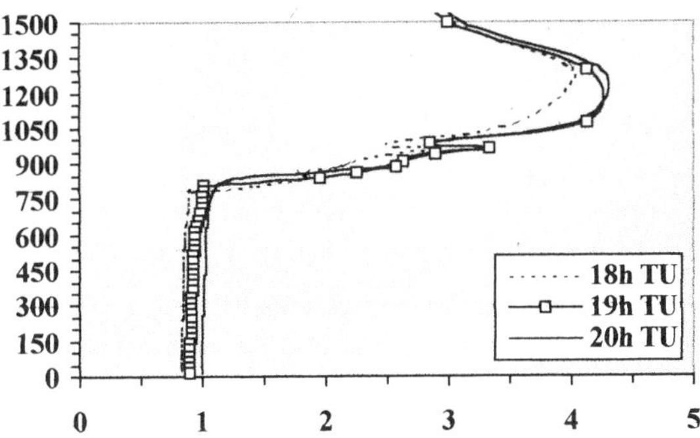

Figure 4. Temporal evolution of H_2O_2 gaseous concentration (in ppbv) when only the gas-phase chemistry and the photolysis disturbed by the cloud layer are accounted for.

of H_2O_2 by liquid water involves a strong decrease of H_2O_2 concentration in gas-phase which remains after the passage of the cloud.

A last simulation was driven in order to assess the relative contributions of radiative effects of the stratiform cloud layer. This will be done by comparing to a simplified run where only gas-phase reactions are present and photolysis rates are calculated in clear sky (run 1 in box model). Above the inversion level, the impact of the cloud layer has its origin in the modification of photolysis rates. At 1000 m AMSL, the H_2O_2 concentration is about 3.5 ppbv in clear sky, 25% less in comparison with the cloudy sky case. This shows that the cloud layer has a significant effect on the gaseous chemistry 500 m above by disturbing the photodissociation of the different chemical species. Below the inversion, the differences between the run 1 and run 2 point out the weak impact of the photolysis relative to the aqueous phase. Numerical results give concentrations less than 10% if the photolysis rates disturbed by the cloud layer is considered.

CONCLUSION

Different simulations were driven to assess the impact of a stratiform cloud layer on the hydrogen peroxide gaseous concentration framework of the 1993 N.A.R.E. campaign. First, we watched only the chemical aspect by using a box model, including the photochemistry of ozone and its precussors and the chemistry of sulfur in both gas and aqueous phases. Mass transfer is included. The photolysis rates vary during the entire simulation and are disturbed by the evolutive cloud layer. Numerical results point out that the cloud modifies the H_2O_2 concentration at cloud level but also after its passage.

Secondly, the coupling between the meteorological code R.A.M.S. and the chemical module used latter allows to account for the dynamical, microphysical radiative and chemical interactions at cloud scale and mesoscale too. Ozone and sulfur fluxes have been released at west edge of the grid in order to characterize the air mass coming from the United States and arriving over Nova Scotia. Different runs hightlighted the radiative and chemical impacts of the stratiform cloud layer on the H_2O_2 concentration. Above the inversion layer, the photolysis is the driving process in the modification of the gaseous concentration. Below the inversion level, the effect of the cloud is mainly due to the aqueous phase. The direct impact of the aqueous phase is two-fold: the reaction with HSO_3^- and the removal from aqueous phase. The indirect effect is the weakening of HO_2 gas concentration and subsequent gas-phase production of H_2O_2. However, too much H_2O_2 is destroyed by the cloud layer and it will be necessary to consider the evaporation of drops in our next simulations.

REFERENCES

Banic, C.M., Leaitch W.R., G. A. Isaac, M.D. Couture, 1996: Transport of ozone and sulfur to the North Atlantic atmosphere during NARE, *J. Gepohys. Res.*, **101**, 29091-29104.

Beattie, B.L. and D.M. Whepdale, 1989: Meteorological characteristics of large acidic deposition events at Kejimkujik, Nova Scotia. *Water Air Soil Pollut.*, **46**, 45-59.

Brice, K.A., J.W. Bottenheim, K.G. Anlauf and H.A. Wiebe, 1988: Long-term measurements of atmospheric peroxyacetylnitrate (PAN) at rural sites in Ontario and Nova Scotia: Seasonal variations and long-range transport. *Tellus*, **40B**, 408-425.

Fehsenfeld, F.C., S. Penkett, M. Trainer and D.D Parrish, 1996a: NARE 1993 Summer Intensive: Forward. *J. Gepohys. Res.*, **101**, 28869-28876.

Fehsenfeld, F.C., S. Penkett, M. Trainer and D.D Parrish, 1996b: Transport of O_3 and O_3-precursors from Anthropogenic Sources to the North Atlantic, *J. Gepohys. Res.*, **101**, 28877-28892.

Grégoire P.J., N. Chaumerliac and E.C. Nickerson, 1994: Impact of cloud dynamics on tropospheric chemistry: Advances in modeling the interactions between microphysical and chemical processes., *J. Atmos. Chem.*, 18, 247-266,.

Heestvedt, E., O. Hov and I.S.A. Isaksen, 1978: Quasi-steady-state approximations in air pollution modeling: comparison of two numerical schemes for oxydant prediction. *Int. J. Chem. Kin.*, **10**, 971-994.

Kelleher, T.J. and W.A. Feder, 1978: Phytotoxic concentrations of ozone on Nantucket Island: Long range transport from Middle Atlantic States over the open ocean confirmed by bioassay with ozone-sensitive tobacco plants. *Environ. Pollut.*, **17**, 187-194.

Kleinman, L.I., P.H. Daum, Y. Lee, S.R. Springston, L. Newman, W.R. Leaitch, C.M. Banic, G.A. Isaac andJ.I. MacPherson, 1996: Measurement of O3 and related compounds over the southern Nova Scotia, 1, Vertical distributions. , *J. Gepohys. Res.*, **101**, 29043-29060.

Logan, J.A., 1985: Tropospheric ozone: Seasonal behavior, trends, and anthropogenic influence. *J. Gepohys. Res.*, **90**, 10463-10482.

Pielke, R.A., W.R. Cotton, R.L. Walko, C.J. Tremback, M.E. Nicholls, M.D. Moran, D.A. Wesley, T.J. Leeand J.H. Copeland, 1992: A comprehensive meteorological modeling system - RAMS. *Meteor. Atmos. Phys.*, **49**, 69-91.

Spicer, S.W., 1982: Nitrogen oxide reactions in the urban plume of Boston. *Science*, **215**, 1095-1097.

Schwartz, S.E., 1986: Mass-transport considerations pertinent to aqueous phasereactions of gases in liquid water clouds. in W. Jaeschke (ed.). Chemistry ofMultiphase Atmospheric Systems. *Springer-Verlaag, Berlin*, 415-471.

Vukovich, F.M. and J. Fishman, 1986: The climatology of summertime O_3 and SO_2 (1977-1981). *Atmos. Env.*, **20**, 2423-2433.

Zeller, K.F., R.B. Evans, C.K. Fitzimmons and G.W. Siple, 1977: Mesoscale analysis of ozone measurements in the Boston Environs. *J. Gepohys. Res.*, **82**, 5879-5888.

INFLUENCE OF THE RESOLUTION OF EMISSIONS AND TOPOGRAPHY ON THE AIR POLLUTION DISTRIBUTION IN A MESOSCALE AREA

Klaus Nester, Hans-Jürgen Panitz, Franz Fiedler, and Walburga Wilms

Institut für Meteorologie und Klimaforschung
Forschungszentrum Karlsruhe/Universität Karlsruhe
D-76021 Karlsruhe

INTRODUCTION

The air pollution in a region is influenced by all atmospheric scales. Because it is not possible to resolve all scales in a single model, different models have been developed which describe the air pollution in certain regions, like the European scale model EURAD (EURopean Acid Deposition model (Ebel et al., 1989)) and the mesoscale model system KAMM/DRAIS (KArlsruher Meteorologisches Modell (Adrian and Fiedler, 1991)/DReidimensionales Ausbreitungs- und Immissions-Simulationsmodell (Schwartz, 1996)). Those effects which are smaller than the grid resolution of a model (subgrid effects) have to be considered by appropriate parameterizations or by nesting procedures. Effects resulting from scales which are even larger than the model domain are usually introduced into the model by the boundary conditions.

In the frame of the EUMAC (EUropean Modelling of Atmospheric Constituents) project simulations of the air pollution over Europe and in subareas have been carried out for special episodes using the EURAD and the KAMM/DRAIS models, respectively Especially, in mountainous areas or in highly industrialized regions, it can be expected, that the European scale model results in too uniform distributions of the air pollutants. Therefore nesting procedures have been developed which allow to simulate the air pollution in subareas with higher spatial resolution. One of the nesting procedures used in the EUMAC project is the coupling of the model system KAMM/DRAIS to the EURAD model (Nester et al., 1995).

In order to demonstrate the performance of the coupling procedure and to analyse the effects of the better resolution of the topography and the emissions on the air pollution distribution, different simulations have been carried out with the KAMM/DRAIS model. The simulations comprise the four possibilities to vary the resolution of the topography and the emissions. The model area selected is South-west Germany and parts of Alsace. This is a mountainous area with highly industrialized regions. Although more than 40 species are simulated by the model, only NO_x and O_3 are considered here. They represent the emitted and non emitted species, respectively.

THE MODEL SYSTEM KAMM/DRAIS

The model system KAMM/DRAIS consists of the meteorological model KAMM and the dispersion model DRAIS. The meteorological model KAMM is forced by the large scale meteorological conditions, which are introduced into the KAMM model by the definition of a basic state. The basic state variables are the geostrophic wind, the pressure, and the potential temperature.

The non hydrostatic model KAMM solves the Navier-Stokes equations of motion, the continuity equation, and the heat equation in a terrain following coordinate system. This transformation allows a better resolution close to the ground as compared to the upper levels.

The interaction between the soil and the atmosphere is described by a soil-vegetation model (Schaedler et al., 1990). This model calculates the fluxes of heat and humidity in the soil, between the soil and the vegetation and between the vegetation and the atmosphere as well as the momentum flux to the ground.

The dispersion model DRAIS solves the diffusion equation on an Eulerian grid. Like in the KAMM model, the equations are transformed into a terrain following coordinate system.

The dry deposition of the different species is described by their dry deposition velocity. It involves a complex linkage between turbulent diffusion in the surface boundary layer, molecular scale motion at the air-ground interface and the interaction of the material with the surface The details of the deposition model are described in Bär and Nester (1992).

In this study the gas phase chemical reaction mechanism of the RADM2 model (Regional Acid Deposition Model (Stockwell et al., 1990)) is applied. This mechanism takes into account 158 reactions among 63 species. It explicitly integrates 41 species concentrations. Twenty-one photochemical reactions are considered. Methan is held constant.

In the stable and neutral boundary layer a local first order approach for the diffusion coefficients is chosen. In the convective boundary layer a non-local approach is used (Degrazia, 1988).

SIMULATIONS

The simulations with the model system KAMM/DRAIS have been performed for 15 July, 1986, a day of the EUMAC Joint Dry Case (JDC). The model domain encloses the south-west part of Germany including the Upper Rhine Valley and the Black Forest as well as parts of Alsace (France) including the Vosges Mountains (Figure 2). The model area covers 260 km * 240 km with a horizontal resolution of 5 km. In the vertical the model domain is divided by 35 levels between the ground and 8000 m.

In order to carry out the dispersion simulations three-dimensional flow fields, temperature fields and turbulence fields are necessary. These data have been provided by simulations with the mesoscale model KAMM. The basic state variables, which describe the large scale conditions are derived from the EURAD model simulation, which is available for the same day.

The initial and boundary data for all chemical species considered are also calculated from the results of the EURAD model by applying the coupling interface.

Two emission inventories are available. The first, having a horizontal resolution of 80 km, is the inventory used by the EURAD model. It is interpolated to the grid points of the DRAIS model. The second inventory is the one for the Federal State of Baden-Württemberg which has been inquired for the period of the TULLA-Experiment (Fiedler et al., 1991). This emission data set has been combined with the inventory of the EURAD model by modifying the TULLA data in such a manner, that the better resolution of point and area sources is retained and the total release rates in both emission inventories are the same. This modification of the fine grid data is necessary if the results of both models are compared, which has been done in the EUROTRAC project. The data for the topography are the original EURAD data (80 km resolution) interpolated to the DRAIS grid and the data, based on a 5 km resolution, respectively.

Four different simulations have been carried out with the model system KAMM/DRAIS. Table 1 contains the selected resolutions for topography and emissions. Coarse means, that the data are based on a 80 km resolution and fine means, that the resolution is 5 km. All simulations are carried out in the mesoscale domain of the DRAIS model using a horizontal grid size of 5 km.

RESULTS

The flow field has been simulated from 01:00 UTC until 24:00 UTC for 15 July, 1986. The general flow direction during the day is from north-east with an average wind speed of about 5 m/s at a height of 1000 m above ground. The wind field with the EURAD topography (Fig.1) is quite uniform during the whole day.

Table 1. Compilation of the model runs

	Topography + Meteorology		**Emissions**	
	(fine resol.)	**(coarse resol.)**	**(fine resol.)**	**(coarse resol.)**
case 1		×		×
case 2		×	×	
case 3	×			×
case 4	×		×	

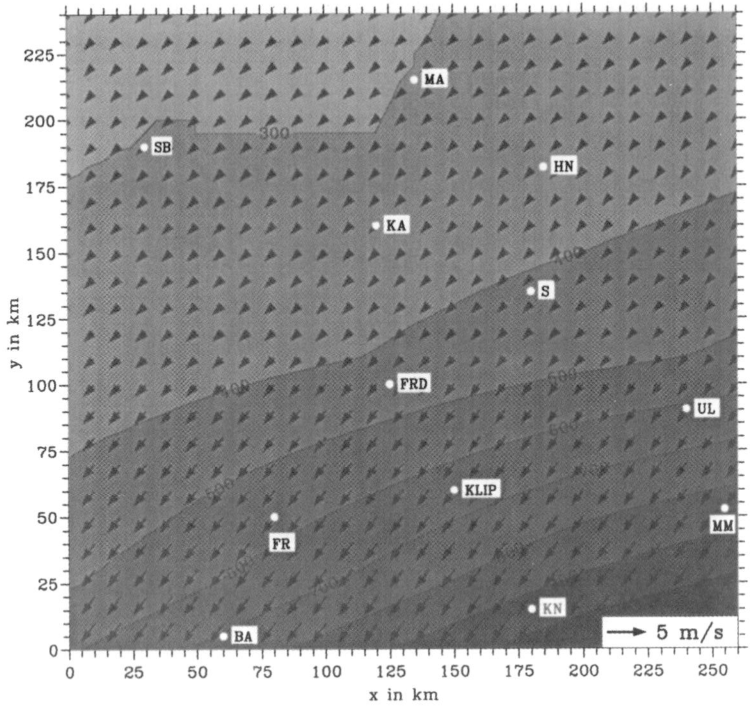

Figure 1. Wind field in 28 m above ground, 8:00 UTC, coarse topography

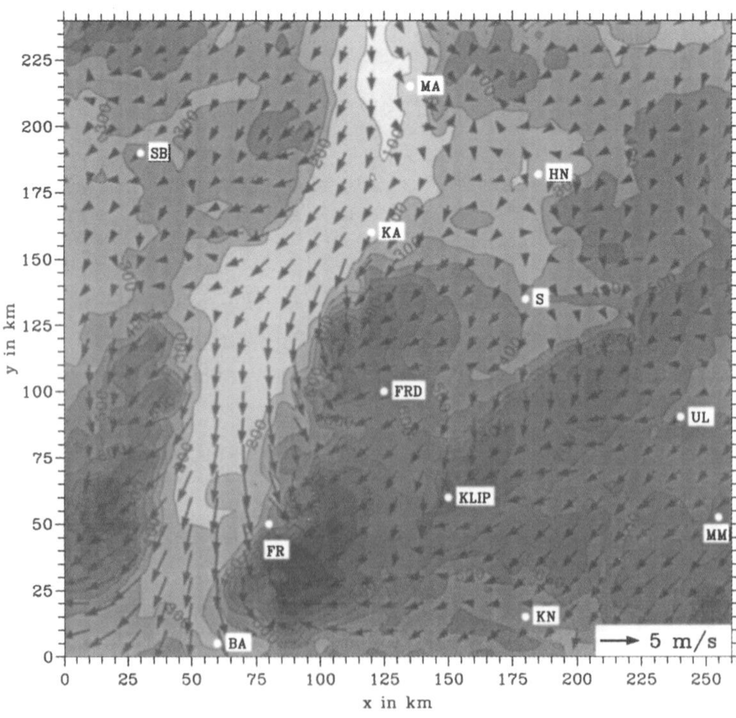

Figure 2. Wind field in 28 m above ground, 8:00 UTC, fine topography

The dominating wind direction near the ground does not differ very much from that of the general flow. A quite different wind field is simulated with the fine resolution of the topography (Fig.2). Especially, close to the ground it is mainly influenced by the orography. The most striking effect is the channelling of the flow along the Upper Rhine Valley. In the afternoon the strong valley wind in the region of Freiburg is higher than the wind speed of the general flow. This is confirmed by the observations as well as the simulated downslope winds in the evening hours.

The DRAIS simulations started at 01:00 UTC. The concentration distributions have been calculated for each hour till 22:00 UTC. Figure 3 shows the distributions of the NO_x concentrations at 08:00 UTC, representing the results of the four simulations. They show that the resolution of the emissions mainly determines the amount and the location of the peak concentrations (cases 1 and 2). The resolution of the topography alters the distribution remarkably (case 3) as compared to the reference case 1. This modification is mainly caused by the flow field, which differs considerably from that of case 1. The flow along the valleys and around the mountains cause the differences in the concentrations between elevated and valley areas. This influence of the topography via the flow field is the reason why the structure of the distribution of case 3 and case 4 are more similar than between case 2 and case 4. In case 4 the better resolution of the emissions mainly cause an amplification of the extreme values.

A comparison of the four simulations for ozone is shown in Figure 4. The influence of the emissions on the O_3 concentration distribution can be seen in the figure of case 2. Here the ozone concentration is reduced in the areas of the high NO_x concentrations and in the plumes starting from these areas. The figure for case 3 is already completely different to that of case 1 or case 2. The modification of the ozone concentration distribution due to the better resolution of the topography is caused by the flow field. The structure of the valleys and the mountains are illustrated by the ozone distribution close to the ground. The additional effect of the better resolution of the emissions alters this distribution only slightly (case 4). At the locations of the highest NO_x concentrations a reduction of the ozone concentration can be recognized. It is obvious that the distributions of case 3 and case 4 are much more similar for ozone than for NO_x.

Due to the intensified mixing the differeces between the four simulations are less pronounced in the afternoon. But the general statements are still valid. In the evening the results are comparable to the morning. Because of the higher NO_x concentrations calculated in the fine-grid emission cases the simulated O_3 concentrations of case 3 and 4 agree less than in the morning hours.

CONCLUSIONS

From the comparison of the simulations it can be concluded that the resolution of the emission inventory plays an important role for the concentration distribution of the emitted species like NO_x, especially in the morning and evening hours. In those areas where the main sources are located the NO_x concentrations may be much higher than in the case of a coarse emission resolution.

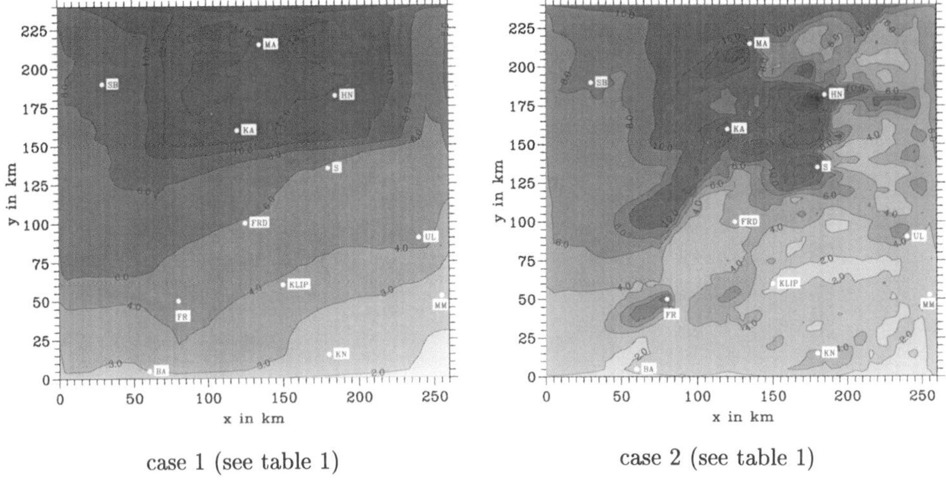

case 1 (see table 1)

case 2 (see table 1)

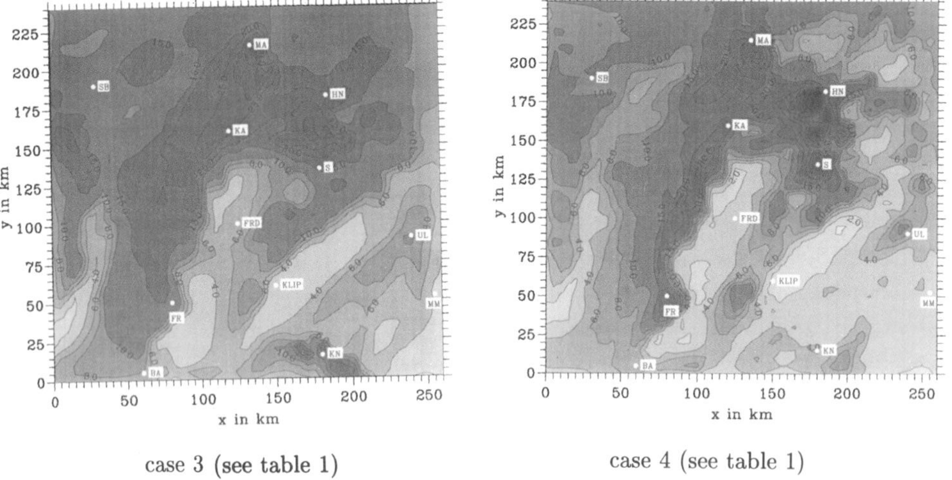

case 3 (see table 1)

case 4 (see table 1)

Figure 3. NO_x Concentration distributions at ground level in ppb, 8:00 UTC

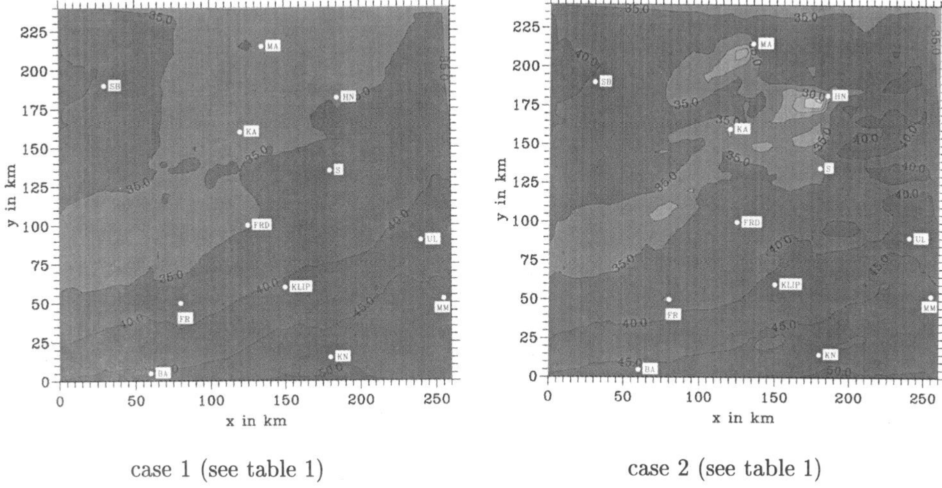

case 1 (see table 1)

case 2 (see table 1)

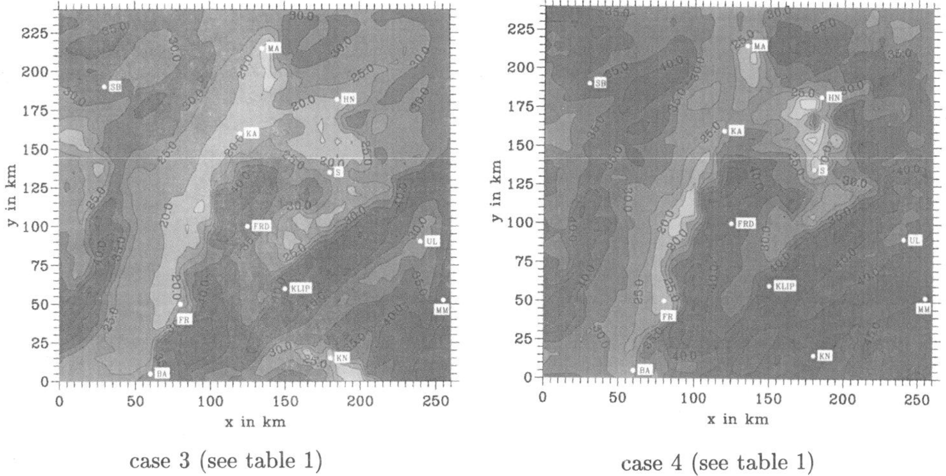

case 3 (see table 1)

case 4 (see table 1)

Figure 4. O_3 Concentration distributions at ground level in ppb, 8:00 UTC

In areas of low emission rates the usage of coarse-grid emissions leads to higher NO_x concentrations. But the simulations show also that the influence of the resolution of the topography is remarkable. The topography influences the flow field, which mainly determines the structure of the concentration distribution. In the afternoon, intensive vertical mixing causes a good dilution. Therefore only in areas where high NO emissions occur, the results are still remarkably different.

The secondary species ozon shows a different behaviour. During the morning and evening hours lower ozone concentrations are simulated with the fine-grid emissions in those areas where the NO_x emission rates are highest. Nevertheless, the general features of the ozone distributions mainly result from the influence of the topography. The emissions play a minor role. In the afternoon, the size of the areas with high ozone concentrations is larger for the coarse-grid emission cases as compared to the fine-grid cases.

REFERENCES

Adrian, G., Fiedler, F., 1991, Simulation of unstationary wind and temperature fields over complex terrain and comparison with observations, *Beitr. Phys. Atmos. 64*, 27:48

Bär, M., Nester, K., 1992, Parameterization of trace gas dry deposition velocities for a regional mesoscale diffusion model, *Ann. Geophysicae 10*, 912:923

Degrazia, G. A., 1988, Anwendung von Ähnlichkeitsverfahren auf die turbulente Diffusion in der konvektiven und stabilen Grenzschicht, *Dissertation am Institut für Meteorology und Klimaforschung, Universität Karlsruhe*, 99 pp.

Ebel, A., Neubauer, F. U., Raschke, E., Speth, P., 1989, Das EURAD Modell, Aufbau und erste Ergebnisse, *Mitteilungen aus dem Institut für Geophysik und Meteorologie der Universität zu Köln Heft 61*, 161 pp.

Fiedler, F. et al., 1991, *Transport und Umwandlung von Luftschadstoffen im Lande Baden-Württemberg und aus Anrainerstaaten (TULLA). Forschungsbericht KfK-PEF 88*, 225 pp.

Nester, K., Panitz, H.-J., Fiedler, F., 1995, Comparison of the DRAIS and EURAD Model Simulations of Air Pollution in a Mesoscale Area, *Meteorol. Atmos. Phys. 57*, 135:158

Schädler, G., Kalthoff, N., Fiedler, F., 1990, Validation of a model for heat, mass and monumentum exchange over vegetated surfaces using LOTREX-10E/HIBE88 data, *Contr. Phys. Atmosph. 63*, 85:100

Stockwell, W. R., Middleton, P., Chang, J. S., 1990, The second generation Regional Acid Deposition Model; chemical mechanism for regional air quality modelling, *J. Geophys. Res. 95*, 16,343:16,367

Schwartz, A., 1996, Numerische Simulationen zur Massenbilanz chemisch reaktiver Substanzen im mesoskaligen Bereich, *Dissertation am Institut für Meteorologie und Klimaforschung, Universität Karlsruhe*, 200 pp.

DISCUSSION

D. W. BYUN: I enjoyed your discussion on effects of resolution in topography and emission data on air quality. You said that your model is linked with the entirely (dynamically) different model EURAD. How did you resolve the potential conflict existing in the dynamic description of regional & local (urban) scale problems? Our failure of linking a global scale model with a mesoscale model may be due to the difference in chemical mechanisms in the models for two different scales. For your case, at least you used the same chemical mechanism in both models.

K. NESTER: All simulations are carried out with the mesoscale model system KAMM/DRAIS with a resolution of 5 km. The effect of coarse grid resolution is simulated by using the emissions and the topography interpolated from the data of the coarse grid distribution. Only the large scale information is taken from the EURAD model. In the meteorological model KAMM these are the geostrophic wind, the large scale pressure and potential temperature fields. A local dynamic link does nor exist. In the dispersion model DRAIS the initial and boundary conditions are taken from the EURAD model results. Therefore, only conflicts at the boundaries are possible. Because we use the same chemical mechanism in both models these conflicts are of minor importance. I cannot say what happens if the chemical mechanisms are different in both models.

R. ROMANOWICZ: What were the results of comparing model predictions with observations for the pollutants concentrations?

K. NESTER: Your question gives me the opportunity to show such a comparison for the city station Rastatt, which is located in the Upper Rhine Valley about 15 km south of Karlsruhe. The comparison between the measurements and the simulations for NO, NO_2 and O_3 shows that the simulation with the fine grid resolution of both emissions and topography give the best agreement. The lowest agreement is found for the coarse grid resolution of both parameters. The peak NO concentration is underestimated by about a factor of two, which means that a resolution of 5 km is still too coarse. All simulations show that the ozone concentrations during the afternoon do not differ very much from each other. This result confirms my statement that during this time period the effect of the resolution on the species concentrations is lowest.

A COMPARISON OF CALPUFF MODELING RESULTS WITH 1977 INEL FIELD DATA RESULTS

John S. Irwin*

Atmospheric Sciences Modeling Division
Air Resources Laboratory
National Oceanic Atmospheric Administration
Research Triangle Park, NC 27711
U.S.A.

ABSTRACT

CALPUFF is a non-steady-state, multi-layer puff dispersion model that simulates the effects of time and space varying meteorological conditions. The puff modeling simulation results are compared with data obtained following a single 3-hour late-afternoon tracer release conducted on April 19, 1977 near Idaho Falls, Idaho, USA. Samplers were positioned on arcs at downwind distances of 3.2, 48 and 90 km. Low-level instrumented masts provided hourly values of wind and temperature at 17 sites within the experimental area. Hourly rawindsondes were available 600 m northwest of the release point. And hourly pibals provided winds aloft at three other sites within the area. In this discussion, alternative combinations of the available meteorological data were used to assess the differences to be seen in the simulation results. Analysis of the results suggests that the simulated lateral dispersion along each arc was best characterized when all of the surface and upper wind observations were used, and that the position of the simulated maximum on each arc was poorly characterized regardless of data used.

INTRODUCTION

This paper provides a summary of results from a series of analyses in which puff dispersion modeling results were compared with data obtained following a single 3-hour late afternoon tracer release, lasting from 1240 to 1540 Mountain Standard Time (MST), conducted on April 19, 1977 near Idaho Falls, Idaho. The puff modeling results were obtained using the CALPUFF dispersion modeling system (EPA, 1995a,b). This modeling system consists of a meteorological processor called CALMET, which is capable of developing time-dependent multi-layered wind fields using a diagnostic wind model; and a puff dispersion model called CALPUFF, which is capable of simulating the hour-by-hour variations in transport and dispersion. The tracer release results (Clements, 1979) were obtained as a consequence of an

---

* On assignment to the Office of Air Quality Planning and Standards, U.S. Environmental Protection Agency

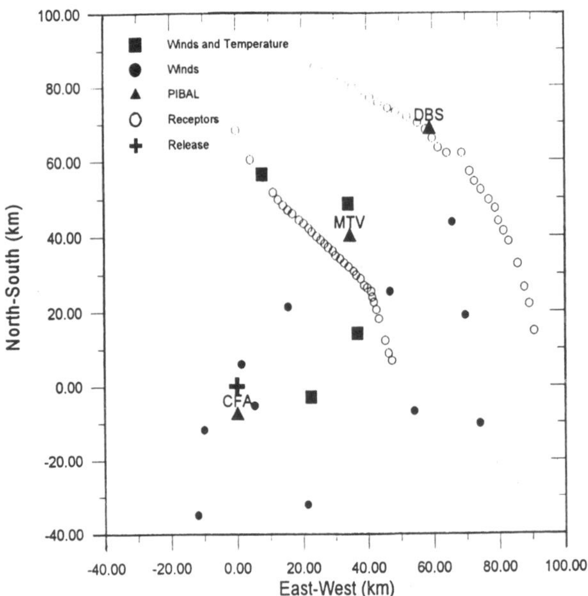

Figure 1. The Idaho tracer experiment sampling arcs and meteorological data collection network. The sampling arcs at 48 km and 90 km are shown. The receptor arc at 3.2 km downwind of the release is omitted for clarity.

investigation into the feasibility of using certain perfluorocarbons and heavy methanes as alternative tracers in place of sulfur hexafluoride (SF6). Hence, although the results have found use for testing alternative characterizations of dispersion and transport, this was not a primary purpose in the original design of the investigation. Draxler (1979) included this experiment in an assessment of the effects of alternative methods of processing wind data for characterization of the mesoscale trajectory and dispersion. He concluded that a network of wind observations having a spacing on the order of 25 kilometers might be needed to simulate mesoscale transport associated with variable-flow situations, and that spacing of order 100 kilometers might prove adequate for stationary and homogeneous flow situations.

METEOROLOGICAL DATA PREPARATION

The design for meteorological data collection and sampling locations relative to the release location is shown in Figure 1. Since locations of towers and sites were extracted from data volume figures, the relative positions are likely accurate but the absolute positions are no better than 0.5 km. The receptor arcs at 48 and 90 km downwind from the release are shown in Figure 1. Meteorological data were available from eleven sites providing hourly-averaged winds; four sites providing hourly-averaged winds and temperatures, three sites providing hourly pibal observations of winds aloft (CFA, MTV, DBS). Two of the pibal sites (CFA and DBS) also provided hourly-averaged winds and temperatures. Hourly rawindsonde observations were taken at about 600 m northwest of the release location. The meteorological masts ranged in height above ground with two at 6.1 m, eleven at 15.2 m, three at 22.8 m, and two at 30 m. The pibal observations taken at Billings, Montana (well past the farthest sampling arc downwind) were not used in this investigation. The skies were clear of clouds and no precipitation occurred during the experiment. The National Weather Service observations taken at Pocatello, Idaho (approximately 75 km southeast of the release location) were included to provide station pressure (required input for CALMET).

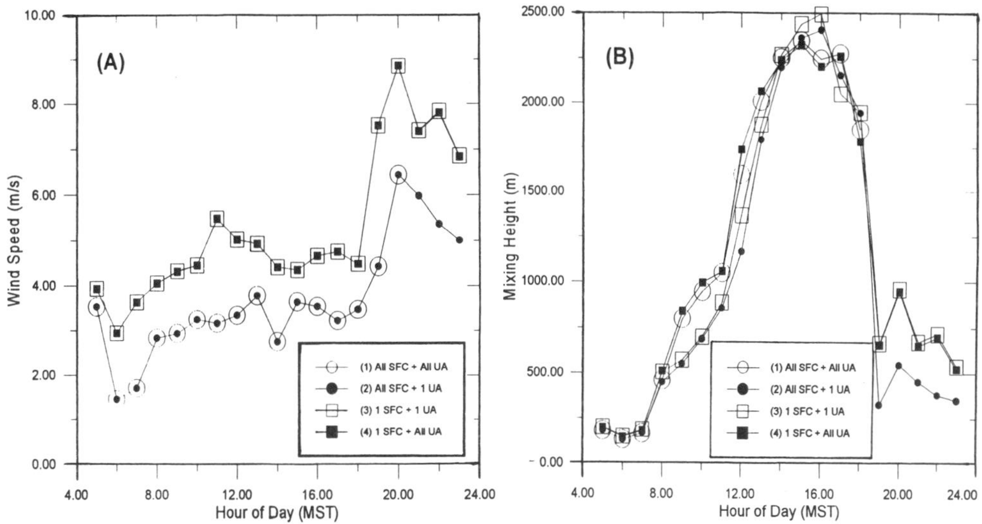

Figure 2. Variation of horizontal wind speed at 10 m (A) and mixing height (B) for April 19, 1977. Different symbols are used to indicate results obtained by various methods used to process the meteorological data.

To estimate the effects of drainage flow on the near-surface wind field, gridded values of land use and terrain heights are needed. The land use data are used as surrogates for typical values of surface roughness, albedo, soil heat flux, anthropogenic heat flux and leaf area index. These surface parameters are used in estimating the surface energy balance. For this analysis, U.S. Geological Service land use and terrain height data were extracted from data bases included in U.S. EPA (1996). The basic grid size for these data is approximately 900 m. They were processed into a 20 by 20 grid with a grid resolution of 10 km. Default values, as defined in U.S. EPA (1996), for the surface parameters to be associated with the land use data were used. The southwest corner of this grid was approximately 50 km southwest of the release. The area depicted in Figure 1 is fairly flat, but the terrain sharply increases in height to the west and north of the area depicted. The dominant landuse was rangeland; and the surface roughness was estimated based on landuse to be on the order of 10 centimeters.

Hourly-averaged winds and temperature were available from midnight April 18 through midnight April 19. To mitigate the effects of not having surface data beyond midnight of April 19, the surface meteorological tower data were duplicated to form two 24-hour periods, having identical meteorology. The assumption being made is that conditions were steady-state. The pibal and rawindsonde data, which were available from 0700 MST to 1900 MST, were treated in a similar manner.

CALMET assumes all upper-air observations are from rawindsondes, and thus expects upper-air observations to provide winds, dry-bulb temperature and pressure with height. CALMET interpolates in height for missing data values at intermediate heights in an observation; but CALMET will not extrapolate upper air data. Thus observations are rejected that fail to reach the user-prescribed top of the modeling domain (3300 m for this analysis), or have missing data values at the surface. To make use of the hourly pibal observed winds, temperature and pressure values were added by linearly interpolating in time and height between available rawindsonde observations, which were available every 1 to 3 hours. The pibal wind directions were consistent with those from the one rawindsonde, but the wind speeds were generally less in magnitude.

The CALMET wind field module is based on the Diagnostic Wind Model (DWM), (Douglas and Kessler, 1988). A two-step procedure is involved in the computation of the gridded wind fields. In the first step, an initial guess field is adjusted for kinematic effects of terrain, slope flows, terrain blocking effects, and three-dimensional divergence minimization. In this analysis, the initial guess wind field varies spatially from the available upper air observations using a $1/r^2$ weighting, where r is the distance from the observation to the grid point. The second step includes four substeps: inverse distance interpolation of observations into the Step 1 field, smoothing to reduce sharp gradients in the field, adjustments of vertical velocities using the O'Brien procedure (O'Brien, 1970), and divergence minimization. In this analysis the O'Brien procedure was not used, hence the vertical velocities were not constrained to be zero at the top of the computation grid.

A purpose of this investigation was to assess the effects of having different amounts of meteorological data for use in the development of the time varying field of meteorological data. For this purpose four separate runs were made: Case 1 using all available upper-air and surface mast observations, Case 2 using all surface mast observations but only the one onsite rawindsonde upper-air observation, Case 3 using only the CFA wind and temperature observations with the one onsite rawindsonde upper-air observations, and Case 4 using only the CFA wind and temperature observations with all upper-air observations. In Cases 1 and 2, all the onsite hourly wind and temperature data are employed but different amounts of upper-air observations are used. In Cases 3 and 4, hourly winds and temperatures taken close to the release are used with different amounts of upper-air observations. For all the CALMET simulations, winds and temperatures were computed for six layers in the vertical, the midpoints of which were: 10 m, 35 m, 75 m, 300 m, 1250 m, and 2650 m.

Figures 2a and 2b illustrate the major differences to be seen in the lowest-level wind speed and mixing heights, for the grid square containing the release. The winds at CFA were higher than those generally seen throughout the network. Hence in Cases 1 and 2 when all the onsite winds were employed (open and closed circles in the Figure 2), the low-level winds were lower than when only CFA data were used. In Cases 1 and 2, the afternoon stability was Pasquill category B/C (Monin Obukhov lengths of order -30 m). As a consequence of higher winds in Cases 3 and 4, the surface friction velocities were higher, and the Monin Obukhov lengths were larger (in magnitude), thus closer to neutral stability. The afternoon mixing heights, shown in Figure 2b, are similar regardless of data used. This results because the "upper-air" temperatures all have a common source, namely the rawindsonde observations taken 600 m northwest of the release. The nighttime mixing heights are mostly a function of the magnitude of the friction velocity. Hence, where estimated friction velocities were largest and differ most among the various processing methods, differences were to be seen in the nighttime mixing height values (see hours 1800 to 2300 MST in Figure 2b).

DISPERSION MODEL CALCULATIONS

Each of the four analyses of meteorology was used to produce two CALPUFF simulations of ground-level concentrations for each of the three sampling arcs. In the first simulation, the dispersion was described using Pasquill-Gifford dispersion parameters. In the second simulation, the dispersion was described using dispersion parameters suggested by Draxler (1976), which require values of the standard deviation of the vertical and lateral wind fluctuations (referred to hereafter as "similarity dispersion"). The wind fluctuation standard deviations estimated within CALMET are primarily dependent on the surface friction velocity. The surface friction velocity is a strong function of stability (largest during unstable conditions), roughness length and wind speed (increases as roughness length or wind speed increase). The CALPUFF user's guide (EPA, 1995b) provides a complete listing of the various equations, which is not possible to provide in this limited discussion.

CALPUFF options were set as follows: the maximum puff travel distance during one sampling step (controls the puff generation rate) was set to 1 km, maximum puff separation was set to 1 km, Gaussian vertical distribution was assumed, concentrations were determined as if over flat terrain, no wet or dry deposition, and no transition to Heffter long-range dispersion parameters was made.

CALPUFF internally computes for each sampling step, a transport wind averaged over the depth of the puff from the multi-layer winds provided to it from CALMET. As a surface release puff grows in the vertical, the depth through which the wind is averaged increases. The SF6 tracer emission was reported to be steady at 25.37 gs^{-1} over the three hour period, and was simulated within CALPUFF as a 3-hour point-source release at 10-m starting at 1300 MST. The release height was set at the midpoint of the lowest CALMET layer, to insure that the internally computed standard deviations of lateral and vertical velocity fluctuations (for use in the similarity dispersion parameter characterizations) at the specified release height, were in accord with the wind speed used by CALPUFF for the lowest layer.

MEASUREMENT UNCERTAINTIES

The primary sampler used in this experiment was a bag sampler, consisting of a 50-liter Saran bag enclosed in a plastic barrel. The bag was inflated with a small battery-powered pump, which was turned on and off manually In general, 6-hour samples were available with approximate start times of 1200 MST on the 3.2-km arc, 1300 MST on the 48-km arc, and 1500 MST on the 90-km arc. Two aliquot samples were taken from each bag, one into a 2-liter bag and the other into a 1-liter steel cylinder. Of the 68 valid SF6 concentration values reported, 48 were from cylinder samples. For the 17 occasions where SF6 concentrations from both the bag and cylinder samples could be compared, the bag samples were consistently 37% lower than the cylinder samples. No discussion was present in the data volume regarding these differences between the bag and cylinder samples, but all plots were shown using the cylinder samples. In light of this and the differences seen between the bag and cylinder samples, only the cylinder concentrations were used in the following analyses. To estimate the sampling uncertainty in the SF6 cylinder samples, comparisons where made with the three perfluorocarbons which were released simultaneously at the same location with the SF6: PDCH (C_8F_{16}), PDCB (C_6F_{12}), and PMCH (C_7F_{14}). Comparison of SF6 concentration values with concentrations from cylinder samples from all three perfluorocarbons was possible at 33 sites along the three sampling arcs. The perfluorocarbon concentration values were found to be consistently 14% greater than the SF6 concentrations. For SF6 concentrations greater than 20 ppt, the standard deviation of the percentage differences was 11%; and for SF6 concentration less than 20 ppt, the standard deviation of the percentage differences was 77%, implying greater uncertainty for the lower concentration values. A background of 0.5 ppt was subtracted from all SF6 concentration values as suggested in the data volume.

MODEL RESULT COMPARISONS

For each 6-hour period, the second moment (lateral dispersion, Sy) of SF6 concentration values about its centroid position along the arc was computed. The crosswind integrated concentration, CWIC, was computed by trapezoidal integration. By assuming the concentration profile along the arc is Gaussian, the central maximum, Cmax, was computed as, $Cmax = CWIC/(\sqrt{2\pi}\ Sy)$.

A goal of this investigation was to assess the sensitivity of the modeling results to different treatments of processing the meteorology, as well as to assess the performance of CALPUFF in characterizing dispersion for transport distances beyond 50 km. Results are summarized in Table 1 for the different wind field and dispersion treatments. Figure 3 depicts the observed SF6 concentrations with the simulation results where all the surface and upper-air observations

Table 1. Summary of observed and estimated centerline maximum concentration, Cmax, lateral dispersion, Sy, and crosswind integrated concentration values, CWIC.

	Pasquill Dispersion			Similarity Dispersion		
Cmax (ppt)	3.2 km	48 km	90 km	3.2 km	48 km	90 km
Observed	103	16.6	6.4	103	16.6	6.4
All SFC + All UA	793	11.4	4.0	945	21.0	9.4
All SFC + 1 UA	808	9.8	2.5	955	20.0	5.4
1 SFC + 1 UA	1903	25.2	12.5	896	10.0	5.9
1 SFC + All UA	1712	27.2	12.0	854	17.1	8.6
Sy (km)	3.2 km	48 km	90 km	3.2 km	48 km	90 km
Observed	0.70	9.66	13.06	0.70	9.66	13.06
All SFC + All UA	0.46	5.22	18.93	0.66	5.31	12.27
All SFC + 1 UA	0.46	5.15	13.66	0.66	5.27	8.43
1 SFC + 1 UA	0.30	4.79	5.76	0.55	4.38	5.38
1 SFC + All UA	0.37	4.67	9.21	0.60	4.35	9.31
CWIC (ppt-m x 10^5)	3.2 km	48 km	90 km	3.2 km	48 km	90 km
Observed	1.82	4.02	2.09	1.82	4.02	2.09
All SFC + All UA	9.17	1.50	1.88	15.63	2.79	2.88
All SFC + 1 UA	9.21	1.27	0.87	15.67	2.58	1.15
1 SFC + 1 UA	14.22	3.03	1.81	12.31	1.10	0.79
1 SFC + All UA	15.89	3.18	2.78	12.91	1.86	2.01

were used to generate the hourly wind fields. For the observed values, there were from 14 to 17 receptors along each arc with valid data for analysis. For analysis of the simulation results, receptors were spaced at each arc distance at 2 degree intervals, over the 90 degree sector northeast of the release location. The second moment, Sy, represents a measure of the puff horizontal dispersion. For these 6-hour periods, the observed lateral dispersion ranged from roughly 22% to 15% of the travel distance downwind. The crosswind integrated concentration values characterizes the amount of pollutant mass seen at the surface. From Figure 2, assuming a mixed layer wind speed of 4 m/s, a mixed layer depth of 2500 m, a sample duration of 6 hours, we would anticipate CWIC values of approximately $2\text{x}10^5$ ppt-m, if the puff was well mixed. The observed CWIC values at 3.2-km and 90-km arcs are close to $2\text{x}10^5$ ppt-m. The CWIC value at 48 km, shown in Table 1 as $4\text{x}10^5$ ppt-m, would be $2\text{x}10^5$ ppt-m, if we assumed the observed concentrations beyond receptor 425 rapidly approached zero. There are indications of such a falloff in concentration in the bag sampling results, but only the cylinder results shown in Figure 3 where analyzed in developing Table 1.

As shown in Figure 3 (which is typical for all of the simulations), the simulated transport was somewhat south of the observed position along the first two arcs. It is also apparent that the concentrations simulated for the first arc are at least a factor of 5 higher than observed. There is no obvious reason to dismiss the 3.2-km observations, as the other tracers at this arc were within 11% of those reported for the SF6 when adjusted for differences in release rates.

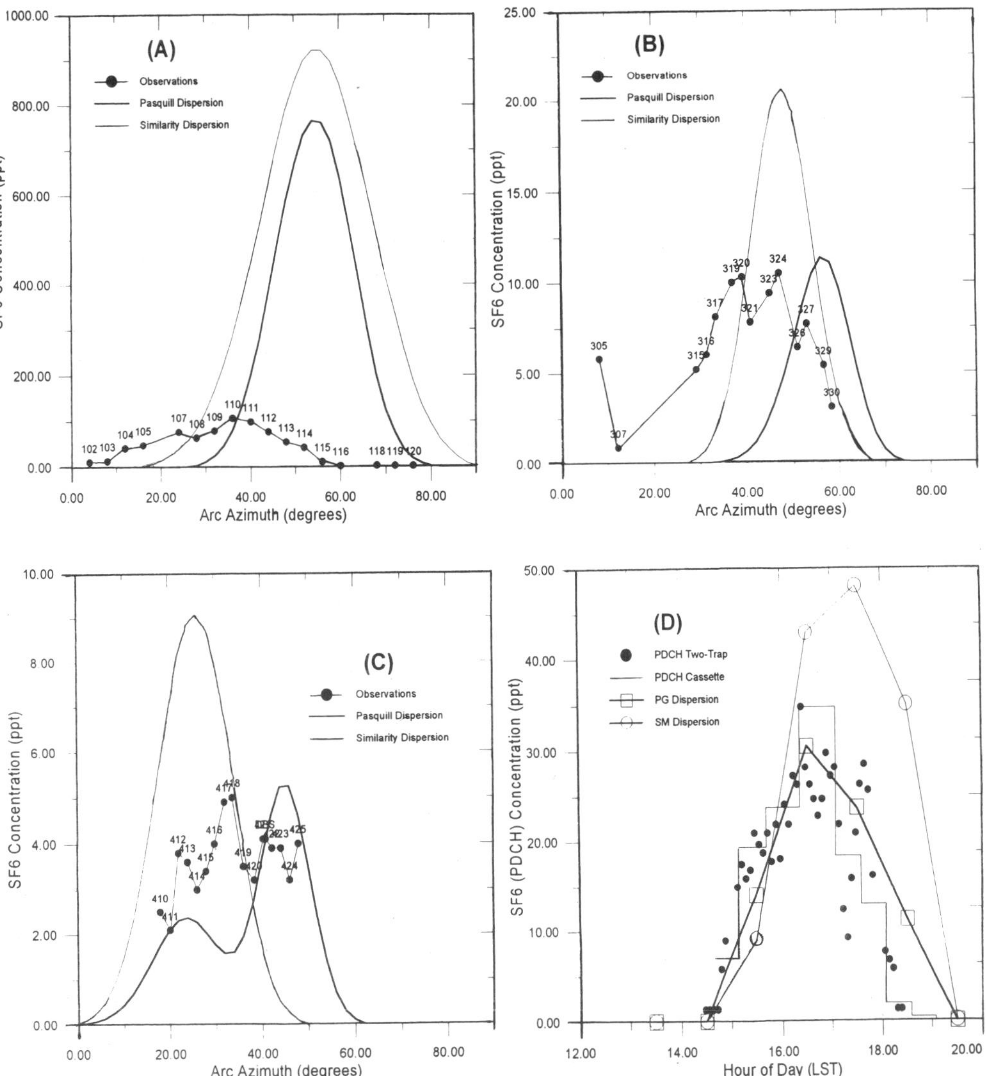

Figure 3. Six-hour average SF6 concentration values observed and estimated for April 19, 1977, (A) 3.2-km arc, 1300-1900 MST; (B) 48-km arc, 1400-2000 MST; (C) 90-km arc, 1600-2200 MST. Azimuth is defined as viewed from the release position with 0 due North and 90 due East (see Figure 1). Receptor numbers are shown just above each observed concentration value. (D) Time history of observed PDCH and estimated SF6 concentrations along the 48-km arc for April 19, 1977. Observed PDCH values were multiplied by 3.16 for comparison with estimated SF6 values (volume of SF6 divided by volume of PDCH released equals 3.16).

The differences seen in the different treatments to develop the wind fields can largely be explained by the increase in the lower level winds by a factor of 2 when only the winds at CFA are used to develop the wind fields. As the transport winds increase, not only was the transport to the arc decreased, but also the variability in the wind directions within the wind field were greatly reduced. The transport time to the 3.2 and 48 km arcs varied from 1.5 to 3 hours, depending on meteorology used. With these transport times and 10-km meteorology, the lateral dispersion is relatively insensitive to the meteorology used for the 3.2 and 48 km arcs.

The CWIC values vary inversely to the product of the simulated vertical dispersion and the mean transport speed. For the Pasquill results, an increase in wind speeds results in neutral stability, which caused the vertical dispersion to be nearly a factor of 4 less. This more than compensated for the increased dilution, hence higher CWIC values. For similarity dispersion, the vertical dispersion was never well-mixed and was nearly the same, regardless of meteorology used. Beyond 75 km, an increase in transport winds tended to increase the simulated vertical dispersion by 30% or so, which tended to reduce CWIC values for the 90-km arc

Figure 3d provides a comparison of the time history of the puff, as it passed by the 48-km arc. Sampling results are shown for the two-trap sampler which provided 5-minute samples, and a cassette sampler which provided approximately 15-minute samples. These samplers were quite close to the observed position of the 6-hour SF6 maximum along this arc. The dispersion results are for the simulated position of the maximum, which was somewhat displaced from that observed. The Pasquill dispersion results are in remarkable accord with the tracer results. The similarity results arrive and depart slightly later than observed. The slower transport for the similarity dispersion results because the vertical dispersion was less than that simulated by Pasquill dispersion, hence the transport speed was computed over a more shallow layer for the similarity results. These results and those discussed above suggest that the similarity dispersion was underestimating the vertical dispersion for this case.

CONCLUSIONS

A goal of this investigation was to assess whether the CALPUFF simulations were in reasonable accord with the observed concentrations, and the sensitivity of the simulation results to different methods of processing the available meteorological observations. The comparison results presented reveal as yet unexplained differences for the nearest arc, 3.2 km downwind from the release. Possible speculations are that the puff became well-mixed sooner than we would otherwise expect, or that the puff lifted somewhat off the surface at the 3.2-km arc. The simulated pattern of dispersion was displaced as much as 40 degrees from that observed, regardless of how the wind fields were characterized. For all arcs, the lateral dispersion along the sampling arcs was best characterized by both dispersion characterizations when all the surface tower winds were used. Except for the first sampling arc, the simulated maxima along the arcs were typically within a factor of 2 of that observed. The Pasquill simulations were most sensitive to how the wind fields were characterized, showing the most variability between the various wind field results. Having only one puff release limits conclusions to be reached. For this one realization, it would appear that simulations by both dispersion characterizations were in best accord overall with observations when all the low-level winds and upper-air observations were used. And for this case, the similarity dispersion simulations may have underestimated the vertical dispersion.

DISCLAIMER

The information in this document has been funded wholly or in part by the United States Environmental Protection Agency under an Interagency Agreement (DW13937039-01-0) to NOAA. It has been reviewed in accordance with the Agency's peer and administrative review policies for approval for presentation and publication. Mention of trade names or commercial products does not constitute endorsement or recommendation for use.

REFERENCES

Clements, W.E (ed.)., 1979: Experimental Design and Data of the April 1977 Multitracer Atmospheric Experiment at the Idaho National Engineering Laboratory, Los Alamos Scientific Laboratory Informal Report, LA-7795-MS/UC-11, Los Alamos, NM 87545, U.S.A., 100 pp.

Douglas, S., and R. Kessler, 1988: User's guide to the diagnostic wind field model (Version 1.0). Systems Applications, Inc., San Rafael, CA, 48 pp.

Draxler, R.R., 1979: Modeling the results of two recent mesoscale dispersion experiments. Atmos. Environ., 13:1523-1533.

Draxler, R.R., 1976: Determination of atmospheric diffusion parameters. Atmos. Environ., 10: 99-105.

O'Brien, J.J., 1970: A note on the vertical structure of the eddy exchange coefficient in the planetary boundary layer. J. Atmos. Sci., 27:1213-1215.

U.S. Environmental Protection Agency, 1995a: A User's Guide for the CALMET Meteorological Model. EPA-454/B-95-002, Office of Air Quality Planning and Standards, Research Triangle Park, NC 273 pp.

U.S. Environmental Protection Agency, 1995b: A User's Guide for the CALPUFF Dispersion Model. EPA-454/B-95-006, Office of Air Quality Planning and Standards, Research Triangle Park, NC 338 pp.

U.S. Environmental Protection Agency, 1996: CALMET, CALPUFF and CALPOST Modeling System, PB 96-502-083INC (CD ROM and diskette), National Technical Information Service, U.S. Department of Commerce, Springfield, VA, 22161.

DISCUSSION

R. PIELKE: Are you also planning to use the observational data to test the accuracy of Lagrangian particle model techniques to represent the dispersion?

J. IRWIN: Not at this time. We are currently focused on assessing the strengths and weaknesses of puff dispersion modeling, in anticipation of possibly suggesting its use in routine regulatory assessments. I believe that use of Lagrangian particle dispersion modeling (routinely) will naturally follow, once puff dispersion modeling is considered acceptable for routine applications.

R. YAMARTINO: The narrower plumes seen when using CALPUFF with puff growth driven by the micrometeorological formulations versus the P-G curves, suggests that the micrometeorological determined lateral turbulence intensities are missing some of the lateral meander energy. Some of this meander might be captured by allowing CALMET to update the winds more frequently than the once per hour rate currently being used.

J. IRWIN: The turbulence intensities are a function of Monin-Obukhov length, surface roughness length and mixing height. All of these were estimated. The surface roughness length was estimated by using typical values assigned to various land-use types. The heat flux (and thereby the Monin-Obukhov length) was estimated through estimates of insolation, low-level wind speed, and estimates of the Bowen ratio. Given the chain of estimates needed to ultimately estimate the turbulence intensities, I suspect they are likely no better than within 50% of reality. This is easily sufficient to explain the differences seen in the simulated lateral dispersion when using the Draxler lateral dispersion versus using P-G lateral dispersion.

What is more disconcerting and impossible to address through more frequent updating of the meteorology, is the substantial over predictions seen at the first arc at 3.2 km downwind of the source. From discussions with various participants at the conference, there does not appear to be any ready explanation. The most reasonable suggestion offered, is to go to the experimental site, and to visually inspect the path between the release point and this arc, to see if there is some physical obstructions, terrain variations, etc. that might possibly offer some new insights.

Short of some physical terrain induced explanation, the other possibility is that the plume became well-mixed prior to reaching this arc. This is not a satisfying explanation in that we are speaking of 6-hour average concentrations resulting from a 3-hour release that could easily reach 3.2 km downwind in less than 20 minutes. Thus the mixing could not have been caused by some passing intermittent convective eddy. The process causing the mixing would have had to have been consistently effective for the entire 3-hour release period. This seems questionable.

GLOBAL AND LONG-RANGE TRANSPORT

chairmen: G. Carmichael
A. Flossmann

rapporteur: S. Rafailidis
M. Rotach

ATMOSPHERIC TRANSPORT - CHEMISTRY MODELLING — SOME IDEAS FOR THE FUTURE

Prof.dr.ir. Peter J.H. Builtjes

TNO-MEP
P.O. Box 342
7300 AH Apeldoorn
The Netherlands

INTRODUCTION

The purpose of developing atmospheric transport-chemistry models is to create a tool, an instrument to investigate the underlying processes which lead to the chemical composition of the atmosphere. Modelling, in the sense both of developing, using and applying models is a scientific activity as it is focused on answering fundamental questions in the field of atmospheric chemistry.

There are different levels and methods of modelling. At least four levels can be distinguished. In the first place, modelling can be a form of data analysis of observed concentrations and associated meteorological parameters. Second, modelling can be a tool to investigate the relative importance of the different processes described in the model which lead to the calculated concentrations at a specific point and time.

A measured concentration, for example ozone, is by itself insufficient to determine the processes which have led to that concentration. Modelling can be used to reveal these processes. In the third place, modelling can be used to try to find phenomena, as for example a specific correlation between a concentration and a meteorological factor which is found as a result of extensive model exercises, which have not been observed yet in the field. Subsequently, these mathematically found phenomena can be tested by performing dedicated field experiments.

Finally, modelling can be used in a restricted sense to study one specific aspect, the effectivity of abatement strategies. In that case, the emission input to the model is modified, and resulting concentrations are determined. Or, in the other direction, concentration guidelines are given and the emission reduction required to reach these guidelines is calculated. It is obvious that this fourth kind of modelling is of high political relevance, whereas the first three modelling activities are more science oriented.

The basis of all forms of atmospheric transport chemistry modelling is the diffusion equation:

$$\frac{\partial c_i}{\partial t}+u\frac{\partial c_i}{\partial x}+v\frac{\partial c_i}{\partial y}+w\frac{\partial c_i}{\partial x}=\frac{\partial}{\partial x}\left(K_h\frac{\partial c_i}{\partial x}\right)+\frac{\partial}{\partial y}\left(K_h\frac{\partial c_i}{\partial y}\right)+\frac{\partial}{\partial z}\left(K_z\frac{\partial c_i}{\partial z}\right)+$$

chemistry + emissions - dry deposition - wet deposition

Here, c_i is the concentration of a specific species, which is changing in time, and is transported by the mean wind in three directions, and is dispersed by atmospheric turbulence both horizontally and vertically.

The four types of modelling mentioned, all have there specific way of using this equation.
Data analysis of observed concentrations often takes the form of chemical box modelling, restricting the full equation to $\partial c_i/\partial t$ = chemistry by neglecting the meteorological processes, or by considering them in a purely empirical way.

By studying the relative importance of the different terms contained in the model, the full equation is used. Often, also the impact of different parametrizations and descriptions of the different terms on the resulting concentrations is studied. In this way improved descriptions can be tested in their contribution to the overall process.

By trying to detect new phenomena the results of the full equation are used. As an example, correlations and relations between chemical species can be determined as has been done by Sillman (1995) to define indicator species like NO_y, HCHO, H_2O_2, HNO_3 to determine the NO_x- and VOC-limited regimes. In the EUROTRAC subproject LOOP (Neftel, 1997) an attempt will be made to verify this approach in the field. Also, relations between chemical species and meteorological parameters can be found, like dependencies with wind direction and -speed, and tested in the field.

Also in analysing the effect of abatement strategies the full equation is used, and numerous sensitivity calculations directed to emissions are performed.

DIFFERENT ELEMENTS OF FUTURE RESEARCH

In the following, the more general discussion in the introduction will be restricted to modelling of photo-oxidant formation in the troposphere. This restriction enables to describe more concrete directions of future research.

The direction of future research will depend on the four kinds of modelling considered in the introduction. Also, the direction will depend on the weak parts in the overall process of modelling. The consideration here will also be based on the current research trends as they can be found in the EUROTRAC subproject GLOREAM (Global and Regional Atmospheric Modelling, Builtjes, 1997), and, which is definitely not the least important, on the personal taste and impression of the author. The discussion will be structured according to the different terms of the full diffusion equation.

Considering photo-oxidant modelling in the troposphere, the following elements of future research can be mentioned.

$$\frac{\partial c_i}{\partial t}$$

What is still unknown to a large extend is the variability of ozone and related species over time and space, and the real reasons of it. It is a well known fact that the year to year variation in ozone, as they are for example depicted in exceedances of thresholds, is substantial. It is also known that this variation is influenced by temperature ('good' summers will lead to 'high' ozone values), but a quantitative explanation is lacking. Performing a 15 year run with a 3-D model covering Europe, and, which is even a larger task than performing the run itself, analysing its results in detail, could contribute substantially to investigate this year to year variability.

Also changes in time as they show up in trends over the years are an item of future research. Again, the fundamental question is in this case the underlying processes and reasons which lead to this trend, see also Roemer, 1996. Extending the trend to pre-industrial times brings the item of natural background into focus. Research has been carried out considering the natural background of ozone, although there are still doubts about the exact natural level of ozone. Knowledge about the natural background of related species and precursors is nearly lacking, and dedicated modelling studies could be used in this respect.

$$w\frac{\partial c_i}{\partial t} + \frac{\partial}{\partial z}\left(K_z\frac{\partial c_i}{\partial z}\right)$$

By considering elements of future research and focusing on the mean transport and turbulent diffusion the clear impression is that the weak part is formed by the description of the vertical exchange processes, the horizontal processes are known more better. In general, a reliable description of the vertical exchange between the boundary layer and the free troposphere, and between the free troposphere and the lower stratosphere is still lacking, and consequently poorly described in models.

The mixing height description is an essential element in the exchange between boundary layer and free troposphere. The knowledge of mixing height is still limited, caused by the fact that mixing height is not an important element in weather forecast, and has not received much attention from meteorologists. The magnitude of the mixing height, the 'strength' of the inversion, its behaviour in time, especially at sunset, its horizontal gradients, especially at land-sea interfaces, are all important elements of research. It is important that a dedicated EURASAP workshop on mixing height will be held at Risø, October 1-3, 1997.

The description of the behaviour of convective clouds is an essential element in the exchange between the free troposphere and the lower stratosphere. Locally, convective clouds will cause vertical upward motions of considerable strength, bringing polluted, low level air to much higher elevation. The real occurring overall process, and its subgrid scale character and description in models is still inadequately known.

The results of the ETEX-project, to which a special session is devoted in these 22e ITM on Air Pollution Modelling and this Application, do also indicate the importance of vertical motion, and the adequate description of the amount of released trace gas which is transported below, or above the inversion.

Budget calculations using the results of 3-D Eulerian grid models also reveal the importance of vertical motion, showing that the vertical motion term is often of similar magnitude as the chemical production term in budget calculations, Builtjes, 1997.

Chemistry

In trying to formulate future areas in the field of the chemistry of photo-oxidant formation it should be noted that the main fundamental processes leading to ozone formation are reasonably well known, and have not changed substantially over the last decade.

What is true for ozone, is not true for other trace gases which play a role in photo-oxidant formation. Ambient validation of the different chemical mechanisms which are currently in use in modelling studies requires ambient measurements of for example PAN, HO_2, RO_2, OH, species of which hardly any measurements in the field exists at the moment.

Also, ambient measurements of NO_y, H_2O_2, HCHO, HNO_3 would contribute to increase our knowledge of photo-oxidant formation and the relative importance of VOC- and NO_x-emissions. Finally, speciated ambient measurements of VOC could contribute to assess the accuracy of both anthropogenic and biogenic emissions, see below.

Emissions

The knowledge of anthropogenic and biogenic emissions might well form the weakest part in the overall process of photo-oxidant modelling, because their uncertainty is large, and to a large extent unknown. When calculated O_3-concentrations differ substantially from measured O_3-concentrations, the reason of the deviation is very often errors in the emission input data. The determination of the accuracy of emissions, their validation, is of utmost importance. Because the accuracy of ambient measurements, and the reliability of dispersion models is often better than the accuracy of emission estimates themselves, inverse modelling and data-assimilation can be used to improve the accuracy of emissions. Using reliable ambient measurements, which are also representative for the grid size of the model used, inverse dispersion modelling will lead to estimates of the precursor emissions. Although inverse modelling and data-assimilation techniques are valuable tools, it is expected that it will not be possible by this method to come to emission estimates which have an accuracy better than $\pm$ 30%, see also Pulles, 1997.

Dry deposition

The dry deposition of ozone, and especially the large difference of its deposition over sea, which is nearly zero due to the low solubility of ozone into water, and over land, has been studied extensively, and is rather well known. However, there might still be a weak part in the description of the dry deposition of ozone, especially in the southern part of Europe. The discrepancies between modelled and measured ozone concentrations are often higher in the south of Europe than in the rest of Europe. The relatively dry atmosphere in the south of Europe and the fact that stomata are closed in that case can have an impact on the dry deposition process which might have a larger influence than is currently considered in modelling studies.

Wet deposition

Clouds are essential in ozone formation, as they have a determining influence on the photolysis rates, both below, above and between clouds. Improvements have been made recently in the description of this process (van Weele, 1996).

A matter of debate is still the chemical influence of clouds and aqueous phase chemistry on the ozone formation process. Recent results point into the direction that this influence is less than previously expected (Mathijssen, 1996).

THREE AREAS OF FUTURE RESEARCH

There are three important areas of future research that can not be categorized using the different terms of the full atmospheric transport-chemistry equation. These areas are the use of tropospheric satellite data, the improvements in numerical methods and the increase in computer capacity and the area of overall model validation. These areas will be described below.

Tropospheric satellite data. Measurements of trace gases, most notably stratospheric ozone, from satellites is a well established technique. The determination of tropospheric ozone from satellites is in principle possible, but much more complicated than in the case of the stratosphere. This is due to the fact that 90% of the ozone, and consequently 90% of the strength of the signal, is present in the stratosphere.

The GOME - instrument (Global Ozone Monitoring Experiment) has been launched in April 1995 on board the ERS-II. This instrument is developed to measure at least the tropospheric ozone column, but hopefully also two points in the troposphere, from 0-5.5 km and from 5.5-11 km can be detected. The horizontal grid size is 40x320 km^2, the measurement takes place once a day, at 9.30 AM, and returns to that position in 3 days. The signal is strongly disturbed in case of cloud cover and elevated aerosol loading. Currently (June 1997) there is a severe delay in the ESA-program in developing and applying the retrieval algorithm for tropospheric ozone. However, it is still expected, and hoped, that tropospheric ozone data from GOME will become available in the near future. It is likely, because of the associated large uncertainty, and the lacking data in case of clouds, that only by integrating the experimental data by data-assimilation techniques for chemical reactive species with 3-D models, useful results can be obtained from GOME.

It should be noted that, because of the very valuable information which is in principle contained in the GOME-data, much effort should be devoted to extract all available information from GOME. However, it is also clear that it will indeed take much effort.

The Sciamachy-instrument, which is scheduled to be launched in 1999, is an improved version of GOME and is planned to deliver at least 4 tropospheric vertical points upto 11 km. Also in this case, data-assimilation will be required to extract the available information.

Numerical aspects and improvements. New developments in the field of high performance computing and networking, and faster and more accurate numerical methods will in the near future substantially increase the capabilities of photo-oxidant modelling. At the moment, only condensed chemical mechanisms can be handled in the models because of the restrictions in computer capacity. The chemical condensation is focused on the VOC-part of the mechanism. Hundreds of VOCs are known to exist in the atmosphere, and their chemical behaviour is known, from laboratory studies, tens of VOCs are know to exist in the emissions, but only a very limited number (order 10) is used, in a condensed manner, in the modelling itself. A lot of research effort is devoted to condensation and simplification of chemical mechanisms (Schurath, 1997) which would not be required in case larger and faster computers would be available.

Other aspects of modelling that are currently limited due to the computational restrictions are a full statistical uncertainty analysis, real time ozone forecasts, interactive data processing and 'virtual reality' visualisation and optimal use of all relevant satellite data.

Model validation. At the moment, the validation of photo-oxidant models is nearly always restricted to comparing measured ozone concentrations at a certain station with calculated ozone concentrations in the corresponding grid, followed by claiming 'reasonable' agreement.

In case that models are used to study scientific questions, and are applied to perform abatement strategy calculations, it is of utmost importance that the models are reliable and of proven accuracy.

The validation of photo-oxidant models is hampered by two fundamental problems. First, point measurements are compared with volume averaged calculated concentrations. Consequently such a comparison only has meaning in case the measurement is representative for that volume. Ideally, this should be shown by having numerous measuring points inside that volume. In reality this is hardly ever the case, and the word 'representative' attached to a measuring station is often more intuition, or just wishful thinking.

The second fundamental problem is due to the highly non-linear character of the ozone formation process. This means that in case a calculated ozone concentration is in agreement with a measurement, it is fundamentally impossible to prove that the agreement is a result of an in all aspects correct modelling. The possibility exists that the result is right for the wrong, compensating, reasons.

Model validation covers also three different evaluation areas. First, the evaluation of all input data, emissions, meteorology, land use, initial and boundary conditions should be performed to the extent possible.

Secondly, all processes described in the overall model should be evaluated separately like the chemical scheme, the photolysis description, the dry deposition mode etc.

Thirdly, the ambient measurements themselves against which the validation takes place should be evaluated concerning their accuracy and representativeness.

A complete model validation study for photo-oxidants should cover all the following aspects (Builtjes, 1997).

- A comparison between calculated and measured, representative, O_3-concentrations.
- Multi-component comparison, taking into account next to O_3-concentrations also measurements of PAN, NO, NO_2, speciated VOC, etc. This aspect would mainly test the chemical mechanism.
- A comparison of observed and modelled phenomena. The measured and calculated concentrations could for example be analysed as a function of wind speed, or wind direction, temperature etc. This would mainly test the relation between concentrations and meteorology.

These three aspects are the most important ones. Information concerning model validation can also be obtained from

- Detailed sensitivity analysis of the models.
- A comparison between observed and calculated trends.
- The intercomparison between models.

A complete and thorough model validation study is considered to be of high priority. Not only to be able to garantee the scientific reliability of model results, but also because further emission reductions, which will in general be very expensive and/or have a large impact on society, will require very strict proofs that the answers given by models are reliable and accurate.

CONCLUSIONS CONCERNING FUTURE RESEARCH

In conclusion, an attempt will be made to describe the situation concerning photo-oxidant modelling in the year 2014, at the 30th ITM (which is also the year of my retirement at the age of 65).

- It is clear that more will be known concerning all items mentioned in this paper, and there is hope that funding will still be available and that we will still be active in this fascinating field.

- However, the impression is that no major breakthroughs will have been made. At the moment, in 1997, we might know 60-80% of the subject of photo-oxidant formation. This is the result of an x amount of money and effort, spend from the discovery of this subject in about 1945 until now, about 50 years later. It can be expected that we might know about 80-90% of the subject in the year 2014, but that the cost of this increase of 10-20% is about 3 to 5 times x, the amount spend until now. There will be always restrictions in our knowledge, and there will be areas of uncertainty that can not be decreased, independent of the amount of money spend.

- In 2014, there will still be weak points in our knowledge. It might be expected that model validation is still limited. It might well be that troposphere satellite data are available, but of limited reliability. It might also well be that the improvement in numerical techniques and computer capacity is not in balance with, for example, the validation of emission input data.

- And most serious, in view of the current political and economical developments, tropospheric ozone levels will still, all over the world, exceed guidelines and limit values.

REFERENCES

Builtjes, P.J.H., *Eurotrac subproject description GLOREAM (Global and Regional Atmospheric Modelling)*, 1997.

Builtjes, P.J.H., P. Esser, G. Boerssen, M. Roemer, *To a methodology for model validation.* Proceedings First Int. Conf. Measurement and Modelling in Environmental Pollution (MMEP97). Madrid, Spain, April 1997.

Builtjes, P.J.H., P. Esser and M.G.M. Roemer, *An analysis of regional differences in tropospheric ozone over Europe*. Proceedings 22e ITM on Air Pollution Modelling end its Application, Clermont-Ferrand, France, June 1997.

Mathijsen, J., *Modelling of tropospheric ozone and clouds.* Ph-D thesis, Univ. Utrecht, The Netherlands, June 1995.

Neftel, A., *Eurotrac subproject description LOOP (Limitation of Oxidant Production)*, 1997.

Pulles, M.P.J. and P.J.H. Builtjes, *Validation and verification of emission inventory data*. Wessex Inst. of Technology. Air Pollution IV, 1997, in press.

Roemer, M.G.M., *Trends of tropospheric ozone over Europe.* Ph-D thesis, Univ. Utrecht, The Netherlands, May 1996.

Schurath, U., *Eurotrac subproject description CMD (Chemical Mechanism Development),* 1997.

Sillman, S., *The use of NO_y, H_2O_2 and HNO_3 as indicators for ozone-NO_x-hydrocarbon sensitivity in urban locations*. J. Geophys. Res 100, 14175-14188, 1995.

Weele, M. van, *Effect of clouds on ultra violet radiation - Photodissociation rates of chemical species in the troposphere.* Ph-D thesis, Univ. Utrecht, The Netherlands, January 1996.

ACKNOWLEDGMENT

The author would like to thank the organisers of the 22th ITM to invite him -as replacement - to present a key-note talk. However, due to this fact, only very limited time was availabe to present, and write, this contribution. As a consequence, the references are limited to the inner circle of the author. The author apologizes for this - unavoidable - restriction.

DISCUSSION

H. VAN JAARSVELD: You mentioned the quality and the availability of emission data as (one of the) limitations to future model improvements. Remarks on the quality of emission data are made on almost every workshop on modelling. What in your opinion, should be done in the future to improve the situation?

P. BUILTJES: In my opinion we should try to combine the measurements and modelling capabilities to address this. Inverse modelling, and data assimilation, using representative ambient measurements will become a strong method in improving the accuracy of emission data, or at least in determining the accuracy. A recent study we performed at TNO funded by the EEA was already reasonale successful.

A. HANSEN: In displaying the plot of the comparison of simulated and observed PAN-concentrations you suggested the source of large overprediction may be associated with emission errors. Since you are using the CBM-IV chemical mechanism, I would suggest you also consider errors in the mechanism as the source of discrepancy.

P. BUILTJES: You might be well correct in this. At the moment, I use in the LOTOS-model not yet the latest CBM-version. On the other hand, changing the VOC-speciation to some extent, and leaving the mass still untouched, has a substantial influence on calculated concentrations of ozone and PAN.

D. W. BYUN: In your discussion, and several presentations yesterday, there have been many suggestions of using inverse method or data assimilation for photochemical modelling. I would like to challenge the audience if anybody can show that the inverse problem we have (highly complex non-linear, multi-species, with 3D spatial and temporal variation) can form a closed system without employing extraneuos simplifying assumptions. Especially expecting we can assimilate ozone data with the air quality model may not help our understanding as much. Although research in this area should continue, our expectation on this method should be dampened. It might be more profitable to master current forward integration method as much.

P. BUILTJES: Thank you for your remark. Maybe my expectations are too high, but I have recently seen promising progress in the EU-DG-12 funded RIFTOZ-project, also including satellite data from the GOME-instrument for ozone in the troposphere.

B. BORNSTEIN Reasonable, expected results are not as interesting as unexpected results, which could lead to either discovery of a model error or a new phenomena not previously seen in the observation.

P. BUILTJES: I agree fully. We should be also careful in claiming, or even wanting such a vague thing as reasonable agreement. However, we as modellers have to come up with Quality assurance and quality control, to show that our models are adequate.

D. FISH: Do you see a role for simple models that include chemistry and meteorology, rather than just chemical box models.

P. BUILTJES: Yes, of course also simple models are very useful tools to address specific questions. However, in my opinion they should have a balanced approach, and not pay 90 % of the attention to chemistry, and only 10% to meteorology. Also, the UK-community could use some 3-D photo-oxidant modelling in my opinion

D. FISH: Do you see an important role for on-line modelling?

P. BUILTJES: Ideally, yes, this might be the future, having on-line ECMWF with all the chemistry you want. However, again this should be in balance. When the uncertainty in emissions is still large, or even determining your final result, on-line modelling will not improve your overall result that much.

A MODEL FOR AIRBORNE POLI-DISPERSIVE PARTICLE TRANSPORT AND DEPOSITION

Michail V.Galperin,[1] Dimiter E.Syrakov[2]

[1]Hydrometeorological Research Centre of Russian Federation, Moscow 123242, Russia
[2]National Institute of Meteorology and Hydrology, Sofia 1784, Bulgaria

INTRODUCTION

According to Stokes law which is well fulfilled within the range of 0.1-30 μm, the velocity of gravitational deposition (sedimentation) and scavenging with precipitation for particles is proportional to their surface area, i.e. to D^2 (D - particle size). With D less than 0.1 μm, their behavior is similar to that of "weightless" gas molecules and the velocities of surface dry deposition and the efficiency of particle captures by precipitation increasing.

While particles settle down, their distribution with size distorts and the mean sedimentation velocity changes. Hence, modeling of scavenging of substances, sorbed on particles, requires an adequate description of size spectrum variation. Usually, the continuous distribution $f(w)$ of sedimentation velocity w is substituted by a discrete one :

$$f(w) \cong \sum_{i=1}^{N} \bar{f}(w_i),$$

where w_i is the mean w at the i-th interval of discretization, $\bar{f}(w_i)$ is distribution density value at this interval, N is the number of sections of $f(w)$ approximation. Thus, N similar problems are solved, since independent calculations are made for each i-th interval. At the same time, one has to compromise between approximation accuracy and computer resources.

Another approach is used in this paper. It is based on the fact that the calculation of a precise form $f(w)$ is practically impossible and not obligatory.

GENERAL DESCRIPTION

Let us assume that at height z at moment t there is a poli-dispersive mixture of particles. The concentration can be presented as:

$$C(z,t) = \int f(w,z,t)\, dw, \tag{1}$$

where $f(w,z,t)$ is particle distribution with sedimentation velocities w, i.e. concentration of particles whose sedimentation velocity is within the range $(w, w+dw)$. Particle settling can be described by the equation:

$$\frac{\partial C}{\partial t} = \frac{\partial Q}{\partial z} + \frac{\partial}{\partial z} K_z \frac{\partial C}{\partial z}, \tag{2}$$

where K_z is the vertical exchange coefficient and $Q = \int wf.dw$ is momentum (impulse). During sedimentation, a change of momentum Q takes place leading to particle spectrum variations at all levels. Therefore, Eq.(2) should be supplemented with an equation for Q:

$$\frac{\partial Q}{\partial t}=\frac{\partial \sigma_w^2 C}{\partial z}+\frac{\partial \overline{w}^2 C}{\partial z}+\frac{\partial}{\partial z}K_z\frac{\partial Q}{\partial z}, \tag{3}$$

where $\overline{W}$ is the mathematical expectation and σ_w^2 is the variance of w.

Equations (2) and (3) describe fully sedimentation if $\sigma_w^2(z,t)$ and $\partial\sigma_w^2/\partial z$ are known. Thus, there is a problem of closure similar to that existing in the turbulence theory. Designating the central moments of distribution $f(w)$ through μ_n (μ_1=0, $\mu_2=\sigma_w^2$ etc.) one can obtain the general formula:

$$\frac{d\mu_n}{dt}=n\mu_2\mu_{n-1}-\mu_{n+1},\ n=2,3,........ \tag{4}$$

The problem can be closed supposing a definite distribution. For important real cases, the situation can be greatly simplified approximating $f(w)$ by Γ-distribution with the only parameter $\overline{W}$:

$$f(w)=\frac{m(w)}{M}=\frac{w^{p-1}\exp(-pw/\overline{w})}{(\overline{w}/p)^p\Gamma(p)}, \tag{5}$$

where $p>0$ is a factor of the distribution shape, $\Gamma(p)$ is the gamma-function.

It is easy to show that $dp/dt\equiv 0$, i.e. shape factor p does not change during the sedimentation process and only the parameter $\overline{W}$ is varied. All equations for the moments are reduced to the following one:

$$d\overline{w}/dt=-\overline{w}^2/p \qquad \text{or} \qquad w(t)=w(0)/(1+t.w(0)/p) \tag{6}$$

If there is no information on particle distribution, it is reasonable to assume $p=1$.

During precipitation, coarse particles are easily captured and small particles with low sedimentation velocity are captured less frequently. Therefore, washout velocity like dry settling velocity depends on the particles mass and the dependence of particle washout efficiency on size is similar to that of dry sedimentation. The processes of particle washout in clouds are very complicated. Particles themselves can be center of droplets coagulation and enter the air again through evaporation.

The free fall of heavy particles in the atmosphere and the efficiency of their capture by precipitation depend on appropriate Stokes numbers and they are functions of D^2. The bulk of information on particle washout was accumulated in the 50-ies in connection with ground nuclear weapon tests (Stade, 1968). It is obtained that the washout process is described by the following equation:

$$C(t+\Delta t)=C(t)\exp(-\Lambda_T\Delta t), \tag{7}$$

where Λ_T is washout coefficient and Δt - precipitation duration. Adopting the hypothesis for similarity of gravitational sedimentation and washout (in the sense explained above) it is possible to express Λ_T as:

$$\Lambda_T=\Lambda_0+\Lambda=(\gamma_0+\gamma)R, \tag{8}$$

where R is precipitation intensity; γ_0 reflects diffusion mechanism of capture and does not depend on D (and consequently on $\overline{W}$); γ reflects inertial processes and is proportional to $\overline{W}$:

$$\gamma=\gamma_{\max}\overline{w}/w_{\max}, \tag{9}$$

where $w_{\max}$ is sedimentation velocity of particles >20 μm (Chemberlain, 1953, 1960). Another coefficient must be introduced in (9) accounting for the different efficiency of capturing in clouds and in precipitation.

The particle spectrum variations during washout will be much faster than during dry settling. Concentration and impulse vary according to the following equations:

$$\partial C/\partial t=-[\Lambda_0+\Lambda(\overline{w})]C; \tag{10}$$

$$\partial Q/\partial t=-[\Lambda_0-2\Lambda(\overline{w})]Q. \tag{11}$$

In Eq.(11), the multiplier 2 in the second term on the right is raised due to simultaneous decrease of both the mass and the settling velocity.

In Galperin et al.(1995) and Galperin and Masljaev (1996) the proposed approach for handling of aerosol specific processes is presented in more details, and one-layer algorithm is created and tested.

ONE-DIMENSIONAL MULTI-LAYER GRAVITATIONAL SETTLING ALGORITHMS

Two dependent variables are considered - the concentration C of the pollutant of interest, adsorbed on aerosols, and the mean gravitational settling velocity w of the carrying particles. In fact, instead of w, the impulse $Q=wC$ is introduced as a second dependent variable. This provides the possibility to describe easily the settling velocity changes due to the mixing during release, advection and diffusion. The impulse is a conservative quantity with respect to the mentioned processes. The boundary conditions for the impulse during diffusion are the same as for concentration. The settling velocity can be determined, when necessary, dividing impulse by concentration. Two kinds of impulses are involved here - C-impulse $Q=wC$ and M-impulse $Q_m=wM$, where M is the mass in the layer. C-impulse is the dependent variable. M-impulse is calculated when necessary, multiplying C-impulse by layer depth.

Let us suppose that at moment t the concentration and impulse profiles are C_k and Q_k, $k = \overline{1,N_z}$, N_z - number of layers. Calculations start from the highest level and continue downwards layer by layer.

Dry gravitational settling

Let us consider layer k. The first step is to calculate the mass and M-impulse in the layer

$$M_k(t) = C_k(t)Dz_k, \qquad Q_{mk}(t) = Q_k(t)Dz_k, \qquad (Dz_k \text{ - layer depth}). \tag{12}$$

Due to the gravitational settling, some mass DM_{k+1} and impulse DQ_{mk+1} come from the upper layer to the k-layer and mix there, yielding

$$M'_k(t) = M_k(t) + DM_{k+1}, \qquad Q'_{mk}(t) = Q_{mk}(t) + DQ_{mk+1}. \tag{13}$$

The mean settling velocity of the mixture in the k-layer is

$$w_k(t) = Q'_{mk}(t) / M'_{mk}(t). \tag{14}$$

During time interval Δt, part of the mass goes downwards and seeds the lower layer, removing M_k. It is

$$DM_k = w_k(t)C_k\Delta t. \tag{15}$$

The remaining mass is

$$M_k(t+\Delta t) = M'_k(t) - DM_k. \tag{16}$$

M-impulse is also changed. At first, it changes because the down-going mass takes some impulse $(w_k DM_k)$. This mass consists mainly of larger particles and the spectrum of the remaining aerosol is changed. This will reflect on the mean settling velocity by decreasing it according to Eq.(6). It can be shown easily that in such a case $[w_k(t+dt) - w_k(t)]M_k \cong w_k(t)DM_k$. Finally,

$$Q_{mk}(t+dt) = Q'_{mk}(t) - 2w_k(t)DM_k. \tag{17}$$

The final operation is to calculate concentration and C-impulse, dividing $Q_k(t+dt)$ and $Q_{mk}(t+dt)$ by the k-layer's depth. The quantities DM_k and DQ_{mk} are used in the lower layer's calculations.

Wet removal

During precipitation, the larger particles are removed faster than the smaller ones. Wet removal velocity depends on particle size in the same way as dry settling velocity, i.e. (Galperin et al., 1995; Galperin and Masljaev, 1996)

$$\Lambda = k_w w = E_f \gamma(w) R, \tag{18}$$

The already mentioned coefficient E_f accounts for the different capturing efficiency of droplets in clouds and in rain. This coefficient is set from geometrical considerations, keeping in mind that in clouds the capturing surface of the droplet is its whole area, while in rain this surface is the cross-section of the falling droplets (E_f=1 in clouds and E_f=0.25 below clouds). The trapping coefficient $\gamma(w)$ can be presented as

$$\gamma(w) = \gamma_{\max}(0.2 + 0.8w / w_{\max}), \tag{19}$$

where γ_{max} is a constant and w_{max} is the gravitational velocity of the largest particles. Expression (19) accounts for the inertial (80%) and diffusion (20%) mechanisms of aerosol trapping by water droplets.

The wash-out algorithm is like the gravitational one but for three differences. The first one is that the removed mass from each layer is calculated from Eq.(7). Second, it is not added to the mass in the lower layer, but is accumulated to give the one time step wet removal. The third difference is that w evolution is calculated after Eq.(6), using $k_w w$ instead of w in the denominator of Eq.(6).

Testing of gravitational settling and wet removal blocks

To test the ability of these algorithms to describe adequately the physical processes, some one dimensional experiments are performed. A vertical diffusion block, realized in PBL on a six-layer log-linear grid, is conjugated with a proper shell, providing for the flexible testing of the new built schemes. The bottom boundary condition is the dry deposition flux. An open boundary condition is used at the model domain top. An additional layer is introduced over the last one where the up-going mass is accumulated, accounting parametrically for free atmosphere. In all cases, a point source at height 200 m is assumed. The instantaneous and continuous action of this source is modeled with strength 10^3 units per time step. The stratification is supposed to be neutral, with geostrophic wind of 10 m and roughness of 0.037 m. The dry deposition velocity is taken V_d=0.001 m/s - typical for aerosols. The time step is 500 s. There is an obvious lack of information about aerosol spectra near the sources, i.e. the initial mean settling velocity. The available reference sources display great divergence of estimates. Here a value of w_0 = 0.005 m/s is mainly used.

The first experiments test the dry sedimentation algorithm. In Fig.1, the effect of introducing settling velocity on concentration profile evolution in the case of an instantaneous source can be seen. Initial w is set to 0 m/s (gas) and to 0.005 m/s (aerosol). In aerosol profiles, there is a tendency of increasing the concentration in the lower layers and decreasing it in the upper ones. This tendency is better observed with time increase, resulting the shift of maximum from the source height to the lowest level after 100 time steps.

In Fig.2, the evolution of some accumulated quantities (called totals hereafter) - the up-going mass, airborne mass and deposited masses (diffusion dry deposition and gravitational settling) are shown as percents of the released mass. It is clearly seen that gravitational settling is more effective than dry deposition, especially at the early stage of dispersion, when there are enough large particles in the air. The mass in the air is decreasing faster. Small decrease of the up-going mass is noticeable, too.

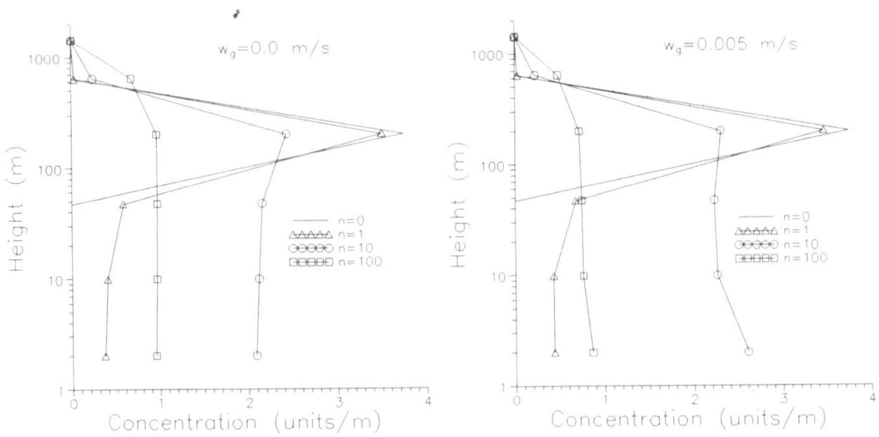

Figure 1. Effect of gravitational settling velocity on concentration profiles evolution. An instantaneous point source.

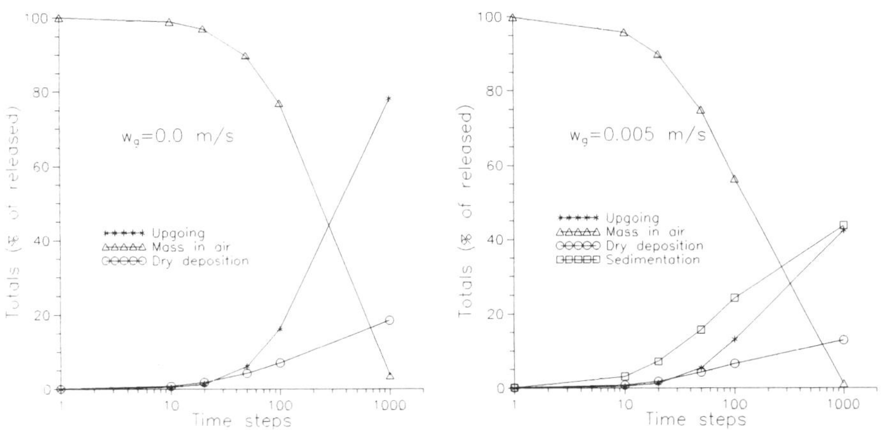

Figure 2. Effect of gravitational settling velocity on time behavior of some total quantities

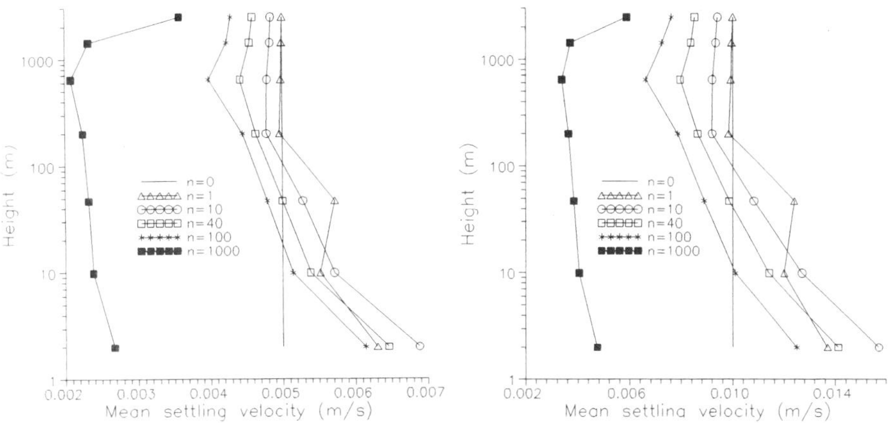

Figure 3. Settling velocity evolution at different initial values.

Fig.3 shows the evolution of mean settling velocity profile in two cases: initial values of 0.005 and of 0.01 m/s. Both figures show similar behavior of settling velocity. The self-decreasing effect of this quantity is clearly seen. This decrease is much faster at initial velocity of 0.01 m/s. The decrease is not regular. It is combined with the influence of mass and impulse redistribution (due to diffusion and sedimentation). Sedimentation concentrates the larger particles in the lower layers, so the curves slope is negative. The settling velocity of aerosol in free atmosphere is also shown as an additional level at height 2500 m. The great thickness of the highest layers leads to slower decrease of settling velocity. The gravitational mass flux, from free atmosphere to the highest model level, makes the minimum to be placed at level 5. There is a small irregularity at the first time step because of the discontinuity in the initial mass distribution (point source), but the further behavior of the profiles can easily be explained by physical reasoning.

In Fig.4, the concentration and mean settling velocity profiles evolution is presented as well as the evolution of totals in the case of a continuous source. The profiles show an increase of concentration mainly in the lower levels - at and under the source height, where the typical concentration profile, influenced by sedimentation, can be seen. This increase is not proportional to the released quantity increase, because a great deal of mass is deposited and gone out through the upper boundary. The highest layer concentration increases, too, and may be faster than in the source layer. This cannot be noticed in the figure, because of the scale.

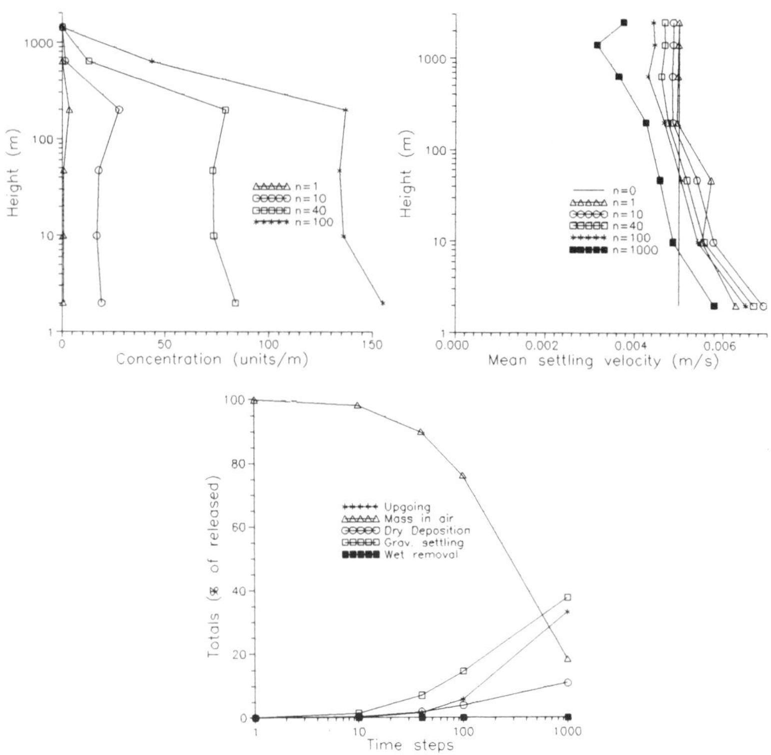

Figure 4. Concentrations, settling velocities and totals. A continuous point source.

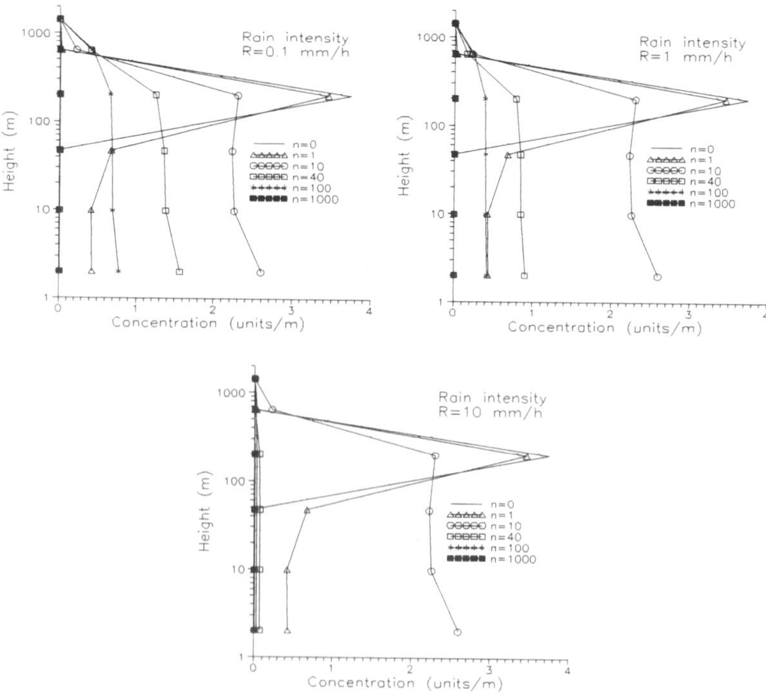

Figure 5. Concentration profiles at different rain intensities. An instantaneous point source.

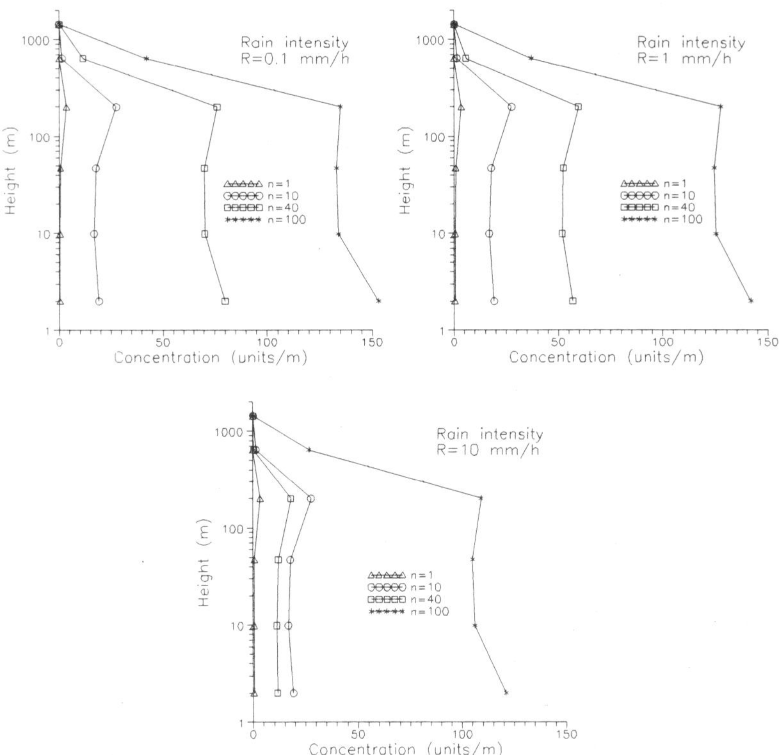

Figure 6. Concentration profiles at different rain intensities. Continuous point source.

The experiments for testing the wet removal algorithm consist in prescribing different rain intensity (0.1, 1 and 10 mm/h) while the other parameters remain fixed. Rain duration is set to 40 steps (about 6 hours) - from 11th to 40th time step. Levels 5 and 6 are set to be cloud levels. The initial settling velocity is 0.005 m/s. In Fig.5, the concentration profiles evolution for an instantaneous source is displayed. The concentration profile behavior at R=0.1 m/s is similar to the dry case.

In Fig.6, the same characteristics are presented but for the case of a continuous source - a case close to reality. It can be seen that the profile at n=40 (end of the rain) is the one decreased most of all. The continuing release following that moment tends to converge the curves to those for the dry case. The gravitational settling profiles evolution for both cases displays very fast decrease of the wash-out ability, due to the loss of the bigger particles, which increase with the increase of precipitation intensity.

The presented results, as well as the results from many others tests (see Syrakov, 1996) show that, from a physical point of view, the presented algorithms seem to provide an adequate description of the behavior of the different characteristics of interest during the dispersion of harmful materials, carried by aerosol particles. They give parametrical accounting for spectrum changes as a result of gravitational settling and washing-out by precipitation. The test results are quite realistic and the improved schemes can be built in three-dimensional models for calculating long range transport of specified pollutants.

3-DIMENSIONAL MULTI-LAYER AEROSOL MODEL - CALCULATION RESULTS AND MODEL VALIDATION

The developed approach is built in the 3D model EMAP (Eulerian Model for Air Pollution), described in BC-EMEP (1994) and Syrakov (1995). The model is realized using the time splitting approach in maintaining the various processes as well as the different directions. The model domain in vertical is the Planetary Boundary Layer (PBL), approximated by a log-linear terrain-following staggered Z-coordinate system. A proper transformation makes it regular and the vertical grid parameters can be easily changed by the user. A staggered, Arakawa's C-type, grid is used to describe the variables in the horizontal plain. A flux-type explicit scheme, called TRAP (BC-EMEP, 1994; Syrakov, 1995; Syrakov and Galperin, 1997), is

Table 1. Lead concentrations in the air [in ng/m^3] and in precipitation in μg/l. The best run. Comparison of calculated values with measurements in the PARCOM/ATMOS stations for 1990.

	concentration in air		concentration in precipitation	
station	meas.	calc.	meas.	calc
B2b	104.0	71.57		
D1	17.30	13.98	2.43	2.22
DK1	17.00	13.67	2.73	2.58
F1			1.53	1.69
GB1	21.00	48.52	4.93	7.26
GB2			11.17	9.19
GB3	6.76	9.72	3.03	3.31
GB4			.88	1.12
N2			.20	.36
N3			3.60	.41
NL2			5.91	7.71
NL3	27.00	27.62	2.78	2.83
S2			2.05	2.26
mean	32.18	30.85	3.44	3.41
correl.		.864		.879

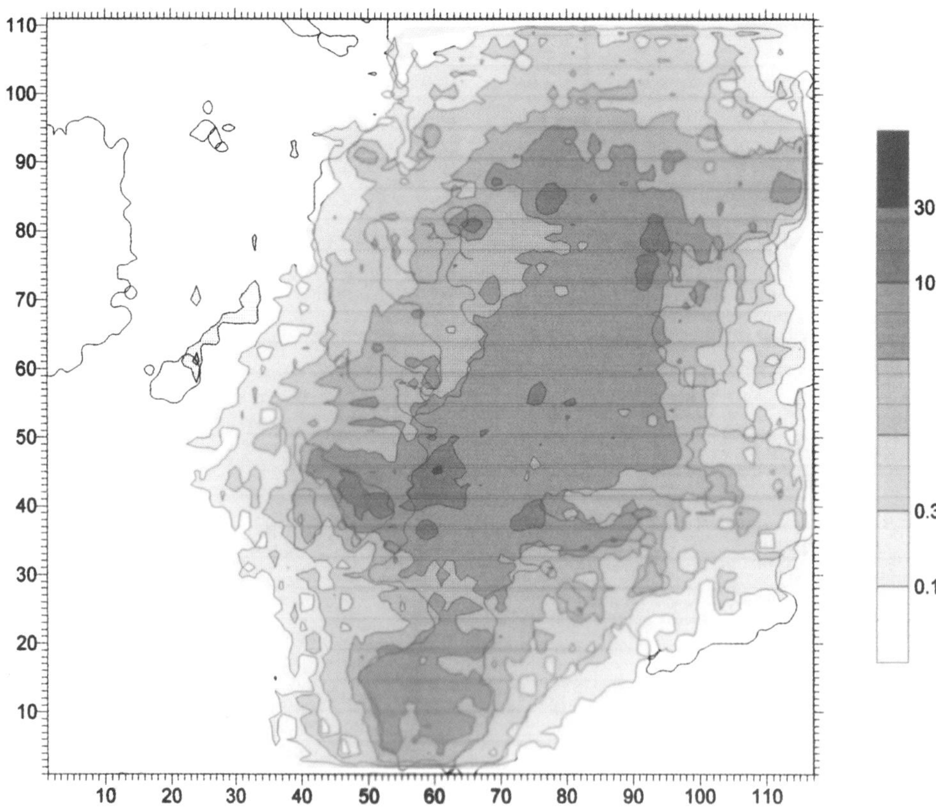

Figure 7. Total deposition of lead for 1990 [mg/m^2].

used for simulating horizontal advection. The scheme is a Bott type one (Bott, 1989), elaborated especially for the EMAP model. The simplest 2-order time forward explicit scheme is used for description of horizontal diffusion. The vertical diffusion block uses the simplest implicit scheme. It approximates the diffusion equation, taking into account the variability of the diffusion coefficient and the changeable vertical grid steps. The obtained difference equation is solved using Thomas algorithm. The dry deposition flux is set as a bottom boundary condition. An open top boundary condition is used, introducing an additional upper layer, where the outgoing masses are accumulated, horizontally advected and diffused. This provides the possibility to account parametrically for the mass exchange between the boundary layer and the free atmosphere. A proper surface layer parametrization allows to have the first computational level at the top of this layer (Syrakov and Yordanov, 1996). Its improved version can account for continuous sources on the ground level, too.

The lead emission data for 1990 over Europe are received from MSC-E/EMEP (Sofiev et al., 1996), together with meteorology for 1990 and some topography data (orography and sea-land mask). The meteorology data consist of wind field at 850 hPa level, ground level temperature and precipitation amounts, time discretization of 6 hours.

Observational data have been taken from the data base accumulated within the scope of the Comprehensive Atmospheric Monitoring Program of Paris Commission (PARCOM). Only concentrations in air and concentrations in precipitation are handled. Model estimates are obtained by bilinear interpolation between the respective grid point values.

A great number of annual integrations are made over the EMEP 50 km grid (117×111 points), varying the control parameters in the model. Each time, the computed parameters were compared with the measured ones. The comparison of model concentration in the air and in precipitation with the observed data is displayed on Table 1.

The annual total deposition of lead over Europe for 1990 is presented in Fig.7

The model shows a rather good performance on the whole 50 km EMEP grid, in spite of the poor meteorological information, coarse land-use data and rough vertical resolution. Improvement of some characteristics is possible and has to be done. It is quite obvious that the experiments have to continue.

CONCLUSIONS

The presented results from application of the new approach for describing aerosol specific processes in the atmosphere show that this scheme reflects adequately the process of dispersion of particle-carried pollutants. The EMAP_A model may be used for estimating the emitter-receiver relationships in regional and local scales.

REFERENCES

BC-EMEP, 1994: Bulgarian Contribution to EMEP. Annual Report for 1994., NIMH, Sofia-Moscow, March 1995.

Bott A., 1989: A positive definite advection scheme obtained by nonlinear renormalization of the advective fluxes, *Mon.Wea.Rev.*, 117, pp.1006-1015.

Chamberlain A.C. ,1953: Aspects of travel and deposition of aerosol and vapor clouds, *British Report AERE-HP/R-1261*.

Chamberlain A.C. ,1960: Aspects of deposition of radioactive and other gases and particles, *Int.J. Air Pollution*, v.3, pp.63-88.

Galperin M, A.Masljaev, 1996: Pilot version of the EMEP/MSC-E model for transport of airborne persistent organic pollutants. Description and preliminary sensitivity analysis, EMEP/MSC-E Technical report 3/96, June, 1996, MSC-E, Moscow

Galperin M, M.Sofiev, A.Gusev, O.Afinogenova, 1995: The approaches to modelling of heavy metals transboundary and long-range airborn transport and deposition in Europe, *EMEP/MSC-E Technical report 7/95*, June, 1995, MSC-E, Moscaw

Sofiev M., A.Masljaev, A.Gusev, 1996: Heavy metal intercomparison. Methodology and results for PB in 1990., , EMEP/MSC-E report 2/96, March 1996, MSC-E, Moscow

Stade D.H., ed.,1968: *Meteorology and Atomic Energy*, NY, Pergamon Press.

Syrakov D., 1995: On a PC-oriented Eulerian multi-level model for long-term calculations of the regional sulphur deposition, *Proc. 21st NATO International Technical Meeting on Air-Pollution Modeling and its Applications*, November 6-10, 1995, Baltimore, Maryland, USA

Syrakov D., 1996: Parameterization of gravitational settling and wet removal of particle matter in a multi-level eulerian dispersion model, in: *Bulgarian contribution to EMEP. Annual Report for 1995.*, NIMH, Sofia-Moscow, January 1996.

Syrakov D. and M.Galperin, 1997: Improvement and validation of EMAP_A model. Adaptation to benzo(a)pyrene calculations, *in*: *Bulgarian contribution to EMEP. Annual Report for 1996.*, NIMH, Sofia-Moscow, January 1997.

Syrakov D. and D.Yordanov, 1996: On the surface layer parametrization in an Eulerian multi-level model, *Proceedings of the 4th Workshop on Harmonisation within Atmospheric Dispersion Modelling for Regulatory Purposes*, v.1, 6-9 May 1996, Oostende, Belgium.

DISCUSSION

A. HANSEN: In order to understand your results, would you clarify for me the assumed particle size distribution? I am used to seeing bi- or tri-modal size distributions for ambient aerosol particles. But your results appear to be based on ensemble averages for aerosol properties.

D. SYRAKOV: The approach, used here is similar to the turbulence description - to solve the equations for the statistical moments of the aerosol distribution. In our case this is the distribution of the aerosol particles over their sedimentation velocities. A problem of closure appear, like in turbulence theory. Here for closure, a Gamma-distribution is assumed, which has only one parameter - mean settling velocity. The evolution equation for this quantity (mainly for the mean impulse = concentration times mean settling velocity) is build and solved together with the equation for the concentration.

M. KAASIK: Have you tried to estimate the wet deposition by snow? It is suggested in the literature, that the scavenging rate may be several tens of times higher for the aerosols by snow than by rain. For a number of gases the scavenging by snow is suggested to be about zero.

D. SYRAKOV: Different scavenging rates are used for rain and snow, but due to the lack of information about the type of precipitation, during the calculations a surface temperature of -4 degree C is used as a threshold, dividing these two events.

NUMERICAL STUDY OF REGIONAL AIR POLLUTION TRANSPORT AND PHOTOCHEMISTRY IN GREECE

Ioannis C. Ziomas[1], Paraskevi Tzoumaka[1], Dimitrios Balis[1], Dimitrios Melas[1], Dionisis Asimakopoulos[2], Georgia Sanida[1], Panagiotis Simeonidis[1], Ioannis Kioutsioukis[1] and Christos S. Zerefos[1]

[1]Laboratory of Atmospheric Physics, Aristotle University of Thessaloniki, 54006 - Thessaloniki, Greece.
[2]Department of Chemical Engineering, National Technical University of Athens, Greece

ABSTRACT

The aim of the present project is to study the transport and photochemistry of regional air pollution in Greece using numerical models, a higher-order turbulence closure dynamic model coupled with an Eulerian photochemical dispersion model. The model used for the simulation of the three-dimensional wind field and the boundary layer structure is developed at the Department of Meteorology, Uppsala University (MIUU) while the calculations of pollutant concentrations were performed using the Urban Airshed Model (UAM).

For the purposes of the present study, a new inventory for the biogenic emissions is prepared comprising hourly emissions from forest canopies and other vegetation as a function of temperature, sunlight (cloud cover), and the coverage of each vegetation category. The VOC speciation of the inventory is suitable as input to the UAM model. In addition, the industrial and traffic emissions were estimated for the whole region with a resolution of 10 x 10 km and used as input for the UAM.

One summer day was selected for simulation during the period of the PAUR project, when the conditions were characterized by a moderate-to-strong synoptic forcing. The resulting meteorological conditions reveal a northerly flow prevailing over the Aegean sea which is known as Etesians. In the areas that are influenced by the Etesians, the development of local circulations is not favored and the flow field is rather homogeneous. The predicted concentrations of air pollutants are in good agreement with observations at rural sites.

Air Pollution Modeling and Its Application XII
Edited by Sven-Erik Gryning and Nadine Chaumerliac, Plenum Press, New York, 1998

INTRODUCTION

The study area is Greece which is characterised by high levels of solar irradiation and temperature throughout the year stimulating the photochemical activity and a geographical topography facilitating the accumulation of primary/secondary pollutants locally.

During the summer months synoptic north-eastern winds prevail, known as Etesians. Previous experimental studies (Ziomas, 1997, Kallos et al., 1996) have shown high ozone background levels (greater than 50 ppb) over remote areas and even above the Aegean Sea. Similar results have been obtained during the ongoing PAUR (Photochemical Activity and Solar Ultraviolet Radiation) project. The mechanisms leading to these high background ozone concentrations have not been completely determined yet.

Many investigations have suggested that naturally emitted non-methane hydrocarbons play an important role in tropospheric chemistry. It has been demonstrated that biogenic hydrocarbons can contribute to the photochemical production of ozone in rural and urban areas (Chameides et al, 1988,1992). The amount of biogenic emissions is equal or even greater from the anthropogenic non-methane emissions, in regional, continental and global scale. The biogenic VOCs are emitted during natural processes of plant growth. Most common biogenic VOCs are isoprene, mainly emitted by deciduous trees and monoterpenes, mainly emitted by coniferous trees.

In the present study an attempt is made to investigate the transport and formation of ozone over the region of Greece by means of a 3-d photochemical model taking into account both anthropogenic and biogenic emissions. One summer day was selected for simulation during the period of the PAUR project, when the conditions were characterized by a moderate-to-strong synoptic forcing.

The Photochemical Activity and Solar Ultraviolet Radiation Experiment

The PAUR experiment took place in Greece during June 1996. PAUR was sponsored by the EU and participants were from the Universities of Thessaloniki (coordinator), of Munich and Oslo and from the Polytechnic School of Lausanne. The purpose of the experiment was to study the complex relationships between changes in spectral solar Ultraviolet-B and tropospheric ozone formation. Observations were performed at a number of sites including both rural sites in north Aegean sea and in the mainland and two large urban centres namely Athens (37°N) and Thessaloniki (~ 40°N).

The instrumentation deployed consisted of remote sensing instruments (LIDAR, DOAS, Double and single monochromators) and in situ measurements from conventional gas analysers, ozonesondes/radiosondes and gas chromatographers. In addition meteorological observations were performed at most of the observing sites and meteorological maps were prepared by the National Meteorological Service of Greece.

The photochemical model

The UAM is a three dimensional photochemical grid model which calculates the concentrations of inert and chemical reactive pollutants by simulating the physical and chemical processes in the atmosphere that affect pollutant concentrations (Scheffe and Morris 1993). The basis for the UAM is the atmospheric diffusion equation that represents a mass balance in which emissions, transport, diffusion, chemical reactions and removal processes are expressed in mathematical terms. The UAM accounts for spatial and temporal variations as well as differences in the reactivity of emissions.

The UAM simulates the emission, advection and dispersion of precursors and the formation of ozone within every grid cell of the modelling domain. For this purpose the following steps have taken place for the preparation of the input data :

a) The modelling domain has been divided to a 50 x 50 grid (grid cell: 10 x 10 km^2) and the topographical and land use data have been prepared for this area.

b) An emission inventory has been also prepared according to the methodology described below, for the main primary pollutants (NOx, and VOC). The main categories of sources considered for the preparation of the inventory are traffic and industrial installations, as well as natural sources (biogenic emissions of isoprene and terpenes). However, it is noted that no emission data were available for Turkey, thus causing some uncertainty to the model results near the eastern boundary.

c) The meteorological input data have been prepared by interpolating the results of MIUU, which is a three-dimensional mesoscale model with a terrain influenced coordinate system. A detailed description for MIUU can be found in other literature references (Enger 1986, Tjenstrom 1987). In the horizontal a telescoping grid is employed, the grid distance being 9 km in the central parts of the model. The total amount of grid points on each vertical level is 87 x 79, covering a horizontal domain with dimensions 1200 x 1300 km^2. The vertical coordinate contains 18 levels which are log linearly spaced, with mean and turbulent quantities vertically staggered. The model top is at 8000. The time step was chosen at 15 s. The initial temperature and humidity profiles were taken from the 00 UTC radiosonde measurements at Athens airport.

Industrial and traffic emissions

The calculation of traffic and industrial emissions for the entire domain over Greece was based on fuel consumption data from the respective activities and population density, obtained by a G.I.S. data base.

More specifically industrial emissions were calculated using fuel consumption per prefecture. These emissions were then distributed to the municipalities according to population density. Fuel consumption is converted to emission values by using the following conversion factors:

- For NO_x: from mazout and diesel 5.59 kg/tn of fuel, from natural gas 5.02 kg/tn of fuel, from coal 5.63 kg/tn of fuel.
- For VOC: from mazout 0.18 kg/tn of fuel, from natural gas and diesel 0.063 kg/tn fuel, from coal 0.77 kg/tn of fuel.

The same approach was applied for the calculation of traffic emissions. The conversion of fuel consumption to emissions was made according to CORINAIR methodology (Eggleston et al., 1992), which expresses the traffic emissions per km as a function of velocity and vehicle category. The vehicles are deviated into the following categories:

- Gasoline Vehicles (Conventional 58.06%,Catalytic 41.94%)
- Diesel Vehicles (Light Trucks (<3.5 tn) 77.9%, Trucks 22.1%)

The number of km per vehicle category and municipality was calculated based on the mean fuel consumption of the corresponding vehicle categories, the percentage of the total number of vehicles for each category and the fuel consumption per municipality.

The calculation of emissions from power production plants was based on the produced electrical energy. The following conversion factors are used to convert the energy production to pollutant emissions:

Fuel type	NO_x [kg/TJ]	VOC [kg/TJ]
Coal	260	3,4
Mazout	190	7

The spatial distribution of the anthropogenic NOx and VOC emissions are presented in Figure 1 a, b respectively.

Biogenic Emissions

A biogenic VOCs inventory for Greece has been constructed, based on a G.I.S data base, containing the spatial distribution of the vegetation type as well as climatological data from 86 meteorological stations operated by the Hellenic National Meteorological Service. The corresponding biomass density and the emission rate were then calculated with a resolution of 5x5 km^2.

The emitted amounts of isoprene, depending on radiation and temperature, were calculated by the following equation: $E_i = S_i$ BLT, where E_i is isoprene emission rate in $\mu g/m^2h$, S_i is the mean isoprene emission rate ($\mu g/g*h$) at 30 °C and light intensity of 800

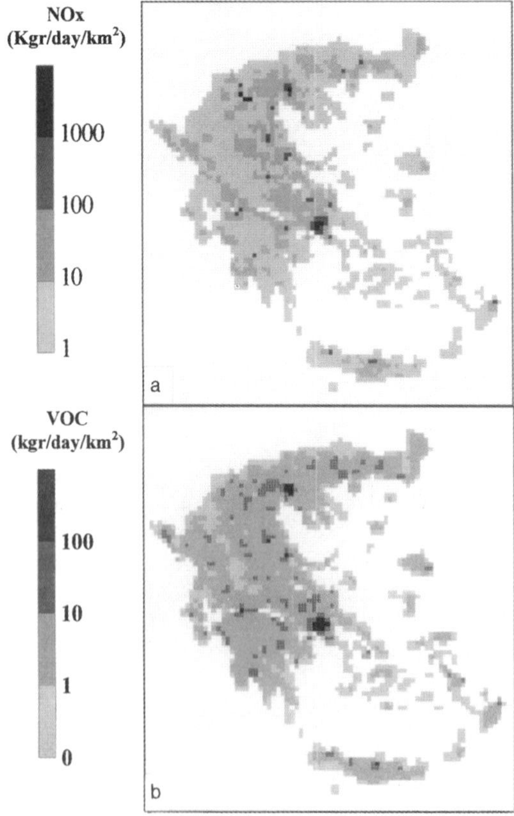

Figure 1. (a) Anthropogenic NOx emissions and (b) VOC emissions for Greece.

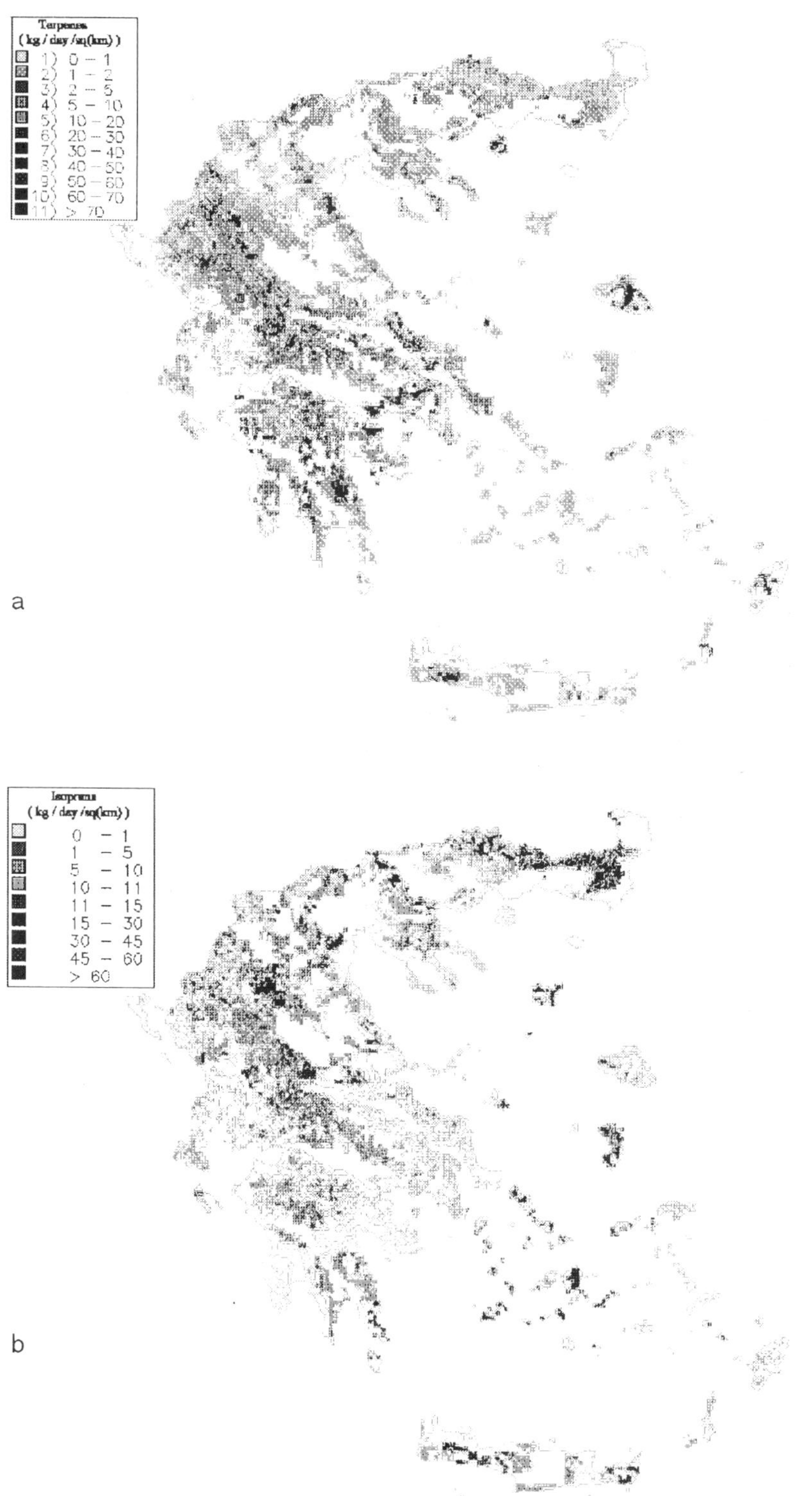

Figure 2. (a) Terpene emissions and (b) isoprene emissions for Greece.

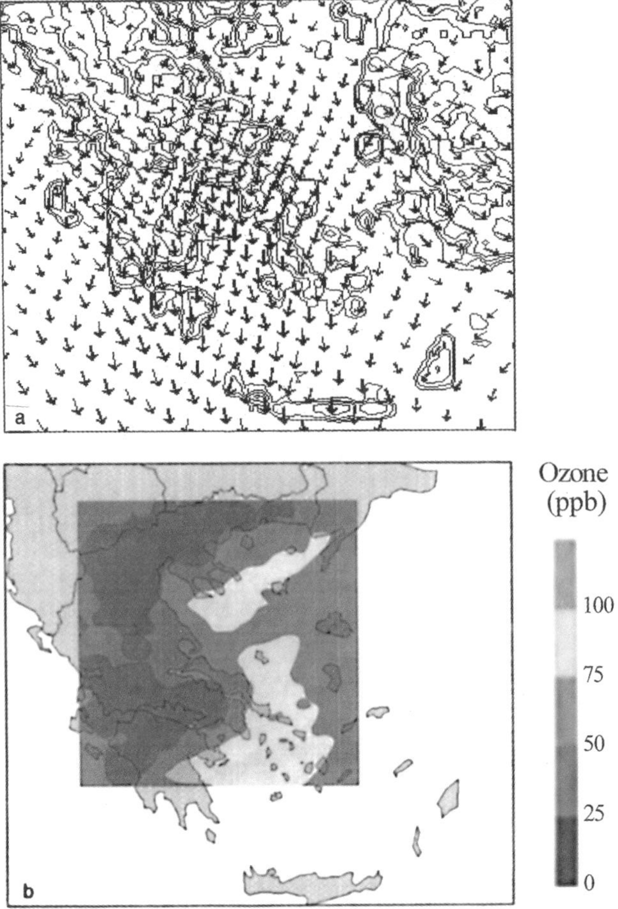

Figure 3. (a) The simulated wind field at 15:00 LST at 10 m AGL and (b) maximum simulated ozone concentartions.

μmole/m^{2}sec, B is the biomass density in g/m^2, L and T are the correction factors for irradiance and leaf temperature, respectively (Farquhar et al., 1980, SCAQMD, 1991)

The monoterpene emissions are temperature depended. The following equation describes the monoterpene emission rate E_t in μg/m^{2}h: $E_t = S_t\ B\ C_t$ where S_t is the mean monoterpene emission rate (μg/g*h), B is the biomass density in g/m^2 and C_t is the correction factor for temperature. C_t was calculated by the equation $C_t = \exp[0.0739*(T-30)]$, where T is the leaf temperature in K (Guenther, 1991).

The spatial distribution of the biogenic terpene and isoprene emissions are presented in Figure 2 a, b respectively.

RESULTS AND CONCLUSIONS

During the simulation day of the 10th of June, 1996, the area was influenced by a broad low situated to the east of Greece, over Turkey, while an anticyclone was located over the central and northern Balkan area up to Black sea. The synoptic meteorological conditions were characterized by clear skies. The surface geostrophic wind, estimated from ECMWF analyses was from the NE with speeds of ~9.0 ms^{-1}. The pressure distribution favored a strong north-easterly over the Aegean sea.

The main entrance of air to the Aegean sea is via the Dardanelles gap. Figure 3a shows the simulated horizontal wind fields at 15.00 LST for the day of simulation. The corresponding height is approximately 10 m above the surface and for clarity of presentation, a fraction of the computational domain near the lateral boundaries is omitted.

In Figure 3b, the spatial distribution of the maximum hourly ozone concentrations, as calculated by UAM, over the whole modelling domain is presented. As it appears from this figure, the highest ozone concentrations, about 80 ppb, are observed at the northern and central part of the Aegean Sea. The lowest ozone concentrations, about 40-50 ppb, are calculated for the mountainous areas of north Peloponese, central and northern Greece. These results are confirmed by ground measurements performed during the PAUR experiment at the remote stations of Pertouli (central Greece) and Agios Efstratios island (north Aegean), as well as at urban and suburban stations at Athens and Thessaloniki.

In order to examine the role of biogenic VOC emissions a second simulation of the photochemical model was performed excluding this time these emissions. It appears that the calculated ozone concentrations were slightly lower, 3 to 9 ppb downwind of the significant biogenic emission sources (see Figure 2)

Acknowledgments

The authors wish to thank professor M. Karteris, Department of Forestry, Aristotle University of Thessaloniki and his staff for their significant help for the construction of the biogenic emissions inventory.

REFERENCES

Chameides W. R., Fehsenfeld F., Rodgers M. O., Cardelino C., Martinez J., Parrish D., Lonneman W., Lawson D. R., Rasmussen R. A., Zimmerman P., Greenberg J., Middleton P., Wang T. (1992) Ozone precursor relationships in the ambient atmosphere. J. Geophys. Res. 97, 6037-6055.

Chameides W.R., Lindsay R., Richardson J., Kiang C. (1988) The role of biogenic hydrocarbons in urban photochemical smog: Atlanta as a case study. Science 241, 1473-1475.

Eggleston H. S, D. Gaudioso, N. Gorissen, R. Joumard, R.C. Rijkeboer, Z. Samaras, K.-H. Zierock, 1992, "CORINAIR Working Group on Emission Factors for Calculating 1990 Emissions from Road Traffic", Volume 1: Methodology and Emission Factors, Final Report CEC ISBN 92-826-5771-X.

Enger L., 1986, A higher order closure model applied to dispersion in a convective PBL. Atmos. Environ. 20, 879-894.

Farquhar G. S., Caemmer S., Berry J. (1980) A biochemical model of photosynthetic CO_2 assimilation in leaves of C_3 species Planta, 149, 78-90.

Guenther A. B., Monson R. K., Fall R. (1991) Isoprene and monoterpene emission rate variability:observations with Eucalyptus and emission rate algorithm development. J. Geophys. Res. 96, 10799-10808.

Kallos G. et al. (1996) Transport and transformation of air pollutants from Europe to the east Mediteranean region. Program AVICENNE, Final Report for the DGXII of the E.C. 352 pp

South Coast Air Quality Managment District. (1991) A data base managment system for spatialy and tempraly resolved inventories of hydrocarbon emissions from vegetation in the South Coast Air Basin. Final Technical Report III-D.

Scheffe, R.D. and R.E. Morris, 1993, A review of the development and application of the Urban Airshed Model. Atmos. Environ. 27B, 23-39.

Tjenstrom M., 1987, A study of flow over complex terrain using a three dimensional model. A preliminary model evaluation focusing on stratus and fog. Annales Geophysicae 5B, 469-486

Ziomas I., 1997, The MEDCAPHOT-TRACE project: An outline. Atmos. Environ. (in press)

SIMULATED ACIDIC AEROSOL LONG-RANGE TRANSPORT AND DEPOSITION OVER EAST ASIA - ROLE OF SYNOPTIC SCALE WEATHER SYSTEMS

Itsushi Uno[1], Toshimasa Ohara[2] and Kentaro Murano[1]

[1] National Institute for Environmental Studies
Onogawa 16-2, Tsukuba 305, JAPAN
[2] Institute for Behavior Sciences
Honmura-cho 2-9, Ichigaya, Shinjuku, Tokyo 162, JAPAN

INTRODUCTION

Acid deposition is widely recognized as one of the most serious global atmospheric pollution problems. Regional scale, international actions to tackle this problem have been taken in Europe and North America. East Asian countries also face a potential regional scale, international acid deposition problem, and have recently started to expand their monitoring activities as a result.

Numerical modeling is one important method to assess transboundary long-range transport (LRT) and chemical transformation processes. The first work in East Asia was reported by Kotamarthi and Carmichael (1990). They applied a two-dimensional version of the STEM (Sulfate Transport Eulerian Model) model, which includes a detailed chemical transformation module, to a straight line region traversing Mongolia, China, Korea and Japan.

LRT models have also been developed in Japan. Ichikawa *et al.* (1994) and Sato *et al.* (1993) have developed Lagrangian-based models. Katatani *et al.* (1991) have reported simulated results based on an Eulerian model. Except for the results reported by Kitada *et al.* (1993), all modeling results are based on simple linear chemical reactions (i.e., assuming a linear conversion rate of SO_2 to SO_4; which for some of the above models is season dependent).

In this study, a three dimensional numerical model which includes detailed atmospheric chemistry (the STEM-II model of Carmichael *et al.*, 1986) was used with precise meteorological input data. Simulated atmospheric aerosol concentrations (e.g., sulfate and nitrate) were then compared with the intensive measurement data reported by Wakamatsu *et al.* (1996). It is important to point out that the present model results are compared with short time interval measurements of atmospheric aerosol concentrations whereas previous model applications in Japan used only accumulated monthly deposition data.

Air Pollution Modeling and Its Application XII
Edited by Sven-Erik Gryning and Nadine Chaumerliac, Plenum Press, New York, 1998

NUMERICAL MODEL

Numerical Model (STEM-II) and Model Domain

To understand the observed characteristics of long-range transport of pollutants, an atmospheric transport model which includes detailed chemical reactions, the STEM-II (Carmichael *et al*, 1986), was applied. STEM-II is a three-dimensional Eulerian numerical model which accounts for transport, chemical conversion, and deposition of atmospheric pollutants. The major features of STEM-II include: emission of pollutants from point and area sources; transport by advection, convection, and turbulent diffusion; and spatially and temporally varying wind, temperature, pressure, water vapor, precipitation, and cloud fields.

The model treats chemical species in the gas, cloud, rain and snow phases. It includes mechanisms related to both cloudy and cloud-free environments, and in- and below-cloud wet removal and chemistry. For the gas phase chemical processes, the reaction mechanism of Lurmann *et al.* (1986) was used.

STEM-II uses chemical, dynamic and thermodynamic parameterization, which allows for a wide variety of applications such as meso-, regional- and global scale air pollution studies. The code is modular in structure and can be driven by observed or modeled meteorological data. The dry deposition model proposed by Wesely (1987) was used in the current application. A detailed description of STEM-II can be found in Carmichael *et al.* (1986). In the present application, no precipitation or cloud processes were considered.

The East Asia model domain was a 1° latitude x 1° longitude grid system. The model domain covers from 110° E to 150° E and 20° N to 50° N. This horizontal domain is divided into a 41 x 31 mesh. The vertical model domain consists of 10 layers from ground level to 10,000 m. The vertical grid levels are 200, 400, 700, 1000, 2000, 3000, 4000, 5000, 7000, 10000 m. The model uses a terrain-following z_* coordinate system. Figure 1 shows the model domain and field measurement sites.

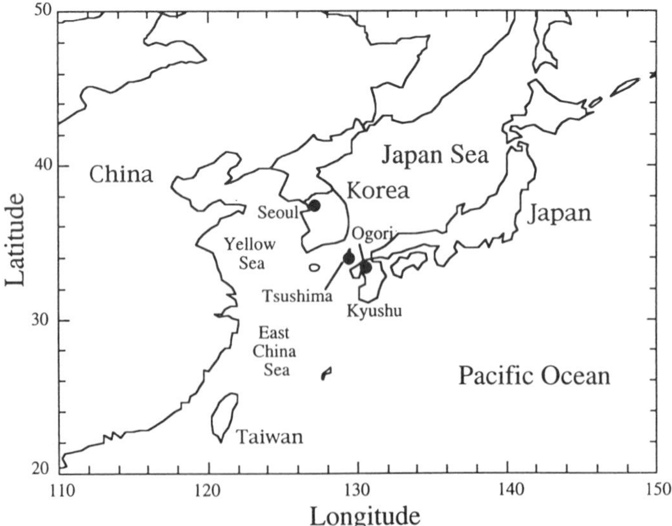

Figure 1. Calculation domain and field observation sites

Model Chemistry and Emission Data

The gas phase reaction mechanism proposed by Lurmann *et al.* (1986) was used. This scheme contains 53 species and 112 reactions. The present version of STEM also includes additional biogenic volatile organic carbon reactions. The model includes the detailed NH_3 - NO_3 - NH_4NO_3 thermodynamical equilibrium mechanism and the heterogeneous reaction from SO_2 (gas) to aerosol sulfate (SO_4^{2-}). The model chemistry also assumes that the formation of $(NH_4)_2SO_4$ occurs simultaneously after the generation of aerosol sulfate.

The 1° latitude x 1° longitude gridded NO_x and SO_2 emission inventory reported by Akimoto and Narita (1994) was used. Emissions for far eastern Russia were not included. Anmunia emission was taken from Zhao and Wang (1994) and Murano *et al.*(1995).

Piccot *et al.* (1992) have estimated global VOC (volatile organic carbon) emissions due to anthropogenic activities on a grid scale of 10° longitude by 10° latitude. Their data was used to prepare a regional scale VOC emissions inventory for the 1° x 1° grid domain shown in Figure 1. These 1° x 1° estimates were obtained by breaking down the larger scale emissions using population density (on a 1/3 degree scale) as the weighting factor. VOC category classes was re-classified into Lurmann *et at's* chemical reaction mechanism scheme. NO_x, SO_2, NH_3 and VOC were emitted into the 1st and 2nd vertical levels.

Meteorological Data Set and Numerical Simulation Setting

All meteorological datasets were obtained from the Japan Meteorological Agency (JMA) Global Objective Analysis Data (GANAL) . The GANAL data consists of 12 hour interval 1.875° spatially gridded meteorological data points (wind speed and direction, temperature, dew-point temperature) at specified pressure levels. GANAL wind speed, temperature and humidity data were all interpolated into the 1° x 1 ° degree grid system shown in Figure 1. The 12 hour interval GANAL data were interpolated linearly in time space to generate an hourly interval meteorological data set.

High concentration episodes (as reported in Wakamatsu *et al.*, 1996) were simulated. The episode occurred from February 4 to February 28, 1992. The STEM numerical calculations were started with zero initial values except for O_3 (which was set 35 ppb throughout the calculation domain).

Table 1 shows the numerical experiment design. Experiments were conducted by changing the heterogeneous reaction rate (SO_2-gas → SO_4^{2-}-aerosol) from 0.0 %/h to 2.0 %/h.

Table 1. Sensitivity experiments configuration

Heterogeneous Reaction Rate*) (% / hour)	Experiments
0.0	Case 00
0.5	Case 05
1.0	Case 10
2.0	Case 20

*) Heterogeneous reaction rate of SO_2(gas) → SO_4(aerosol)

RESULTS

Synoptic Weather Pattern and Observation Data

The field observation details and the results of a basic analysis of the observed aerosol data have already been reported in Wakamatsu *et al.* (1996). Therefore only a brief outline is given here. Aerosols were collected at three locations in Japan and Korea - two urban sites (Ogori in Fukuoka prefecture, Japan and Seoul, Korea) and one rural site (Tsushima in Nagasaki prefecture, Japan). Aerosol samples were collected over periods of one to two weeks every two months from August 1990 to February 1992 (with four or eight hour intervals). Figure 1 shows the monitoring locations. The highest average concentrations of F^-, Cl^-, NO_3^-, SO_4^{2-}, Ca^{2+} and Mg^{2+} were detected in February 1992. The maximum aerosol sulfate concentration was 38.6 $\mu g/m^3$ on February 1992.

The observation in Feb. 1992 was conducted during the winter monsoon season. A low pressure system moved from near the Taiwan area to the east of the Japan Islands within the latitude range 25° N - 30° N on the 10th-11th, 12th-13th and 22th-24th. After the passage of this low pressure system, a high pressure system slowly moved from the Yellow Sea to the Kyushu area. Total precipitation of 22 mm was recorded on the 15th and 23th February at Tsushima. The concentration peaks of aerosol sulfate were observed on the 6th, 11- 13th and 23- 24th of February.

Figure 2 shows the isentropic backward trajectories calculated from Tsuhsima from Feb. 4 to 27, 1992 at 1500 m level. Trajectory analysis clearly indicates the understanding of the transboundary pollutant transport is extremely important during this field observation.

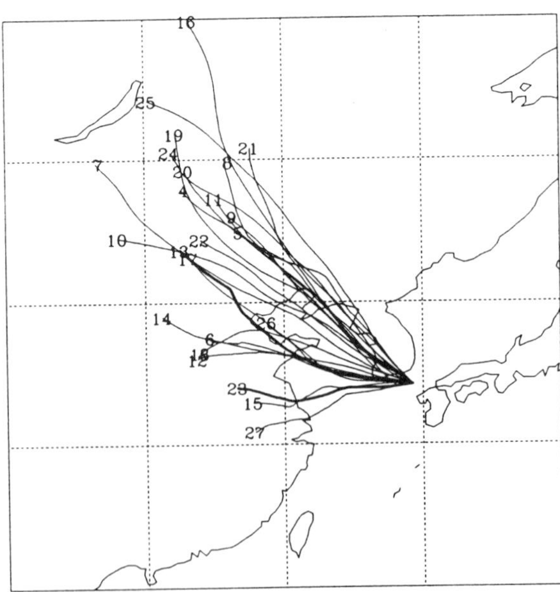

Figure 2. Isentropic backward trajectories calculated at Tsushima 0900 JST from Feb 4 to 27, 1992. The numbers located at the termination of the trajectory lines indicate the starting date of the backward trajectory calculation. Trajectory analysis clearly indicates the transboundary pollutant transport is extreamly important during this field observation period.

Transport Pattern and Time Variation Comparison

3-D visualization of SO_4^{2-} isosurface are shown in Figure 3 from Feb. 22 to Feb. 24 with 24 hour interval (for Case 10). Figures 4 and 5 show the time variation of SO_4^{2-} and NO_3^- observed in these three sites, respectively (Open circles in figures shows the observation), and the model results (legends in Figures indicate the heterogeneous reaction rate experiments). Arrows in figures indicates the GANAL interpolated wind speed and direction at Tsushima at z=1000 m.

A sharp peak of aerosol SO_4^{2-} was observed on the 13th and 24th (see Figure 4), and these periods were related to the passage of the low pressure system. The low pressure system and meso-front line movements (typical wind vectors shown in Figure 3) clearly indicate the transport of pollutants from the Asian mainland to the Yellow Sea and the western part of Japan direction, which occurred after the passage of the low pressure system and meso-front line. This fact indicates that the synoptic scale pressure system change play an important role in LRT in the East Asia region.

Simulated SO_4^{2-} aerosol concentration generally trace the observed SO_4^{2-} concentration variation. There are obvious diurnal variation of SO_4^{2-} concentration in Seoul, while no obvious variation at Tsushima and Ogori. This is because Seoul is located the strong emission area and controlled by the diurnal development of mixing layer height. On the other hand, two Japanese sites are away from the strong emission area, so only the synoptic scale long-range transport plays an important role for the sulfate concentration level.

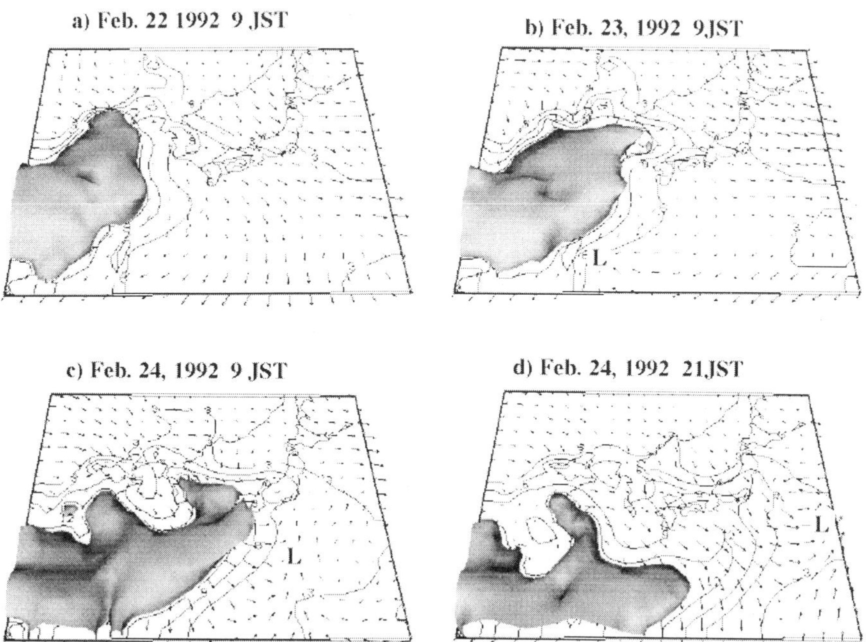

Figure 3. Simulated sulfate isosurfaces for 13.0 μg/m^3 and contours (3 μg/m^3 interval) . Wind vectors from Feb. 22 to 24, 1992 at 500m level are also plotted. Low pressure system (denoted by L) located east of Taiwan (Feb. 23) moved to south of Tokyo (Feb. 24). Transport of pollutants from the Asian mainland (a) to the Yellow Sea (b) and the western part of Japan (c), which occurred after the passage of the low pressure system.

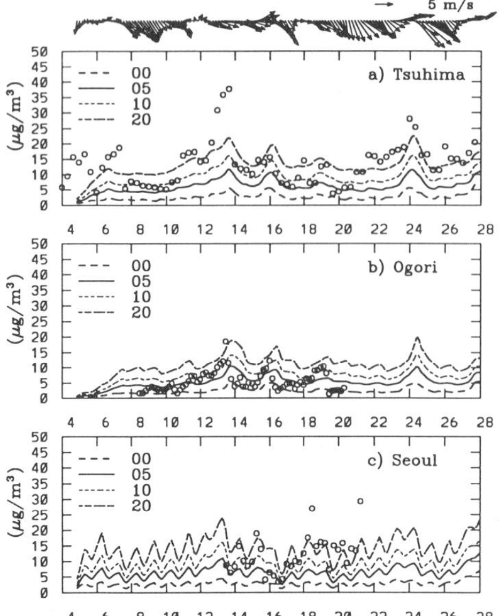

Figure 4. Comparison of observed nss-SO_4^{2-} data (open circle) and model simulation results at Tsushima, Ogori and Seoul for Feb. 1992. Symbols indicate observed data and lines indicate calculated results(Each lines denoted by the legends represent the Case 00 - 20 described in Table 1). Arrows at the top of the figure indicate the wind speed and direction at Tsushima at z=1000m.

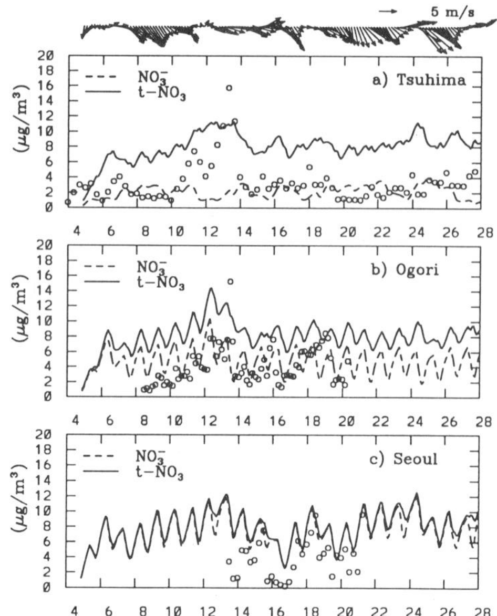

Figure 5. Comparison of observed NO_3^- data (open circle) and model simulation (NO_3^- and t-NO_3^-) results at Tsushima, Ogori and Seoul for Feb. 1992. Symbols indicate NO_3^- observed data and lines indicate calculated results for Case 10 experiment in Table 1. Arrows at the top of the figure indicate the wind speed and direction at Tsushima at z=1000m.

Most of the time variation of observation data for sulfate were located within the heterogeneous reaction rate of 0.5 - 2.0 %/h (Figure 4). The rate of 0.5 %/h corresponds to the lower limit and 2.0 %/h becomes to the upper limit. In general, the higher heterogeneous reaction rate agreed with time period when having a high concentration (vice versa). This fact indicate that the precise modeling of the heterogeneous sulfate formation rate is very important for the prediction of sulfate concentration.

The time variation of t-NO_3 (total nitrate) and NO_3^- shown in Figure 5 indicate that the simulated NO_3^- level agreed with observation at the three sites. At Ogori and Seoul (inland area), the particulate ratio was reached upto 60% and 95%, respectively. On the other hand, the particulate ratio at Tsushima was 28 %, which indicates the existence of nitric acid (HNO_3) at Tsushima island. These phenomena can be understand in view of the fact that HNO_3 and NO_3^- are in an equilibrium relation dependent on atmospheric temperature and relative humidity. These relation are also controlled by the ambient ammonia (NH_3) concentration by (Stelson and Seinfeld, 1982; Chang *et al.*, 1986),

$$NH_4NO_3 \text{ (aerosol)} \Leftrightarrow NH_3 \text{ (gas)} + HNO_3 \text{ (gas)}$$

Tsushima is an isolated island and the ambient anmonia concentration is lower than that of Ogori and Seoul. Furthermore because of the rapid decrease of NH_3 concentration over the ocean (no emission source and the formation of $(NH_4)_2SO_4$), the model predicts a very high rate of the conversion of aerosol nitrate to gaseous nitric acid. Therefore, even in cold temperature conditions, the NH_3 - NO_3 - NH_4NO_3 thermodynamical equilibrium moves to the gas phase and predicts the gas phase HNO_3. The existence of gas phase HNO_3 plays an significant role in the calculation of nitrate deposition because the dry deposition velocity of nitric acid is one order faster than the velocity of aerosol nitrate.

CONCLUSIONS

A numerical simulation of the long-range transport of pollutants in East Asia using STEM was conducted from Feb. 4 to Feb. 28, 1992 and the numerical results were compared with detailed high-time resolution (4-8 hour interval) aerosol monitoring data set at Tsushima, Ogori and Seoul. The STEM model successfully simulated the typical time variation of observed concentrations (both SO_4^{2-} and NO_3^-). It is found that wind pattern variations associated with a winter monsoon synoptic scale pressure system are extremely important for the transport of pollutants. Sensitivity analysis shows that most of the observation data are located within the heterogeneous reaction rate 0.5 - 2.0 % /h. Model predicted the large fraction of HNO_3 in Tsushima island even in winter, which is because of the low concentration of NH_3 during the long-range transport over the ocean (the NH_3 - NO_3 - NH_4NO_3 thermodynamical equilibrium moves to the gas phase condition).

The present model application reveals that the heterogeneous sulfate formation rate can range between 0.5 and 2.0 %/h, and precise modeling of this heterogeneous reaction rate plays significant role in the calculation of sulfate deposition. Model results also indicates that the NH_3 - NO_3 - NH_4NO_3 thermodynamical equilibrium mechanism is important to evaluate the nitrate deposition, and a much more complete aerosol formation model including the interaction of NH_3 - HNO_3 - SO_4 is important.

Acknowledgments The authors also wish to sincerely thank Prof. G.R. Carmichael of Iowa University and Dr. Yang Zhang of Battle Pacific Northwest Laboratory in the United States for their kind permission to use the STEM code and for their help in its installation.

REFERENCES

Akimoto H. and Narita H., 1994: Distribution of SO_2, NO_x and CO_2 emissions from fuel combustion and industrial activities in Asia with $1^\circ \times 1^\circ$ resolution, *Atmos. Environ.*, **28**, 213-225.

Carmichael G.R.,Peters L.K. and Kitada T., 1986: A second generation model for regional-scale transport/ chemistry/ deposition, *Atmos. Environ.*, **20**, 173-188.

Chang, J.S., Brost, R.A., Isaksen, I.S.A., Mardonich,S., Middelton,P., Stockwell, W.R. and Walcek, C.J., 1987: A three-dimensional Eulerian acid deposition model: Physical concepts and formulation, *J. Geophy. Res.*, **92**, 14681-14700.

Ichikawa, Y., Fujita, S. and Ikeda, Y. , 1994: An analysis of wet deposition of sulfate using a trajectory model for east Asia, Journal of Japan Society of Civil Engineering, ***II-28***, 127-136 in Japanese).

Katatani, N., Murao, N., Okamoto, S. and Kobayashi, K., 1991: A modeling study on acid deposition in eastern Asia, Proceeding of the 2nd IUAPPA Regional Conference on Air Pollution, Seoul, Vol. II, p. 59-64.

Kitada, T., Lee, P.C.S. and Ueda, H., 1993: Numerical Modeling of Long-Range Transport of Acidic Species in Association with Meso-B-Convective-Clouds Across the Japan Sea

Resulting in Acid Snow over Coastal Japan--I. Model Description and Qualitative Verifications, II. Results and Discussion, *Atmos. Environ.*, **27A**, 1061-1090.

Kotamarthi, V.R. and Carmichael, G.R., 1990: The long range transport of pollutants in the Pacific rim region, *Atmospheric Environment*, **24A**, 1521-1534.

Lurmann,F.W., Lloyd, A.C., Atkinson, R., 1986: A chemical mechanism for use in long-range transport/acid deposition computer modeling, *J. Geophy. Res.*, **91**, 10905-10936.

Murano, J., Hatakeyama, S, Mizuguchi,T. and Kuba, N., 1995: Gridded ammonia fluxes in Japan , *Water Air and Soil Pollution*, **85**, 1915-1920.

Piccot S., Watson S.D. and Jones J.W., 1992: A global inventory of volatile organic compound emissions from anthropogenic sources, *J. Geophys. Res.*, **97**, D9, 9897-9912.

Sato, J., Satomura, Y., Sasaki, H., 1993: The long-range transport model of air pollutants in east Asia, Meteorological Research Institute Report. (in Japanese)

Stelson, A. and Seinfeld, J.H., 1982: Relative humidity and temperature dependence on the ammonium nitrate dissociation constant, *Atmos. Environ.*, **16**, 983-992.

Wakamatsu, S., A. Utsunomiya, J.-S. Han, A. Mori, I.Uno and K. Uehara, 1996: Seasonal variation in atmospheric aerosol concentration covering northern Kyushu, Japan and Seoul, Korea, *Atmos. Environ.*, **30**, 2343-2354.

Wesely,M.L., 1988: Improved parametrizations for surfaces resistance to gaseous dry deposition in regional scale model, *EPA/6003-86/037* (PB86-218104).

Zhao, D. and Wang,A., 1994: Estimation of anthropogenic ammonia emissions in Asai, *Atmos. Environ.*, **28**, 689-694.

DISCUSSION

S. PRYOR: Are wet deposition concentration available for the site you are evaluate your model performance?

I. UNO: Yes, Japan Environment Agency has a wet/dry deposition monitoring network, and bi-monthly deposition data are available to evaluate model performance.

A. KEIKO: How you make an attempt of simultaneous modelling of NO_x and SO_x , I mean NO_x-SO_x aerosol? If not, do you plan to tackle this problem?

I. UNO: My model study does not include an interaction between NH_4NO_3-aerosol and $(NH_4)_2SO_4$-aerosol processes. I understand these interaction is very important to include model study. I hope that US-EPA MODELS-3 projects will provide such aerosol chemistry module to tackle this problem.

LONG-RANGE TRANSPORT OF LEAD AND CADMIUM IN EUROPE RESULTS OF THE MODEL SIMULATION FOR THE 1985 - 1995 PERIOD

Jerzy Bartnicki[1], Krzysztof Olendrzynski[1], Marek Uliasz[2]

[1]Norwegian Meteorological Institute,
P.O. Box 43 Blindern, N-0313 Oslo, Norway
[2]Mission Research Corporation, ASTER Division,
P.O. Box 466, Fort Collins, CO 80526-0466, USA.

INTRODUCTION

In 1994, the UN Economic Commission for Europe (ECE) proposed that the future work on heavy metals should concentrate on priority elements: Pb, Hg, Cd, Cr, Ni, Zn, Cu, As and Se with particular focus on the first three metals. Concerning environmental impact of heavy metals, their concentrations, especially in remote areas, are usually too low to cause any serious adverse effects. However, the concentrations in remote parts of Europe can increase significantly during episodes of long range transport from anthropogenic sources. In addition, even low deposition fluxes accumulate in the soil and can reach the level at which heavy metals become mobile in the environment. Therefore, depositions and especially long-term cumulative depositions of heavy metals have to be monitored.

The existing models developed for simulating heavy metals transport and deposition in Europe have been reviewed by Petersen and Iverfeldt (1994). Among all models discussed by Petersen and Iverfeldt, only several can be used for long-term simulations. This group is further limited for practical applications by the lack of meteorological input data necessary for such simulations. For the transport period longer than one year, three-dimensional models cannot be used directly because specific, for these models, meteorological input data are not directly available for routine computations at present. For two-dimensional models, a unique set of meteorological input data has been developed at the Norwegian Meteorological Institute (DNMI) in the frame of the EMEP (Co-operative Programme for Monitoring and Evaluation of the Long Range Transmission of Air Pollutants in Europe) program (Jakobsen, 1996).

The latest version of the Heavy Metals Eulerian Transport (HMET) model has been also developed at DNMI, especially for the long-term simulations. The HMET model is using the 1985 - 1995 EMEP data set as a meteorological input and its structure is a compromise between fully three-dimensional approach and a flat, two-dimensional model.

Emission inventories for two priority metals, Pb and Cd, have been developed at the International Institute for Applied Systems Analysis (IIASA) within the Project on Regional Material Balance Approaches to Long-Term Environmental Policy Planning. These inventories include annual emissions of Pb and Cd for all European countries during the period 1985-1995. The gridded spatial distributions of Pb and Cd emissions are available in the EMEP grid system (39 x 37 nodes, 150 x 150 km mesh size) for Europe, exactly the same as used by the HMET model.

Using meteorological input data from DNMI and emission data developed at IIASA, long-term simulations of Pb and Cd transport and deposition in Europe have been performed with the HMET model and the results are presented in this paper.

THE HMET MODEL

The previous version of the HMET model (Bartnicki *et al.*, 1993, Bartnicki, 1994) has been improved by adding one vertical layer above the mixing height (Bartnicki, 1997). In this way, diurnal variation of the mixing height and the vertical flux of heavy metals between two layers are taken into account in the new version.

Model Structure

Atmosphere is represented by two dynamic layers in the HMET model: the mixing layer and the residual layer. The thickness of the mixing layer is defined by the mixing height which is variable in space and time. The residence layer is located above the mixing layer up to 3 km. The top of the residence layer was chosen in such a way that the maximum of the mixing height is still below the top of the residence layer.

A homogenous vertical distribution of heavy metals is assumed in the mixing layer immediately after emission. Horizontal transport in both layers is calculated using the same wind from the σ-level = 0.925, which corresponds to approximately 570 m. Diurnal variation of the mixing height is the main mechanism for the vertical exchange of heavy metals between the mixing layer and the residence layer.

Metals are removed from the mixing layer by dry and wet deposition processes. Efficiency of the dry deposition is a function of the mixing height, terrain type and atmospheric stability. Wet deposition depends on the mixing height, precipitation intensity and the scavenging ratio, which is different for each metal. During transport, metals in the residual layer are only affected by wet deposition process which is parameterized in a similar way like in the mixing layer. However, the scavenging ratio is higher in this layer because of the presence of clouds.

Transport of heavy metals in the HMET model is described by a set of two transport equations (one for each layer):

$$\frac{\partial c}{\partial t} + \frac{\partial uc}{\partial x} + \frac{\partial vc}{\partial y} = S \qquad (1)$$

$$\frac{\partial c_r}{\partial t} + \frac{\partial uc_r}{\partial x} + \frac{\partial vc_r}{\partial y} = S_r \qquad (2)$$

where $c(x,y,t)$ and $c_r(x,y,t)$ are the ambient concentrations of Cd and Pb in the mixing layer and in the residence layer respectively; $u = u(x,y,t)$ and $v = v(x,y,t)$ are components of the wind field; S and S_r represent all sources and sinks of heavy metals in the mixing layer and in the residence layer respectively. The local deposition coefficients are introduced in the model equations to account for more efficient deposition close to the sources. For Cd, the local deposition coefficient $\alpha = 0.10$, and for Pb, $\alpha = 0.15$. A full description of other model assumptions and discussion of parameters values can be found in Bartnicki (1997) and (Bartnicki *et al.*, 1993).

The advection part of the transport equations is solved according to Botts method (Bott 1989a, 1989b), which compared to other algorithms gives more accurate solutions with relatively low numerical diffusion.

Heavy metals emissions are estimated at each node and they represent the sum of all sources within the grid square. Meteorological data (e.g. transport wind, precipitation and mixing height) are also assigned to each node. All meteorological data (except mixing height) are updated every six hours during the model run. New values of the mixing height are introduced every 12 hours at 12:00 UTC and 0:00 UTC.

Model Validation

A new measurement data base for heavy metals which includes observations from approximately 70 stations in Europe has been recently developed at the Norwegian Institute for air Research (Berg *et al.*, 1996). This database, which covers the period 1987-1995, has been used for the validation of the HMET model (Bartnicki, 1997). Comparison of model results and measurements for Pb and Cd were compared based on annual average concentrations in air and precipitation. Number of samples, measured and computed annual average concentrations, standard deviations and correlation coefficients, for all metals, are shown in Table 1.

The results presented in Table 1 indicate a good (within 30%) agreement between measured and computed annual concentrations in air and precipitation. This agreement is slightly better for air concentrations than for concentrations in precipitation.

EMISSION DATA

The anthropogenic emission databases for Pb and Cd have been developed at IIASA using various production and consumption statistics from all European countries. The details about the methodology can be found in Olendrzynski *et al.* (1996). When developing emission inventories in the HMET grid system, the following source categories were considered: (1) stationary fuel combustion (including industrial, commercial and residential boilers), (2) non-ferrous metal manufacturing, (3) road transport, (4) iron and steel production, (5) waste disposal and (6) other sources (primarily cement production). On country by country basis, the most

Table 1. Computed vs. measured concentrations of Pb and Cd in air and precipitation.

Metal/Media	Air concentration ($ng\ m^{-3}$)		Conc. in prec. ($mg\ l^{-1}$)	
	Pb	Cd	Pb	Cd
No. of observations	67	84	154	188
Measured mean	21.89	0.47	2.97	0.10
Computed mean	23.16	0.44	2.59	0.07
Standard deviation	11.10	0.29	2.63	0.09
Correlation coefficient	0.75	0.64	0.60	0.56

important contributors to European anthropogenic emissions are: in Central and East Europe - the European part of the former Soviet Union (Pb and Cd), former Yugoslavia (Cd); in Western Europe - United Kingdom, France (Pb) and Spain (Cd).

Annual Totals for the Period 1985 - 1995

Total emissions from the model domain were calculated as a sum of emissions from each individual grid-square for each year of the simulation. Trends for Pb and Cd emissions, are shown in Figure 1.

There is clear decline in Pb and Cd emission trends during the period 1985 - 1995. Pb emissions are reduced by 29% and Cd emissions by 14% during this 11-year period. Since information about emissions in 1993 - 1995 was not available, inventories developed at IIASA assumed constant emission level of Pb and Cd beyond 1992.

Spatial Distribution of Pb and Cd Emissions

Spatial distribution of Pb and Cd does not change much during the 11-year period. As an example, Pb and Cd emission maps for Europe in 1989 are presented in Figures 2 and 3. Distribution of Pb emissions in Europe is more uniform than Cd emissions. Both for Pb and Cd, absolute maxima of emissions can be found in Central and Eastern Europe. Maximum of Pb emissions - 969 tonnes is located in the model grid square (26,28) close to Moscow in Russia. Maximum of Cd emissions - 75.5 tonnes is located in Poland in the grid square (25,19).

RESULTS OF THE SIMULATION FOR THE 11-YEAR PERIOD

Atmospheric transport and deposition of Pb and Cd in Europe has been simulated with the HMET model for the period 1985 - 1995, using as model input emission data developed at IIASA and routine meteorological data compiled at DNMI. Annual mean (over entire period) concentration maps for Pb and Cd are shown in Figures 4 and 5 respectively.

Concentration of Pb is quite uniform over entire Europe with the maximum located close to Moscow in Russia. Concentrations of Cd is slightly higher in Central Europe (Black Triangle region) than in other parts of the continent with the maximum close to Katowice in Southern Poland.

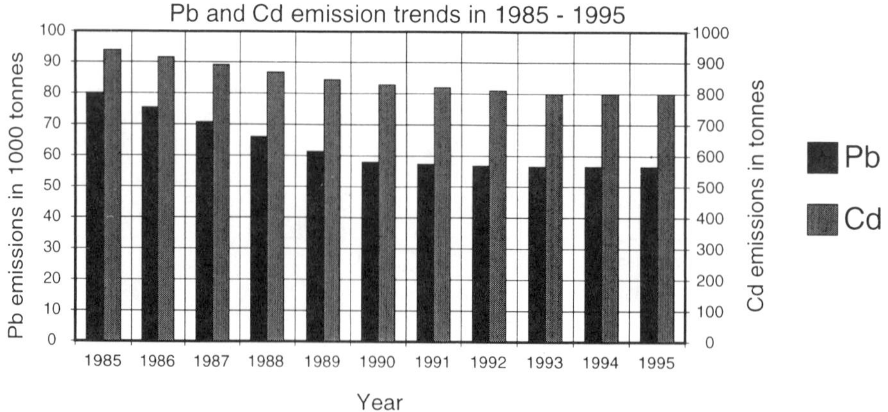

Figure 1. Annual total emissions of Pb and Cd in the period 1985 - 1995.

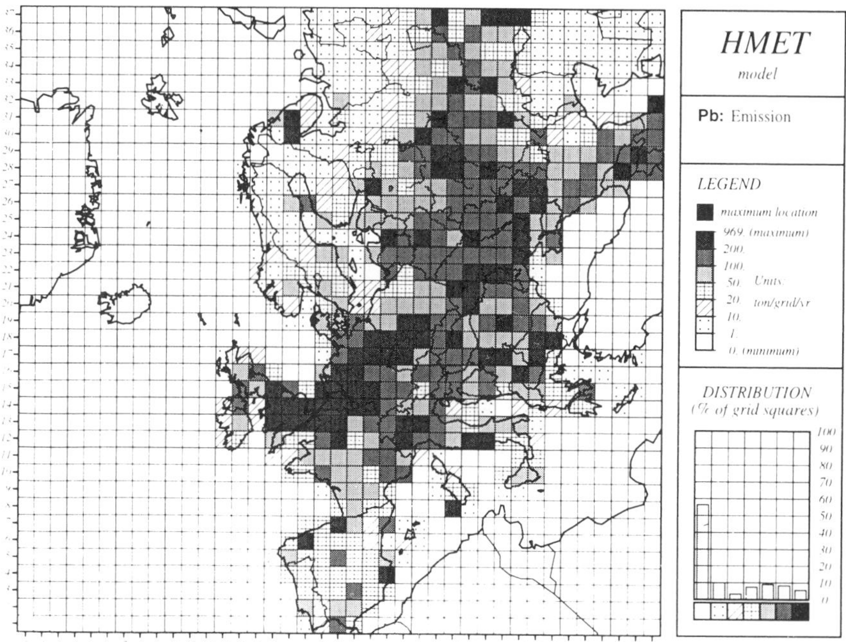

Figure 2. Spatial distribution of Pb emissions in Europe in 1989. Units: 10^3 kg grid^{-1} yr^{-1}.

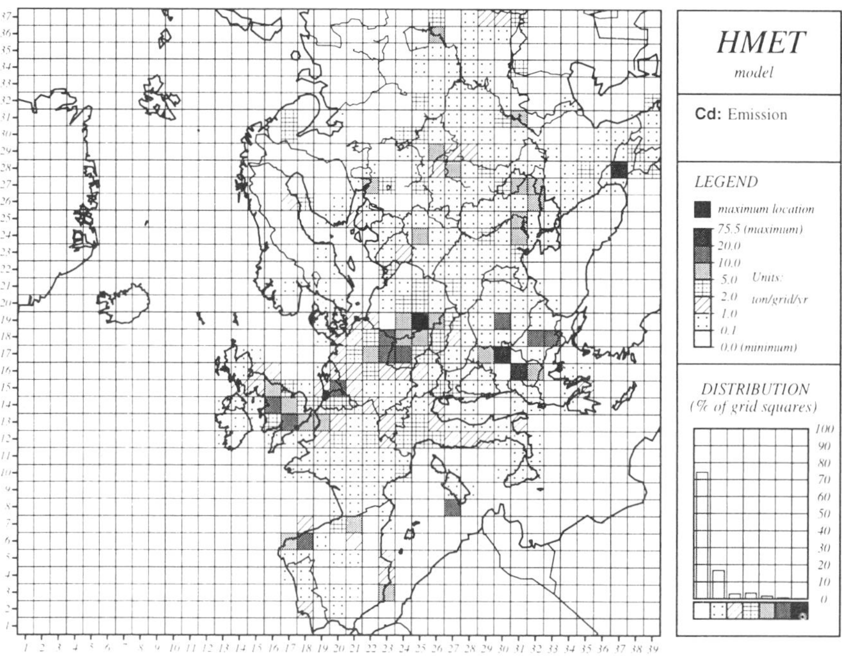

Figure 3. Spatial distribution of Cd emissions in Europe in 1989. Units: 10^3 kg grid^{-1} yr^{-1}.

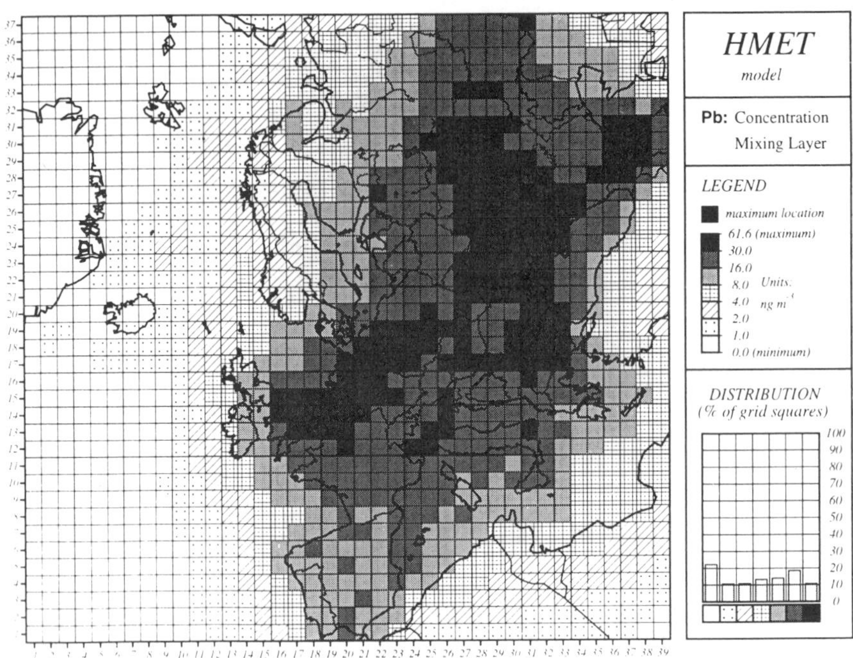

Figure 4. Annual average (over 11-year period) Pb concentrations in Europe. Units: ng m^{-3}.

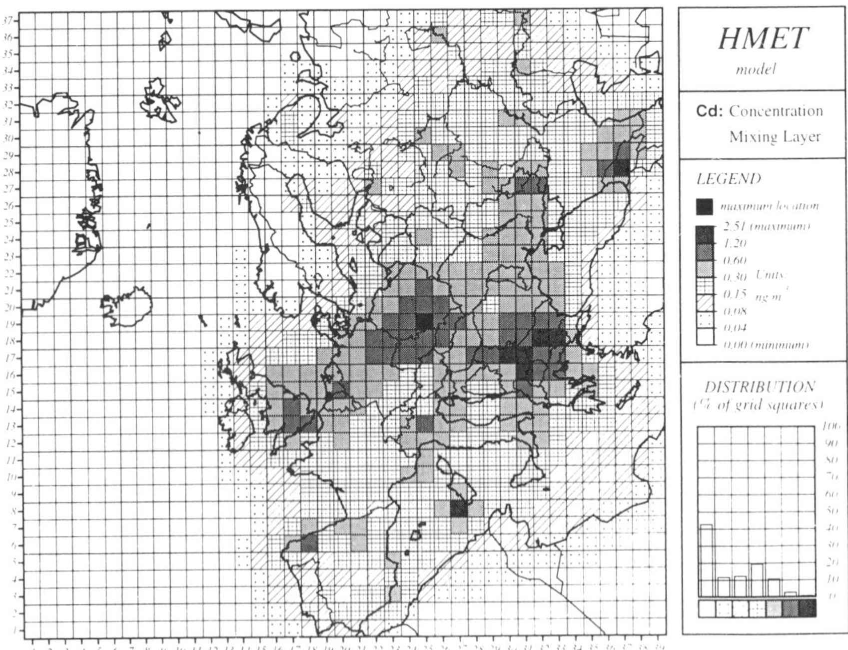

Figure 5. Annual average (over 11-year period) Cd concentrations in Europe. Units: ng m^{-3}.

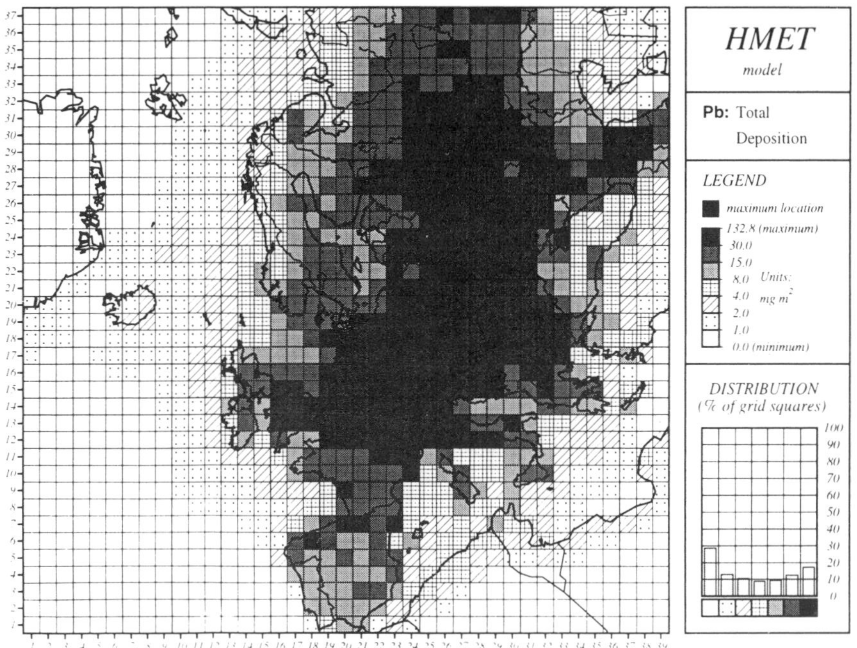

Figure 6. Cumulative (over 11-year period) Pb deposition in Europe. Units: mg m^{-2}.

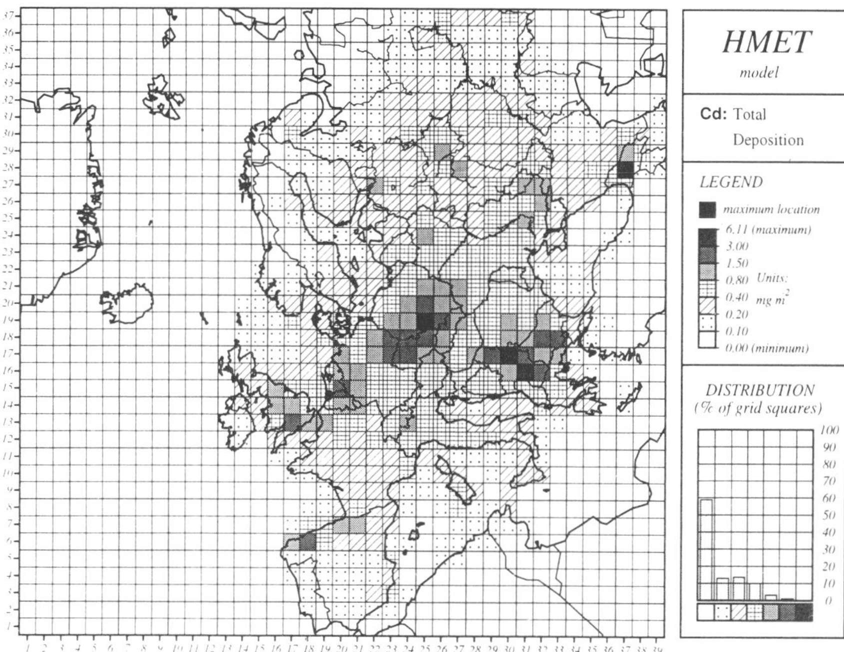

Figure 7. Cumulative (over 11-year period) Cd deposition in Europe. Units: mg m^{-2}.

Cumulative deposition map for Pb is even more uniform than the concentration field. For Cd, again the cumulative deposition in Central Europe is higher than elsewhere.

Both for annual average concentrations and cumulative depositions computed maxima are located in the same model grid squares as maxima of Pb and Cd emissions.

CONCLUSIONS

The results presented in the paper show that the HMET model is fully operational and can be used for simulating long-range, long-term transport of heavy metals in Europe. There is a good agreement between the computed and measured annual concentrations, in air and precipitation, for selected stations in Europe. Computed annual average concentration maps for Pb and Cd indicate that, distribution of Pb is rather uniform over entire Europe, whereas for Cd higher concentrations are observed in Central Europe. Distribution of the total cumulative depositions of Pb and Cd in Europe is similar to the distribution of the concentration fields for these metals.

REFERENCES

Bartnicki J. (1994) An Eulerian model for atmospheric transport of heavy metals over Europe: Model description and preliminary results. *Water, Air and Soil Pollution* **75**, 227-263.

Bartnicki J. (1997) Comparison of long-term concentrations of heavy metals in air and precipitation computed by the HMET model and measured at the EMEP stations in Europe. Proceedings of the First international Conference on Measurements and Modelling in Environmental Pollution. April 22-24, Madrid, Spain (in press).

Bartnicki J., Modzelewski H., Szewczyk-Bartnicka H., Saltbones J., Berge E. and A. Bott (1993) An Eulerian model for atmospheric transport of heavy metals over Europe: Model description and testing. Technical Report no. 117. Norwegian Meteorological Institute. Oslo - Norway.

Berg T, A.G. Hjellbrekke, J.E. Skjelmoen (1996). Heavy Metals and POPs within the ECE region. EMEP/ CCC Report 8/96. Norwegian Institute for Air Research, Kjeller, Norway

Bott A. (1989a) A positive definite advection scheme obtained by nonlinear renormalization of the advective fluxes. *Monthly Weather Review* **117**, 1006-1015.

Bott A. (1989b) Reply. *Monthly Weather Review* **117**, 2633-2636.

Jakobsen H. (1996) Preparation of meteorological input data to the 150 km Lagrangian air-pollution models. DNMI Research Report No. 40. Norwegian Meteorological Institute, Oslo, Norway.

Olendrzynski K., Anderberg S., Bartnicki J., Pacyna J. and W. Stigliani (1996) Atmospheric emissions and depositions of cadmium lead and zinc in Europe during the period 1955 - 1987. *Environmental Reviews* **4** (4), 300-320.

Petersen G. and A. Iverfeldt (1994) Numerical modelling of atmospheric transport and deposition of heavy metals. A review of results from models applied in Europe. In: *Proceedings of the EMEP Workshop on European Monitoring, Modelling and Assessment of Heavy Metals and Persistent Organic Compounds* (R.M. van Aalst ed.) Beekbergen, The Netherlands.

DISCUSSION

A. V. KEIKO: How complete is your emission database? How detailed is it?

J. BARTNICKI: Our emission data base consists of annual emission inventories for Pb and Cd for the 1985 - 1995 period. These emissions, developed at the International Institute for Applied Systems Analysis in Austria, are available in the model grid system with 150 km x 150 km resolution for Europe.

R. YAMARTINO: Studies in the US suggest that a significant fraction of lead emissions are deposited close to roadways. How do you account for this in the model with a 150 km horizontal resolution?

J. BARTNICKI: We have introduced so called local deposition coefficient (15% for lead) in the HMET model. This coefficient specifies the amount of emission which is deposited directly at the same grid where emission takes place and is not a subject of the long range transport. In this way we take into account a correction for large subgrid concentrations.

LONG RANGE TRANSPORT OF NO_X, NO_Y, O_3 AND SO_X OVER EAST ASIA AND THE WESTERN PACIFIC OCEAN IN WINTER SEASON–A NUMERICAL ANALYSIS

Toshihiro Kitada, [1] Masato Nishizawa, [1] and Yutaka Kondo [2]

[1]Dept. of Ecological Engineering, Toyohashi University of Technology
Tempaku-cho, Toyohashi, Japan 441
[2] Solar-Terrestrial Environment Lab., Nagoya University
Honohara, Toyokawa, Japan 442.

INTRODUCTION

The PEM West-B, Pacific Exploratory Mission West: phase B, experiment was carried out from the beginning of February to the mid of March, 1994. The experiment explored chemical composition of the atmosphere over the tropical Pacific Ocean and the Pacific rim area of Asia. The experiment is valuable for understanding of long range transport/transformation of air pollutants released over Asian industrial area in late winter and early spring, and is useful for model validation in semi-global scale.

In this paper, results of numerical simulation for the PEM West-B period, using our 3-D Eulerian semi-global scale model, are reported. The calculated concentrations showed good agreement with observation for most of the chemical species. The agreement was better in emission source area than in remote area where no single strong source exists. Spatial distribution of the depositions of SO_x and NO_y were also calculated with and without background concentrations of SO_2, SO_4^{2-} and HNO_3, the background concentrations which were estimated using PEM-B observation in remote area. The results suggested that NO_y deposition in the tropical area such as Borneo and Java islands may be largely affected by the natural NO_x emissions due to lightning and soil microbiological activity, and that relative contribution of the background- to the total-depositions was less than 10 % for both N- and S-compounds in anthropogenic emission-dominated area. Additionally, contributions of the emission sources in the Asian continent to distributions of SO_x and NO_y depositions over the Pacific rim area are evaluated.

TRANSPORT/TRANSFORMATION/DEPOSITION MODEL

Governing Equations

The semi-global scale model consists of a set of partial differential equations that account for advection, diffusion and chemical reactions of trace chemical species. The model also includes processes of vertical mass-transport by subgrid scale cumulus convection and wet- and dry-depositions. The previous version of this model was applied to east Asia and the northwestern Pacific ocean for the analysis of data from PEM West-A campaign (Kitada et al., 1996). The governing equations are as follows, where index for chemical species, i = 1, 2, ..., 38:

$$C\frac{\partial X_i}{\partial t} + CU\frac{\partial X_i}{\partial x} + Cv\frac{\partial X_i}{\partial y} + C\dot{\sigma}\frac{\partial X_i}{\partial \sigma} = \frac{\partial}{\partial x}\left(CE_\phi\frac{\partial X_i}{\partial x}\right)$$

$$+ \frac{1}{cos\theta}\frac{\partial}{\partial y}\left(Ccos\theta E_\theta\frac{\partial X_i}{\partial y}\right) + \frac{\rho g^2}{\pi^2\, r^2}\frac{\partial}{\partial \sigma}\left(C\rho r^2 E_\sigma\frac{\partial X_i}{\partial \sigma}\right) + R_i \tag{1}$$

where $dx = rcos\theta d\phi$, $dy = rd\theta$, $\sigma = (P - P_T)/\pi$, $\pi = P_S - P_T$, $\dot{\sigma} = \{(\frac{\partial z}{\partial t})_\sigma + \vec{V}\cdot\nabla_\sigma z - W\}\frac{\rho g}{\pi}$. X_i is the non-dimensional concentration of the ith chemical species; ρ the air density in kg m^{-3}; C the air density in kmol m^{-3}; θ and ϕ denote latitude and longitude, respectively; r is the distance from the center of the Earth (approximated by the averaged radius of the earth); P_S and P_T denote the atmospheric pressures at the earth's surface and the top boundary, respectively: P_T is set at 10 hPa; z is the altitude of a σ-surface; U and V are the horizontal wind velocities for ϕ and θ directions, respectively; W is the vertical wind velocity; g is the gravitational acceleration; E_ϕ and E_θ denotes horizontal eddy diffusivity; and E_σ denotes vertical eddy diffusivity and R_i represents the chemical reaction term for the ith species. Here, $E_\phi = E_\theta = 0.02(r\triangle\theta)^2\sqrt{D_T^2 + D_S^2},\ where\ D_T = \left(\frac{\partial U}{\partial \phi} - \frac{\partial(Vcos\theta)}{\partial\theta}\right)\frac{1}{rcos\theta}\ and\ D_s = \left(\frac{\partial V}{\partial\phi} + \frac{\partial(Ucos\theta)}{\partial\theta}\right)\frac{1}{rcos\theta}$, have been used (Washington and Williamson, 1977). To determine the vertical diffusivity E_σ in the atmospheric boundary layer (i.e., ABL), outputs from a $k-\epsilon$ turbulence model for ABL (Kitada, 1987) has been used. The reaction terms, R_i, accounts for 25 advected species (NO, NO_2, HNO_3, PAN, $RONO_2$, HNO_2, NO_3, N_2O_5, HNO_4, CO, SO_2, SO_4^{2-}, O_3, H_2O_2, CH_4, C_2H_6, C_3H_8, C_2H_4, $ALKENEs$, $HCHO$, $ALDEHYDEs$, $HCOOH$, $ACETIC\ ACID$, $ROOH$, $and\ LUMPED\ KETONEs$) and 13 steady-state assumption applied species (e.g., OH, HO_2,...,etc.), forming a system of 90 chemical reactions (an adaptation from the model in Lurmann et al., 1986).

Dry and Wet Deposition Processes

Dry deposition onto the earth's surface was expressed using the dry deposition velocity. Chemical species subjected to the dry deposition were SO_2, SO_4^{2-}, NO_2, HNO_3, H_2O_2, O_3, PAN, $HCHO$, $HCOOH$, etc. Typical velocity values used in the simulations were, for example, 0.5 $cm\ s^{-1}$ for land surface and 0.1 $cm\ s^{-1}$ for sea surface for SO_2; and 0.1 $cm\ s^{-1}$ for both land and sea surfaces for SO_4^{2-}.

For wet removal processes, semi-empirical equations were used for SO_2, SO_4^{2-}, HNO_3, H_2O_2, $and\ O_3$; the equations were derived on the basis of capture of aerosol particle by cloud and rain drops and gas absorption into the rain drops (Kitada, 1994). Meteorological parameters used in the wet deposition equation were the heights of cloud-top and -base, and precipitation intensity. Scavenging coefficients of sulfate and nitrate particles by rain and snow were expressed as follows:
For rain,

$$\Lambda_{p,\,rain} = 6\times 10^{-4}\,\eta\, P^{0.75} \tag{2}$$

where $\Lambda_{p,\,rain}$ denotes the scavenging coefficient for particle due to rain in s^{-1}, η the collection efficiency of aerosol by rain and was assumed to be 0.05 $\sim$ 0.5, and P the precipitation intensity in $mm\, hr^{-1}$.
For snow, following Slinn (1974),

$$\Lambda_{p,\,snow} = \frac{\rho_w\, g\, \eta\, (3.6 \times 10^{-6}\, P)}{\rho_a\, V_t^2} \tag{3}$$

where ρ_w denotes the density of water ($= 1000\ kg\, m^{-3}$), g the gravitational acceralation ($= 9.8\ m\, s^{-2}$), ρ_a the air density ($\sim 1kg\, m^{-3}$), V_t the average settling velocity of the snow flakes in $m\, s^{-1}$ and is expressed by the following equation recommended by Knutson et al. (1976): $V_t = (102 + 51\, log_{10}\, d_c)/100$, where d_c denotes the diameter of the circle circumscribed about the average snowflake in cm. $\Lambda_{p,\,snow}$ in Eq.(3) was approximated as $5.6 \times 10^{-4} P$ by assuming $d_c = 500\, \mu m$ and $\eta = 0.002$ (judged from figure in Slinn, 1977).

Coefficients (s^{-1}) for scavenging, due to rain, of gaseous species were given in the model, for example, for SO_2,

$$\Lambda_{SO2} = \beta \frac{\alpha\, P}{3.6\, H} \tag{4}$$

$$\alpha = 10^{-6} R\, T\, H_{eff} \tag{5}$$

where H denotes the height of cloud-top in m, R the universal gas constant ($= 0.082\, \ell\, atm\, K^{-1}\, mol^{-1}$), T the air temperature in K, and H_{eff} the effective Henry's law constant for SO_2 in $mole\, \ell^{-1}\, atm^{-1}$, which is a function of the hydrogen ion concentration in rain drop. For example, the estimated value of α is 0.672 for pH = 5.6 at 273 K, and 0.089 for pH = 5.0 at 298 K. Equation (4) has been derived by assuming that S(IV) concentration in precipitation can be expressed with the hypothetical S(IV) concentration, which is in equilibrium with the atmospheric SO_2 concentration averaged over the height from ground to cloud-top, multiplied by a factor β, which represents the ratio of the real S(IV)- to the hypothetical equilibrium S(IV)-concentrations in precipitation, and was assumed unity for the $SO_2 - S(IV)$ system. Expression similar to Eq. (4) was used also for HNO_3 and H_2O_2, where the "equilibrium" ratio β was given as 0.38 x 10^{-8} for HNO_3 and 0.055 for H_2O_2; these values were determined by a series of numerical experiments in which temporal development of various chemical species in a water drop, falling through pollutants-containing-atmosphere, was calculated using a sub-module of a comprehensive acid rain model described in Kitada et al. (1993a,b). The scavenging coefficients Λ in Eqs. (2), (3), and (4) were used in the following first order removal equation, which was incorporated into the model:

$$\frac{\partial\, C\, X_i}{\partial\, t} = -\, \Lambda\, C\, X_i \tag{6}$$

Vertical Mass Transport in Sub-grid Scale

Vertical mass transport due to sub-grid scale cumulus convection was included in the model. Height of the cumulus convection and the cloud air mass subjected to the convection were determined using vertical profile of water vapor and resolvable scale vertical wind velocity of input meteorological data (i.e., ECMWF data). The assumptions made for the process, which are similar to those used in Strand and Hov (1993), were: (1) cumulus convection draws air mass out of boundary layer (i.e., approximately below 1 km high above ground) and then redistributes it to upper layers; (2) continuity of air mass is maintained within a column over each horizontal grid cell; and (3) the compensating subsidence occurs from one layer to the one below.

METEOROLOGICAL INPUTS, EMISSION SOURCES AND SIMULATION CASES

For input meteorological fields of air flow, temperature, humidity and pressure, the ECMWF basic consolidated data set was used. The data has longitudinal and latitudinal resolutions of 2.5 degree, and time resolution of 12 hours, and is defined at 15 vertical layers from the earth's surface to 10 hPa. These meteorological data were interpolated and modified vertically and temporally so that the data give information at two additional vertical grid points within atmospheric boundary layer and at intervals of 30 minutes, i.e. the time step for transport processes in the simulation. Diurnal variation of boundary layer activity was modelled using the outputs of a k-ϵ model (Kitada, 1987 and Kitada et al., 1997). Precipitation fields were estimated from the NCAR archive for those of monthly values (Xie and Arkin, 1995). Figure 1 shows precipitation field in March, 1994.

In the present simulations, five types of NO_x emission sources are considered: (1) source distribution due to fuel combustion at surface level is shown in Fig. 2, where original 1° x 1° distribution by Akimoto and Narita (1994) was converted into a 2.5^o x 2.5^o distribution; (2) NO_x emission due to soil micro-biological activity in March is displayed in Fig. 3, which was also adapted from that by Yienger and Levy (1995); (3) NO lightning emission is shown in Fig. 4 for February and March, which was estimated from that by Kumar et al. (1995); (4) aircraft NO_x emission was taken from an ICAO committee report on aviation environmental protection (ICAO, 1995); and (5) biomass burning NO_x emission was tentatively assumed to have the same distribution as the soil NO_x emission. Table 1 lists source strength of various NO_x emissions in calculation domain (see, for example, Fig. 2). Non methane hydrocarbons were grouped into C_2H_6 (ethane), C_3H_8 for propane and benzene, C_2H_4 (ethene) and ALKE for $\geq$ C_3 alkenes. Emissions for these species were determined so as to be proportional to the NO_x emissions by fuel combustion, aircraft emission, and biomass-burning. The proportional coefficients were estimated using data in Aichi Prefecture, Japan (Nakanishi, 1996) as follows: 0.25 for C_2H_6, 0.45 for C_3H_8, 0.4 for C_2H_4, and 0.2 for $ALKE$ when NO_x emission on molar basis is assumed to be 1. Methane emissions were collected from the work by Matthews and Fung (1987) and Lerner et al. (1988). SO_2 emission due to fuel combustion was again adapted from that by Akimoto and Narita (1994).

Table 1. NOx emissions in calculation domain

Source type	Emission strength(kg-/month)	Relative strength
Fuel Combustion	3.59×10^8	1
Aircraft	6.92×10^6	0.019
Lightning	3.40×10^7	0.097
Soil	3.87×10^7	0.108
Biomass Burning*	3.87×10^7	0.108

* Assumed

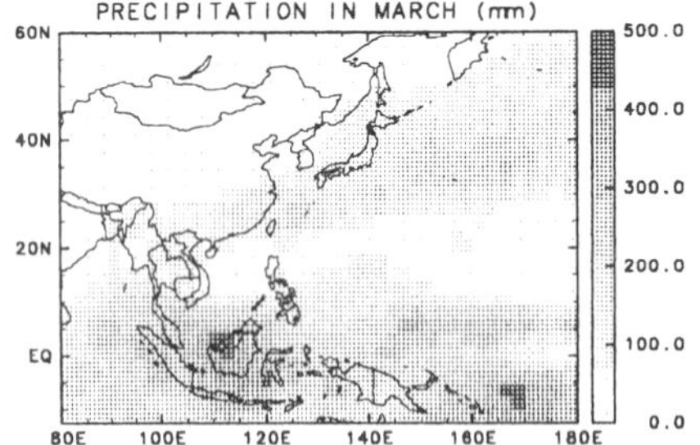

Figure 1. Precipitation in March, 1994 in mm month^{-1}.

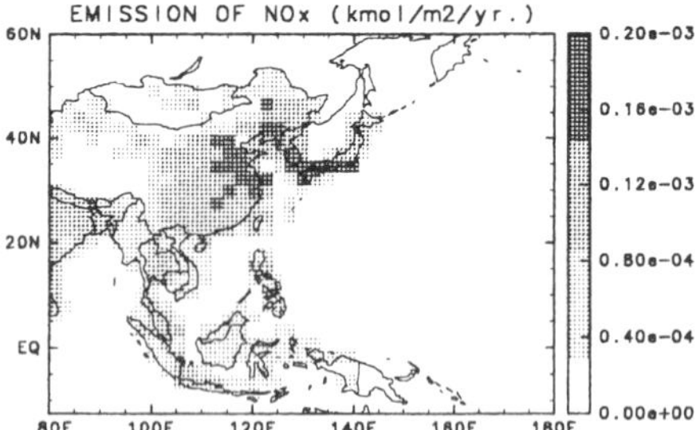

Figure 2. NOx emission by fuel combustion in 1987 in kmol m^{-2} yr^{-1}.

The simulation domain was 80 $^oE \sim 180^oE$ and $12.5^oS \sim 60^oN$, with grid size of 2.5 degree for horizontal directions. In the vertical direction, 17 variable grids were used. Time step was 30 minutes for transport calculation and 12 seconds for chemistry. Appropriate minimum background concentrations, which were estimated from PEM West-B data, were assumed for all species throughout the entire simulation domain and period. Under these conditions, two cases of simulations were performed;

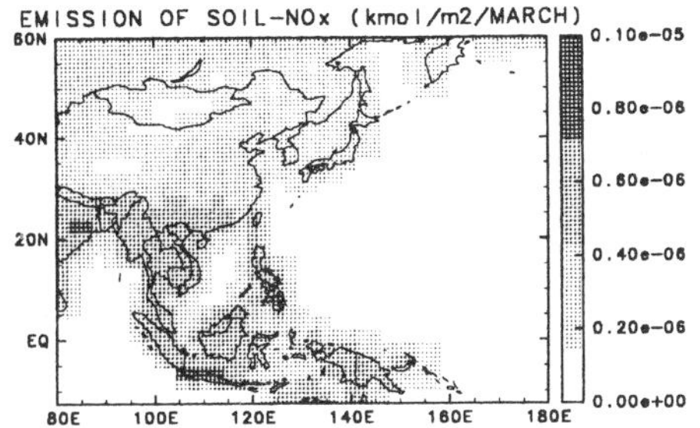

Figure 3. Soil-NOx emission in March, 1990 in kmol m^{-2} $month^{-1}$.

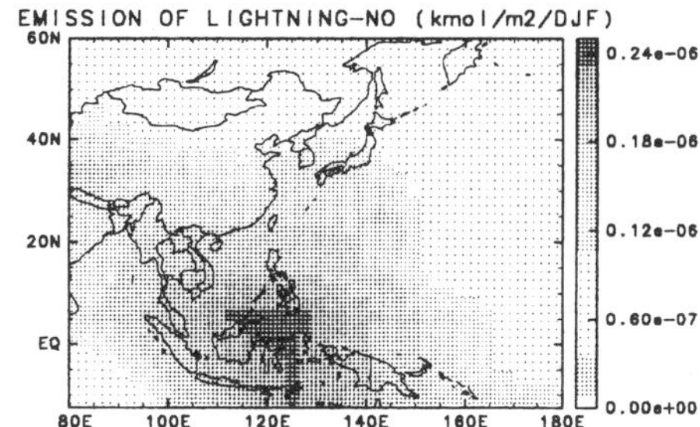

Figure 4. NOx emission, within vertical column, by lightning during 3months from Dec., 1979 to Feb., 1980 in kmol m^{-2} $3months^{-1}$.

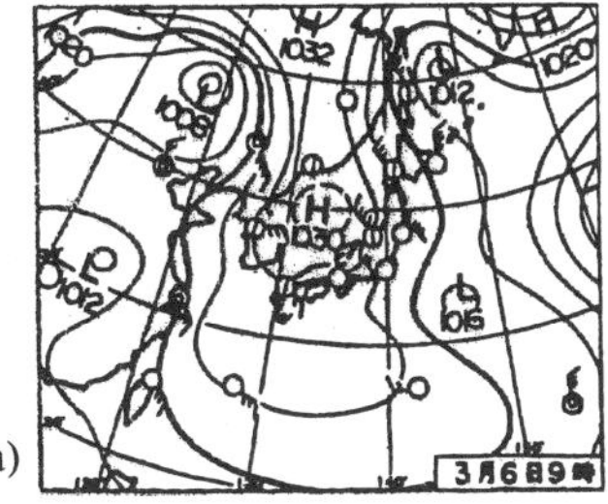

Figure 5a. Surface weather chart for East Asia at 00GMT on 6 March, 1994.

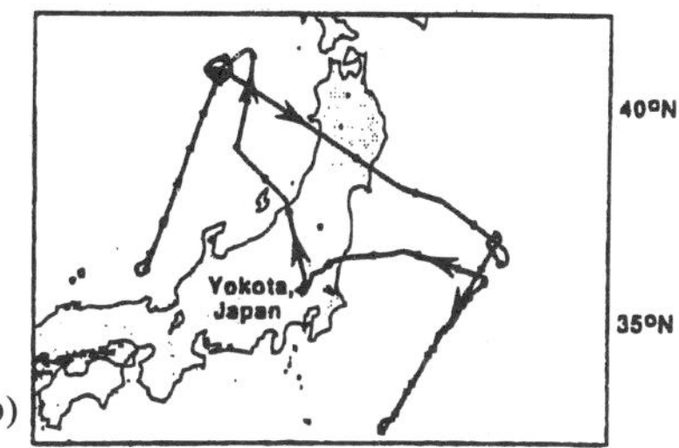

Figure 5b. Flight route of Mission15 on 6 March, 1994 (PEM-WEST:B).

the simulated time for each case was from 00GMT on 21 Feb. to 00GMT on 15 March, 1994: Case 1 with- and Case 2 without-wet deposition process.

RESULTS AND DISCUSSION

Comparison between Calculated and Observed Concentrations

During the first two weeks in March, the PEM West-B campaign launched six missions over Japan area, i.e. Flight 13 to 18. Here we will discuss results from Flight 15, which left Yokota Base at 23:30GMT on 5 March (i.e. 08:30JST on 6 March; JST, Japanese Standard Time) and returned at 07:00GMT (i.e. 16:00JST) on 6 March. Figure 5a shows the surface weather chart at 00GMT (i.e. 09JST) on 6 March, and Fig. 5b the route of Flight 15. Figure 5a indicates that the Japan area was under a typical winter weather system, i.e. an anticyclone was over the western part of the Sea of Japan with a cyclone over the northern Sea of Okhotsk. Stably stratified layer associated with the anticyclone, thus, existed over and around the Japanese islands at relatively low altitudes such as 2 km over the Sea of Japan and 3 km over the Pacific Ocean according to the temperature profile observed during the flight, though it was cloudy and weakly snowing below the stable layer on the Japan Sea side of the islands.

Figure 6 compares calculated concentrations with the observed along the flight route (see Fig. 5b): (a) NO, (b) HNO_3, (c) C_2H_4, (d) O_3, (e) SO_2, and (f) SO_4^{2-}; in Fig. 6, open square (□) denotes the DRY case (Case 2), open circle (○) the WET case (Case 1), and solid lines without symbol indicate the flight height which varied from 300 m to 11 km. Figure 6 demonstrates that the calculated concentrations well simulate those observed. It can be found in Fig. 6 that there are two clear peaks in both of observed and calculated concentrations of NO, C_2H_4, SO_2, and SO_4^{2-}, which represent chemical species with anthropogenic emission sources at surface or those chemically produced from the primary pollutants; the first peak around 26 and 27 GMT, i.e. 2 and 3 GMT on 6 March, was observed in the boundary layer over the Sea of Japan, and the second peak around 29 hr also in that over the Pacific Ocean, thus indicating the first peak was largely affected by the continental sources and the second peak similarly by both continental and Japanese sources. Observed HNO_3 in Fig. 6b does not show so remarkable peaks as SO_4^{2-} (see Fig. 6f). This observed HNO_3 might be too small since sum of the observed NO, PAN, HNO_3, and NO_2, assumed to be twice of NO concentration, was 400 to 800 pptv less than the observed NO_y, which was about 1700 ppt for the first- and 2000 ppt for the second-peaks.

The ratio of the calculated $[SO_4^{2-}]$ to $[SO_x]$ $(= [SO_2] + [SO_4^{2-}])$ was about 24 % for the two peaks in the WET-case (Case 1), while in the DRY-case (Case 2) the ratio was 24 % for the first peak at 26 hr (Fig. 6e) and 33 % for the second at 29 hr, suggesting complex effects of chemical conversion of SO_2 to SO_4^{2-} and wet deposition. The same ratios in observed data were about 40 % for those two peaks, and 53 % at altitude of 2300 m over the Japan Sea and 73 % at about 9 km high over both of the Japan Sea and the Pacific Ocean.

The O_3 concentration in Fig. 6d also has two peaks, but in contrast to other chemical species those peaks were observed at altitudes of about 9 km when the airplane was flying in lower stratosphere,

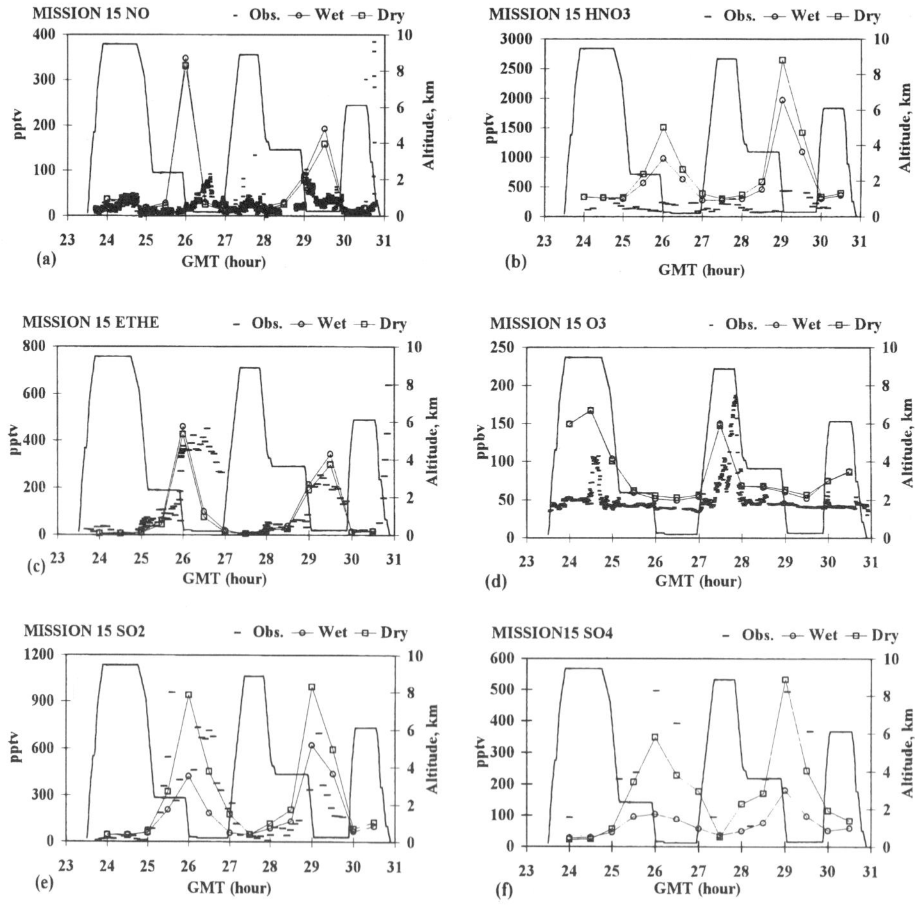

Figure 6. Comparison of the calculated concentration with the observed along Mission 15 flight route : (a) NO, (b) HNO_3, (c) C_2H_4, (d) O_3, (e) SO_2, and (f) SO_4^{2-}.

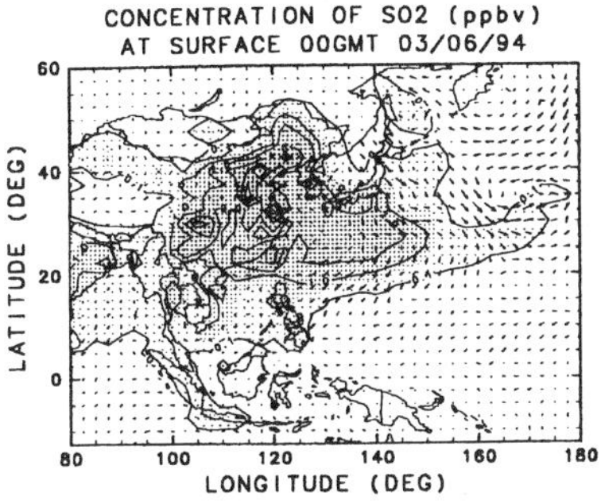

Figure 7. Horizontal distribution of calculated SO_2 in ppbv at surface level at 00GMT on 6 March, 1994.

indicating influence of stratospheric air mass. As a whole, it can be seen that the observed and also calculated concentrations tend to be high/low, when the flight height is low/high, on all species except for O_3 which does not have direct source at surface level. This suggests that there was no significant air-mass exchange between boundary layer and upper troposphere over Japan area. That was due to the stably stratified layer associated with the above-mentioned synoptic-scale weather condition, i.e., temperature inversion at and above the top of anticyclone which consists of relatively shallow cold air mass formed by radiative cooling in Siberia.

Calculated HNO_3 in Fig. 6b and SO_2 in Fig. 6e demonstrate that inclusion of the wet deposition process improves agreement between simulation and observation.

Figure 7 shows an example of the calculated SO_2 at surface level at 00GMT on 6 March, 1994.

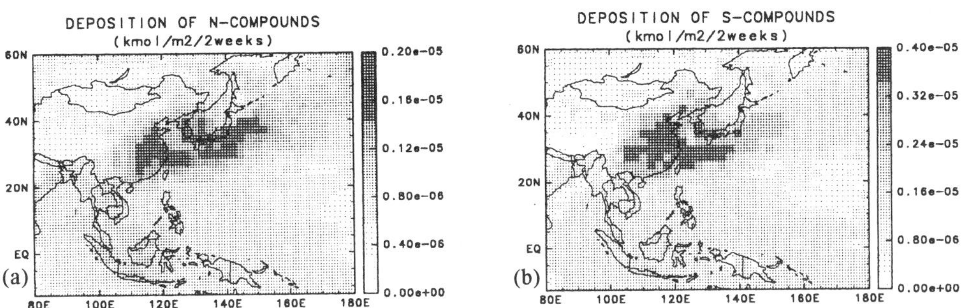

Figure 8. Calculated dry+wet deposition of (a) N- and (b) S-compounds in kmol m^{-2} $(14days)^{-1}$.

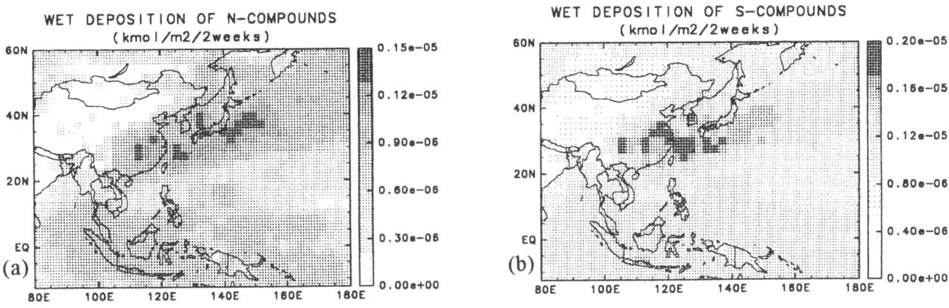

Figure 9. Same as in Fig. 8 but for wet deposition ; (a) N- and (b) S-compounds.

Spatial distributions of the deposition of N- and S-compounds

Figures 8a and b show calculated distributions of the "total" dry- plus wet-depositions of N- and S-compounds for 14 days from 00GMT on 1 March to 00GMT on 15 March, in $kmol\ m^{-2}\ (14\,days)^{-1}$, respectively. The "total" deposition includes those due to background concentrations, which were assumed, using PEM-B observation, to be 50, 25, and 10 pptv for HNO_3, SO_2, and SO_4^{2-} in the troposphere, respectively. Figures 9a and b show the "total" wet depositions of N- and S-compounds. Both N- and S-compounds depositions in Fig. 8 show extremely large values in and near anthropogenic emission source area of East Asia. However, also some difference between N- and S-depositions can be found, i.e. N-deposition in Japan area is relatively large as compared with that in China area, which is different from S-deposition map (Fig. 8b) and apparently reflects larger relative strength of Japanese N-emissions; ratio of Japanese emission to Chinese is 0.24 for NOx and about 0.05 for SOx. Wet depositions in Figs. 9a, b show similar tendency to those, i.e. dry + wet depositions, in Figs. 8a, b. N-deposition also shows its large values over tropical Southeast Asia, suggesting effects of NOx sources due to soil-microbiological activity and lightning. To see effect of deposition due to the "background" HNO_3, SO_4^{2-}, and SO_2 concentrations, an additional simulation without emission sources and with only background concentrations was performed. The simulation results indicated that contribution of "background" deposition to the "total" deposition was less than 10 % in China and Japan emission source area in both N- and S-depositions, while in the tropical Southeast Asia such as Borneo and Java islands and Malay peninsula, the background contribution in N-deposition was about 40 %, which is small compared with that in S-deposition, i.e. more than 90 % over Borneo,

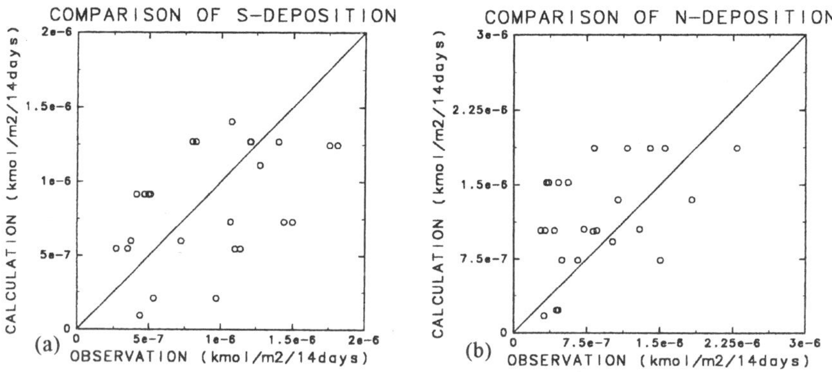

Figure 10. Calculated vs. observed depositions of (a) S-compounds and (b) N-compounds over Japan.

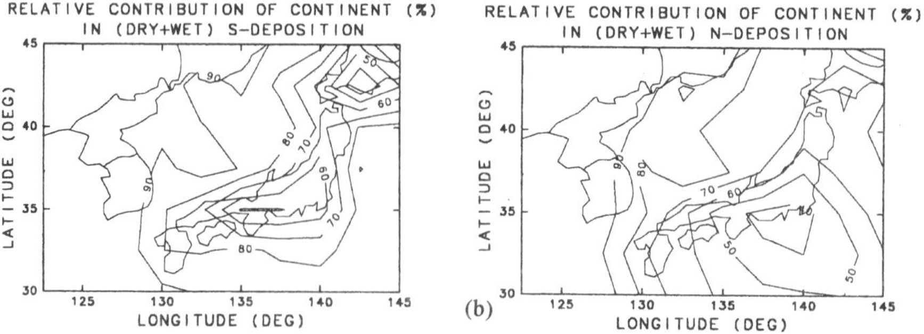

Figure 11. Contribution of the emission sources in the East Asia continent to (a) S- and (b) N-depositions, unit: %.

indicating that the N-deposition in this area is largely affected by NO_x sources other than fossil fuel combustion, i.e. lightning and soil microbiological activity.

Calculated deposition was compared with observed data (Suuri Keikaku Co., 1997) over Japan. Figures 10a and b show relations of calculated versus observed dry + wet depositions for S- and N-compounds, respectively, where the calculated depositions denote those accumulated during the simulated two weeks from 1 to 15 March, and the observed are those derived by linear interpolation of the original monthly data. Correlations between the calculated and the observed, shown in Figs. 10a, b, are acceptable though a little scattering of the points is found. This scattering may be partly due to a relatively coarse horizontal resolution in the simulation, i.e. 2.5° × 2.5°, compared with spatial scales of natural topography and change of land use type.

To investigate relative contribution of emission sources in the Asian continent to acidic deposition in Japan area, another simulation without Japanese sources was also carried out. Figures 11a and b illustrate estimated contribution, in %, of the continental sources to (dry + wet) S- and N-depositions, respectively. The figures show that the contribution of the continental sources in Japan area is larger for S-compounds than for N-compounds. For example, Fig. 11 shows that in Tokyo area less than 40 % of N-deposition is due to HNO_3 etc. produced during long range transport of air mass from emission source area in the continent while around 60 % of the deposited S-compounds was due to those transported from the continent. This reflects relatively large NOx emission in Japan as described above and also shorter life time of HNO_3 in the atmosphere compared with SO_4^{2-}.

Relative contributions of various chemical species to the dry and wet depositions of N- and S-compounds, whose horizontal distributions are shown in Figs. 8a and b, were calculated over Japan area as follows: N-deposition (50.2 % by wet deposition of HNO_3, hereafter abbreviated as "wet-HNO_3", 39.4 % by dry-HNO_3, 9.7 % by dry-NO_2, and 0.7 % by dry-PAN), and S-deposition (41.2 % by wet-SO_2, 41.9 % by dry-SO_2, 14.5 % by wet-SO_4^{2-}, and 2.4 % by dry-SO_4^{2-}). These ratios and the source contributions in Figs. 11a and b, of course, can vary depending on how dry and wet deposition processes are modeled. To examine the model processes related to acid deposition, it is necessary that observed deposition data collected over not only Japan but also large Pacific rim area are compared with simulation results simultaneously.

SUMMARY

Long range transport/transformation/deposition of various trace chemical species has been simulated over east- and southeast-Asia and the west Pacific ocean during February and March, 1994. Emission sources considered in the simulation include natural sources of NOx due to lightning, soil micro-biological activity and biomass-burning, and of SOx from volcano as well as anthropogenic sources of NOx, SOx and hydrocarbons; the anthropogenic NOx sources also include aircraft emissions. Comparison of calculated concentrations with PEM-West-B (Pacific Exploratory Mission-West, Phase B) campaign data showed good agreement for O_3, CO, SO_2 and hydrocarbons such as C_2H_4, C_2H_6 and C_3H_8, and acceptable agreement for NOx, HNO_3 and PAN. Dry and wet depositions of both S- and N-compounds were calculated and the results are discussed in terms of relative contributions of local sources to the depositions in Japan area. That deposition of N-compounds in the tropical southeast Asia is probably largely affected by NOx produced by lightning etc. was suggested.

REFERENCES

Akimoto, H., and Narita, H., 1994, Distribution of SO_2, NO_x and CO_2 emissions from fuel combustion and industrial activities in Asia with 1° x 1° resolution. *Atmos. Environ.*, 28:213.

ICAO, Committee on Aviation Environmental Protection: Working Group 3 1995, Report on the Emissions Inventory Sub-group. Bonn, Germany, June.

Kitada, T., 1987, Turbulence structure of sea breeze front and its implication in air pollution transport - Application of $k - \epsilon$ turbulence model. *Boundary-Layer Meteor.*, 41:217.

Kitada, T., 1994, Transport, transformation and deposition model for acidic species. *Meteorological Research Notes*, 182:95 (in Japanese).

Kitada, T., Isogawa, S., and Kondo, Y., 1996, Long range transport of NO_x, SO_x and O_3 over east Asia and the northern Pacific ocean caused by typhoons. *Air Pollution Modelling and Its Application XI*, Plenum Pub. Co., 191.

Kitada, T., and Lee, P.C.-S., 1993, Numerical modeling of long-range transport of acidic species in association with meso- β -convective clouds across the Japan sea resulting in acid snow over coastal Japan -II. Results and discussion. *Atmos. Environ.*, 27A:1077.

Kitada, T., Lee, P.C.-S., and Ueda, H., 1993, Numerical modeling of long-range transport of acidic species in association with meso- β -convective clouds across the Japan sea resulting in acid snow over coastal Japan -I. Model description and qualitative verifications. *Atmos. Environ.*, 27A:1061.

Kitada, T., Okamura, K., and Tanaka, S., 1997, Effects of topography and urbanization on local winds and thermal environment in Nohbi plain, coastal region of central Japan -A numerical analysis by meso-scale meteorological model with k- ε turbulence model. *J. Appl. Meteor.*, 36, to appear.

Knutson, E. O., and Stockham, J.D., 1976, Aerosol collection by snow and ice crystals. *Atmos. Environ.*, 10:395.

Kumar, P.P., Manohar, G.K., and Kandalgaonkar, S.S., 1995, Global distribution of nitric oxide produced by lightning and its seasonal variation. *J. Geophys. Res.*, 100:11,203.

Lerner, J., Matthews, E., and Fung, I., 1988, Methane emission from animals: a global high-resolution data base. *Global Biogeochemical Cycles*, 2:139.

Lurmann, F.W., Lloyd, A.C., and Atkinson, R., 1986, A chemical mechanism for use in long-range transport/acid deposition computer modeling. *J. Geophys. Res.*, 91:10905.

Matthews. E., and Fung, I., 1987, Methane emission from natural wetlands: global distribution, area, and environmental characteristics of sources. *Global Biogeo-chemical Cycles*, 1:61.

Nakanishi, H., 1996, Emission sources and concentrations of non-methane hydrocarbons over the Nohbi plain in central Japan. *Toyohashi Univ. Technology, B.S. thesis*, 42 (in Japanese).

Slinn, W.G.N., 1974, Dry deposition and resuspension of aerosol particles—a new look at some old problems. In R.J. Engelmann and G.A. Sehmel (coord.) *Atmosphere-Surface Exchange of Particulate and Gaseous Pollutants*, 1-40. Avail. NTIS, Springfield, Virginia as CONF-740921.

Slinn, W.G.N., 1977, Some approximations for the wet and dry removal of particles and gases from the atmosphere. *Water, Air, and Soil Poll.*, 7:513.

Strand, A., and Hov, O., 1993, A two-dimensional zonally averaged transport model including convective motions and a new strategy for the numerical solution. *J. Geophys. Res.*, 98:9023.

Washington, W.M., and Williamson, D.L., 1977, A description of the NCAR global circulation models. In *Methods in Computational Physics, Vol. 17: General Circulation Models of the Atmosphere*, Academic Press, 111-265.

Xie, P., and Arkin, P.A., 1995, An intercomparison of gauge observations and satellite estimates of monthly precipitation. *J. Appl. Meteor.*, 34:1143.

Yienger, J.J., and Levy II, H., 1995, Empirical model of global soil-biogenic NO_x emissions. *J. Geophys. Res.*, 100:11,447.

DISCUSSION

A. KEIKO: How precise are the data on NO_x soil fluxes you use?

T. KITADA: Although global estimates on the nitric oxide emissions from soils include potentially large uncertainties, especially on those from tropical savanna and agricultural soils receiving N fertilizers, recent global modeling exercises have shown a global soil source of 5-10 TgN yr^{-1} (Davidson and and Kingerlee, 1996). In our study one of the recent estimates by Yienger and Levy (1995) has been used, which gives 5.5 TgN as best estimate for the annual emissions. We are, of course, ready for sensitivity simulations to look at effects of soil-NO_x emission on tropospheric chemistry.

Reference: Davidson, E.A., and Kingerlee, W., 1996, Global inventory of emissions of nitric oxide from soils: A literature review. In abstract volume: Inter. Workshop on NO_x Emission from Soils and its Influence on Atmos. Chemistry, Tsukuba, Japan, March 4-6, 1996.

P. SEIBERT: Why did you use such a coarse resolution, not only in space but also in time? With 12 h time resolution it is difficult to capture the diurnal cycle of the boundary layer.

T. KITADA: We have used ECMWF Basic Consolidated Data Set, which has relatively coarse resolutions, because of two resources-related reasons: first, the data is much cheaper than the other ECMWF data with finer resolutions in space and time; secondly, available CPU resources for us had been limited when we did the simulations.

In the simulations diurnal boundary layer activity was taken into account by using formula which describe diurnal variations of vertical profiles of vertical eddy diffusivity, and was derived from outputs of our k-epsilon turbulence model.

THE TRANSPORT OF DUST AND SO_X IN EAST ASIA DURING THE PEM–B EXPERIMENT

Hui Xiao[1], Gregory R. Carmichael, and James Durchenwald

Center for Global & Regional Environmental Research and the Department of Chemical & Biochemical Engineering, University of Iowa, Iowa City, IA 52240 USA
[1]Present Address, Institute of Atmospheric Physics, Chinese Academy of Sciences, Beijing, 100029 China

INTRODUCTION

The subject of long range transport (LRT) of pollutants is in the early stages of study in Asia (Merrill et al., 1985; Kotamarthi and Carmichael, 1990; Arndt et al., 1996). The results from the PEM West A & B experiments have shown the widespread impact that long range transport of materials from the continental regions of east Asia has on the chemical composition of the troposphere over the western Pacific (Hoell et al., 1996). The projected growth in emissions for the region suggests that the long range transport of pollutants in east Asia will grow in importance over the next several decades.

There is growing evidence that chemical reactions on mineral aerosol surfaces may be an important pathway for sulfate and nitrate formation (Zhang et al., 1994). Observational data support the formation of sulfate and nitrate on mineral aerosol. For example, single particle studies on Asian aerosols using spot tests and TEM clearly show the mineral aerosol to be coated with sulfates and nitrates (Purungo et al., 1995). Recent measurements in the eastern Mediterranean also indicate that mineral aerosol is coated with sulfate and that this coating makes the aerosol effective CCN (Levin et al., 1996). In addition, analysis of measurements in the plumes of the Kuwait oil fires found SO_2 oxidation rates of 6 to 8%/hr, with the highest rates associated with reactions on mineral aerosol (Herring et al., 1996). The importance of dust as a reaction surface has recently been evaluated using a global three-dimensional model of the troposphere by Dentener et al. (1996). They found the reaction of SO_2 with calcium-rich mineral aerosol to play an important role down-wind of arid source regions. This is especially important for regions in Asia, which are important emitters of sulfur compounds, and where future emissions are expected to increase dramatically.

The PEM-WEST-B experiment provided an opportunity to investigate further the interactions between dust and the sulfur cycle in east Asia. In this paper we apply a three-dimensional tropospheric chemistry model to analyze the long-range transport of sulfur oxides and dust in east Asia during the period of March 1994.

MODEL DESCRIPTION

The long range transport of SO_x (SO_2 + sulfate) and dust in east Asia was investigated by use of the STEM-II regional-scale model. The STEM-II model is a three-dimensional, Eulerian numerical model which accounts for the transport, chemical transformation and deposition of atmospheric pollutants (Carmichael et al., 1991). The model has been used quite extensively for scientific studies and policy evaluations in the Eastern United States and the Pacific Rim region (Chang et al., 1989; Kotamarthi and Carmichael, 1990, 1993).

Model Domain

The study domain was 100°E to 160°E and 20°N to 55°N (shown in **Figure 1**) and included all of Japan, North and South Korea, most of China, and parts of Russia and Mongolia. This region has dramatic variations of topography, land type, and mixtures of industrial/urban centers and agricultural/rural regions. In addition, the interactions between continental and marine influences play a prominent role in determining the effects of pollutant production and transport in this region. The horizontal grid spacing was 1° by 1°, and the vertical domain covered from the surface to 10 km with a vertical resolution of 500 m. In total ~46,000 spatial grid points were used for the calculations.

Input Data

Emissions. The SO_x emissions utilized in this study were those developed under the RAINS-ASIA (Regional Air Pollution Integration Study) project (Arndt et al., 1996; Arndt and Carmichael, 1996; Foell et al., 1996). Emissions from both natural and anthropogenic sources were included. Natural sources consist of the active volcanic sources in the region. Fujita and Takahashi (1994) studied volcanic emissions in and around Japan by compiling information on the number of small releases and explosions each year from 1977 through 1989, and measuring SO_2 releases at the 12 largest volcanoes for a period of one year. For the largest volcano, Sakurajima (the one most relevant to this study), the number of explosions averaged 20 to 30 per month, with an annual variation in emissions of only 25%. They estimate that the annual SO_2 emission from volcanoes equal those arising from Japan's anthropogenic activities, and account for ~3% of the total SO_x emissions in the region. These estimates suggest that active volcanoes in Japan provide a persistent contribution to ambient sulfur levels in and around Japan. In this study, we used the estimates of Fujita and Takahashi (1994) to provide daily emission rates. Anthropogenic sources are comprised of regional shipping activity emissions, area surface emissions, and elevated sources. Shipping emissions include both emissions from regional shipping lanes and port activities. Surface sources include industrial activity, domestic activity and transportation emissions. Elevated source emissions associated with large point sources (referred to as Large Point Sources or LPSs) were also included. Approximately 150 LPSs were included in the analysis, and account for ~25% of the total SO_x emissions in the study domain. The area and shipping emissions were emitted at the surface, while the LPS emissions were released at 0.5 km. It was assumed that 5% of the SO_x emitted was in the form of sulfate. Volcanic emissions were released at levels 0.5,1 and 1.5 km.

Meteorological Data. The three-dimensional meteorological fields needed by the model were obtained from ECMWF analyzed winds. This database provides the horizontal winds in both the zonal and meridonal directions and the vertical values at designated atmospheric pressure intervals. The data were available on a 2.5^o resolution, and were interpolated to 1° grids by first interpolating the horizontal fields, and then recalculating the vertical velocities using the divergence theorem in order to avoid mass balance inconsistencies that are likely when interpolating the wind velocities from the original pressure grid to our constant-spacing vertical grid system. These data are available every 12 hours.

Dust Source and Transport. Two large dust source regions were considered in this study: the Gobi desert and the loess areas in the upper stream basin of the Huang river (Chang et al., 1996). The Gobi desert is located from 39°- 45° N and 100°-110° E with a total estimated area of 1 x 10^6 square kilometers. The loess region extends from 34° to 41° N and 102° to 114° E with an area of 8 x 10^5 square kilometers.

One major problem regarding long range transport of dust is how to calculate the emission rate from the sources into the atmosphere. In this work, we utilized a method

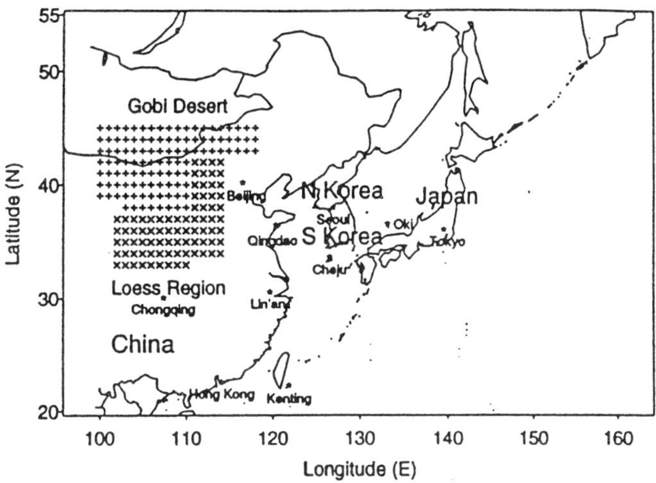

Figure 1. Study domain and location of the major dust emission regions.

which assumes a threshold velocity at which dust is initially produced by the wind erosion. The introduction of the threshold velocity establishes a relationship between dust production and the wind field data. Based on the actual wind field, dust emission can be estimated using the following formula (Gillette and Passi, 1988; Gillette et al., 1992), and discribed in detail in Xiao et al., 1997.

The transport scheme for dust is composed of convection and turbulent diffusion which are set to be the same as the trace gas species. The transport equation for dust can be described as follows:

$$\frac{\partial Q_i}{\partial t}+\nabla\bullet(\mathbf{u}\, Q_i)=\nabla\bullet(\mathbf{K}\bullet\nabla Q_i)+\left(\frac{\partial Q_i}{\partial t}\right)_{cond./evap.}+\left(\frac{\partial Q_i}{\partial t}\right)_{coag.}+\left(\frac{\partial Q_i}{\partial t}\right)_{reac.}+\left(\frac{\partial Q_i}{\partial t}\right)_{source/sinks} \quad (1)$$

$$i=1,\ldots,5$$

where **u** is the wind velocity vector, **K** is the atmospheric eddy diffusivity tensor, $(\partial Q_i/\partial t)$source/sinks is the rate of change of the aerosol mass of size i, Q_i, due to nucleation production, primary aerosol emission and removal, and $(\partial Q_i/\partial t)_{cond./evap.}$, $(\partial Q_i/\partial t)_{coag.}$ and $(\partial Q_i/\partial t)_{reac}$ .are the rates of change due to condensation/evaporation, coagulation and surface aqueous reactions, respectively. In this study, five size fractions of dust were modeled ranging in size from submicron to 20 microns. The coagulation and condensation parameters were calculated as discussed in Zhang et al. (1994).

SO_x Chemistry on Dust. The interaction of SO_x with dust is complicated and involves the sorption of SO_2, followed by an oxidation reaction of SO_2 or associated anions. In this study we assumed that the reaction of SO_2 with dust is diffusion-limited. Under this assumption the pseudo-first order rate coefficient k_{SO_2} [s-1] which describes the net removal rate of SO_2 on the aerosol surface is given by:

$$k_{SO_2} = \int_{r1}^{r2} k_{dSO_2}(r)n(r)dr \quad (2)$$

where n(r)dr [cm-3] is the number density of particles between r and r + dr, and k_{SO_2} is the size dependent mass transfer coefficient [cm^3s^{-1}] calculated using the Fuchs and Sutugin (1970) interpolation equation.

The reaction of the sorbed SO_2 may proceed via metal-sulfito-complexes, with HO_2 and SO_3^- as chain carriers (Beilke and Gravenhorst,1978). In addition, SO_2 may be oxidized by ozone (Maahs, 1983). This reaction is strongly pH dependent, and for pH>8, in the presence of water, the oxidation of SO_2 is sufficiently fast to make heterogeneous reaction of SO_2 on mineral aerosol essentially diffusion limited.

The reaction rate is also strongly dependent on the reaction probability, g(SO_2). Judeikes et al. (1978) measured reaction probabilites on various materials like Fe_2O_3, fly ash and soot, ranging from 10^{-6} to 10^{-3}. Reaction probabilites decreased during prolonged exposure, but much less so at high relative humidities (RH). In the atmosphere, at least in areas removed from the dust source areas, relative humidities in the boundary layer are generally higher than 50%. Hanel (1976) showed that Saharan dust at RH>50% takes up significant amounts of water, making the SO_2 oxidation by O_3 or radical reactions likely.

The uptake of SO_2 is favored when the alkalinity from the dust aerosol exceeds the acidity from the dust-associated H_2SO_4 and HNO_3 aerosol. The calcium content of soils in the arid regions of Asia range from 4-8 % (by weight) (Wang and Wang, 1995; Wang et al., 1990). In Dentener et al. (1996) the impact of the calcium content of the aerosol on the sulfate conversion rate was evaluated and found not to be a limiting quantity. In this study we used a constant reaction probability of 0.005 (and sensitivity studies using 0.001).

In addition to the heterogeneous oxidation of SO_2 on mineral aerosol, we included the gas phase reaction with the OH radical. The gas phase reaction rates were paramaterized from previous model studies (Dentener et al., 1996). The values averaged 0.65%/hr at latitudes below 25°N, 0.25%/hr at 40°N and 0.1%/hr above 50°N. We assumed that the H_2SO_4 produced in the gas phase was either forming new particles, which can go to the dust via coagulation, or was directly condensing onto the mineral surface. In addition to the gas phase and mineral aerosol pathways, sulfate can also be formed by reactions in cloud droplets. In this study we used a parameterization which is first order in the fractional cloud cover and the ambient SO_2 concentration, with a maximum value of 1%/hr at 100% cloud cover.

The implications of the above chemical mechanism for sulfate formation on mineral aerosol should be noted. The mechanism provides for both direct reactions on the surface as well as condensation and coagulation of sulfate formed in the gas phase. This allows for sulfate to be formed on both the small and large particles. Size resolved data in Asia (Horai et al., 1993) supports the fact that sulfate is found in both the accumulation mode and on the mineral aerosol surfaces. They found the fraction of sulfate associated with the mineral aerosol to vary from 10 to 90%. Recent analysis of the aerosol at Cheju, shows 10 to 20% of the sulfate mass to be associated with the coarse particle mode (Chen et al., 1996).

The PEM-WEST-B Period. The focus of this study is the period of March 1 through March 14, 1994, during which the PEM-WEST-B aircraft missions were performed around Japan. This period was characterized by strong continental outflows associated with cold fronts which moved through the study domain from west to east. On March 3, low pressure systems passed over the Gobi Desert and Loess region, bringing strong surface winds which resulted in the first dust emission. On March 5, a new and strong low pressure system formed over the northwest of Gobi Desert and moved first southeasterly and then turned to the northeast. This low pressure system resulted in the dust release from the Gobi Desert on March 6. In front of the low pressure system, there existed a high pressure system located over the Japan Sea, which blocked westerly flows. On March 8 through March 11, a stronger low pressure system, accompanied by a stationary front, dominated most areas in east Asia. The center of this system was located to the north of Japan and wind speeds of ~ 18 m/s, were observed on the backside of the front. On March 10, a high pressure center formed over east China and moved to the east, bringing with it fair conditions. Further meteorological details during this time period are discussed in Merrill et al., (1996).

RESULTS AND DISCUSSIONS

Only one representative result is presented in this paper. Mission 16 took place on March 7, with a flight out of Tokyo along a path to the east-southeast. This is a most interesting case as this is the time period during which the long range transport of dust occurred. The dominant meteorological feature on this day was the high pressure system centered south of Tokyo. During this period the flows in the lower and mid- to upper troposphere were distinctly different. The SO_X in the lower layers in the southeast corner of the domain were influenced by the flows associated with the bottom side of high pressure system. SO_X rich air had been transported to this region during the previous days, and now was being transported to the west by the high pressure system. Air in the mid- to upper-troposphere in this region resulted from transport around the high pressure system, and had elevated levels of dust and SO_X. Dust and SO_X are transported into the mid troposphere in the frontal region, and subsequently transport around Japan. Through these processes dust and SO2 are mixed together, enabling chemical interactions .

The impact of the reactions with the dust on the sulfate distribution is presented in Figure 2 . Shown are the results for simulations without dust, with dust (dust-events only), and the percent influence of the dust reactions. The largest impacts are calculated over northern China. It is in this region that the dust-rich air and the SO_2-rich air come together in the frontal region on the backside of the high pressure system. The dust interactions account for greater than 50% of the sulfate production. In the region to the southeast of Japan the dust reaction accounts for 30 to 50% of the SO_2 conversion in the middle troposphere, and 10 to 20% in the lower troposphere.

The calculated vertical distribution of SO_2 and sulfate were compared with the aircraft observations. The outbound leg of the flight at an altitude of ~9 km measured SO_2 and sulfate levels less than ~100 ppt. On the return leg of the flight, from 160°E to 140°E, the

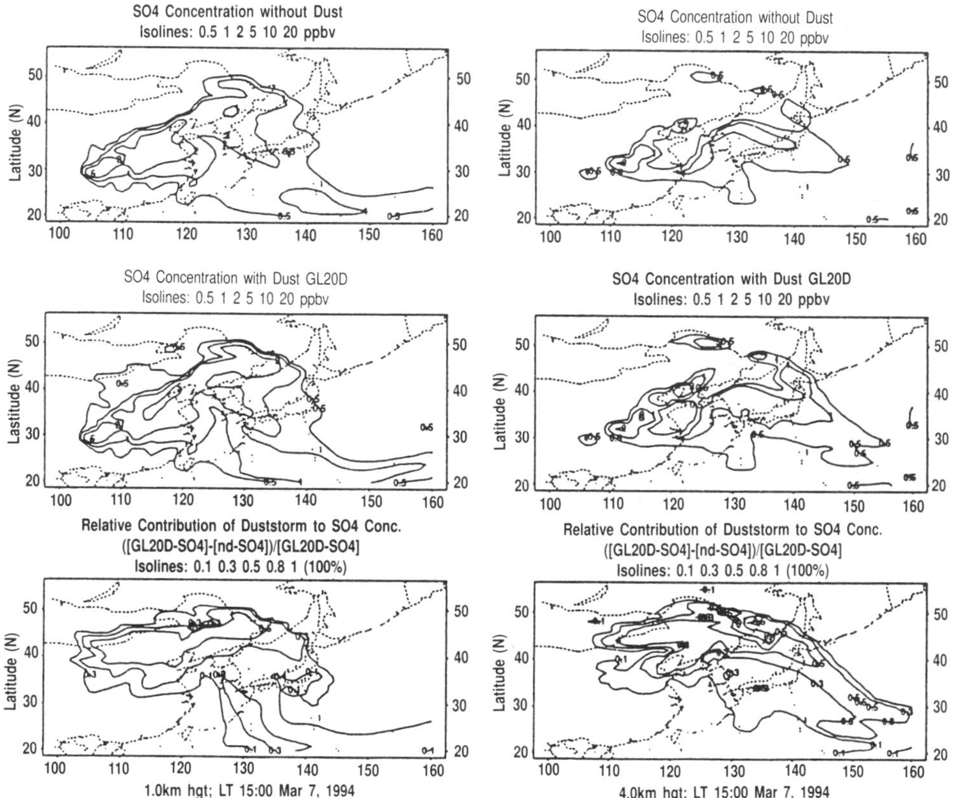

Figure 2. Calculated sulfate concentrations at 1 and 4 km, on March 7, 1994. Shown are values calculated with no-dust, with reactions on the dust-event aerosols (no background), and the fraction of total sulfate due to the dust reactions.

aircraft sampled air at altitudes of 0.5 to 4 km. At the lowest altitudes the aircraft was sampling air characteristic of the low-level continental outflow, discussed above. This air mass had been in the region for approximately 1 day, and the SO_2 levels were depleted largely by dry deposition to the sea surface.

The air sampled above ~2 km was the result of the air transported at higher altitudes around Japan, and then subsiding in this region. The measurements and the calculations show this air to contain elevated levels of aerosol and SO_2. Dust concentrations in excess 15$\mu g/m^3$ were calculated in the region of the flightpath, with the highest levels at altitudes between 2 to 5 km, and between 150°E and 160°E. These predictions are consistent with the aircraft observations. Elevated levels of aerosol calcium were sampled by the aircraft (Pueschel et al., 1996), with calcium levels higher than the sulfate levels, and ranged from 300 to 600 ppt at altitudes of 3 to 6 km. Assuming a calcium content of dust of 4 to 8% percent by weight, implies a mineral aerosol values of ~ 10 to 20 $\mu g/m^3$. Furthermore, the lidar measurements of Browell et al.(1996) showed an elevated layer of aerosol between 2 to 6 km, consistent with the location of the calculated layer of elevated sulfate and dust. Elevated particle levels were also detected by Pueschel et al. (1996).

The general influence of dust on the distribution of SO_x throughout the study period is shown in Figure 3 where boundary-layer, 14 day average quantities of sulfate are plotted for conditions with and without dust. These results are for a threshold friction velocity of 20 cm/s and an accommodation coefficient of 0.005. One difficulty in accessing the role of dust reactions is deciding what to do with background aerosols. Asia is characterized by high concentrations of mineral aerosols, and a background level of 5 mg/m^3 to 10 mg/m^3, is consistent with observations in the region (Chen et al., 1996). These levels provide sufficient surface area for significant reaction rates. To illustrate this point, results are presented for conditions where the sulfur reactions were allowed to occur on both the background dust and the dust due to the dust-events, and the case where reactions occurred only on the dust-event surfaces. As shown in Figure 3, the dust reactions account for a significant fraction of the conversion of SO_2 to sulfate. Over China sulfate levels are doubled in the presence of dust. Throughout eastern Asia the dust pathway accounts for more than 40% of the sulfate production. The fraction of sulfate due to the dust events accounts for the bulk of the sulfate production over China, while over east Asia, these events have a 20% effect or more over broad regions. Reactions on the background aerosol caused an increase in sulfate formation by 40% throughout the eastern parts of the study domain. The fraction of sulfate calculated on the background aerosol is perhaps too high, and this suggests that the sulfur reactions on the mineral aerosol may be regulated. We discussed previously the possibility that the reactions take place most rapidly at high pH, and slow down after the strong acids in the aerosol exceed the calcium content as discussed in Dentener et al., 1996. This subject requires further study. In the remainder of the paper we focus our attention on the reactions on the dust-released surfaces.

SUMMARY

The transport of SO_x and mineral aerosol in east Asia during March 1 through 14, 1994 was studied using the STEM-II regional scale atmospheric chemistry model. This period was characterized by continental outflow associated with a series of cold fronts which passed through the region. During this period the regional sulfur cycle is dominated by transport out of the model domain, predominately through the western boundary, and by removal by dry deposition. This transport is largely limited to the lower ~4km of the atmosphere and the maximum flux occurs in the latitude band of 30 to 40°N, coincident with the major source regions in east Asia. Approximately 65% of the sulfur emitted in east Asia is transported out of the domain, with approximately equal amounts of SO_2 and sulfate.

The model results were compared with measurements made during the PEM-WEST-B experiment. The predicted values were found to capture much of the observed horizontal and spatial variability. The calculated results showed that Mission 14 may have sampled a volcanic plume from Sakurajima. Volcanic emissions from Japan while accounting for ~3% of the total sulfur emissions in the region can have a significant influence in and around Japan.

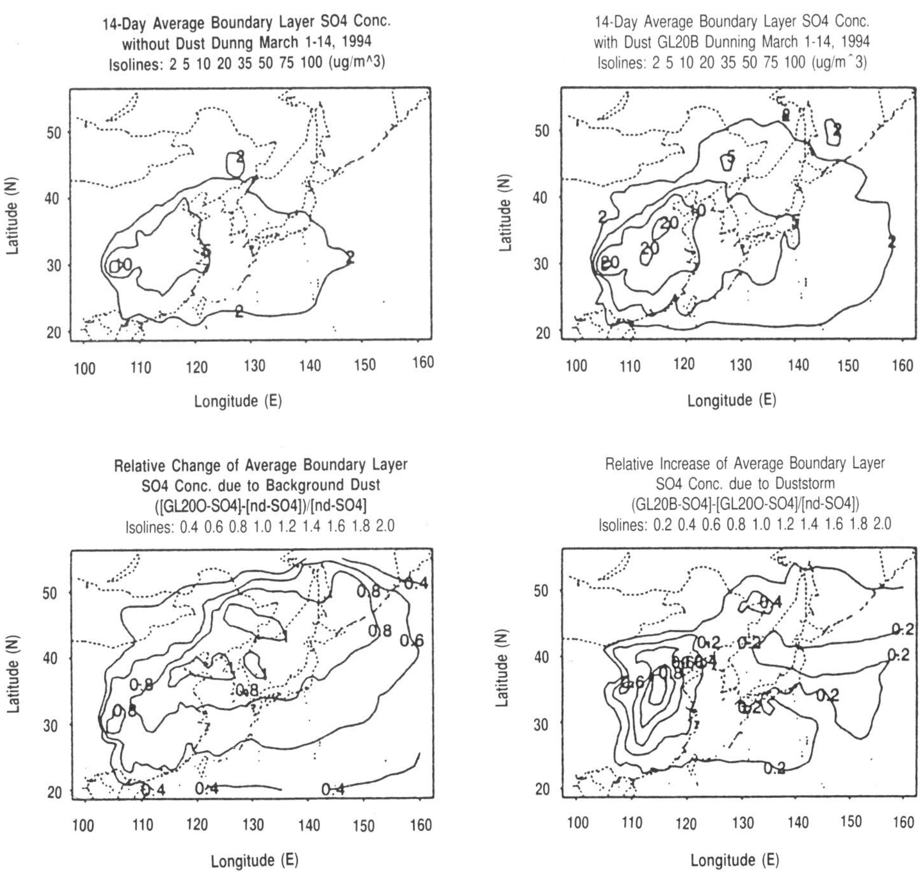

Figure 3. The effect of dust reactions on the predicted sulfate concentrations. Shown are the 14-day averaged boundary-layer values for simulations with and without dust reactions. The with-dust values include the effect of reactions on background aerosol. The fraction of increase in sulfate (relative to the no-dust case) for the simulations which include background dust, and when only the dust-release cases are also shown.

The role of mineral aerosol as a reactive surface for sulfate formation was also studied. The mineral aerosol, rich in calcium and iron oxides, may be an important reaction surface. Model results indicate that this pathway may account for 20 to 40% of the sulfate formed in the region during this period. The limited surface observations in the region support these findings. Further work is necessary to better quantify the importance of sulfate production in the presence of mineral aerosol, and to evaluate the importance of these surfaces in nitrate formation, and NO_X - O_3 chemistry.

ACKNOWLEDGMENTS

This research was supported in part by NASA (grant # NAGW-2428) and NOAA (grant # NA46GP0121). Special thanks to the European Center for Medium-Range Weather Forecasting for use of the meteorological fields, and to the Royal Observatory of Hong Kong for the preparation of the visibility data.

REFERENCES

Arndt, R. and G. Carmichael, Long range transport of sulfur in Asia, Water, Air and Soil Pollution, 85: 2283-2288, 1996.

Arndt, R., G. Carmichael, N. Bhatthi and D. Streets, Transport and deposition of sulfur in Asia , in press to *Atmos. Environ.* , 1996.

Beilke, S., and G. Gravenhorst, Heterogeneous SO2 oxidation in the droplet phase, *Atmos. Environ. 12*, 171-177, 1978.

Carmichael, G.R., L.K. Peters, and R.D. Saylor. The STEM-II regional scale acid deposition and photochemical oxidant model: I. An overview of model development and applications , *Atmos. Environ., 25A*, 2077-2090, 1991.

Chang, Y. S., R. Arndt, and G. R. Carmichael. Mineral base-cation deposition in Asia, *Atmos. Environ.*, 30: 2417-2427, 1996Chang, Y.S., G.R. Carmichael, H. Kurita, and H. Ueda, The transport and formation of photochemical oxidants in Central Japan", *Atmos. Environ., 23,* 363-393 ,1989.

Chen, L.L., G. R. Carmichael, M-S Hong, H. Ueda, S. Shim, C. Song, Y. Kim, R. Arimoto, J. Prospero, D. Savoie, K. Murano, J. Park, H. Lee, and C. Kang., Analysis of ground measurements at Cheju Island, South Korea, *J. Geophys. Res. PEM-B special issue*, submitted, 1996.

Denterner, F. , G. Carmichael, Y. Zhang, P. Crutzen, J. Lelifeld, The role of mineral aerosol as a reactive surface in the global troposphere, *J. Geophys. Res.*, in press, 1996.

Foell W., C. Green, M. Amann, G. Carmichael, J..Hettelingh, L. Hordick, D. Streets, Energy use, emissions, and air pollution reduction strategies in Asia, *Water, Air and Soil Pollution, 85*, 2277-2282, 1996.

Fujita, S., and A. Takahashi, *Acidic Deposition in East Asia*, Central Institute of Electric Power Industry, report # T93091, 35 pages, 1994.

Gillette, D., and R. Passi, Modeling dust emission caused by wind erosion, *J. Geophys. Res., 93*, 14233-14242, 1988.

Gillette, D.A., G.J. Stensland, A.L. Williams, W. Barnard, D. Gatz, P.C. Sinclair, and T.C. Johnson, Emissions of alkaline elements calcium, magnesium, potassium, and sodium from open sources in the contigeous United States, *Global Biogeochem. Cycl., 6*, 437-457, 1992.

Hanel, G., The properties of atmospheric aerosol particles as functions of relative humidity at thermodynamic equilibrium with the surrounding moist air, *Adv. Geophys., 19*, 73-188, 1976.

Herring, J. , R. Ferek, and P. Hobbs, Heterogeneous chemistry in the smoke plume from the 1991 Kuwait oil fires, *J. Geophys. Res.*, in press, 1996.

Hoell, J., D. Davis, S. Lui, R. Newell, H. Akimoto, R. McNeal, and R. Bendura, The Pacific Exploratory Mision -West Phase B. *J. Geophys. Res., PEM-West-B first issue*, submitted, 1996.

Horai, S., T. Minari, and Y. Migita, Aerosols Coposition in Kagoshima, Annual Report of the Kagoshima Prefectural Institute, Vol. 9, 1993.

Judeikes, H.S., T.B. Stewart, and A.G. Wren, Laboratory studies of heterogeneous reactions of SO2, *Atmos. Environ., 12*, 1633-1641, 1978.

Kotamarthi, V. and G. R. Carmichael, The Long Range Transport of Pollutants in the Pacific Rim Region, *Atmos. Environ., 24A*, 1521-1534, 1990.

Kotamarthi, V., and G. R. Carmichael, A modeling study of the long range transport of Kosa using particle trajectory analysis, *Tellus, 45B*, 426-441, 1993.

Levin, Z., E. Ganor, and V. Gladstein, The effect of desert particles coated with sulfur on rain formation in the eastern Mediteranean, *J. Atmos. Sci.*, in press 1996.

Maahs, H.G., Kinetics and mechanisms of the oxidation of S(IV) by ozone in aqueous solution with particular reference to SO2 conversion in nonurban clouds, *J. Geophys. Res., 88*, 10721-10732, 1983.

Merrill, J. R. Bleck and L. Avita, Modeling atmospheric transport to Marshall Island,*J. Geophys. Res., 90*, 12927-12936, 1985.

Merrill, J., R. E. Newell and A. S. Bachmeier. A metereological overview for the Pacific Exploratory Mission - West, Phase B, *J. Geophys. Res.,* submitted, 1996.

Parungo, F., Y. Kim, C-J Zhu, J. Harris, R. Schnell, X-S Li, D-Z Yang, M-Y Zhou, Z. Chen and K. Park, Asian dust storms and their effects on radiation and climate, STC report 2906, 1995.

Pueschel, R., et al., Physical, chemical and optical properties of western Pacific free tropospheric aerosols: effects of continental outflow, *J. Geophys. Res., PEM-West-B first issue,* submitted, 1996.

Wang, W., and T. Wang, On the origin and the trend of acid precipation in China, *Water Air & Soil Pollution*, 85, 2295-2300, 1995.

Wang, X., G. Zhu, and X. Shen, Some characteristics of the aerosol in Beijing, *China J. Atmos. Sci., 19*, 211-215, 1990.

Xiao, H., G. carmichael, J. Durchenwald, D. Thornton, and A. Bandy, Long-range transport of Sox and dust in east Asia during the PEM-WEST-B experiment, *J. Geophys. Res.*, in press 1997.

Zhang, Y., Y. Sunwoo, V. Kotamarthi, and G. Carmichael, Photochemical oxidant processes in the presence of dust: an evaluation of the impact of dust on particulate nitrate and ozone formation, *J. Applied Met.*, 33, 1994.

DISCUSSION

H. VAN JAARSVELD: How can you be sure that the sulphate and nitrate you found on mineral aerosol is not simply associated with the dust material in the source regions?

G. CARMICHAEL: We have a variety of evidence. Indeed the soil material contains some sulphate and lesser amounts of nitrates. However when we calculate enrichment factors based on the chemical composition of the soils, we find that the aerosol in east Asia (e.g. at Cheju Island), is several orders of magnitude higher than that in the soils. In addition when these aerosols are put under a electronmicroscope we see that they are coated in a film of sulfate and nitrate.

R. YAMARTINO: Given that a large fraction of sulphate is tied up with Ca rich mineral aerosol doesn't that imply a buffering of acidic deposition?

G. CARMICHAEL: Yes, that is an important point. This is readily apparent in the pH values in the high dust regions in Asia. The precipitation pH is other greater than 6 eventhough the sulphate levels are as high as those measured in N. America and Europe. This not only effects precipitation pH but also soil chemistry, where the strong mineral bases can neutralise the strong mineral acids. However, our model results and measurements both indicate that sulphur and nitrogen emissions are increasing at a sufficient rate to "break-through" the buffering capacity of the mineral aerosol. Many regions have already passed this point. In our analysis of impacts we take these considerations into effect.

A SIMULATION OF LONG-RANGE TRANSPORT OF CFCS IN THE TROPOSPHERE USING A 3-D GLOBAL LAGRANGIAN MODEL WITH 6-HOURLY METEOROLOGICAL FIELDS

W.J. Collins, D.S. Stevenson, C.E. Johnson
and R.G. Derwent

Meteorological Office,
Bracknell,
Berkshire,
RG12 2SZ, U.K.

INTRODUCTION

Throughout this century the composition of the troposphere has been perturbed by anthropogenic emissions. These emissions are not uniformly distributed over the globe but are concentrated in the northern mid-latitudes in industrial regions. Biogenic emissions are largely emitted from the land and so too are biased towards the northern hemisphere. A modelling study of the global chemistry of the troposphere requires knowledge of how pollutants are carried away from their mainly continental source regions to the remote atmosphere and the southern hemisphere.

CFCs are excellent tracers to study and validate the transport characteristics of chemistry models as their sources and sinks are well known. Their atmospheric lifetimes are tens to hundreds of years which makes them ideal to study intercontinental and interhemispheric transport particularly as there are comprehensive records of observational data (Cunnold *et al.* 1994) from all over the globe for the last two decades.

We have developed an offline Lagrangian global chemistry model STOCHEM which advects 50,000 constant mass parcels in three dimensions within the troposphere, driven by 6-hourly global meteorological fields output from the UK Meteorological Office operational forecast model. This enables full simulation of transport processes down to the synoptic scale in our chemistry model, whereas physical processes such as boundary layer depth and convective mixing are parameterised.

The STOCHEM model has been developed to simulate global tropospheric photochemistry but in this work we use our model to simulate the emission and transport of CFC11 ($CFCl_3$), and by comparison with observational data to obtain characteristic timescales for interhemispheric transport.

MODEL DESCRIPTION

So far the main approach to modelling three-dimensional tropospheric chemistry has been Eulerian where the accurate representation of the advection of trace gases is not straightforward if numerical dispersion and short timesteps are to be avoided (Chock and Winkler 1994; Dabdub and Seinfeld 1994). Pseudospectral techniques offer a formally accurate alternative to the conventional finite difference approach in models of atmospheric dynamics. However, when applied to atmospheric trace gas transport, they may generate negative concentrations and spurious oscillations. Lagrangian models guarantee tracer conservation by definition, and eliminate negative concentrations and unwanted numerical diffusion (Walton *et al.* 1988; Taylor 1989). The disadvantages are that species concentrations are defined on parcel centroids but output is generally required on an Eulerian grid and this may be over- or under-determined, also distortions due to wind shear can render the notion of a distinct air parcel meaningless.

The STOCHEM model has been described elsewhere (Collins *et al.* 1997; Stevenson *et al.* 1997; Stevenson *et al.* submitted). Since then there have been improvements to the time resolution of the meteorological data and to the advection scheme. In this study 50,000 constant mass parcels of air are advected by interpolated winds from the United Kingdom Meteorological Office Unified Model (UM) general circulation forecast model (Cullen 1993). As well as the 3-D wind fields, we also use temperature, surface pressure and convective cloud information (fractional cover, height of base, height of top, precipitation rate).

The global UM is run 4 times a day at the Meteorological Office to provide weather forecasts. For use in our chemistry model we have been archiving the data from the assimilation step (T+0 forecast), this is based on the T+6 hour forecast from the previous UM run but constrained to give the best agreement with all the global observations made in a specified time period, thus the data input to our chemistry model should be a realistic representation of the state of the atmosphere. In particular, unlike climate model runs which require a GCM integration of a few years, this operational data should not have any climatological biases or drifts and therefore any transport characteristics derived should pertain to the real atmosphere rather than to a simulated one.

The meteorological data to drive our transport model are stored on a grid of 1.25° in longitude and 0.833° in latitude on 12 unevenly spaced levels up to 5 hPa with a time resolution of 6 hours. The height coordinate (η) used in the UM and our transport model is a hybrid coordinate. It is terrain-following at the surface $\eta = P/P_s$, and follows pressure surfaces above 30 hPa $\eta = P/(1000\text{hPa})$ where P is the pressure and P_s the surface pressure.

The meteorological data at the centroids of the parcels are found by linear interpolation in time and in the horizontal, and cubic interpolation in the vertical. The advection of the parcels is by 4th order Runge-Kutta with a 3-hour timestep.

Sub-grid scale processes

Two sub-grid scale processes (convection and diffusion) are parameterised in the chemistry transport model as described in Collins *et al.* (1997). Convection is a very important means of transporting pollutants out of the boundary layer, where they are emitted, into the free troposphere where the strongest winds and greatest horizontal transport occurs. Diffusion itself is not a very efficient process in terms of long range transport but it can be important in transporting material across barriers to the large scale flow such as the ITCZ or the tropopause. Prather *et al.* (1987) found that increasing diffusive mixing near convergence zones in the tropics was necessary to get agreement between their model and data. In our model, diffusion is parameterised by exchange of material between nearby parcels and by random Gaussian increments to the parcel positions. For this experiment we have turned off the random increments.

Sources and sinks of CFC11

CFC11 is emitted into our model on a 5° × 5° grid using an anthropogenic NO_X inventory from Benkovitz *et al.* (1996) scaled to give an emission of 250 kTonnes of CFC11 per year (more appropriate for the mid-1980s than the mid-1990s). After each advection timestep the emissions for a grid square are distributed equally over all the Lagrangian parcels that are with the boundary layer in that grid square. If there are no cells within the boundary layer for a particular grid square then the emissions are stored until a cell does pass through.

The major loss of CFC11 is destruction in the stratosphere, but we have decided not to include this in our model in order to simplify calculations of transport times. We are interested in the concentration fluctuations brought about by tracer transport rather than by destruction processes.

MODEL RESULTS

The model was set up to run for over two years, from 1st October 1994 until 31st December 1996. This was the entire meteorological data set that was available at the time that the model was run. Since this model in the CFC configuration has only one species and no chemistry, the CPU requirements are small and most of the time to run the model is taken up in reading the large amounts of meteorological data. The CFC11 mixing ratios were output every day as 24 hour averages on a grid of $5° \times 5° \times \Delta\eta = 0.1(\sim 100\text{hPa})$, except the highest level which has $\Delta\eta = 0.95$ between $\eta = 0.1$ and $\eta = 0.005$). With this resolution there should be on average two Lagrangian parcels in each grid volume at mid-latitudes. The great advantage of a Lagrangian model is that the trajectories of the parcels can be followed. To do this we output the positions of all 50,000 parcels every 10 days.

As the global distribution of CFC11 is not well known, we have initialised the model with zero concentrations. The emissions were held constant at 250 kTonnes per year with no attempt to take into account recent emission reductions under the Montreal Protocol. Figure 1 shows the surface mixing ratio of CFC11 on the last day of the run. As expected, there are maxima in concentration over the industrial areas which have spread out giving high concentrations over most of the mid-latitude northern hemisphere. CFC11 has certainly been transported to the remote high latitudes in the southern hemisphere, there seem to be tongues of higher concentration pushing into the southern ocean. The strongest latitudinal gradient in CFC11 is found in the tropics since they act as a barrier to tracer transport.

Interhemispheric exchange of CFC11

In equilibrium, with an constant interhemispheric exchange rate, the difference in mass of CFC11 between the two hemispheres should be constant. An exchange time τ is given by

$$\tau = \frac{M_N - M_S}{\frac{1}{2}(E_N - E_S)}$$

where M_N, E_N and M_S, E_S are the masses and emissions of CFC11 in the northern and southern hemispheres respectively. Figure 2 shows the exchange time as a function of

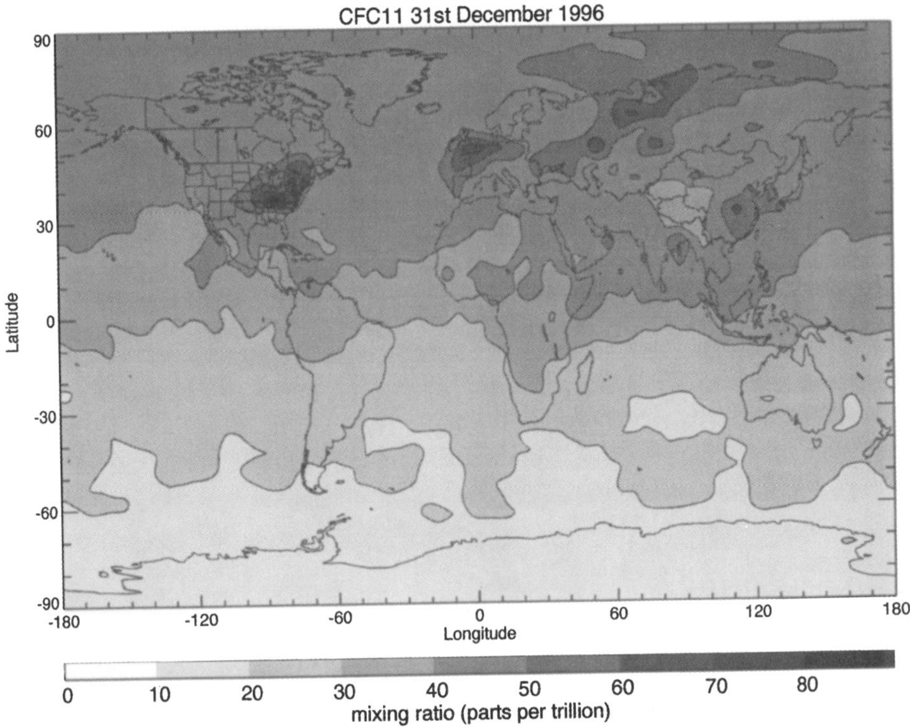

Figure 1. The model calculated surface mixing ratio of CFC11 on 31 December 1996 after 822 days run.

the date. For most of the run the system is not in equilibrium as the masses in both hemispheres are increasing from zero, but by the end of the run the graph is flattening off with an exchange time of between 1.0 and 1.2 years which is significantly longer than calculated by Prather *et al.* (1987). This may be due to insufficient diffusivity globally or possibly the need to account for sub gridscale horizontal transport in convective clouds. The minima in the exchange times occur in the late winter, due the position of the ITCZ which separates the regions of high and low concentrations. In the late winter the ITCZ is south of the equator, so polluted air that is still north of the ITCZ is temporarily classified as being in the southern hemisphere.

CFC11 Comparison with observations

We have obtained CFC11 data from Mace Head on the West Coast of Ireland up to the end of 1995 which gives us a year's overlap with our model run. Figure 3 shows time series from both the model and the observations. Both have had a baseline subtracted to show deviations from the background, in the case of the model this baseline increases linearly from the start of the model run. The observations are plotted upside down for ease of comparison. The comparison appears very good in terms of positions of the peaks but the sizes of the peaks are a factor of 2 or 3 too large. This is largely because the European emissions have substantially decreased since the mid 1980s which we have not taken account of. This comparison does not constitute a test of the long-range transport of the model since the significant pollution peaks are due to European emissions sources a thousand km or less from Mace Head.

The effects of longer range transport can be observed in Samoa. We do not yet have the CFC11 observations from Samoa for 1995 and 1996 so cannot compare these with our detailed meteorology but we can look at the seasonal variation. The CFC11

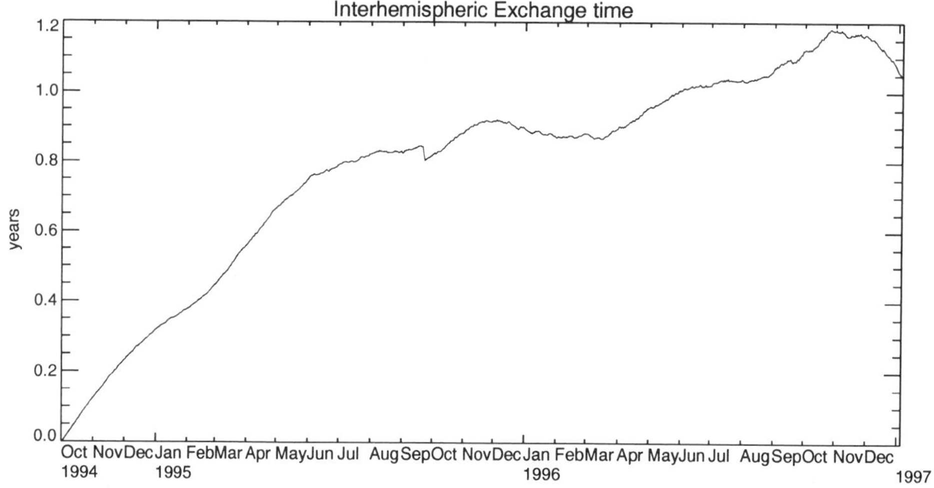

Figure 2. Interhemispheric exchange time for the atmosphere up to ~5 hPa. Concentrations started from zero on 10th October 1994

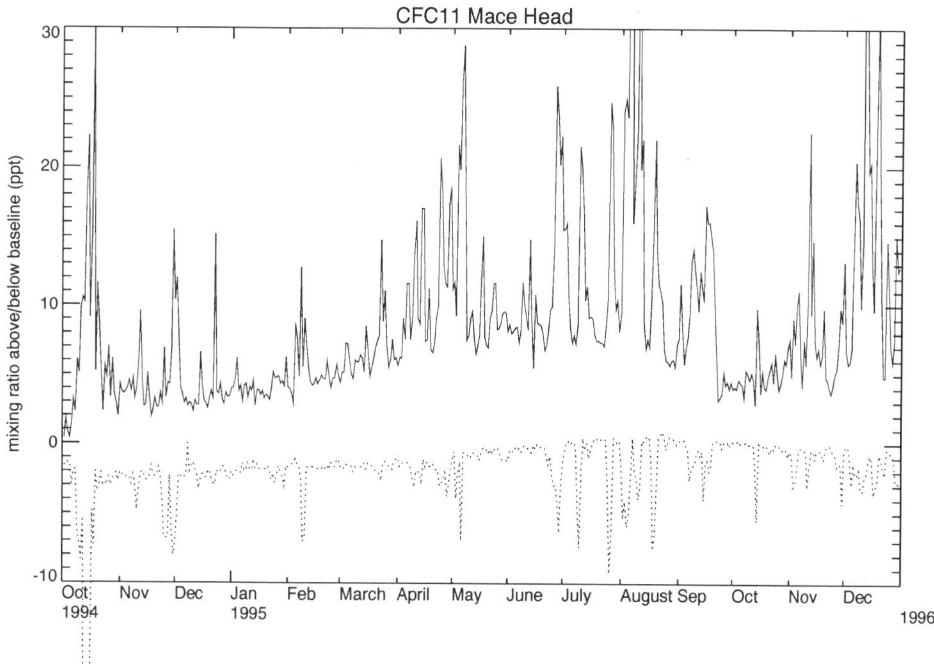

Figure 3. CFC11 concentrations at Mace Head Ireland. The solid line is model results, the dotted line is observations. Both have had baselines subtracted, the observational data have been plotted upside down for clarity.

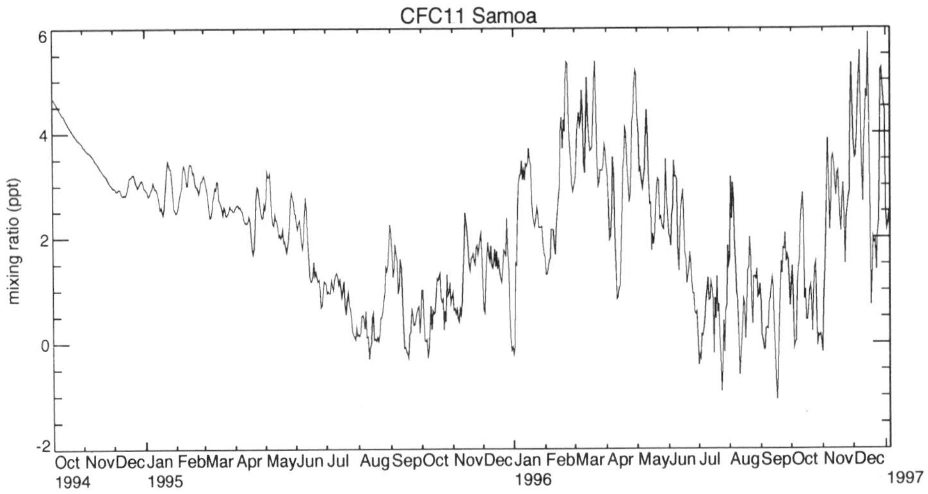

Figure 4. Model CFC11 concentrations at Samoa. The data have had a baseline subtracted.

concentrations in the observations show maxima in the boreal winter (e.g. Hartley *et al.* 1994) with a peak to trough amplitude of about 5 ppt in the 1980s. Our model results are shown in figure 4, again a linearly increasing background has been subtracted. This graph shows clear minima in the boreal summer though the maxima vary from winter to spring. The peak to trough amplitude is also about 5 ppt suggesting that with comparable emission strengths we are modelling the climatological variations in transport well.

TRAJECTORY CALCULATIONS

The strict definition of interhemispheric exchange is based on the transport across the equator. This is not necessarily the most useful definition from the point of view of tropospheric chemistry since the boundary between northern polluted air and the cleaner southern air occurs near the ITCZ rather than at 0° latitude. Of more interest is the timescale for exchange between the industrial northern mid-latitudes and the remote southern mid-latitudes.

As mentioned previously we kept track of the parcel positions during our two year run. We have selected those which travelled from near the surface ($\eta > 0.9$) in the extratropics of one hemisphere (latitude> 30°N or 30°S) to near the surface in the extratropics of the opposite hemisphere. The number of trajectories crossing between the two regions is plotted against the date at which the trajectory crossed the equator in figure 5 (a and b). The total number crossing from north to south in the 822 day run was 5354 out of 50,000 parcels, the number crossing from south to north was 6655. The figure shows that more parcels crossed from north to south in the boreal winter than in the summer whereas the maximum transport from south to north was in the boreal summer. This seasonality cannot just be due to the position of the ITCZ, as the trajectories are from extratropics to extratropics, but must be due to the seasonality of the flow across the ITCZ as described by Feichter *et al.* (1991).

We can also look at the characteristic time for trajectories to cross between the two extratropical surface regions. This is shown in figure 6 (a and b). For the north

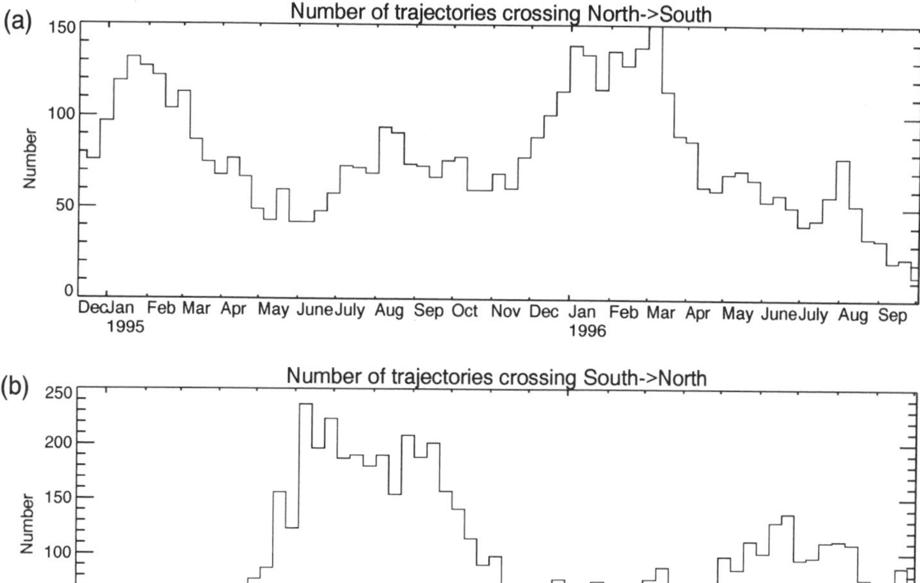

Figure 5. Number of trajectories crossing between the near-surface extratropics in opposite hemispheres.

to south transport, the modal transport time is about 180 days with a mean of 280 days. The south to north transport is quicker, with a modal value about 150 days and a mean of 260 days. Analysis of individual trajectories projected onto a latitude-height plane shows that for the quicker transport times (around the modal value), the parcels have been circulating several times in Hadley cell type patterns before escaping to the higher latitudes. Those taking significantly longer than the modal value tend to have got caught in the stratospheric circulation before finding their way down to the surface once more, however as the vertical resolution of our meteorological data is poor in the stratosphere and does not resolve the tropopause we cannot be confident that this feature is representative of the true atmospheric circulation.

CONCLUSIONS

Using meteorological data from 1995 and 1996 we have shown that our Lagrangian tracer transport model can reproduce the observed patterns in CFC timeseries, giving some confidence in our ability to simulate short and long range transport, although we need more observational data from 1995 and 1996 to confirm this more rigorously. Our interhemispheric exchange time of 1.0–1.2 years is quite long and may prove to be a function of subgrid-scale diffusion.

Looking at the statistical properties of Lagrangian parcel trajectories has demonstrated the seasonality of exchange between the extratropical regions. Whereas the study of individual trajectories is starting to give insight into the mechanisms for the exchange.

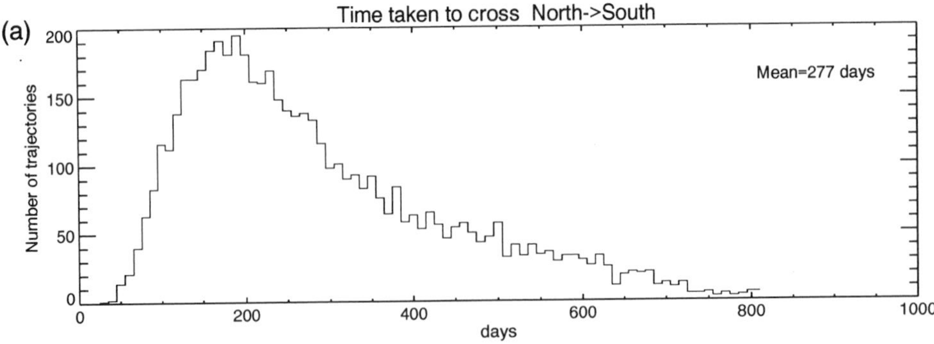

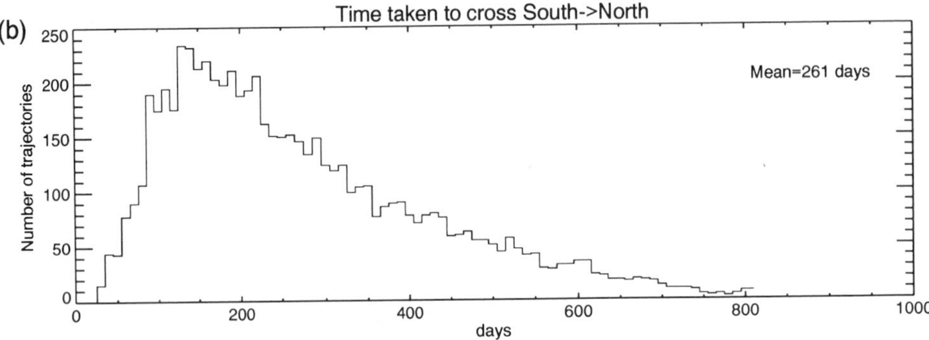

Figure 6. Histogram of the time taken for trajectories to cross between the near-surface extratropics in opposite hemispheres

REFERENCES

Benkovitz, C.M., Scholtz, M.T., Pacyna, J., Tarrasón, L., Dignon, J., Voldner, E.C., Spiro, P.A., Logan, J.A., and Graedel, T.E., 1996, Global gridded inventories of anthropogenic emissions of sulfur and nitrogen, *J. Geophys. Res.*, **101**, 29239–29253.

Chock, D.P., and Winkler, S.L., 1994, A comparison of advection algorithms coupled with chemistry. *Atmos. Environ.*, **28**, 2659–2675.

Collins, W.J., Stevenson, D.S., Johnson, C.E., and Derwent, R.G., 1997, Tropospheric ozone in a global-scale three-dimensional Lagrangian model and its response to NO_X emission controls, *J. Atmos. Chem. in press.*

Cullen, M.J.P., 1993, The unified forecast/climate model, *Meteorological Magazine*, **122**, 81–94, London, U.K.

Cunnold, D.M., Fraser, P.J., Weiss, R.F., Prinn, R.G., Simmonds, P.G., Miller, B.R., Alyea, F.N., and Crawford, A.J., 1994, Global trends and annual releases of CCl_3F and CCl_2F_2 estimated from ALE/GAGE and other measurements from July 1978 to June 1991, *J. Geophys. Res.*, **99**, 1107–1126.

Dabdub, D., and Seinfeld, J.H., Numerical advective schemes used in air quality models — sequential and parallel implementation, *Atmos. Environ.*, **28**, 3369–3385.

Feichter, J., Roeckner, E., Schlese, U and Windelband, M., 1991, Tracer transport in the Hamburg climate model, in: *Air Pollution Modeling and its Application VIII,* H. van Dop and D.G. Steyn, ed., Plenum Press, New York.

Hartley, D.E., Williamson, D.L., Rasch, P.J., and Prinn, R.G., 1994, Examination of tracer transport in the NCAR CCM2 by comparison of $CFCl_3$ simulations with ALE/GAGE observations, *J. Geophys. Res.*, **99**, 12885–12896.

Prather, M., McElroy, M., Wofsy, S., Russel, G., and Rind, D., 1987, Chemistry of the global troposphere, Fluorocarbons as tracers of air motion, *J. Geophys. Res.*, **92**, 6579–6613.

Stevenson, D.S., Johnson, C.E., Collins, W.J. and Derwent, R.G., 1997a, Changes to tropospheric oxidants from aircraft nitrogen oxide emissions studied with a 3-D Lagrangian model. *Atmospheric Environment* (in press)

Stevenson, D.S., Johnson, C.E., Collins, W.J. and Derwent, R.G., Intercomparison and evaluation of atmospheric transport in a Lagrangian model (STOCHEM), and a Eulerian model (UM), using ^{222}Rn as a short-lived tracer. *submitted to the Quarterly Journal of the Royal Meteorological Society.*

Taylor, J.A., 1989, A stochastic Lagrangian atmospheric transport model to determine global CO_2 sources and sinks — a preliminary discussion. *Tellus*, **41B**, 272–285.

Walton, J., MacCracken, M., Ghan, S., 1988, A global-scale Lagrangian trace species model of transport, transformation, and removal processes, *J. Geophys. Res.*, **93**, 8339–8354.

DISCUSSION

G. COSEMANS: Another source of information on interhemispheric exchange could be the radioactive fallout after the atomic bomb explosions in the atmosphere in the 1950s - 1960s. Have these data been analysed and do they lead to similar conclusions?

W.J. COLLINS: Yes, atomic bomb fallout data have been used by many authors to study interhemispheric exchange, and the work I have presented here is consistent with their findings.

PRODUCTION AND LONG-RANGE TRANSPORT OF DESERT DUST IN THE MEDITERRANEAN REGION: ETA MODEL SIMULATIONS

Slobodan Ničković[1], Dusan Jović[2], Olga Kakaliagou, George Kallos

University of Athens, Department of Physics,
Laboratory of Meteorology, Athens, Greece

INTRODUCTION

Saharan dust storms are the main source of the atmospheric dust in the Mediterranean region. Once injected to the atmosphere, dust may pass long distances under favourable meteorological conditions before it deposits to the ground or sea surfaces. Typically, several hundreds millions of tonnes of dust is transported away from sources annually (D'Almeida, 1986). Continuous presence of dust in the atmosphere causes diverse climatic and environmental effects. For example, dust modifies radiation properties of the air through absorption and scattering of the solar energy on dust particles (e.g. Chen *et al.*, 1995). Recent estimate of Tegen and Fung (1994) shows that mineral dust may decrease the net radiation for about 1 Wm^{-2}, revealing thus the fact that it could be a significant climate forcing factor. Another environmental effect of the dust process is dust deposition on the sea surface, which may significantly change the marine biochemical properties (e.g. Martin and Fitzwater, 1988; Kubilay and Saydam, 1995). Also, the atmospheric dust may significantly influence human activities: for example, it reduces the visibility, causing thus problems in the air and ground traffic; during dust storms, increased number of eye and respiratory organs infections is recorded, too.

Modelling of the atmospheric dust cycle is one of the approaches which may provide more insight to the process. Several recent studies are performed in this direction (e.g. Genthon, 1992; Tegen and Fung, 1994; Ničković and Dobričić, 1996). In this study, the Eta step-mountain model is used as a driving tool for dust concentration calculations. A viscous sublayer model is incorporated into the dust production parametrization scheme in order to perform more appropriate surface concentration flux calculations. Other dust model components used in this study

[1] On leave from Institute of Physics, Belgrade, Yugoslavia (e-mail: nicko@etesian.meteolab.ariadne-t.gr)
[2] On leave from Institute for Meteorology, University of Belgrade, Belgrade, Yugoslavia

are described in more details in Ničković and Dobričić (1996). In this study, the case of July 1988 dust storm was analysed. The model performance was in a fairly agreement with the observational data.

MODEL DESCRIPTION

Atmospheric component

The dust concentration continuity equation is integrated on-line in the frame of the Eta limited area atmospheric model. The Eta model is originally developed at the University of Belgrade and the Yugoslav Federal Hydrometeorological Institute. Recent model development is performed by the National Centers for Environmental Prediction, Washington and the University of Athens (through the projects SKIRON and MEDUSE).

The Eta model is based on the primitive equations with the hydrostatic equation applied. The model uses several sophisticated numerical methods and paramertization techniques, as briefly described and listed bellow:

- The semi-staggered horizontal Arakawa E grid, combined with a technique for preventing separation of the gravity wave solutions, is applied (Mesinger, 1973; Janjić, 1979; Janjić and Mesinger, 1989).
- The explicit forward-backward time integration scheme is applied to the pressure gradient force and continuity equation terms; the splitting time scheme is used for the advection and 'physics' terms.
- The horizontal advection scheme of Janjić (1984) provides strict control of the nonlinear energy cascade towards smaller scales.
- In the vertical, the eta (generalised terrain following sigma) system is developed, generating step-like mountains in the model (Mesinger, 1984; Mesinger *at al.,* 1988); this technique provides more realistic simulation of effects such as flow splitting, channelling and blocking in the presence of complex topography.
- Mellor-Yamada turbulent mixing scheme (Mellor and Yamada, 1982), improved by Janjić (Janjić,1990; Janjić, 1994) and combined with the viscous sublayer scheme for the surface mixing, is used.
- The deep and shallow convection parametrized by the Betts and Miller (1986) scheme, is modified by Janjić (1994).
- The surface processes, including the surface hydrology are represented by a simple bucket scheme (Janjić, 1990).
- The GLA radiation scheme with a random overlap clouds, is applied.
- The nonlinear lateral diffusion scheme is used to control small scale grid noise.

Most of these model components are developed and/or applied in order to successfully resolve the synoptic and mesoscale features of the atmospheric processes. Such a model design in principle represents an appropriate frame for simulating the atmospheric dust cycle, too.

Dust component

Dust concentration is considered as a passive substance with a unique particle size. The dust concentration continuity equation is applied in the following form

$$\frac{\partial C}{\partial t} = -u\frac{\partial C}{\partial x} - v\frac{\partial C}{\partial y} - \dot{\eta}\frac{\partial C}{\partial \eta} - \nabla K_L \nabla C - \frac{\partial}{\partial \eta}\left(K_C \frac{\partial C}{\partial \eta}\right) + S \tag{1}$$

where, C is the dust concentration, u and v are the horizontal velocity components, $\dot{\eta}$ is the vertical velocity in the η coordinate system, K_L is the lateral diffusion coefficient, K_C is the turbulence exchange coefficient and S is the source/sink term. The dust source points in the model are specified according to the Wilson and Henderson-Sellers (1984) vegetation data set. Numerical schemes for horizontal and vertical advection, turbulent mixing in the free atmosphere and horizontal diffusion schemes for dust concentration are applied in analogy to the schemes for the other scalar model variables (Ničković and Dobričić, 1996).

The viscous sublayer (VSL) scheme of the Eta model (Janjić, 1994) is developed in order to control efficiently heat and momentum surface fluxes over the sea surface, covering broad range of different flow regimes. In this study, this scheme is also applied to the dust concentration field over the desert areas, following the idea of the physical similarity of transport processes between the lower atmosphere and the mobile surfaces such as sea, dust and snow surfaces (Chamberlain, 1983; Segal, 1990). Under smooth flow conditions, a thin viscous sublayer is created just above the ground, in which viscous mixing dominates. The transition toward the rough flow is characterised by increased turbulent mixing. Under fully developed turbulent mixing conditions, the viscous sublayer vanishes; instead, braking waves over sea surfaces and mobilisation and saltation of particles over dust/snow surfaces appear.

Calculation of the dust surface fluxes is performed following Janjić (1994). The dust concentration at the top of the VSL C_1 is defined as

$$C_1 = C_S + \left(\frac{z_{1S}}{\omega}\right) F_S \tag{2}$$

where, C_S is the surface concentration, ω is the viscous diffusivity and F_S is the surface turbulent flux above the VSL. C_S is parametrised as a function of the friction velocity in the following form:

$$C_S = \lambda u_*^3 \left(1 - \frac{u_{*t}}{u_*}\right) \tag{3}$$

where, $u_{*t} = 0.15ms^{-1}$ and λ is an empirical dimensional constant.

The depth of the VSL z_{1S} is specified by

$$z_{1S} = \zeta M \frac{\omega Rr^{1/4} Sc^{1/2}}{u_*} \tag{4}$$

where, $\zeta = 0.35$ and M are empirical constants. The other symbols used in (4) have the following meaning:

- Reynolds number: $Rr = \frac{z_0 u_*}{\nu}$
- Schmidt number: $Sc = \frac{\nu}{\omega} = 10$
- momentum molecular diffusivity ν=0.000015 m^{2}s^{-1}
- roughness height $z_0 = \max\left(0.018\frac{u_*}{g}, 1.59 \times 10^{-5} m\right)$
- model calculated friction velocity u_*

The dust turbulent flux is determined by

$$F_S = \left(\frac{K_s}{\Delta z}\right)(C_{LM} - C_1) \tag{5}$$

where, K_S is the turbulent dust exchange coefficient, considered to be the same as for the sensible heat, and C_{LM} is the concentration at the lowest model level. Assuming that the continuity of fluxes exists across the VSL interface, with the aid of (2) and (5), the concentration at the top of VSL is calculated as

$$C_1 = \frac{C_S + \alpha C_{LM}}{1+\alpha}; \quad \alpha = \frac{\left(\frac{K_S}{\Delta z}\right)}{\left(\frac{\omega}{z_{1S}}\right)} \tag{6}$$

According to (6), C_1 may be considered as a weighted average of C_S and C_{LM}. C_1 represent the lower boundary condition in the Mellor-Yamada turbulent mixing scheme of the model.

The two threshold friction velocities $u_{*1} = 0.225 ms^{-1}$ and $u_{*1} = 0.7 ms^{-1}$ separate different flow regimes. At the first threshold, the empirical constant M in (4) switches from 30 to 10 (Janjić, 1994; Liu *et al.,* 1979).

Details about the other dust concentration calculations (ground wetness effects, wet/dry deposition and advection processes) are presented in Ničković and Dobričić (1996).

MODEL EXPERIMENTS

Several model simulation were performed. In this presentation, the case of 1-7 July 1988 will be discussed. During this period, a substantial transport of dust transport was observed over the Mediterranean region (Dulac *et al.,* 1994). The model was initialised using the objective analyses of the European Centre for Medium-Range Forecasts (ECMWF). The same fields were used in order to specify the model lateral boundary conditions.

The considered period is characterised by coexistence of two systems in the region: a cyclone over North Europe and a high pressure system located between Libya, Italy and Greece, as can be seen in Fig. 1. Strong temperature discontinuity extending over the north-eastern African coast and very high temperatures over the Moroccan, Tunisian and Libyan desert areas can also be noticed. Between 3 and 4 July, the flow suddenly changes from south-east to south direction at the 850 hPa. A pattern with relatively strong southern winds that extends from the Algerian and Tunisian Sahara toward the northern Italy, is positioned along the periphery of the anticyclone. Therefore, one necessary condition for long range dust transport from the Saharan area towards the Mediterranean was satisfied. The other condition - higher dust productivity in the Sahara on 4 July is simulated by the model (Fig. 2). In this figure, the existence of several source patterns in the northern Africa can be observed. Not all of them seem to be important for the current case: probably the region with dust storms over Algeria and Tunisia is the most responsible for dust transport towards Europe, since it is accompanied with stronger southern upper-air winds. On the other side, the simulated sources in Libya and Egypt are related to the northern upper-air winds, and thus do not contribute to the dust transport towards the Mediterranean Sea.

Fig. 3 shows the dust load field (the vertically integrated dust concentration) over the Mediterranean Sea, as simulated by the model. The fields are displayed every six hours during 6 July 1988, showing relatively slow modification of the shape and quantity of dust load during this period. It can be noticed that the origin of the dust cloud corresponds well to the sources mentioned above. The dust pattern extends in the SW-NE direction, reaching the Adriatic Sea and the Black Sea. These results are generally in a good agreement with the dust load calculated from METEOSAT IR images of 6 July (Dulac *et al.,* 1994). It is worth mentioning the agreement between the two methods of estimating the dust cloud and the positioning of it over the area. Very good agreement is also in the positioning of the maxima. The dust load pattern over the Aegean and Black Seas shows good agreement with the observations (Fig. 3). However, it seems that the model is slightly slower in moving the dust cloud (for about 6 hours), compared again with the satellite images. These differences should be due to various reasons, such as model forecasting errors, uncertainties in specifying the dust source areas, etc. Although, differences in the simulated and observed dust quantities seem to be inside the range of the observation accuracy.

CONCLUSION

The atmospheric Eta model with the included dust concentration formulation was used in order to simulate the mechanisms of generation and evolution of a dust storm over Mediterranean. The model contains all the major components of the atmospheric dust cycle. Despite the fact that, some of these

(a)

(b)

(c)

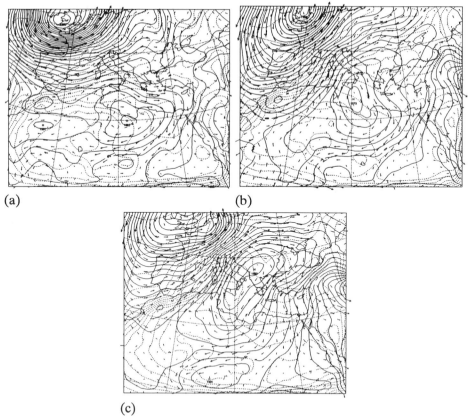

Figure 1. Wind, temperature (dashed lines) and geopotential (full lines) 850 hPa on 12 UTC 3 July (a), 4 July (b) and 5 July (c) 1988.

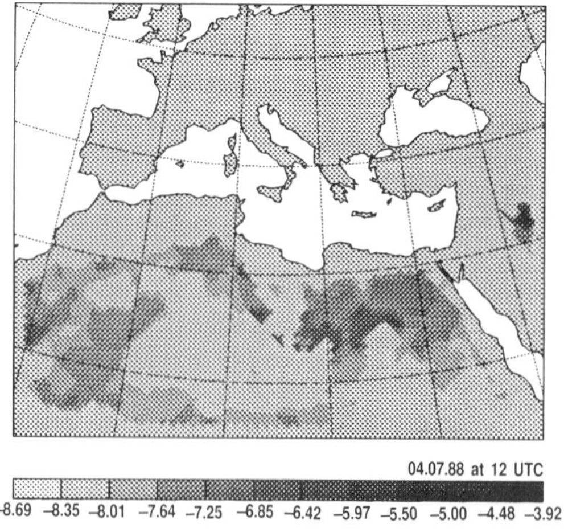

Figure 2. Simulated dust surface fluxes at 12 UTC 4 July 1988. Units: Log($kgm^{-2}s^{-1}$).

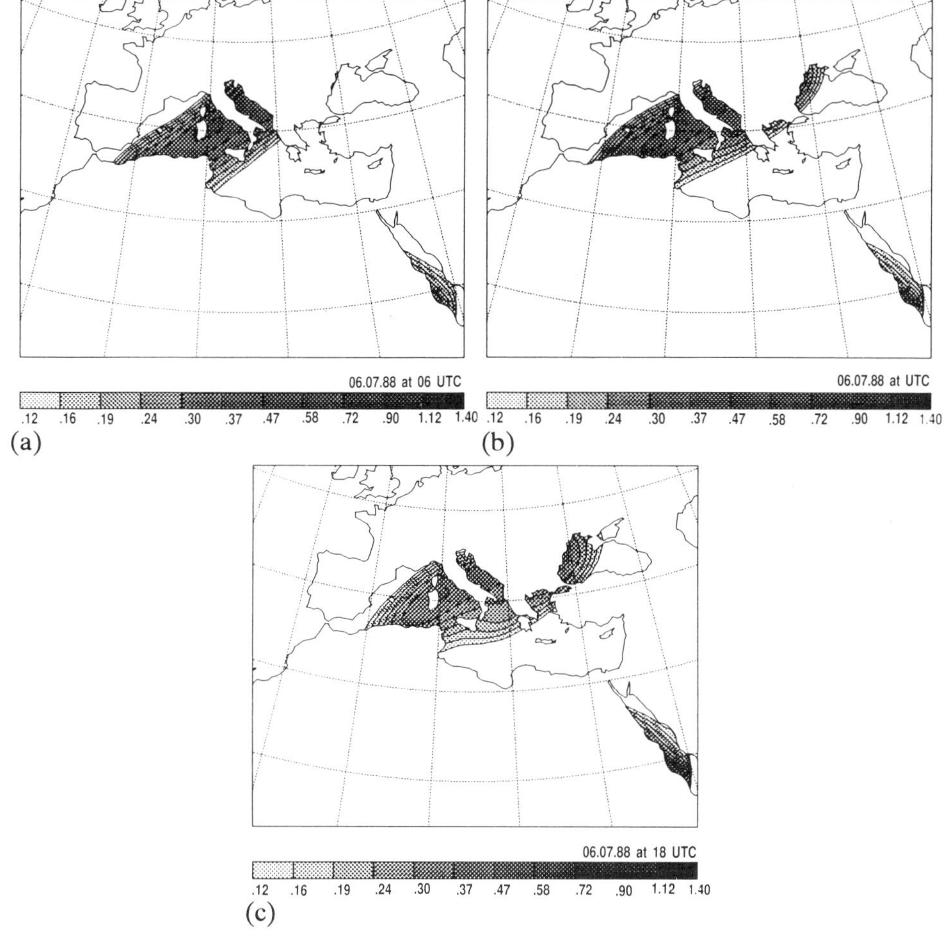

Figure 3. Simulated dust load (vertically integrated dust concentration) at (a) 06 UTC, (b) 12 UTC and (c) 18 UTC, 06 July 1988. Units: kgm^{-2}.

components are used in a simplified form (for example, use of a unique particle size and treatment of dust particles as a passive substance), the model is able to reproduce basic features of the dust uptake and transport. The model could be used as an efficient tool for better understanding of the atmospheric dust cycle, as well as for forecasting of the dust process.

Acknowledgments

This research was supported by the research projects MEDUSE (contract number ENV4-CT95-0036) funded by the DG-XII of EU and SKIRON (project ΕΠΕΤ-II No 322) funded by the Greek Government and EU.

REFERENCES

Betts, A.K., and Miller, M.J., 1986, A new convective adjustment scheme. Part II: Single column tests using GATE wave, BOMEX and arctic air-mass data sets, *Quart. J. Meteor. Soc.*, **112**, 693-709.

Chamberlain, A. C. 1983: Roughness length of sea, sand and snow. *Boundary-Layer Meteor.*, **25**, 405-409.

Chen, S.J., Kuo, Y.H., Ming, W., and Ying, H., 1995, The effect of dust radiative heating on low-level frontogenesis, *J. Atmos. Sci.*, **52**, 1414-1420.

D'Almeida, G.A., 1986, A model for Saharan dust, *J. Clim. Appl. Meteor.*, **25**, 903-916.

Dulac, F., Jankowiak, I., Legrand, M., Tanre, D., N'Doume, C.T., Guillard, F., Lardieri, D., Guelle, F., and Poitou, J., 1994, Meteosat imagery for quantitative studies of Saharan dust transport, *Proc. Meteosat Scientific Users Meeting, Cascais, Portugal, 5-9 Sept. 1994*

Genthon, C., 1992: Simulations of desert dust and sea-salt aerosols in Antarctica with a general circulation model of the atmosphere. *Tellus*, **44B**, 371-389.

Janjić, Z.I., 1979, Forward-backward scheme modified to prevent two-grid-interval noise and its application in sigma coordinate models, *Contrib. Atmos. Phys.*, **52**, 69-84.

Janjić, Z.I., 1984, Nonlinear advection schemes and energy cascade on semistaggered grids, *Mon. Wea. Rev.*, **112**, 1234-1245.

Janjić, Z.I., 1990, The step-mountain coordinate: Physical package, *Mon. Wea. Rev.*, **118**, 1429-1443.

Janjić,Z.I., 1994, The step-mountain eta coordinate model: Further developments of the convection, viscous sublayer, and turbulence closure schemes, *Mon. Wea. Rev.*, **122**, 927-945.

Janjić, Z.I., and Mesinger, F.,1989, Response to small-scale forcing on two staggered grids used in finite-difference models of the atmosphere, *Quart. J. Roy. Meteor. Soc.* **115**, 1167-1176.

Kubilay, N. and Saydam, A.C., 1995, Trace elements in atmospheric particulates over the Eastern Mediterranean: Concentrations, Sources, and temporal variability, *Atmos. Environ.*, **29**, 2289-2300.

Liu, W.T., Katsaros, K.B., and Businger, J.A., 1979, Bulk parametrization of air-sea exchanges of heat and water vapour including the molecular constraints at the interface, *J. Atmos. Sci.*, **36**, 1722-1735.

Martin, J.H., and Fitzwater, S.F., 1988, Iron deficiency limits phytoplancton growth in the North-East Pacific subarctic, *Nature*, **331**, 341-343.

Mellor, G. L., and Yamada, T., 1982, Development of a turbulence closure model for geophysical fluid problems, *Rev. Geophys. Space Phys.*, **20**, 851-875.

Mesinger, F., 1973, A method for construction of second-order accuracy difference schemes permitting no false two-grid-interval wave in the height field, *Tellus*, **25**, 444-458.

Mesinger, F., 1984, A blocking technique for representation of mountains in atmospheric models, *Rivista di Meteorologia Aeronautica*, **44**, 195-202.

Mesinger F., Janjić, Z.I., Ničković, S., Gavrilov, D., and Deaven, D.G., 1988, The step-mountain coordinate: Model description and performance for cases of Alpine lee cyclogenesis and for a case of Appalachian redevelopment, *Mon. Wea. Rev.*, **116**, 1493-1518.

Ničković, S., and Dobričić, S. 1996, A model for long-range transport of desert dust, *Mon. Wea. Rev.*, **124**, 2537-2544.

Segal, M., 1990, On the impact of thermal stability on some rough flow effects over mobile surfaces, *Boundary-Layer Meteor.*, **52**, 193-198.

Tegen, I., and I. Fung, 1994, Modelling of mineral dust in the atmosphere: sources, transport, and optical thickness. *J. Geophys. Res.*, **99**, 22897-22914.

Wilson, M.F., and Henderson-Sellers, A., 1984, Land cover and soils data sets for use in general circulation climate models, *J. Climatol.*, **5**, 119-143.

DISCUSSION

S. ZILITINKEVICH: It is known that dust involved in the air alters the density stratification of the air flow. As a result, the surface friction decreases, and the flow accelerates. Is this "acceleration effect" reproduced by your model?

G. KALLOS: In the current version of the model, dust concentration is considered as a passive substance having no influence to the atmospheric fields at all. Thus, 'acceleration effect' or similar model responses induced by dust are not existing

J. LANGNER: Do you have plans to use satellite data to initialize the dust distribution in your dust forecast model?

G. KALLOS: As far as the authors know, satellite (or another) dust measurements do not still provide three-dimensional distribution of dust concentration. When such data will be available, it will be used for the model initialization.

A GLOBAL SCALE INVERSION OF THE TRANSPORT OF CO_2 BASED ON A MATRIX REPRESENTATION OF AN ATMOSPHERIC TRANSPORT MODEL DERIVED BY ITS ADJOINT

Thomas Kaminski, Martin Heimann, and Ralf Giering

Max Planck Institut für Meteorologie
Hamburg, Germany

1. INTRODUCTION

The atmosphere contains a number of radiatively and chemically important trace gases (a.o. carbon dioxide (CO_2), carbon monoxide (CO), nitrous oxide (N_2O) and methane (CH_4)) whose concentrations are changing in the atmosphere, primarily due to human activities. These changing concentrations affect the radiative balance of our atmosphere and may thus lead to climate change. In order to compute reliable projections of the future evolution of the concentrations of these gases their natural and anthropogenic sources and sinks have to be known. Using a direct approach one can extrapolate locally measured fluxes to the entire globe. Because of the many necessary assumptions in the extrapolation, this direct approach, however, is subject to very large uncertainties. In contrast, one can apply an inverse approach, in which ambient observations of the atmospheric trace gas concentrations are used to constrain the surface fluxes. This requires a model of the atmospheric transport which provides the link between the surface fluxes and the concentrations at the monitoring sites.

The trace gases of interest are being monitored at a global network of atmospheric observing stations. However, since this network is spatially sparse, the determination of the surface fluxes constitutes a highly underdetermined inverse problem. Hence, additional information on the fluxes has to be included in the inversion in order to obtain a well-determined problem.

Several inversion studies have been performed on the basis of both, two-dimensional and three-dimensional models of the atmospheric transport [*Enting and Mansbridge*, 1989; *Brown*, 1995; *Hein and Heimann*, 1994; *Hein et al.*, 1996]. In all these studies, the surface flux field is decomposed into prescribed spatio-temporal patterns ("source" or "flux" components) with unknown scaling coefficients. The transport model is run with

Air Pollution Modeling and Its Application XII
Edited by Sven-Erik Gryning and Nadine Chaumerliac, Plenum Press, New York, 1998

each of the source components separately and the contributions to the concentration signal at each of the monitoring sites and times are recorded. These contributions can be interpreted as a discretized impulse response or Greens function which quantifies the response of the modeled concentration at the observation sites and dates to unit changes in the magnitude of each source component.

Formally this impulse response or Greens function is the Jacobian matrix representing the first derivative of the modeled concentration at the observational sites and dates with respect to the coefficients of the source components. Computationally, for nf source components, nf model runs have to be performed to compute the nf differential quotients which constitute the columns of the Jacobian matrix. The complexity of the transport model thus essentially limits the number of source components that can be considered.

Here we present an alternative approach employing the adjoint of the three-dimensional transport model TM2 for the efficient determination of the Jacobian in a "reverse mode". The exact Jacobian is computed line by line, for which the cost is proportional to the number of observations and nearly independent of the number of flux components.

This concept is illustrated by a computation of the Jacobian for about 10000 flux components and about 300 observations of the atmospheric CO_2 concentration. Combined to a priori information on the surface fluxes derived from output of high resolution models of both, the terrestrial biosphere and the ocean, and fossil fuel burning statistics we infer the seasonal cycle and mean annual source and sink distribution on the approximately 8 degree by 10 degree horizontal TM2 grid. Thereby we use the observations in the Globalview data set from the NOAA/CMDL program [*Globalview - CO2*, 1996].

The outline is as follows: In Sect. (2) we briefly introduce the transport model followed by the description of the Jacobian matrix in Sect. (3). We sketch the inversion in Sect. (4). Finally, Sect. (5) contains some perspectives.

2. THE TRANSPORT MODEL

TM2 is a three-dimensional atmospheric transport model which solves the continuity equation for an arbitrary number of atmospheric tracers on an Eulerian grid spanning the entire globe [*Heimann*, 1995]. It is driven by stored meteorological fields derived from analyses of a weather forecast model or from output of an atmospheric general circulation model. Tracer advection is calculated using the "slopes scheme" of *Russel and Lerner* [1981]. Vertical transport due to convective clouds is computed using the cloud mass flux scheme of *Tiedtke* [1989]. Turbulent vertical transport is calculated by stability dependent vertical diffusion according to the scheme by *Louis* [1979]. Numerically, in each base time step the model calculates the source and sink processes affecting each tracer, followed by the calculation of the transport processes.

The spatial structure of the model is a regular latitude-longitude grid and a sigma coordinate system in the vertical dimension. The base "coarse grid" version of the model uses a horizontal resolution of approximately 8° latitude by 10° longitude (i.e. the horizontal dimension of the grid is $ng = 36 \times 24$) and 9 layers in the vertical dimension. The numerical timestep of this model version is four hours. In the present application the TM2 is forced with the meteorological fields of the year 1987, derived from analyses of the European Center for Medium Range Weather Forecast (ECMWF) which are available to the model every 12 hours.

Prescribing the same monthly surface CO_2 source flux fields every year, and starting from zero initial concentration, TM2 is run for four years. At a particular site S the concentration c_S is computed from the simulated concentration fields of the fourth year by first computing monthly means and then performing a bilinear interpolation in the horizontal from the TM2 grid to the exact location of S. This setup of TM2 in the following will be denoted as standard setup.

In the standard setup TM2 has as input $f \in I\!R^{nf}$, a vector of $nf = 12 \times ng$ real numbers characterizing the 12 monthly fluxes into each surface grid cell and as output $c \in I\!R^{nc}$, a vector of $nc = 12 \times ns$ real numbers for the modeled monthly mean concentration at ns observational sites. Since the transport of a passive atmospheric tracer in TM2 is linear, TM2 can be represented by a real $nc \times nf$ matrix T and the application of TM2 can be written as

$$c = Tf \quad . \tag{1}$$

3. JACOBIAN MATRIX

As explained in Sect. (2), for the standard setup, TM2 can be represented by a $nc \times nf$ matrix T. For given surface fluxes f by a model run we are able to compute the resulting concentrations at the station locations $c_{mod} = Tf$. However, the matrix T itself is yet to be determined.

By applying TM2 subsequently to the each of the nf standard basis vectors $e_1 = (1, 0, ..., 0)^T, ..., e_{nf} = (0, ..., 0, 1)^T$ spanning $I\!R^{nf}$, the matrix T could be computed column by column. This requires nf runs of TM2 and is thus only computationally feasible for a small number of flux components. Instead we apply an alternative, much more efficient approach: Using the model adjoint to TM2 in the standard setup the Jacobian matrix is computed line by line in reverse mode. In this case both, the computational cost and the storage requirements, depend on the number of rows, i.e. on nc, rather than on the number of columns, i.e. on nf.

For the construction of an adjoint model there exist several strategies [*Marchuk*, 1995; *Talagrand and Courtier*, 1987; *Talagrand*, 1991]. The adjoint of TM2 has been derived directly from the model code [*Giering and Kaminski*, 1996] based on the concept of differentiation of algorithms [*Iri*, 1991]. Thereby the Tangent linear and Adjoint Model Compiler [TAMC *Giering*, 1996] has been applied to automatically generate the adjoint code. By means of the adjoint of TM2 in the standard setup the matrix representation T for $ns = 27$ NOAA/CMDL stations (see Fig. (1)) has been computed. The rows of T consist of the sensitivity of the modeled concentration at a particular station and month to the fluxes into each of the $ng = 36 \times 24$ TM2 surface layer grid cells at each month. Conversely the rows of T contain the impact of a particular flux component on the modeled concentration at all stations and months.

We run TM2 and its adjoint on a Cray C916 supercomputer: For our standard setup with $nc = 27 \times 12$ the adjoint model run takes the cost of less than 100 TM2 single tracer runs. The computing time scales with the number of observations nc.

4. INVERSION

The inverse problem associated with Eq. (1) consists in the determination of a flux field f, so that for prescribed concentrations c the equation is satisfied. For our matrix T the problem consists of $nc \approx 300$ equations for $nf \approx 10000$ unknowns, and thus is highly underdetermined, i.e. there will be many flux fields which yield the same modeled concentration.

by using the so-called Bayesian approach we include additional information in the inversion procedure to obtain a unique solution: Both, atmospheric observations and a

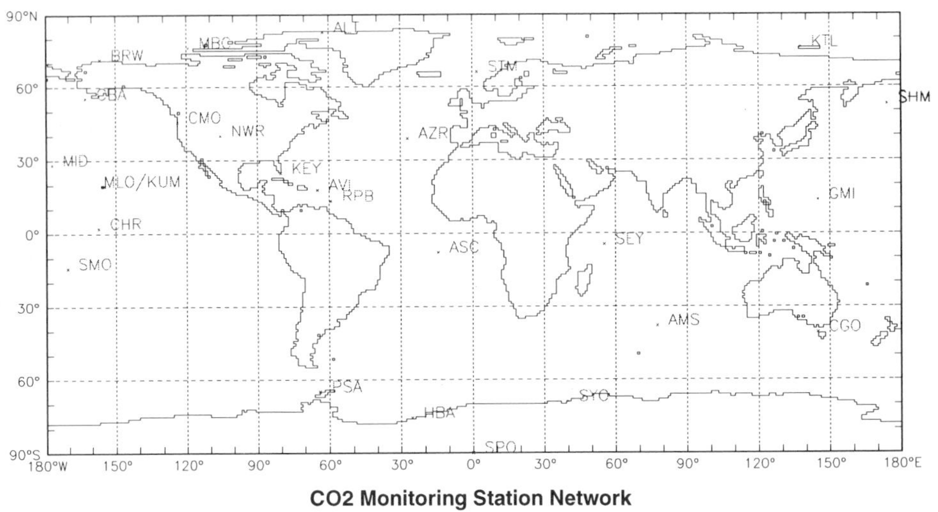

Figure 1. Monitoring Station Network of the NOAA/CMDL

priori information are described in terms of probability densities. Employing the transport as constraint, consistent probability densities are derived. A detailed introduction to the Bayesian approach is given in the textbooks of *Menke* [1989] and *Tarantola* [1987].

For our inversion, we assume fluxes and concentrations to have a Gaussian distribution with diagonal covariance matrices. Hence, the definition of the probability distributions consists in assigning standard deviations for each component. This special type of inverse problem is known as least squares problem. It results in the minimization of the cost function

$$J(\tilde{f}) = {}^1/_2 \left\{ \sum_{i=1}^{nf} \left(\frac{f - \tilde{f}}{\sigma_f} \right)^2 + \sum_{i=1}^{nc} \left(\frac{c_{obs} - T\tilde{f}}{\sigma_c} \right)^2 \right\} \quad , \tag{2}$$

where f is the a priori estimate of the fluxes with uncertainty σ_f and c_{obs} is the observed concentration with uncertainty σ_c.

We compose our a priori estimate of the net fluxes into the atmosphere from three components: the terrestrial biosphere, the ocean, and fossil fuel burning. From fossil fuel burning statistics of *Andres et al.* [1997] on a 1 ^{o} grid, annual mean fluxes on the TM2 grid have been interpolated. The global annual net source is 5.3 Gigatons of Carbon (GtC). Compared to the biospheric and oceanic component, the uncertainty in this source is rather small. Hence we exclude this component by subtracting prior to the inversion from the observations the modeled seasonal cycle, linear trend and mean annual spatial gradient generated by the fossil fuel source alone.

In Eq. (2), as a priori estimate for f we employed the sum of the biospheric and oceanic components computed from high resolution models of both, the terrestrial biosphere [*Knorr and Heimann*, 1995] and the ocean[*Six and Maier-Reimer*, 1996; *Maier-Reimer*, 1993]. The global annual means of the surface flux fields from both models are zero.

The uncertainties assigned to the a priori estimates of the fluxes are crucial parameters in the inversion, because they constitute the weights in the cost function (Eq. (2)). In general, assuming large uncertainties on the a priori flux estimates (we choose values as large as 100% of the respective flux component) results in a solution that fits more closely the observations.

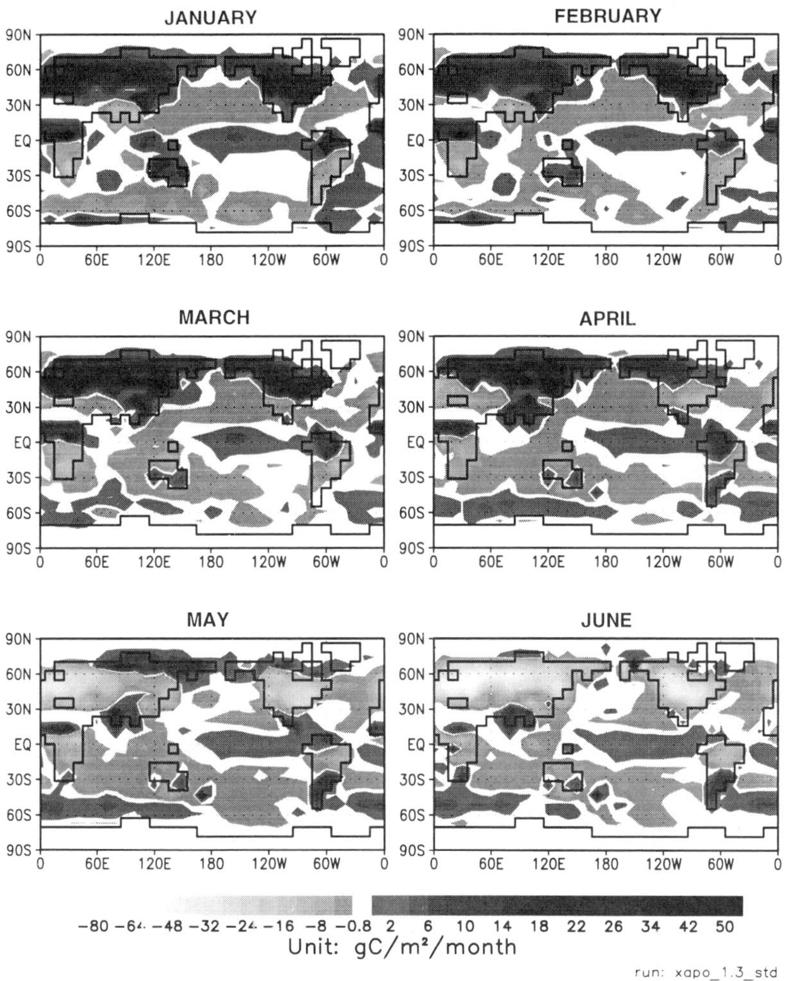

Figure 2. A posteriori estimate of the seasonal cycle of the sum of the biospheric and oceanic flux components

Globalview-CO_2 is a database of high quality atmospheric measurements coordinated by the NOAA/CMDL [*Globalview - CO2*, 1996]. This monitoring network comprises more than 60 sites for which weekly data together with standard deviations are available. have been prepared. For our inversion we use the data from the 26 sites (all stations are displayed in Fig. (1) except for KTL). Similar to *Tans et al.* [1990] we choose a target period of 6 years from January 1981 to January 1987 and extract the seasonal cycle, linear trend and spatial gradient from the observations.

The a posteriori estimate of the fluxes from January to June is displayed in Fig. (2). The cost function of Eq. (2) is the sum of two terms: The contribution of the misfit between modeled and observed concentrations is 10.9^2, while the contribution of the correction of the estimate of the a priori fluxes is $(11.9)^2$. This means that with a relatively small mean square correction of the a priori estimate of 0.001 standard deviations for the flux component a relatively small misfit of 0.35 standard deviations is obtained. Essentially this has two reasons: The a priori source estimate provides already a good fit to the observed concentrations. Furthermore, of course, the assumed uncertainty for the a priori estimate is large. Primarily the fluxes are corrected near several stations like BRW or CGO and also in some areas with large a priori values and thus large uncertainties. The uncertainties of the a posteriori flux estimate is not reduced significantly, except close to a few observing sites [see also *Enting*, 1993]. This is also reflected by the low model resolution [see e.g. *Menke*, 1989; *Tarantola*, 1987] of the present inverse problem [see also *Enting*, 1993].

5. CONCLUSIONS AND PERSPECTIVES

We demonstrated the benefit of the adjoint approach for the computation of the Jacobian matrix for an atmospheric transport model in a particular setup. The computational cost is as low as 1 per cent compared to conventional forward modeling. For a higher spatial or temporal resolution of the fluxes, this percentage is even smaller.

In our example we employed the Jacobian to derive an estimate of the sources and sinks of CO_2 . However, the technique can be efficiently applied to other tracers in the same manner, as long as the number of observations is small compared to the number of flux components of interest.

If both, the number of observations and the number of flux components is large, or if a linearization around the a priori estimate induces a too large error, we recommend a different approach: The adjoint model can be used to provide the gradient of the cost function in Eq. (2) which is required by a class of powerful minimization algorithms. In an iterative procedure the cost function is minimized by variation of the fluxes [*Talagrand and Courtier*, 1987; *Giering and Kaminski*, 1996].

In the present study we characterized the sources and sinks by their net exchange fluxes with the atmosphere, rather than the processes causing the fluxes. After coupling the transport model (or its Jacobian) to process models such as the SDBM [*Knorr and Heimann*, 1995], the corresponding adjoint can be applied to estimate the internal parameters of the process models.

Furthermore an adjoint model constitutes a valuable tool for sensitivity studies. E.g. the adjoint of TM2 has been applied to efficiently decompose features such as the magnitude of the modeled seasonal cycle of the atmospheric CO_2 concentration with respect to the contributions of prescribed fluxes from all TM surface layer grid cells [*Kaminski et al.*, 1996].

6. ACKNOWLEDGMENT

The authors thank Wolfgang Knorr and Katharina Six for providing the flux fields from their models as well as Michael Voßbeck and Walter Sauf for preparing the figures. This work was supported in part by the Commission of the European Communities under contract EV5V-CT92-0120, European Study of Carbon in the Oceans Biosphere and Atmosphere (ESCOBA): Atmosphere Section. Computing support was provided by the Deutsches Klimarechenzentrum (DKRZ) in Hamburg.

References

Andres, R. J., G. Marland, T. Boden, and S. Bischoff, Carbon dioxide emissions from fossil fuel consumption and cement manufacture 1751 to 1991 and an estimate for their isotopic composition and latitudinal distribution, in *The Carbon Cycle*, edited by T. M. L. Wigley, and D. Schimel, Cambridge University Press, 1997.

Brown, M., The singular value decomposition methof applied to the deduction of the emissions and the isotopic composition of atmospheric methane, *J.Geophys.Res.*, (100), 425–446, 1995.

Enting, I. G., Inverse problems in atmospheric constituent studies.iii: Estimating errors in surface sources atmospheric CO_2., *Inverse Problems*, (9), 649–665, 1993.

Enting, I. G., and J. V. Mansbridge, Seasonal sources and sinks of atmospheric CO_2 direct inversion of filtered data, *Tellus*, (41B), 111–126, 1989.

Giering, R., *Tangent linear and Adjoint Model Compiler, users manual*, MPI, Bundesstr. 55, 20251 Hamburg, Germany, 1996.

Giering, R., and T. Kaminski, Recipes for Adjoint Code Construction, *Submitted to ACM Transactions on Mathematical Software*, 1996.

Globalview - CO2, *Cooperative Atmospheric Data Integration Project - Carbon Dioxide , CD-ROM*, NOAA/CMDL, Boulder, Colorado, 1996.

Heimann, M., The global atmospheric tracer model TM2, Technical report no. 10, Max-Planck-Institut für Meteorologie, Bundesstr. 55, 20251 Hamburg, Germany, 1995.

Hein, R., and M. Heimann, Determination of global scale emissions of atmospheric methane using an inverse modelling method, in *Non-CO_2 Greenhouse Gases*, edited by J. van Ham et all, Kluver, 1994.

Hein, R., P. Crutzen, and M. Heimann, An inverse modeling approach to investigate the global atmospheric methane cycle, *Global Biogeochemical Cycles*, 1996.

Iri, M., History of automatic differentiation and error estimation, in *Automatic Differentiation of Algorithms: Theory, Implementation, and Application*, edited by A. Griewank, and G. F. Corliss, SIAM, Philadelphia, PA, 1991.

Kaminski, T., R. Giering, and M. Heimann, Sensitivity of the seasonal cycle of CO_2 at remote monitoring stations with respect to seasonal surface exchange fluxes determined with the adjoint of an atmospheric transport model, *Submitted to Physics and Chemistry of the Earth*, 1996.

Knorr, W., and M. Heimann, Impact of drought stress and other factors on seasonal land biosphere CO_2 exchange studied through an atmospheric tracer transport model, *Tellus*, (47B), 471–489, 1995.

Louis, J. F., A parametric model of vertical eddy fluxes in the atmosphere, *Boundary Layer Meteorology*, (17), 187–202, 1979.

Maier-Reimer, E., Geochemical cycles in an ocean general circulation model. preindustrial tracer distributions, *Global Biogeochemical Cycles*, (7), 645–677, 1993.

Marchuk, G. I., *Adjoint Equations and Analysis of Complex systems*, Kluwer Academic Publishers, Dordrecht, The Netherlands, 1995.

Menke, W., *Geophysical Data Analysis*, Academic Press, San Diego, CA, 1989.

Russel, G. L., and J. A. Lerner, A new finite-differencing scheme for the tracer transport equation, *J. Appl. Met.*, pp. 1483–1498, 1981.

Six, K. D., and E. Maier-Reimer, Effects of plankton dynamics on seasonal carbon fluxes in an ocean general circulation model, *Global Biogeochemical Cycles*, *10*(4), 559–583, 1996.

Talagrand, O., The use of adjoint equations in numerical modelling of the atmospheric circulation, in *Automatic Differentiation of Algorithms: Theory, Implementation, and Application*, edited by A. Griewank, and G. F. Corliss, SIAM, Philadelphia, PA, 1991.

Talagrand, O., and P. Courtier, Variational assimilation of meteorological observations with the adjoint vorticity equation – Part I. Theory, *Q. J. R. Meteorol. Soc.*, *113*, 1311 – 1328, 1987.

Tans, P. P., I. Y. Fung, and T. Takahashi, Observational constraints on the global atmospheric CO_2 budged, *Science*, (247), 1431–1438, 1990.

Tarantola, A., *Inverse Problem Theory - Methods for Data Fitting and Model Parameter Estimation*, Elsevier, Amsterdam, 1987.

Tiedtke, M., A comprehensive mass flux scheme for cumulus parameterization in large-scale models, *Mon. Weath. Rev*, (117), 1779–1800, 1989.

DISCUSSION

R. ROMANOWICZ: What do you mean by posterior estimate of parameters of your model? Is it a single set of values representing a mean set? In my understanding, you should derivate randomly your 10368 parameter values, derive the error model (e.g. assuming Gaussian distribution) and evaluate a posterior Likelihood function, where from the mean estimates for parameters could be derived. However, with this number of parameters this task is impossible numerically.

T. KAMINSKI: In a Bayesian framework the system under consideration is characterized by a joint probability distribution. A priori, i.e. without taking our model of the atmospheric transport into account, the fluxes and the concentrations are independent. For simplicity both are assumed to have Gaussian distributions. The distribution of the fluxes is derived from atmospheric measurements. The transport model quantifies the concentrations as a linear function of the fluxes. It can be shown that under this constraint the resulting joint probability distribution is Gaussian as well. Numerically both, mean and covariance, are computed by means of a singular value decomposition of an 10368*312 matrix (e.g. about 100 cpu seconds on a Cray C90 using about 7MW of core memory).

ETEX session

THE EUROPEAN TRACER EXPERIMENT EXPERIMENTAL RESULTS AND DATABASE

Katrin Nodop, Richard Connolly, and Francesco Girardi

Joint Research Centre, Environment Institute, I-21020 ISPRA (Va), Italy

SUMMARY

As part of the European Tracer Experiment (ETEX) two successful atmospheric experiments were carried out in October and November, 1994. Perfluorocarbon (PFC) tracers were released into the atmosphere in Monterfil, Brittany, and air samples were taken at 168 stations in 17 European countries for 72 hours after the release. Upper air tracer measurements were made from three aircraft. During the first experiment a westerly air flow transported the tracer plume north-eastwards across Europe. During the second release the flow was eastwards. The results from the ground sampling network allowed the determination of the cloud evolution as far as Sweden, Poland and Bulgaria. Typical background concentrations of the tracer used are around 5 to 7 fl/l in ambient air. Concentrations in the plume ranged from 10 to above 200 fl/l.

The tracer release characteristics, the tracer concentrations at the ground and in upper air, the routine and additional meteorological observations at the ground level and in upper air, trajectories derived from constant-level balloons and the meteorological input fields for long-range transport (LRT) models are assembled in the ETEX database. The ETEX database is accessible via the Internet.

INTRODUCTION

ETEX is jointly organised by the European Commission's (EC) Joint Research Centre (JRC), the International Atomic Energy Agency (IAEA) and the World Meteorological Organization (WMO). It was established to evaluate the validity of long-range transport models for real time application in emergency management and to assemble a database which will allow the evaluation of long-range atmospheric dispersion models in general (Klug et al, 1993; Girardi et al., 1997). The objectives of ETEX are: first, to conduct a long-range atmospheric tracer experiment with controlled releases under well-defined conditions; second, to test the capabilities of institutes responsible for producing rapid forecasts of atmospheric dispersion to produce such forecasts in real time; and third, to evaluate the validity of their forecasts by comparison with the experimental data. The results of the tracer experiment are reported in this paper. The results of the model intercomparison studies and on emergency response are also presented in these proceedings.

EXPERIMENTAL DESIGN

The planning of the very large ETEX experiments took nearly two years. The ground-level stations had to be found and equipped with samplers, the meteorological conditions had to be defined, the amount and duration of the tracer release set. This was followed by the preparation of the sampling tubes, as well as the establishment of analytical procedure and of data management. At the moment of the suitable weather conditions, the release activities and the start of sampling (at the stations and by aircraft) had to be co-ordinated with short notice and within a tight schedule. In several "dry runs" the communication links between the headquarters at Ispra, the weather services, the national representatives and station officers were tested.

To conduct a tracer experiment over a distance of up to 2000 km in Europe, a release in the western part of Europe, in meteorological conditions with prevailing westerly-south-westerly air flow, was required.

The release site used was approximately 35 km west of Rennes, at Monterfil, in Brittany, France. The Monterfil site is the highest flat point in the area, with no obstacles in the vicinity. It is located at 2° 00'30"W and 48° 03'30"N, 90 m above sea level.

Perfluorocarbon compounds are suitable tracer substances for experiments over long distances (Dietz, 1986). They are non-toxic, non-depositing, non-water-soluble, inert and environ-mentally safe. At ambient pressure and temperature they are odourless and clear liquids. They are released into the atmosphere by spraying the liquid into a hot air stream, causing it to evaporate. During the first experiment perfluoromethylcyclohexane (PMCH), C_7F_{14}, was released. In the second release perfluoromethylcyclopentane (PMCP), C_6F_{12}, was used to avoid cross contamination.

The weather services in 17 European countries made their ground-level stations and personnel available for taking air samples during the experiments. A total of 168 stations were equipped with sequential air samplers. The air samples were collected in metal tubes filled with absorption material. They were distributed shortly prior to the experiment and sent back to Ispra for chemical analysis immediately after the experiment. Sampling time per sample was set to three hours. The starting time of the first of the 24 samples was calculated according to expected arrival time of the tracer plume. The last measurement was taken 96 hours after the start of the release in the most eastern countries.

During each experiment over 5000 air samples were taken. The chemical analysis was performed at the Environment Institute, Ispra. The samples were thermally desorbed and analysed by gas chromatography with an electron capture detector.

Prior to the ETEX experiment, two studies were carried out to determine the ambient levels of the PFC tracers in Europe. The first took place in early 1994 and the second in October, 1994, just before the first tracer release. At all ETEX stations, passive capillary adsorption tubes (CATs) were exposed for 14 days and their contents analysed to determine PFC levels at Brookhaven National Laboratory (BNL), USA. In parallel with the second tracer release, another set of CATs was exposed. For quality control reasons, at a couple of measuring sites different samplers were co-located and duplicate sampling was performed. A high number of field and laboratory blanks were distributed. Also PFC standards in use at BNL and Ispra were compared.

First Tracer Release

The first tracer release started on 23 October, 16:00 UTC, and ended on 24 October, 3:50 UTC. During these 12 hours a total of 340 kg of PMCH were emitted, resulting in an average release rate of 7.95 g/s. The air stream (67 m^3/h) at the top of the chimney (8 m above ground) had an average temperature of 84 °C and a velocity of about 45 m/s.

The weather situation on 23 October showed a deep low, with its centre east of Scotland, moving north. The cold front had passed the day before and in western Europe south-westerly air flow was prevailing. Also in the release area the air flow was of south-westerly direction. In the surface layer the wind velocity was 6 m/s and the temperature was 11 °C. The air was unstable and no clear mixed layer height was observed.

Second Tracer Release

The second tracer release started on 14 November, 15:00 UTC. and ended on 15 November, 2:45 UTC. During these 12 hours a total of 490 kg of PMCP were emitted, resulting in an average release rate of 11.56 g/s. The air stream (71 m^3/h) at the top of the chimney had an average temperature of 73 °C and a velocity of about 50 m/s.

The weather situation on 14 November showed a deep low, with its centre between Iceland and Norway, moving slowly eastwards. The associated cold front curved from Denmark towards the Channel and the Azores. At the release site there was a very strong south-westerly wind (12 m/s) of very stable and mild (14 °C) warm sector air. The stable layer extended to 400 m. During the release there was a slight drizzle. The cold front passed the site towards the end of the release. After the passage of the front the wind at ground level decreased rapidly.

EXPERIMENTAL RESULTS

Both of the tracer experiments were very successful. All preparations were finished in time, the operation of samplers, shipment of samples for tracer determination, and the GC analysis at Ispra went according to plans, although it took a much longer time than anticipated.

During the first experiment (Nodop et al., 1997) a westerly flow transported the tracer plume from its release point in Brittany north-eastwards across Europe. The results from the ground sampling network allowed the determination of the cloud evolution as it travelled across Europe. In Figure 1 the location of the tracer is shown for 24 and 48 hours after the release on 23 October, 1994.

The large black dots indicate the stations where tracer was found above background level. The concentrations in the plume ranged from 10 to above 200 fl/l in ambient air. From a total of over 4000 samples taken at ground level, 1300 values showed tracer plume concentrations clearly above background. That the tracer plume had not reached some stations at certain times can be seen from 1900 samples giving concentrations within the background noise level. Due to quality control procedures 800, measurement values were discarded.

Preliminary results from the second release (Fig. 2) indicate an eastwards tracer transport, at a higher speed compared to the first release, though not all stations are analysed yet.

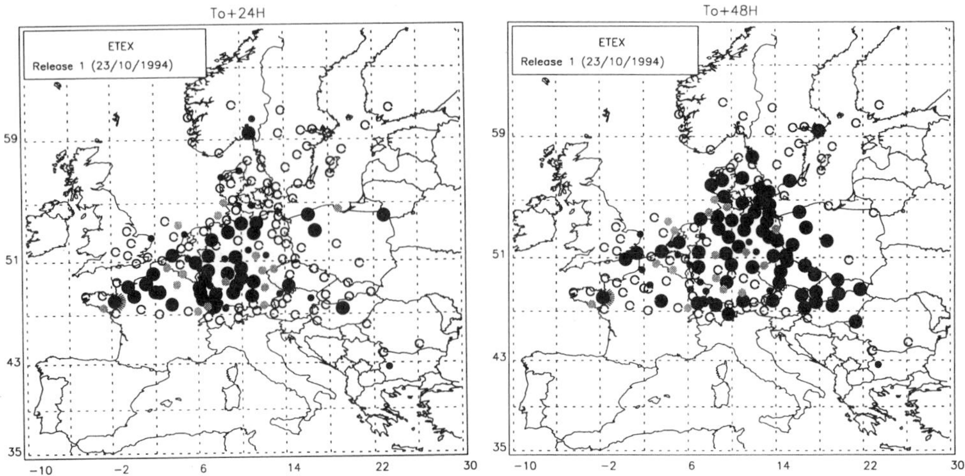

Figure 1. PMCH detected at ground-level stations in Europe, 24 and 48 hours after the first release. Tracer found (O), no tracer found (o), location only (o), tracer found but uncertain (o).

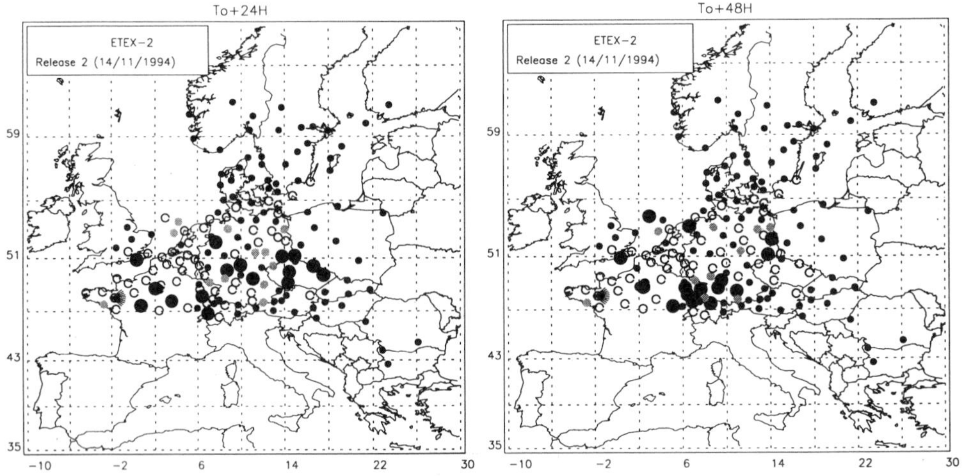

Figure 2. PMCP detected at ground-level stations in Europe, 24 and 48 hours after the second release. Tracer found (O), no tracer found (o), location only (o), tracer found but uncertain (o).

The PFC concentration profiles during the first release show clearly the decrease of the maximum concentration and the increase of the plume extension with distance (Nürnberg, Malmö, Gyor, and Bucharest) from the release point (Fig. 3).The plume shapes are more or less regular. However for the second release the plume evolution shows at some locations (Belgium, The Netherlands and also France - not shown) an interesting secondary peak after 60 hours(Fig. 4). The explanations are not clear yet.

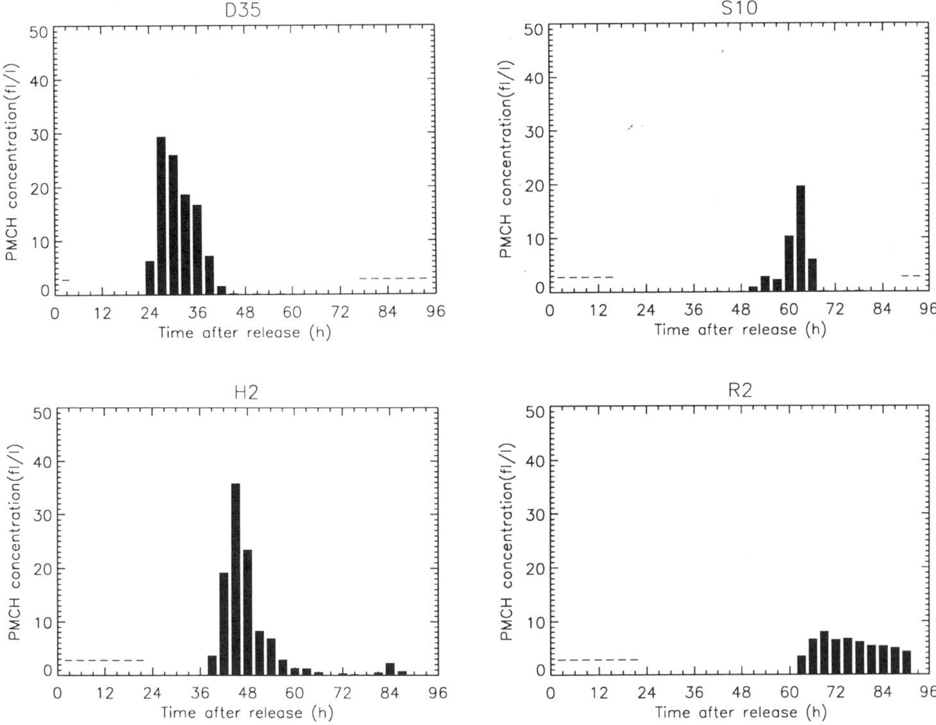

Figure 3. PMCH concentration profile at D35 (Nürnberg, Germany), S10 (Malmö, Sweden), H2 (Gyor, Hungary) and R2 (Bucharest, Romania), during the first release.

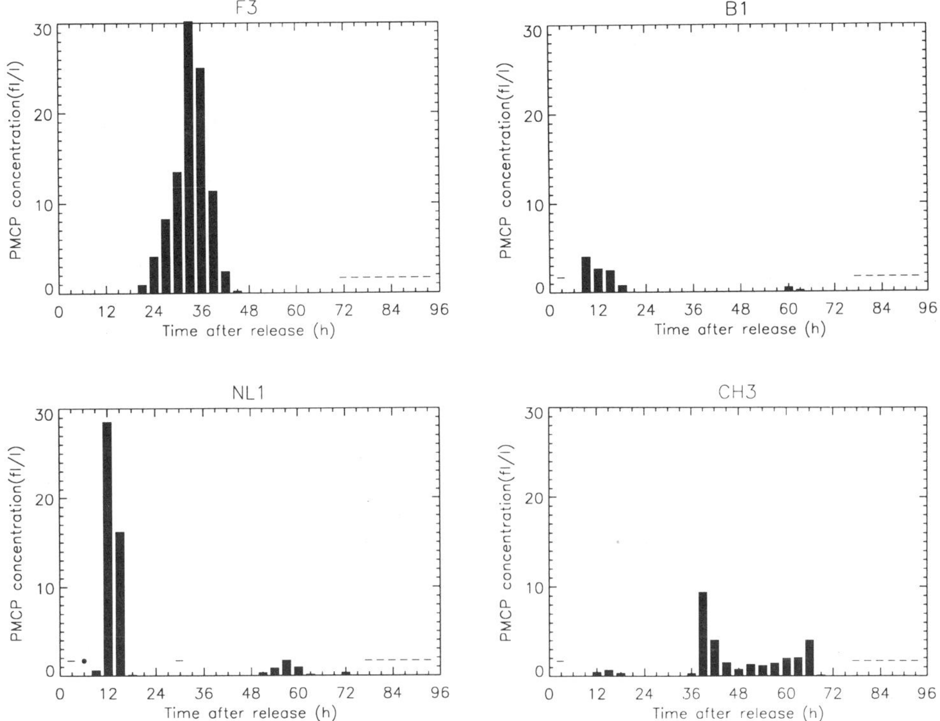

Figure 4. PMCP concentration profile at F3 (Auxerre, France), B1 (Dourbes, Belgium), NL1 (Beek, The Netherlands) and CH3 (Zürich, Switzerland), during the second release.

CONCLUSIONS

ETEX provides a unique experimental database for validating long-range atmospheric dispersion models. Further statistical analysis of the great amount of information collected is still to be done.

ACKNOWLEDGMENTS

ETEX is sponsored by the EC, the IAEA and the WMO. The project is managed by the JRC, Environment Institute, and which has responsibility for the experiments and the evaluation of the models' performances. ETEX was made possible by the enthusiastic participation of the national weather services and responsible institutes inside and outside Europe. Special thanks go to the site personnel operating the samplers. The release site was made available by the University of Rennes, Radio Communications Faculty. Their contributions are gratefully acknowledged.

REFERENCES

R. N. Dietz, Perfluorocarbon Tracer Technology, in: Sandroni (ed.), Regional and Long-range Transport of Air Pollution, 215-247 Elsevier Science Pub. (1986)

F. Girardi, G. Graziani, W. Klug, K. Nodop, European Tracer Experiment, Description of the ETEX Project, EUR Report in prep., EC JRC Ispra 1997

W. Klug, F. Girardi, G. Graziani,, K. Nodop, ETEX, a European Atmospheric Tracer Experiment, in: Proc. Top. Meeting on Environmental Transport and Dosimetry, ANS, Charleston SC. 1993, pp. 147-149

K. Nodop, R. Connolly, F. Girardi, European Tracer Experiment, First Release in October 1994, EUR Report in prep., EC JRC Ispra 1997

DISCUSSION

R. ROMANOWICS: What percentage of tracer mass was found in both experiments in comparison with total released?

A. G. STRAUME: It is very uncertain to calculate the total mass from ground level measured concentrations only. Therefore it has not been done.

INTERCOMPARISON OF TWO LONG-RANGE LAGRANGIAN PARTICLE MODELS WITH ETEX TRACER DATA

F. Desiato[1], D.Anfossi[2], S. Trini Castelli[3], E. Ferrero[4], G. Tinarelli[5]

[1]Agenzia Nazionale per la Protezione dell'Ambiente (ANPA), Roma, Italy
[2]C.N.R., Istituto di Cosmogeofisica, Torino, Italy
[3]Universita', Dipartimento di Fisica Generale, Torino, Italy
[4]Universita', Dip. Scienze Tecn. Avanz., Alessandria, Italy
[5]ENEL/CRAM, Servizio Ambiente, Milano, Italy

INTRODUCTION

The first European long-range tracer experiment (ETEX), jointly organised by the European Commission, the World Meteorological Organisation and the International Atomic Energy Agency, took place on October 23, 1994. The aim of the experiment was to simulate an emergency situation following a release of harmful material into the atmosphere, and to test both real-time and a-posteriori modeling capabilities of reproducing in space and time the tracer concentration field. An inert tracer (perfluoromethylcyclohexane, a perfluorocarbon compound) was released near Rennes, in Northwest France, for twelve hours starting form 16 h UTC, and sampled at three-hourly intervals by 168 ground sites up to a distance of about 2000 km. The synoptic situation at the beginning of the release was characterised by a deep low east of Scotland slowly moving North, maintaining a strong flow from west-south-west in the lower layers over the release site. This condition, together with the correct forecast of its evolution for the following three days, yielded a large number of sampling stations, sparse over Central and Northern Europe, detecting concentrations above background, and so determined the success of the experiment. The concentration data set constitutes the base for both real-time and a-posteriori model evaluations. We present here an a-posteriori intercomparison between two Lagrangian particle models (APOLLO, developed at ANPA, and MILORD, developed at CNR-ICG) simulations of ETEX.

MODELS DESCRIPTION

APOLLO (Atmospheric POLlutant LOng range dispersion) and MILORD (Model for the Investigation of LOng Range Dispersion) are Lagrangian particle models for simulating the long range transport and dispersion of either chemical pollutants or radionuclides. Both were validated against the Cs-137 air concentrations of the Chernobyl accident (Klug et al., 1992; Anfossi et al., 1995; Desiato, 1992).

In both models particle positions X_i (i=1,2,3) are computed at each time step Δt as follows:

$$X_i(t+\Delta t) = X_i(t) + \overline{U}_i(t)\,\Delta t + U'_i(t)\,\Delta t \qquad (1)$$

where X_i is the i-th component of particle position, $\overline{U}_i$ represents the transport due to the mean wind and U'_i refers to the atmospheric turbulence contribution. The two models use similar schemes, based upon an algorithm developed by Reap (1972), for the pollutant transport ($\overline{U}_i$), even if they slightly differ in the horizontal and time interpolation of the ECMWF gridded wind data: APOLLO makes use of a bi-linear interpolation on standard pressure levels and a linear one in time, whereas MILORD interpolates in space by means of a bi-cubic formula (making use of 12 points) and in time with a parabola (three consecutive analysed fields are taken into account).

In both models the horizontal components of U'_i are computed by:

$$U'_i(t)\,\Delta t = \sqrt{2K_i\Delta t}\;\mu_i \qquad (i=1,2) \qquad (3)$$

where K_i are the horizontal diffusion coefficients and μ_i are random numbers picked up from a Gaussian distribution having zero mean and unit standard deviation.

For the vertical component, in APOLLO a drift correction for inhomogeneous turbulence, $(dK_z/dz)\Delta t$, is added to the r.h.s. of eq. (3). K_z is parameterised in the stable, neutral and unstable boundary layer as a function of meteorological scaling parameters derived as in Hanna and Chang (1993). In MILORD, two cases are considered : i) if particles move above the mixing layer, i.e. if $X_3(t) > Z_g + H$ (where Z_g is the height of the orography below the particles), a relationship like eq. (3) with $K_z = 1\ m^2s^{-1}$ is used; ii) if particles are within the mixing layer, their vertical position is randomly re-assigned within H.

In order to calculate air concentrations, a kernel density distribution is assigned to each particle. In the vertical the distribution is uniform and stepwise below the mixing height. In the horizontal, a Gaussian distribution is assumed. In APOLLO, the standard deviation is assigned dynamically as a function of the average distance between the particle and the surrounding particles inside a grid cell. In MILORD the following expression (Gifford, 1982) is used for the kernel standard deviation:

$$\log(\sigma) = -2.81524 + 0.118807\log(t) + 0.274799[\log(t)]^2 - 0.0241565[\log(t)]^3, \qquad (4)$$

where t is the particle age, i.e. the time elapsed from the emission at the source.

APOLLO and MILORD were run for a period of 60 hours after the release using the ECMWF analyses with a temporal resolution of 6 hours. In APOLLO time step was set equal to 60 minutes and 120 particles per Δt were emitted during the release period. The similar figures for MILORD are 36 minutes and 500 particles.

SENSITIVITY TO K_H AND H VARIATIONS

The scope of the ETEX modeling exercise was to test different dispersion models and to try to improve their performance through the comparison between calculated and measured tracer concentrations. To investigate on the APOLLO and MILORD behaviours, a number of model runs were performed with different model versions. In particular, the attention was focused on the horizontal diffusivity K_H and mixing height H parameterizations, which appear to play a major role in determining the shape, size and position of the computed concentration pattern. In the following, the most noteworthy model runs are described.

In APOLLO, K_H was firstly derived by the empirical fit (eq. 4) of a large number of experimental data on horizontal atmospheric diffusion at time scales ranging form a few seconds to a few days (Gifford, 1982). In this version, K_H depends on travel time, as discussed by Zannetti (1990). Alternative runs were performed with fixed values of K_H. Among the others, $K_H = 2.5x10^4\ m^2s^{-1}$ gave the best results. For the parameterization of H, a space and time varying mixing depth is firstly computed by choosing the largest value between a mechanical mixing depth proportional to friction velocity, and, for daytime hours, a convective one based on the integration scheme proposed by Batchvarova and Gryning (1990). Input data are cloud cover, temperature and wind speed. This scheme was preferred to the temperature profile "dry parcel method", since a recent study showed that H is underestimated using ECMWF gridded temperature profiles (Verver and Holtslag, 1991). The following values are chosen for the roughness height: 0,01 m over sea, 1 m over topography higher than 1000 m, 0.1 m elsewhere. Alternative runs were performed with fixed values of H. Among the others, H = 1000 m gave the best results.

In MILORD K_H may assume fixed values only. Different runs were performed and the one with $K_H = 4.5x10^4\ m^2s^{-1}$ gave the best results. Different runs were also performed using different options for prescribing the space and time variations of H: i) - following a daily modulation.(unstable by day and stable by night) deduced at each grid point by a meteorological pre-processor on the basis of wind and temperature information at 10 m, cloud cover and land use information; ii) - deducing it as in the previous case but considering the mechanical mixing depth (proportional to friction velocity); iii) H constant in time and space. Particles are not allowed to cross the three vertical boundaries: the top of the computational domain $H_T(x,y)$ (to which a constant value is assigned), the height of the mixing height H(x,y,t) and the bottom of the domain $Z_g(x,y)$. The only condition that allows particles to move between the mixing layer and the free atmosphere is related to the height variation of H along the trajectory. Perfect reflection is imposed at the boundaries.

A model evaluation based on the MODIA post processing package (Morselli and Brusasca, 1991) allowed to compare predicted and observed three-hourly concentrations (paired in time and space). According to the ETEX protocol, measured zeroes were discarded. The statistical indexes considered are: maximum value (Max), top ten average (Top_10), fractional bias (FB), normalised mean square error (NMSE), correlation coefficient (CC), percentage of model predictions falling within a factor of two (FA2) or five (FA5). Results are indicated in Table I. They put in evidence some interesting features of the various simulations. Particularly interesting are the values of Max, Top_10 and NMSE indexes in run 4, in which a typical PBL value of K_H was used. Max and Top_10 are larger than the observed values by one order of magnitude. This means that the crosswind size of the computed tracer cloud is too small, so that a large number of samplers are not reached by the plume which is not correctly diluted. On the other hand FB is good and the same can be said for the number of observed/calculated pairs that are foreseen within a specified factor. However this result may be fortuitous and due to the fact that the large overestimation at the nearest samplers during the first hours is counterbalanced by a large underestimation at the farthest samplers during the last hours. This causes the very large NMSE value. Looking at Table I allows to conclude that the results improve with fixed H and increased K_H values. For the interpretation of these results a critical discussion of the meaning of K_H and H in our models might be useful. Mixing height plays a very specific role in APOLLO and MILORD. In both models, all particles located inside the mixing layer contribute in the same way to ground level air concentration, as they have a uniform density distribution in the vertical,

Table I - Statistical indexes for three-hour concentrations

		Max (ng/m^3)	Top_10 (ng/m^3)	FB	NMSE	CC	FA2 (%)	FA5 (%)
	observed	12.57	4.87	-	-	-	-	-
1	APOLLO K_H var H var	16.82	12.07	0.39	8.68	0.62	25	57
2	APOLLO $K_H=2.5x10^4$ H var	12.89	8.52	0.32	4.79	0.63	13	60
3	APOLLO $K_H=2.5x10^4$ H = 1000	7.60	5.22	0.12	2.92	0.62	35	71
4	MILORD H var/mech $K_H = 50$	128.9	19.15	0.13	142.8	0.23	40	66
5	MILORD H var/mech $K_H = 1.5x10^4$	13.53	5.40	- 0.48	6.89	0.53	34	62
6	MILORD H var/mech $K_H = 4.5x10^4$	29.99	9.93	- 0.24	15.81	0.66	34	65
7	MILORD H var $K_H = 4.5x10^4$	12.78	8.13	- 0.16	7.95	0.48	33	62
8	MILORD H = 1500 $K_H = 4.5x10^4$	13.51	4.17	- 0.29	2.94	0.73	44	74

between the terrain and the mixing top. This assumption relies on the consideration that particles are rapidly mixed in the vertical, with respect to the long travel times. On the other hand, an inaccurate estimate of H has the consequence that a number of upper air particles may be erroneously included to, or excluded from, contributing to ground concentration. Due to the effect of wind shear, this may result in a wrong position and size on the calculated air concentration pattern. Through the ETEX simulation, we have some evidence of this effect at least in two cases. During the first hours, when strong winds near the release point result in a mechanical H > 1000 m, the calculated plume is transported a few degrees south of the observed. Since the ECMWF winds show a clockwise rotation with height, this might be an indication of overestimation of H. After 36-48 hours, when low winds in central Europe result in a mechanical H < 1000 m, the calculated concentration pattern is somewhat backward and smaller than the observed. This might be an indication of underestimation of H. Both cases represent a possible explanation of the better performance of model runs with H = constant with respect to the original versions. In any case, from our sensitivity analysis it can be concluded that the estimate of a space and time varying mixing height, at least in the sense we use it in our models, based on low-resolution meteorological fields coming from a global circulation model, is not reliable.

As far as K_H is concerned, the large values needed by both models to give reliable results, confirm that it is necessary to increase the horizontal diffusion in long range dispersion models, in order to include the advection fluctuations non-resolved by large scale meteorological models (Zannetti, 1990). The same result was also obtained by Ishikawa (1995), who found that his best simulation of the Chernobyl dispersion data required a fixed value of horizontal diffusion lying in the range $3.3x10^4$-$1.0x10^5$ m^2s^{-1}. He also found that using a variable K_H, derived from an empirical formula for the horizontal mean square displacement as the ones suggested by Gifford (1982), did not improve the simulations. The present results from APOLLO and MILORD confirm such results. A deeper insight on this problem may derive from the following considerations. Gifford (1982), looking for what mechanism could maintain the observed plume growth at times greater than the typical time scale of PBL turbulence suggested that this growth is due to substantial energy contained in the mesoscale (motions in the range 1-48 hours), at difference (on the basis of some experimental evidence) with what generally assumed, i.e. that in this region there is the presence of a spectral gap. On the other hand, McNider et al. (1988) attributed such growth "to the vertical shear in the horizontal wind produced by diurnal and/or inertial oscillations in conjunction with or followed by vertical PBL mixing".

Table II - Three-hour concentrations: best simulations

	Max (ng/m^3)	Top_10 (ng/m^3)	FB	NMSE	CC	FA2 (%)	FA5 (%)
observed	12.57	4.87	-	-	-	-	-
APOLLO $K_H=2.5x10^4$ H = 1000	7.60	5.29	0.23	3.78	0.61	28	64
MILORD H = 1500 $K_H= 4.5x10^4$	13.52	4.85	- 0.06	4.64	0.66	40	69

Table III - Time integrated concentrations: best simulations

	Max (ng/m^3)	Top_10 (ng/m^3)	FB	NMSE	CC	FA2 (%)	FA5 (%)
observed	27.24	9.65	-	-	-	-	-
APOLLO $K_H=2.5x10^4$ H = 1000	33.12	13.00	0.23	1.11	0.80	33	62
MILORD H = 1500 $KH= 4.5x10^4$	35.17	9.27	- 0.06	1.76	0.73	34	64

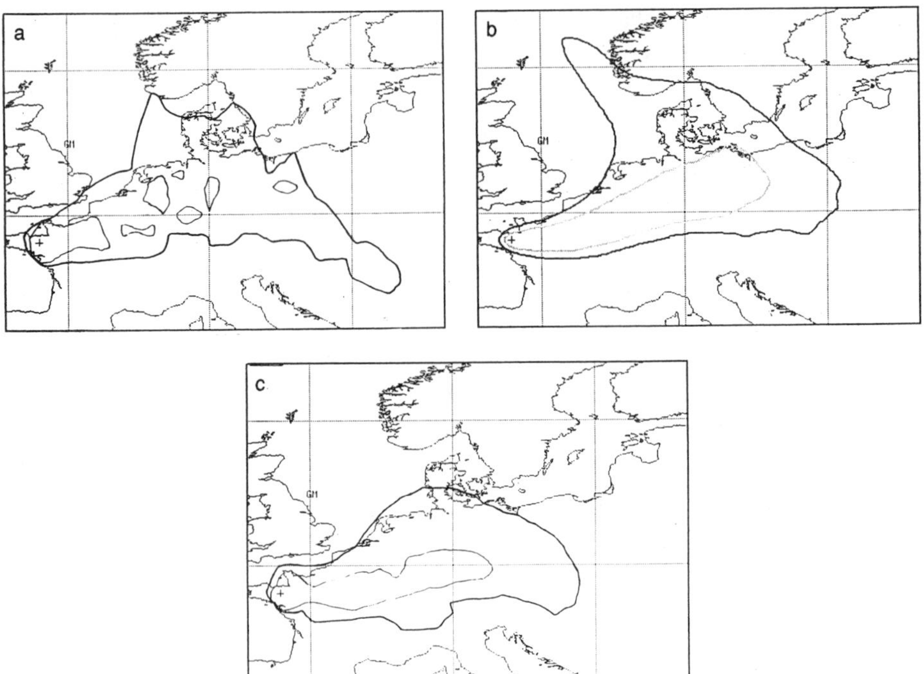

Figure 1. (a) Measured time integrated concentration isolines. The two contour levels correspond to 10 and 1 ng/m^3 h. (b) Time integrated concentration isolines produced by APOLLO. (c) Time integrated concentration isolines produced by MILFORD.

BEST SIMULATION RESULTS

Runs 3 and 8 of Table I are the best simulations obtained by APOLLO and MILORD, respectively. Their model evaluation indexes are reported again in Table II. However in this case only the pairs in which both observed and predicted were zeroes were discarded. Comparing Tables I and II gives an insight on the importance of retaining in the statistics measured zeroes when the corresponding computed values are not zero.

Table III shows the results of the best model simulations in the case of time integrated concentrations. These results show a good correspondence to the experimental reality. The same holds true looking at Figures 1 in which the observed time integrated isolines (Figure 1a) are compared to those produced by APOLLO (Figure 1b) and Milord (Figure 1c).

CONCLUSIONS

The simulations of ETEX long range dispersion through the APOLLO and MILORD have been shown. A brief survey of their basic characteristics and differences have been described. The influence of the more important input parameters (mixing height and horizontal diffusion coefficient), was studied. It was found that the best simulations were obtained by setting a fixed value for both H and K_H. In particular, it was found that a rather large K_H value is needed to correctly simulate the size and dilution of the tracer cloud. This fact confirms previous work on the same subject by Gifford (1982), McNider at al. (1988) and Ishikawa (1995). The accuracy of both models, stated through a model evaluation against measured tracer concentrations, appeared to be good and comparable to each other.

ACKNOWLEDGMENTS

The authors acknowledge Dr. Sonia Mosca and JRC - Ispra for having allowed the publication of the graphs shown in Figures 1 a-c.

REFERENCES

Anfossi D., Sacchetti D. and Trini Castelli S., 1995, Development and sensitivity analysis of a Lagrangian particle model for long range dispersion, Environmental Software, 10, 263-287

Batchvarova E. and Gryning S.E., 1990, Applied model for the growth of the daytime mixed layer, Boundary Layer Meteorology, 56, 261-274.

Desiato F.,1992, A long range dispersion model evaluation study with Chernobyl data, Atmospheric Environment, 26 A, 2805-2820

Desiato F. and Bider M., 1994, ARIES-I: A computer system for the real-time modeling of the atmospheric dispersion at different space and time scales, Environmental Software 9, 201-212

Gifford F.A., 1982, Horizontal diffusion in the atmosphere: a Lagrangian-dynamical theory, Atmospheric Environment, 16, 505-512

Hanna S.R. and Chang J.C., 1993, Hybrid plume dispersion model (HPDM) improvements and testing at three field sites, Atmospheric Environment, 27 A, 1491-1508

Ishikawa H., 1995, Evaluation of the effect of horizontal diffusion on the long-range atmospheric transport simulation with Chernobyl data, Journal of Applied Meteorology, 34, 1653-1665

Klug W., Graziani G., Grippa G., Pierce D. & Tassone C. (Eds), 1992, Evaluation of Long Range Atmospheric Transport Models using Environmental Radioctivity Data from the Chernobyl Accident: the ATMES Report, Elsevier Applied Sciences

McNider R.T., Moran M.D. and Pielke R.A., 1988, Influence of diurnal and inertial boundary-layer oscillations on long-range dispersion, Atmospheric Environment, 22, 2445-2462.

Morselli M.G. and Brusasca G., 1991, MODIA: Pollution dispersion model in the atmosphere, Environmental Software Guide, 211-216.

Reap R.M., 1972, An operational three-dimensional trajectory model, Journal of Applied Meteorology, 11, 1193-1201

Verver G.H.L. and Holtslag A.A.M., 1992, Sensitivity of an operational puff dispersion model to alternative estimates of mixed-layer depth, in Air Pollution Modeling and its Application IX, H. van Dopp and G. Kallos, ed., NATO, Challenges of Modern Society 17.

Zannetti P.,1990, Air Pollution Modelling, Computational Mechanics Publications

DISCUSSION

J. H. SØRENSEN: If my memory serves, radiosoundings for the area covered by the ETEX-1 plume show that the boundary-layer height varied between 500 and 1000 m. How do you justify the use of a constant boundary-layer height of 1000 m, and even 1500 m ?

F. DESIATO: The mixing height is used in both our models as the top of the layer containing particles which contribute to ground level air concentrations. In our opinion, our results indicate that, for the purpose of its estimate, punctual, in space and time, PBL height derived from radiosoundings or from ECMWF gridded temperature profiles are not adequate.

J. H. SØRENSEN: You have used different values of the horizontal eddy diffusivity coefficient for the two Lagrangian long-range transport models, which used the same meteorological data (ECMWF). In my understanding, this coefficient is used to describe sub-grid scale diffusion. Thus the value depends mainly on the horizontal resolution of the underlying meteorological model, and the value should be the same for the two dispersion models. Can you comment on this?

F. DESIATO: We have tried several values of the horizontal eddy diffusivity coefficient K in order to test the sensitivity of our models to this parameter. It turns out that the best results are obtained with K=25000 and K=45000, respectively, which are of the same order of magnitude and both fit satisfactorily with the interpolation of several experimental data (Gifford, 1982, in the list of references). Both models are not very sensitive to changes of K values of the order of 20000 around their "best" estimate.

A STUDY OF THE TRACER DISPERSION MODEL SNAP USING ENSEMBLE FORECASTS AND ITS EVALUATION BY USING EUROPEAN TRACER EXPERIMENT DATA

Anne Grete Straume[1] Ernest N'Dri Koffi,[2] and Katrin Nodop[1]

[1]Joint Research Centre
Environment Institute
21020 Ispra (VA), Italy
[2]Laboratoire d'Aėrologie
(UMR CNRS/UPS 5560) Universitė Paul Sabatier
3100 Toulouse, France

INTRODUCTION

Numerous numerical dispersion models have been developed to predict long range transport of hazardous air pollution in connection with accidental releases. When developing and evaluating such a model, it is important to detect uncertainties connected to insufficient characterization of the accidental release, the meteorological input data, errors in the formulation of the Numerical Weather Prediction (NWP) model, and errors in the formulation of the dispersion model.

A tracer dispersion model, The Severe Nuclear Accident Program (SNAP), developed and operated at the Norwegian Meteorological Institute (DNMI), is here used to investigate the effect of errors in the meteorological input data. 32 ensemble forecasts produced by the European Centre for Medium-Range Weather Forecasts are then used as input to the SNAP model. The resulting 32 puff predictions are compared, and we expect to find a spread in the predicted puff evolutions. This will indicate the importance of the quality of the meteorological input data for the success of the dispersion model.

In order to evaluate the SNAP model, its calculations are compared with measurements from the European Tracer Experiment (ETEX) (Klug et al, 1993; Archer et al, 1996). The aim of the ETEX project was to evaluate the ability of dispersion models to predict puff dispersion from an accidental release. It was also performed to test the ability of dispersion modelers to predict the outspread from the accidental release in real time. The ETEX experiment contained two tracer releases. We will here consider data only from the first release. The modelled dispersion will be compared with the measurements up to 60 hours after the start of the release. Reasons for deviations between model and measurements are then to be examined.

A description of the SNAP model will first be presented, together with a description of the ensemble forecasts. Results from the SNAP model calculations using ensemble forecasts as input data are then analyzed and discussed. At last, an inter-comparison of the SNAP results and the measurements from the first ETEX experiment is presented.

DESCRIPTION OF THE SNAP MODEL AND THE ENSEMBLE FORECASTS

The SNAP model is described in detail by Saltbones et al (1995). Only a short overview of the main model features will be given here.

The SNAP model is a Monte Carlo Lagrangian particle model. It has 14 layers in the vertical, of which seven are located below approximately 1800 meters. The vertical coordinate is terrain-following at the surface, and pressure-surface-following in the upper atmosphere. The height of the atmospheric boundary level (ABL) is calculated by the critical gradient Richardson number (Ri_c). The diffusion processes within the ABL are described by the "random walk" technique (Physick and Maryon, 1995). In this technique, the vertical position of the particles within the whole ABL is changed randomly within every time step (which is 15 minutes in the SNAP model).

As mentioned above, the meteorological data used to study the influence of forecast errors in dispersion modelling, are ensemble forecasts collected from the ECMWF. In order to obtain the necessary spatial and time resolution, the forecasts were processed through the NWP model; the High Resolution Limited Area Model (HIRLAM), described by Haugen et al (1995) and Källen (1996).

Several papers, e.g. Molteni et al (1996), describe the methodology and validation of the ECMWF ensemble prediction system. Only a brief summary is given here. Ensemble prediction is a method to predict the evolution of the atmospheric probability density function beyond the range in which error growth can be described by linearized dynamics. The ensemble forecasts are generated for the same period of time by perturbing the initial meteorological fields of the weather forecast. The 32 perturbations are calculated from 16 singular vectors (SV), meant to represent possible forecast developments due to instabilities in the atmospheric flow during the early part of the forecast. The instabilities are generated by errors in the analyzed fields used to define the initial conditions for the forecast. One forecast, the Control Forecast, is generated by using unperturbated operational analysis input. The SVs are chosen to be located in the Northern extratropical Hemisphere, above 30° N, since instabilities in this area will influence the weather over Europe. The ensemble forecasts are organized in 3 to 6 clusters, and classified by comparing the root mean square deviation (rmsd) between the height of the 500 hPa fields for each perturbed forecast and the Control Forecast from day 5 to day 7 after forecast start. The clusters are meant to represent the main phase space directions the ensemble forecasts can develop into.

PUFF DISPERSION USING ENSEMBLE FORECASTS

All ensemble forecasts together with the Control Forecast were used as input to the SNAP model, and their respective tracer dispersions plotted. The results from the Control Forecast and two of the ensemble forecasts, are shown in Figure 1 together with the ETEX tracer measurements. The two ensemble members are picked out by looking at all puff dispersions, and choosing the two that differed the most from the Control Forecast dispersion. As can be seen from Figure 1, the puff locations and shapes are slightly different already after 24 hours but the differences are more visible after 48 hours.

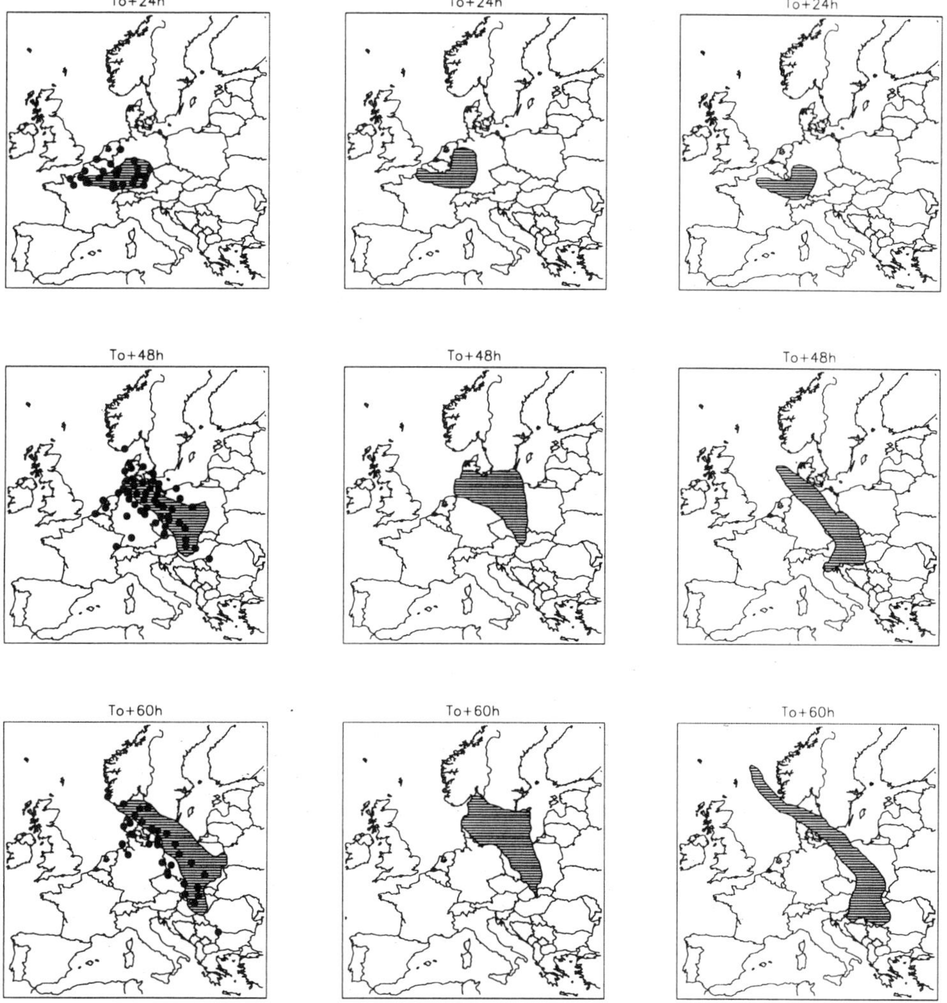

Figure 1. Puff dispersion using meteorological input data from the Control Forecast, ensemble 29, and ensemble 30. The shaded area indicate a tracer concentration greater than 0.1 ng/m^3. The ● indicate ETEX stations with a measured concentration above 0.1 ng/m^3.

The spread among all ensemble forecasts, among the forecasts within each cluster and in-between the clusters will now be quantified through different statistical parameters.

According to Molteni et al (1996), ensemble spread appears to be connected to forecast skill in the way that the less spread, the more predictable is the weather situation. We now examine the ensemble spread for this forecast period, comparing the root mean square deviations (rmsd) between the total ABL tracer concentrations using input from the Control Forecast, and the total ABL concentrations from each ensemble member. The results are presented in Figure 2. The deviations are varying between 0.031 and 0.596 ng/m^3, which is a measure for the uncertainties connected to the meteorological input data. The limit of

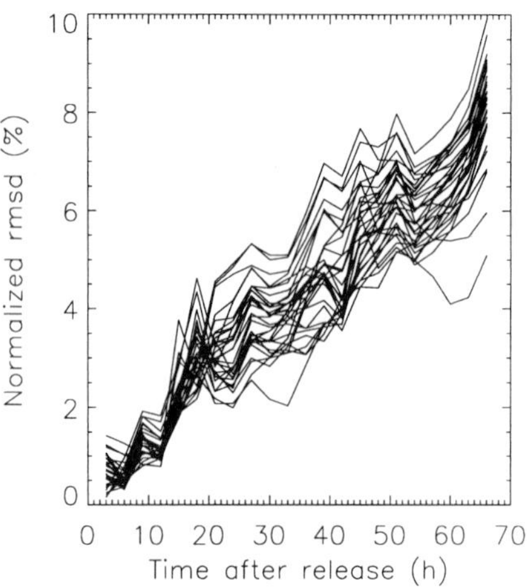

Figure 2. Root mean square deviation (rmsd) in tracer concentration (ng/m^3) between each ensemble member and the Control with time.

detection of the ETEX tracer, above the background concentration, is 0.1 ng/m^3. The uncertainties connected to the measurements are on average of the order of 20%, which should correspond to a variation of the order of 0.05 to approximately 0.320 ng/m^3, depending on measurement location and time. The meteorological uncertainties are strongly dependent upon the geographical area chosen, due to the number of gridpoints containing 0 tracer concentration. They are therefore assumed to be greater than indicated above, and greater than the uncertainties connected to the measurement procedure. The spread between the ensemble members is large during the time period 15 to 45 hours after the release. A continuously increasing spread between the ensemble forecasts is normally detected after 3-4 days (Molteni et al, 1996). The changes in ensemble spread with time for the first 3 days is therefore normally found to be small, due to the definition of the initial perturbations.

The classification of the ensemble forecasts into clusters, done at the ECMWF, is here checked to see whether the members within a cluster are providing similar weather forecasts (a small cluster spread). The mean value and standard deviation of the rmsd ABL concentration between each cluster member and the Control Forecast is calculated. The larger spread within a cluster, the less accurate is the cluster classification for our time period, concerning the ABL. Figure 3 shows the mean value plus and minus one standard deviation for each cluster. The variations between the three are not large, but cluster 2 has a larger mean value and standard deviation than the other two before 50 hours after the release. Cluster 1 and cluster 3 seems therefore to be better classified in this time period. Cluster 1 has 17 members, cluster 2, 8 members, and cluster 3, 7 members. Cluster 1 has still a smaller standard deviation than cluster 2.

The spread in phase space between the three clusters is compared, to see whether the clusters develop in different direction. The spread is defined as the difference in mean ABL concentration between two clusters in each grid point, and is denoted by $d_{n,m}$. If $d_{n,m}$ is small, the spread among the clusters is small, which again implies a small ensemble spread and a quite predictable weather situation. Figure 4 shows the development of the difference between the three clusters with time. As can be seen from the figure, the spread within the

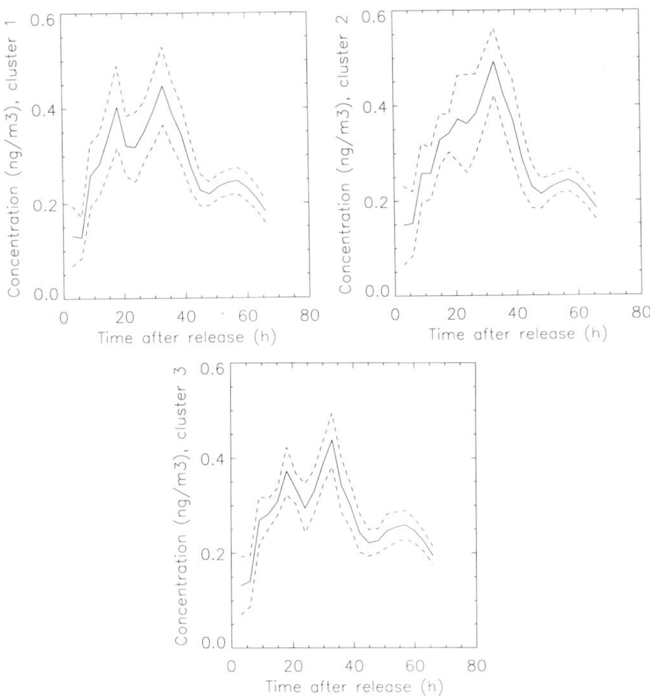

Figure 3. Mean value (__) of rmsd between ensemble members and the Control within each cluster. One standard deviation is added to and subtracted from the mean value (- - -).

pairs of clusters is increasing with time until approximately 40 hours after the release started. The differences are then stabilizing on three levels. Comparing this result with the study of the rmsd between the ensemble members and the Control Forecast, we see that the ensemble forecasts differed most from the Control Forecast at 33 hours after release, and then the difference decreased. Here we see that the difference in-between the clusters still increase after 33 hours, but at a much slower rate than before. We therefore conclude that the clusters develop differently throughout the whole forecast period, even though at a slower rate in the second half. Cluster 2 is differing most from the two other clusters.

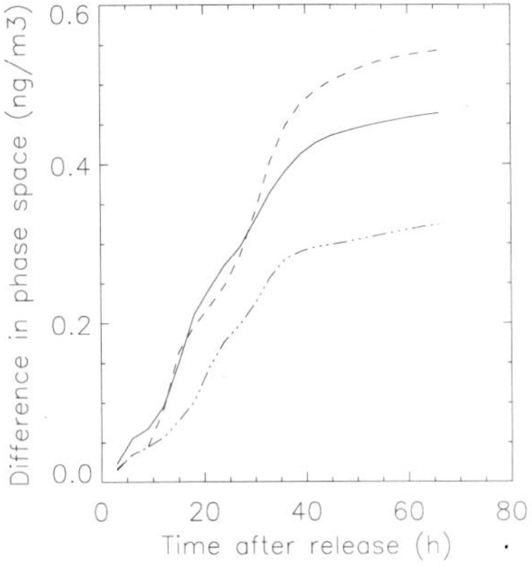

Figure 4. Difference in phase space ($d_{n,m}$) between cluster one and two (__), cluster two and three (- - -), and cluster one and three (-...-...-). The $d_{n,m}$ is represented by difference in tracer concentration (ng/m^3).

EVALUATION OF DISPERSION BY ETEX MEASUREMENTS

The SNAP model is now evaluated by using the Control Forecast as meteorological input data. The calculated puff dispersion is compared with the measurements from the ETEX experiment. The experiment contained two tracer releases. The tracers, both inert gases of the perfluorocarbon family, were released in France during South-Westerly winds in October and November 1994. The outspread of the tracers were measured over a period of 72 hours after each release. The measured and calculated puff evolutions from the first release are here presented (Figure 1). As can be seen from the figure, the model predict the puff evolution to a great extent. Still, it seems to predict a slower tracer dispersion until 60 hours after the release. Note that the modelled tracer concentrations are calculated as the total concentrations in the whole ABL. The puff is therefore expected to cover a larger area than the surface measurements show. The concentrations should also be higher. This is biasing the following inter-comparison study between the measurements and the model.

Several statistical measures can be used to quantify the differences between the measured and the modelled puff dispersion. Here the following parameters have been calculated; bias, sample correlation coefficient, time of arrival of the tracer at each station together with the corresponding puff duration.

The bias between the measurements and the modelled data is found by calculating the mean difference between the measured and modelled tracer concentrations at each station, summed over all time steps. Figure 5 shows the number of stations with a bias within a given interval. A positive value for the bias means that the sum of the measured values at the station is higher than the sum of the modelled values. 69% of the bias values are located between -0.2 and 0.2 ng/m^3 which shows that the overall agreement between the measurements and the model result is good.

Another important parameter for data set inter-comparison is the sample correlation coefficient, also called Pearsons Product Moment Correlation Coefficient (PPMCC). The PPMCC is averaged in the same way as the bias. Figure 6 shows the number of stations within given PPMMC-intervals. Note that when the concentration values are 0 at a station during all time steps for both measurements and model calculations, the correlation is optimal (PPMCC equal 1). The results from the calculations show that the mean PPMCC for all stations is 0.33, while a large number of locations have a PPMCC above 0.8.

The time of arrival of the tracer puff at locations down-wind from the source is an important output from dispersion models designed for emergency management. Differences

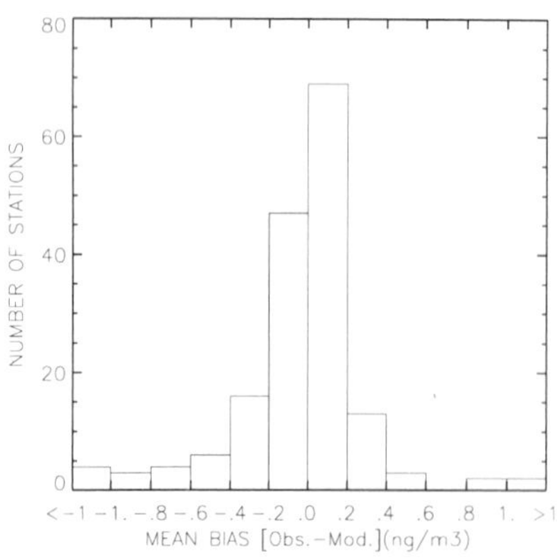

Figure 5. Bias between observed and modelled tracer concentration at each ETEX station, averaged over all time steps.

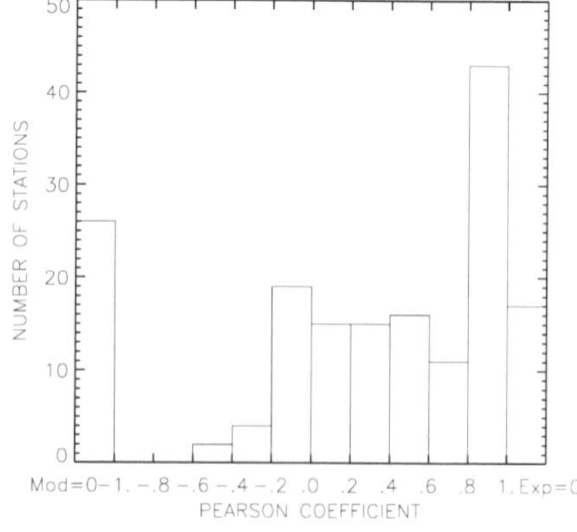

Figure 6. PPMCC between observed and modelled tracer concentration at each ETEX station, averaged over all time steps.

CONCLUSIONS

In this article, ensemble forecasts were used in order to show the necessity of having as precise meteorological input to a dispersion model as possible. Modelled and measured tracer dispersion from the European Tracer Experiment (ETEX) was then compared to evaluate the dispersion model.

The quality of meteorological input data was studied by using ensemble forecasts from the ECMWF. A difference in the ensemble tracer spreads appeared already after 24 hours, which shows the dispersion models dependence upon the meteorological input. The spread between the ensemble forecasts was on the order of 0.031 and 0.596 ng/m^3. The spread is in time of arrival for the modelled data and the measurements are plotted in a global scatter diagram (Figure 7). As can be seen from the figure, the model seems to give a slower advection of the puff, 15 to 60 hours after the release. This may be due to the modelled wind speed, wind direction, vertical diffusion, topography or that the modelled concentration is from the whole ABL. Meteorological input from ensemble 29, gave a faster puff advection than the Control Forecast.

A puff duration scatter plot (Figure 8) gives information about the time the tracer is present at each station. The puff duration measured at the stations is much longer than the modelled one. A possible explanation of this discrepancy is that the tracer might be trapped between obstacles at the surface, which means that the assumption of full relocation of the tracer within the ABL every 15 minutes is not describing the vertical dispersion processes correctly. The scheme for vertical dispersion therefore has to be looked into in more detail. a measure of uncertainty due to meteorological input, and is assumed to be larger due to the influence of gridpoints with 0 tracer concentration. The uncertainty is somewhat larger than the uncertainty connected to the tracer measurements. The classification of the ensemble forecasts, done at the ECMWF, was checked by comparing the internal spread of the forecasts within each cluster. The differences in spread between the three clusters are not large, but cluster 2 has more often a larger standard deviation than the other two for the time period from 0 to 50 hours after the release. The difference in phase space between the three clusters shows that the clusters develop in different phase space directions during the 3 days forecast period, but at a much slower rate after 33 hours.

The inter-comparison of the SNAP model and the measurements show a good agreement. The mean bias between the two, at 69% of the stations, is between -0.2 and 0.2 ng/m^3. This shows that the model manage to simulate the tracer dispersion to a high extent.

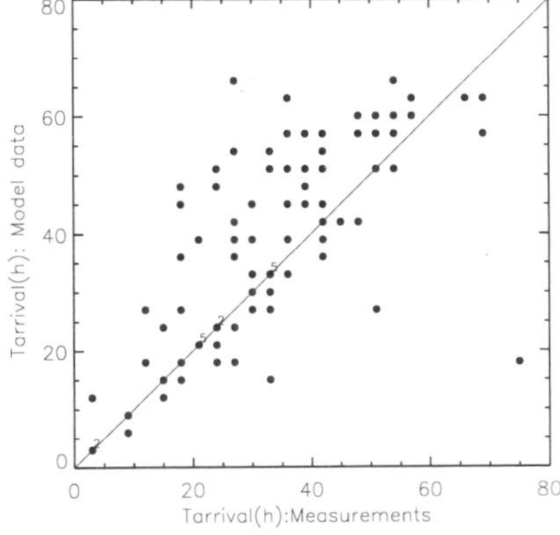

Figure 7. Global scatter diagram, comparing the time of arrival (h) between measured and modelled data at each station.

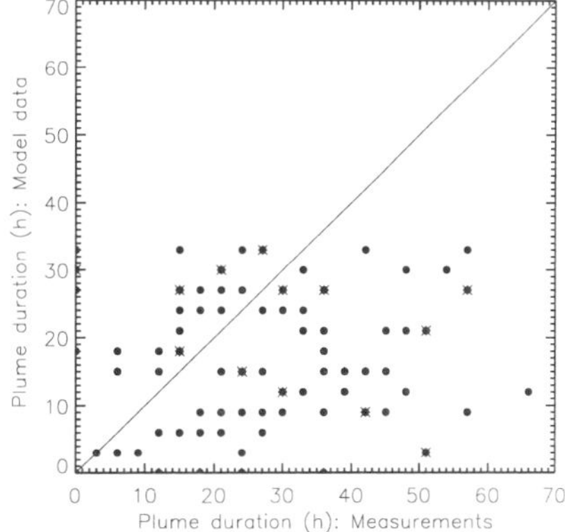

Figure 8. Global scatter diagram comparing plume duration (h) between measured and modelled data at each station.

The correlation between the model and the measurements is, by calculating the PPMCC, shown to be positive on average. A large number of stations had PPMCC values larger than 0.8. The model seems to have a slower puff advection than shown by the measurements between 15 and 60 hours. This may be due to the modelled wind speed, wind direction, vertical dispersion, topography or that the modelled concentration is from the whole ABL. The model seems to transport the tracer too fast away from the stations after the puff has arrived. A reason for this phenomenon is assumed to be that the model mixes the tracer too fast upwards from the ground.

ACKNOWLEDGMENT

First of all thanks to Jørgen Saltbones and his colleagues at the Norwegian Meteorological Institute for lending out the SNAP model, and for the training and guiding of the work with the SNAP model. Special thanks goes to Jan Erik Haugen and Anstein Foss (at the DNMI) for their help in running the HIRLAM and SNAP model with ECMWF ensemble data. Thanks to the ECMWF for providing the ensemble data. Thanks to Professor Øystein Hov at the University of Bergen and Professor Trond Iversen at the University of Oslo for supervision and comments to the article.

REFERENCES

Archer G., Girardi, F., Graziani, G., Klug, W., Mosca, S., and Nodop, K., 1996, The European Long Range Tracer Experiment (ETEX): Preliminary Evaluation of Model Inter-comparison Exercise, *Proceedings from the 21st NATO/CCMS International Technical Meeting on Air Pollution Modelling and Its Application XI* :181-190

Haugen J.E., and Midtbø, K.H., 1995, Det operasjonelle HIRLAM systemet ved DNMI, *Available from The Norwegian Meteorological Institute, Po.Box 47, N-0313 Oslo, Norway.*

Källen E. (ed.), 1996, HIRLAM Documentation Manual, System 2.', *Available from SMHI, S-60176 Norrköping, Sweden.*

Klug W., Girardi, F., Graziani, G., and Nodop, K., 1993, ETEX, a European Atmospheric Tracer Experiment, *Proc. Top. Meeting on Environmental Transport and Dosimetry, ANS, Charleston sc.*.:147-149.

Molteni F., Buizza R., Palmer T.N., and Petrolagis T., 1996, The ECMWF Ensemble Prediction System: Methodology and Validation, *Q.J.R.Meteorol. Soc.* 122:73-119.

Physick W.L., and Maryon, R.H., 1995, Near-source turbulence parameterization in the NAME model, *Met O(APR) TDN No. 218, Available from Meteorological Office, London Road, Bracknell, Berkshire, RG12 2SZ, UK.*

Saltbones J., Foss, A., and Bartnick, J., 1995, The Norwegian MEMbrain application. SNAP: Sever Nuclear Accident Program, A real time dispersion model for major emergency management, *DNMI-report: Project NORMEM(175), Available from The Norwegian Meteorological Institute, Po.Box 47, N-0313 Oslo, Norway.*

DISCUSSION

M. KAASIK: What is the main advantage of your approach (random walk method) compared to methods calculating the turbulent diffusion from given wind fields?

A. G. STRAUME: In the Lagrangian particle approach, polluted air is represented by a cloud of individual particles. Dispersion of this cloud is computed using random walk method with dispersion coefficients calculated from the available wind fields and atmospheric stability. There are two main advantages of using Random Walk method: (1) it is a convenient numerical solution of the problem and (2) it is an efficient method for parameterizing subscale processes leading to horizontal and vertical diffusion.

R. YAMARTINO: You indicated that the modelled tracer arrived later than observed, and had a receptor exposure duration shorter than observed. Couldn't both these effects be accounted for by increasing the rate of models along-wind dispersion (i.e. increase σ_u)?

A. G. STRAUME: In the present SNAP model version horizontal diffusion is isotropic. Increased diffusion along-wind could partly improve the model results compared to the measurements. However, as has been pointed out in the article, the same effect could be achieved by substituting the random relocation of the particles within the whole atmospheric boundary layer (ABL), by a random relocation within several ABL sublayers. Regarding discrepancies in puff arrival time, the reasons can also be related to the meteorological input data, the grid resolution or that surface measurements are compared with a calculated concentration average within the ABL. The two are not necessarily comparable, since the surface values are not measured at the puff centre line, nor are they necessarily representative for the ABL average value.

THE EUROPEAN TRACER EXPERIMENT ETEX: A COMPARISON OF LONG RANGE ATMOSPHERIC DISPERSION MODELS IN DIFFERENT WEATHER CONDITIONS

G. Graziani [1], W. Klug[2], S. Mosca[3]

[1] JRC Ispra, Environment Institute,I-21020
[2] Mittermayerweg 21, Darmstadt,D-64289
[3] Idea, viaVolturno 80, Brugherio, I-20047

INTRODUCTION

In fall 1994 two long range tracer experiments were conducted in Europe over a distance of 1,800 km. For 12 hours an inert, non-depositing tracer was released at Rennes, Brittany, France. The releases took place at the surface and the tracer was sampled at 168 stations throughout Europe. The sampling stations were run by the National Meteorological Services and the whole experiment was sponsored by the EC, the WMO and the IAEA. 24 Institutions took part in the real-time forecasting of the cloud evolution with 28 long range dispersion models. They simulated surface concentration evolution at the locations where the tracer was sampled. The results of these calculations were compared for both experiments with the measurements, using a number of statistical parameters. The results of this comparison (see Archer et al.,1995 for details) indicated for the first release that a limited group of models (7-8) were capable to obtain a good reproduction of the cloud displacement throughout Europe for any time intervals between 24 to 60 hours after the release start. Large differences were however found when examining the predicted tracer concentration at a particular location. In this paper some results for the second release are also presented, similarities and differences with the first are evidenced and tentative explanations are proposed for the differences in model performance.

PARTICIPANTS AND REAL TIME MODELS

The ETEX participants were asked to forecast both the flow and diffusion field to calculate concentration evolution in real time. Therefore, discrepancies between concentration forecasts of different models are due to the initialization of flow models, to the forecasted flow and to the type of dispersion model used. The main characteristics of the models employed can be found in Archer et al.(1995).

Air Pollution Modeling and Its Application XII
Edited by Sven-Erik Gryning and Nadine Chaumerliac, Plenum Press, New York, 1998

DESCRIPTION OF THE WEATHER SITUATION DURING THE RELEASES

First release

On Sunday 23 October (date chosen for the first release), the following weather criteria were matched:

- a rather strong West to South-westerly flow, advecting the tracer during the experiment over many stations.
- no centre of high pressure or low pressure, no extending ridges or troughs passing the release site during the release or being very close to it.
- no frontal system passing the release site short before or after the period of release.

Synopsis 24 October:
A deep Low, 975 hPa East of Scotland was slowly moving North, maintaining a strong Southwesterly flow over release site.

Synopsis 25 October:
The wind was backing more to the south in North-West France while the front approached during the day, and after its passage the wind was veering to the south-west.

Synopsis 26 October:
There was still a complex Low over the North Sea and Scotland. Showery weather with a wind tending to veer a little bit over western Europe towards west-south-west.

Possible problems for the dispersion models caused by the meteorological situation were identified as:

Advection Speed: Due to the rather strong surface winds during the first twelve hours, the mixing layer depth was probably rather large during the night following the emission. Therefore it is possible that some models which consider a fixed stable layer during the night, have kept the tracer cloud to lower levels, then using too low wind speeds to advect it.

Advection Direction: It is expected then that only marginal differences in dispersion direction solutions are likely to occur. However, there was a definite shear of the wind with height and those models that used higher wind speeds could show a more veered direction positioning of the cloud.

Second release

The synoptic situation on November 14 (date of the second release) was characterised by a deep Low, between Iceland and Norway, slowly moving East and filling. An associated coldfront over Denmark and Holland bent towards the Channel, with a minor frontal wave along the front. On the release site, a very strong south-westerly wind was blowing, advecting stable and mild warm air. The coldfront passed at Rennes at about 0300 UTC on November 15. After the frontal passage, the surface wind decreased in strength and veered to West.

Synopsis on November 15
The westerly wind decreased at the release site, where the weather became rather stable. The weather was more unstable, showery and windy over eastern France, with wind directions mainly from West.

Synopsis on November 16

West-North-westerly winds advecting unstable, fresh air with showers were observed over northern part of Central Europe. More stable weather and less strong winds were present over central France and southern Germany.

Synopsis on November 17

Weather over Central Europe was rather quiet, with weak westerly winds at the surface. A Low over Oslo was still giving rise to rather strong NW winds and showery weather, with a trough passage over Denmark, North Germany and Holland.

Possible problems for the dispersion models caused by the meteorological situation were identified as:

Advection speed: A forecast of an earlier passage of the front would also influence the cloud speed, since at the end of the dispersing cloud , the surface windspeed was steadily decreasing. Due to the strong shear in windspeed with height, some models might split the cloud into two parts, while a deeper boundary layer is building up during 15 November daytime.

Advection direction: The frontal position as evaluated by various meteorological models is critical for the dispersion, since immediately after the wind veered and dropped. During the release, contrary to the first one, the boundary layer was about 600 m deep and the temperature profile in the lower atmosphere was stable and wind direction was about constant. These conditions may have produced a rather narrow plume during the first 12 hrs after the release start. The NW movement of the cloud during the first period of time had to change on the day after when the cloud should bent more to the East.

STATISTICS USED IN THE COMPARISON

Similar to what has been done in ATMES (Klug et al., 1992), statistical comparison has been performed globally, pairing model results with observations independently of time and space, or with a time-dependent analysis where the results were examined for a number of selected sites, or with a space-dependent analysis for given time intervals. Among the indicators used,

- Time Series, the simplest aids to model evaluation, and also some of the most effective. The time series plots of model predictions over time help to examine the qualitative behaviour of the models, as much as anything else. Associated with the time series is the **Figure of Merit in Time(FMT)**, which is the total percentage agreement between observed and predicted values over the twenty time points.

- **Figure of Merit in Space** (**FMS**) , that is evaluated for a fixed time. It is defined as the percentage of common area between two spatial distributions. The **FMS** is useful to evaluate the percentage of agreement between two sets of data and for a graphical estimate of the geometrical shape and extension of the two areas.

RESULTS

Participants provided in real-time forecasted and analyzed data. Here, only the full forecasted results will be discussed.

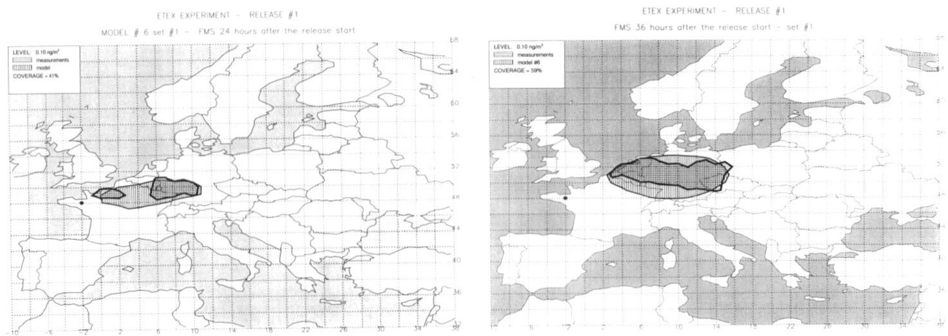

Figure 1. Release#1, FMS for Model 6 at 24 hrs (left)and 36 hrs(right) after start 0.10 ng/m^3 contour values are 41%(left) and 59%(right)

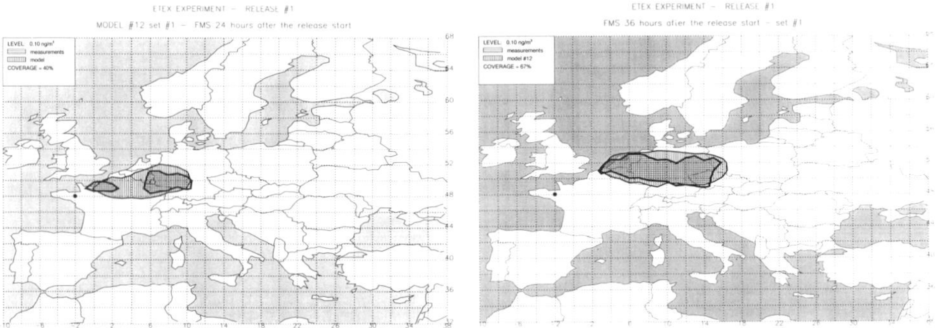

Figure 2 . Release#1, FMS for Model 12 at 24 hrs (left)and 36 hrs(right) after start 0.10 ng/m^3 contour values are 40%(left) and 67%(right)

First release

Figures 1 and 2 show the calculated versus measured cloud projection on the map of Europe for two of the most successful models (Model#6 and #12) at 24 and 36 hrs after the start of the release. These models are capable to reproduce the average shape and dimensions of the tracer concentration at ground at the selected time intervals. Similar graphs can be produced for 48 and 60 hrs. At the 12 hours interval, however, the measured cloud position was still over France, where the number of sites is limited, their average spacing is comparable with the cloud horizontal dimensions: therefore the interpolation errors can lead to inconsistencies. It is interesting to analyze the time evolution of observed and calculated concentration at some of those stations. The results are shown in Figure 3a for a site close to the release site (F21). There, two models (selected among the best) reproduce rather well the tracer evolution as the location is rather at the centreline of the cloud trajectory. Some other sites are placed on the edge of plume trajectory, like F3 (displaced to the South) or F9 (to the North). For those sites, even a small error in the evaluation of the lateral spread can therefore result in large variation of the calculated concentrations. In fact, at these locations, even the most successful models fail to reproduce correctly the cloud passage (Figures 3b and 3c). Simulation is even worst at F25, since both models forecasted similar tracer evolution where no tracer was measured. The results shown at those sites are generally worst than those at the sites located far away and touched later by the cloud. One can then conclude that the "best" model ability in reproducing the tracer

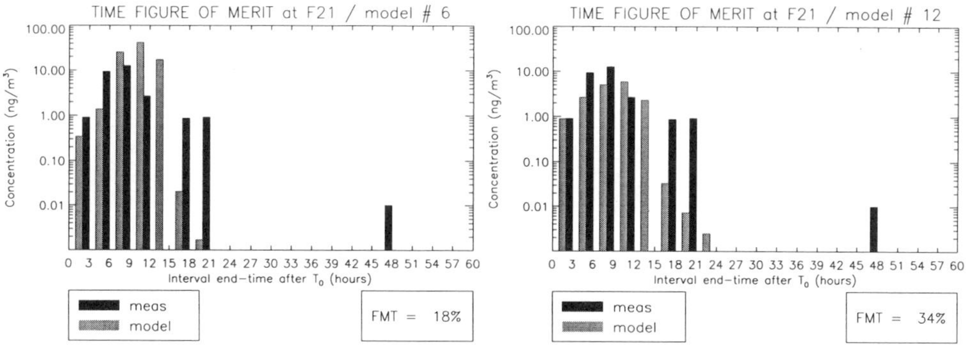

Figure 3a. Release#1; FMT at F21 for model #6 (left, FMT=18%) and model#12 (right, FMT=34%)

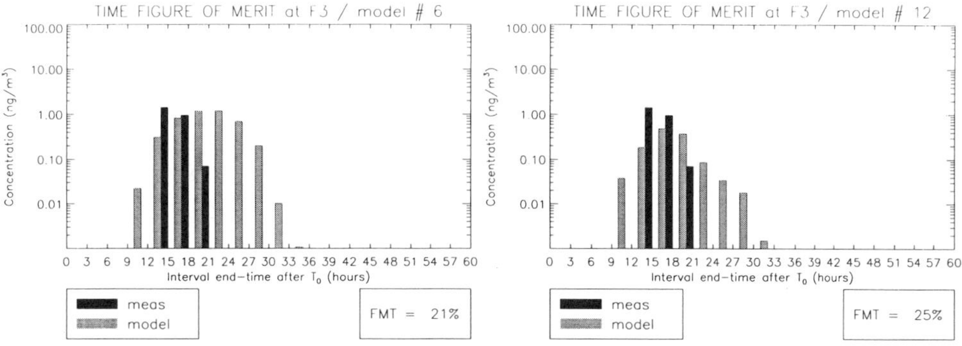

Figure 3b. Release#1; FMT at F3 for model #6 (left, FMT=21%) and model#12 (right, FMT=25%)

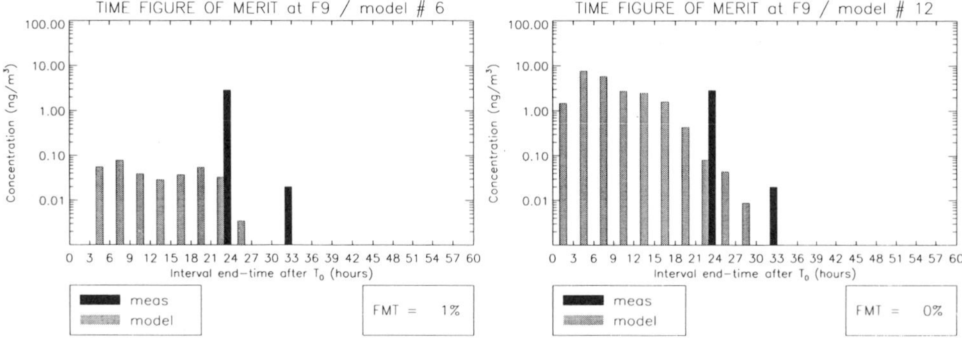

Figure 3c. Release#1; FMT at F9 for model #6 (left, FMT=1%) and model#12 (right, FMT=0%)

concentrations improves after the first period, when the sparseness of the network does not allow one to describe correctly the limited cloud dimensions. The improvement can be attributed to the steadiness of the meteorological situation which was such to maintain the bulk of the emission within the planned network.

Second release

Similarly to what has been done for the first, the tracer cloud projection over the map of Europe is compared with the measured one for two of the best models in the evaluation (Figures 3 and 4). This time, however, due to the much stronger wind field

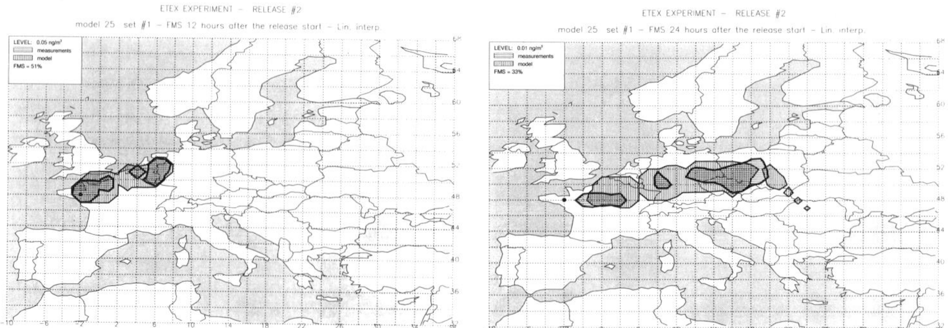

Figure 4. Release#2, FMS for Model 25 at 12 hrs (left)and 24 hrs(right) after start 0.05 (left) and 0.01 (right) ng/m^3 contour values are 51%(left) and 33%(right)

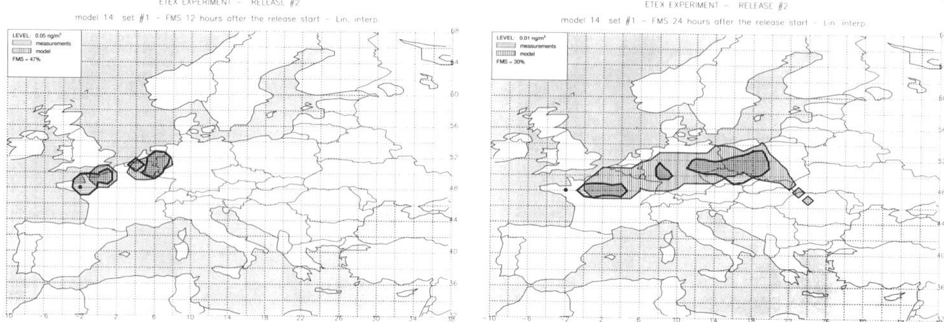

Figure 5. Release#2, FMS for Model 14 at 12 hrs (left)and 24 hrs(right) after start 0.05 (left) and 0.01 (right) ng/m^3 contour values are 47%(left) and 30%(right)

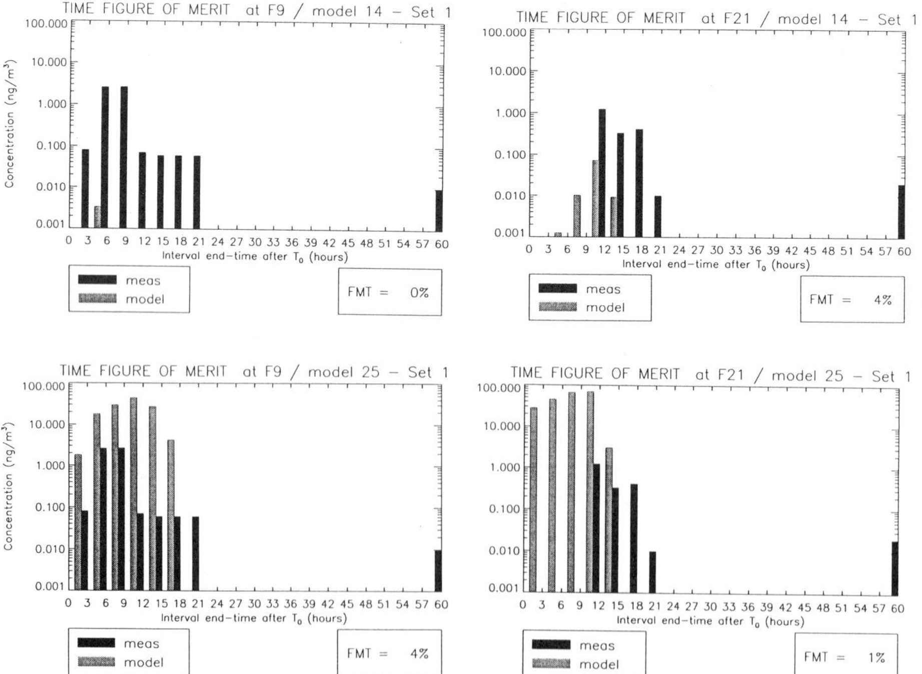

Figure 6. Release #2, FMT at F9 (left) anf F21 (right) for Model#14 (top, FMT values 0% and 4%) and Model#25 (bottom, FMT values 4% and 1%)

during the first period, shorter time intervals are considered, 12 and 24 hours. For these intervals, the models reproduce correctly the north-westerly cloud displacement during the first 12 hours and so does for the next period, when the cloud elongates more towards East.

For this interval, however, measured concentration values are much lower than the calculated ones. The contours shown in the Figures are at the lowest concentration level (0.01 ng/m3). For higher levels, the comparison worsen considerably. This result indicates that the general features of the flow were correctly taken into account; however, due to the sudden variations of the meteorological conditions during the release, large part of the cloud escaped the sampling network and could not be measured. In fact the tracer evolution at the sites close-to-source presents similar features to the first release (Figure 5). Due to the low wind direction shear, the cloud might have been rather narrow and differences in concentration values might have occurred because some sites are just at its edge. The later arrival of the tracer at site F21 can be explained by the sudden decrease of wind direction at the end of the release period. Some preliminary diagnostic mesoscale calculations have confirmed this hypothesis.

CONCLUSION

The thorough analysis and understanding of the results of the two ETEX experiments will take much more time than was available till now. Preliminary conclusions as presented here, however, indicate that the initial cloud development in the mesoscale region tends to determine its future evolution. The importance of correct determination of the tracer in the mesoscale region is therefore enhanced, which could suggest the necessity of increasing the density of the measurement network and of nesting mesoscale and long range dispersion models. This is particularly true for specific meteorological conditions as those of release #2.

References

Archer G., Girardi F., Graziani G., Klug W., Mosca S. and Nodop K.: “The European long range tracer experiment (ETEX): preliminary evaluation of model intercomparison exercise”, presented at the 21st NATO/CCMS International Technical Meeting on Air Pollution Modelling and its Application, Baltimore (USA), November 6-10,1995.

Klug W., Graziani G., Grippa G., Pierce D. and Tassone C.: “Evaluation of Long Range Atmospheric Models using Environmental Radioactivity Data from the Chernobyl Accident: ATMES Report”; 366 p. Elsevier Science Publishers, Barking, England, 1992.

DISCUSSION

R. YAMARTINO: For release 1 you mentioned factor of three disagreement between modeled and predicted peak ground level concentration. Was this a factor-of-three over - or under-prediction?

W. KLUG: The mentioned factor of three was typical for the *range* of under- or over-prediction of the models considered.

R. YAMARTINO: Was there any additional evidence (e.g. balloon trajectories, aircraft measurements) to indicate what happened to the release 2 tracer material?

W. KLUG: There was no additional evidence whatsoever what happened to the tracer.

D. FISH: Were both Eulerian and Lagrangian models used and if so were there any differences in their ability to simulate trace concentrations?

W. KLUG: Both types of models were used by participants. We did not detect any differences in their behaviour.

NEW DEVELOPMENTS

chairmen:	D. Carruthers D. Steyn
rapporteur:	R. Salvador D. Melas

RECENT DEVELOPMENTS IN CLOSURE AND BOUNDARY CONDITIONS FOR LAGRANGIAN STOCHASTIC DISPERSION MODELS

William L. Physick

CSIRO Division of Atmospheric Research
PB 1, Aspendale, Victoria, 3195
Australia

INTRODUCTION

Until the definitive paper of Thomson (1987), arguably the most pressing problem in Lagrangian particle modelling was which form of the Langevin equation should be used for inhomogeneous, non-stationary, non Gaussian turbulence to ensure well-mixedness, i.e. to ensure that particle accumulations did not occur in regions of low turbulence. Since that time, the major topics of research in Lagrangian dispersion modelling have been associated with convectively unstable conditions. This review paper focuses on two recent topics: (1) the closure problem associated with specification of the probability density function (PDF) for vertical turbulent fluctuations, and (2) the most appropriate boundary conditions to apply at the ground and at the top of the convectively mixed layer, particularly when simulating the fumigation[1] process. The reviewed studies emphasise the importance of turbulence and concentration data from a laboratory saline water tank (Hibberd and Sawford, 1994) for testing the closure schemes and the various methods for incorporating entrainment processes into stochastic models.

CONVECTIVE TURBULENCE IN LAGRANGIAN MODELS

In Lagrangian particle models which simulate dispersion in convective conditions, the turbulence in the vertical direction is often represented by a probability distribution function (PDF) which is the weighted sum of two Gaussian distributions (Baerentsen and Berkowicz, 1984). The general form of this bi-Gaussian PDF, which accounts for the skewness of the turbulence, is

$$P_E(w,z) = A(z)P_A(w,z) + B(z)P_B(w,z), \tag{1}$$

[1] Fumigation refers to the process whereby an elevated plume trapped in a stable (or neutral) flow is entrained into a developing convective boundary layer and undergoes rapid vertical mixing.

Air Pollution Modeling and Its Application XII
Edited by Sven-Erik Gryning and Nadine Chaumerliac, Plenum Press, New York, 1998

where

$$P_A = [2(\pi)^{1/2}\sigma_A]^{-1}\exp[-(w-\overline{w}_A)^2/(2\sigma_A^2)], \tag{2a}$$
$$P_B = [2(\pi)^{1/2}\sigma_B]^{-1}\exp[-(w+\overline{w}_B)^2/(2\sigma_B^2)], \tag{2b}$$

w is the vertical turbulent velocity and z the height above the ground. The two Gaussian functions P_A and P_B can be considered to represent the updrafts (A) and downdrafts (B) of the convective boundary layer, with A being the probability of a particle being in an updraft (or the proportion of area occupied by an updraft), $\overline{w}_A$ the mean velocity in an updraft and σ_A the velocity standard deviation in an updraft, and similarly for the downdraft terms.

The PDF equation (1) can be solved for the six unknown variables A, σ_A, $\overline{w}_A$, B, σ_B, and $\overline{w}_B$ by solving the six moment equations:

$$A + B = 1, \tag{3a}$$
$$A\overline{w}_A - B\overline{w}_B = \overline{w} = 0, \tag{3b}$$
$$A(\overline{w}_A^2 + \sigma_A^2) + B(\overline{w}_B^2 + \sigma_B^2) = \sigma_w^2, \tag{3c}$$
$$A(\overline{w}_A^3 + 3\overline{w}_A\sigma_A^2) - B(\overline{w}_B^3 + 3\overline{w}_B\sigma_B^2) = \overline{w^3}, \tag{3d}$$
$$A(3\sigma_A^4 + 6\overline{w}_A^2\sigma_A^2 + \overline{w}_A^4) + B(3\sigma_B^4 + 6\overline{w}_B^2\sigma_B^2 + \overline{w}_B^4) = \overline{w^4} \tag{3e}$$
$$A(10\sigma_A^2\overline{w}_A^3 + 15\overline{w}_A\sigma_A^4 + \overline{w}_A^5) - B(10\sigma_B^2\overline{w}_B^3 + 15\overline{w}_B\sigma_B^4 + \overline{w}_B^5) = \overline{w^5}, \tag{3f}$$

assuming known values of the Eulerian moments $\sigma_w^2(=\overline{w^2}), \overline{w^3}, \overline{w^4}$ and $\overline{w^5}$. However, the presence of the fourth and fifth moments $\overline{w^4}$ and $\overline{w^5}$, which are not usually known to any great accuracy anyway, means that the equation set (3) must be solved numerically. This is undesirable in Lagrangian dispersion models, which typically use small timesteps and large numbers of particles. Analytical solutions can be obtained by making closure assumptions which reduce the number of equations, and correspondingly the number of unknowns. In a recent paper, Luhar *et al.* (1996) hereafter referred to as LHH, examined some of the closures currently in use. They compared the resulting vertical velocity PDFs (equation (1)) with PDFs obtained from laboratory experiments with a saline tank, and derived a new closure with more realistic limiting behaviour. The schemes were evaluated further by comparing dispersion predictions with additional data from the saline tank experiments. A summary of their experiments and findings is presented in the following pages.

Various closures

LHH considered the following closures, which have been used by earlier authors:

- Closure 1. Baerentsen and Berkowicz (1984): $\overline{w}_A = \sigma_A$, $\overline{w}_B = \sigma_B$;
- Closure 2. Weil (1990): $\overline{w}_A = (2/3)\sigma_A$, $\overline{w}_B = (2/3)\sigma_B$;
- Closure 3. Du *et al.* (1994) (referred to as DWY): $A = 0.4$.

The closure models 1 and 2 solved the first four moment equations (3a)-(3d), while Closure 3 solved the first five moment equations (3a)-(3e) assuming the kurtosis $(K = \overline{w^4}/\sigma_w^4)$ to have a value of 3 (the value for a Gaussian distribution), independent of the skewness $(S = \overline{w^3}/\sigma_w^3)$. It is desirable that forms of the bi-Gaussian PDF (equation (1)) collapse smoothly to a single Gaussian PDF with $K = 3$ as the skewness tends to zero, as in neutral conditions. Closures 1 and 2 do not satisfy this requirement, and although Closure 3 does, its kurtosis value of 3 in convective conditions is not supported by observations (Fig.1).

A New Closure. LHH state that the problem with Closures 1 and 2, which are of the form

$$\overline{w}_A = m\sigma_A, \tag{4a}$$
$$\overline{w}_B = m\sigma_B, \tag{4b}$$

arises because m is a constant, which means that the PDF is not Gaussian for zero skewness. If m was a suitable function of the skewness, then the bi-Gaussian would collapse to a simple Gaussian in the zero skewness limit. Hence they proposed a new closure scheme with

$$m = \frac{2}{3}S^{1/3}. \tag{5}$$

If the first four moment equations (3a)-(3d) are solved with equations (4) and (5), the following solution is obtained for the bi-Gaussian PDF:

$$\sigma_A = \sigma_w \left[\frac{B}{A(1+m^2)}\right]^{1/2}, \tag{6a}$$

$$\sigma_B = \sigma_w \left[\frac{B}{B(1+m^2)}\right]^{1/2}, \tag{6b}$$

$$A = \frac{1}{2}\left[1 - \left(\frac{r}{4+r}\right)^{1/2}\right], \tag{6c}$$

$$B = 1 - A, \tag{6d}$$

where

$$r = [(1+m^2)^3 S^2]/[(3+m^2)^2 m^2] \tag{6e}$$

and the kurtosis is given by

$$K = (1+r)(3+6m^2+m^4)/(1+m^2)^2. \tag{7}$$

The value of 1/3 for the exponent in equation (5), while necessarily less than 1, was chosen to give the simplest forms for equation (4), while the value of 2/3 chosen for the constant was obtained from equation (6e) and equation (7) when $S = 0.8$ and $K = 3.9$. These values for skewness and kurtosis are the average values from LHH's saline tank experiments on the convective boundary layer (CBL), which are plotted in Fig. 1. Also shown are aircraft data from Lenschow *et al.* (1994) and the relation between skewness and kurtosis (from equations (6e) and (7)) for the various closure schemes. The new closure scheme and Closure 2 agree quite closely, with the other two closures having significantly lower values of K. From a survey of field and laboratory measurements, LHH conclude that $S = 0.7 \pm 0.2$ and $K = 3.7 \pm 0.5$ in the bulk of the convective boundary layer.

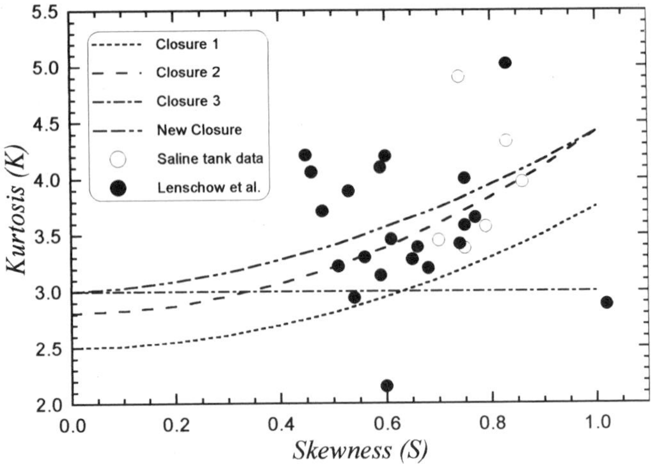

Figure 1. Variation of kurtosis with skewness given by the various closure assumptions. The open circles are the water tank data while the filled ones are the data from Lenschow *et al.* (1994). From Luhar *et al.* (1996).

COMPARISON OF PDF PARAMETERS WITH LABORATORY DATA

Vertical velocities were measured at eight equidistant levels within the CBL in a 3.2 m x 1.6 m x 0.8 m saline water tank (see Hibberd and Sawford (1994) for a description of the tank facility). Approximately 20,000 samples were obtained at each level from data spanning about 200t_* of experimental time, where $t_* = z_i/w_*$. Using a least squares approach, corresponding PDFs were fitted using the bi-Gaussian form given by equation (1) and equation (2). The filled circles in Fig. 2a-2e represent values of $\overline{w}_A / w_*$, $\overline{w}_B / w_*$, σ_A / w_*, σ_B / w_* and A, respectively, as determined from the least-squares method, at the eight levels. The solid lines denote relatively simple expressions fitted to the data points (see LHH). The broken lines represent the values of the bi-Gaussian parameters calculated for the various closure schemes using equations (3a)-(3d) and the second and third moments obtained directly from the water tank PDFs, as shown in Figs. 3a and 3b.

Figure 2 shows that both Closure 2 and the new closure closely approximate the direct fits to the data. Although the new closure curves agree only slightly better than those from Closure 2, the new closure has the advantage of giving the correct PDF behaviour as the skewness approaches zero. The figure also shows that Closures 1 and 3 overestimate $\overline{w}_A / w_*$ and $\overline{w}_B / w_*$, but underestimate σ_A / w_* and σ_B / w_*, a result which has implications for maximum ground-level concentrations (GLCs) from elevated point sources.

The $\overline{w^4} / w_*^4$ profiles obtained using equation (3e) for the various closure schemes are compared with the laboratory data in Fig. 3(c). Closure 3, which assumes $\overline{w^4} = 3\sigma_w^4$, deviates most from the data whereas Closure 2 and the new closure agree well.

DISPERSION RESULTS

LHH evaluated the effects of the differences in the closure schemes on GLCs by simulating dispersion from four source heights z_s with the one-dimensional Lagrangian stochastic model of Luhar and Britter (1989). The downwind variation of dimensionless cross-wind integrated GLC is shown in Fig. 4 for the four source heights non-dimensionalised by the mixed-layer height z_i (z_s/z_i = 0.067, 0.24, 0.49 and 0.99). The solid line in Fig. 4 is a benchmark calculation using a PDF calculated directly from the tank data,

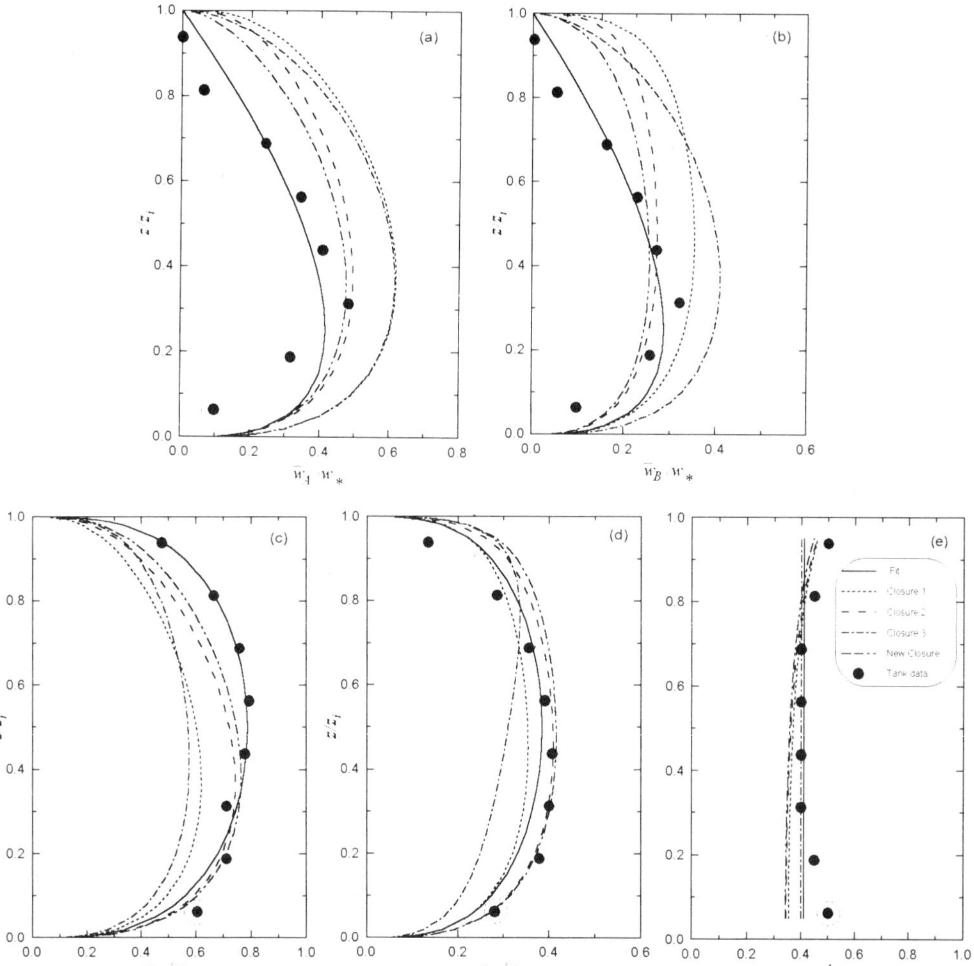

Figure 2. Vertical profiles of (a) $\overline{w}_A / w_*$, (b) $\overline{w}_B / w_*$, (c) σ_A / w_*, (d) σ_B / w_* and (e) A. Filled circles — derived from the water tank PDFs; solid line — fits to the filled circles; dotted line — calculated using Closure 1; dashed line — calculated using Closure 2; dot-dash line — calculated using Closure 3; dot-dot-dash-dash line — calculated using the new closure. The circled data points were not considered in calculating the solid line fits. From Luhar *et al.* (1996).

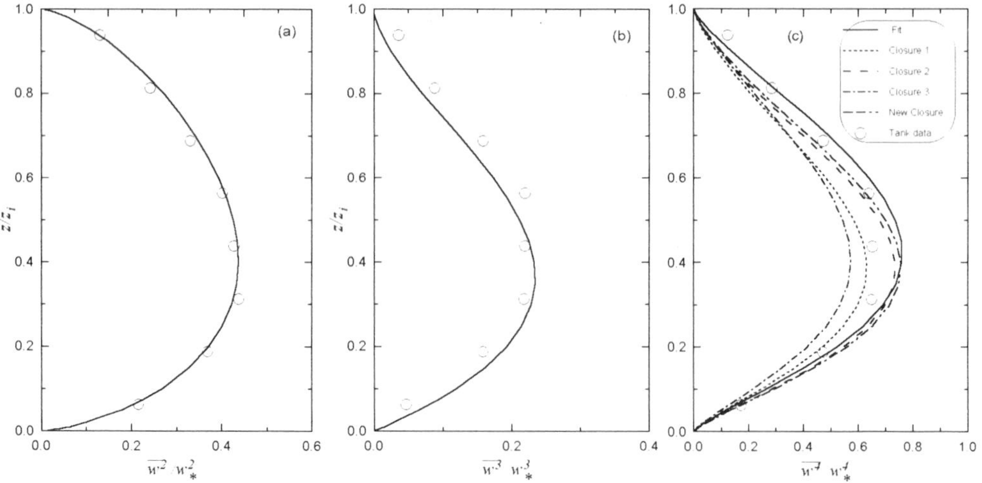

Figure 3. Vertical profiles of (a) $\overline{w^2} / w_*^2$, (b) $\overline{w^3} / w_*^3$, and (c) $\overline{w^4} / w_*^4$. Open circles — water tank data; solid line — determined using equations (3) and solid fitted curves of Fig. 2. The other line types in (c) are as described for Fig. 2. Note that the variance and skewness are the same for all four closure schemes. From Luhar *et al.* (1996).

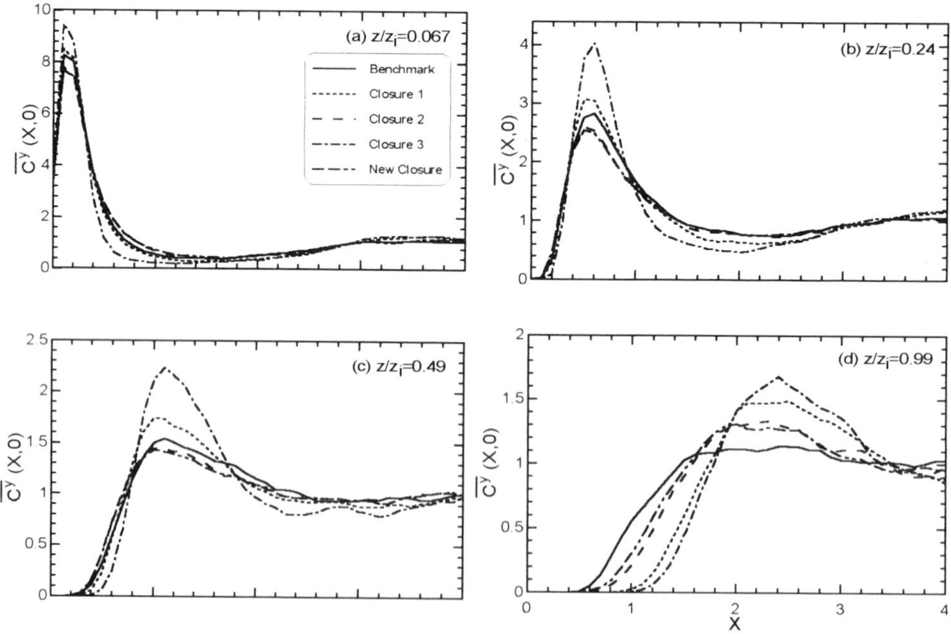

Figure 4. Variation of the dimensionless ground-level cross-wind-integrated concentration $(\overline{C}^y)$ with the dimensionless downwind distance (X) predicted by the Lagrangian stochastic model for the source height (z_s/z_i) (a) 0.067, (b) 0.24, (c) 0.49, and (d) 0.99. Solid line — benchmark calculation using direct fits to tank velocity data . Other line types as in Fig. 2 caption. From Luhar *et al.* (1996).

i.e. Fig. 2. For all heights, the GLCs predicted using Closure 2 and the new closure are very similar and give results closer to the benchmark than Closures 1 and 3. As the source height increases, the differences between the GLCs computed using different closures increase. Note that all curves approach the well-mixed GLC value of 1 far downwind, a consequence of using the correct form of the Langevin equation. Closure 3 gives the largest maximum GLC, and also the smallest minimum downwind of the peak value. This behaviour is a consequence of Closure 3 having the smallest values of standard deviation of velocities in updrafts and downdrafts in the bulk of the boundary layer (Fig. 2). Similarly, the GLC variations of the other closure schemes can be interpreted from the curves of Fig. 2.

The largest differences between closures occur for the case z_s/z_i = 0.99, which is relevant to fumigation. It is not only the peak value that is strongly influenced by the choice of closure scheme, but also the distance downwind at which the plume first reaches the ground. The non-dimensional difference between this location for Closure 3 and the benchmark case corresponds to a distance of around 3 km. Interestingly, the benchmark curve for this height shows no distinguishable peak and is flat for a distance of about 1.5X, a characteristic which is reproduced to some extent by Closure 2 and the new closure and is also observed for low entrainment rates in direct measurements of GLCs in the fumigation tank experiments of Hibberd and Luhar (1996).

CONDITIONS AT THE BOUNDARIES OF LAGRANGIAN STOCHASTIC MODELS

Discussion of fumigation leads into the second topic of this paper: boundary conditions. We reserve an evaluation of the mathematical boundary conditions until later sections, and begin by examining how downward entrainment of a plume into a growing mixed layer is affected by physical conditions at the top of the layer. These include the horizontal variation of the mixed layer depth and the thickness of the entrainment zone, and we discuss how these characteristics can be accounted for in a Lagrangian stochastic model. Once again, laboratory experiments provide the data and insight into the entrainment process.

Fumigation Experiments

The laboratory fumigation experiments described by Hibberd and Luhar (1996), hereafter referred to as HL, were designed to determine the influence of CBL growth rate (also referred to as entrainment rate) on concentration levels, both at the surface and throughout the CBL, for a wider range of growth rates than previously examined by Deardorff and Willis (1982). The rates studied by the latter researchers are smaller than those commonly found in coastal areas, where intersection of an elevated plume with a growing thermal internal boundary layer can produce steady fumigation over a number of hours.

A dominant visual feature of HL's experiments was the horizontal variability in mixed-layer height across the tank at any instant, leading to patchy rather than uniform entrainment in the early stages of fumigation. The variability arises not only from the updraught/downdraught nature of convection, but also from gravity wave undulations, caused by the impingement of thermals on the overlying stable layer. There has been little recognition of the importance on plume dispersion of this variability of the local mixed-layer height since it was identified by Deardorff and Willis in the early eighties. For all experiments, HL obtained a quantitative estimate of the variability (Δz_i) from the scatter of the individual conductivity probe measurements about the best-fit curve for z_i. These are plotted in Fig. 5 as a function of non-dimensional entrainment rate. Fig. 5 also contains data points from detailed measurements of entrainment zone thickness from the heated water tank experiments of Deardorff *et al.* (1980). The entrainment zone is usually defined as the region

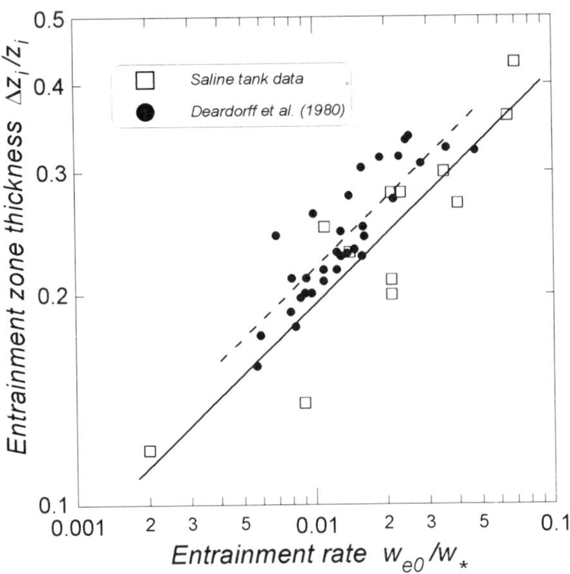

Figure 5. Variation of entrainment zone thickness with entrainment rate. Saline tank estimates of $\Delta z_i/z_i$ obtained from scatter in conductivity probe measurements of z_i. One-third power law fits to the saline tank data (solid line) and the Deardorff *et al.* data (dashed line) are also shown. From Hibberd and Luhar (1996).

of negative buoyancy flux at the top of the CBL. The good agreement between the two data sets suggests that the HL z_i variability technique yields good estimates of the entrainment zone thickness. A power-law fit to the HL data yields the relation

$$\frac{\Delta z_i}{z_i} = 0.9\left(\frac{w_{e0}}{w_*}\right)^{1/3} . \tag{8}$$

HL found that the value of the maximum crosswind-integrated GLC was only weakly dependent on the entrainment rate, but that it occurred later with decreasing growth rate. The duration of the fumigation episodes was found to be proportional to the sum of the entrainment zone thickness and the initial plume spread, and inversely proportional to the entrainment rate. These findings were used to modify a PDF model of fumigation and to evaluate the model over a range of boundary-layer growth rates.

Fumigation Modelling

HL showed that in order to obtain good agreement with crosswind-integrated GLCs from the tank experiments, the analytical PDF fumigation model of Luhar and Sawford (1996) needed modifying so that it accounted for the spatial variability of the local mixed-layer height. The first technique tested, originally employed in a Lagrangian stochastic fumigation model by Luhar and Sawford (1995), weighted GLC predictions from eleven simulations with different values of z_i, equally spaced between $z_i + \Delta z_i / 2$ and $z_i - \Delta z_i / 2$. It was found that a Gaussian distribution of weights provided much better agreement with the tank concentrations than the procedure of Luhar and Sawford (1995), which used equal weights (top-hat distribution) for all mixed-layer heights.

In an effort to reduce the computing load of the above technique, HL used the experimental finding that mixed-layer-height variability and the initial plume spread (in the stable layer above the mixed layer) both act to reduce the magnitude of the GLCs. By increasing the initial plume spread from σ_{z0} to $z_{i0}\left((\sigma_{z0}/z_{i0})^2+(\Delta z_{i0}/4z_{i0})^2\right)^{1/2}$, where $\Delta z_{i0}/4z_{i0}$ is the dimensionless standard deviation of mixed-layer heights (obtained from equation (8)), and by specifying a mixed layer with no z_i variability, they were able to obtain predictions very similar to those from the previous more time-consuming technique. A comparison of crosswind-integrated concentrations from the various methods is shown in Fig. 6 for one of the experiments. With z_i variability not considered at all, the GLC increases rapidly and shows a pronounced peak because all the fumigant is entrained over a shorter time. In contrast, the effective duration over which the other three methods entrain the fumigant is much larger, causing the GLC curve to be flatter. The Gaussian and enhanced plume spread approaches give much better agreement with the data than does the top-hat method.

Upper and Lower Boundary Conditions

The mathematical boundary condition on turbulent vertical velocity which is commonly applied at the upper and lower boundaries of a Lagrangian stochastic model, is that of perfect reflection, i.e. a particle impinging on a boundary leaves the boundary in the opposite direction but at the same speed. This condition is appropriate for Gaussian turbulence or skewed inhomogeneous turbulence (where the Lagrangian time scale is normally very small near the boundaries), but if used for skewed homogeneous turbulence will lead to an accumulation or deficit of particles at the boundaries. Modifications of the perfect reflection condition for non-zero skewness by Weil (1990) and Hurley and Physick (1993) were shown by Thomson and Montgomery (1994) (TM) to provide acceptable solutions for small values of the Lagrangian time-scale τ, but the departure from a uniformly-mixed profile became greater as τ increased. Considering now only the lower boundary, TM proposed that the correct boundary condition is

$$\int_{w_r}^{\infty} wP_E(w, z_b)dw = -\int_{-\infty}^{w_i} wP_E(w, z_b)dw \tag{9}$$

where w_r is the reflected velocity, w_i is the incident velocity and P_E is the assumed vertical velocity distribution of particles at the boundary z_b (a corresponding equation can be derived for the upper boundary). The basis of this equation was their assertion that the relevant

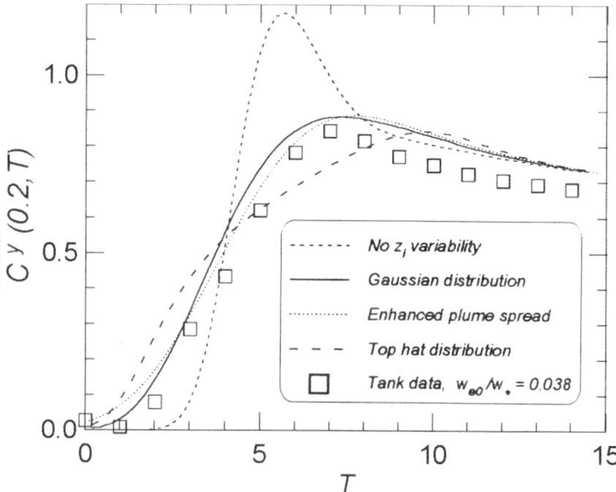

Figure 6. Comparison of time dependence of crosswind-integrated concentrations at $0.2z_{i0}$ from saline convection tank ($w_{e0}/w_* = 0.038$, $\sigma_{z0}/z_{i0} = 0.026$) with PDF fumigation model predictions using four different methods for including the z_i variability. From Hibberd and Luhar (1996).

quantity to be considered is the PDF of the velocities of particles which leave z_b during a fixed time interval, rather than just the PDF of particles leaving the boundary at a particular time t, $P_E(w,z_b)/\int_0^\infty P_E(w,z_b)dw$ ($w \geq 0$), as used by Weil (1990). The former PDF is $wP_E(w,z_b)/\int_0^\infty wP_E(w,z_b)dw$ ($w \geq 0$) since more of the faster-moving particles will leave $z = z_b$ in a given time interval.

Knowing the incident velocity w_i of a particle, equation (9) is used to obtain w_r. When P_E is the commonly-used bi-Gaussian expression for skewed convective turbulence (equations (1 and 2)), the solution to equation (9) consists of numerical integrals and error functions and is obtained for each particle from prepared look-up tables (TM). The time-consuming nature of this process in a three-dimensional Lagrangian stochastic model has recently been addressed by Anfossi *et al.* (1997) who proposed two approximate analytical solutions to equation (9). The first one made use of a Taylor series expansion and still involved error functions, but the second one involved a regression curve between w_i and w_r as a function of skewness and $w_i/(\overline{w^2})^{1/2}$. Curve coefficients were obtained from many "exact" (look-up table) solutions to equation (9) over a range of variance and skewness values. Although both solutions of Anfossi *et al.* satisfy the well-mixed condition and do not appreciably depart from the correct or "exact" solution, the regression method uses considerably less computing resources and seems a suitable approach to applying boundary conditions in three-dimensional particle models. Note that although Anfossi *et al.* used Closure 1, their solutions can be modified for other closure schemes.

An Interface Condition

Perfect reflection was applied at the lower and upper boundaries of the mixed layer by Luhar and Sawford (1995) in their Lagrangian stochastic model of fumigation. Although they specified non-Gaussian turbulence, a perfect reflection condition was valid because the skewness went to zero at the boundaries. When applied at the top of the CBL, this condition and that of equation (9) do not allow any interchange of particles/material between the CBL and the overlying stable layer - in the models, material enters the mixed layer as the rising z_i jumps from below the material to above it over the course of a time step. Thus the CBL top is effectively a rigid boundary rather than as a porous interface across which there is exchange of material.

This limitation has been addressed in some recent work by Thomson *et al.* (1997) (TPM) who examined the more general problem of random-walk modelling of diffusion across an infinitesimally-thin interface, at which the turbulence statistics change discontinuously. TPM argued that if the Lagrangian time scale τ on which particles forget

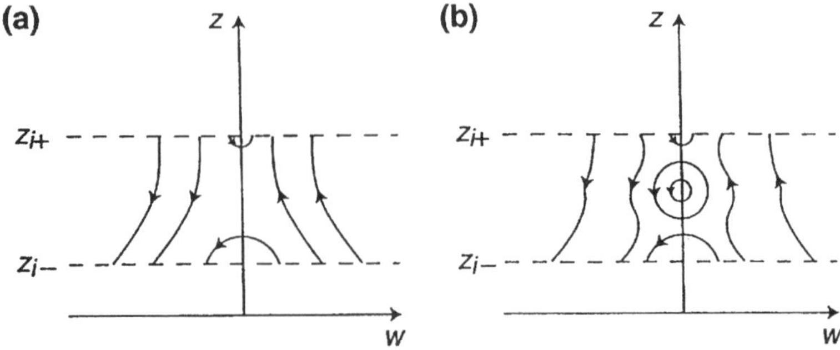

Figure 7. Illustration of some possible flows in (z,w)-space. From Thomson *et al.* (1997).

their velocity is much larger than the time particles spend within the interface, then particle trajectories in (z,w)-space within the interface are deterministic and do not cross each other. As a result, the trajectories will generally take the form illustrated in Fig. 7a , although cut-off circulations (Fig.7b) and other configurations are possible.

By using the fact that the flux between two streamlines in (z,w)-space must be conserved, and considering a particle entering the interface (lower and upper boundaries z_{i-} and z_{i+}) from below with incident velocity w_i, they show that its velocity w at a height z within the interface can be obtained from

$$\int_w^\infty wp_a(z,w)dw = \int_{w_i}^\infty wp_a(z_{i-},w)dw \tag{10}$$

where p_a is the density of well-mixed tracer particles. Equation (10) is equally applicable if $w < 0$. From here on, the integrals will be denoted by F, e.g. the left-hand side of equation (10) is $F_w^\infty(z)$. If $F_0^\infty(z) < F_{w_i}^\infty(z_{i-})$ at any height in the interface then the particle will be reflected, i.e. the particle will be reflected if w_i is less than the critical value w_c which is defined for the CBL case of $F_0^\infty(z)$ decreasing monotonically within the interface by

$$F_{w_c}^\infty(z_{i-}) = F_0^\infty(z_{i+}). \tag{11}$$

Otherwise the particle will be transmitted. The reflection and transmission velocities w_r and w_t are given by

$$F_{w_r}^\infty(z_{i-}) = F_{w_i}^\infty(z_{i-}) \quad (w_r < 0) \tag{12}$$

and

$$F_{w_t}^\infty(z_{i+}) = F_{w_i}^\infty(z_{i-}) \quad (w_t > 0). \tag{13}$$

Similar relations can be derived for particles entering the interface from above. Note that particles approaching the interface from above (weaker turbulence) are always transmitted. When implementing the interface condition in a random walk model, the particle velocity should be changed at the instant the particle reaches the interface, with $z\ (t+\Delta t)$ being calculated in a way that accounts for the change in velocity during the time step.

Using a (z,w) stochastic model based on the general form of the Langevin equation (Thomson, 1987), TPM carried out a number of one-dimensional simulations, with the geometry of Fig. 8, to illustrate that the interface conditions (equations (11-13)) preserve the correct well-mixed state. Figure 9 illustrates how the model is able to simulate the entrainment of a plume from a less-turbulent layer into a more-turbulent one, i.e. fumigation for the case of the convective boundary layer in the atmosphere. Note that z_i is constant throughout the simulation. Skewed homogeneous turbulence (S = 0.6) was specified for the boundary layer and Gaussian homogeneous for the layer above, with a time step of 0.05τ. Boundary conditions were perfect reflection at the top of the domain and the TM approach (equation (9)) at the ground. The initial concentration field was uniform in the region above the boundary layer top, and zero below. Twenty thousand particles were followed and concentration profiles were evaluated by counting particles in boxes of height 30 m. The instantaneous profiles in Fig. 9 show quite rapid entrainment in the early stages followed by an approach towards a well-mixed profile. A comparison is underway of this approach to fumigation modelling with the Lagrangian stochastic and PDF models of Luhar and Sawford (1995, 1996) described earlier, and all models will be applied to data from a recent field study of coastal fumigation at the Kwinana industrial complex south of Perth, Australia.

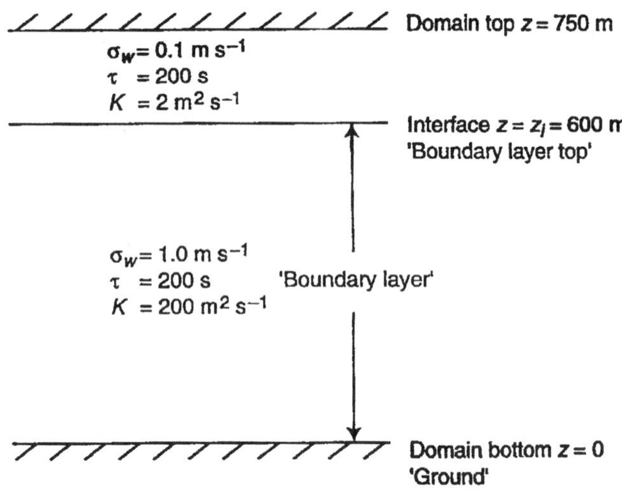

Figure 8. Illustration of the geometry and parameter values used in the entrainment simulation. From Thomson *et al.* (1997).

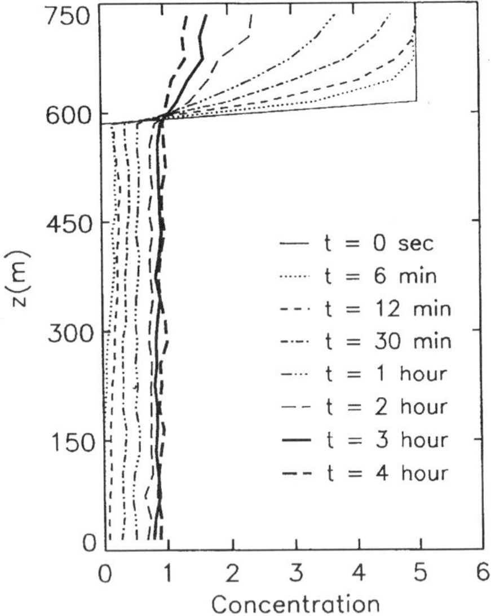

Figure 9. Evolution of the concentration profile for a situation where material which is initially distributed uniformly above the boundary layer is entrained into the boundary later. Concentrations are normalised to equal unity when vertically well mixed throughout the whole domain. From Thomson *et al.* (1997).

CONCLUDING REMARKS

Detailed saline water tank experiments have allowed a comparison of various closure schemes used for PDFs in convective conditions. When the system of equations is closed by specifying the mean vertical velocities in the bi-Gausssian PDF as a function of associated variances, the best results are obtained when the relationship involves the skewness of the turbulence, specified in such a way that allows the bi-Gaussian PDF to collapse to a simple Gaussian in the zero skewness limit.

Similarly, we have shown how tank experiments have provided insight into those physical factors which are important for the entrainment of a plume into a growing mixed layer. The detailed nature of the data enabled development of a technique to incorporate the spatial variability of the mixed-layer height into Lagrangian stochastic and PDF fumigation models.

Mathematical boundary conditions were also discussed, including a new formulation to allow exchange of material across an interface with a discontinuity in turbulence statistics.

Acknowledgments

I am pleased to acknowledge the permission granted by my colleagues Ashok Luhar, Mark Hibberd and Peter Hurley to quote extensively from their work for this review and also to David Thomson and Roy Maryon from the U.K. Meteorological Office for their agreement in allowing part of our upcoming paper to be presented here.

REFERENCES

Anfossi, D., Ferrero, E., Tinarelli, G., and Alessandrini, S., 1997, A simplified version of the correct boundary conditions for skewed turbulence in Lagrangain particle models. *Atmos. Environ.*, 31:301-308.

Baerentsen, J.H., and Berkowicz, R., 1984, Monte Carlo simulation of plume dispersion in the convective boundary layer. *Atmos. Environ.* 18:701-712.

Deardorff, J.W. and Willis, G.E., 1982, Ground-level concentrations due to fumigation into an entraining mixed layer. *Atmos. Environ.* 16:1159-1170.

Deardorff, J.W., Willis, G.E., and Stockton, B.H., 1980, Laboratory studies of the entrainment zone of a convectively mixed layer. *J. Fluid Mech.* 100:41-64.

Du, S., Wilson, J.D., and Yee, E., 1994, Probability density functions for velocity in the convective boundary layer, and implied trajectory models. *Atmos. Environ.* 28:1211-1217.

Hibberd, M.F., and Luhar, A.K., 1996, A laboratory study and improved PDF model of fumigation into a growing convective boundary layer. *Atmos. Environ.* 30:3633-3649.

Hibberd, M.F., and Sawford, B.L., 1994, A saline laboratory model of the planetary convective boundary layer. *Boundary-Layer Met.* 67:229-250.

Hurley, P.J., and Physick, W.L., 1993, A skewed homogeneous Lagrangian particle model for convective conditions. *Atmos. Environ.* 27A:619-624.

Lenschow, D.H., Mann, J, and Kristensen, L., 1994, How long is long enough when measuring fluxes and other turbulence statistics? *J. Atmos. Oceanic Technol.* 11:661-673.

Luhar, A.K., and Britter, R.E., 1989, A random walk model for dispersion in inhomogeneous turbulence in a convective boundary layer. *Atmos. Environ.* 23:1911-1924.

Luhar, A.K., and Sawford, B.L., 1995, Lagrangian stochastic modelling of the coastal fumigation phenomenon. *J. Appl. Meteor.* 34:2259-2277.

Luhar, A.K., and Sawford, B.L., 1996, An examination of existing shoreline fumigation models and formulation of an improved model. *Atmos. Environ.* 30:609-620.

Luhar, A.K., Hibberd, M.F., and Hurley, P.J., 1996, Comparison of closure schemes used to specify the velocity pdf in Lagrangian stochastic dispersion models for convective conditions. *Atmos. Environ.* 30:1407-1418.

Thomson, D.J., 1987, Criteria for the selection of stochastic models of particle trajectories in turbulent flows. *J. Fluid. Mech.*, 180:529-556.

Thomson, D.J., and Montgomery, M.R., 1994, Reflection boundary conditions for random walk models of dispersion in non-Gaussian turbulence. *Atmos. Environ.*, 28:1981-1987.

Thomson, D.J., Physick, W.L., and Maryon, R.H., 1997, Treatment of interfaces in random walk dispersion models. *J. Appl. Meteor.*, in press.

Weil, J.C., 1990, A diagnosis of the asymmetry in top-down and bottom-up diffusion using a Lagrangian stochastic model. *J. Atmos. Sci.* 47:501-515.

DISCUSSION

R. BORNSTEIN: How do you use PBL model results to obtain values of T_L? Do you account for breaking gravity waves?

W. PHYSICK: The Lagrangian timescale T_L is not used explicitly anywhere in the formulation of our particle model, although a value can be obtained through the relation $T_L = 2\sigma_w^2/(C_0\varepsilon)$. The dissipation rate ε is obtained from the PBL model as a function of the convective velocity scale w_* and the PBL height. The breaking of gravity waves is roughly parameterised by specifying σ_w in stable conditions to be a function of a critical Richardson number at each model level.

D. CARRUTHERS: How is plume buoyancy considered in your particle modelling algorithms, and in particular how does it affect plume fumigation? Does the buoyancy inhibit downward mixing?

W. PHYSICK Effective stack height (*esh*) is calculated by integrating the Briggs' equations for a bent-over plume and using wind and stability at each model level. Initial buoyancy is a specified source characteristic. At each timestep, particles are distributed about the *esh* in a Gaussian manner, on occasions resulting in some particles within and some above the mixed layer, i.e. partial penetration. There is no mechanism in the model for buoyancy to inhibit downward mixing, and thus fumigation.

OBSERVATION AND MODELLING OF BOUNDARY LAYER DEPTH IN A REGION WITH COMPLEX TERRAIN AND COASTLINE

Douw Steyn[1], Ekaterina Batchvarova[2], Marina Baldi[3], Bob Banta[4], Xiaoming Cai[5], Sven-Erik Gryning[2] and Ray Hoff[6]

[1] Atmospheric Science Programme, Department of Geography
The University of British Columbia, Vancouver, Canada
[2] Risoe National Laboratory, Roskilde, Denmark
[3] Institute for Atmospheric Physics, IFA-CNR, Rome, Italy
[4] ERL, NOAA, Boulder, USA
[5] School of Geography, University of Birmingham, Edgbaston, U.K.
[6] Air Quality and Inter-Environmental Research Branch, Atmospheric Environment Service, Egbert, Canada

INTRODUCTION

The purpose of this paper is to test the ability of two quite different models to simulate the combined spatial and temporal variability of Thermal Internal Boundary Layer (TIBL) depth and Mixed Layer Depth (MLD) in the complex terrain and coastline of the Lower Fraser Valley (LFV) of British Columbia, Canada during the course of one day. The models (the simple applied model of Gryning and Batchvarova (1996), and the Colorado State University Regional Atmospheric Modelling System (CSU-RAMS) described by Pielke *et al.* (1992)). will be tested by comparison with data gathered during a field study (called Pacific '93) of photochemical pollution in the LFV. The data utilized here are drawn from tethered balloon flights, free flying balloon ascents, and downlooking lidar operated from an aircraft flown at roughly 3500m ASL.

OBSERVATIONAL DATA

In the summer of 1993, an intensive field study was conducted in the LFV to investigate processes leading to frequent episodic ground-level ozone. Steyn *et al.* (1997) and Pottier *et al,*.(1994) provide an overview of the study and details of all data collected. On 5 August 1993, during the height of the study, the instruments were operated in a particularly intense effort. This day was characterized by warm, dry conditions as the

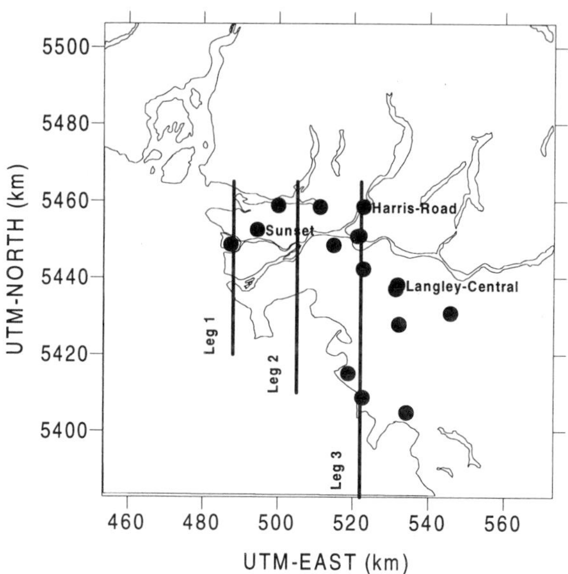

Figure 1. Map of the LFV modelling domain showing coastline (solid line), aircraft flight tracks (NS lines labeled leg 1, leg 2, leg 3), tethered balloon and free flying balloon launch sites. Coordinates are UTM.

regional weather was dominated by a stationary anticyclone. Data from this day will be used for analysis in this study. Figure 1 indicates the location of fixed sensors and aircraft tracks.

Information on the vertical structure of wind direction and speed, temperature and humidity at Harris Road were obtained from measurements by tethered balloon flights up to 1000 m AGL. The instrument was operated continuously from 0816 to 1837 PST on 1993.08.05, resulting in eight ascending or descending flights. Free flying balloons released from Langley-Central returned data up to 10 km AGL during four flights (0400, 1000, 1300 and 1600 PST) on 1993.08.05. Profiles of temperature and specific humidity from all flights are interpreted to provide information on MLD and the warming rate in the air above it.

Spatial variability of the TIBL in the region were estimated from measurements performed with a downlooking airborne lidar. The platform for this program was a Convair 580 carrying a downward looking Nd-Yag lidar emitting at 1.064 μm. Two flight sequences, each forming a grid of nine NS tracks at roughly 3500 m ASL were flown on 1993.08.05. The flight sequences lasted from 0611 to 0803 and 1317 to 1448. The horizontal spatial resolution was approximately 200 meters. Indicated on Figure 1 are three legs from the afternoon flight that cover the transition from coastline to the valley floor. MLD can be diagnosed from these aircraft lidar images by defining an absolute value of backscatter intensity (BSI) which distinguishes aerosol laden lower boundary layer air from relatively cleaner upper boundary layer air. The cutoff value of BSI is chosen based on trial and error analysis of the data set. This method will not work if regional differences in surface emissions of aerosols produce horizontal differences in ML-top BSI that span all possible BSI cutoff values. The same data were utilized by Hoff *et al.* (1997) and Hayden *et al.* (1997), who employed an analysis scheme for ML depth based in vertical gradients of BSI.

Measurements of turbulence fluxes of heat and momentum were carried out using 20 m tower mounted eddy correlation instrumentation at a location in the residential area of Vancouver about 10 km from the coastline (Sunset observational site, Figure 1).

Wind speed and direction were measured on an hourly basis at a large number of routine meteorological stations scattered over the area. In the present study we used data from 13 stations situated on the valley floor (Figure 1).

TWO MODELS OF MIXED LAYER DEPTH

A Simple Applied Model

This model is based on a zero-order model for the development of the TIBL during near neutral and unstable atmospheric conditions (Gryning and Batchvarova, 1996). Potential temperature jump at the inversion base and energy balance within the internal boundary layer are parameterized (Batchvarova and Gryning, 1994). Initially, the internal boundary layer depth growth is proportional to the friction velocity, with mechanical production of TKE being the controlling mechanism. The importance of mechanical production diminishes gradually as the production of convective turbulence becomes increasingly important with a deepening layer and increasing turbulent sensible heat flux density (Gryning and Batchvarova, 1990). The equation for the height of the internal boundary layer includes coastline curvature and spatially varying winds (Gryning and Batchvarova, 1996), and is solved using a second order Lax-Wendroff explicit differential scheme.

Model Simulation The spatial and temporal evolution of the internal boundary layer over LFV on 5 August 1993 was simulated, starting at 0700 PST when the heat flux at the Sunset site became positive. Simulations were based on hourly averages of the model input parameters. The model domain shown in Figure 2 extends 120 km in the west-east direction and 124 km in the south-north direction with a 0.5 km grid resolution. The interpolated u and v wind components in each grid point are derived by inverse distance (to the power 2) interpolation among the 13 stations. The kinematic heat flux measured at the Sunset observational site was taken as basis for the simulation with a correction applied in each grid point for the actual land use. The land use distribution is shown on Figure 2, while heat flux and friction velocity corrections are given in Table 1. Over the water the height of the boundary layer was kept constant, in the actual simulation arbitrarily chosen as 10 meters. Radiosoundings performed throughout the day at Langley-Central indicate negligible warming of the air in the free atmosphere, suggesting that subsidence need not be included in the model simulation for this day. A potential temperature gradient of 0.007 K/m was applied in this study.

Table 1. Land use corrections to heat flux and friction velocity.

Land use type	Fraction between friction velocity and interpolated wind speed in the grid point	Fraction between the kinematic heat flux in a grid point and at the Sunset observational site
Urban	0.13	1
Agricultural and horticultural	0.1	0.75
Parks and forest	0.13	1
Bogs and swamps	0.1	0
Lakes and rivers	0	0
Mountainous forest	0.1	1.25

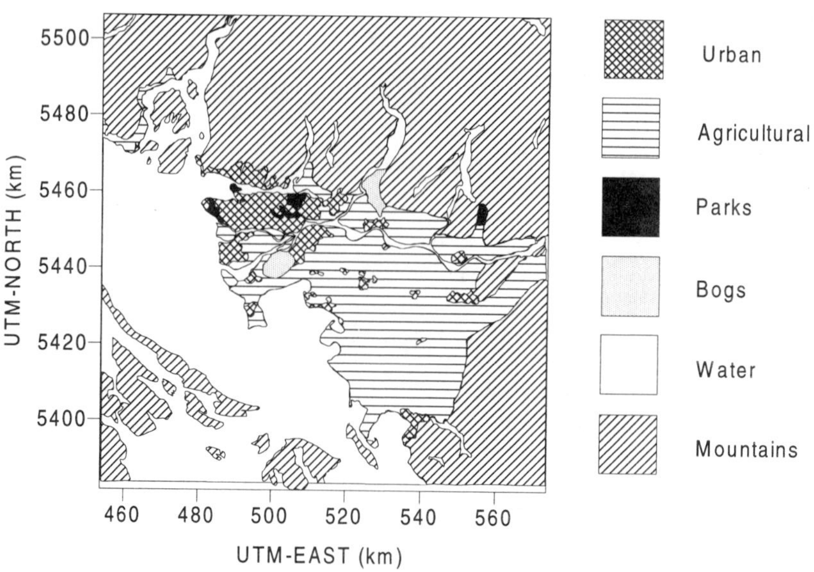

Figure 2. Map of the LFV modelling domain showing land use.

Fully Three Dimensional Mesoscale Model

The Colorado State University Regional Atmospheric Modelling System (CSU-RAMS) described by Pielke *et at.* (1992) can be used to simulate atmospheric phenomena ranging from large eddy simulation of the atmospheric surface layer to mesoscale thunderstorms. When applied to a mesoscale simulations, CSU-RAMS allows nesting of model grids. In the present study, Smagorinsky's deformation turbulence closure scheme is used in the horizontal direction and the Mellor-Yamada (1974) second order closure scheme is used in the vertical direction. The Mellor-Yamada's closure scheme allows the simulation of realistic TKE fields. In the present application, MLD is diagnosed from the vertical profile of TKE.

Model Simulation In the current simulation, three nested grids are used: the largest grid covers most of British Columbia, while the smallest one covers the region shown in Figure 1. Horizontal mesh spacings are 40 km, 10 km, and 2.5 km for the three grids. Vertical mesh configuration is the same for all three grids. The first vertical mesh point for u and v is at 56.35 m AGL, and the stretching ratio of the vertical mesh spacings is 1.2, with the maximum grid spacing being 2000 m. The top of the domain is at 18.3 km, and a lid condition is adopted here for all variables. To prevent vertically propagating gravity waves from being reflected by this lid, a nudging layer is imposed from 4 km above with a time scale of 600 sec. At the surface, vertical velocity is specified zero, horizontal velocity components and temperature are given by the Businger-Dyer surface layer similarity relationship, through Louis' scheme (Businger *et al.*, 1971; Louis, 1979). Five soil levels are adopted and their depths are 0.0, 0.1, 0.3, 0.6, 1.0 m. The simulation starts at 1600 PST on August 4, 1993, with 12 extra hours to allow non-physical solutions to be diffused or dissipated. Simulation ends at 0400 PST on August 6 so that a total of 36 hours' results are available. We are only interested in the results in the daytime of 5 August during which observational data are available for comparison.

TKE profiles produced by the model generally have simple shapes. Analysis of a large number of such profiles shows that ML depth can be detected by choosing a critical value of TKE (denoted by E_{cr}). Since the typical maximum simulated value of TKE in the

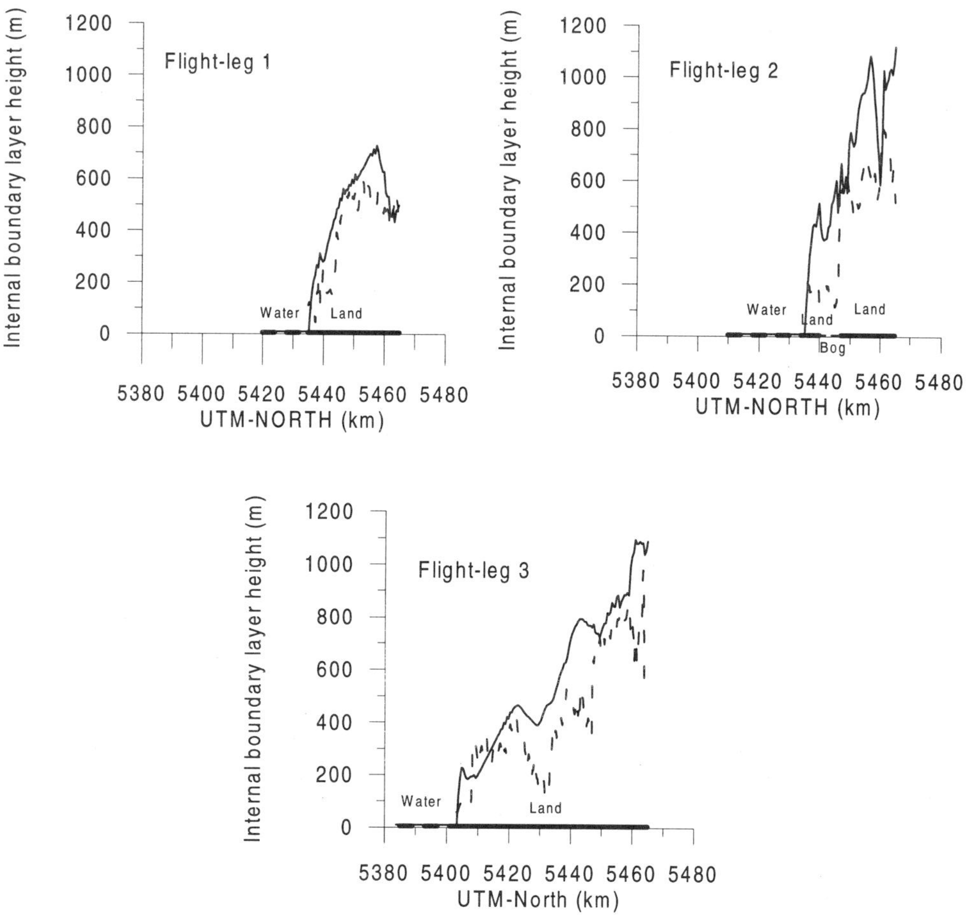

Figure 4. Spatial variation of Simple Applied Model simulation (full line) and measurement (dashed line) of TIBL height along flight legs 1, 2 and 3.

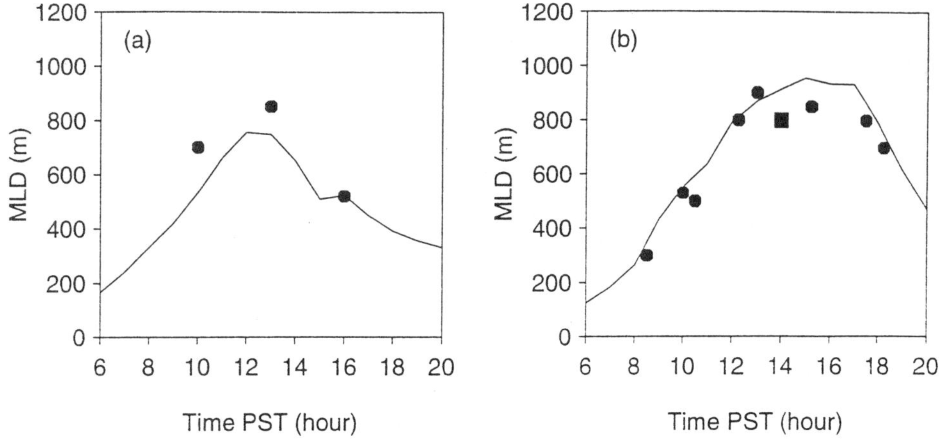

Figure 5. Temporal variation of CSU-RAMS simulation (full line) and measurement (•) of TIBL height at Langley Central and Harris Road.

domain is 0.5 $(m/s)^2$ (in kinematic units), we choose E_{cr}=0.03 $(m/s)^2$. For simplicity, we determine the ML depth as the first level above the ground at which TKE is smaller than the critical value, E_{cr}.

MODELLING RESULTS AND DISCUSSION

A Simple Applied Model

During the morning hours on 5th August the wind in the LFV was calm. At around 1000 PST, the wind speed gradually increased in a south-westerly direction, but with substantial variability in both direction and speed over the area. An hour later a westerly wind was established over the whole LFV, with a northerly component over central Vancouver, and a southerly component to the south and in the valley. Wind continued to increase during the next hours becoming more uniform in direction and speed over the LFV. The afternoon the wind was fairly strong with a rather complicated pattern.

Figure 3 shows the simulated and measured temporal behaviour of TIBL depth at Langley Central and Harris Road. The 10 PST sounding at Langley Central shows a significantly greater TIBL depth than simulated, but the next sounding (at 13 PST) indicates good agreement between model simulations and measurements. At Harris Road the TIBL depth was deduced from the tethersoundings carried out throughout the whole day. The agreement between measurements and model simulations at this site is very good.

Observed and modelled spatial structure of TIBL depth along three flight legs are shown in Figure 4. Flight leg 1 is the most westerly, crossing over the western part of the Vancouver area the agreement between observed and modelled depth is generally fair. Flight leg 2 crosses an area of complicated landuse, with tidal mud flats, swamps and a very poorly defined coastline; the agreement between model predictions and measurements is not very good. Attempts to correct this weakness by assuming an evaporation inversion over the bog areas were partially successful. Flight leg 3 has a very long land fetch covering a multitude of land use types with an overland fetch of more than 50 km; agreement between model and measurement is fair.

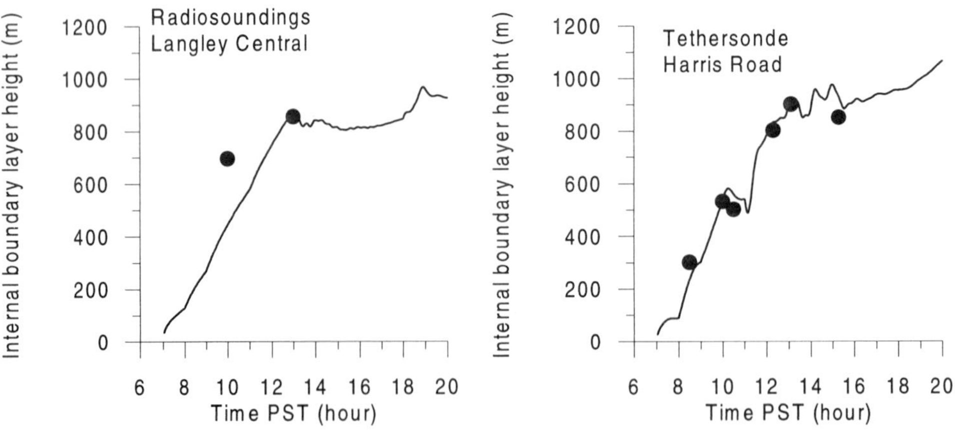

Figure 3. Temporal variation of Simple Applied Model simulation (full line) and measurement (•) of TIBL height at Langley Central and Harris Road.

Fully Three Dimensional Mesoscale Model

Using a range of statistical techniques, Cai *et al.* (1996) show that the modelled wind on this day matches very closely the wind measurements made using a superset of the stations used to provide interpolated wind fields for the simple applied model.

Figure 5 shows the simulated and measured temporal behaviour of TIBL depth at Langley Central and Harris Road. For the Langley-Central site, the modelled MLD closely matches the measurements, except for a slight underestimate around mid-day. For the Harris Road site, we see very good agreement between the modelled and the observed MLD not only during the morning with a developing mixed layer, but also during the afternoon with a decaying mixed layer. Note that one data point denoted by a solid square is derived from the aircraft lidar scanning.

Observed and modelled spatial structure of TIBL depth along three flight legs are shown in Figure 6. Notice that the left panel is a geographical map of the modelling domain rotated clockwise by 90°. The time of flight for these three legs is 1330 PST to 1406 PST. The modelled MLD is derived from the hourly averaged TKE field for the hour from 1300 PST to 1400 PST. The overall agreement between modelled and measured MLD is very good. For leg 1 which is very close to the western coastline of the basin, the modelled results badly underestimate the MLD in the city of Vancouver, but capture the overwater values well. This is due to the fact that the model produces too strong a westerly sea breeze in this area. In contrast, the observed wind near leg 1 had a stronger southerly component thus increasing overland fetch and MLD. For leg 2, the modelled MLD has a slight overestimation over the land, but spatial structure is well reproduced by the model. Near the root of the mountains, both modelled and observed MLD show a decrease. These good results can be associated with the wind field near leg 2 which agrees with the observed wind fairly well. Most of leg 3 is over land where

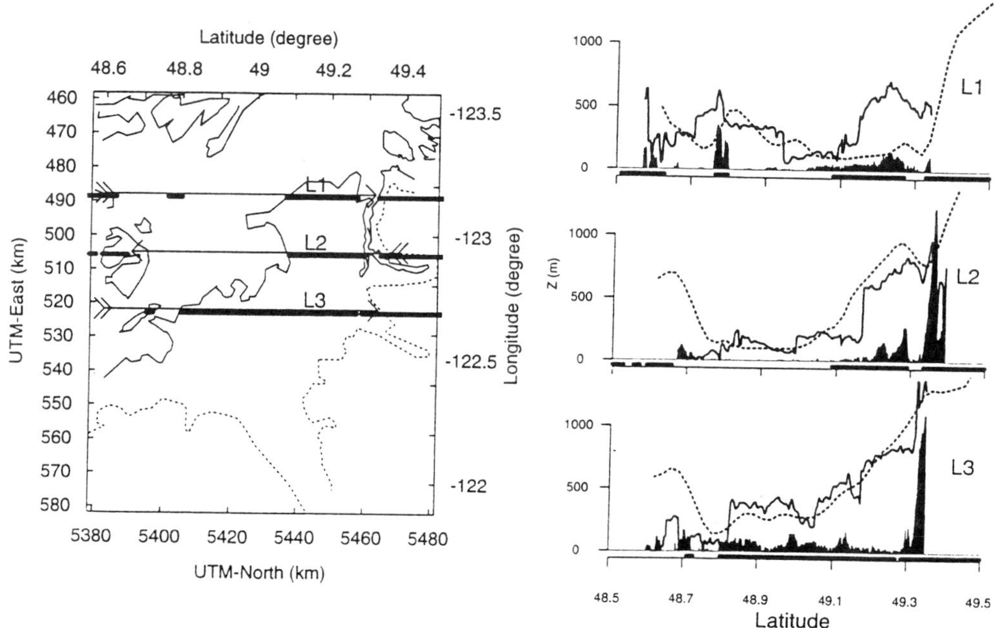

Figure 6. Spatial variation of CSU-RAMS simulation (full line) and measurement (dashed line) of TIBL height along flight legs 1, 2 and 3.

modelled MLD agrees very well with that detected from aircraft lidar signals. Two reductions in MLD correspond to two bays close to leg 3, the reduction being in response to stronger advection from the shoreline in these bays.

CONCLUSIONS

Given the complexity of topography, coastline and landuse in the LFV region, both models perform remarkably well. The simple applied model performs extremely well, given its simplicity. It is clear that correct specification of spatially resolved surface sensible heat flux and wind field are crucial to the success of this model which can be operated at very fine spatial resolution. The 3D model performs extremely well, though it too must capture the local wind field correctly for complete success. Its limited horizontal resolution result in strongly smoothed MLD fields.

ACKNOWLEDGMENTS

Data was drawn from parts of the Pacific '93 field study funded by Environment Canada, The U.S. National Oceanographic and Oceanic Administration, and was aprtiaslly funded by grants awarded by the Atmospheric Environment Service of Environment Canada and the Natural Science and Engineering Research Council of Canada. Kate Stephens and Roy Hourston analysed some of the data.

References

Batcharova E. and Gryning S.E., 1994: Applied model for the height of the daytime mixed layer and the entrainment zone. *Bound Layer Met*, **71**, 311-323.

Cai, X., D.G. Steyn, and R. Hourston, 1996: *Mesoscale meteorological modelling study of the Lower Fraser Valley, B.C., Canada from July 31 to August 05, 1993*. Report to Environment Canada. August, 1996. 32 pp.

Gryning S.E. and Batcharova E. 1990: Analytical model for the growth of the coastal internal boundary layer during onshore flow. *Quart. J. Roy. Met. Soc.*, **116**, 187-203.

Gryning S.E. and Batcharova E. 1996: Modelling the coastal internal boundary layer height over the Vancouver area. Presented to 9th Conference on the Applications of Air Pollution Meteorology. American Meteorological Society with the Air and Waste Management Association. Atlanta, 28 Jan-2 Feb., 1996.

Hayden, K.L., K.G. Anlauf, R.M. Hoff, W.J. Strapp, J.W. Bottenheim, H.A. Wiebe, F.A. Froude, J.B. Martin, D.G. Steyn and I.G. McKendry, 1997: The vertical chemical and meteorological structure of the boundary layer in the Lower Fraser Valley during PACIFIC '93. *Atmospheric Environment*, In Press.

Hoff, R.M., M. Harwood, A. Sheppard, F. Froude, B.A. Martin and W. Strapp, 1997: Use of airborne lidar to determine aerosol sources and movement in the Lower Fraser Valley (LFV), British Columbia. *Atmospheric Environment*, In Press.

Louis, J.-F., 1979: A parametric model of vertical eddy fluxes in the atmosphere. *Boundary Layer Meteorology*, **17**, 187-202.

Mellor, G. and Yamada, T., 1974: A hierarchy of turbulence closure models for planetary boundary layer. J. Atmos. Sci.., **31**, 1791-1806.

Pielke, R., Cotton, W., Walko, R., Tremback, C., Lyons, W., Grasso, L., Nicholls, M., Moran, M., Wesley, D., Lee, T. and Copeland, J., 1992: A comprehensive meteorological modelling system - RAMS. *Meteorology and Atmospheric Physics*, **49**, 69-91.

Pottier, J., B. Thomson, J. Bottenheim and D.G. Steyn, 1994: *Lower Fraser Valley oxidants study and Pacific '93 meta data report*. Canadian Institute for Research in Atmospheric Chemistry. 50p.

Steyn, D.G., J.W. Bottenheim, and R.B. Thomson, 1997: Overview of tropospheric ozone in the Lower Fraser Valley, and the PACIFIC '93 field study. *Atmos. Env.* In Press.

DISCUSSION

M. KAASIK: Did you estimate how much NO_x and other air pollution was released by your aircraft during measurements? I think for such experiments the effect of new knowledge should be compared to the effects of pollution.

D. G. STEYN: Of course the aircraft emitted NO_x and VOCs, but the amount would have been a small increment on the amount emitted by the very busy summertime air traffic into local airports, let alone the vast number of automobiles in the valley. With regard to cost/benefit, the estimated health costs of ozone and fine particulates are on the order of $100 million per year. It is difficult to estimate the dollar value of our data, but I am sure it is much greater then the incremental effect of the aircraft emissions.

R. BORNSTEIN: Have you thought of using a smaller grid spacing in the area of the elevated inversion base?

D. G. STEYN: I have indeed thought of that, and the approach is the central topic of an M.Sc. thesis by my student Magdalena Rucker. She employed a number of grid generation techniques to achieve high resolution in the entrainment layer, all with no improvement in the simulation of mixed layer height and entrainment energetics. The reason for this is that the problem lies in TKE parameterization weaknesses, rather than resolution. Present closure schemes do no better than the simple slab models.

NON-LOCAL VERTICAL TRANSPORT IN THE SHEAR-FREE CONVECTIVE SURFACE LAYER: NEW THEORY AND IMPROVED PARAMETERIZATION OF TURBULENT FLUXES

Sergej Zilitinkevich[1,2], A.A. Grachev[2], and J.C.R. Hunt[3]

[1]Institute for Hydrophysics, GKSS Research Centre, 21502 Geesthacht, Germany
[2]A.M. Obukhov Institute of Atmospheric Physics, 109017 Moscow, Russia
[3]DAMTP, University of Cambridge, Cambridge, CB3 9EW, UK

During the last several decades the surface frictional processes in the shear-free convective boundary layer (CBL) are considered conceptually in the spirit of the Prandtl (1932) theory of free convection, implying the ideas of (i) universal chaotic turbulence and (ii) local correspondence between turbulent fluxes and mean gradients. Accordingly the fluxes of heat and water vapour in the atmospheric surface layer are parameterized disregarding gross features of the CBL. Conventional practical tools are either the Monin-Obukhov similarity theory or simple downgradient turbulence closure models.

In strong convection regimes the concepts of perfectly chaotic turbulence and local transport, underlying conventional theories, break down. Essentially the problem is as follows. At very high Rayleigh numbers buoyancy-driven large-scale coherent (semi-organised) structures develop embracing the entire CBL. They consist of comparatively narrow but strong uprising plumes surrounded by wider but weaker downdraughts. The surface layer plays the role of a feeder layer for plumes, being filled with the CBL-scale convergence (towards plume axis) flow patterns. The latter can approximately be treated as internal boundary layers of radial geometry strongly affected by the buoyancy forces.

Generally convergence flows superimpose on (and interact with) mean wind. In the shear-free regime they yield their own velocity shears whose absolute values are characterised by the "minimum friction velocity", U_*. The shear-free convection turbulence turns out to be dependent on the surface friction and can be characterised by the "minimum Monin-Obukhov length", $L = L_* \equiv U_*^3 / B_s$, where B_s is the buoyancy flux at the surface (in contrast to classical formulation that implies $L \to 0$). Then turbulent mixing and heat/mass transfer facilitate depending on gross features of the CBL, such as the CBL depth, h, and the surface roughness length, z_{0u}.

Air Pollution Modeling and Its Application XII
Edited by Sven-Erik Gryning and Nadine Chaumerliac, Plenum Press, New York, 1998

Additional shears due to coherent structures are oriented against each other. They annihilate after the vector averaging and do not immediately contribute to the mean momentum transfer (hence the term "inactive turbulence").

Only the first step has been made in analysing the above mechanism theoretically (Businger, 1973; Schumann, 1988; Sykes et al., 1993; Siggia, 1994, Zilitinkevich, 1997; Grachev et al., 1997). In the present paper early theoretical models are shown to be not sufficiently advanced.

A revised model of the non-local friction processes is proposed for strong convection regimes over not too rough surfaces, $z_{0u}/h < 10^{-4}$. Theoretical predictions are compared with new experimental data on the minimum friction velocity and the heat/mass transfer.

$$\ln\frac{h}{z_{0u}} = C_{u0}\frac{W_*}{U_*} + 3\ln\frac{W_*}{U_*} - C_{u1} + C_{u2}\left(\frac{W_*}{U_*}\right)^{-1},$$

agrees well with data from the SCOPE, COARE and BOREX field experiments. Results from large-eddy simulation (LES) strongly underestimate U_*/W_*.

$$\frac{1}{C_{AH}} \equiv \frac{\Delta\theta U_*}{Q_s} - k_T\ln\left(\frac{z_{0u}}{z_{0T}}\right) = C_{T0}\frac{W_*}{U_*} - C_{T1} + C_{T2}\left(\frac{W_*}{U_*}\right)^{-1},$$

agrees well with data from the SCOPE, COARE and BOREX field experiments. Results from LES strongly underestimate C_{AH}.

It is shown that data from large-eddy simulation (LES) studies (Schmidt & Schumann, 1989; Sykes et al., 1993), served as "empirical" basis in the early models, diverge drastically from atmospheric data (Figures 1 and 2). It is conceivable that LES is not an appropriate tool for the investigation of fine features of the flow-surface interaction as long as the surface fluxes within LES are parameterized using traditional local models.

The proposed theoretical model is in excellent agreement with atmospheric data (Figures 1-3). It results in reliable parameterization of surface fluxes in atmospheric problems.

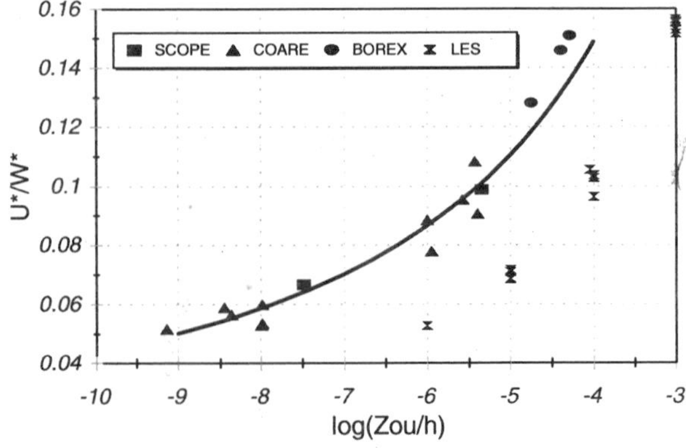

Figure 1. Dimensionless minimum friction velocity, U_*/W_*, versus the roughness length to the CBL depth ratio, z_{0u}/h. Theoretical curve, after the equation

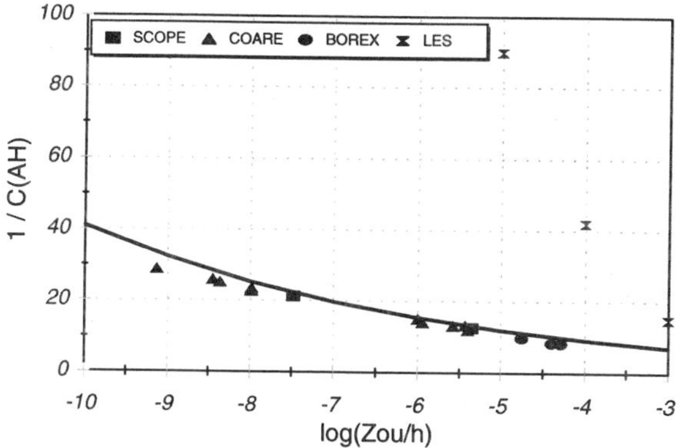

Figure 2. Reciprocal of the aerodynamic heat transfer coefficient, $1/C_{AH}$, versus the roughness length to the CBL depth ratio, z_{0u}/h. Theoretical curve, after the equation

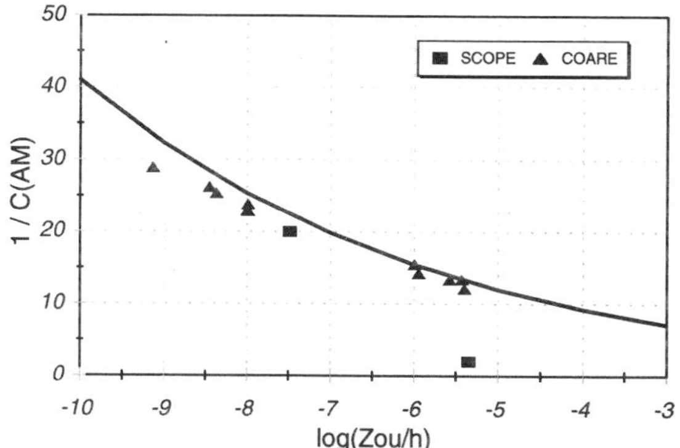

Figure 3. As for Fig. 2, but for reciprocal of the aerodynamic mass transfer coefficient, $1/C_{AM}$. Theoretical curve, after the equation

The model is shown to be valid when the roughens length to the CBL depth ratio, z_{0u}/h, is less than 10^{-4}. The nature and the theory of the purely convective atmospheric surface layer over very rough surfaces is a subject of further investigation. Another challenging problem playing a key role in the convective heat and mass transfer is proper determination of the roughness lengths for scalars.

$$\frac{1}{C_{AM}} \equiv \frac{\Delta q U_*}{E_s} - k_q \ln\left(\frac{z_{0u}}{z_{0q}}\right) = C_{q0}\frac{W_*}{U_*} - C_{q1} + C_{q2}\left(\frac{W_*}{U_*}\right)^{-1},$$

agrees well with data from the SCOPE and COARE field experiments.

REFERENCES

Businger, J.A., 1973: A note on free convection. *Boundary-Layer Meteorol.*, 4, 323-326.

Grachev, A.A., C.W. Fairall, and S.S. Zilitinkevich, 1997: Surface-layer scaling for the convection-induced stress regime. To appear in *Boundary-Layer Meteorol.*

Prandtl, L., 1932: Meteorologische Anvendungen der Strömungslehre. *Beit. Phys. Fr. Atmos.*, **19**, 188-202.
Schmidt, H., and U. Schumann, 1989: Coherent structure of the convective boundary layer derived from large-eddy simulations. *J. Fluid Mech.*, **200**, 511-562.
Schumann, U., 1988: Minimum friction velocity and heat transfer in the rough surface layer of a convective boundary layer. *Boundary-Layer Meteorol.*, **44**, 311-326.
Siggia, E.D., 1994: High Rayleigh number convection. *Ann Rev. Fluid Mech.*, **26**, 137-168.
Sykes, R.I., and D.S. Henn, 1989: Large-eddy simulation of turbulent sheared convection. *J. Atmos. Sci.*, **46**, 1106-1118.
Sykes, R.I., D.S. Henn, and W.S. Lewellen, 1993: Surface-layer description under free-convection conditions. *Quart. J. Roy. Meteorol. Soc.*, **119**, 409-421.
Zilitinkevich, S., 1997: *Heat/Mass Transfer in the Convective Surface Layer: Towards Improved Parameterization of Surface Fluxes in Climate Models.* Alfred-Wegener-Institut für Polar- und Meeresforschung. Berichte aus dem Fachbereich Physik. Report 76, 34 pp.

DISCUSSION

R. BORNSTEIN: What is the basis of your equation which states that the log law is valid when $z > 30z_{0u}$, not, say, $4z_{0u}$?

S. ZILITINKEVICH: In classical laboratory studies of logarithmic boundary layers over rough surfaces with roughness elements made of sand particles of height h_0 (so-called "sand roughness") the roughness length with respect to momentum, z_{0u}, is shown to be equal to $\sim 1/30\, h_0$. For natural land surfaces with typical height of roughness elements h_0, the roughness length, z_{0u}, lies usually between $1/30\, h_0$ and $1/10\, h_0$. Clearly, within the roughness-element layer, $0<z<h_0$, the velocity profile is immediately affected by the flow-obstacle interactions, and nothing similar to the logarithmic profile is observed. The latter is theoretically grounded at $z >> h_0$. Factually it is observed already at $z > (2 \div 3)h_0$. Anyhow, $z>30\ z_{0u}$ is a necessary condition for the applicability of the logarithmic profile.

R. BORNSTEIN: Is it correct to have $z_{0u} \neq z_{0(for\ scalars)}$?

S. ZILITINKEVICH: Yes, over dynamically rough surfaces, $z_{0u} \neq z_{0(for\ scalars)}$. The reason is that the momentum is transferred from flowing fluid to an adjacent rough surface basically through the pressure forces, which are much more efficient than the molecular-viscosity forces. On the contrary, the heat (mass) transfer at the very surface is always controlled by the molecular conductivity (diffusivity), no matter how rough is the surface. As a result the resistance of the roughness layer for scalars is much higher than for momentum. In other words, the roughness length for scalars is much smaller than for momentum.

STUDY OF THE SPACIAL AND TEMPORAL STRUCTURE OF TURBULENCE IN THE NOCTURNAL RESIDUAL LAYER

Frank R. Freedman,[1,2] and Robert D. Bornstein[2]

[1]Department of Civil Engineering
Stanford University
Stanford, CA 94305-4020

[2]Department of Meteorology
San Jose State University
San Jose, CA 95192-0104

INTRODUCTION

Of the planetary boundary layer (PBL) sublayers defined by Stull (1988, pp. 10-11), the nighttime residual layer (RL) may be the least understood. One reason for this is that, unlike the daytime convective boundary layer and surface based nocturnal boundary layer, the RL is in general decoupled from the surface. Consequently, turbulence in the RL does not scale with surface turbulent quantities, and derivation of accurate similarity relationships is difficult, if not impossible. Although qualitative success has been achieved in relating RL turbulence to the gradient Richardson number (Ri) (Mahrt et al., 1979; Lenschow et al., 1987; Kim and Mahrt, 1992), oftentimes these scalings quantitatively fail due to overly coarse vertical differencing in computing Ri (Padman and Jones, 1985). Study of RL turbulence is further complicated by the wide variety of atmospheric forcing mechanisms present during the night. Since nighttime turbulent intensities are small, forcing by radiation, baroclinity, topographic drainage flows, and gravity waves can become important. Variability in each of these leads to a seemingly infinite number of possible RL turbulent time/height structures (e.g., André et al., 1978; Garratt, 1985; Lenschow et al., 1978; Heilman and Takle, 1991; Weber and Kurzeja, 1991).

One unresolved issue resulting from the sparsity of knowledge on RL turbulence is the origin of the turbulence, i.e. whether it is truly "residual" daytime turbulence, "new" turbulence generated after dissipation of daytime turbulence, or both. The following addresses this and the spacial/temporal structure of RL turbulence for the Wangara Day 33 case (Clarke et al., 1971) through 1-D numerical simulation. Turbulence is modeled by use of a prognostic equation for turbulent kinetic energy (TKE). By investigating the magnitude of the terms in this equation, the structure and temporal dynamics of the turbulence are studied. Information obtained, in addition to serving academic interest, is useful to air pollution meteorologists interested in the transfer of pollutants stored in the RL to the surface (e.g., Neu et al., 1994).

Air Pollution Modeling and Its Application XII
Edited by Sven-Erik Gryning and Nadine Chaumerliac, Plenum Press, New York, 1998

MODEL DESCRIPTION AND SETUP

The current study employs a 1-D formulation of the TVM mesoscale model of Schayes et al. (1996). The model is governed by the basic conservation equations for momentum (formulated in vorticity mode), potential temperature, and specific humidity. The large-scale pressure gradient force is represented by a specified geostrophic wind profile. Cooling by radiative flux divergence is included largely following the parameterization used by Garratt and Brost (1981). Turbulent fluxes are computed from eddy diffusivity theory (a.k.a. K-theory), with eddy diffusivity calculated as a function of TKE (computed prognostically) and a turbulent length scale. Computation of turbulent length scales is achieved by algebraic equations of Therry and Lacarrère (1983) modified for nighttime conditions. These modifications involve the assurance of continuity of mixing lengths across the surface layer/PBL interface, and the use of an asymptotically approached (with height) free-atmospheric length scale so that turbulence is more properly modeled within the RL. Further details of the model are given in Freedman (1996).

The model domain extends to 2 km. The simulation is initialized at 0600 LST on Day 33 with measured profiles of wind velocity, potential temperature, and specific humidity interpolated to the model grid. By starting the simulation at 0600 LST, and hence modeling the previous daytime convective boundary layer, accurate conditions at the evening transition are more easily achieved than would be possible by initializing at sunset. Time and height varying geostrophic winds are imposed from interpolation of observed surface geostrophic and 0-1 km and 1-2 km thermal winds, as described by Yamada and Mellor (1975). At 0600 LST Day 33, geostrophic speeds are ~ 5 ms^{-1} at the surface decreasing to small values at 2 km. At 0600 LST Day 34, the geostrophic wind speed increases to ~ 10 ms^{-1} at the surface decreasing to ~ 4 ms^{-1} at 2 km. Directionally, the geostrophic wind shifts from southeasterly to northeasterly throughout the PBL.

RESULTS

Simulated nighttime profiles of mean variables were in fairly close agreement with observations Discrepancies were mainly due to horizontal advective effects, which cannot be included in the 1-D

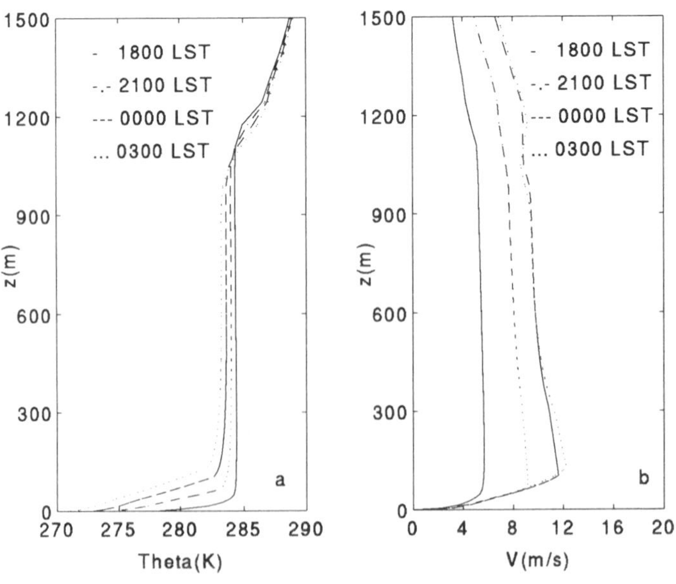

Fig. 1. Vertical profiles of simulated a) potential temperature and b) wind speed.

model formulation. Profiles, however, were in excellent agreement with third-order turbulence closure model results of André et al. (1978).

Simulated profiles of mean potential temperature and wind speed (Fig. 1a,b) show typical nighttime structure. A nocturnal boundary layer (NBL), characterized by strong stratification and wind shear, is present in the lowest ~ 150 m. This region is capped by a wind speed maximum resulting from an inertial oscillation. The region of both weak stratification and wind shear above the speed maximum and below the capping elevated stable layer (at ~ 1200 m) is the RL. Although RL mean profiles result largely from residual effects of daytime mixing, important non-steady behavior occurs during the evening: potential temperature decreases with time due to clear-sky radiative flux divergence and wind speed increases with time due to an inertial oscillation.

The influence of mean potential temperature and wind profiles on RL TKE generation can be described by Ri, proportional to the ratio of buoyancy destruction to shear production of TKE. As such, "small" Ri values correspond to regions of relatively high TKE generation, and "large" values correspond to regions where relatively low or zero TKE generation. Turbulent production ceases when Ri exceeds a critical value, which ranges from 0.25-0.50 (Miles et al., 1961; Padman and Jones, 1985). Vertical profiles of simulated Ri (Fig. 2) for early and late evening hours show regions of sub-critical Ri (< 0.50) above 400 m within the RL, signifying that TKE generation is expected in the RL.

It is interesting that Ri < 0 in much of the RL during the early evening (see the 2000 LST profile in Fig. 2). The implied positive heat flux results from a superadiabatic lapse rate produced during the day. The prediction of a superadiabatic rather than adiabatic/near-adiabatic lapse rate in the middle to upper region of the PBL is erroneous, and results from application of local-gradient based K-theory to the convective boundary layer, where fluxes are not determined by the local gradient (Stull, 1988). While the generation of positive heat fluxes *from superadiabatic lapse rates* is therefore

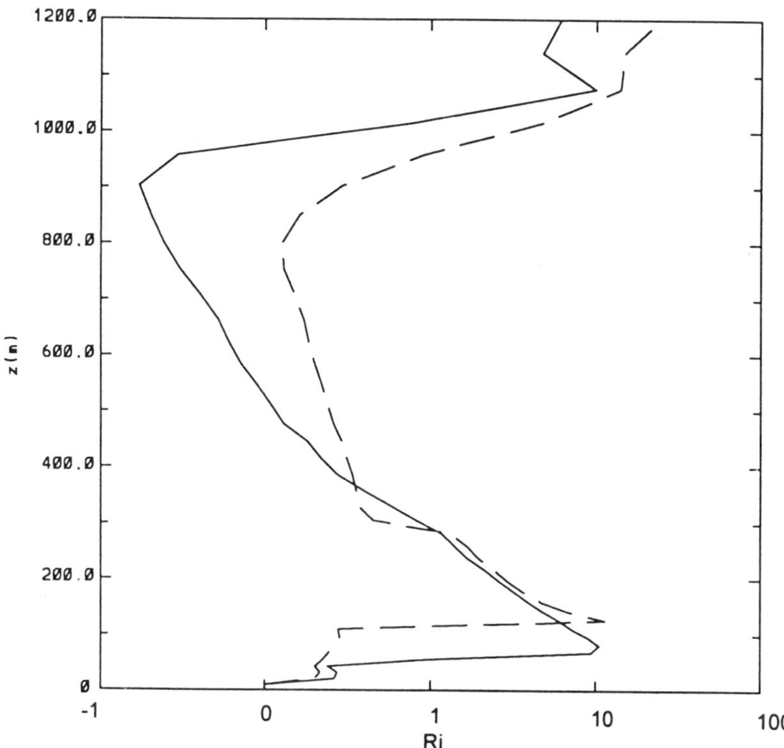

Fig. 2. Vertical profiles of simulated Ri: —— 2000 LST, ----- 0200 LST; abscissa changes from linear to logarithmic at Ri = 1.

erroneous in this case, it is not clear whether positive heat fluxes can persist in the RL during early evening from, for example, large residual values of temperature variance (which also lead to a positive heat flux). The answer to this question is reserved for future work; in the meantime the simulated positive values of heat flux in the early evening are viewed as merely plausible rather than as fact.

Simulated TKE profiles (Fig. 3a) show clear distinction between the daytime well-mixed layer (evident from the 1600 LST profile), the nocturnal boundary layer (below 150 m), and the RL (from 300-1000 m). RL TKE is ~ 1% of that within the daytime mixed layer, and is ~ 10% of that within the NBL. In contrary, RL eddy diffusivity (Fig. 3b) is higher than that within the NBL by an order of magnitude. This feature, due to larger turbulent length scales within the RL (as eddy size is not limited by proximity to the ground), has important implications for pollutant mixing within the nighttime PBL.

Explanation of the height and time structure of RL TKE is deduced from Figs. 4-6, which show the time tendency of TKE and its production budget at 450, 600, and 850 m. Turbulent diffusion is small at all three locations; the TKE tendency is thus primarily governed by the shear, buoyancy, and dissipation terms. Net TKE production takes place within a narrow time interval (for ~ two hours in the early evening). The magnitude of this production is equal at 600 and 850 m (Figs. 5a and 6a), and is smaller and occurs later in the evening at 450 m (Fig. 4a). The narrow time region of net production results from competing effects of shear production, buoyancy destruction and molecular dissipation. Immediately after sunset, wind shear has not developed to the point where its TKE production can outweigh rapid daytime dissipation. However, as wind speed and shear increase in the early evening, net production takes place. Later in evening (~ after hour 16; 10 PM) the RL has stabilized to the point where buoyancy destruction causes net TKE destruction, although wind shear and shear production increase for a part of this time. The narrow time region of net TKE production thus coincides with the time period in which wind shear is strong enough to outweigh both molecular dissipation (strong immediately after sunset) and buoyancy destruction (strong later in the evening as the RL stabilizes). The net TKE production that takes place during this time leads to a significant portion of "new" turbulence within the RL.

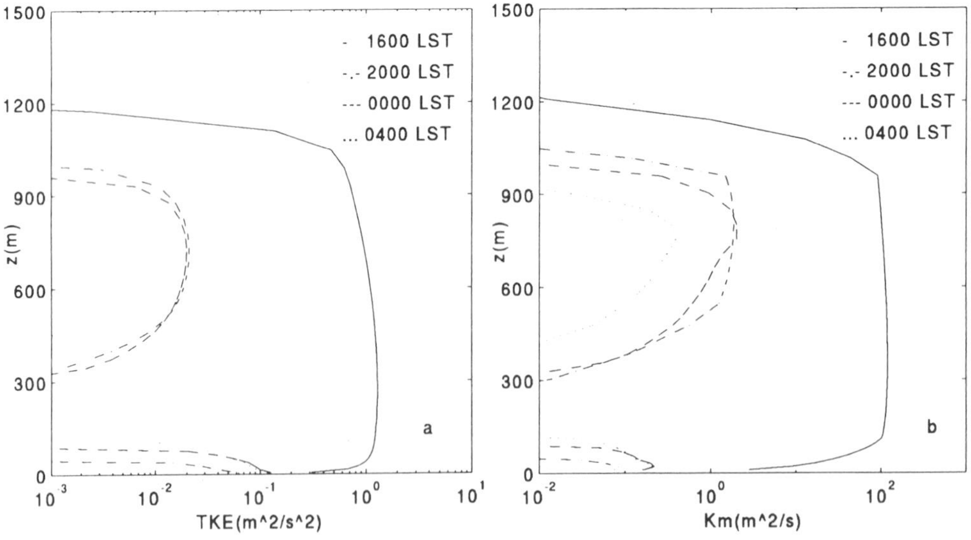

Fig. 3. Vertical profiles of simulated a) turbulent kinetic energy and b) eddy diffusivity.

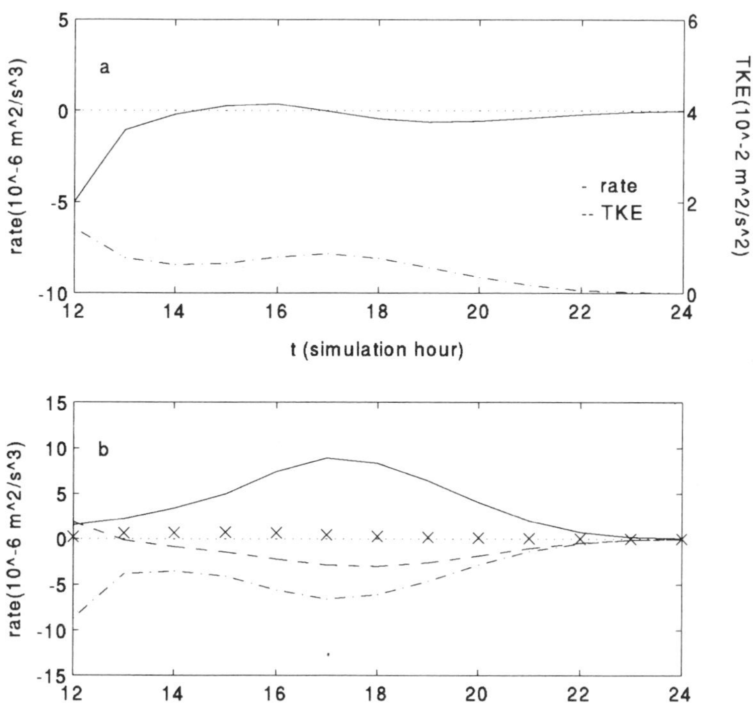

Fig. 4. Time variation of a) TKE and its total production rate and b) TKE production terms at 450 m. Simulation hour 12 corresponds to 1800 LST. In b), symbols represent production rates due to: —— wind shear, ----- buoyancy, -.-.- dissipation, ××× turbulent diffusion.

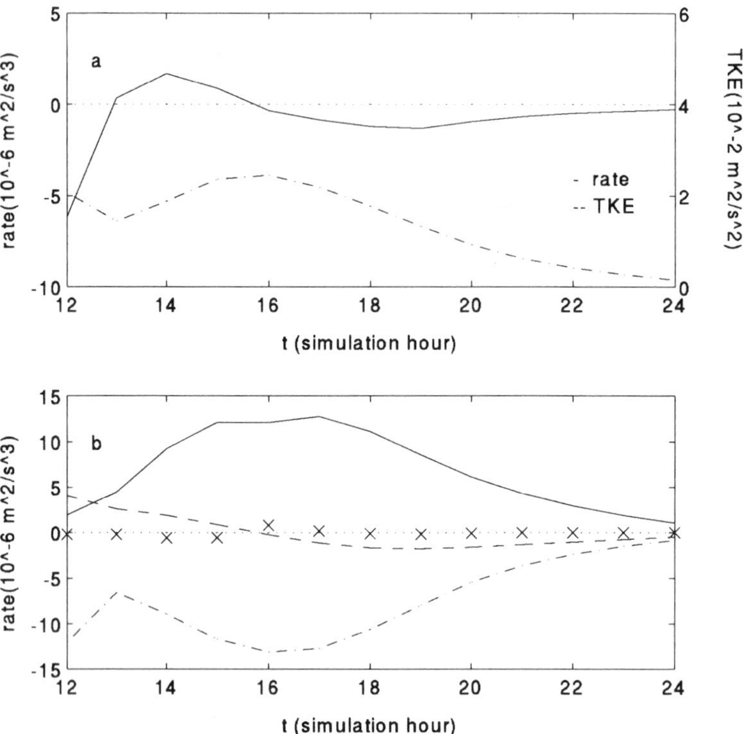

Fig. 5. Same as Fig. 4 except at 600 m.

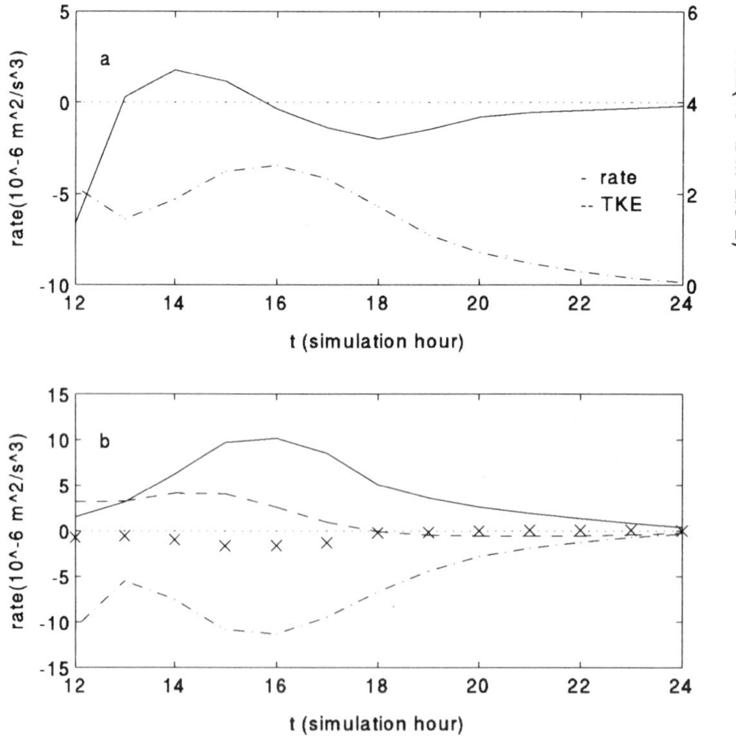

Fig. 6. Same as Fig. 4 except at 850 m.

RL stabilization take place from "bottom-to-top", evident from the sequentially later time at which the buoyancy term changes sign from positive to negative at 450, 600, and 850 m (Figs. 4-6). One explanation for this is growth of the top of the surface-based stable layer into the base of the RL, in turn stabilizing the RL from the bottom upwards. However, an equally if not more important factor was found to be the role of upper level moisture on the height dependency of radiative cooling. During the Day 33 period, dryer air was present within the elevated stable layer than that within the PBL. Largely as a result of this decreased amount of infrared emitting gas, substantially less longwave radiation is emitted downwards from the elevated stable layer into the top of the RL than that emitted from the top of the RL downwards into the middle of the RL. This difference between radiative input to and output from the upper RL leads to enhanced cooling in this region, in turn inhibiting stabilization in the upper RL. Weaker RL stability in the upper RL enables TKE at 850 m (Fig. 6) to be produced at the same rate as at 600 m (Fig. 5), although wind shear (Fig. 1b) is weaker at 850 m than at 600 m.

CONCLUSION

Simulation of Wangara Day 33 was performed to study the spacial and temporal structure of turbulence within the residual layer (RL). A 1-D formulation of the TVM mesoscale model was used with modified turbulence closure to allow for improved nighttime simulation of TKE.

Inspection of net and component TKE production showed that net TKE production took place within a narrow time region (~ 2 hours in duration) in the early evening. The major source of TKE was shear production, which is only able to outweigh dissipation (important immediately after sunset) and buoyant destruction (important later in the evening as the RL stabilizes) during this narrow time interval. Production during this period provided a significant portion of "new" TKE to the RL. Moisture content within the elevated stable layer, through effects on radiative cooling, was found to be an important factor in the determination of RL stabilization. Significantly dryer air in the elevated stable layer than that present in the PBL led to enhanced cooling in the upper RL, allowing weak stability to be maintained, and TKE to be produced in this region in the presence of relatively weak wind shear.

The effects of the mean moisture profile on RL stabilization and TKE generation encourage further study of the roles of mean thermodynamic profiles on the spacial and temporal structure of RL TKE. In addition, since wind shear is in large part determined by the degree and type of baroclinity in the PBL, different geostrophic wind profiles than those used in the current study are hypothesized to give different TKE structures within the RL. Current work is being carried out to examine the roles of mean thermodynamic and geostrophic wind profiles on RL TKE.

ACKNOWLEDGMENTS

The authors would like to thank Dr. Ted Yamada for making available the Clarke et al. (1971) Wangara report. Appreciation is also addressed to Dr. B. B. Hicks for his help in locating weather maps for the Wangara Day 33 period.

REFERENCES

André, J.C., De Moor, G., Lacarrère, P., Therry, G., and Du Vachat, R., 1978, Modeling the 24-hour evolution of the mean and turbulent structures of the planetary boundary layer, *J. Atmos. Sci.* 35: 1861-1883.

Clarke, R.H., Dyer, A.J., Brook, R.R., Reid, D.G.,Troup, A.J., 1971, *The Wangara Experiment: Boundary Layer Data*, Commonwealth Scientific and Industrial Research Organization, Australia.

Freedman, F.R., 1996, Reevaluation of lower order turbulence closure prediction the PBL, M.S. thesis, San Jose State University.

Garratt, J.R., 1985, The inland boundary layer at low altitudes, *Bound. Layer Meteor.* 42: 307-327.

——, and Brost, R.A., 1981, Radiative cooling effects within and above the nocturnal boundary layer, *J. Atmos. Sci.* 38: 2730-2746.

Heilman, W.E., Takle, E.S., 1991, Numerical simulation of the nocturnal turbulence characteristics over Rattlesnake Mountain, *J. Appl. Meteor.* 30: 1106-1116.

Kim, J., and Mahrt, L., 1992, Simple formulation of turbulent mixing in the stable free atmosphere and nocturnal boundary layer, *Tellus* 44A: 381-394.

Lenschow, D.H., Li, X.S., Zhu, C.J., and Stankov, B.B., The stably stratified boundary layer over the Great Plains, *Bound. Layer Meteor.* 42: 95-121.

Mahrt, L., Heald, R.C., Lenschow, D.H., and Stankov, B.B., 1979, An observational study of the structure of the nocturnal boundary layer , *Bound. Layer Meteor.* 17: 247-264.

Miles, J.W., 1961, On the stability of heterogeneous shear flows, *J. Fluid Mech.* 10: 496-508.

Neu, U., Künzle, T., and Wanner, H., 1994, On the relation between ozone storage in the residual layer and daily variation in near-surface concentration - a case study, *Bound. Layer Meteor.* 69: 221-247.

Padman, L., and Jones, I.S.F., 1985, Richardson number statistics in the seasonal thermocline, *J. Phys. Oceanogr.* 15(7): 844-854.

Schayes, G., Thunis, P., and Bornstein, R.D., 1996, Topographic vorticity-mode mesoscale-β (TVM) model: Part I- formulation, *J. Appl. Meteor.* 35: 1815-1823.

Stull, R. B., 1988, *An Introduction to Boundary Layer Meterology*, Kluwer Academic Pub., Dordrecht.

Therry, G., and Lacarrère, P., 1983, Improving the eddy kinetic energy model for planetary boundary layer description, *Bound Layer Meteor.* 25: 63-88.

Weber, A.H., and Kurzeja, R.J., 1991, Nocturnal planetary boundary layer structure and turbulence episodes during the project STABLE field program, *J. Appl. Meteor.* 30: 1117-1133.

Yamada, T., and Mellor, G., 1975, A simulation of the Wangara atmospheric boundary layer data, *J. Atmos. Sci.* 32: 2309-2329.

DISCUSSION

D. STEYN: In your nocturnal “mixed layer”, you invoke shear generated turbulence to maintain the mixing. If the layer is truly mixed, there should be no shear. Is this a contradiction?

R. BORNSTEIN: The term residual/shear production layer (R/SPL) is a better term for this layer. In the early evening, its turbulence is that which is left over from the daytime CBL, while later on new turbulence forms in the layer due to shear production.

D. STEYN: We have observed elevated polluted layers in what you will call the nocturnal “mixed layer”. Radiative divergence in these layers can induce superadiabatic lapse rates which will result in buoyancy generated turbulence, thus keeping the layer mixed.

R. BORNSTEIN: I assume that you are referring to long wave radiative flux divergence. Our simulations have shown that this effect can also be a source of turbulence.

MODELING SIZE-DISTRIBUTED SEA SALT AEROSOLS IN THE ATMOSPHERE: AN APPLICATION USING CANADIAN CLIMATE MODELS

S.L. Gong [1], L.A. Barrie[1], J.-P. Blanchet[2] and L. Spacek[2]

[1]Atmospheric Environment Service, 4905 Dufferin Street
Downsview, Ontario M3H 5T4 Canada
[2]Earth Sciences Department
University of Quebec at Montreal (UQAM)
P.O. Box 8888, Stn "Downtown"
515 Ste-Catherine St, Montreal, QC, H3C 3P8 Canada

ABSTRACT

An algorithm for size-distributed atmospheric aerosols designed for the Northern Aerosol Regional Climate Model [NARCM] is applied to three versions of the Canadian climate models: GCM, RCM and FIZ-C/LCM. It incorporates the processes of aerosol generation, diffusive transport, transformation and removal as a function of particle size to simulate global and regional sea-salt aerosol spatial and temporal distributions. A comparison was made between observations and model predictions of sea-salt. Size-resolved aerosol properties such as transport patterns, fluxes and removals are obtained from the simulations. Since the sea-salt generation term is relatively well quantified, the comparison ensures that a reasonable parameterization of removal and transport schemes is used.

1 INTRODUCTION

The effect of aerosol particles on radiative forcing occurs by two processes: (1) directly scattering and absorption of solar radiation and (2) indirectly altering cloud droplet size distribution and concentration, and thereby the albedo of optically thin clouds. In addition, aerosols can alter precipitation formation processes and change the life time of clouds in the atmosphere. Natural sources, such as sea salt and desert dust and biomass burning are important due to their influence on aerosol background concentrations, and because of the possibility that the source strength of these particles may respond to future changes in climate. Together with biogenic sulphur aerosols, sea-salt aerosols form the natural background aerosol surface area and concentration of cloud condensation nuclei (CCN) in marine areas upon which atmospheric sulphur aerosols are superimposed. In addition, sea-salt aerosol particles are chemical carriers of species containing Cl, Br, I and S and therefore play a role in the atmospheric cycles of these important elements. The halogens Br and Cl, once mobilized by heterogeneous reactions from sea-salt inorganic forms to reactive gaseous forms (e.g. Br_2, Cl_2) [e.g., Mozurkewich 1995, Behnke *et al.* 1997], can play a role in atmospheric ozone depletion and destruction of light hydrocarbons [Jobson *et al* 1994].

In the following sections of this paper, the NARCM aerosol algorithm is presented with up-to-date built-in parameterizations of micro-physics and chemistry of aerosols. As an initial step, the algorithm was used to predict large-scale sea-salt aerosol distributions.

2 NARCM Aerosol Algorithm

The aerosols are represented by a size-segregated prognostic equation in such a way that for aerosol particles with a dry size range [or section] i, the mass balance equation can be written as

$$\frac{\partial \rho \chi_i}{\partial t} + div \cdot \rho \chi_i U = S_i + I_i \tag{1a}$$

where χ_i is the mass mixing ratio (mmr) in the size bin i [kg/kg$_{air}$], U horizontal wind velocity vector [m s^{-1}], ρ air density [kg m^{-3}] and S_i the source and sink terms which may include following contributions: coagulation, chemical transformation, cloud interaction, gas-to-particle (GTP) conversion, precipitation scavenging, dry deposition and gravitational settling. The term I_i represents the intersectional transfer rate for all the source and sink terms which modify the size distribution of the aerosols. Except for the dry and wet removal processes, all the source terms contribute to the intersectional transfer of aerosols from one size bin to another. In accordance with the GCM convention, equation (1a) can be re-written in terms of tendencies which govern the aerosol concentrations

$$\frac{\partial \chi_i}{\partial t} = \left.\frac{\partial \chi_i}{\partial t}\right|_{DYNAMICS} + \left.\frac{\partial \chi_i}{\partial t}\right|_{SURFACE} + \left.\frac{\partial \chi_i}{\partial t}\right|_{CLEAR\ AIR} + \left.\frac{\partial \chi_i}{\partial t}\right|_{DRY} + \left.\frac{\partial \chi_i}{\partial t}\right|_{IN-CLOUD} + \left.\frac{\partial \chi_i}{\partial t}\right|_{BELOW-CLOUDS} \tag{1b}$$

In equation (1b), the aerosol concentration change has been divided into tendencies for dynamics, surface, clear air, dry deposition, in-cloud and below-cloud processes. The dynamics includes resolved motion as well as sub-grid turbulent diffusion and convection. The surface processes include surface emission rate of both natural and anthropogenic aerosols and serve as boundary conditions for the model. Particle nucleation, coagulation and chemical transformation are included in clear-air process. These tendency calculations constitute the aerosol module within NARCM.

Three versions of Canadian climate model are now run with the NARCM aerosol algorithm: FIZ-C/LCM, RCM and GCM. FIZ-C/LCM is a column version [Therrien 1993] of the Canadian global GCM while RCM is a regional climate model developed by Robert *et al.* [1985]. Both of them share the same physics as the Canadian General Circulation Model (CGCM). In FIZ-C/LCM the horizontal advection of aerosols is neglected. Time integration for aerosols in the RCM and GCM is done by a semi-Lagrangian and semi-implicit scheme. The prognostic equations take the following form

$$d_t \psi = \mathbf{L}(\psi) + \mathbf{R}(\psi) + \mathbf{E}(\psi) + \mathbf{P}(\psi) + \mathbf{H}(\psi) + \mathbf{T}(\psi) \tag{2}$$

where ψ represents one of the variables such as temperature, velocity, pressure, humidity liquid water amount as well as aerosol concentrations. The right-hand side of the prognostic equation has been broken down into six separate parts. The first term, **L**(ψ), represents the linear part of the gravity and elastic waves. The second term, **R**(ψ), represents all the remaining dynamic terms. The **P**(ψ) terms contains the physical parameterization acting in the vertical. This is the term where aerosol physics and chemistry is built into. The **H**(ψ) term represents horizontal diffusion, while **E**(ψ) and **T**(ψ) terms are directly linked to the numerical treatment. The **E**(ψ) term represents the nesting done to blend the model's variables with the driving model values near the lateral boundaries. **T**(ψ) term is Robert's time filter [Robert 1966]. The numerical method applied to solve for ψ is termed "successive corrections". By this method, the solution for ψ is achieved by firstly considering only the dynamic terms of the prognostic equations [**L**(ψ) and **R**(ψ)] and then correcting the value of ψ by successively adding the terms [**E**(ψ), **P**(ψ), **H**(ψ) and **T**(ψ)]. Comparing (1) and (2), it is apparent that:

$$\mathbf{L}(\chi)+\mathbf{R}(\chi)=\left.\frac{\partial\chi_i}{\partial t}\right|_{DYNAMICS} \tag{3}$$

$$\mathbf{P}(\chi)=\left.\frac{\partial\chi_i}{\partial t}\right|_{SURFACE}+\left.\frac{\partial\chi_i}{\partial t}\right|_{CLEAR\,AIR}+\left.\frac{\partial\chi_i}{\partial t}\right|_{DRY}+\left.\frac{\partial\chi_i}{\partial t}\right|_{IN-CLOUD}+\left.\frac{\partial\chi_i}{\partial t}\right|_{BELOW-CLOUDS} \tag{4}$$

The tendencies for aerosol processes are computed according to equation (4). By solving equation (2) with (4), the spatial and temporal distributions of aerosols are obtained.

2.1 Aerosol Sources

A semi-empirical formulation [Monahan et al. 1986] is used to relate the size-segregated surface emission rates of sea-salt aerosols to the surface wind speed. It has shown reasonably good results in sea-salt aerosol modelings [Fitzgerald, 1992; Stramska, 1987; Gong *et al.*; 1997a]. For the indirect production of sea-salt aeroosls through bubbles, the density function dF_0/dr [particles m^{-2} s^{-1} μm^{-1}], which expresses the rate of sea-salt droplet generation per unit area of sea surface, per increment of droplet radius, is given by:

$$\frac{dF_0}{dr}=1.373U_{10}^{3.41}r^{-3}\left(1+0.057r^{1.05}\right)\times10^{1.19e^{-B^2}} \tag{5}$$

where B=(0.380-*log r*)/0.650, r the particle radius and U_{10} the 10-m wind speed. The emission rate for a size bin is obtained by integrating equation (5) over the size range in the bin.

2.2 Physical Processes

A detailed description of the physical processes affecting atmospheric sea-salt aerosols has been given by Gong *et al.* [1997a]. They include: (1) dry deposition; (2) below-cloud scavenging; (3) in-cloud processes and (4) coagulation and equilibrium growth. For sea-salt aerosols, the coagulation process has been neglected because large particles dominate the total mass. It should be pointed out that all the processes are size-dependent and are computed for each size bin based on the averaged radius of particles in the bin. This ensures that a reasonable size distribution is achieved to be used later in NARCM for climate effect assessment.

3 SEA-SALT MODELING RESULTS

The results from applying NARCM aerosol algorithm with three scale climate models: FIZ-C/LCM, RCM and GCM are used in the discussion. The global sea-salt results were obtained using Canadian GCMIII [version 11] at a horizontal resolution of 96×48 Gaussian grid (~370 km), vertical resolution of 22 levels (from surface to 12 hPa pressure level) and 20 minute integration time step. With the same integration time and vertical resolution, a higher resolution result was obtained with RCM run on a domain centered at the north pole (>35° North) with a resolution of 1°×1° and driven by GCM outputs at the boundary. The FIZ-C/LCM is basically one-grid run of the global GCM without advection. The size spectrum of sea-salt aerosols (dry) from 0.03 μm to 8 μm in radius was divided into 8 size bins: Bin 1: 0.03-0.06, Bin 2: 0.06-0.13, Bin 3: 0.13-0.25, Bin 4: 0.25-0.50, Bin 5: 0.50 -1, Bin 6: 1 - 2, Bin 7: 2 - 4 and Bin 8: 4 - 8.

3.1 Global Transport Pattern and Fluxes of Sea-salt Aerosols

Sea-salt aerosols are generated at the open ocean by mechanically breaking waves by wind. The sea-salt aerosol particles span a large size spectrum from 0.03 to 10 μm in radius. Due to the large gravitational settling velocity and dry deposition velocity, large particles will not participate in long range transport. Gong et al [1997a] show that the average atmospheric residence time of these particles is about 30 minutes. The lack of long range transport is seen in Figure 1a for the global spatial distribution of large particles in Bin 8 [r=4-8 μm] which is governed largely by the spatial distribution of oceanic surface wind speeds.

In contrast to the large sea-salt particles, the smaller sea-salt aerosols [r=0.13-0.25 μm] show clearly long range transport patterns [Figure 1b] in the surface layer [1000 mb]. Particles produced in the north Pacific are transported eastward across the north American continent in the westerlies. Particles originating in the north Atlantic ocean are also transported eastward to Europe and Asia. Poleward transport of sea-salt aerosols from the northern middle latitudes to the high latitude also occurs. Once the aerosols are carried over to the high latitude, they reverse the direction and move westward. This is consistent with the global general circulation patterns.

A two year run using the GCM was used to investigate the global sea-salt aerosol budget. The total global annual sea-salt flux from ocean to atmosphere is estimated to be 3.33×10^{12} kg. This agrees well with estimate of Erickson and Duce [1988] of $1\text{-}3\times10^{13}$ kg using an empirical relationship between atmospheric sea-salt aerosol concentrations and 10 m wind speed. The current methodology employs the model generated surface wind speed and a size-distributed source function to dynamically compute the sea-salt flux of each size bin. Therefore, relative contributions to the total sea-salt budget from different size range of sea-salt aerosols are resolved.

3.2 Long Range Transport of Sea-salt Aerosols to the Arctic

Running the aerosol algorithm in the high resolution-NARCM mode yielded a more detailed picture of the transport of sea-salt aerosols from open oceans to the remote area such as Alert on the Arctic ocean. During the competing processes of source region production, removal and transport, small sea-salt aerosols are gradually moved to the arctic regions while big sea-salt aerosols are always constrained in the source regions. If the simulated sea-salt concentration at Alert of Canada is monitored as a function of starting time, it needs about 20 days for sea-salt aerosols to reach Alert from source regions via

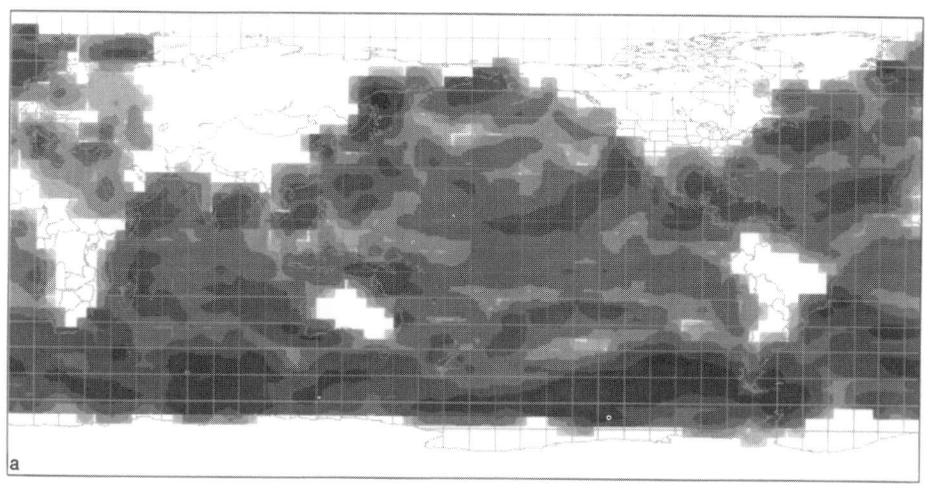

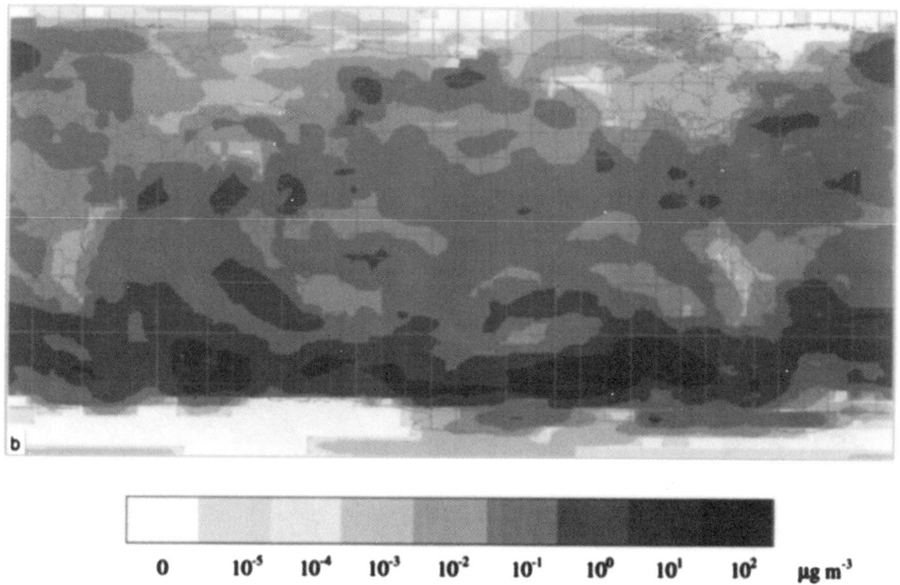

Figure 1 Global distributions of sea-salt aerosol for two size bins in June: (a) Bin 8 and (b) Bin 3.

Northern Atlantic, Europe and Asia [Figure 2]. This is consistent with the transport time of other contaminants to the arctic oceans [Patterson and Husar 1981].

3.3 Sea-salt Dependence on Wind Speed

The wind dependence of sea-salt aerosols was studied at two locations with the column version of the Canadian climate model: FIZ-C/LCM. The simulation results for Mace Head and Heimaey were compared with some observations in Figure 3. A general agreement for the wind effect is obtained between the predictions and experiments. The relationship for the dependence can be generally expressed as

$$\ln \chi = \ln(b) + aU_{10} \qquad (6)$$

One can conclude from the curves in Figure 3 that the dependence of total sea-salt aerosols, i.e., a and b, varies with geographic locations. It seems that the climate pattern in a specific site may regulate the removal processes and hence the dependence of χ on wind speed.

4 CONCLUSIONS

Size-segregated sea-salt aerosols have been incorporated as dynamic constituents in a global, regional and column climate models. With two year global runs, the annual global sea-salt flux is estimated to be 3.33×10^{12} kg. In the northern hemisphere, it takes about 20 days for sea-salt to be transported from open oceans to Alert in the Arctic. The relationship

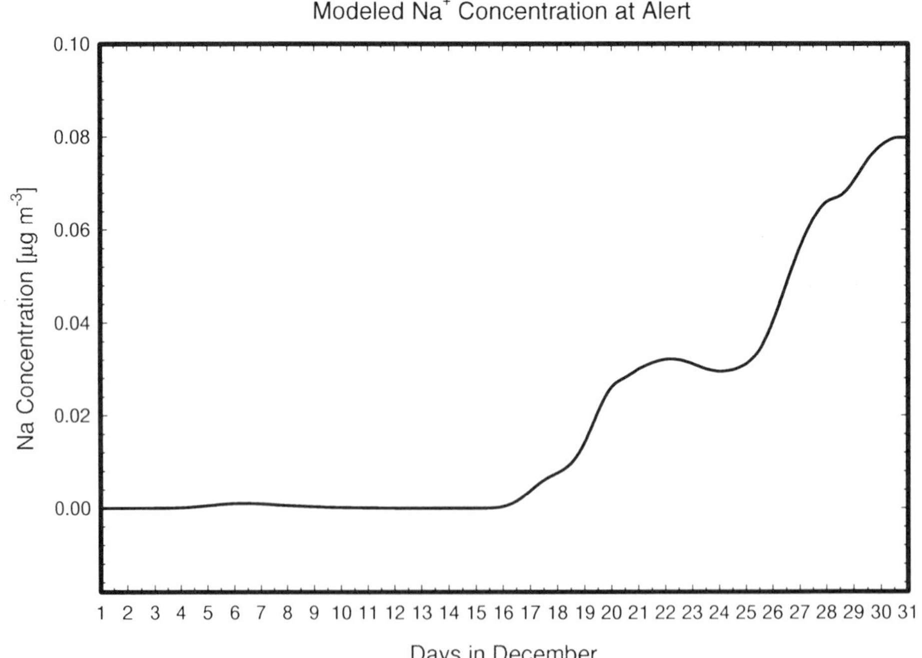

Figure 2 Simulated Sea-salt concentration at Alert monitored as a function of time. The initial concentration was set to zero.

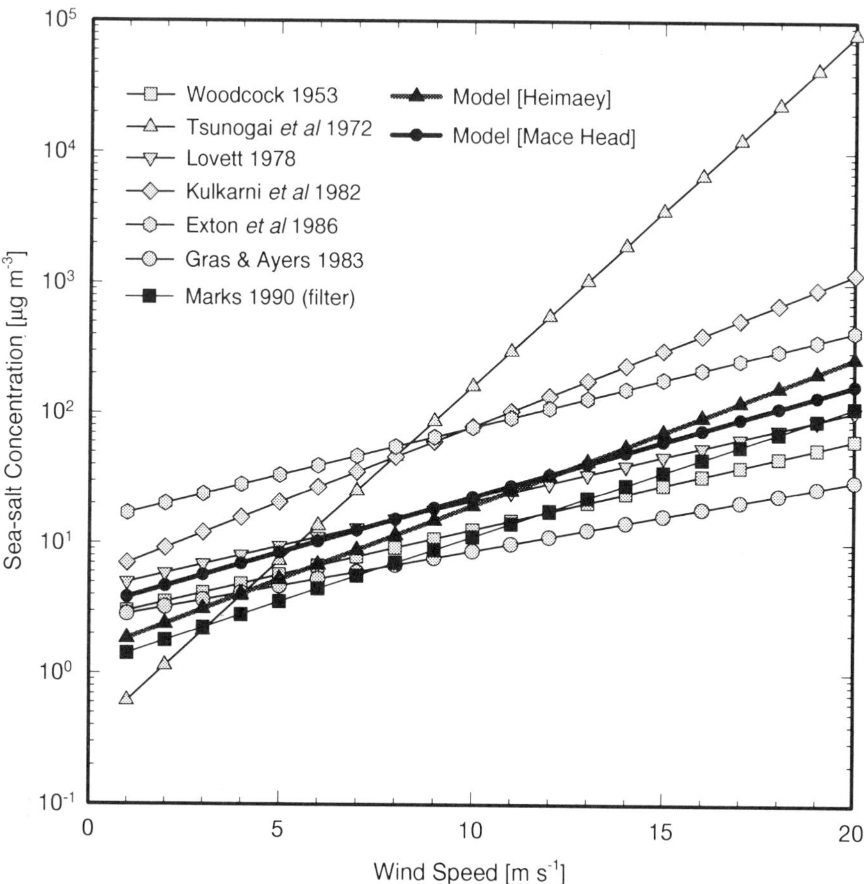

Figure 3 Comparison of sea-salt mass concentration as a function of wind speed between observations and the model predictions at two sites

between sea-salt concentration and surface wind speed depends on precipitation and wind frequency distributions and therefore varies with geographic locations.

5 REFERENCE

Behnke, W., George, C., Scheer, V. and Zetzsch, C. (1997), Production and decay of $ClNO_2$ from the reaction of gaseous N_2O_5 with NaCl solution: Bulk and aerosol experiments, J. Geophys. Res., 102(D3), 3795-3804.

Exton, H.J., Latham, J., Park, P.M., Smith, M.H. and Allan R.R. (1986), The production and dispersal of maritime aerosol, in *Oceanic Whitecaps*, E.C. Monahan and G. Mac Niocaill (eds.), D. Redeil Publishing, Dordrecht, Holland, 175-193.

Fitzgerald J.M. (1992), Numerical simulation of the evolution of the aerosol size distribution in a remote marine boundary layer, in Nucleation and Atmospheric Aerosols, Fukuta and Wagner (eds.), 157-160.

Gong, S.L, Barrie, L.A., and Blanchet, J.-P. (1997a), Modeling Sea-salt Aerosols in the Atmosphere --- I, Model Development, J. Geophys. Res., **102**(D3), 3805-3818.

Gong, S.L., Barrie, L.A., Prospero, J.M., Savoie, D.L., Ayers, G.P., Blanchet, J.-P. and Lubos, S. (1997b), Modeling Sea-salt Aerosols in the Atmosphere --- II, Atmospheric concentrations and fluxes, J. Geophys. Res., **102**(D3), 3819-3830.

Gras, J.L. and Ayers, G.P. (1983), Marine aerosol at southern mid-latitudes, J. Geophys. Res., 88, 10661-10666.

Jobson, B.T., Niki, H., Yokouchi, Y., Bottenheim, J., Hopper, F. and Leaitch R. (1994), Measurements of C_2-C_6 hydrocarbons during the Polar Sunrise 1992 Experiment: Evidence for Cl atom and Br atom chemistry, J. Geophys. Res., 99(D12), 25355-25368.

Kulkarni, M.R., Adiga, B.B., Kapoor, R.K. and Shirvaikar V.V. (1982), Sea salt in coastal air and its deposition on porcelain insulators, J. Appl. Meteor, 21, 350-355.

Lovett, R.F. (1978), Quantitative measurement of airborne sea-salt in the North Atlantic. Tellus, 30, 358-363.

Marks, R. (1990), Preliminary investigation on the influence of rain on the production, concentration, and vertical distribution of sea salt aerosol, J. Geophys. Res., 95(C12), 22299-22304.

McFarlane, N.A., Boer, G.J., Blanchet, J.-P. and Lazare, M. (1992), The Canadian climate centre second-generation general circulation model ant its equilibrium climate, J. Climate, 5, 1013, 1044.

Monahan, E.C., Spiel, D.E. and Davidson, K.L. (1986), A model of marine aerosol generation via whitecaps and wave disruption, in *Oceanic Whitecaps*, Monahan and Mac Niocaill (eds.), 167-174.

Mozurkewich, M. (1995), Mechanisms for the release of halogens from sea salt particles by free radical reactions, J. Geophys. Res., 100(D7),14199-14207.

Patterson, D.E. and Husar, R.B. (1981), A direct simulation of hemispherical transport of pollutants, Atmospheric Environment, 15(8), 1479-1482.

Robert, A. (1966). The integration of a low order spectral form of the primitive meteorological equations. *J. Meteor. Soc. Japan*, Ser. 2, 44, 237-245.

Robert, A., Yee, T.L. and Ritchie H. (1985). A semi-Lagrangian and semi-implicit numerical integration scheme for multilevel atmospheric models. *Mon. Wea. Rev.*, **133**, 388-394.

Stramska, M. (1987), Vertical profiles of sea-salt aerosol in the atmospheric surface layer: A numerical model, ACTA Geophysica Polonica, Vol. xxxv (1).

Therrien,D. (1993), Le modèle de circulation générale atmosphérique Canadien en version colomne: FIZ-C, MSc Thesis, University of Quebec at Montreal, 123pp.

Woodcock, A.H. (1953), Salt nuclei in marine air as a function of altitude and wind force,. J. Met., 10, 362-371.

DISCUSSION

D. CARRUTHERS: The parameterizations of the sink process (wet and dry deposition) are a crucial part of the sea salt aerosol model. What is the sensitivity of the model to the parameterization of these processes?

S. GONG: Some limited sensitivity studies have been done. It has been found that dry deposition accounts for more than 50% of the total sea-salt mass removed from the atmosphere while the rest of it attributes to the wet removals. Since the parameterization depends on size of the particles, the model results are also size-dependent. For example, for small (r=0.03-0.25μm) and large (r=1-8μm) particles, particle dry deposition is the dominant process while for medium (r=0.25-1μm) particles, below-cloud scavenging dominates (Gong et al. 1997b).

A COMPREHENSIVE EULERIAN MODELING FRAMEWORK FOR AIRBORNE MERCURY SPECIES: DEVELOPMENT AND TESTING OF THE TROPOSPHERIC CHEMISTRY MODULE (TCM)

G.Petersen [1] , J.Munthe [2] , K.Pleijel [2] , R.Bloxam [3] and A.Vinod Kumar [4]

[1] GKSS Research Centre,Institute of Hydrophysics
Max-Planck-Strasse , D-21502 Geesthacht , Germany

[2] Swedish Environmental Research Institute (IVL)
P.O. Box 47086 , S-402 58 Göteborg , Sweden

[3] Ontario Ministry of Environment and Energy , Science and Technology Branch
2 St.Clair Ave W ,Toronto , ON M4V 1L5 , Canada

[4] Environmental Assessment Division, Bhabha Atomic Research Centre
Mumbai 400085 , India

INTRODUCTION

In recent years there has been a revival of interest in mercury as an atmospheric pollutant on local,regional and global geographical scales.An important indicator of increasing interest regarding mercury species in Europe's atmosphere is the recent decision of the UN-ECE to prepare a protocol for heavy metals and persistent organic pollutants under its co-operative programme for monitoring and evaluation of the long range transmission of air pollutants (EMEP) with mercury as a priority substance.In North America,the 1990 U.S. Clean Air Act Amendments have identified mercury as one of the trace substances listed in the legislation as "hazardous air pollutants" because of its potentially significant effects on ecosystems and human health.

An assessment of the potential ecological and health risks associated with atmospheric mercury species requires an understanding of the relationships between sources of emission to the atmosphere and the levels of concentrations measured in ambient air and precipitation in a given receptor area.However,the complexity of the physico-chemical processes of atmospheric mercury species makes results from measurement programs difficult to interprete without a clear conceptual model of the workings of the atmosphere.Further,measurements alone cannot be used directly by policy-makers to form balanced and cost-effective strategies for dealing with this problem;an understanding of individual processes within the atmosphere does not automatically imply an understanding of the entire system.A complete picture of individual mercury processes and their interactions with the atmospheric system as a whole- can only be obtained by means of numerical modeling.As a first step in this direction the cloud mixing,scavenging,chemistry and wet deposition components of the comprehensive three-dimensional Eulerian Acid Deposition and Oxidants Model (ADOM),originally designed for regional-scale acid precipitation and photochemical oxidants studies in North America and Europe (Venkatram et al.,1988) are restructured to accomodate the most recent developments in atmospheric mercury chemistry.A stand-alone version of these components referred to as the Tropospheric Chemistry Module (TCM) has been developed and tested under different environmental conditions allowing changes to the chemistry and meteorology alone to be evaluated at different vertical TCM levels.

In the subsequent sections,the TCM together with results from sensitivity runs are briefly described.Emphasis is given to individual physical and chemical mercury processes that are incorporated into the module.Finally,conclusions are summarized and an outlook with respect to future application of the TCM as a major component of the full ADOM mercury version for North America and Europe is given.More in depth information on the design and application of the TCM is provided in Petersen et al.,1997.

DESCRIPTION OF THE TROPOSPHERIC CHEMISTRY MODULE

The module is designed to simulate the meteorology and chemistry of the entire depth of the troposphere to study cloud mixing,scavenging and chemical reactions associated with precipitation systems that generate wet deposition fluxes of atmospheric mercury species.The scavenging and chemistry components of the module are based on the mercury chemistry scheme depicted in figure 1.It incorporates 14 mercury species and 21 reactions including mass transfer reactions (R1-R5),aqueous-phase (R6-R17) and gas-phase (R20-R21) chemical reactions and equilibrium reactions for adsorption on soot particles (R18-R19).

In view of their relevance to the results from sensitivity studies presented in this paper the TCM components for cloud mixing and chemistry are described in the next two sub-sections.Special emphasis is given to chemical reaction mechanisms,which have been derived from a far more complex series of chemical reactions in the aqueous phase (Pleijel and Munthe,1995a,b)

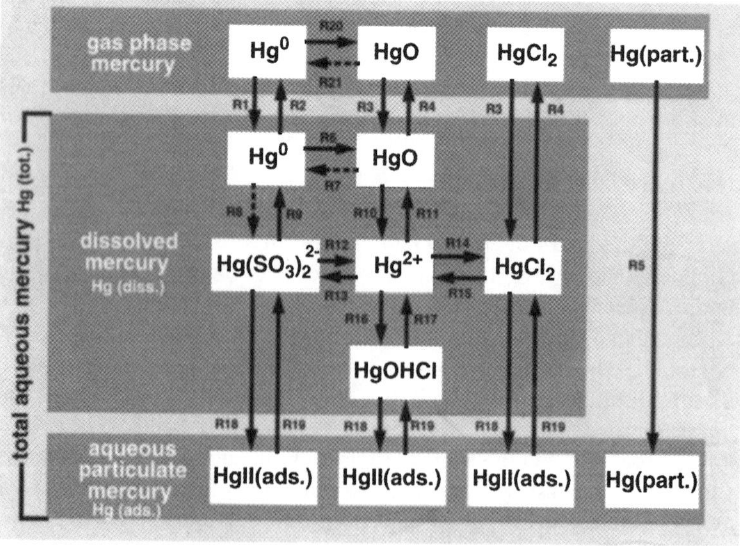

Figure 1. The mercury chemistry scheme used with the Tropospheric Chemistry Module (TCM).

Cloud mixing

The cloud mixing submodule simulates the vertical distribution of mercury species in clouds. Two different modules are incorporated: one describes stratus (layer) clouds and other simulates cumulus (convective) type clouds. One or the other or a combination (cumulus deck embedded in a stratus cloud) is used in the calculations depending upon the characteristics of the precipitation observed.

The stratiform cloud module solves the conservation equations for cloud water, rain water and snow. The 'Bulk Water' technique (Kessler,1969) is used for the microphysical formulation. A continuous supply of water vapour is assumed to be available at all times for conversion to cloud water (via condensation) or snow (via vapour deposition), depending upon the temperature. The precipitating field concentrations, cloud water concentrations and the updraft velocities are determined by working backwards from surface precipitation data. The following microphysical process rates are included in the model:vapor condensation to cloud water,cloud water autoconversion to rain,cloud water collection by rain,vapor deposition to snow,cloud water collection by snow (riming),evaporation of rain water below cloud base and sublimation of snow below cloud base.Vapour condensation rates are calculated based on the vertical temperature profile and the vertical velocity.

For cumulus clouds a detailed cloud mixing model is used (Raymond and Blyth,1986). This model allows subparcels in each parcel of ascending cloud air to mix in different proportion with the environmental air and eventually to settle at its level of neutral buoyancy. The life cylce of a cumulus cloud is modelled in three stages. First, an active region is formed by ingestion of air (also pollutants) from the cloud surroundings with 50% of the cloud air assumed to enter through the cloud base. Cloud water concentration profiles are generated,derived from observations made in cumulus clouds (Warner,1970). Second, chemical reactions take place in the cloud water formed in the active region. This stage referred to as the dwell phase is assumed to make up most of the life cycle of the cloud. At the end of the dwell phase, scavenging occurs. Finally, in the third stage, the cloud dissipates and the remaining pollutants are ejected into the cloudy region. Redistribution of pollutants occurs during both the cloud formation and the cloud dissipation stages. The final concentration profile is the weighted average of the initial concentration in the inactive cloud air,the final concentration in the active cloud air and the well mixed concentrations corresponding to displaced cloud air.

Chemistry

One of the critical steps in the atmospheric modeling of mercury is the description of the chemical processes involved in the transformations. Atmospheric mercury consists of three major forms; elemental mercury (Hg^0), oxidised species (e.g. $HgCl_2$, HgO) and particulate mercury (Hg(part.)). Methods capable of identifying individual species such as $HgCl_2$ or HgO in ambient air are not available but operationally defined methods have been used to sample and measure these forms at different locations (e.g. Bloom et. al, 1996; Xiao et al., 1996).For a correct parameterization, conversions between these species need to be described in terms of mathematical expressions. In Pleijel and Munthe (1995a,b) an extensive chemistry was tested and evaluated in a sensitivity analysis aimed at determining the most important parameters and identifying the critical processes in the atmospheric cycling of mercury. The model employed in those studies was considerably simplifyed in terms of physical processes and meteorology and only calculated cloud or fog droplet concentrations of mercury species.

Due to the constraints in computer resources it was necessary to condense the chemistry described in Pleijel and Munthe (1995a).The resulting condensed scheme depicted in figure 1 consists of four gas phase species,six dissolved aqueous species and four aqueous particulate phase species. The reaction rates are derived from published data and from assumptions of the rates of complex formation.

For each process in the gas and aqueous phase, the stoichiometric reaction rate expressions have been transformed into first order expressions where the first order rate constants are functions of the reactant species.Only one gas phase reaction of mercury is treated in the model; the oxidation of elemental mercury by ozone: $Hg^0 + O_3 \Rightarrow HgO + O_2$

$$k_{20} = 0.739*10^{-9}\ ppb^{-1}\ s^{-1} \qquad \text{(Hall, 1995) (1)}$$

This reaction is treated as a first order process with a constant ozone concentration so that $R20 = k_{20}*[O_3]$. For reasons of calculation procedure, all reactions in the TCM must have a reverse process. In the gas phase, no reduction of Hg(II) is known so the rate constant for this process was given a very low value (R21=R20/1000) making it insignificant in the overall process.The same principle is used for the aqueous oxidation of Hg^0 by ozone.The reaction is written as:

$$Hg^0 + O_3 \Rightarrow HgO + O_2 \qquad k_6 = 0.47*10^{-8} \text{ (Munthe, 1992) (2)}$$

but is treated as a first order reaction in the model. The aqueous concentration of ozone is given by the gas phase concentration and the Henrys law constant. The product of both the gas and aqueous phase oxidations is assumed to be HgO. A low value for the reverse reaction is set for the same reasons as for the gas phase process. The HgO species is not very likely to be long lived in the aqueous phase and a rapid transformation to Hg^{2+} is assumed ($HgO + H^+ \Rightarrow Hg^{2+} + OH^-$).In the subsequent calculation of oxidised mercury speciation in the aqueous phase, Hg^{2+} takes part in a number of reactions leading to the formation of three separate complexes: $HgCl_2$, $Hg(SO_3)_2^{2-}$ and HgOHCl.

The formation of $HgCl_2$ is described as a hypothetical third order reaction:

$$Hg^{2+} + 2Cl^- \Rightarrow HgCl_2 \text{ (3)}$$

for which the rate expression is

$$d[HgCl_2]/dt = k * [Hg^{2+}]*[Cl^-]^2 \text{ (4)}$$

where k is a third order reaction rate constant.

Ligand formation reactions of mercury are well known to be rapid so that the hypothetical third order rate constant can be given a value corresponding to the diffusion controlled limit, i.e. about $10^{15}\ M^{-2}s^{-1}$. Thus R14 is given the value $10^{15}*[Cl^-]^{-2}\ s^{-1}$ where the chloride ion concentration is set in the model run. For HgOHCl a similar process is assumed except that the OH^- is given by the pH set before the model run. The rate expression for formation of $Hg(SO_3)_2^{2-}$ is similar but slightly more complicated since the concentration of SO_3^{2-} is calculated from the air concentration of SO_2 and the cloudwater pH:

$$[SO_3^{2-}] = K_2*[HSO_3^-]/[H^+] = K_2*K_1*[SO_2(aq)]/[H^+]^2 = K_2*K_1*H_{SO2}*[SO_2(g)]/10^{-2pH} \text{ (5)}$$

where K_2 and K_1 are equilibrium constants for the dissociation of aqueous SO_2 (H_2SO_3) and HSO_3^- and H_{SO2} is the Henrys law constant:$K_2 = 6.42*10^{-8}$, $K_1 = 1.32*10^{-2}$,$H_{SO2} = 1.24$ mol $L^{-1}\ atm^{-1} = 1.24*10^{-9}$ mol $L^{-1}\ ppb^{-1}$.With these values inserted, the expression becomes

$$[SO_3^{2-}] = 1.05*10^{-18} *[SO_2(g)]/10^{-2pH} \text{ (6)}$$

For a hypothetical third order formation reaction, $Hg^{2+} + 2[SO_3^{2-}] => Hg(SO_3)_2^{2-}$ the rate expression is given by

$$d[Hg(SO_3)_2^{2-}] /dt = k *[Hg^{2+}] * [SO_3^{2-}]^2 \quad (7)$$

In our case we assume a first order reaction which means that:

$$R13 = k * [SO_3^{2-}]^2 \quad (8)$$

Our assumed reaction is third order reaction for which the diffusion controlled limit is about 10^{15} M^{-2} s^{-1} and R13 can be simplified to:

$$R13 = 1.1 * 10^{-21} * ([SO_2(g)] / 10^{-2pH})^2 \quad (9)$$

For all the three major dissolved complexes $HgCl_2$, $Hg(SO_3)_2^{2-}$ and HgOHCl,equilibrium constants are well known and can be applied for the calculation of the reverse reaction rate constants R15, R12 and R17.The whole procedure described above for these complexes is meant to allow kinetic treatment of these processes without disturbing the known equilibrium properties.

Of the three major complexes nearly all the oxidised mercury exists in the chloride form. This is due to the relatively high chloride concentration found in the droplets (Pleijel and Munthe,1995a). HgOHCl is the second most important complex and $Hg(SO_3)_2^{2-}$ is added due to the possibility of back reduction of oxidised Hg to Hg^0 via reaction R9. The stoichiometry of this reaction is written as:

$$Hg(SO_3)_2^{2-} <=> HgSO_3 + SO_3^{2-} \qquad k_f = 4.4*10^{-4}\ s^{-1}, \qquad k_b = 1.1*10^8\ M^{-1}\ s^{-1} \quad (10)$$

and

$$HgSO_3 =>Hg^0 \qquad k = 0.6\ s^{-1} \text{ (Munthe et al., 1991) } (11)$$

The rate limiting step in this reaction scheme is the dissociation of $Hg(SO_3)_2^{2-}$ and this rate constant has been used for reaction R9 in the condensed scheme. The reverse process is not considered to be a realistic process and R8 is given a sufficiently low value in order not to affect the process.

The parameterization of adsorption of dissolved mercury species on particles is based on an observed empirical equilibrium relation between mercury concentrations (Petersen et al.,1995) and the occurence of soot particles in precipitation samples (Iverfeldt, 1991).The adsorption of airborne mercury onto carbon based surfaces (i.e. activated carbon) is well known and used in many applications for trapping vapour phase mercury. The reversibility of the aqueous adsorption of different mercury species, as described in the TCM, is to some extent an assumption necessary for computing reasons.

In the TCM it is assumed that the adsorption rate is diffusion controlled and the desorption rate is calculated from the adsorption rate and the empirical equilibrium relation mentioned above. The adsorption rate is calculated as the typical time for diffusion in an aqueous phase to be about 0.02 s^{-1} which is used for the forward rate constant R18 in the condensed scheme. Thus the reverse desorption rate constant R19 will depend on the soot particle concentration and the soot particle radius.For the reference case a value of $4*10^{-3}$ s^{-1} is used.

TEST RESULTS

Tests of the chemical scheme have been performed by investigating the formation of total mercury concentration in the aqueous phase (Hg(tot.)).This species represents the sum of all dissolved (Hg(diss.))and adsorbed (Hg(ads.)) species originating from Hg^0 in ambient air,i.e. for these tests $HgCl_2$ and (Hg(part.) in ambient air have been decoupled from the system depicted in figure 1, thus allowing to determine the formation of Hg(tot.) in cloud and rainwater droplets as a function of the Hg^0 depletion.The development of Hg(tot.) concentrations during a 48 h-simulation is shown in figure 2.The solid line represents the formation of Hg(tot.) in a non-precipitating cumulus in a relatively warm environment.The dashed and the chaindashed line show the build up in this deep warm and in a shallow cold cloud with precipitation rate of 0.3 mm h^{-1} in each of the two clouds.Since aqueous phase chemistry occurs up to about 10 000 m height,the deep cloud processes a much larger mass of the cloudy region air in an hour resulting in a more effective build up of Hg(tot.) concentration.The warm cloud with 2 mm h^{-1} precipitation (dotted line) shows a less effective Hg(tot.) build up than the same cloud with a precipitation rate of 0.3 mm h^{-1}(dashed line),as expected due to more effective scavenging.The significant influence of the bigger cloud volume on Hg(tot.) formation becomes evident when the curves for the deep warm cloud (precipitation rate 2 mm h^{-1}) and the shallow cold cloud (precipitation rate 0.3 mm h^{-1}) are compared;in the case of the warm cloud,enhanced production of Hg(tot.) by aqueous phase chemistry overcompensates the enhanced depletion by a higher precipitation rate resulting in a slightly higher build up of Hg(tot.) than in the shallow cloud with a smaller precipitation rate.

Comparisons of calculated Hg(tot.) concentrations as a function of Hg^0 in ambient air and simultaneous measurements of Total Gaseous Mercury (TGM) and Hg(tot.) obtained at four different sites in Sweden (Iverfeldt,1991) and in Ireland and Germany (Ebinghaus et al.,1995) are presented in figure 3.This comparison is between TCM calculations of Hg(tot.) in cloud and rainwater ranging from a non-precipitating cloud to a deep cumulus associated with a precipitation rate of 2 mm h^{-1} and measurements of Hg(tot.) at ground level associated with precipitation rates up to 2 mm h^{-1}.In general,TCM predicted Hg(tot.) concentrations compare satisfactorily with observations from all the stations,indicating that the module is based on an adequate parameterization of cloud mixing,scavenging and chemistry.Compared to the linear fit of observed Hg(tot.) concentrations the TCM seems to underpredict Hg(tot.) at the upper end of the range shown in figure 3.This is most probably due to two factors which could adversely affect the skill of the model to reproduce the observed Hg(tot.) concentrations.First,slight differences between Hg^0 model input and measured TGM could affect the comparison,at least at sites like Langenbrügge which is located in the vicinity of industrial sources with a potential of additional emissions of oxidised gaseous mercury species such as $HgCl_2$ resulting in higher Hg(tot.) concentrations at the measurement site.Second,the upper end of the range of observed TGM concentrations shown in figure 3 is frequently associated with soot concentrations greater than $1.0*10^{-6}$ g m^{-3}.This value was used with the entire range of TCM simulations,thus suggesting that Hg(tot.) concentrations at the upper end of the calculated range are underestimated.However,the limited data material does not allow to draw any firm conclusions and a larger data set is needed to evaluate the TCM performance with respect to adsorption of dissolved mercury species on soot particles and the impact of gaseous mercury species other than Hg^0 on aqueous phase concentrations of both dissolved and adsorbed mercury species.

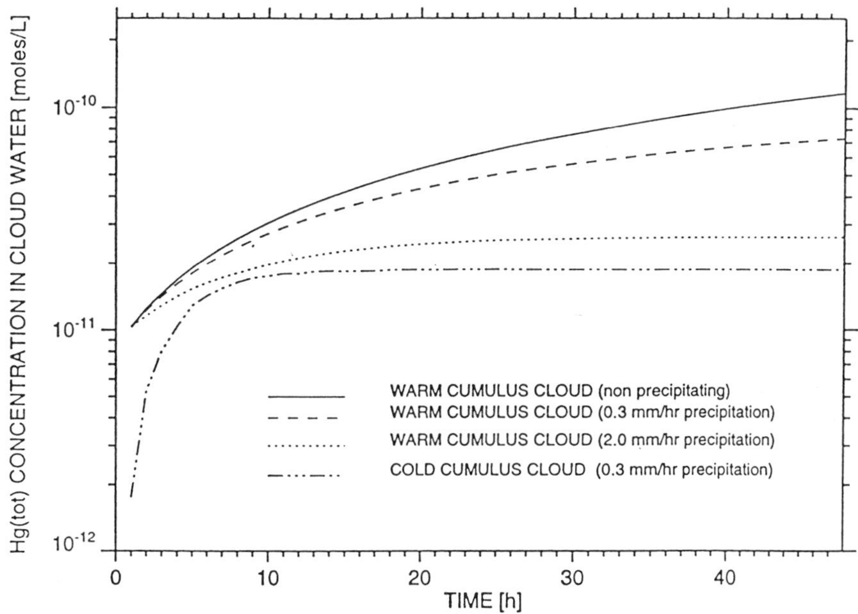

Figure 2. Total mercury concentration in cloudwater and rain for different cloud conditions.

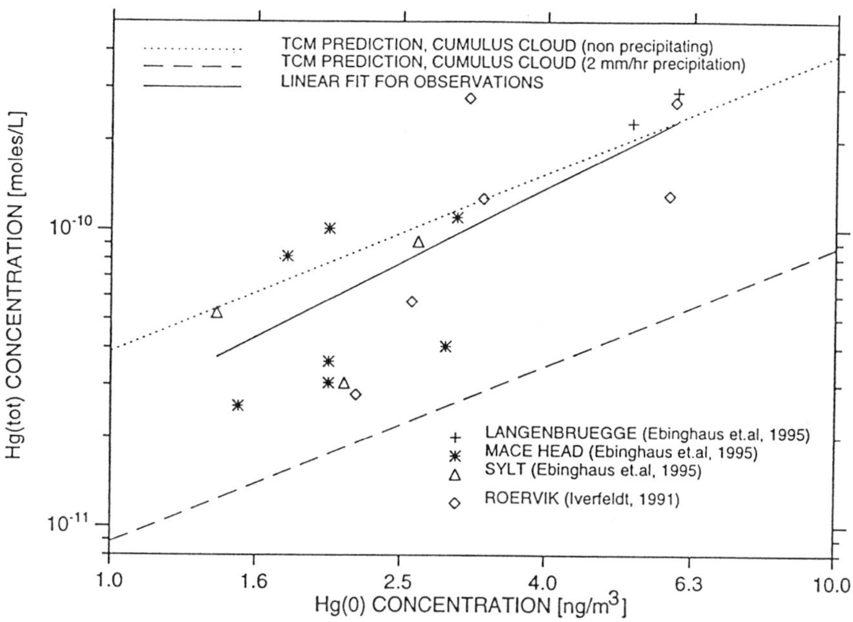

Figure 3. Comparison of TCM predicted and observed total mercury concentrations in cloudwater and rain.

SUMMARY AND CONCLUSIONS

The stand-alone version of a Tropospheric Chemistry Module (TCM) containing sub-modules for cloud mixing,scavenging and chemistry of atmospheric mercury species is presented.The TCM is consistent with current understanding of cloud processes and treats both cumulus and stratus clouds .The module is computationally simple enough to be readily incorporated into comprehensive Eulerian models for atmospheric mercury species without significant increase of total computational time.Sensitivity studies have been conducted with the TCM to investigate its cloud mixing, scavenging and chemical transformation characteristics.In general,results from these studies indicate that the TCM has capabilities to gain scientific insights of tropospheric mercury transport,transformation and deposition problems which cannot be obtained through field measurements or experiments in the laboratory.The evaluation of the module performance led to three main issues:

(1) TCM predictions of mercury concentrations in rainwater compare satisfactorily with observations at four European sites thus indicating that the module is based on an adequate parameterization of atmospheric mercury processes.However,the available data material is scarce and more simultaneous measurements of mercury concentrations in ambient air and precipitation are necessary to evaluate the ultimate TCM performance.

(2) Due to lack of analytical methods capable of identifying individual oxidised mercury species,the TCM performance with respect to the impacts of $HgCl_2$ and HgO on the atmospheric cycling of mercury cannot be evaluated at present.

(3) TCM test runs have shown that adsorption of mercury species to particulate matter is very important to the atmospheric fate of mercury.Currently,the TCM chemical equation set is restricted to adsorption of dissolved mercury species on soot particles based on an empirical adsorption coefficient.For further TCM development,a more accurate parameterization of adsorption processes is clearly needed.

Overall,the present development level of the TCM is such that its implementation in full ADOM mercury versions for Europe and North America is justified.These models with the TCM as their major component will be developed further according to advancements in the knowledge of atmospheric mercury processes to ensure that they maintain their capability to address effectively the scientific and policy questions that may arise over the next decade.

REFERENCES

Bloom N.S.,Prestbo E.M. and Von der Geest E. (1996) Determination of atmospheric gaseous Hg(II) at the pg/m3 level by collection onto cation exchange membranes,followed by dual amalgamation/cold vapor atomic fluorescence spectrometry,*Fourth International Conference on Mercury as a Global Pollutant,Book of Abstracts,* GKSS Forschungszentrum,Max-Planck-Strasse,D-21502 Geesthacht,Germany.

Ebinghaus R.,Kock H.H.,Jennings S.G.,McCartin P. and Orren M.J. (1995) Measurements of atmospheric mercury concentrations in northwestern and central Europe - Comparison of experimental data and model results, *Atmospheric Environment* 29,3333-3344.

Hall B. (1995) The gas phase oxidation of elemental mercury by ozone, *Water,Air and Soil Pollution* 80,301-315.

Iverfeldt A. (1991) Occurrence and turnover of atmospheric mercury over the nordic countries, *Water,Air and Soil Pollution* 56,251-265.

Kessler E. (1969) On the Distribution and Continuity of Water Substances in Atmospheric Circulation, *Meteorological Monographs* 10, No. 32,American Meteorological Society,Boston,Mass.02108,U.S.A.

Munthe J.,Xiao Z.F. and Lindqvist O. (1991) The aqueous reduction of divalent mercury by sulfite,*Water, Air and Soil Pollution* 56,621-630.

Munthe J. (1992) The aqueous oxidation of elemental Mercury by ozone,*Atmospheric Environment* 26A, 1461 - 1468.

Petersen G.,Iverfeldt A. and Munthe J. (1995) Atmospheric mercury species over Central and Northern Europe. Model calculations and comparison with observations from the nordic air and precipitation network for 1987 and 1988,*Atmospheric Environment* 29, No.1,47-67.

Petersen G.,Munthe J.,Bloxam R. and Vinod Kumar A. (1997) A comprehensive Eulerian modelling framework for airborne mercury species: development and testing of a Tropospheric Chemistry Module (TCM), *Atmospheric Environment - Special Issue on Atmospheric Transport,Chemistry,and Deposition of Mercury* (edited by S.E.Lindberg,G.Petersen and G.Keeler) (in press).

Pleijel K. and Munthe J. (1995a) Modelling the atmospheric mercury cycle - chemistry in fog droplets, *Atmospheric Environment* 29, No.12,1441-1457.

Pleijel K. and Munthe J. (1995b) Modeling the Atmospheric Chemistry of Mercury - The importance of a detailed description of the chemistry in cloud water.*Water,Air and Soil Pollution* 80,317-324.

Raymond D.J. and Blyth A.M. (1986) A stocastic mixing model for nonprecipitating cumulus clouds,*J. Atmos. Sci.* 43,2708-2718.

Venkatram A.,Karamchandani P.K. and Misra P.K. (1988) Testing a comprehensive acid deposition model, *Atmospheric Environment* 22,737-747.

Xiao Z.F.,Sommar J.,Wei S. and Lindqvist O. (1996) Sampling and determination of gas phase divalent mercury with KCl coated denuders,*Fourth International Conference on Mercury as a Global Pollutant,Book of Abstracts*, GKSS Forschungszentrum,Max-Planck-Strasse,D-21502 Geesthacht,Germany).

DISCUSSION

S. ZILITINKEVICH: What about mercury pollution in the River Elbe? What is more important the air pollution, you spoke about, or the water pollution in the river?

G. PETERSEN: In case of the Elbe, the pollution of the river is still more important, as the Elbe was significantly polluted by direct mercury discharges of the chemical industry in the former German Democratic Republic during the last 40 years. However, in most of the other areas affected by high mercury pollution such as inland waters in Scandinavia and Canada, and the Everglades in Florida, the mercury load originates from atmospheric long-range transport and subsequent mercury accumulation in the water bodies and their drainage basins. This has been proven by several researchers in North America and Europe.

DEVELOPMENT AND IMPLEMENTATION OF THE EPA'S MODELS-3 INITIAL OPERATING VERSION: COMMUNITY MULTI-SCALE AIR QUALITY (CMAQ) MODEL

Daewon W. Byun[§], Jason K.S. Ching[§], Joan Novak[§], and Jeff Young[§]

Atmospheric Sciences Modeling Division,
Air Resources Laboratory,
National Oceanic and Atmospheric Administration,
Research Triangle Park, NC 27711, USA.

INTRODUCTION

For the last fifteen years, the Office of Research and Development (ORD) of the U.S. Environmental Protection Agency (EPA) has been developing three-dimensional Eulerian based air quality models (AQMs) to study air quality problems, such as urban and regional tropospheric ozone and regional acid deposition. These AQMs simulate comprehensively atmospheric processes such as chemical transformations, transports, and removal of pollutants and their precursors. Model application experience with second generation air quality modeling systems has revealed several shortcomings such as slow execution speed, difficulty in implementing improved science algorithms in the model, and complexity in data exchange among system submodels. Byun et al. (1995) listed some of the shortcomings of the present AQM modeling systems in detail.

The U.S. EPA's next generation air quality modeling project, Models-3, is intended to alleviate these problems (Dennis et al., 1996). The project combines two distinct activities; development of the computational framework and integration of related science components into a state-of-the-art air quality modeling system. The framework provides a sophisticated environmental modeling platform for science and regulatory communities. It incorporates recent advances in software engineering, communication, and computer hardware technology. Models-3 is a system framework that helps users build, apply, and study regulatory or customized environmental models. The Community Multi-scale Air Quality (CMAQ) model is the initial operating science model under the Models-3 framework. It integrates developments in physical and chemical science process algorithms together with efficient numerical algorithms and data intercommunication methods. The CMAQ system consists of emissions processing, meteorological modeling, and chemistry-transport modeling. A significant effort has been made to maintain consistent scientific formulations throughout the modeling process. In the present paper, we describe the high-level architecture, the modular design concept, and key implementation features of CMAQ.

ARCHITECTURE OF SYSTEM FRAMEWORK

The system framework provides a user interface and facilitates management of data, codes, and computer resources. Framework, in the context of the Models-3 system, is the mechanism necessary to manage the scheduling and execution of computational models, the data produced by the computational models, the analysis and visualization tools necessary to

[§]On assignment to the National Exposure Research Laboratory, U.S. Environmental Protection Agency

provide understanding of their results for decision making, and the interfaces to all these capabilities. Models-3 initial implementation is as an air quality modeling system. Future extension as a generalized computational paradigm for integrated multi-media environmental modeling, such as those that model water quality and ecosystems, is possible and desirable. For the development of the framework, we have applied software engineering methods to define requirements, to design the system and its components, to manage the development efforts, to standardize the implementation and documentation efforts, and to improve the model verification and evaluation processes. Software engineering provides not only a disciplined method of defining large-scale environmental software, but also a rigorous approach to managing the development process. Key components of the framework have been identified through the rapid prototyping process and the relations among objects, their attributes, and their operations through Object Oriented Analysis (OOA).

Because the Models-3 framework must be able to adapt to advances not only in science but also in the computer software and hardware technology, it is structured with isolated architectural layers, including user interface, system manager, UNIX environment, computational program, data access, and data storage. Science models reside in the computational program layer. The system management layer, implemented as a three-tier client-server architecture, provides user functionalities necessary for environmental modeling. Component software in each layer can be replaced independently as individual computer technology advances and standards change. The implementation specifics of the key framework subsystems (servers) are described below:

Dataset Manager: A dataset is made up of one or more physical files. These datasets are inputs/outputs of scientific models or data needed for analysis and decision making. The Dataset Manager manipulates the datasets. It allows users to register, search, update, archive and restore datasets, to work with a data directory structure, and move data transparently without data conversion across different platforms. Users can access any registered data on a network through the use of browsers and keyword selection, and track the metadata (information about the source, content, and location of the data, including temporal, spatial, variable, and coordinate attributes) of datasets. Models-3 facilitates management of metadata, datasets, and globally shared information such as grid and coordinate system definitions, chemical mechanisms, and scenario case dates through the use of an object-oriented database management system.

Study Planner: Models-3 provides a very special tool that allows users to set up a complicated sequence of processor runs without having to deal with the complexity behind them. The Study Planner allows users to graphically set up an execution sequence of a set of programs, to link data files, to enter how they connect to make a plan, and to execute its processes. With the Study Planner, users can build a study from the ground up or manipulate (save, edit, delete, etc.) an existing study plan. Several plans can be linked in a study to allow for different processing pathways. The Study Planner, together with the Dataset Manager, provides interactive and automatic batch processing capability for the Models-3 system.

Source Code Manager and Model Builder: Comprehensive modeling involving many scientists developing different process modules requires careful code management. Many modelers may work concurrently on different computer platforms. Because model development is a continuously evolving process, there must be a well-defined procedure to coordinate groups of modelers and to allow access to various versions of the related codes. The Source Code Manager allows users to retrieve a version of a source code, change it, and return it to the code archive after the change has been tested. Once this code has been returned to the archive, it can be accessed by other users. The Source Code Manager tracks historical information on the source code as well. The Model Builder embodies the key framework function for constructing models and processors. It allows users to pick and choose from different science modules to compose a complete model. Without having to know details about the code structure or the locations of physical files, users can build models or replace science modules in a model. It also simplifies the complicated module assembling process required to build a nested CMAQ version. The Model Builder uses the Source Code Manager to retrieve science codes and scripts for compilation. It also allows a user to modify the modeling domain, horizontal or vertical grid resolution, and chemistry mechanism without any user coding changes.

Program Manager: A program is defined here as any executable file. The Program Manager allows users to manage executable programs (object binary code) and scripts. It lets users register executables and enter characteristics of programs into the framework. For the registration, users are expected to provide command-line parameters, input/output (I/O) files (logical names), the grid information, and default run-time options of the programs. These programs then become available for the Study Planner to define studies.

Science Manager: The Science Manager helps users to register the persistent science objects (coherent sets of information for science components in the system) that are needed for model simulations. The Models-3 objects capture spatial (grid and domain definitions), temporal (episode definitions), and chemical (mechanism and science process setup) information. Thereafter, other parts of the system can use the registered objects in a consistent manner. This saves users from having to enter spatial, temporal, or chemical specifications more than once, and ensures internal consistency throughout system science components.

Analysis and Visualization (A&V) Tools: Models-3 employs a multi-faceted approach for visualization and analysis. It supports many popular analysis and visualization tools such as; Vis5D, PAVE, SAS, NCAR Graphics, and Data Explorer. These tools can be used to perform two- and three- dimensional data analyses. The Models-3 A&V toolset includes translators between different data formats used in the supported visualization packages and Models-3 I/O Applications Programming Interface (API) NetCDF data format. Data probe functions available in these packages can be used to perform investigatory analysis of model outputs. Two- and three-dimensional animation capabilities are also supported. In addition, we provide several air quality specific exploration methods by utilizing the A&V tools listed above. They include simultaneous visualization of measurements and model outputs following aircraft trajectories, analysis of processor budget, and presentation of chemistry cycles and pathways. Visualization tools for plume-in-grid model outputs are under development.

DESIGN AND IMPLEMENTATION OF THE CMAQ MODEL

The Models-3 system framework provides a generalized computational environment that enables integrated multi-media environmental modeling. The CMAQ modeling system is the first science implementation of the Models-3 system framework for a single media application (i.e., air quality simulation). The science components of CMAQ consist of a meteorological modeling system, an emissions processing system, chemical-transport models (CTMs), and analysis and visualization tools. Figure 1 shows the key components of the Models-3 CMAQ modeling system. Development of a new generation of Eulerian CTMs to take advantage of the high performance computing and communication (HPCC) technology required redesigning of the modeling framework and science algorithms in the second-generation AQMs. In the following we describe key design characteristics of CMAQ.

Prototyping: Because of the complexity of comprehensive AQMs and because of the rapidly changing computer technology, we have employed an evolutionary prototyping method. Evolutionary prototyping is a combination of research prototyping and incremental development. The detailed data flow of each prototype model is analyzed to improve the requirements and structure of the target science models. It allows for the building of more complex models based on the simpler prototypes. Through the prototyping process, we have tested science and framework concepts such as; modular and interchangeable science implementations, extensibility of science models and system framework, use of standard input/output interfaces, and consistent modeling practices over heterogeneous computational environments.

Community Model: One of the main objectives of Models-3 CMAQ is to develop a flexible community modeling paradigm upon which the atmospheric and air quality community (scientists and federal/state/local managers) can build and continue to add, in a unified fashion, a one-atmosphere modeling capability. The concept is beyond the one-group/one-model practice and will entail more than incremental changes in the current modeling culture. There are impending needs for ever increasing complexity in the air quality models to fully understand processes involved in determining air quality. It is difficult to expect any one group of scientists to possess and to maintain all the expertise in diverse disciplinary areas incorporated into state-of-the-science comprehensive AQMs. The community modeling practice will provide a foundation for continuous developments in all modeling aspects and furnish a mechanism for rigorous peer reviews of modeling results.

Modular concept: We have found that it is critical to define the levels of modularity in the Models-3 CMAQ system. The levels of modularity in science models are classified based on the granularity of the modeling components. The coarsest level of modularity is the distinction between system framework and science models. The second level is the division of science models into submodels such as meteorology, emissions, chemistry-transport models, and analysis and visualization subsystem. The modularity within the CTM is based on the operator-splitting concept of science processes. The third level of modularity involves distinction of driver, processor modules, data provider modules, and utility subroutines in a CTM. The fourth level of the modularity is the division based on the computational functionality in a processor module, i.e., science parameterization, numerical solver, processor analysis, and input/output routines. The last meaningful modularization level is the isolation of sections of code that can benefit from machine architecture dependent optimization. Currently, the CMAQ is structured based on the third level of modularity taking advantage of the state-of-the-art Models-3 I/O API utilities that provide selective random access to machine portable data files.

Implementation: The initial operating version (IOV) of the CMAQ system offers a fully functional multi-scale and multi-pollutant air quality modeling capability. The Models-3 system framework provides a graphical user interface and facilitates management of data, codes, and computer resources. CMAQ's science subsystems are composed of the MM5 meteorological model, Models-3 Emissions Processing and Projection System (MEPPS), several interface processors, and analysis and visualization tools. MEPPS is an upgraded and integrated version of GEMAP (Wilkinson and Emigh, 1994) for CMAQ. The interface processors, which provide consistent linkage between preprocessor submodels and the CTM, include the meteorology-chemistry interface processor (MCIP), initial condition and boundary condition processors, photolytic rate (J-value) processor, emissions-chemistry interface processor (ECIP), and a mechanism reader. The Models-3 CTM can be configured with all or some of the critical science processes, such as atmospheric transport, deposition, cloud mixing, emissions, gas- and aqueous-phase chemical transformation processes, and aerosol dynamics and chemistry. The machine portable science codes are written in FORTRAN-77 and the I/O API routines are accessible both in FORTRAN and C languages.

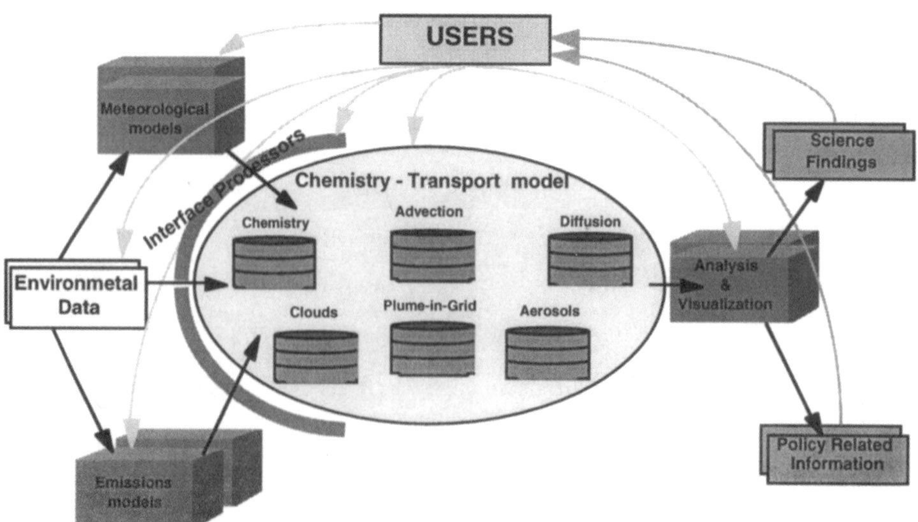

Figure 1. Models-3 CMAQ components showing science adaptability. the meteorology and emissions submodels and chemistry-transport model's science process modules are interchangeable to adapt to user's air quality problem.

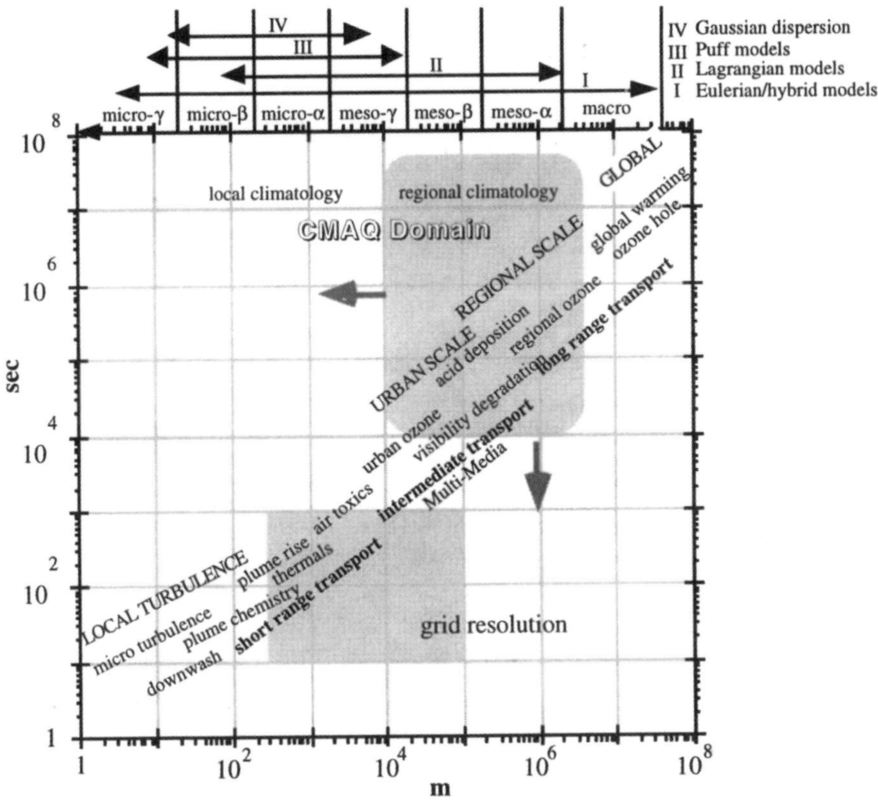

Figure 2. Ranges of temporal and spatial grid resolutions and modeling domains covered by CMAQ.

MULTI-SCALE CHARACTERISTICS OF THE CMAQ MODELING SYSTEM

As shown in Figure 2, the target grid resolutions and modeling domain sizes for Models-3 CMAQ cover several decades of temporal and spatial scales. We have achieved the multi-scale capability by considering several distinct aspects of scalability issues, such as scalability of atmospheric dynamics and physical parameterization, grid-resolution scalability, sub-grid-scale plume modeling, and input data scalability. Although not described below, an aggregation technique for providing seasonal to annual air quality data is being developed.

Scale Dependent Atmospheric Dynamics: For regional scale simulations, whose problem size is on the order of a few thousand kilometers, hydrostatic meteorological models can be used to provide atmospheric characterization. The hydrostatic assumption relates not only the static components of pressure to the density, but also the dynamic pressure field to the perturbation density field. The vertical acceleration is much less than the pressure gradient force for the large-scale atmospheric motions, so that it is impractical to use the vertical momentum equation. Instead, we use a diagnostic relation obtained from the mass continuity equation to estimate the vertical velocity field. This practice enforces mass consistency in the meteorological data. However, the hydrostatic approximation is not valid for small scale motions of on the order of 1-5 km, i.e., deep convection flows or complex terrain flows, where the vertical acceleration can be comparable to gravity. For this situation, a set of governing equations for the compressible non-hydrostatic atmosphere should be used. Under non-hydrostatic flows, which would be prevalent for the fine grid nest, it is essential to ensure mass consistency in meteorological fields because errors in air density and wind fields act as first order reaction terms in CTMs causing enormous errors in the computation of trace gas concentrations. CMAQ addresses this mass conservation issue and is capable of handling a full range of atmospheric dynamics from hydrostatic to fully compressible atmospheres.

Generalized Coordinates: The CMAQ's governing equations are formulated with a generalized coordinate system to benefit from its wide adaptability to many different scales of atmospheric dynamics. The generalized coordinate system allows CMAQ to be configured consistently with the algorithms used in the preprocessor meteorological model. The consistency among meteorological parameters and the way they are utilized in a CTM address the mass conservation issues of pollutants that are crucial for the success of air quality model simulations. In addition, the science process parameterizations are expected to be implemented with the generic coordinates, i.e., utilizing the physical coordinates to best represent the specific physical and chemical processes. This facilitates communication with research communities specializing in science process research while furthering the community modeling concept.

The governing atmospheric diffusion equation in the generalized coordinate system is given as

$$\underset{(a)}{\frac{\partial(\lambda\bar{c}_i)}{\partial t}} + \underset{(b)}{\nabla_s \bullet \left(\lambda\bar{c}_i\bar{\hat{\mathbf{V}}}_s\right)} + \underset{(c)}{\frac{\partial(\lambda\bar{c}_i\bar{\dot{s}})}{\partial s}} + \underset{(d)}{\frac{\partial}{\partial\hat{x}^1}\left[\bar{\rho}\lambda\hat{F}^1_{q_i}\right] + \frac{\partial}{\partial\hat{x}^2}\left[\bar{\rho}\lambda\hat{F}^2_{q_i}\right]} + \underset{(e)}{\frac{\partial}{\partial s}\left[\bar{\rho}\lambda\hat{F}^3_{q_i}\right]}$$

$$= \underset{(f)}{\lambda R_{c_i}(\bar{c}_1,\ldots,\bar{c}_N)} + \underset{(g)}{\lambda S_{c_i}} + \underset{(h)}{\left.\frac{\partial(\lambda\bar{c}_i)}{\partial t}\right|_{cld}} + \underset{(i)}{\left.\frac{\partial(\lambda\bar{c}_i)}{\partial t}\right|_{ping}} + \underset{(j)}{\left.\frac{\partial(\lambda\bar{c}_i)}{\partial t}\right|_{aero}} \quad (1)$$

where λ is the three-dimensional Jacobian that determines grid and coordinate transformations, s is the generalized vertical coordinate, $\bar{c}_i$ is the mean concentration of trace gas species i, $\hat{F}^l_{q_i}$ are the turbulence flux components expressed in terms of the mixing ratio q, and $\bar{\hat{\mathbf{V}}}_s$ and $\bar{\dot{s}}$ are horizontal and vertical contravariant wind components, respectively. Each term in Equation (1) represents a science process for CMAQ: (a) time rate change of pollutant concentration; (b) horizontal advection; (c) vertical advection; (d) horizontal turbulent diffusion; (e) vertical turbulent mixing and deposition; (f) production or loss from chemical reactions; (g) emissions; (h) cloud mixing and aqueous-phase chemical production or loss; (i) plume-in-grid; and (j) aerosol process.

Models-3 CMAQ accommodates various vertical coordinates (e.g., geometric height, terrain following Sigma-z, Sigma-p, Sigma-p_o) and map projections (e.g., rectangular, Lambert, Mercator, polar stereographic) by simple changes in a few scaling parameters, boundary conditions, map origin, and orientation. Therefore, each component in CMAQ can adapt to the coordinate system used in meteorological modeling.

Nesting Technique: To resolve fine scale features such as those found in urban and complex terrain areas, CMAQ provides a grid nesting capability. Nested CTMs can be constructed either one-way or two-way, to update concentration predictions. The choice depends on the target air quality problems users need to solve. CMAQ accommodates several different methods for passing the coarse-grid information to the nested fine grid: the one-way nesting can be implemented by explicitly linking a coarse model output with the nest run through the use of the Study Planner, or within a CTM by concurrently processing the coarse and nest domains. A "set of cooperating processors" concept (running two or more different scales of models independently while letting the I/O API take care of the coordination of inter-processor communications) can be used for two-way nested modeling.

Subgrid scale plume modeling: Anthropogenic precursors of the lower tropospheric loading of ozone, aerosols and acidic species are largely emitted from major point sources, mobile sources, and urban-industrial area sources. Inadequate representations of these emissions can cause inaccurate predictions in regional and urban Eulerian AQMs especially for power plant plumes with characteristically high NO_X. A plume-in-grid (P-in-G) approach, in combination with the nesting capability of the modeling framework, enhances overall accuracy of the model outputs. Key processes in the preprocessor plume dynamics submodels include plume rise, plume transport (trajectory), and plume growth in vertical and cross sectional directions. The corresponding plume-in-grid module in CMAQ deals with chemical reactions inside the plume including plume-background interactions, emissions input to the plume, and surface deposition.

MULTI-POLLUTANT CHARACTERISTICS OF THE CMAQ MODELING SYSTEM

Many different air pollution problems challenge scientists and the public. They are acid deposition, nutrient loading, fine particles, visibility degradation, tropospheric ozone, and toxics, etc. However, as our understanding of the relationships among the key pollutants has improved, it has become more evident that the air quality problems should be studied from a "one atmosphere" perspective. To transcend the old practices of developing an air quality model for each separate air pollution issue, CMAQ is designed to be a "one atmosphere" air quality model. Some of the key processes that give CMAQ multi-pollutant capability are described below.

Photolysis Rate Computation: Almost all chemical reactions in the atmosphere are initiated by the photodissociation of a number of trace gases. In order to accurately model atmospheric chemical reactions, good estimates of the photolytic rates of trace gases must be made. The CMAQ's photolysis rate processor is largely based on the model for Tropospheric Ultraviolet and Visible radiation (Madronich, 1987), with added capabilities incorporating temperature and pressure profiles from the meteorological model, modeled cloud fields, gridded surface albedo, and the total ozone column data (TOMS). The processor utilizes a generalized interpolation algorithm for specifying different sets of absorption cross section and quantum yield data to allow generation of photolysis rates for three chemical mechanisms (i.e., RADM, CBM-IV, and SAPRC-90) (Roselle et al., 1997).

Base Chemistry Mechanisms: An important component of any CTM is its treatment of the chemical transformations among the modeled species. The ability of a model to accurately describe the distributions of various pollutants, especially secondary pollutants, is highly dependent on the various chemical processes represented in a mechanism. Although CMAQ is capable of incorporating many different mechanisms easily, currently we have implemented two important atmospheric chemical mechanisms, the Carbon Bond Mechanism IV (CBM-IV) and the RADM mechanism. The SAPRC-90 mechanism will be implemented in later CMAQ versions. The limited choice is due to the limitation of the Models-3 emissions processing system that can provide emissions data only for species in the three base chemistry mechanisms.

Generalized Chemistry Reader and Solvers: The chemistry solver components of second-generation air quality models are usually hand-coded for a specific reaction mechanism, thereby requiring considerable re-coding when the mechanism must be replaced or enhanced. In the Models-3 CMAQ system we have emphasized development of a generalized representation of the chemistry. Here, we wanted to have a finer-grained modularity beyond the science process level. The numerical ordinary differential equation (ODE) solver is isolated from the chemical mechanism in order to achieve a high-level of modularity. Currently two generalized numerical solvers based on QSSA (a quasi-steady state approximation) (Young et al., 1993) and SMVGEAR (Sparse Matrix Vectorized GEAR) (Jacobson and Turco,1994) techniques are implemented in CMAQ. Also, we incorporated a generalized mechanism reader to allow experimentation with chemical mechanisms in CMAQ. Users can input mechanisms using a simple syntax for describing chemical reactions. Then, the mechanism reader compiles the list of modeled species, sets up reaction rates, and the CMAQ computes production and loss terms for each species automatically.

Particulate Submodule: Atmospheric aerosols are both emitted and formed from complex photochemical processes. Such aerosols are typically distributed in several size modes based upon their derivative processes. Gas-phase pollutants can nucleate to produce ultra fine particles; additional growth, chemical transformations and aggregation of such particles into sub-micron particles occur in an accumulation mode. Coarse particles come from some industrial processes or appear as fugitive or wind blown dust. Particles are transportable, enter into clouds (fog), deposit differentially with size and chemical composition including water content, and are chemically active in heterogeneous-phase reactions, and can serve as transport sites for other pollutants such as airborne semi-volatile toxic compounds. Models-3 CMAQ incorporates an aerosol modeling approach using two interacting lognormal modes (nucleation and accumulation). Unlike the previous implementation in the Regional Particulate Model (RPM) (Binkowski and Shankar, 1995), the CMAQ version provides interaction with the photochemistry at the synchronization time step, which is set by a Courant condition determined by wind speed and grid cell size. The particulate sub-module includes aerosol water, nitrates and organics, in addition to sulfate and ammonium.

Table 1. Summary of science implementation for CMAQ Initial Operating Version.

Process	Interim Science	Target Science+
Meteorology	MM5 Hydrostatic, Kuo/Betts-Miller Cloud, Standard Surface Exchange & PBL algorithms	MM5 Hydro/Nonhydrostatic Kuo/Betts-Miller/Kain-Fritsch Cloud Enhanced Pleim-Xiu Surface Exchange & PBL schemes
Meteorology-Chemistry Interface Processor	MM4/MM5 coordinates only RADM deposition velocity algorithm	Generalized coordinates for other meteorological models (RAMS, ETA). Models-3/CMAQ deposition velocity algorithm
Other Interface Processors	Table form photolysis rates for RADM & CB-IV mechanisms Boundary conditions: constant/dynamic Emissions-Chem. Interface Processor (ECIP) and Plume Rise algorithm	4-D dynamic photolysis processor for RADM/CB-IV/SAPRC-90 mechanisms Initial and boundary conditions for multi-level nesting. ECIP and Plume-Dynamics Model
Emissions	Enhanced GEMAP/EMS-95	Models-3 Emissions Processing and Projection System (MEPPS)
Chemistry-Transport Model	Hydrostatic One-way nesting Advection: Botts Diffusion: K-theory Chemistry: RADM-2/CBM-4 Generalized solvers (QSSA, SMVGEAR) Sulfate and nitrate aerosol Walcek cloud & aqueous chemistry	Generalized coordinates Advection: multiple choice Diffusion: ACM and TTM Chemistry: SAPRC-90 Nest: two-way/multi-level Plume-in-grid Aerosol: nitrates, organics, ammonium Improved cloud & aqueous chemistry
Analysis and Visualization	Process Analysis: mass budget and integrated reaction rates 2-D/3-D analysis with Vis5D and PAVE	Pointer flyer Plume-in-grid visualization

+ Target science includes all the capabilities implemented for the CMAQ interim science version.

CONCLUSIONS

A new framework for comprehensive air quality modeling is emerging. As our knowledge of the environment improves, the linkage of natural processes will be diversified. Models-3 should provide a generalized computational paradigm that enables integrated multi-media environmental modeling. We are at the final stage of the model development and expect to release the Models-3 CMAQ Initial Operating Version (IOV) in 1998. Table 1 summarizes the components being implemented in CMAQ. To expand the depth of science in the system and to improve our understanding of the air quality problems, CMAQ should be nurtured through involvement of the impacted and interested science communities.

DISCLAIMER

This paper has been reviewed in accordance with the U.S. Environmental Protection Agency's peer and administrative review policies and approved for presentation and publication. Mention of trade names or commercial products does not constitute endorsement or recommendation for use.

REFERENCES

Binkowski F. and U. Shankar, 1995. The regional particulate model, Part I: Model description and preliminary results. *J. Geophys. Res.*, 100: 26191-26209.

Byun D.W., A. Hanna, Coats C., and D. Hwang. 1995. Models-3 Air Quality Model Prototype Science and Computational Concept Development. *Transactions of A&WMA Specialty Conference on Regional Photochemical Measurement and Modeling Studies*, Nov. 8-12, San Diego, CA. 1993, pp. 197-212.

Dennis, R.L., D.W. Byun, J. H. Novak,, K.J. Galluppi and C.J. Coats, 1996. The next generation of integrated air quality modeling: EPA's Models-3. *Atmos. Environ.* 30, No. 12, 1925-1938.

Jacobson, M.Z. and R.P. Turco, 1994. SMVGEAR: A sparse-matrix, vectorized Gear code for atmospheric models. *Atmos. Environ.* 28, No. 20, 3369-3385.

Madronich, S.J., 1987. Photodissociation in the atmosphere I. Actinic flux and the effects of ground reflection and cloud. *J. Geophys. Res.*, 92: 9740-9752

Roselle, S.J., et al., 1997. Development and testing of an improved photolysis rate model for regional photochemical model. *Proceedings of the Air & Waste Management Association's 90th Annual Meeting*, June 8-13, 1997, Toronto, Ontario, Canada.

Wilkinson, J.G., and B. Emigh, 1994. The Geocoded Emissions Modeling and Projections (GEMAP) System Training Workshop. Sacramento, CA. Jan 19-21, 1994. 270pp.

Young, J.O., Sills, E., and Jorge, D., 1993. *Optimization of the Regional Oxidant Model for the Cray Y-MP, EPA/600/R-94/065*. U.S. Environmental Protection Agency, Research Triangle Park, NC.

DISCUSSION

D. CARRUTHERS: The Models-3/CMAQ appears very ambitious, perhaps over ambitious.

D. W. BYUN: In spite of many difficulties, we are now at the final stage of development of the initial operating version (IOV). Next month (July 1997) we will convene a beta-user workshop to introduce a preliminary integrated version of the system framework (Models-3) and air quality modeling system (CMAQ) to a select group of regulatory modelers from the EPA regional offices. IOV will be released to the public next year (1998). After that, we will need your help to achieve the goal of community air quality modeling.

D. CARRUTHERS: What users is the model aiming at?

D. W. BYUN: Models-3/CMAQ is not a single model, but a system that allows users to build an air quality model to fit the user's problem. As such, the system is targeting both the regulatory users and the science processor developers.

D. CARRUTHERS: How can you ensure that the users have sufficient technical know-how to use the model correctly? It will be very easy to run the model in an inappropriate manner.

D. W. BYUN: Because Models-3/CMAQ is designed for a wide spectrum of users, it is difficult to say that we can ensure that all the potential users have sufficient technical know-how to use the model. Like all other models and tools available to us, there are always risks of misusing air quality models. As described earlier, we are classifying the Models-3/CMAQ users into two groups; science developers and regulatory users. For science developers, we want to provide as many options as possible. Flexibility, extensibility and modularity of the system were the main design objectives of CMAQ to achieve a community modeling system to exchange new ideas for air quality modeling. For regulatory users, we are planning to define a regulatory study plan (i.e., a set of linkages of science processors such as meteorological model, emissions model, interface processors, chemistry-transport model module options with associated basic data) as a standard procedure of operation for a specific air quality problem. The Study

Manager of the Models-3 system framework then handles the study plan and processes the computational steps automatically. The regulatory users will be advised to change only certain model characteristics, such as modeling domain, resolution, and input data, etc. For the development of initial operating version (IOV), we are not focusing on this issue as much. In the near future, we will work out this issue with the EPA's regulatory air quality modelers.

D. CARRUTHERS: What plume model (PING) will be implemented in the grid based system? How will you integrate the chemistry module within the different schemes (grid and plume)?

D. W. BYUN: For the IOV, we are implementing a Lagrangian plume model proposed by Dr. Noor Gillani of the Univ. Alalbama at Huntsville. Currently, we are using identical chemical mechanism (RADM-2 or CB-IV) and generalized chemistry solver (QSSA or SMVGEAR) for grid and plume. We prefer this way because it is more complete and consistent. If researchers want to experiment with simpler chemical mechanisms in plume, they need to develop customized plume chemistry subroutines and link them with the PING module of CMAQ. The modularity of CMAQ, Models-3 framework's Model Builder and I/O API utility routines greatly simplify this type of work. This can be an example of how we could use the Models-3/CMAQ system as a tool for the community modeling paradigm.

B. PHYSICK: Is the Models-3/CMAQ system being developed for use by scientists only, or is it envisaged that it will be available to the general community, for example lawyers?

D. W. BYUN: Thank you for asking the question. The system has kind of a split personality. As a research scientist myself, definitely science model CMAQ is for scientists. However, many functionality of the system framework, such as the Study Planner, Object Dataset Manager, Tools Manager, Science Manager and the graphical user interfaces are aimed at the regulatory users. Because the system automatically documents modeling procedure and data through the use of the study plan and meta data (they are kept in the database), the quality assurance part is much more rigorous than any other EPA's second-generation regulatory models. In that regard, we can say that the system is lawyer savvy. However, outputs of the model should be interpreted by air quality modelers for the appropriate regulatory applications. The intent of the system is to provide air quality simulation output and analysis that may be relevant for environmental policy decision making. Models-3/CMAQ is not an artificial intelligence modeling system.

TRAFFIC-INDUCED URBAN POLLUTION: A NUMERICAL SIMULATION OF STREET DISPERSION AND NET PRODUCTION

Jean-François Sini and Patrice G. Mestayer

Laboratoire de Mécanique des Fluides, UMR CNRS 6598
and SUB-MESO, Groupement de Recherches CNRS 1102
Ecole Centrale de Nantes
F-44072 NANTES, France

ABSTRACT

The CFD model CHENSI, based on the k-ε two equation closure, including thermal buoyancy and scalar transport-diffusion, is first used to simulate the experiment of Cadle et al. (1976), allowing to deduce the turbulence produced by car motion. Then, a heuristic study of the streets capacity to ventilate the pollutants emitted by traffic, as a function of their geometrical and thermal properties, is presented from simulations of the flow and turbulent fields, the pollutant concentrations within the streets, and their fluxes to the atmosphere. Isothermal and non-isothermal simulations are presented, in the case when the street geometry corresponds in isothermal conditions to Oke's (1988) skimming flow type. The wall heating is shown to induce different types of flow resulting in different pollutant distributions and net fluxes. The influence of the turbulence produced by cars on the pollutant dispersion is shown.

INTRODUCTION

The major source of urban pollution in the large western cities is due to traffic. It is especially important in the densely constructed parts of the city center. The pollutants behavior is closely linked to their dispersion close to the sources. To evaluate the pollutant fluxes and to parameterize the net emissions of city quarters to the atmosphere, it is first necessary to assess the influence of the urban canopy geometry and micro-meteorology, especially in weak wind conditions.

The range of geometrical arrangements of the buildings and streets is extremely wide. Considering the cost and difficulty of running sufficiently detailed on-site experiments during sufficiently long periods, considering the high variability of urban micro-meteorological conditions, and considering the high intermittence of both the flow conditions and the vehicle traffic, it is unlikely that a large understanding of the pollutant dispersion processes within streets can be drawn from on-site measurements alone. On the contrary numerical simulation offers the possibility to multiply the observations ... at the condition that the numerical model be validated by comparison with reliable experimental

Air Pollution Modeling and Its Application XII
Edited by Sven-Erik Gryning and Nadine Chaumerliac, Plenum Press, New York, 1998

data. We think that the two efforts, model validation and heuristic numerical simulations, must be lead in parallel.

The numerical simulation is especially interesting to study the situations of low wind speed when the flow is a mixed convection and the pollutant dispersion depends largely on the coupling between wind dynamics and thermal convection, as shown by Mestayer et al. (1995) and Sini et al. (1996). In these situations the transport by the mean flow and the turbulent diffusivity are very weak in the bottom of the streets, and additional sources of turbulence like the moving vehicles can be of importance. The study briefly presented here aims at evaluating this importance, although the reduced selection of results is more illustrative than demonstrative. It concerns the simulation of the dispersion of CO concentration from vehicle exhaust pipes in one main street of Nantes city center. First we present the numerical model, the evaluation of turbulence production by car motion, the parameterization of CO production by cars. Then simulations are presented for several thermal and traffic conditions. Finally the influence of these conditions on CO concentrations in the street is shown.

CHENSI MODEL

The numerical model used in this study is the finite-difference three-dimensional version of CHENSI (V3.0). This model solves the Reynolds-averaged Navier-Stokes equations on a staggered non-uniform Cartesian grid, using Chorin's method to solve the pressure-velocity coupling. It uses Boussinesq's first gradient assumption to express the stress tensor components; the turbulent viscosity ν_t is computed with the k-ε model, $\nu_t = C_m k^2/\varepsilon$, by solving simultaneously the budget equations for the turbulent kinetic energy k ($= {}^1/_2\ \overline{u_i u_j}$) and its dissipation rate ε. To take into account the kinetic-thermal coupling, an equation for the potential temperature is also solved, allowing to compute the kinetic energy production by buoyancy in the equation of k, and in that of ε. Thus, in the equation of k the production term P_k includes three contributions,

$$P_k = P_m + P_b + P_{car} \qquad (1)$$

where $P_m = -\overline{u_i u_j}\, dU_i/\partial x_j$ is the mechanical production of k, $P_b = \beta\, g_i\, \overline{u_i \theta}$ is the production by buoyancy, and where P_{car}, the production by cars, has been added for this study. These contributions appear also implicitly in the budget equation of ε since $P_\varepsilon = P_k\, \varepsilon/k$. To simulate the dispersion of pollutants an equation of transport-diffusion of a passive contaminant is also solved either in the same time than the equations of dynamics, or off-line using stationary turbulent fields. Here, CO concentration is considered as an inert gas produced by the exhaust pipes.

The model CHENSI has been extensively validated against the results of reference experiments presenting the characteristic features of in-street turbulent flows, including thermal convective coupling (Levi Alvares, 1991) and of experiments in the lower atmospheric surface layer over non-homogeneous terrain (Costes, 1996; Mestayer et al., 1996). In particular the standard k-ε model weaknesses and those of the "improved" alternatives have been analyzed as concerns the urban canopy flows (Bottema and Sini, 1997). In the low wind speed conditions we think that the standard model weaknesses are of little influence, while this model version is best calibrated.

TURBULENT KINETIC ENERGY PRODUCTION BY CARS

Model calibration

The General Motors experiment described by Cadle et al. (1976 and 1977) and Chock (1977) has been mainly conducted to assess sulfate dispersion from a highway, but it also included measurements of wind and turbulence. The experiment was conducted in the GM Proving Ground. The North-South Straightaway is a 6-lane close-loop freeway, straight over 10 km. A fleet of 352 car was split into packs of 22 cars evenly distributed on the two outer lanes in each direction, in an arrangement allowing to control car speed. A line of six 10.5 m high experimental masts extended 3 m upwind and 26 m downwind of the 25.4 m wide, level road bank, at a 72° angle to the track, in the middle of the Straightaway. The length of the test field insured flow and turbulent homogeneity along the direction of the road at the test section.

We have taken advantage of the previous analyses of Eskridge and Hunt (1979) and Sedefian et al. (1981) to model the influence of car motion on the turbulent field by the additional turbulent kinetic energy production term

$$P_{car} = C_{car}\, U_{car}^2\, Q_{car} \tag{2}$$

where U_{car} is the mean car speed and Q_{car} the rate of vehicles per second per lane. C_{car} is a non-dimensional constant.

Since we assume that the main process of kinetic energy production is by the generation of a turbulent wake behind each vehicle, P_{car} is set to the computed value in the computational grid meshes traversed by the cars, and to zero everywhere else. C_{car} has been calibrated by adjustment of a numerical simulation to the experimental case ref. 279-0809, whose main set-up and results are presented in Figure 1. The computational domain is the x-z plane of the experimental masts, 83 m long, 52 m high, with a non-uniform grid of 174 x 60 meshes. Note that the cars are considered as kinetic energy sources in a set of meshes, not as solid obstacles, since their occupation time of the mesh volume is much smaller than the vacancy time. The inflow conditions have been deduced from the measurements at the first mast, 30 m upwind of the road, and the C_{car} constant has been adjusted to the value 0.41 giving the best agreement with the measurement at the first mast 3.8 m downwind of the road (see Figure 1). the values of k were deduced from the measured vertical velocity variance σ_w^2 using the σ_w^2/k ratios deduced from Eskridge and Hunt's (1979) analysis.

Discussion

We must note that we present here a calibration of the model rather than a validation, considering the spatial distribution of k in Figure 1, the small number of available measurements, and the fact that only one experimental set-up is simulated.

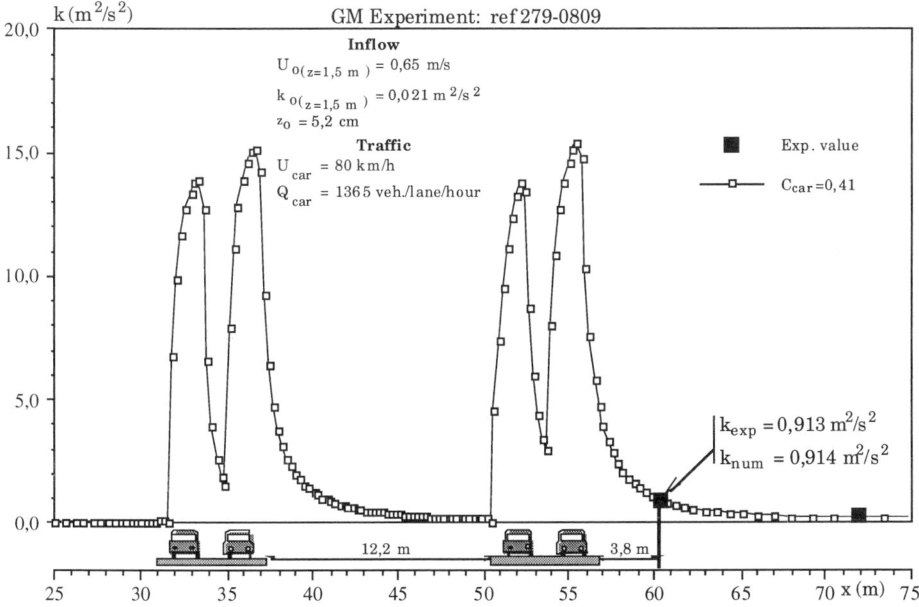

Figure 1. Horizontal profile of k at the level z = 1.5 m.

Another point of importance is that the vehicle motion is also a source of stress in the direction of the track. If the traffic is in both directions, this generates a strong shear over the central bank, but the effect probably vanishes rapidly downwind of the road. If the traffic is in one direction only, it adds to the mean flow increasing the mechanical production of kinetic energy. These points are being explored experimentally by our colleagues of CSTB Nantes (Delaunay et al., 1996; Sacré, 1996).

Here, to stay coherent with the results of the GM experiment we have chosen to simulate the flow in the street when the wind is normal to the street axis, assuming homogeneity in the average, and a zero mean velocity component, in the y-direction.

PRODUCTION OF CO

When the traffic rate varies in a street, it affects 3 parameters simultaneously: the number of pollution sources, the diffusion of the emitted pollutants, and the mixing with the pollutant immission. This last effect influences, e.g., the Ozone-NO_x mixing and the NO-NO_2 conversion rate. To identify the dynamical effect separately, we simulate here the dispersion of CO which is essentially emitted by the cars and chemically inert.

The simulated street being in the city center forbidden to truck circulation, in first approximation buses and lorries are ignored and the CO production is that of the cars with gas engines. In the CO concentration transport-diffusion equation, a production term P_{CO} is added in the only grid meshes occupied by exhaust pipes, i.e., one mesh per lane, in the left lower corner of the set of meshes traversed by the cars. This source term is

$$P_{co} = (E_{co}\ Q_{car}) / (\Delta x_p\ \Delta z_p) \qquad (3)$$

where P_{co} is in g m^{-3} s^{-1}, Δx_p and Δz_p are the sizes of the exhaust pipe meshes in the simulation plane, their size in the y-direction being 1 m (see below), and E_{co} is the unit CO emission rate per car.

This last factor depends on the engine regime. We empirically parameterize it as a function of U_{car} by a power law best fit numerical adjustment to the results of Joumard & Lambert (1991) for the gas engine cars:

$$E_{co} = 0.083\ U_{car}^{-.558} \qquad (r = 0.977) \qquad (4)$$

The car speed and the traffic rate are linked by relationships that are strongly dependent on the type of street, the cross-roads, inlets, and outlets, the local driving practices... Rather than using a "theoretical" function we have derived an empirical relationship by linear fit to the counting furnished by the road system service of the city of Nantes. In the range of interest from 0_+ to 0.15 vehicle per second per lane, the car speed decreases about linearly and we obtain car speeds of 5, 10, 20, and 50 km/h for traffic rates of 3860, 3470, 2690, and 349 vehicles per hour, respectively, for the three lane street.

FLOW AND DISPERSION SIMULATIONS

The rue de Strasbourg is a 3-lane, 1-way, heavily trafficked, straight street of the city center. It is 14.8 m wide and bordered by quite regular buildings whose heights average at 22 m. The simulated domain is a vertical plane normal to the street axis, 150 m in length and 112 m in height. The computational domain is a non-uniform grid of 70 x 81 x 3 meshes. The 20 x 40 meshes covering the canyon are 0.74 m wide and 0.55 m high (and 1 m thick), while outside the canyon the meshes expand regularly. It must be noted that these simulations, like that of the GM experiment, are fully three-dimensional and the computational grid has three 1 m meshes in the y-direction, while the mean wind component in this direction remains null. The data presented here are the results in the grid central plane.

The inflow is an atmospheric surface layer with U = 3 m/s at z = 100 m (U ≈ 1 m/s at roof level in the street center) and k = 0.45 m^2s^{-2}. The roughness length of all surfaces is z_o = 1 cm. The temperature profile is adiabatic. Two thermal conditions are presented: with either all walls at air temperature T = 20°C, or one of them, the wind-facing wall, at 30°C.

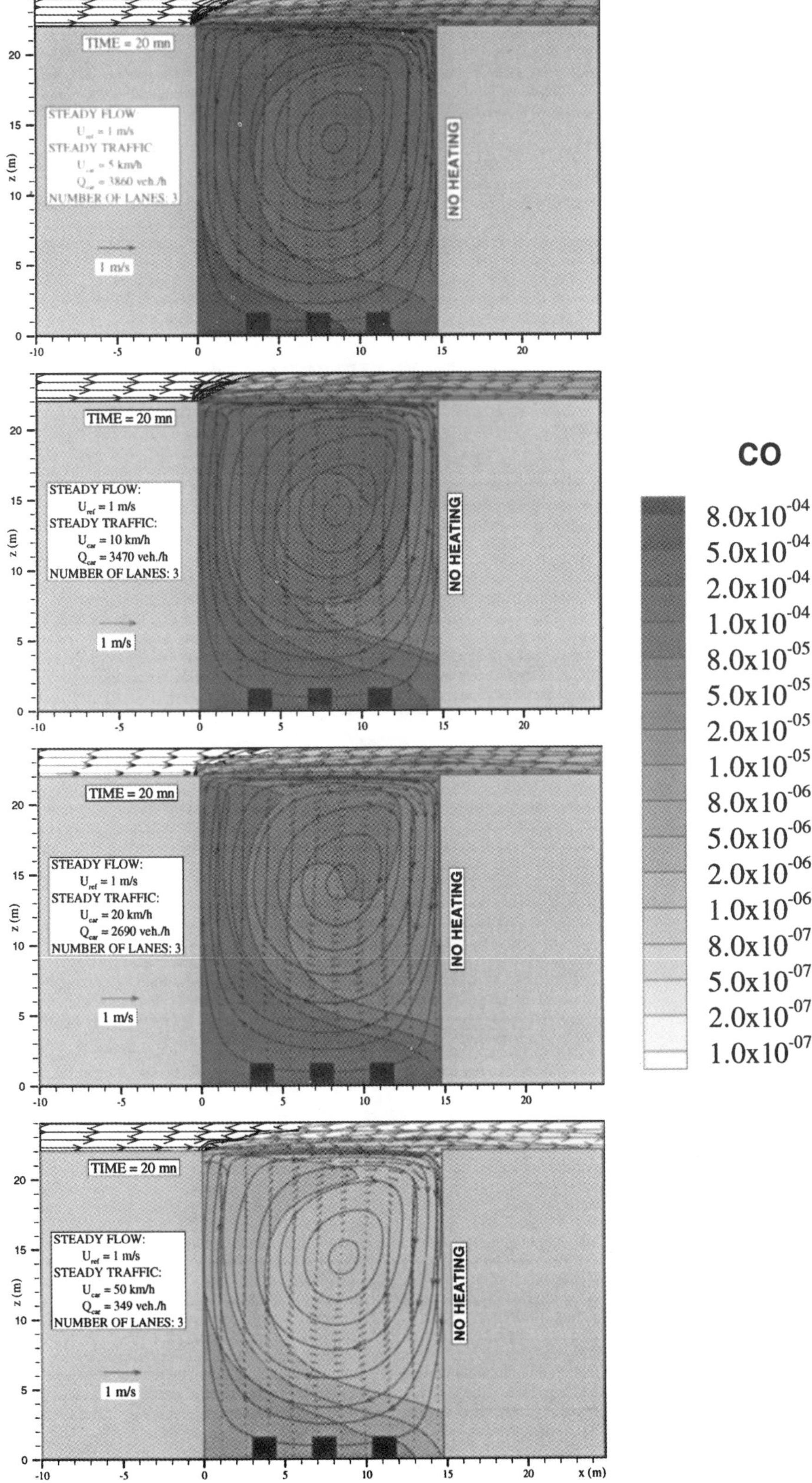

Figure 2a. Flow lines and CO concentrations in the street in isothermal conditions

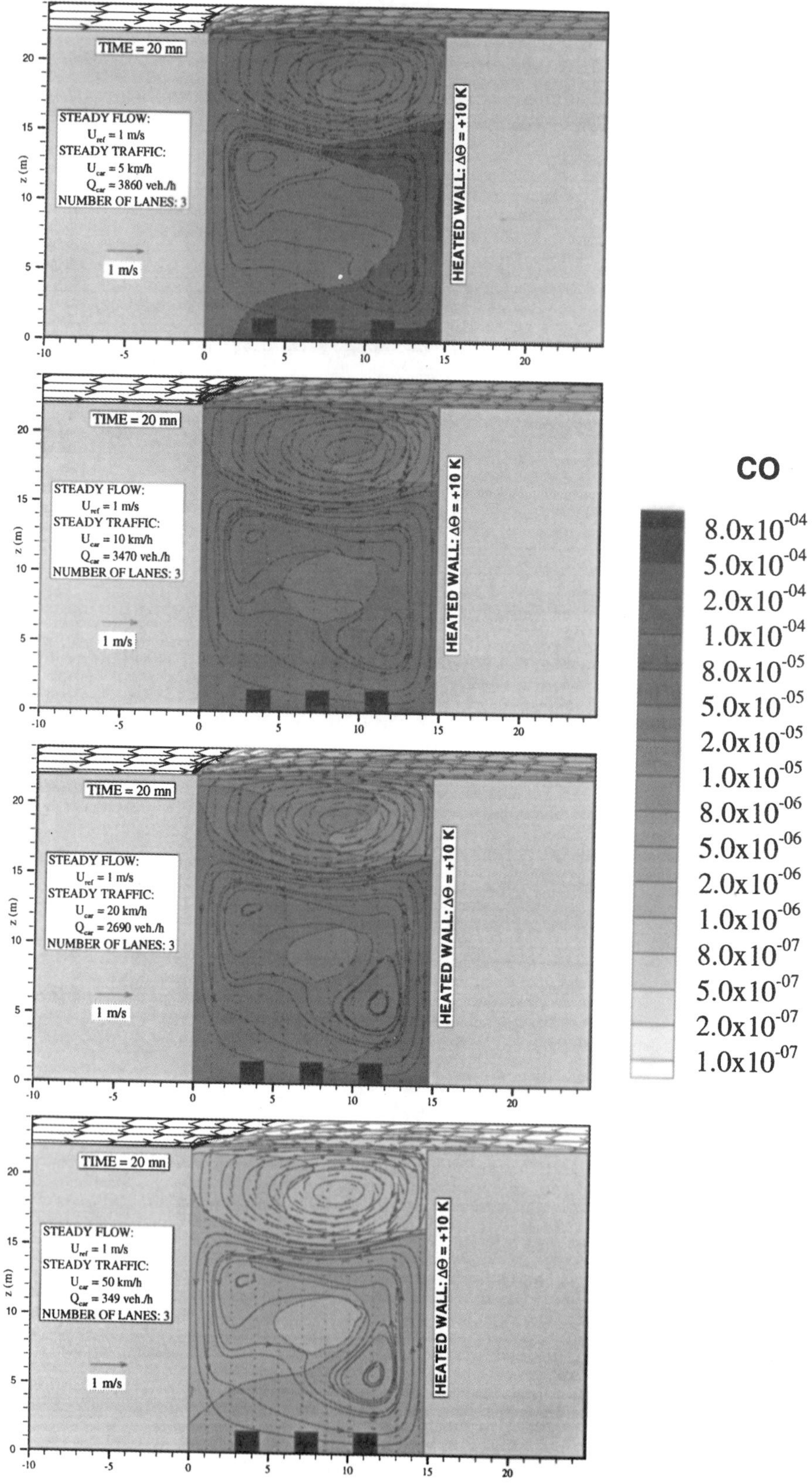

Figure 2b. Flow lines and CO concentrations when the wind-facing wall is heated by 10 K

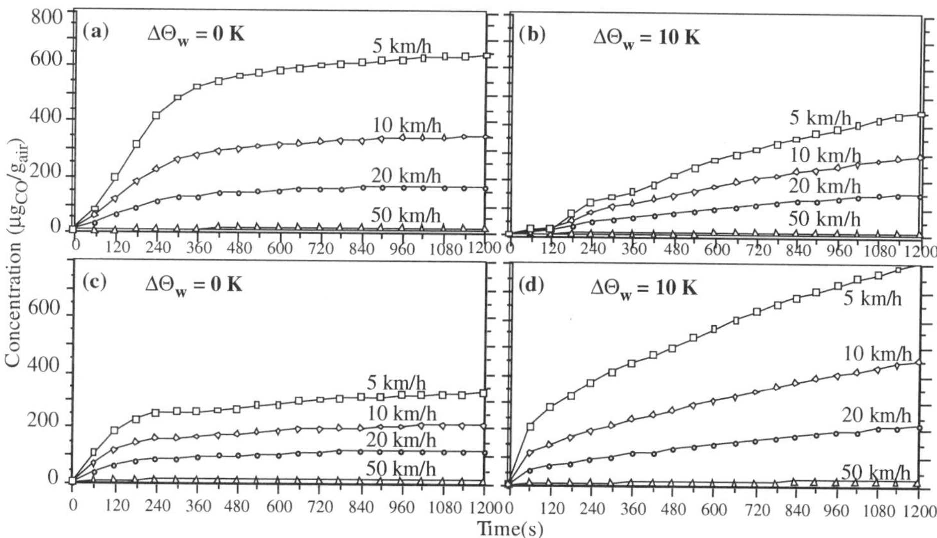

Figure 3. Time evolution of CO concentration over the left sidewalk (a and b) and over the right sidewalk (c and d) in isothermal condition (a and c) and with the right-side wind-facing wall heated by + 10 K (b and d).

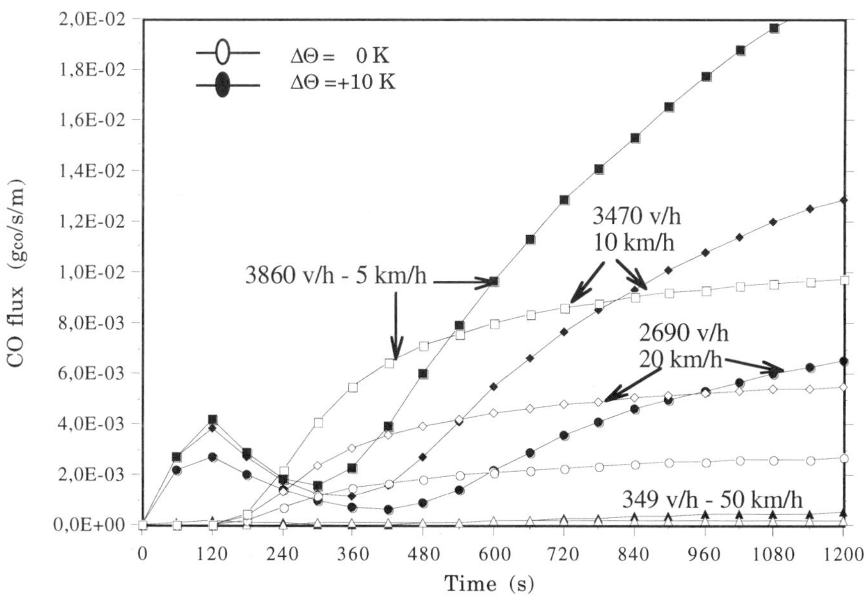

Figure 4. Time evolution of the CO flux from the street to the atmosphere. The corresponding CO production rates are $T_5 = 74$, $T_{10} = 45$, $T_{20} = 24$, $T_{50} = 2$ $mg_{CO}/s/m$ (mg of CO per second per lane per linear m).

The results are obtained after a simulated time of 20 minutes, all equations (dynamic, temperature, and concentration) being computed simultaneously. The time step is 2.10^{-2} s. For the CO concentration the initial condition is always an empty canyon. After 20 minutes the concentration field is not fully stationary and the flux at the top of the street is still smaller than the production rate.

The Figure 2 shows that the heating of the wind-facing wall changes largely the flow structure, creating a flow blockage and a secondary slow vortex, nearly a stagnation zone, in the lower part of the canyon. The traffic rate does not appear to change much the flow structure in the canyon, and the turbulent diffusion is increased only in the bottom part of the street. In the average, the CO concentration varies about as the traffic rate, but the influence of the traffic rate is very different for the different parts of the street.

The Figure 3 presents the CO concentrations, integrated from 0 to 2 m over each of the pedestrian sidewalks. It shows a contradictory influence of the wall heating on the concentration at pedestrian level, due to the generation of the secondary rotor that inverses the direction of the transverse flow in the bottom of the street.

The time evolution of the CO flux from the street to the atmosphere (Figure 4) reflects the complex structure of the flow, hence the pollutant transport from the bottom to the top of the atmosphere. The bulk response times of the street to a change in pollutant emission appear to be rather large, on the order of several tens of minutes.

Acknowledgments

This work has benefited of computing time allocation by the CNRS computing center IDRIS. We also want to thank the "Service de la voirie" of the City of Nantes.

REFERENCES

Bottema, M. & J.-F. Sini, 1997, High-Schmidt number mass transfer through turbulent gas-liquid interfaces, 11th Symposium on Turbulent Shear Flows, Grenoble, France.

Cadle, S.H., D.P. Chock, J.M. Heuss and P.R. Monson, 1976, Results of the General Motors sulfate dispersion experiment, GM Research Publication GMR-2107, GM Research Laboratories, Warren, MI.

Cadle, S.H., D.P. Chock, P.R. Monson and J.M. Heuss, 1977, General Motors sulfate dispersion experiment: Experimental procedures and results, J. Air Pollution Control Assoc. 27: 33.

Chock, D., 1977, General Motors sulfate dispersion experiment - An overview of the wind, temperature, and concentration fields, Atmospheric Environment 11: 553

Costes, J.-P., 1996, Simulations numériques des écoulements atmosphériques sur sols fortement hétérogènes, Thèse de Doctorat, Ecole Centrale de Nantes, France.

Delaunay, D., J.P. Flori, C. Sacre, 1996, Numerical Modelling of gas dispersion from road tunnels in urban environments: comparison with field experiment data, 4th Workshop on harmonisation within atmospheric dispersion modelling for regulatory purposes, Oostende, Belgium: 361.

Eskridge, R.E. and J.C.R. Hunt, 1979, Highway modeling. Part I: Prediction of velocity and turbulence fields in the wake of vehicles, J. Applied Meteorology 18: 387.

Joumard, R. and J. Lambert, 1991, Evolution des emmissions de polluants par les transports en France de 1970 a 2010, Rapport INRETS n° 143.

Levi Alvares, S., 1991, Simulation numérique des écoulements urbains à l'échelle d'une rue à l'aide d'un modèle k-ε., Thèse de Doctorat, Université de Nantes - ENSM, France.

Mestayer, P.G., J.-F. Sini and M. Jobert, 1995, Simulation of the wall temperature influence on flows and dispersion within street canyons, 3rd Int.Conference on Air Pollution, Porto Carras, Greece, 1: 109.

Mestayer, P.G., M. Bottema, J.-P. Costes, J.-F. Sini, 1996, Modelling urban canopy and terrains for transport-diffusion simulations at sub-mesoscales, 4th Workshop on harmonisation within atmospheric dispersion modelling for regulatory purposes, Oostende, Belgium: 337, to appear in Int. J. Environm. Pollution 6.

Sacré, C., 1996, Modélisation de la turbulence induite par le traffic automobile, CSTB report EN-CLI 96.21 C, CSTB Nantes, France.

Sedefian, L., S. Trivikrama Rao and U. Czapski, 1981, Effects of traffic-generated turbulence on near-field dispersion, Atmospheric Environment 15: 527.

Sini, J.-F., S. Anquetin and P.G. Mestayer, 1996, Pollutant dispersion and thermal effects in urban street canyons, Atmospheric Environment, Urban Atmosphere, 30: 2659.

DISCUSSION

S. RAFAILIDIS: By looking into your present comparison between the isothermal and the Dt = 10°C modelling results, we see that in the latter condition two vortices develop. But, if we compare your conclusions from the Porto Carras (1995) paper, then you reported also for non-isothermal cases a single vortice only.

P. G. MESTAYER: In the Porto-Carras paper we reported isothermal simulations showing in narrow streets one, two and even three piling-up vortices, depending on the street geometry (narrowness). Non-isothermal simulations showed that the flow regime induced by geometry can be largely changed by the induced thermal convection, the resulting flow regime being very sensitive to the air-wall temperature difference. Especially, some of our non-isothermal simulations showed shifts from a one-vortex flow regime to a two-vortex regime where the vortices were slanted with a diagonal separation.

DEVELOPMENT AND VALIDATION OF THE MULTILAYER MODEL MUSE - THE IMPACT OF THE CHEMICAL REACTION MECHANISM ON AIR QUALITY PREDICTIONS

P. Sahm, F. Kirchner*, N. Moussiopoulos

Laboratory of Heat Transfer and Environmental Engineering, Aristotle University Thessaloniki, 54006 Thessaloniki, Greece

ABSTRACT

As a new constituent of the European Zooming Model (EZM) system, the multilayer model MUSE is designed to serve as an efficient tool for simulating transport and transformation of air pollutants in the urban scale and thereby in supporting local scale air quality management in the most cost effective way.

Comparison of simulation results achieved with MUSE with corresponding results of the validated three-dimensional photochemical dispersion model MARS reveals that the model MUSE is capable of reproducing the spatial and diurnal variation of the major photochemical air pollutants.

In order to investigate the effect of the chemical reaction mechanism on air quality predictions, three different reaction mechanisms ranging from the compact mechanism KOREM to the comprehensive mechanisms EMEP and RACM are compared. The latter mechanism is a revised version of the RADM2 mechanism, the improvement mainly focussing on the description of the RO_2 chemistry and biogenic emissions.

The intercomparison reveals that despite of similar predicted ozone concentrations, the chemical mechanisms are still performing differently in many aspects. Thus, the choice of a suitable chemical reaction mechanism is mainly depending on the accuracy of the emission inventory as well as on the available computer memory and CPU time.

INTRODUCTION

The European Zooming Model system (EZM) is a complete system of models for simulating the wind flow and pollutant transport and transformation in the mesoscale. Main modules of the EZM system are the nonhydrostatic mesoscale model MEMO (Kunz and

---

* present address: Ecole Polytechnique Fédéral de Lausanne, DGR - LPAS, CH-1015 Lausanne, Switzerland

Moussiopoulos, 1995) and the photochemical dispersion model MARS (Moussiopoulos, 1989; Moussiopoulos et al., 1995).

In spite of the dramatic increase in computing power in the last decade, the application of the above or even more complex combined transport/chemistry models are still limited to high end computing facilities. However, there is an increasing demand on efficient tools supporting local scale air quality management in the most cost effective way. Consequently, the reasons for the design of the model MUSE are twofold. On the one hand, MUSE should serve as a basis for an efficient air quality management system. On the other hand it should allow to serve as a tool to address effectively many of the important scientific questions, e.g. through comparing individual chemical reaction mechanisms, which vary extensively in their complexity.

There are many intercomparisons of chemical schemes based on box model calculations (e.g. Poppe et al., 1996). Although such exercises are absolutely necessary to improve the scientific understanding of the photochemical oxidant formation mechanisms, they do not account for the complex interactions between the processes of transport, transformation and deposition of pollutants, which may amplify or dampen the influences of the differences in the details of a chemical scheme (cf. Hass et al., 1996).

The present paper describes the application of the EZM system to the Greater Athens area to study both, the capability of the newest constituent MUSE and the impact of the choice of the chemical reaction mechanism on air quality predictions.

MODEL OVERVIEW

MUSE is a multilayer Eulerian photochemical dispersion model for reactive species in the local-to-regional scale. The conceptual basis of the MUSE model is the division of the boundary layer into individual layers; the thickness of each layer is allowed to vary in the course of the day in order to adequately simulate the dynamics of the atmospheric boundary layer. In this study five vertical layers have been used. A shallow layer adjacent to the ground is used for simulating dry deposition and other sub-grid phenomena. This layer practically corresponds to the surface layer. The upper limit of the second layer is defined as the half height of the mixing height. The latter is described by Deardorff's prognostic equation (Deardorff, 1974). The upper limits of the third and fourth layer are defined by the lower and the upper limit of the entrainment zone. The parameterisation of the latter is based on the entraiment zone depth proposed by Batchvarova and Gryning (1994). In this study the top of the upper layer, which serves as a reservoir layer, is set at 3000 m, chosen so that it is well above the maximum mixing layer height.

The mathematical analysis is based on the coupled, three-dimensional advection-diffusion equations for the ensemble averaged quantitites of reactive species. The equations are solved by operator splitting according to the method of lines, that is by solving the advection dominated terms seperately from the diffusion dominated terms (in vertical direction) and the chemical reaction terms. The derivation of the advection, the vertical diffusion and the entraiment operator is then added as a source term to the chemical reaction equation system.

Advective transport is described with the scheme of Smolarkiewicz (1984). The description of vertical transport due to turbulent diffusion is based on the turbulent kinetic energy defined at the interface of the layers. To avoid unrealistic vertical diffusion rates associated with the growth of the mixing height, one-sided concentration gradients are calculated at these interfaces. Turbulent diffusion in the upper limit of the fourth layer is neglected. The chemical reaction rate equation system in MUSE is solved with a backward difference solution procedure, i.e. by applying the Gauss-Seidel iteration scheme (Kessler, 1995). The dry deposition process is calculated following the resistance model concept.

Due to the modular structure of MUSE, chemical transformations can be treated by any suitable chemical reaction mechanism.

THE PHOTOCHEMICAL REACTION MECHANISMS

All three reaction mechanisms in this study are so-called condensed chemical mechanisms, in which a necessarily limited number of reactions are used to represent the many thousands of reactions known to influence the atmospheric chemistry.

The KOREM mechanism is a modified version of the Bottenheim-Strausz mechanism (Bottenheim and Strausz, 1982). The chemistry has been described in detail in Moussiopoulos (1989). The condensation method used is lumped molecule. In this intercomparison study, emissions of VOC are mapped onto the lumped model species alkanes, olefines, o-xylene and aldehydes.

The EMEP MSC-W chemistry has been described in detail in Simpson et al. (1993) and Simpson (1995). The condensation method used is lumped molecule. In this study, emissions of VOC are mapped onto the molecules ethane, n-butane, ethene, propene, isoprene, o-xylene, formaldehyde, acetaldehyde, acetone and acrolein.

The RACM mechanism is a revised version of the RADM2 mechanism, the improvement mainly focussing on the description of the RO_2 chemistry and biogenic emissions (Stockwell et al., 1997). Emissions are split into a number of representative species (C2H6, HC3, HC5, HC8, C2H4, OLT, OLI, ISO, TOL, XYL, HCHO, ALD, KET and MACR).

There are important differences between these mechanisms in the hydrocarbon split, the formation of carbonyl species and nitrates, and formation and reaction of peroxy radicals: KOREM includes only 4 non oxidized hydrocarbons: methane, higher alkanes, alkenes and aromatic species. Compared to KOREM in EMEP three species are removed from the lumped groups and treated separately: ethane has been separated from the higher alkanes and ethene and isoprene from the alkene group. The remaining lumped groups alkanes, alkenes and aromatics are represented by n-butane, propene and o-xylene, respectively. n-butane and propene have significantly lower reactivities than most of the alkanes and alkenes they should represent, whereas the reactivity of o-xylenes is above the average of aromatic species. RACM includes the same single species as EMEP, but the lumped groups of alkanes, alkenes and aromatics are divided due to different reactivities into 3, 2, and 2 model species, respectively, and two terpene groups are added.

The most important carbonyl products from the hydrocarbon decay in EMEP and RACM are HCHO, higher aldehydes and ketones, whereas KOREM produces only aldehydes. EMEP and RACM treat the higher aldehydes as acetaldehyde and therefore underestimate the chain lengths of the aldehydes whereas the treatment of the aldehydes in KOREM correspond to an infinite chain length reproducing continuously aldehydes from the aldehyde decay.

All mechanisms include the reactions of peroxy radicals with NO, HO_2 and other RO_2, RACM also the reactions of RO_2 with NO_3. But they differ concerning the products and their reactions. KOREM produces no organic nitrates, EMEP considers only the small amount of organic nitrates resulting from the isoprene decay, whereas in RACM the formation of organic nitrates is considered for all reactions. Organic hydroperoxides resulting from the RO_2 + HO_2 reaction are treated as stable species in KOREM and are lumped into two groups in RACM, whereas EMEP produces one hydroperoxide for each

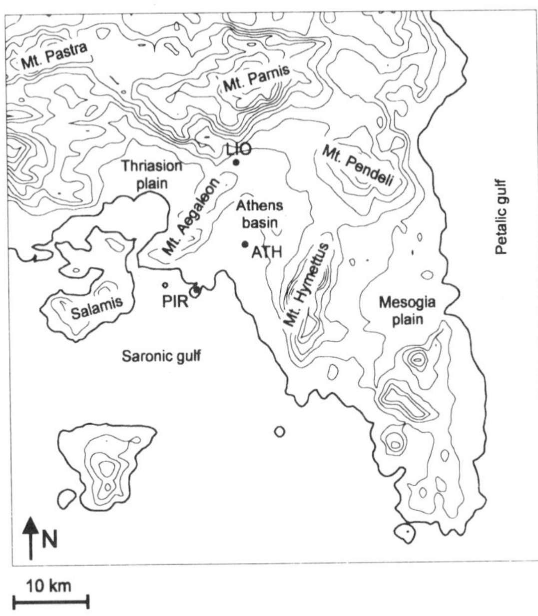

Figure 1. Topography of the Greater Athens area. Altitude isopleths are contoured at 100 m intervals. The dots denote the location of stations for which simulation results are presented (LIO: Liosia, ATH: Athinas, PIR: Piraeus).

peroxy radical enabling a correct description of the kinetic and the products of the ROOH + OH reactions.

CASE SPECIFICATION

Figure 1 shows the model domain as employed in this study. The city of Athens is located in the centre of the domain, in a basin surrounded at three sides by fairly high mountains, while to the SW the basin is open to the sea. The complex topographical and meteorological features of the Greater Athens area (GAA) in conjunction with high anthropogenic emissions result in alarmingly high air pollution levels.

The meteorological conditions prevailing on 25 May 1990 can be considered as the worst possible regarding the potential for the occurence of an air pollution episode in Athens. The reason for choosing this particular time period for this study was that this episode has been selected to be studied in the APSIS initiative[1] (Moussiopoulos, 1993; Moussiopoulos, 1995). In some ways, the following intercomparison between the models (MUSE and MARS) and the chemical mechanisms is a follow-up to the APSIS initiative. The model MARS in conjunction with the reaction mechanism KOREM was in fact a part of the earlier study (Moussiopoulos et al., 1995).

The necessary meteorological input for the dispersion simulations presented in this paper was derived from results of the nonhydrostatic mesoscale model MEMO and are discussed elsewhere (Moussiopoulos and Papagrigoriou, 1997). All simulations are based on the emission inventory specifically compiled for the needs of the Athens 2004 Air Quality study (ibid.). Due to the lack of more detailed information, monthly average

[1] The Athenian Photochemical Smog - Intercomparison of Simulations

concentration values for Athens derived from EMEP MSC-W model results have been used as initial and lateral boundary concentrations at inflow (Simpson, 1996). Though defined as boundary layer averages, these values have been used for all model layers. Photolysis rates were computed as functions of the solar zenith angle.

EVALUATION OF THE MODEL MUSE

Figure 2 shows the spatial distribution of the maximum hourly ozone concentrations (ppb) in the GAA at approx. 10 m above ground on 25 May 1990, predicted with MUSE (left) and MARS (right), both in conjunction with the KOREM reaction mechanism. This figure confirms that the model MUSE yields ozone concentration patterns which are in general similar to those obtained with MARS both in the qualitative and the quantitative point of view. According to the predictions of both models ozone concentrations remain low in the centre of Athens even during a severe air pollution episode. This is due to the fact that in the city centre ozone formation is counterbalanced by depletion reactions with near-ground emissions, primarily with NO.

As a main exception to the generally good agreement of model results, the multilayer model MUSE is found to overestimate the ozone concentration in the northwestern part of the study area. This deviation makes apparent the limitations of a multilayer model over non-homogeneous terrain: The complex flow field at the narrow strip at the foot of Mt. Parnis, where the sea breeze converges with katabatic winds, cannot be adequately resolved within few vertical layers. Thus the differences in advection and vertical diffusion compared to the model MARS lead to deviations in predicted ozone concentrations in this area.

ANALYSIS OF THE EFFECT OF THE CHEMICAL REACTION MECHANISM ON AIR QUALITY PREDICTIONS

Figure 3 shows the spatial distribution of the maximum hourly ozone concentrations (ppb) at approx. 10 m above ground on the 25 May 1990, predicted with MUSE in

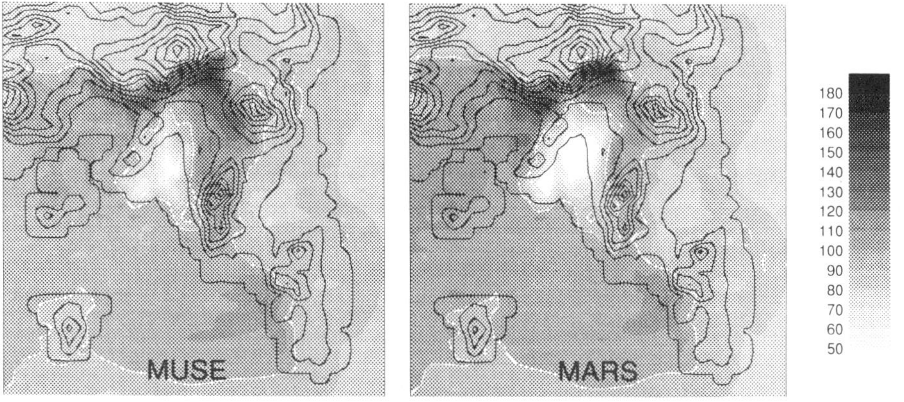

Figure 2. Spatial distribution of the maximum hourly ozone concentrations (ppb) at approx. 10 m above ground on 25 May 1990, predicted with MUSE (left) and MARS (right), both in conjunction with the KOREM reaction mechanism.

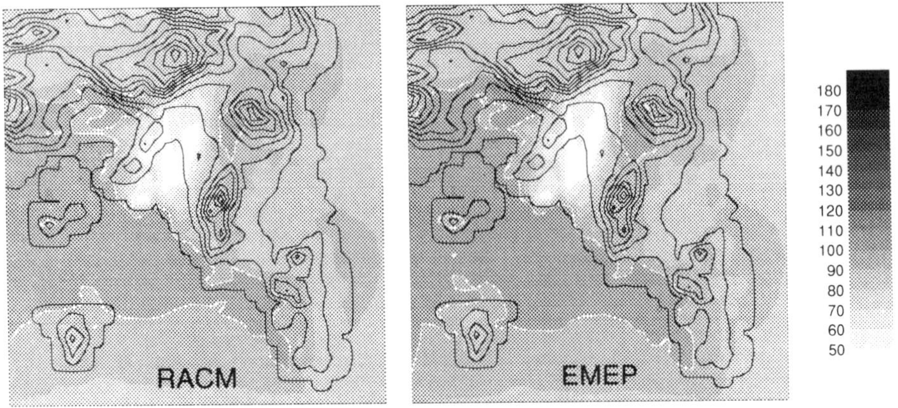

Figure 3. Spatial distribution of the maximum hourly ozone concentrations (ppb) at approx. 10 m above ground on 25 May 1990, predicted with MUSE in conjunction with the RACM reaction mechanism (left) and the EMEP reaction mechanism (right).

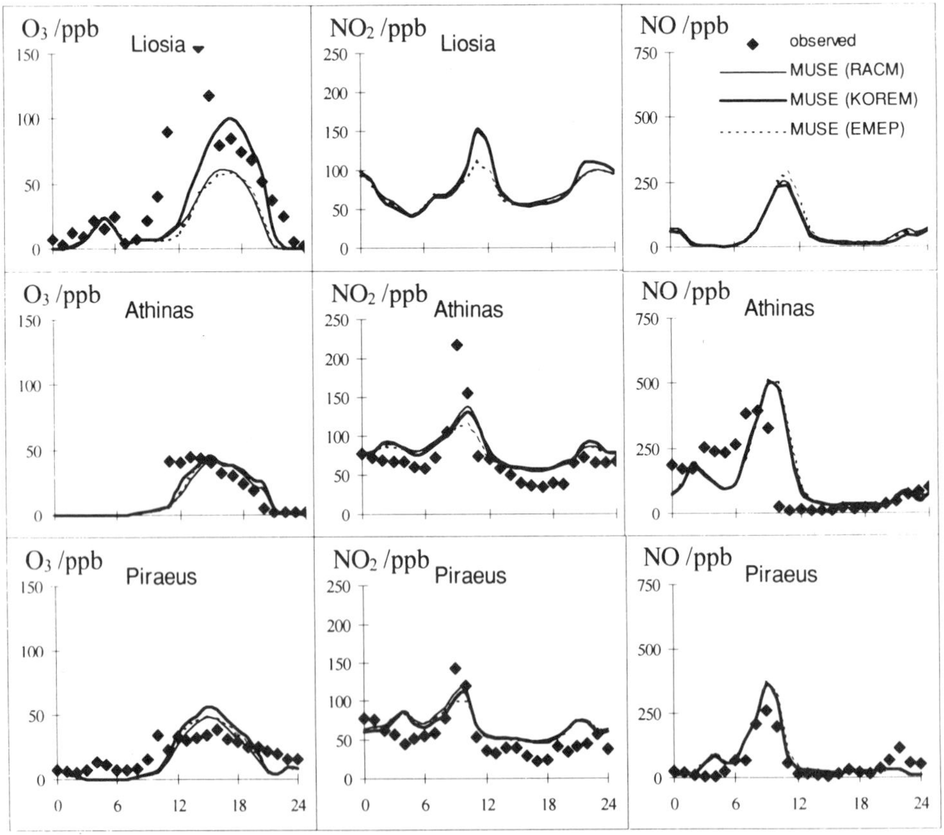

Figure 4. Ozone, NO_2 and NO concentrations predicted with MUSE in conjunction with the chemical reaction mechanisms KOREM (thick line), RACM (thin line) and EMEP (thin dotted line) compared with observations.

conjunction with the RACM reaction mechanism (left) and in conjunction with the EMEP reaction mechanism (right). The predicted ozone patterns as well as the peak ozone concentrations are similar for both mechanisms. These peak ozone concentrations differ, however, considerably from those predicted with the KOREM mechanism (cf. Figure 2, left).

In Figure 4 predictions of MUSE in conjunction with three different reaction mechanisms are compared with observed diurnal variations of the ozone, NO_2 and NO concentrations at three measuring stations in the Athens basin (the locations of these stations are indicated in Figure 1). It can be stated that the concentrations of ozone predicted with the mechanisms RACM and EMEP are remarkably similar. There is also a good agreement between these two mechanisms and KOREM for the stations Athinas and Piraeus where the NO_x concentrations are high. For Liosia where the consumption of ozone by NO is lower and the ozone production higher KOREM predicts much higher ozone concentrations than the other two mechanisms.

The higher ozone concentrations predicted with the KOREM mechanism in comparison with the mechanims EMEP and RACM might well be attributed to the cycle:

$$RO_2 + NO \rightarrow NO_2 + RCHO + HO_2$$
$$HO_2 + NO \rightarrow HO + NO_2$$
$$RCHO + HO \rightarrow 0.3 \cdot HO_2 + 0.3 \cdot RO_2 + 0.7 \cdot RCO_3 + 0.3 \cdot CO$$
$$RCO_3 + NO \rightarrow RO_2 + NO_2$$

After the initial step (i.e. $VOC + OH \rightarrow RO_2$), this cycle keeps the VOC in the system in their highest reactive form (RO_2 and aldehydes). In RACM and EMEP the aldehyde concentrations are lowered by considering the loss path:

$$ALD + HO \rightarrow RCO_3$$
$$RCO_3 + 2 \cdot NO \rightarrow HCHO + 2 \cdot NO_2 + HO_2$$
$$HCHO + HO \rightarrow CO + HO_2$$

Indeed, a comparison of the aldehydes predicted with the three reaction mechanisms shows significantly higher aldehyde concentrations for KOREM than for RACM and EMEP (cf. Figure 5). The differences between RACM and EMEP can be explained by the higher reactivity of the aldehyde producing precursors alkanes and alkenes in RACM.

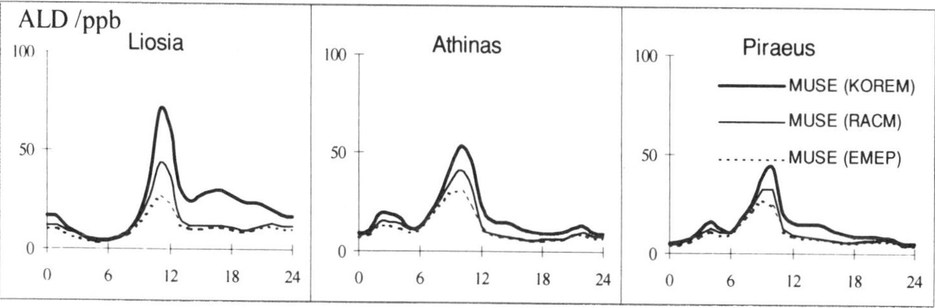

Figure 5. Aldehyde concentrations predicted with MUSE in conjunction with the chemical reaction mechanisms KOREM (thick line), RACM (thin line) and EMEP (thin dotted line).

In contrast to the good agreement in ozone and in the sum of NO_x between EMEP and RACM there are considerably differences in the NO_2/NO ratio, especially in the case of Liosia. At the other stations this effect is less pronounced but also existent. Differences in the photolysis rates and in the RO_2 chemistry may cause these deviations.

However, in conclusion it must be stated that the similarities between the reaction mechanisms with regard to the major air pollutants are greater than their differences, with the spatial distribution being rather similar in all. The response of the reaction mechanisms to changes in boundary conditions, however, has not been investigated yet.

CONCLUSIONS

The multilayer model MUSE has been applied to the Greater Athens area for meteorological conditions which favour the occurrence of air pollution episodes. The comparison of the simulation results with corresponding results of the full 3D photochemical dispersion model MARS shows that the model MUSE is capable of reproducing most features of the spatial and diurnal variation of the major photochemical air pollutants in the Greater Athens area. Hence, the model MUSE is a reasonable extension of the EZM system, as it may be used in conjunction with the prognostic mesoscale model MEMO for air quality studies on small scale computers (e. g. workstations).

In order to investigate the effect of the chemical reaction mechanism on air quality predictions, three different reaction mechanisms ranging from the compact mechanism KOREM to the comprehensive mechanisms EMEP and RACM have been compared. It was found that the pattern of the spatial distribution of the ozone values is rather similar for all three mechanisms but KOREM privides higher ozone maxima than RACM and EMEP. All mechanisms are in reasonable agreement to the avaiable field measurement results.

REFERENCES

Batchvarova E. and Gryning S.E. (1994), An applied model for the height of the daytime mixed layer and the entrainment zone, *Bound.-Layer Meterol.* **71**, 311-323.

Bottenheim J.W. and Strausz O.P. (1982), Modelling study of a chemically reactive power plant plume, *Atmos. Environ.* **16**, 85-97.

Deardorff J.W. (1974), Three-dimensional numerical study of the height and mean structure of the heated planetary boundary layer, *Bound.-Layer Meterol.* **7**, 81-106.

Hass H., Builtjes P.J.H., Simpson D., Stern R., Jakobs H.J., Memmesheimer M., Piekorz G., Roemer M., Esser P. and Reimer E. (1996), Comparison of Photo-oxidant Dispersion Model Results, EUROTRAC Special Report, ISS Garmisch-Partenkirchen.

Kessler Chr. (1995), Entwicklung eines effizienten Lösungsverfahrens zur Beschreibung der Ausbreitung und chemischen Umwandlung reaktiver Luftschadstoffe, Verlag Shaker, Aachen, pp. 148.

Kunz R. and Moussiopoulos N. (1995), Simulation of the Wind field in Athens Using Refined Boundary Conditions, *Atmos. Environ.* **29**, 3375-3591.

Moussiopoulos N. (1989), Mathematische Modellierung mesoskaliger Ausbreitung in der Atmosphäre, Fortschr.-Ber. VDI Reihe 15 Nr. 64, VDI-Verlag Düsseldorf, pp. 316.

Moussiopoulos N., ed. (1993), Special issue on APSIS, *Environ. Software* **8**, 1-71.

Moussiopoulos N., ed. (1995), Special issue on APSIS, *Atmos. Environ.* **29**, 3573-3728.

Moussiopoulos N. and Papagrigoriou S., eds. (1997), The Athens 2004 Air Quality Study, Proceedings of the International Workshop, Athens, 18-19 February 1997, in press.

Moussiopoulos N., Sahm P. and Kessler Ch. (1995), Numerical simulation of photochemical smog formation in Athens, Greece - A case study, *Atmos. Environ.* **29**, 3619-3632.

Poppe D., Andersson-Sköld Y.A., Baart A., Builtjes P.J.H., Das M., Fiedler F., Hov Ø, Kirchner F., Kuhn M., Makar P.A., Milford J.B,. Roemer M.G.M., Ruhnke R., Simpson D., Stockwell W.R., Strand A., Vogel B. and Vogel H. (1996), Intercomparison of the gas-phase chemistry of several chemistry and transport models, EUROTRAC Special Report, ISS Garmisch-Partenkirchen.

Simpson D. (1995), Biogenic emissions in Europe 2. Implications for control strategies, *J. Geophys. Res.* **100**, D11, 22891-22906.

Simpson D. (1996), private communications.

Simpson D., Andersson-Sköld Y.A. and Jenkin M.E. (1993), Updating the chemical scheme for the EMEP MSC-W oxidant model: current status. EMEP MSC-W Note 2/93, The Norwegian Meteorological Institute, Oslo, Norway.

Smolarkiewicz P.K. (1984), A fully multidimensional positive definite advection transport algorithm with small implicit diffusion, *J. comput. Phys.* **54**, 325.

Stockwell W.R., Kirchner F., Kuhn M. and Seefeld S. (1997), A new mechanism for regional atmospheric chemistry modeling, *J. Geophys. Res.*, (accepted for publication).

DISCUSSION

R. BORNSTEIN: The change from UAM-IV to UAM-V overcome two problems:

1. K-fields could now include advective effects
2. late afternoon boundary layer collapse no longer compressed concentrations, as horizontal outflow was allowed.

Does MUSE have the same limitations that UAM-IV had, as they both have the same basic formulations?

F. KIRCHNER:

1. K-fields are taken from previous applications of the wind field model MEMO. MEMO is a full 3D non-hydrostatic mesoscale model and includes advective effects of the turbulent kinetic energy.
2. In MUSE all layers are permeable. With the increase of the mixing height and thus the increase of layer thickness, concentrations are entrained from upper layers. The opposite occurs in the late afternoon: With the collapse of the mixing height and thus the layer thickness, concentrations may remain in its original height and thus in higher layers (Entrainment in upper layers).

D. W. BYUN: I am troubled by the subjective approach you used to evaluate different mechanisms. Chemical mechanisms used in current air quality models are very complex. So, we need rely on quantitative methods that characterize process budgets and integrated production and loss of photochemical species. I think it is time to use new tools available to use to analyse what model is actually computing rather than relying on the subjective comparison method. Basically you can use process Analysis/Intgrated Reaction Rate to objectively analyse differences among different chemical mechanisms.

F. KIRCHNER: The aim of our paper was not a complete mechanism evaluation. For a special ozone episode we investigated the dependence of the results upon the used chemical mechanisms and the used models and determined the causes for the most important differences. Clearly, a comprehensive mechanism evaluation would require additional tools.

THE TREATMENT OF RELATIVE DISPERSION WITHIN A COMBINED PUFF-PARTICLE MODEL (PPM)

Peter de Haan and Mathias W. Rotach

Swiss Federal Institute of Technology, GIETH
Winterthurerstr. 190, 8057 Zurich, Switzerland

Abstract—The Puff-Particle Model (PPM) combines the advantages of both, puff and particle dispersion models. In short, in this approach the centre of mass of each puff is moved along a 'particle trajectory', so trying to mimic the quickly changing turbulent flow field. However, particle models account for dispersion of turbulent eddies of all sizes (1-particle statistics, absolute dispersion) while puff models use relative dispersion to describe the puff growth. Therefore, on combining these two approaches as described above, the dispersing effect of small eddies (smaller than approximately the puff's size) is accounted for twice. A method is therefore presented to correctly simulate the relative dispersion of puffs within the framework of the PPM. It is based on removing the effect of the high-frequency part of the spectrum when using a 'particle trajectory' as the trajectory of the puff centre. It is shown on the basis of tracer data, that the correct treatment and interpretation of the two contributions to the dispersion process is crucial for reproducing experimental results to a good correspondence.

CONCEPT OF THE PPM

One of the major advantages of puff models is their ability to simulate inhomogeneous and instationary conditions. In principle this allows for concentration predictions of a 'sudden release' of pollutants from a source. In such cases relative dispersion must be used to describe the growth of the puff. Relative dispersion only takes into account the dispersing effect of those turbulent eddies which are able to enlarge the size of the puff and do not move the puff as a whole without dispersing it. The use of absolute dispersion (like the widely used parametrizations of Pasquill and Turner), which accounts for the dispersing effect of turbulent eddies of any size, leads to underpredicted concentrations relatively close to the source due to too large a dispersion.

However, to correctly describe the dispersing effect due to the meandering of the released plume, a rather frequent updating of the flow field information is necessary. In principle, the updating of the flow field should resolve all turbulent eddies larger than the 'size' of the puff (characterized by its standard deviations). Since in practice such a frequent updating of

Air Pollution Modeling and Its Application XII
Edited by Sven-Erik Gryning and Nadine Chaumerliac, Plenum Press, New York, 1998

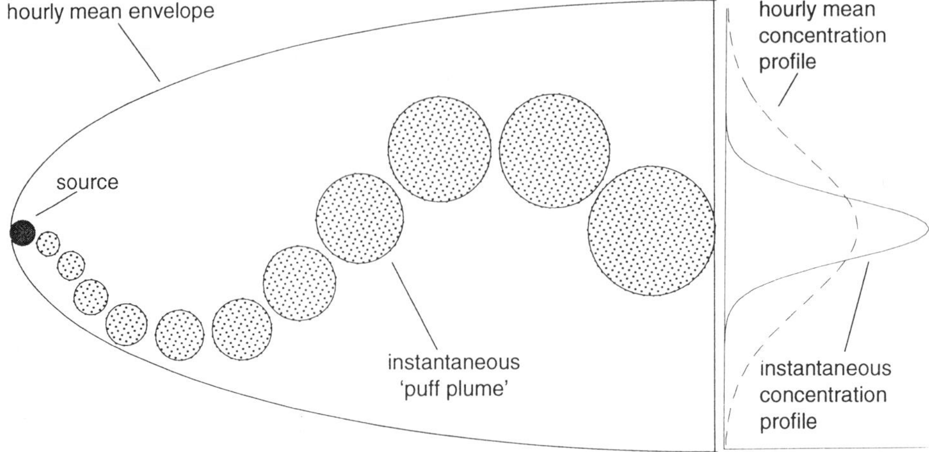

Figure 1. Conceptual sketch of the principle of the PPM. The meandering of the instantaneous 'puff plume' is modelled by moving the centre of mass of the puffs along a particle trajectory.

the flow field and meteorological information is hardly possible, the puff-particle approach (de Haan and Rotach, 1995) aims at simulating the dispersing effect of plume meandering by introducing puff centre trajectories (see Fig. 1). These trajectories are determined by the low-frequency part of the turbulence spectrum, since relative dispersion only describes the effects of the high-frequency part. Clearly, as the puff's size grows, the relative dispersion covers an increasing part of the spectrum. Therefore, the trajectory of the puff's centre of mass has to simulate the effect of a decreasing amount of turbulent eddies, and thus has to become 'smoother' as puff sizes grow.

RELATIVE VS. ABSOLUTE DISPERSION

Relative dispersion corresponds to the expansion of a cluster of particles. The spread σ of an ensemble of marked passive particles from each other is

$$\sigma^2(t) = \overline{\langle (\int_0^t \upsilon(\xi) d\xi)^2 \rangle} \tag{1}$$

where $\upsilon = u - V_{cm}$, u is the absolute velocity of the particles and the velocity of the centre of mass of the cluster is denoted by V_{cm}. The overbar in Eq. (1) and hereafter denotes the average over all the particles within the puff, and the angular brackets refer to an ensemble average. In the concept of absolute dispersion, on the other hand, the spread σ is

$$\sigma^2(t) = \overline{\langle (\int_0^t u(\xi) d\xi)^2 \rangle}. \tag{2}$$

For example, a turbulent eddy larger than the cluster of particles will displace the cluster as a whole. This will increase absolute dispersion, whereas the relative dispersion remains unchanged. From this it becomes clear that when using a particle model to simulate the meandering of a puff, as a surrogate for a frequently updated flow field, part of the spectrum is accounted for twice. This would cause the total dispersion to become overestimated more and more as the travel distance increases, leading to underestimated ground level concentrations far away from the source.

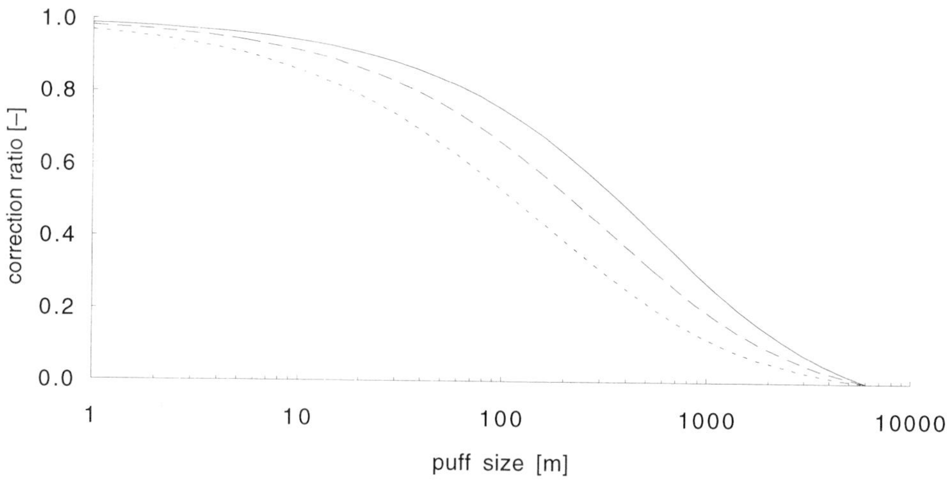

Figure 2. Plot of the correction ratios r_i $(i = u, v, w)$ (u-component: solid line; v-component: dashed line; w-component: dotted line) as a function of puff size (conditions of forced convection; $u_* = 0.4\,\mathrm{m \cdot s^{-1}}$, $z/L = -2$, $z_i = 1000\,\mathrm{m}$, $\bar{u} = 3.4\,\mathrm{m \cdot s^{-1}}$, averaging time of flow field one hour).

To separate the contribution of small eddies (contributing to an increase in puff size) for those of larger eddies (contributing to the meandering of the whole puff) the following procedure is introduced. From the actual puff sizes (i. e., the standard deviations) σ_x, σ_y, σ_z, a threshold frequency $n^*_{x,y,z} = \bar{u}(t)/(2\sigma_{x,y,z})$ is defined for each direction using Taylor's frozen turbulence hypothesis. The integral over the low frequency part of the turbulence spectrum is denoted as $\langle u_i^2 \rangle_{\mathrm{eff}}$ $(i = u, v, w)$, where $\langle u_i^2 \rangle_{\mathrm{eff}} = \int_{n_{\min}}^{n^*} S_i \, \mathrm{d}n$. Then, the ratio $r_i = \langle u_i^2 \rangle_{\mathrm{eff}} / \langle u_i^2 \rangle$ is determined (Fig. 2). The integration of the whole spectrum runs from $n_{\max} = \bar{u}/\eta$, where $\eta = (\upsilon^3/\varepsilon)^{1/4}$ is the Kolmogorov micro-scale, υ is the kinematic molecular viscosity and ε is the dissipation rate, to $n_{\min} = 1/T$, where T is the averaging time for the measurements of the turbulence statistics of the flow field.

In the present work velocity spectra models are taken from Højstrup (1981) for the unstable surface layer and Højstrup (1982) for the unstable planetary boundary layer. The model of Kaimal *et al.* (1972) is used for the neutrally stratified surface layer. For the stable surface layer, the model of Olesen *et al.* (1984) is adopted. For the upper part of the neutral and of the stable boundary layer, the same formulations as for the surface layer are used.

REDUCTION OF THE TRAJECTORY VARIABILITY AS PUFF SIZES GROW

The combination of a puff model with a particle model representing the whole turbulence spectrum asks for the removal of the dispersing effect from the high-frequency part of the energy spectrum (i. e., the small turbulent eddies already covered by the puff model) from the particle trajectories. The procedure to remove this part of the turbulent fluctuations from the 'particle-part' of the PPM is straightforward: The time series of stochastic turbulent velocity components of each particle is smoothed. The trajectory of the centre of mass of a puff is then calculated based on these smoothed turbulent velocities. This leads to an increasingly smooth puff centre trajectory as the puff size grows.

The Kalman filter originates from the need to estimate the true value of an underlying stochastic process which can only be observed with an error, where it is assumed that the observational error is normally distributed with a standard deviation τ. Under such circumstances there is a need not to consider the measured time series of, for example, pressure, but to filter out the noise signal originating from the observational error, thus

obtaining a smoothed time series as an improved, less fluctuating estimation of the quantity which originally had been measured. Often running mean values are used for such purposes, where the smoothed value is calculated as a weighted function of several values at both sides of the position for which a smoothed value has to be estimated. These running mean values have two disadvantages with respect to the present need for smoothing the turbulent velocities calculated by a particle model: first, the future values of the turbulent velocity are not known at the moment where a smoothed value has to be estimated for use for the puff centre trajectory. This way, running mean estimations could only be based on past values of the turbulent velocities, leading to a biased estimation. Second, such a running mean makes it necessary to store an increasing number of 'past values' of the turbulent velocities of each particle, since the running mean is based on these past values.
The Kalman filter procedure, on the other hand, is computed recursively. This way, the filter estimation at the time step t is only based on the filter estimate at the time step $t-1$, the measured value at time step t and the parameters of the underlying stochastic process. Stochastic particle dispersion models generally are modelled as a so-called AR(1)-process, i. e. an auto-regressive process in which the position and velocity of a particle only depend on the velocity and position of the same particle one time-step ago.
If the underlying AR(1)-process is

$$X(t) = \alpha \cdot X(t-1) + E(t) \tag{3}$$

where X denotes the turbulent velocity vector (u', v', w'). The stochastic process from Eq. (3) has a part correlated with the turbulent velocities at the preceding time step and a uncorrelated stochastic distribution $E(t)$ which is normally distributed with zero mean and variance σ^2. The process given by Eq. (3) is our underlying, 'true' process. We assume it to be observed with a normally distributed observational error with variance τ^2. The Kalman filter estimate $\hat{X}$ for the next time step then is

$$\hat{X}_{t+1,t} = \alpha \cdot \hat{X}_{t,t} \tag{4}$$

where the filter probability density has a variance

$$R_{t+1,t} = \alpha^2 R_{t,t} + \sigma^2 \tag{5}$$

Here, the first subscript gives the time step for which the filter estimate is valid, and the second subscript indicates the time step of the last observation on which this filter estimate is based. Given the new observation at time step $t+1$, $X(t+1)$, the filter estimate is corrected to

$$\hat{X}_{t+1,t+1} = \hat{X}_{t+1,t} + \frac{R_{t+1,t}}{R_{t+1,t} + \tau^2}\left(X_{t+1} - \hat{X}_{t+1,t}\right) \tag{6}$$

with the new variance

$$R_{t+1,t+1} = \frac{R_{t+1,t} \cdot \tau^2}{T_{t+1,t} + \tau^2} \tag{7}$$

This recursive algorithm can be started as soon as initial values $\hat{X}_{0,0}$ and $R_{0,0}$ are chosen, the influence of which vanishes after just a few iterations.
The concept used within the puff-particle approach is to consider the turbulent velocity

components computed by the particle-part of the model to be 'observed values', where the 'observational error' is normally distributed with mean zero and standard deviation τ. By choosing τ proportional to the correction ratios r_i (and thus proportional to the growth of the size of the puff), increasingly smoothed time series of turbulent velocities are obtained. These smoothed velocity components are then used to calculate the trajectory of the puff's centre of mass. In the beginning, i. e. when the puff size is small and r_i approximately equals unity, τ is chosen to be zero. As puff sizes increase and the r_i decrease (Fig. 2), the high-frequency fluctuations of the turbulent velocity components are eliminated from the trajectory of the puff's centre (Fig. 3). This corresponds to the concept of relative diffusion where all fluctuations originating from turbulent eddies with sizes smaller than the size of the puff are taken into account. As the puff size further increases, the r_i eventually drop to zero, and τ is chosen in such a way that the smoothed turbulent velocity time series remain constant.

Between the two limiting cases ($\tau_i = 0$ for $r_i = 1$ and τ_i large for $r_i = 0$),

$$\tau_i(r_i) = c\sigma \cdot \text{logit}\{d(0.5 - r_i)\} \tag{8}$$

is chosen in the PPM, where $\text{logit}(x) = \exp(x)/\{1 + \exp(x)\}$. In the present work, $c = 100$ and $d = 10$ are used. The ratios r_i are evaluated for each puff individually. As puff sizes grow, the values of the r_i decrease and eventually approach zero, leading to turbulent movements of the puff centres of very low frequency only. This allows the particle model within the PPM to be switched off as soon as r_i equals unity, and the puff centres are moved by the average flow field only. For far-field concentration predictions, this causes considerable computational savings.

Even when τ is large, the smoothed velocity fluctuation of a stochastic process will not give exactly zero. Therefore, to ensure that the smoothed turbulent velocities approach zero as $r_i \to 0$, the smoothed values are forced towards zero as soon as $r_i > 0.9$.

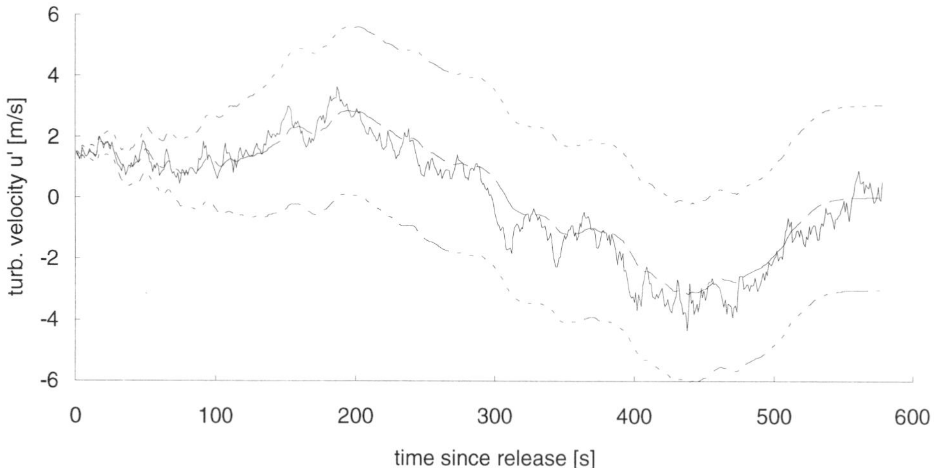

Figure 3. Smoothing of the turbulent velocity components calculated by the particle model (solid line) for use as puff centre velocities in the PPM (dashed line). The smoothing is increased as the correction ratios drop, i. e. the puff size grows. The dotted lines below and above of the smoothed turbulent velocity indicate the range $[u' - \tau, u' + \tau]$. Example for the u-component.

VALIDATION

For the prediction of concentrations averaged over approximately one hour, the puff-particle approach with smoothed puff centre trajectories should yield similar concentrations as a dispersion model based on absolute dispersion (e. g., a particle model). Therefore, the validation in this section aims at comparing the ground-level concentrations of four different models. The first model which is to be compared is a 'pure' particle model after Rotach *et al.* (1996) which fulfils the well-mixed criterion of Thomson (1987). The second and third model in this comparison are the PPM without and with additionally smoothed puff centre trajectories, respectively. The fourth model is a 'pure' puff model, in which the puffs are dispersed with relative dispersion. A second validation of the performance of the PPM with smoothed puff centre trajectories, against the data from three tracer experiments under different atmospheric conditions, can be found in de Haan and Rotach (1997).

The comparison of the predicted ground-level concentrations of these four models allows to validate the treatment of relative dispersion within the PPM. If the averaging time is one hour, the stochastic puff centre trajectories will account for almost the whole dispersing effect of plume meandering. Therefore the total dispersion will equal absolute dispersion, and this leads to comparable ground-level concentrations predictions of both, the 'pure' particle model on the one side and the PPM with the smoothed puff centre trajectories on the other side. The PPM without the smoothing procedure, on the other hand, is expected to give too low concentration predictions, since a increasing part of the energy spectrum is taking into account twice, leading to an overestimation of the total dispersion. At the opposite, the 'pure' puff model does only cover part of the energy spectrum, leading to underestimated dispersion and hence to overpredicted ground-level concentrations.

To validate the predictions of the different models against measurements, the data from the Copenhagen tracer experiment are used. Data from 9 hours of measurements under conditions of forced convection are available. The non-buoyant tracer was released over a suburban surface at a height of 115 m. The receptors were placed on several arcs downwind of the source and hourly average measurements of the tracer concentrations were made at 2 m above ground. The mean wind speed was measured at different levels. The mixing height, the friction velocity and the Obukhov length were measured close to the release point, and measurements of the velocity statistics, $\langle u'^2 \rangle$ and $\langle w'^2 \rangle$, are available. More details about the experiment can be found in Gryning and Lyck (1984).

The simulation of the tracer experiment was performed by rebuilding the experimental set-up within the model and by predicting the concentrations at arcs of receptors at the same locations as in the tracer experiment. This allows the calculation of the cross-wind integrated concentration (CIC), the standard deviation on the arc in meters (SIGY) and of the maximum concentration occurring on an arc (ArcMax). Additionally, the ground-level concentrations in the plume centre line are predicted.

As an example, the resulting concentration profiles down-wind from the source for one of the hours of the tracer experiment are depicted in Fig. 4. The poor performance of the 'pure' puff model is completely due to the underestimated total dispersion, causing the maximum concentration to occur further down-wind and to remain on a high level as compared to the other three models as well as with the experimental data. The other three models all show very similar concentration patterns close to the source. This is due to the fact that in this early stage the relative dispersion of the puffs is only a minor contribution to the total dispersion, and also due to the fact that the 'pure' particle model is identical with the particle-part of the PPM used to calculate the stochastic puff centre trajectories. The removal of part of the dispersing effect of these stochastic trajectories in the third model, the PPM with smoothed puff trajectories, leads to a larger predicted maximum concentration. It is similar to the maximum concentration as predicted by the 'pure' particle model. The PPM without smoothing procedure, on the other hand, predicts an about ten

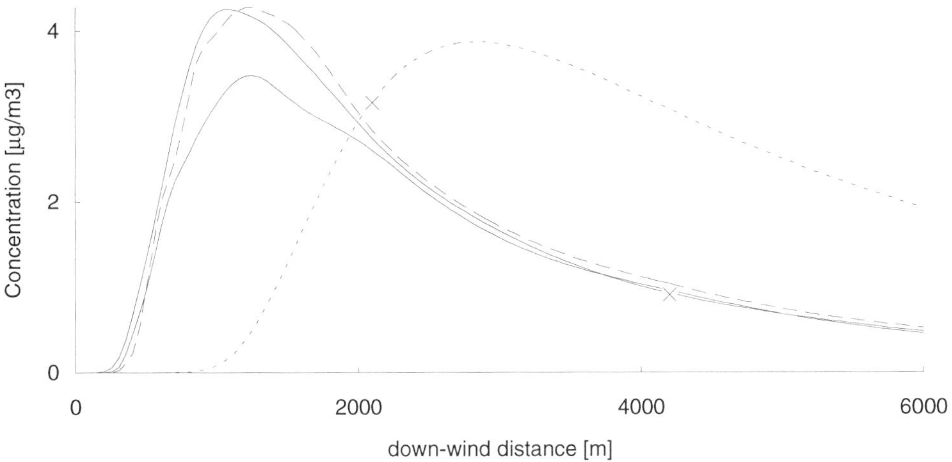

Figure 4. Example of the plume centre line concentration at 2 m above the ground for the particle model (dashed line), the puff-particle model without and with Kalman filter (lower and upper solid line, respectively) as well as for the puff model (dotted line). Crosses depict the measurements (Copenhagen experiment from Sep. 26, 1978).

percent lower maximum concentration, due to the overestimation of total dispersion. Further down-wind, the concentration predictions of the particle model and of the smoothed PPM remain similar, whereas the PPM without smoothing simulates somewhat lower concentrations.

In Table 1 the following statistical measures are compared for the different simulations of the tracer experiments: the fractional bias $FB = (\overline{c}_{obs.} - \overline{c}_{pred.})/\{0.5(\overline{c}_{obs.} + \overline{c}_{pred.})\}$; the normalised mean square error $NMSE = \overline{(c_{obs.} - c_{pred.})^2}/(\overline{c}_{obs.}\overline{c}_{pred.})$; the correlation coefficient $COR = \overline{(c_{obs.} - \overline{c}_{obs.})(c_{pred.} - \overline{c}_{pred.})}/(\sigma_{obs.}\sigma_{pred.})$; the percentage of simulations within a factor of two of the measurement, FAC2. $c_{obs.}$ is the observed, $c_{pred.}$ the simulated concentration. As can be seen, the overall performance of the 'pure' particle model is better than the measures of the other three models. However, the smoothed PPM shows almost identical FB and NMSE. Additionally, the measures of the smoothed PPM are clearly improved as compared with the PPM without smoothing procedure. The puff model, finally, shows a rather poor performance. Even the particle model, however, shows a general underprediction (a positive fractional bias). In Rotach and de Haan (1996) it is shown that taking into account the rough character of the suburban area where the tracer experiment took place

Table 1. Comparison of statistical measures (see text) for the different simulations.

	ArcMax				CIC				SIGY			
	NMSE	COR	FAC2	FB	NMSE	COR	FAC2	FB	NMSE	COR	FAC2	FB
observations	0.000	1.000	100%	0.000	0.000	1.000	100%	0.000	0.000	1.000	100%	0.000
Particle model	0.137	0.871	91%	0.029	0.189	0.700	87%	0.197	0.088	0.836	100%	0.006
PPM (Kalman f.)	0.186	0.858	87%	0.050	0.179	0.644	83%	0.123	0.091	0.879	100%	–0.079
PPM (no filter)	0.218	0.852	78%	0.047	0.180	0.654	83%	0.134	0.103	0.901	95%	–0.104
Puff model	1.687	0.255	30%	–0.470	1.057	0.029	39%	–0.124	0.388	0.947	60%	0.569

leads to the vanishing of this systematic underprediction.
It must be noted that the data from the Copenhagen experiment do not allow to validate one of the most important effects of the smoothing procedure within the PPM. The predicted maximum down-wind concentration closely resembles the predicted values from the 'pure' particle model, whereas the PPM without correction predicts lower maximum concentrations. Unfortunately, all arcs in the Copenhagen experiment were placed at relatively large distances from the source, so that the maximum concentration was not observed.

SUMMARY AND CONCLUSIONS

The puff-particle approach is suited to simulate the effect of plume meandering in absence of frequently updated meteorological and flow field information. However, the combination of a puff and a particle model leads to a double representation of the dispersing effect of part of the turbulence spectrum, dependent on the size of the individual puffs. In the present contribution, a method is proposed which corrects this overestimation of dispersion by filtering out the high-frequency part of the changes of turbulent velocity within the particle model. This filtering is realised using a Kalman filter. The extent of the smoothing depends on the proportion of the velocity spectrum of which the dispersing effect is taken into account by the puff part of the model. This way, the smoothing increases as puffs sizes grow. The validation against data from a tracer experiment and against the results of a pure particle model show than the corrected PPM (i. e., with smoothed puff centre trajectories) does not over- nor underpredict the total dispersion and shows approximately the same results as does the particle model.

Acknowledgments— The present work was partly financed by the Swiss Federal Department of Education and Sciences (BBW) and the Swiss Federal Department of Environment, Forest and Landscape (BUWAL) through a project in the framework of COST 615 (citair), and by the Swiss National Science Foundation through Grant Nr. 21-46849.96.

REFERENCES

de Haan, P., and Rotach, M. W. (1995): 'A puff-particle dispersion model', *Int. J. Environment and Pollution,* **5**, Nos. 4–6, 350–359.

de Haan, P., and Rotach, M. W. (1997): 'A novel approach to atmospheric dispersion modelling: the Puff-Particle Model (PPM)', submitted to the *Quarterly Journal of the Royal Meteorological Society.*

Gryning, S.-E., and Lyck, E. (1984): 'Atmospheric Dispersion from Elevated Sources in an Urban Area: Comparisons between Tracer Experiments and Model Calculations', *J. Clim. Appl. Met.*, **23**, 651–660.

Højstrup, J. (1981): 'A Simple Model for the Adjustment of Velocity Spectra in Unstable Conditions Downstream of an Abrupt Change in Roughness and Heat Flux', *Boundary-Layer Meteorol.,* **21**, 341–356.

Højstrup, J. (1982): 'Velocity spectra in the Unstable Planetary Boundary Layer', *J. Atmos. Sci.,* **39**, 2239–2248.

Kaimal , J. C., Wyngaard, J. C., Coté, O. R. and Izumi, Y. (1972): 'Spectral characteristics of surface layer turbulence', *Quart. J. Roy. Met. Soc.*, **98**, 563–589.

Olesen, H. R., Larsen, S. E. and Højstrup, J. (1984): 'Modelling velocity spectra in the lower part of the planetary boundary layer', *Boundary-Layer Meteorol.,* **29**, 285–312.

Rotach, M. W., Gryning, S.-E. and Tassone, C. (1996): 'A Two-Dimensional Stochastic Lagrangian Dispersion Model for Daytime Conditions', *Quart. J. Roy. Met. Soc.,* **122**, 367–389.

Rotach, M. W. and de Haan, P. (1996): 'On the Urban Aspect of the Copenhagen Data Set', Pre-prints 4th Workshop on Harmonisation within Atm. Dispersion Models, May 5–9, Oostende, Belgium, 1996.

Thomson, D. J. (1987): 'Criteria for the Selection of Stochastic Models of Particle Trajectories in Turbulent Flows', *J. Fluid Mech.*, **180**, 529–556.

DISCUSSION

T. MIKKELSEN: The Copenhagen experiments you refer to are maybe not the best suited data set for evaluation of your model, because these experiments (of twenty minutes to one hour sampling time) are well described by absolute diffusion (G. I. Taylor, 1921)?

P. de HAAN: This is true. However, the use of the Copenhagen experiment allows the comparison with the results given by a Lagrangian particle model. This way, we are able to compare the predicted maximum ground level concentration of PPM and the particle model. In the future, we will validate the PPM using experiments with averaging times shorter than one hour.

T. MIKKELSEN: Have you investigated the sensitivity of your model to your assumed form of the "logit"-function, and can you reproduce the classical $\sigma^2 \sim \varepsilon T^3$-regime?

P. de HAAN: The low-pass filtering effect caused by the Kalman filter with the "smoothing width" described by the "logit"-function should closely correspond to the high-frequency part of the spectrum (all frequencies higher than the frequency f^* attributed to the puff). Therefore, the values of the coefficients c and d depend on the model of the velocity spectrum used. The values of c and d presented here were determined empirically, taking care of the above considerations. A more accurate way would be to determine the cut-off frequency belonging to the filtering with the Kalman filter using a Fourier transformation, and relating this frequency to f^*.

The $\sigma^2 \sim \varepsilon T^3$-regime has to do only with the relative diffusion of a Gaussian puff. The parametrisation of the relative diffusion used is, for intermediate travel times, based on Batchelor's $\sigma^2 \sim \varepsilon T^3$-rule (see de Haan and Rotach, 1997, for details).

R. BORNSTEIN: Is your "absolute dispersion" really the sum of relative diffusion (due to eddies smaller than a puff or plume) plus single-point diffusion (due to eddies about the size of the puff/plume) plus transport (due to eddies larger than the puff/plume), and is your technique the correct removal of the single point diffusion effect from the transport component of the motion?

P. de HAAN: I believe that what we call "meandering of the puff centre" relative to a fixed point and meandering of the plume center line relative to a fixed axis corresponds to the sum of what you refer to as single-point diffusion and transport. The technique presented removes the relative diffusion part from the stochastic velocity components of the puff's center, leaving both the single point diffusion and the transport component.

D. CARRUTHERS: The combined Puff-Particle Model has a stochastic element in it although this is less than in a particle model. How many realisations are necessary to calculate ensemble means? How long does the run take?

P. de HAAN: Typically, the three-dimensional ensemble average is taken over 1000 realisations, compared to 20.000 to 50.000 realisations used in the pure particle model. On a Unix Sparc 5 workstation, the simulation of a tracer experiment such as Copenhagen (1000 realisations) takes about 30 minutes.

M. KAASIK: You successfully performed the comparison with the Copenhagen data set. Did you try to compare your model with the Lillestrøm data set?

P. de HAAN: Yes. The NMSE is 0.63, the correlation coefficient 0.75. The fractional bias is 0.08, and 70% of all predictions are within a factor of two of the measurements.

MODELING THE EFFECTS OF URBAN VEGETATION ON AIR POLLUTION

David J. Nowak,[1] Patrick J. McHale,[2] Myriam Ibarra,[1] Daniel Crane,[1] Jack C. Stevens,[1] and Chris J. Luley[3]

[1]USDA Forest Service, Northeastern Forest Experiment Station, Syracuse, NY 13210
[2]State University of New York, College of Environmental Science and Forestry, Syracuse, NY 13210
[3]ACRT Inc., Cuyahoga Falls, OH 44221

INTRODUCTION

Urban vegetation can directly and indirectly affect local and regional air quality by altering the urban atmospheric environment. Trees affect local air temperature by transpiring water through their leaves, by blocking solar radiation (tree shade), which reduces radiation absorption and heat storage by various anthropogenic surfaces (e.g., buildings, roads), and by altering wind characteristics that affect air dispersion. During the summertime, trees predominantly reduce local air temperatures, but may increase within- and below-canopy air temperature due to reduced turbulent exchange with above-canopy air (Heisler et al., 1995). Reduced air temperature due to trees can improve air quality because the emission of many pollutants and/or precursor chemicals are temperature dependent. Decreased air temperature can also reduce ozone (O_3) formation (Cardelino and Chameides, 1990).

Besides affecting air temperature, the physical mass, water transpiration, and thermal/radiative properties of trees can affect wind speed, relative humidity, turbulence, and surface albedo. In addition, trees affect surface roughness and consequently the evolution of the mixing-layer height, which in turn affects O_3 formation (Berman et al., in press). These changes in local meteorology can alter pollution concentrations in urban areas.

Trees remove gaseous air pollution primarily by uptake via leaf stomata, though some gases are removed by the plant surface (Smith, 1990). Once inside the leaf, gases diffuse into intercellular spaces and may be absorbed by water films to form acids or react with inner-leaf surfaces. Trees also remove pollution by intercepting airborne particles. Some particles can be absorbed into the tree (e.g., Zeigler, 1973), though most particles that are intercepted are retained on the plant surface. The intercepted particle often is resuspended to the atmosphere, washed off by rain, or dropped to the ground with leaf and twig fall. Consequently, vegetation is only a temporary retention site for many atmospheric particles.

Emissions of volatile organic compounds (VOCs) by trees can contribute to the formation of O_3 and carbon monoxide (CO) (Brasseur and Chatfield, 1991). Because VOC emissions are temperature dependent and trees generally lower air temperatures, increased tree cover can lower overall VOC emissions and, consequently, O_3 levels in urban areas (Cardelino and Chameides, 1990).

Trees reduce building energy use by lowering temperatures and shading buildings during the summer, and blocking winds in winter (e.g., Heisler, 1986). However, they also increase energy use by shading buildings in winter, and may increase or decrease energy use by blocking summer breezes. Thus, proper tree placement near buildings is critical to achieve maximum building energy conservation benefits.

When building energy use is lowered, pollutant emissions from power plants are also lowered. While lower pollutant emissions generally improve air quality, lower NO_x emissions, particularly ground-level emissions, may lead to a local increase in O_3 concentrations under certain conditions due to NO_x scavenging of O_3 (Rao and Mount, 1994). The cumulative and interactive effects of trees on meteorology, pollution removal, and VOC and power plant emissions determine the overall impact of trees on air pollution.

Modeling Urban Vegetation Effects on Air Quality

Research integrating the cumulative effects of urban vegetation on air quality, particularly ozone, is limited. Cardelino and Chameides (1990) modeled vegetation effects on ozone concentrations in the Atlanta region using the OZIPM4 model. The study's primary focus was on the interaction of VOC emissions and altered air temperatures, and revealed that a 20 percent loss in the area's forest due to urbanization could have lead to a 14 percent increase in O_3 concentrations for June 4, 1984. Although there were fewer trees to emit VOCs, an increase in Atlanta's air temperatures due to the urban heat island, which occurred concomitantly with tree loss, increased VOC emissions from the remaining trees and anthropogenic sources, and altered O_3 photochemistry such that concentrations of O_3 increased.

A model simulation of California's South Coast Air Basin suggests that the air quality impacts of increased urban tree cover may be locally positive or negative. The net basin-wide effect of increased urban vegetation is a decrease in ozone concentrations if the additional trees are low VOC emitters (Taha, 1996). This study used the Colorado State University Mesoscale (CSUMM) and Urban Airshed Models (UAM-IV), and accounted for vegetation temperature reduction effects on altered chemical reaction rates, reduced temperature dependent biogenic VOC emissions, and changes in the depth of the mixed layer. It also accounted for increased pollution deposition and possible increased VOC emissions due to increased vegetative cover.

This paper is an overview of a current research project, funded by the National Urban and Community Forestry Advisory Council, that is investigating the cumulative and interactive effects of altered meteorology, dry deposition, VOC emissions, and power plant emissions, due to increased and decreased urban tree cover, on local and regional ozone concentrations. This study is focusing on four cities (Baltimore, MD; Boston, MA; New York, NY; and Philadelphia, PA) and the total urban megalopolis of Washington, DC, to Boston, MA. A new Urban Forest Effects (UFORE) model is also detailed, and preliminary results of pollution removal by trees in Philadelphia are presented.

MODELING METHODS

To determine base conditions in each city, approximately 210 stratified (by land use), random, 0.04-ha ground plots were measured to determine vegetative and artificial surface parameters. From these data, leaf-surface area and biomass were calculated (Nowak, 1996). Aerial photographs and satellite imagery were analyzed to determine the amount of tree cover and potential space for tree cover in the future. To model the effects of altered urban tree cover on ozone, three scenarios are being modeled: 1) Base Case: existing vegetation configuration, 2) No Tree: all tree cover removed and replaced with grass, and 3) High Tree: all non-tree cover grass space filled with trees.

To model the effects of altered tree cover on meteorological conditions, the Pennsylvania State University / National Center for Atmospheric Research Mesoscale Model 5 (MM5) is being used (Dudhia, 1993). To model the effects of urban trees on ozone removal (dry deposition), VOC emissions, and power plant emissions an Urban Forest Effects (UFORE) model was developed. Results of changes in meteorological conditions, deposition velocities, VOC emissions and power plant emissions due to altered urban tree cover will be input into the SARMAP air quality model (SAQM) (Pleim et al., 1991) to quantify the overall effect of altered tree cover on local and regional ozone concentrations for the ozone episode period of July 13-15, 1995. Project completion is expected in December 1997. Recently completed results revolve around the completion of the UFORE model, in particular, the dry deposition component.

The UFORE model currently has five modules in development or completed. UFORE-A: Anatomy of the Urban Forest, quantifies urban vegetation and artificial surface characteristics (e.g., species composition, tree condition, leaf surface area and biomass, tree biomass, ground cover distribution, artificial surface characteristics) by land use type based on field data inputs. UFORE-B: Biogenic VOC Emissions, calculates hourly isoprene, monoterpene, and other VOC (OVOC) emissions based on species leaf biomass calculations (from UFORE-A), National Climatic Data Center (NCDC) hourly weather data, base VOC emission factors, and emission temperature and light correction factors (Guenther et al., 1994). UFORE-C: Carbon Storage and Sequestration, determines the amount of carbon currently stored and annually sequestered by urban trees within each land use type (based on UFORE-A inputs). UFORE-D: Dry Deposition of Air Pollutants, calculates the hourly dry deposition of ozone, sulfur dioxide (SO_2), nitrogen dioxide (NO_2), and carbon monoxide, and daily deposition of particulate matter less than 10 microns (PM10) to tree canopies throughout the year based on tree cover data, NCDC weather data, and U.S. Environmental Protection Agency (EPA) pollution concentration monitoring data. UFORE-E: Energy Conservation and Power Plant Emission Effects, computes the effect of trees on local building energy use (based on UFORE-A data and published tree-energy effects data) and the consequent effect on local power plant emissions. The UFORE model is programmed in SAS. One of the most detailed modules is UFORE-D, which accounts for the interaction of the hourly variation in tree transpiration and pollution deposition (based on local meteorological data) and the hourly pollutant concentration data.

UFORE -D: Dry Deposition of Air Pollutants

In UFORE-D, the pollutant flux (F) is calculated as the product of the deposition velocity (V_d) and the pollutant concentration (C):

$$F\ (g\ m^{-2}\ s^{-1}) = V_d\ (m\ s^{-1}) \times C\ (g\ m^{-3}) \quad (1)$$

Deposition velocity is calculated as the inverse of the sum of the aerodynamic (R_a), quasi-laminar boundary layer (R_b) and canopy (R_c) resistances (Baldocchi et al., 1987):

$$V_d = (R_a + R_b + R_c)^{-1} \quad (2)$$

Hourly meteorological data from the NCDC are used in estimating R_a and R_b. The aerodynamic resistance is calculated as (Killus et al., 1984):

$$R_a = u(z)\, u_*^{-2} \quad (3)$$

where u(z) is the mean wind speed at height z (m s^{-1}) and u_* is the friction velocity (m s^{-1}).

$$u_* = (k\, u\, (z\text{-}d))\, [\ln ((z\text{-}d)\, z_o^{-1}) - \psi_M ((z\text{-}d)\, L^{-1}) + \psi_M (z_o\, L^{-1})]^{-1} \quad (4)$$

where k = von Karman constant, d = displacement height (m), z_o = roughness length (m), ψ_M = stability function for momentum, and L = Monin-Obuhkov stability length. L was estimated by classifying hourly local meteorological data into stability classes using Turner classes (Panofsky and Dutton, 1984) and then estimating 1/L as a function of stability class and z_o (Zannetti, 1990). When L<0 (unstable) (van Ulden and Holtslag, 1985):

$$\psi_M = 2 \ln [0.5(1+ X)] + \ln [0.5(1+ X^2)] - 2 \tan^{-1}(X) + 0.5\pi \quad (5)$$

where $X = (1 - 28\, z\, L^{-1})^{0.25}$ (Dyer and Bradley, 1982). When L>0 (stable conditions):

$$u_* = C_{DN} u\{0.5 + 0.5[1-(2u_o/(C_{DN}^{1/2} u))^2]^{1/2}\} \quad (6)$$

where $C_{DN} = k\, (\ln (z/z_o))^{-1}$; $u_o^2 = (4.7\, z\, g\, \theta_*)\, T^{-1}$; g = 9.81 m s^{-2}; $\theta_* = 0.09\, (1 - 0.5\, N^2)$; T = air temperature (K°); and N = fraction of opaque cloud cover (Venkatram, 1980; U.S. EPA, 1995). Under very stable conditions, u_* was calculated by scaling actual wind speed with a calculated minimum wind speed based on methods given in U.S. EPA (1995).

The quasi-laminar boundary-layer resistance was estimated as (Pederson et al., 1995):

$$R_b = 2(Sc)^{2/3}\, (Pr)^{-2/3}\, (k u_*)^{-1} \quad (7)$$

where k = von Karman constant, Sc = Schmidt number, and Pr is the Prandtl number.

In-leaf, hourly tree canopy resistances for ozone, sulfur dioxide, and nitrogen dioxide were calculated based on a hybrid of big-leaf and multi-layer canopy deposition models (Baldocchi et al., 1987; Baldocchi, 1988). Canopy resistance (R_c) has three components: stomatal resistance (r_s), mesophyll resistance (r_m), and cuticular resistance (r_t), such that:

$$1/R_c = 1/(r_s + r_m) + 1/r_t \quad (8)$$

Mesophyll resistance was set to zero for O_3 and SO_2 (Wesely, 1989), and 600 s m^{-1} for NO_2, to account for the difference between transport of water and NO_2 in the leaf interior and to bring the computed deposition velocities in the range typically exhibited for NO_2 (Lovett, 1994). Base cuticular resistances were set at 8,000 m s^{-1} for SO_2, 10,000 m s^{-1} for O_3, and 20,000 m s^{-1} for NO_2 to account for the typical variation in r_t exhibited among the pollutants (Lovett, 1994).

Hourly inputs to calculate canopy resistance are photosynthetic active radiation (PAR; μE m^{-2} s^{-1}), air temperature (K°), wind speed (m s^{-1}), u_* (m s^{-1}), carbon dioxide concentration (set to 360 ppm), and absolute humidity (kg m^{-3}). Air temperature, wind speed, u_*, and absolute humidity are measured directly or calculated from measured hourly NCDC meteorological data. Total solar radiation is calculated based on the National Renewable Energy Laboratory Meteorological / Statistical Solar Radiation Model (METSTAT) with inputs from the NCDC data set (Maxwell, 1994). PAR is calculated as 46 percent of total solar radiation input (Monteith and Unsworth, 1990).

As carbon monoxide and particulate matter removal by vegetation is not directly related to transpiration, R_c for CO was set to a constant for in-leaf season (50,000 s m^{-1}) and leaf-off season (1,000,000 s m^{-1}) based on data from Bidwell and Fraser (1972). For particles, the deposition velocity (based on average V_d from the literature) was set at 0.0064 m s^{-1} for the in-leaf season and 0.0014 m s^{-1} for the leaf-off season, both of which incorporate a 50 percent resuspension rate of particles back to the atmosphere (Zinke, 1967).

The model uses an urban tree leaf area index of 6, and a distribution of 90 percent deciduous and 10 percent coniferous leaf surface area (Nowak, 1994). Local leaf-on and leaf-off dates are input into the model so that deciduous tree transpiration and related pollution deposition are limited to the in-leaf period, and seasonal variation in removal can be illustrated for each pollutant. Particle collection and gaseous deposition on deciduous trees in winter assumed a surface-area index for bark of 1.7 (m^2 of bark per m^2 of ground surface covered by the tree crown) (Whittaker and Woodwell, 1967). To limit deposition estimates to periods of dry deposition, deposition velocities were set to zero during periods of precipitation.

Hourly pollution concentrations (ppm) for gaseous pollutants were obtained from the EPA. Hourly ppm values were converted to $\mu g\ m^{-3}$ based on measured atmospheric temperature and pressure (Seinfeld, 1986). Average daily concentrations of PM10 ($\mu g\ m^{-3}$) also were obtained from the EPA.

Average hourly pollutant flux (g m^{-2} of tree canopy coverage) among the pollutant monitor sites was multiplied by city tree canopy coverage (m^2) to estimate total hourly pollutant removal by trees across the city. Bounds of total tree removal of O_3, NO_2, SO_2, and PM10 were estimated using the typical range of published in-leaf dry deposition velocities (Lovett, 1994).

Monetary value of pollution removal by trees is estimated using the median externality values for the United States for each pollutant. The externality values are: NO_2 = \$6,750 t^{-1}, PM10 = \$4,500 t^{-1}, SO_2 = \$1,650 t^{-1}, and CO = \$950 t^{-1} (Murray et al., 1994). Externality values for O_3 were set to equal the value for NO_2.

To approximate boundary-layer heights in the study area, mixing-height measurements from a nearby station were used. Daily morning and afternoon mixing heights were interpolated to produce hourly values using the EPA's PCRAMMIT program (U.S. EPA, 1995). Minimum boundary-layer heights were set to 150 m during the night and 250 m during the day based on estimated minimum boundary-layer heights in cities. Hourly mixing heights (m) were used in conjunction with pollution concentrations ($\mu g\ m^{-3}$) to calculate the amount of pollution within the mixing layer ($\mu g\ m^{-2}$). This extrapolation from ground-layer concentration to total pollution within the boundary layer assumes a well-mixed boundary layer. The amount of pollution in the air was contrasted with the amount of pollution removed by trees to calculate the relative effect of trees in reducing local pollution concentrations:

$$E = R\,(R+A)^{-1} \qquad (9)$$

where E = relative reduction effect (%); R = amount removed by trees (kg); A = amount of pollution in the atmosphere (kg).

POLLUTION REMOVAL BY TREES IN PHILADELPHIA, PA.

The City of Philadelphia (362 km^2) was analyzed using UFORE-D for 1994. Within Philadelphia there are 5 O_3, 6 SO_2, 3 NO_2, 4 CO, and 7 PM10 monitors. Weather data from the Philadelphia airport and boundary-layer height measurements from Sterling, VA, were input into the model. Overall tree cover in Philadelphia is 21.6 percent. UFORE-D calculations predicted that total air pollution removal by Philadelphia's trees was 1,084 metric tons in 1994 with an estimated value of \$5.4 million (Table 1). Total removal and percent air-quality improvement exhibit diurnal and seasonal patterns based on vegetation and meteorological conditions, and atmospheric pollution concentration (Table 2). Average percent air-quality

Table 1. UFORE-D estimates of total dry deposition to trees, associated monetary value, and average in-leaf daytime deposition velocities (V_d), for carbon monoxide (CO), nitrogen dioxide (NO_2), ozone (O_3), sulfur dioxide (SO_2), and particulate matter less than 10 microns (PM10) in Philadelphia, PA, in 1994. Number in parentheses represent expected range based on the typical range of V_d found in the literature (Lovett, 1994).

Pollutant	Total Deposition (t)	Value ($ x 1000)	V_d (m s^{-1})
PM10[a]	418 (160 - 789)	1,884 (723 - 3,555)	0.0064
O_3	306 (89 - 418)	2,069 (604 - 2,820)	0.0056
NO_2	169 (86 - 209)	1,138 (581 - 1,410)	0.0037
SO_2	163 (82 - 256)	270 (136 - 423)	0.0055
CO	28	27	0.00002
Total	1,084 (445 - 1700)	5,388 (2,071 - 8,235)	

[a] Assumes 50% resuspension of particles

Table 2. Monthly pollution removal (t) attributed to trees in Philadelphia, PA (1994).

	Month											
Pollutant	J	F	M	A	M	J	J	A	S	O	N	D
PM10	12	11	12	28	59	66	62	64	54	21	14	16
O_3	3	4	4	18	48	64	68	49	36	7	3	3
NO_2	5	6	6	12	23	27	22	23	22	9	6	6
SO_2	4	4	4	9	24	28	28	27	22	7	4	4
CO	0[a]	0[a]	0[a]	2	5	5	4	5	5	2	0[a]	0[a]

[a] 0.3 t

improvement due to dry deposition to trees during the in-leaf season for the entire city was: PM10 = 0.72%, O_3 = 0.29%, SO_2 = 0.29%, NO_2 = 0.2%, CO = 0.002%. Average air-quality improvement during the in-leaf season was highest just after sunrise when boundary layer heights are still relatively low. Maximum air-quality improvement for the city was about 3 percent for a one-hour period for SO_2, O_3, and PM10. In areas with complete (100 percent) tree cover, air-quality improvement may occasionally reach 13 percent for an hour period depending on boundary-layer height.

The next phase in UFORE-D development is to integrate all urban surfaces together in an urban deposition model. This model will account for the diurnal and seasonal deposition to trees, shrubs, grasses, and other plants, and will include deposition to the myriad of artificial surfaces encountered in urban areas (e.g., buildings, roads, etc.). Model results in urban areas are being validated through eddy-flux measurements that were made in Chicago, IL (King et al., 1995). Preliminary validation of the model results based on published field measurements reveals that model predictions are within the typical range of deposition values for all pollutants. However, mesophyll resistance was increased for NO_2 to account for the difference between the transport of water and NO_2 in the leaf interior, and to bring the estimated NO_2 deposition values within the measured values found in the literature (Lovett, 1994). Future research needs to investigate mechanisms by which gas deposition is different from the resistance to water efflux (after adjustment for the relative diffusivities of water and the pollutant gas), particularly for NO_2.

ACKNOWLEDGMENTS

We thank D. Baldocchi for use and help with the Big-Leaf / Multi-layer hybrid model, E.L. Maxwell for use and help with the METSTAT model, and S.T. Rao, G. Heisler, and K. Civerolo for reviews of an earlier draft of this manuscript.

REFERENCES

Baldocchi, D., 1988, A multi-layer model for estimating sulfur dioxide deposition to a deciduous oak forest canopy, *Atmos. Environ.* 22:869-884.

Baldocchi, D. D., Hicks, B. B. and Camara, P., 1987, A canopy stomatal resistance model for gaseous deposition to vegetated surfaces, *Atmos. Environ.* 21:91-101.

Berman, S., J.Y. Ku, and S.T. Rao, 1997, Uncertainties in estimating the mixing depth: comparing three mixing-depth models with profiler measurements, *Atmos. Environ.*, (in press).

Bidwell, R.G.S. and Fraser, D.E., 1972, Carbon monoxide uptake and metabolism by leaves, *Can. J. Bot.* 50:1435-1439.

Brasseur, G.P. and R.B. Chatfield, 1991, The fate of biogenic trace gases in the atmosphere, in: *Trace Gas Emissions by Plants.* Sharkey, T.D., Holland, E.A., Mooney, H.A., eds., Academic Press, New York.

Cardelino, C.A. and W.L. Chameides. 1990, Natural hydrocarbons, urbanization, and urban ozone, *J. Geophys. Res.* 95(D9):13,971-13,979.

Dudhia, J., 1993, A nonhydrostatic version of the Penn State - NCAR mesoscale model: validation tests and simulation of an Atlantic cyclone and cold front, *Mon. Wea. Rev.* 121:1493-1513.

Dyer, A.J. and C.F. Bradley, 1982, An alternative analysis of flux gradient relationships, *Boundary-Layer Meteorol.* 22:3-19.

Guenther, A., P. Zimmerman and M. Wildermuth, 1994, Natural volatile organic compound emission rate estimates for U.S. woodland landscapes, *Atmos. Environ.* 28(6):1197-1210.

Heisler, G.M., 1986, Energy savings with trees, *J. Arboric.* 12(5):113-125.

Heisler, G.M., R.H. Grant, S. Grimmond, and C. Souch, 1995, Urban forests--cooling our communities? in: *Inside Urban Ecosystems*, Proc. 7th Nat. Urban Forest Conf., American Forests, Washington, DC.

Killus, J.P., J.P. Meyer, D.R. Durran, G.E. Anderson, T.N. Jerskey, S.D. Reynolds, and J. Ames, 1984, *Continued Research in Mesoscale Air Pollution Simulation Modeling. Volume V: Refinements in Numerical Analysis, Transport, Chemistry, and Pollutant Removal,* EPA/600/3-84/095a. U.S. EPA, Research Triangle Park, NC.

King, T.S, C.S.B. Grimmond, and D.J. Nowak, 1995, Eddy correlation determination of local-scale energy and pollutant fluxes in Chicago, in: *Abstracts of the Association of American Geographers 91st Annual Meeting*, Assoc. Amer. Geog., Washington, DC.

Lovett, G.M., 1994, Atmospheric deposition of nutrients and pollutants in North America: an ecological perspective, *Ecol. Appl.* 4:629-650.

Maxwell, E.L., 1994, A meteorological / statistical solar radiation model, in: *Proc. of the 1994 Annual Conf. Amer. Solar Energy Soc.*, San Jose, CA.

Monteith, J.L. and M.H. Unsworth, 1990, *Principles of Environmental Physics*, Edward Arnold, New York.

Murray, F.J., L. Marsh, and P.A. Bradford, 1994, *New York State Energy Plan, Vol. II: Issue Reports*, New York State Energy Office, Albany, NY.

Nowak, D.J., 1994, Understanding the structure of urban forests, *J. For.* 92(10):42-46.

Nowak, D.J., 1996, Estimating leaf area and leaf biomass of open-grown deciduous urban trees, *For. Sci.* 42(4):504-507.

Panofsky, H.A. and J.A. Dutton, 1984, *Atmospheric Turbulence*, John Wiley, New York.

Pederson, J.R., W.J. Massman, L. Mahrt, A. Delany, S. Oncley, G. den Hartog, H.H. Neumann, R.E. Mickle, R.H. Shaw, K.T. Paw U, D.A. Grantz, J.I. MacPherson, R. Desjardins, P.H. Schuepp, R. Pearson Jr., and T.E. Arcado, 1995, California ozone deposition experiment: methods, results, and opportunities, *Atmos. Environ.* 29(21):3115-3132.

Pleim, J.E. and J.S. Chang, and K. Zhang, 1991, A nested grid mesoscale atmospheric chemistry model, *J. Geophys. Res.* 96:3065-3084.

Rao, S.T. and T. Mount, 1994, Least-cost solutions for ozone attainment in New York State: photochemical modeling analysis, *Project Final Report to Niagara Mohawk Power Corp.*, Syracuse, NY.

Seinfeld, J.H., 1986, *Air Pollution*, John Wiley, New York.

Smith, W.H., 1990, *Air Pollution and Forests*, Springer-Verlag, New York.

Taha, H., 1996, Modeling impacts of increased urban vegetation on ozone air quality in the South Coast Air Basin, *Atmos. Environ.* 30(20):3423-3430.

U.S. EPA, 1995, *PCRAMMIT User's Guide*, U.S. Environmental Protection Agency, Research Triangle Park, NC.

van Ulden, A.P and Holtslag, A.A.M., 1985, Estimation of atmospheric boundary layer parameters for diffusion application, *J. Clim. Appl. Meteorol.* 24:1196-1207.

Venkatram, A., 1980, Estimating the Monin-Obukhov length in the stable boundary layer for dispersion calculations, *Boundary-Layer Met.* 19:481-485.

Wesely, M. L., 1989, Parameterization for surface resistance to gaseous dry deposition in regional-scale numerical models, *Atmos. Environ.* 23:1293-1304.

Whittaker, R.H. and Woodwell, G.M., 1967, Surface area relations of woody plants and forest communities, *Amer. J. Bot.* 54:931-939.

Zannetti, P., 1990, *Air Pollution Modeling*, Van Nostrand Reinhold, New York.

Zeigler, I., 1973, The effect of air-polluting gases on plant metabolism, in: *Environmental Quality and Safety, Volume 2*, Academic Press, New York.

Zinke, P. J., 1967, Forest interception studies in the United States, in: *Forest Hydrology,* Pergamon Press, Oxford.

DISCUSSION

R. BORNSTEIN: Have you included feedback's associated with the effects of trees on surface albedo, etc.?

D. J. NOWAK: Yes, they are included in the Mesoscale Model 5 (MM5) simulations.

D. FISH: Do you know enough about deposition of pollutants onto buildings and roads to be able to say whether trees have a positive or negative effect on air pollution?

D. J. NOWAK: Literature is being compiled on deposition velocities to various artificial surfaces. This information will be incorporated as part of an urban deposition model with results being input into the SARMAP air quality model to help determine whether the overall impact of urban trees is either positive or negative.

R. SAN JOSE: Have you compared the effects due to the ozone removal by deposition and the ozone production due to isoprene and monoterpene emissions due to urban vegetation?

D. J. NOWAK: Though this has not been done yet, project results will allow for a comparison of ozone removal due to deposition with the ozone effects from volatile organic compound emissions and the effect of the trees on local meteorology.

APPLICATION OF THE CART AND NFIS STATISTICAL ANALYSES TO FOGWATER AND THE DEPOSITION OF WET SULPHATE IN MOUNTAINOUS TERRAIN

Natty Urquizo[1,2], John L. Walmsley[2], William R. Burrows[2], Robert S. Schemenauer[2,3] and Jeffrey R. Brook[2]

[1]Contractor
[2]Atmospheric Environment Service
4905 Dufferin Street
Downsview, Ontario M3H 5T4 Canada
[3]Retired

INTRODUCTION

Between 1985 and 1991, the Chemistry of High Elevation Fog (CHEF) experiment was conducted on three mountains in southern Quebec, Canada. The CHEF project was linked with the Mountain Cloud Chemistry Project (MCCP) in the Appalachian Mountains of the USA. Measurements were made of liquid water content (LWC) as well as standard meteorological parameters (temperature, pressure, relative humidity, wind speed, wind direction and precipitation).

Fogwater and precipitation samples were collected and analyzed for the concentrations of various acid ions. Previous studies at two of the high-elevation (almost 1000 m above sea level) sites showed that the acid deposition from fog was at least as important as that from precipitation.

To extrapolate to areas where, and time periods when, chemistry data do not exist, in the present study we estimated LWC and acid deposition on Roundtop Mountain (see Figure 1) from routine meteorological measurements at Sherbrooke weather station and from gridded, objectively analyzed meteorological fields. In these approaches, we used two powerful statistical packages: Classification and Regression Trees (CART) (Breiman et al., 1984; Steinberg and Colla, 1995) and Neuro-Fuzzy Inference Systems (NFIS) (Chiu, 1994). From the routine meteorological data, we obtained estimates of mountain-top LWC and compared them with hourly observations. From the gridded fields, we computed 72-h back trajectories over a period of almost six years to help determine the meteorological conditions that lead to acid deposition, using wet sulfate ($SO_4^=$) as the main ion indicator.

The present paper is a continuation of a project, the earlier stages of which appeared in Bridgman et al. (1994), Walmsley et al. (1995), Walmsley et al. (1996a) and Walmsley et al. (1996b). Here we show sample CART results and preliminary NFIS output. We hope to use CART and NFIS analyses to develop diagnostic techniques for estimating LWC and acid ion deposition on mountain tops. If verification against observations at two mountain sites shows sufficient promise, we plan to apply the techniques to other high-elevation areas in southeastern Canada.

Air Pollution Modeling and Its Application XII
Edited by Sven-Erik Gryning and Nadine Chaumerliac, Plenum Press, New York, 1998

CLASSIFICATION AND REGRESSION TREES (CART)

The CART program is ideally suited to the problem of analyzing a large amount of data consisting of an observed value that one wishes to be able to estimate or predict (*i.e.*, the *predictand*) and a number of possible *predictors*. The predictand and predictors may take on continuous values or may be classified into a finite number of categories. An important input option is the method of error estimation. In the present study, we selected the *test set* method for the analyses of both LWC and $SO_4^=$. In the case of LWC, one year was chosen as a learning set and another as the test set, whereas for $SO_4^=$, a specified percentage of the total data set was reserved for testing purposes. CART processes the data and attempts to find a *tree*, consisting of a number of *nodes*, that minimizes the prediction error. Node 1 of the tree is the entire data set. CART builds a tree by splitting the data, beginning with Node 1, into two new nodes. These nodes may also be split; a node which is not split is referred to as a *terminal node* (TN). Each split is accomplished according to whether a predictor or a linear combination of predictors is above or below a *threshold* value. For simplicity in this paper, linear combinations of predictors were not permitted, so the splitting was done using a single *split-variable*. In each application, we show the tree, a table giving details of the splitting and nodes and a table giving the terminal nodes and their corresponding predicted values. In both applications, we had many small observed values of the predictand, so we tested the logarithmic transformation option in CART to produce a more even distribution of values. In the end, we used the transformation only for the $SO_4^=$ predictions. Results for both predictions, however, are presented in the untransformed units of g m^{-3} for LWC and μeq L^{-1} for $SO_4^=$.

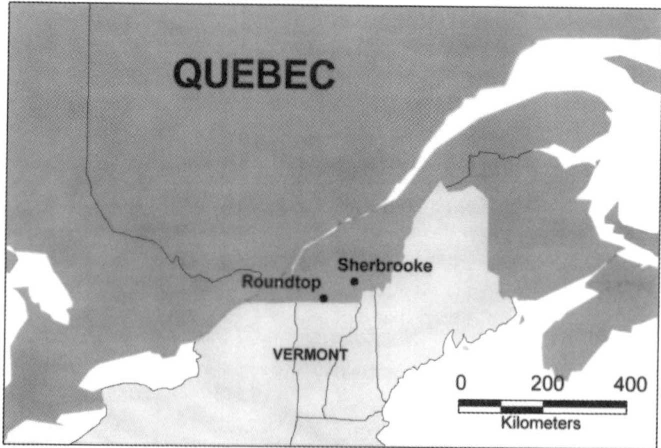

Figure 1. Location map showing the Sherbrooke weather station and the Roundtop Mountain CHEF site.

SURFACE WEATHER DATA

Data from the Sherbrooke weather station used in this study are available in the Canadian Weather Energy and Engineering Data Sets (CWEEDS) compiled by the Atmospheric Environment Service (AES) in 1993.

A number of the CART predictors used in the estimation of both LWC and $SO_4^=$ were obtained directly from the CWEEDS files. These included solar radiation variables, cloud ceiling height, sky condition, visibility, weather elements, station pressure, air temperature, dew point, wind speed and wind direction. Other predictors were derived directly from the CWEEDS data and from date, time, location and elevation. These included dew point depression, relative humidity, air density, lifting condensation level (LCL), solar declination, solar altitude and azimuth angles and a day/night category variable. A preliminary estimate of LWC at the Roundtop observation site (845 m MSL) was made by integrating upward from the LCL assuming, as in Walmsley et al. (1995), that 38% of the condensed liquid water was retained during the ascent. This estimate was then adjusted based on a "probability of cloud" as determined from CWEEDS weather elements, sky condition, cloud ceiling height and an estimate of thermal stability.

ESTIMATIONS OF LIQUID WATER CONTENT

LWC Measurement Program

The fog measuring device [FMD-04R, Advance Electronic Sub-Systems, Inc., Kettleby, Ontario, Canada] was based on a design described by Gerber (1984). It consisted of a planar circular photodiode placed perpendicular to and coaxial with a narrow collimated beam of light. The beam itself was captured by a light trap. The detector sensed the light scattered out of the beam direction through a small angular range. In principle, the scattered flux F (watts) could be related to the beam radiant flux I (watts) and the liquid water content W (g m^{-3}) by Gerber's relation:

$$F = 0.00825 \; I \; W. \qquad (1)$$

In practice, however, the output from the device FMD (mV) was first corrected for zero-drift c (mV) and then used to derive W (g m^{-3}) using a calibration curve (M.A. Wasey, personal communication):

$$W = m \; (FMD - c) + b, \qquad (2)$$

where $m = 0.021$ g m^{-3} mV^{-1}, $b = 0.0029$ g m^{-3} and $c \sim 50\text{-}100$ mV. The zero-drift c depended on temperature and cleanliness of the optics. It therefore varied with time but was readily determined from time-series plots of FMD. The values of m and b were determined from a wind-tunnel comparison with a PMS Forward Scattering Spectrometer Probe (FSSP). Equation (2) is intended for use with drops of size 6 - 10 μm and is valid up to $W \sim 0.8$ g m^{-3}. Hourly LWC data obtained from this device was available for various lengths of time between April and October in the period 1988-91. (Data for 1992 are still being processed at the time of writing.)

In addition to the FMD device, fog detector (FD) data were available for the years 1989-91 from an independent instrument designed to indicate the presence or absence of fog.

Data Preparation

From an examination of the hourly-averaged FD data, it was found that in only about 10% of the cases did the instrument detect fog for part of the hour. The rest of the time, the instrument either detected fog for the whole hour or (in the majority of the cases) did not detect it at all. To simplify the problem, it was decided to eliminate from further consideration the hourly LWC records with a mixture of fog and no fog.

For the sample calculations presented here, all the predictors came from the CWEEDS data set or were derived from the CWEEDS data, as mentioned above. Some tests, not discussed here, were run with additional predictors derived from gridded upper-air data obtained from the Canadian Meteorological Centre (CMC). These were the average of the relative humidities at 925 and 850 hPa and the 925 and 850 hPa temperatures and vertical motions.

Description of CART Runs

A series of 12 CART runs was conducted using hourly data from each of the years 1988-91 as a *learning set* and data from each of the remaining years as a *test set*. Each year had a different number of cases, N, ranging from less than 700 to almost 2900. Input options were chosen to produce the tree with minimum error, with a reasonable number of terminal nodes. (In some cases, there were local minima at two different numbers of terminal nodes. In such instances, one of these optimal solutions was rejected for having too few or too many terminal nodes.)

Results

Figure 2 shows the tree that resulted from using the 1991 data (N = 683) as the learning set and the 1990 data (N = 2377) as the test set. The tree consists of three splits and has four terminal nodes. Table 1 gives details of the splitting of the learning set. At each node, the number of cases to be split is shown, together with the variable used for splitting and its threshold value. When the observed value of the split variable is less than or equal to the threshold, the split goes to the left node; otherwise, it goes to the right node. A bold figure in either the left- or right-node column indicates a terminal node. The split variables in this example are the adjusted estimates of LWC based on Sherbrooke meteorological data (RWTADJ), the wind direction at Sherbrooke (WD) and the sine of the solar declination (SDECL). The "Test Set Relative Error" for this tree was 0.781.

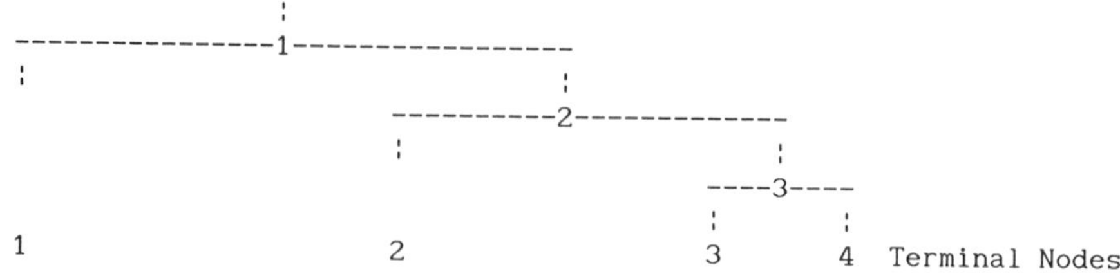

Figure 2. Regression tree diagram for LWC prediction.

Table 1. Node information for LWC prediction. Terminal nodes are shown in **bold** figures.

Node	Cases	Split Variable	Threshold	Units	Left Node	Right Node
1	683	RWTADJ	0.090	g m^{-3}	**1**	2
2	175	WD	215	°	**2**	3
3	57	SDECL	-0.250	-	**3**	**4**

Table 2. Terminal node information for LWC (g m^{-3}) prediction.

	Learning Sample			Test Sample			
Node	Cases	Average	Std.Dev.	Cases	Average	Std.Dev.	Std.Err.
1	508	0.005	0.027	2020	0.008	0.033	0.033
2	118	0.037	0.077	253	0.050	0.079	0.081
3	3	0.509	0.030	0	-	-	-
4	54	0.141	0.127	104	0.121	0.111	0.113

Table 2 shows the terminal node information for this example. In both the learning and test samples, the majority of the cases (74-85%) are in TN 1, with an average LWC near zero, *i.e.*, no-fog events. Only three cases from the learning set and none from the test set fell into TN 3, with an average LWC of 0.5 g m^{-3}. About 17% of the cases in the learning set and 11% in the test set were in TN 2, with an average LWC of 0.04 - 0.05 g m^{-3} and relatively large standard deviation (σ), suggesting that TN 2 contains both fog and no-fog events, but mostly the latter. Finally, TN 4 holds 8% of the learning set and 4% of the test set cases, with an average LWC of 0.12 - 0.14 g m^{-3} and a large σ, suggesting that these are mainly instances of moderate LWC mixed with some no-fog events.

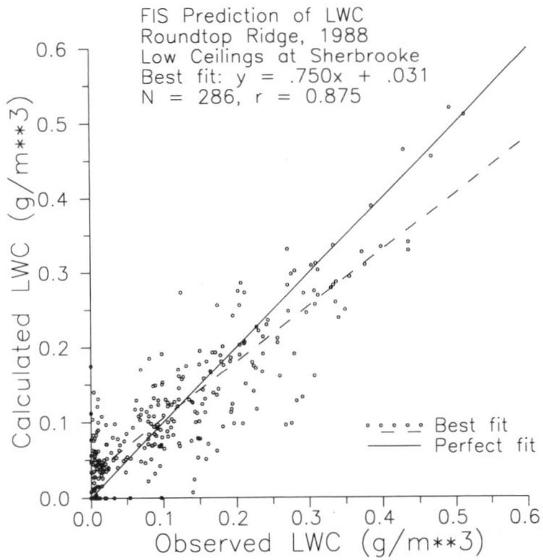

Figure 3. Preliminary NFIS prediction of LWC at Roundtop based on 286 cases in 1988 in which the cloud ceiling at Sherbrooke was ≤ 910 m MSL.

Preliminary NFIS Results for LWC

In Figure 3, we present an example of output derived from an NFIS analysis. In this case, we selected 286 hourly values from the 1988 data set in which the cloud ceiling height at Sherbrooke was observed to be ≤ 910 m MSL. Figure 3 shows the resulting predicted values of LWC compared with the observed LWC at Roundtop Mountain. The agreement is reasonably good. The correlation coefficient was 0.875 and the slope of the best-fit line was 0.75. (As no FD data were available in 1988, cases with a mixture of fog and no fog during the hour could not be eliminated as they were in the previous section. Otherwise, it is possible that some of the low observed values would have been removed before applying NFIS.)

ESTIMATIONS OF WET SULPHATE DEPOSITION

Data Preparation

The chemistry data discussed here were collected during the CHEF project between December 1985 and September 1991. Both fog and precipitation were sampled at the mountain sites, but only precipitation was collected at the valley sites. Details of sampling and field methods are given in Schemenauer et al. (1995). In the present study, we concentrated on the Roundtop database, supplemented by hourly meteorological data from the Sherbrooke weather station (see description above). The hourly meteorological data were averaged into periods corresponding to those of the CHEF sample collection.

Since the MSL elevations of the ridge and valley measurement sites were at 845 and 250 m, respectively, we computed back trajectories at 850 and 925 hPa. (Sample results derived from the 850 hPa trajectories are described below.) Positions in the back trajectories were computed at 6-h intervals for 72 h. To match trajectories to the periods of the chemistry samples, their arrival times at Roundtop were set at 0600 and 1800 UTC. The direction and distance from the trajectory to Roundtop was extracted at 6-h intervals. Additional predictors were derived from the direction and distance, *e.g.*, cosine of the direction, and average distances and directions. The synoptic wind flow category of Brook et al. (1995) was included in two more predictors, one (CAT40) with segregation of dry and wet categories (1-20 and 21-40, respectively) and the other (CAT) with all categories combined. All chemistry data were explicitly *excluded* as predictors.

Description of CART Runs

The sample CART run to be presented here was conducted using 80% of the 1805 data records as a *learning set* and the remainder as a *test set*. As in the LWC runs, linear combinations were not permitted and input options were chosen to produce the tree with minimum error, with a reasonable number of terminal nodes. A logarithmic transformation of the predictand was selected using the relation LSO4 = ln(SO4+0.0001), the constant term being included to permit handling of cases with zero concentration.

Results

Figure 4 shows the tree that resulted from using 80% of the data (N = 1441), randomly selected, as the learning set and 20% (N = 364) as the test set. The tree consists of four splits and has five terminal nodes. Table 3 gives details of the splitting of the learning set (see the above description of Table 1 for details). The split variables in this example

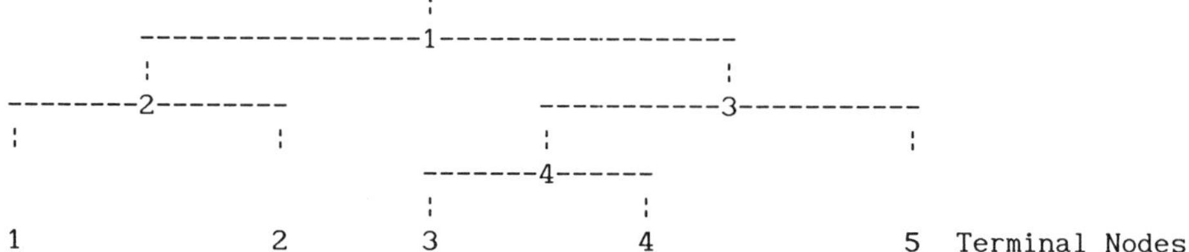

Figure 4. Regression tree diagram for $SO_4^=$ prediction.

Table 3. Node information for $SO_4^=$ prediction. Terminal nodes are shown in **bold** figures.

Node	Cases	Split Variable	Threshold	Units	Left Node	Right Node
1	1441	DIR4	201.6	°	2	3
2	294	TD	17.3	°C	**1**	**2**
3	1147	CAT40	4.5	-	4	**5**
4	766	ISC	2.5	-	**3**	**4**

are the direction from Roundtop of the trajectory location 24 h before arrival over Roundtop (DIR4), the dew point temperature (TD), the synoptic wind flow category with dry-wet segregation (CAT40), and the thermal stability category (ISC) ranging from 1 (very unstable) to 6 (very stable). The "Test Set Relative Error" for this tree was 0.799.

Table 4 shows the terminal node information for this example. It should be noted that CART produced output as the natural logarithm of the concentration, the values of which have been converted back to the original units of μeq L^{-1}. Column headings of "Range" indicate the range within one standard deviation of the average value in the logarithmic scale. In both the learning and test samples, slightly more than 50% of the cases are in TN 4, with the highest average wet sulphate concentrations (106-108 μeq L^{-1}) and ranges from about 26 to over 400 μeq L^{-1}. Only a small number of cases fell into TNs 2 and 3, with the two lowest average concentrations. About 20% of the cases fell into TN 1 and 25% in TN 5, the two nodes with intermediate values of average concentration (30-50 μeq L^{-1}).

Table 4. Terminal node information for $SO_4^=$ (μeq L^{-1}) prediction.

	Learning Sample			Test Sample		
Node	Cases	Average	Range ($\pm\sigma$)	Cases	Average	Range ($\pm\sigma$)
1	285	38.67	5.41-276.17	75	37.41	9.12-153.39
2	9	0.33	0.001-106.91	6	0.08	0.0002-20.68
3	4	0.12	0.001-19.95	1	4.80	4.80-4.80
4	762	108.09	28.53-409.53	191	105.95	26.00-431.82
5	381	50.10	15.00-167.34	91	31.82	2.52-401.02

CONCLUSIONS

We have described the preparation of data, CART runs and sample results for estimates of LWC and $SO_4^=$ deposition. Although the results presented here are encouraging, we are not yet sure that a viable prediction scheme can be achieved with the available predictors and observations. Therefore, we plan to do further CART runs to examine the impact of additional predictors. If satisfactory estimation of LWC proves unfeasible, we shall attempt to predict the occurrence or non-occurrence of cloud at a given elevation. This would seem to be a necessary first step towards the prediction of LWC and wet deposition.

As a post-processing task, we plan to use the best CART trees to predict *nodal* values of LWC, $SO_4^=$, nitrate (NO_3^-) and ammonium (NH_4^+) and then to compute correlation coefficients and perform least-squares fitting to assess skill. Once the most useful predictors have been found, we plan to use NFIS to estimate the predictands on *continuous* scales, an example of which was shown for LWC. The ultimate goal is to develop and verify diagnostic techniques for estimating LWC and acid ion deposition on mountain tops and to apply them to high-elevation areas in southeastern Canada.

Acknowledgments

Thanks to Richard Tanabe for extracting the hourly observed LWC and FD data from the CHEF database, to Mohammed Wasey for information about the fog measuring device and to Dr Yi-Fan Li for Figure 1.

REFERENCES

Breiman, L., Friedman, J.H., Olshen, R.A., and Stone, C.J., 1984, *Classification and Regression Trees*, Wadsworth & Brooks, Monterey, CA.

Bridgman, H.A., Walmsley, J.L., and Schemenauer, R.S., 1994, Modelling the spatial variations of wind speed and direction on Roundtop Mountain, Quebec, *ATMOS.-OCEAN* 32;605.

Brook, J.R., Samson, P.J., and Sillman, S., 1995, Aggregation of selected three-day periods to estimate annual and seasonal wet deposition totals for sulphate, nitrate, and acidity. Part I: A synoptic and chemical climatology for eastern North America, *J. Appl. Meteorol.* 34;297.

Chiu, S., 1994, Fuzzy model identification based on cluster estimation, *J. Intelligent & Fuzzy Sys.* 2.

Gerber, H., 1984, Liquid water content of fogs and hazes from visible light scattering, *J. Clim. Appl. Meteorol.* 23;1247.

Schemenauer, R.S., Banic, C.M., and Urquizo, N., 1995, High elevation fog and precipitation chemistry in southern Quebec, Canada, *Atmos. Environ.* 29;2235.

Steinberg, D. and Colla, P., 1995, *CART: A Supplementary Module for SYSDAT*, Salford Systems, San Diego, CA.

Walmsley, J.L., Bridgman, H.A., and Schemenauer, R.S., 1995, Modelling wet deposition of sulfate on Roundtop Mountain, Quebec, in: *Proc., MODSIM'95 Conf., Newcastle, NSW, Australia, 27-30 November 1995*, 2.

Walmsley, J.L., Schemenauer, R.S., and Bridgman, H.A., 1996a, A method for estimating the hydrologic input from fog in mountainous terrain, *J. Appl. Meteorol.* 35;2237.

Walmsley, J.L., Urquizo, N., Schemenauer, R.S., and Bridgman, H.A., 1996b, Modelling of acid deposition in high-elevation fog, in: *Air Pollution IV. Monitoring, Simulation and Control*, B. Caussade, H. Power, and C.A. Brebbia, eds., Computational Mechanics Publications, Southampton.

DISCUSSION

D.G. STEYN: It seems that you are trying to do something very difficult - forecast a very infrequent event in a large sample. Can you comment?

J.L. WALMSLEY: Your point is exactly the one stressed in a recent article in the New Scientist (R. Matthews, "How Right Can You Be", 8 March 1997, pp 28-31). Out of 3567 hourly observations made during the spring, summer and autumn seasons of 1988-91, we found that fog occurred at the Roundtop CHEF site only 9% of the time, as measured by the AWS fog monitor. By using a multi-step application, we can now use CART to eliminate a large number of cases (6022 out of 7357 or 82%) that with 99% confidence we can label as "no-fog". This makes the fog events less infrequent in the remaining data, which can then be more easily processed, first by CART to find the useful predictors and then by NFIS to predict the LWC. We are hoping that a similar strategy will work for estimating wet sulphate deposition.

D.G. STEYN: How are you able to apply the method under real conditions when observations of LWC and acid ions are not available?

J.L. WALMSLEY: I should have made it more clear in my presentation that we are using the observations of LWC and wet sulphate deposition to let CART and NFIS produce algorithms. Those algorithms can then be applied to the predictors, without the need for observed values of the predictand.

INTRODUCTION OF A FOREST FIRE EFFECT IN A MESOSCALE DISPERSION MODEL

C. Borrego, A. I. Miranda, and M. Nunes

Department of Environment and Planning
University of Aveiro
3810 AVEIRO - PORTUGAL

INTRODUCTION

Burning is a locally, regionally, and globally important biospheric phenomenon. Globally, biomass burning is a major contributor of greenhouse gases and particulate matter to the atmosphere. Andreae (1991) estimated a value of 40% for the annual contribution of biomass burning to the total annual greenhouse gases emissions. In Portugal, the contribution of forest fire's emissions to the total greenhouse gases emissions could reach about 10% (Miranda *et al.*, 1994a).

A wide range of chemically reactive gases is produced during biomass burning, including carbon monoxide (CO), nitric oxide (NO), nitrogen dioxide (NO_2), many hydrocarbons, and ammonia (NH_3). These reactive gases influence strongly the local/regional concentrations of the major atmospheric oxidants ozone (O_3) and the hydroxyl radical (OH). Therefore, in view of the large amounts of hydrocarbons and NO_x emitted from biomass fires, it is not surprising that very high concentrations of O_3 are produced in the plumes, often exceeding values typical of industrialised regions (Andreae, 1991).

In addition, wildfires can produce severe degradation of air quality on a local scale as resulted, for example, from the 1987 and 1988 wildfires that burned in the Western United States (Ward *et al.*, 1993). Regarding the local scale, a fire is a significant source of pollutants, for several hours, or even days, and their effect in the air quality may be of relatively short term duration, but considerable (Miranda *et al.*, 1994b; Borrego and Miranda, 1994).

Air quality studies usually do not take into account this source of pollutants, being fundamental the development of a numerical tool that allow the analysis of this subject. The University of Aveiro is at present studying the effect of forest fire emissions in the air quality and developing this numerical system. Such work implies the combination of several related features: (i) forest fire emissions; (ii) atmospheric flow; (iii) fire progression; and (iv) dispersion of the pollutants.

Taking into account the possible impact of forest fires in the photochemical production, an integrated model, which includes the non-hydrostatic meteorological model MEMO, the

Rothermel fire progression model and the atmospheric dispersion model for reactive species MARS has been developed. The inclusion of the interaction between the fire and the atmospheric flow is a fundamental aspect of this integrated system.

The main purpose of this work is to present the development of the numerical tool, named AIRFIRE, describing the changes introduced into the models and the integrated linking between all the components. Results from two types of applications will also be presented.

MODELS DESCRIPTION

MEMO Model

The mesoscale model MEMO is a non-hydrostatic prognostic model developed to simulate the wind flow over complex terrain (Moussiopoulos, 1989). Within the model, the conservation equations for mass, momentum and scalar quantities like energy, humidity or pollutant concentrations are solved numerically, in terrain following co-ordinates. Radiative transfer is calculated with an efficient scheme based on the emissivity method for longwave radiation and an implicit multilayer method for shortwave radiation. For turbulent parametrization k-theory is applied. Turbulence can either be treated with a zero-, a one- or a two-equation turbulence model. Recently, the University of Aveiro has implemented in MEMO an advanced scheme for the simulation of turbulence (Coutinho and Borrego, 1995), using the transilient theory, a non-local closure for turbulence.

MARS Model

MARS model describes the dispersion and chemical transformation of air pollutants in a three dimensional region, attending to the fact that for the numerical modelling of photochemical oxidant formation the chemical transformation of pollutants should be considered in conjunction with their transport in the atmospheric boundary layer (Moussiopoulos *et al.*, 1995). MARS is a fully vectorized model which solves the concentration parabolic equation system for known meteorological variables (wind velocity, temperature, humidity, eddy diffusivity), i.e. the mass conservation equations are decoupled from the equation of motion of air. In the present work the meteorological information is provided by MEMO.

ROTHERMEL Model

Rothermel developed a semi-empirical fire spread model, based essentially on laboratory experiments (Rothermel, 1972). The model applies to a stationary fire spread in a statistically homogeneous fuel bed and is not suited to be used with high intensity fires, crown fires, or fires where spotting plays a relevant role in fire spread. The model is based in the following equation, which expresses an energy balance within a unit volume of the fuel ahead of the flame

$$R = \frac{I_R \, \pi \, (1 + \phi_w + \phi_s)}{\rho_b \, w \, Q_i}$$

where: R - Rate of spread ($m.s^{-1}$); I_R - Reaction intensity ($J.m^{-2}.s^{-1}$); π - Propagation flux ratio; ϕ_w - Wind factor; ϕ_s - Slope factor; ρ_b - Bulk density ($kg.m^{-3}$); ω - Effective heating number; Q_i - Heat of pre-ignition ($J.kg^{-1}$).

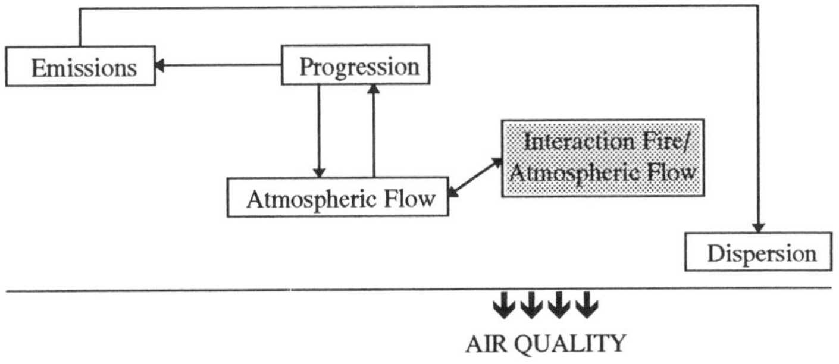

Figure 1. Schema of the AIRFIRE.

This equation illustrates the rate of spread concept, which is a ratio between the rate of heating of the fuel and the energy required to bring the same fuel to ignition. All the quantities are calculated using fuel and environmental characteristics supplied by the user.

In most forest fires, fuel characteristics have large spatial variations. Therefore, for the application of Rothermel model a sub-division of the domain in cells is required. In each of the cells, the input conditions are assumed as locally homogeneous.

DEVELOPMENT OF THE AIRFIRE

As already mentioned, modelling the dispersion of the pollutants emitted during a forest fire implies the combination of several related features: (i) forest fire emissions; (ii) atmospheric flow; (iii) fire progression; and (iv) dispersion of the pollutants. Figure 1 presents a schema of this numerical system, which main purpose is the study of the forest fire effects in the air quality.

The key component of the AIRFIRE is the fire spreading along the time, which is highly related to the wind and fuel conditions and to the topography. So, for a chosen fire progression domain, described by grid cells, one should know for each cell: the altitude, the land cover and the wind (speed and direction). Each type of land cover should be associated to one combustible, clearly described in terms of physical and chemical characteristics, like the heat and mineral content or like the fuel loading. Wind information will be calculated by MEMO linked with Rothermel model to estimate the fire progression.

Since the spatial definition of the fire front requires a more refined grid than the fluid flow calculation, information of wind is transferred to a finer grid, where the fire shape is computed based on the information of fuel characteristics, local topography and wind. Fire grown simulation is performed using Dijkstra's dynamic programming algorithm, which is a procedure designed to find the shortest path between a specified pair of nodes. For each burning cell, the time the fire takes to propagate to all its 'neighbours' is computed with Rothermel model. The instant of time for which each non-burning cell may be ignited is computed, and the next cell to ignite will be the one with the lowest assigned value. This process of successive ignitions determines the advance in time. After each time step defined for the fluid flow calculation, fire front characteristics are transferred to the coarser grid, where the velocity field is recalculated on the basis of a new temperature distribution, assigned to each cell as a function of the percentage of cell area that is burning.

This new temperature distribution calculation has been included in MEMO aiming to consider the effect of the fire, as a source of heat, in the atmospheric flow. According to Van Wagner (1973), the temperature reached at any height in the convection column above a fire

(T_f) depends on the intensity of the heat source (I), the ambient temperature (T_0), and the wind speed (U):

$$T_f = \frac{\sqrt{b}\,(I\ /\ 4187)^{7/6}}{h\sqrt{bI\ /\ 4187\ +\ U^3}} + T_0$$

where: $b = 4187\ g\ (d\ c_p\ T_0)^{-1}$; g is the acceleration due to gravity; d is the air density; c_p is the specific heat of air; T_0 is the ambient absolute temperature.

The heat released by a fire is responsible both for ignition of the fuel ahead of the flame front, due to the heat transfer to the fuel, as well as for the heating of the air and soil. Rothermel model provides a mean of quantifying both the total heat release and the part of this heat absorbed by the fuel, by defining the variable π, the propagation flux ratio. For the remaining energy, Lopes (1993) considers that 15% was absorbed by the soil, and 85% was responsible for air heating.

MEMO model has been modified, chiefly in the calculation of energy fluxes and atmospheric heating rate subroutine, by the introduction of the above equation and taking into account the referred assumptions. All the needed fire variables can be computed through Rothermel model. Therefore, interaction between the fire spread model and flow field calculations is undertaken through a process of sequential computations, in which the velocity field obtained in a certain time step is used as input to the Rothermel fire spread model, and vice-versa.

Rothermel model also allows to estimate the emitted pollutants during the fire, providing the rate of spread. Ward and Radke (1993) use $q_n = (Ef_n)wR$ to determine the fire emissions, where: q_n - source strength of emission n, $g.m^{-1}.s^{-1}$; n - specific emission; Ef_n - ratio of the mass of emission n released to the mass of consumed fuel, $g.kg^{-1}$; w - mass of available fuel per unit area, $kg.m^{-2}$; R - rate of spread of the fire, $m.s^{-1}$.

Each of these variables needs to be evaluated to estimate the smoke emissions. For example, the rate of spread should be known, being advisable the linkage of the fire emissions calculation module with a fire progression module. Another important needed variable is the ratio of the mass of emission n released to the mass of consumed fuel. There are several works which main purpose is to estimate this variable. Borrego *et al.* (1996) summarise a bibliographic research on these emissions factors.

Taking into account that the final purpose of the integrated system is the calculation of air pollutants dispersion, the inclusion of MARS model in the described system is a fundamental component. So, a fire emissions module has been developed in order to be integrated in MARS, considering some data given by the fire spread model. Additionally to this fire spread data, MARS also uses meteorological data supplied by MEMO.

AIRFIRE APPLICATION

AIRFIRE has been applied to two different situations: the first oriented to analyse the effect of a heat source in the atmospheric flow through the definition of an ideal scenario; the second related to the study of the effect of a real fire in the air quality.

Ideal scenario

In order to analyse the performance of the system and to better understand its results in terms of the introduction of a heat source, a simple ideal scenario has been made. This scenario was created taking into account the absence of factors, such as topography, wind, etc., which would also affect the simulation results.

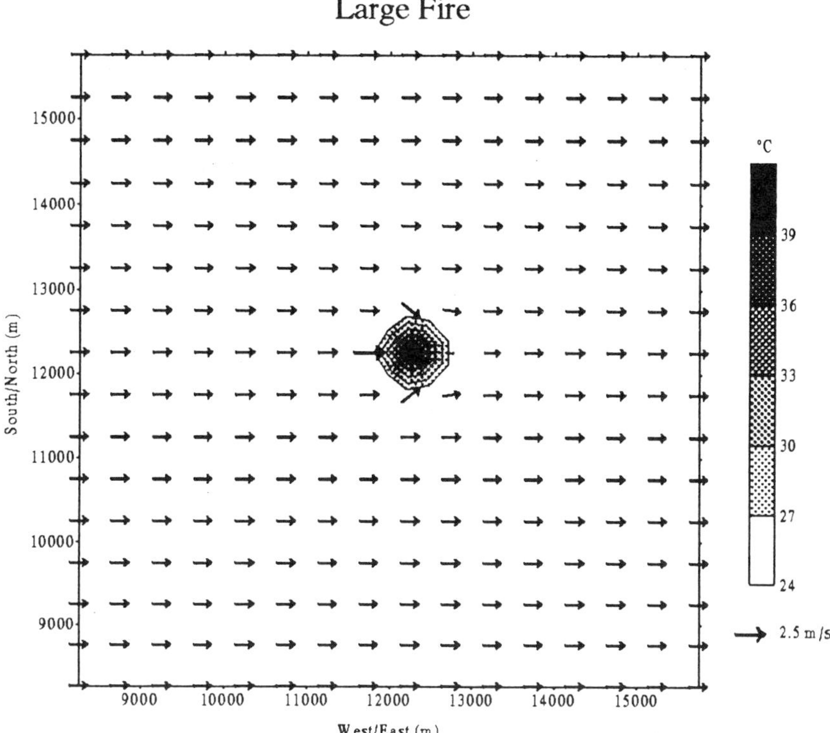

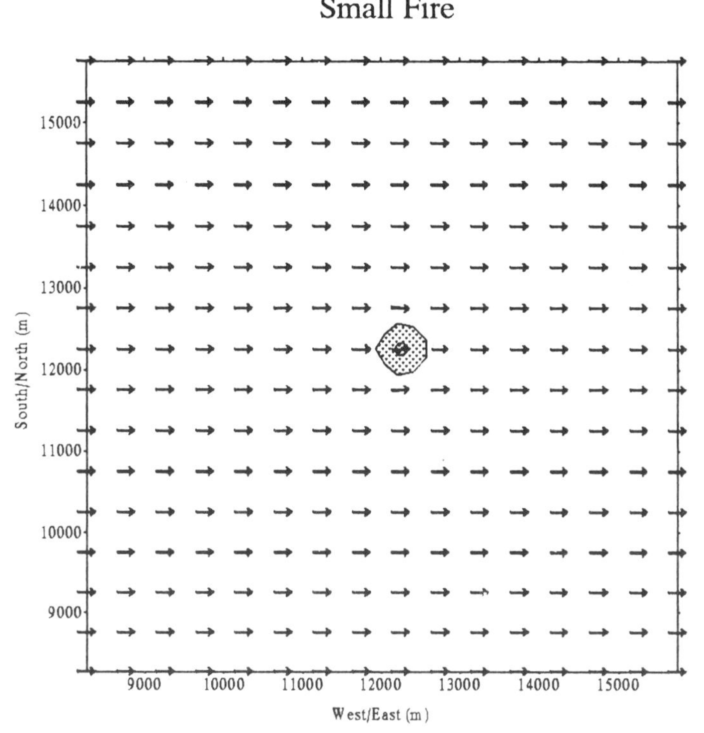

Figure 2. Horizontal and wind temperature fields 50 seconds after the heat source inclusion (horizontal and vertical scales represent only a part of the simulated domain).

It has been considered a flat area, uniformly covered by bush. The simulation domain has an extension of 25 km x 25 km, with a grid spacing of 500 m x 500 m. Meteorological boundary conditions were set as unstable atmospheric conditions with a weak wind blowing to East (Miranda, 1997).

Taking into account that fires can be classified into large and small fires, regarding the vertical height to which the fire affects the normal wind field (Chandler *et al.*, 1991), two different fires have been simulated. A large one (I = 5282 $kW.m^{-1}$), which should affect and be affected by atmospheric processes several hundred to several thousand meters above the fuel bed, and a small one (I = 559 $kW.m^{-1}$), which should be governed by the properties of the fuel bed and air movements within and a few meters above the fuel bed. A heat source, for each fire type, has been included in the middle of the simulation domain for the period of a time step (5 s, which is the recommended value for the chosen grid spacing) and has been removed, being the fire simulated as a thermal.

Figure 2 presents the horizontal (at 5 m above ground level) patterns of temperature and wind, 50 seconds after the inclusion of the heat sources. For the large fire it is clear that the heat source affects the atmospheric flow. The wind orientation changes in the direction of the heat source and there is an acceleration upstream of the flow. The heat source acts as sink, but it is not strong enough to modify the flow direction downstream. The temperature field also reveals the presence of the heat source. In the case of the small fire, this disturbance is not so evident, being the main atmospheric flow almost unchanged.

Figure 3 presents the vertical profiles of wind and temperature (in the same plane than the heat source), 50 seconds after the inclusion of a large heat source. The wind vectors represent the result of the wind components u and w. The orientation of each vector is related to ascendant or descendant movements. So, the region just above the large heat source location presents an ascendant flow and the adjacent regions a flow going down.

As expected, results from both figures point to a clear disturbance of the main flow due to the presence of a heat source representing a large fire. It is also possible to observe in both figures that just a non-significant percentage of the simulation domain has been affected by the heat source.

Arrábida Fire

Once demonstrated the fire effect into the flow, a real fire case will be used to show the fire effect in the air quality results. However, being preliminary results, they are not expected to be compared with real data, they enlighten the good performances of the AIRFIRE system and how far the study can be extended.

Arrábida mountains are part of a very complex topographic region situated approximately 30 km South-East from Lisbon, on the coast with an altitude reaching up to 500 m and an area of approximately 11 x 5 km^2. A fire occurred at the occidental crest of the mountains, above 200 m altitude, and has started on the 17th September 91, at 14.34 Local Standard Time (LST). After 15.00 LST the fire assumed large proportions going quickly upslope and toward East. This behaviour agrees with the time when the sea breeze clearly appeared. Between 17.00 and 18.00 LST the fire progression had its maximum value. The fire was extinct, after a strong shower, on the 20th. An area of about 200 ha has been burned. It was the chosen fire to simulate due to the large amount of information about its behaviour.

As the first day of the fire was the most critical in terms of burned area it has been chosen to apply the AIRFIRE system in order to simulate the atmospheric dispersion of reactive species. The synoptic meteorological situation for this day agrees with the previously defined for a typical Summer day, which favours the formation of coastal breezes.

MEMO was applied to simulate the wind flow during the first day of the fire. In order to simulate meteorological phenomena from different scales, the simulation comprised three

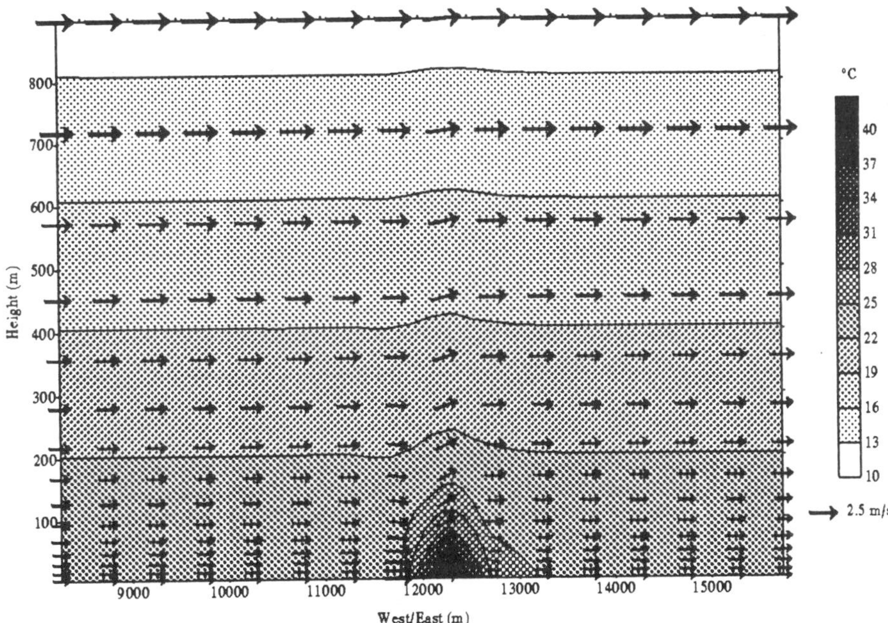

Figure 3. Vertical wind (uw) and temperature profiles 50 seconds after the large heat source inclusion (horizontal and vertical scales represent only a part of the simulated domain).

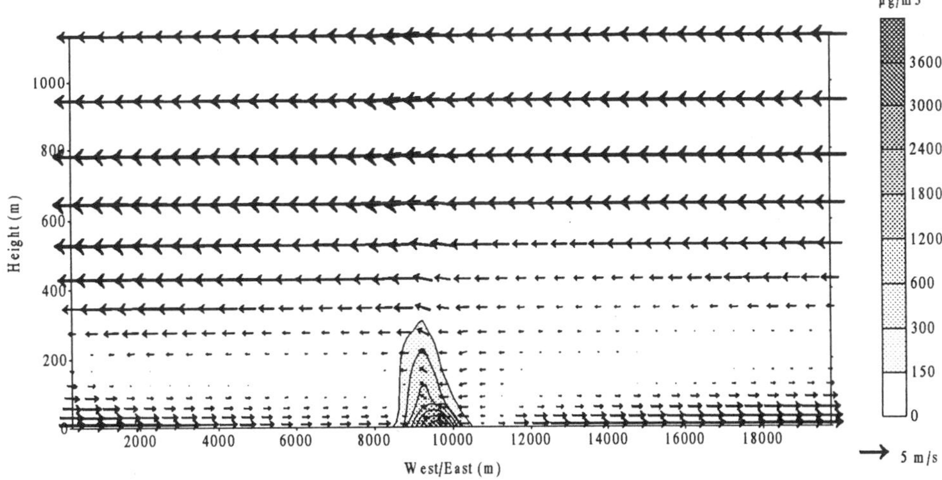

Figure 4. Vertical profile of wind (uw) and CO concentration.

nested model domains, which coincide with three numerical grids: (i) the coarse grid domain has an extension of 120 x 120 km^2 and an horizontal resolution of 3 x 3 km^2; (ii) the 40 x 30 km^2 medium grid domain is given at 1 x 1 km^2 resolution; and (iii) the fine grid domain spreads over 20 x 15 km^2 at a resolution of 0.5 x 0.5 km^2. In the vertical direction the grid consists of 20 layers non-equidistant with a minimum grid spacing of 20 m near the ground.

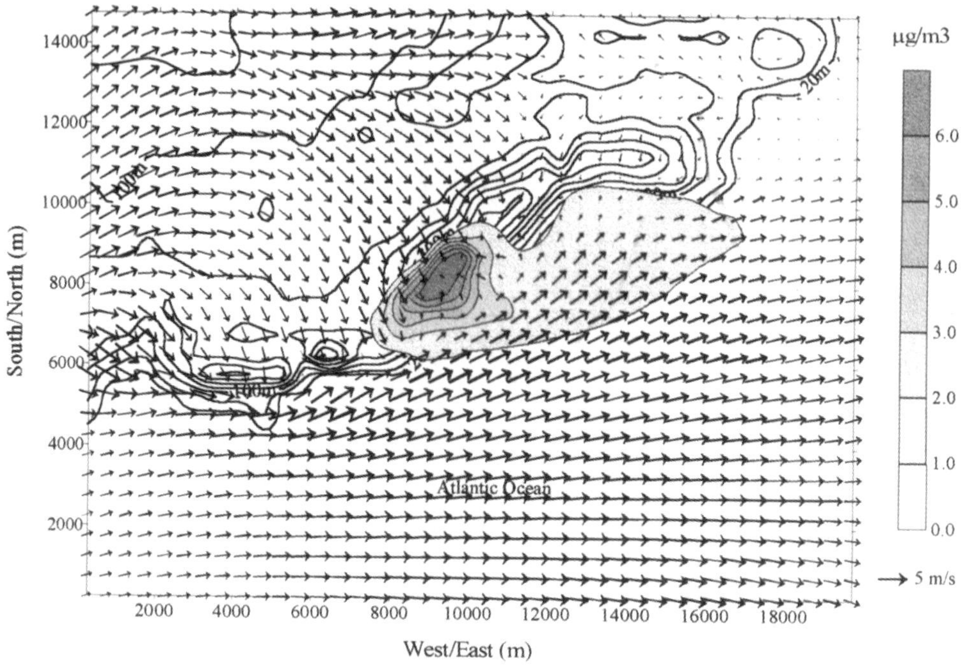

Figure 5. Horizontal ground pattern of wind and O_3 concentration.

The fire, as a source of heat and its spreading, have been included in the fine grid simulation, which has also been used to the dispersion calculations. Modified MEMO, by the linking with the Rothermel model, allowed to estimate the fire progression along the day. Emissions of some reactive species have been calculated by modified MARS (where the fire emissions module has been included) using this information.

Figure 4 presents the vertical profile of wind (components u and w) and CO concentration at 17.00 LST. Figure 5 presents the horizontal ground pattern of wind and O_3 concentration at the same time.

In figure 4 there is a wind intensity increase towards the heat source and an ascendant flow above the heat source. In terms of horizontal patterns, it is also possible to see the small shift of the flow in direction of the heat source. Those changes of wind flow due to the fire, which could be summarised as an acceleration of the horizontal flow towards the fire and an upward flow above it, have not a significant pattern since the simulated fire was a small one.

Wind fields of both horizontal and vertical representations are the result of the sea breeze going into land. The presence of a coastal breezes circulation system in the area and in the day in analysis could provoke a more complex dispersion pattern along the time. So, the pollutants that are being emitted during the day and transported in altitude in direction to the sea, could come back later.

The obtained concentration values can be considered as significant regarding the contribution of the fire, which was a small intensity one, to the air quality of the region.

CONCLUSIONS

Studying the effect of forest fire emissions in the air quality definitively implies an integrated approach between several components, namely emissions, atmospheric flow, progression and dispersion. This integrated approach means that all interactions between the components must be carefully analysed and taken into account. The introduction of the fire as a source of heat in a meteorological model, is a fundamental step for this integrated approach and results from the simulations confirm this assumption. Results also point out an impact of the fire in the air quality, being the obtained CO and O_3 concentration values quite significant.

The fire does affect the atmospheric flow and consequently it will affect the dispersion of the emitted pollutants in the atmosphere. However, these are preliminary results and a more deep study must be done and different situations must be simulated.

Acknowledgments

The authors would like to thank the European Commission, under the framework of the project MINERVE and the Comissão Nacional Especializada de Fogos Florestais, for their support. It is also acknowledge PRAXIS XXI for the PhD. grant of A. I. Miranda and the MSc. grant of M. Nunes.

REFERENCES

Andreae, M. 1991 Biomass Burning: Its History, Use , and Distribution and Its Impact on Environmental Quality and Global Climate. In *Global Biomass Burning. Atmospheric, Climatic, and Biospheric Implications*. Ed. J. Levine. Massachussets Institute of Technology, pp. 3-21.

Borrego, C. and Miranda, A. I. 1994. *Study of the Atmospheric Dispersion of Gases and Aerosols.*Department of Environment and Planning, University of Aveiro. AMB-QA-(3)/94.

Borrego, C., Miranda, A. I. and Carvalho, A. C. 1996 *Modelling Wind and Smoke Dispersion. MINERVE II. Final Report.* Department of Environment and Planning, University of Aveiro. AMB-QA-(3)/96.

Chandler, C., Cheney, P., Thomas, P., Trabaud, L. and Williams, D. 1991 *Fire in Forestry. Volume I. Forest Fire Behavior and Effects.* Krieger Publishing Company. Malabar, Florida.

Coutinho, M. and Borrego, C. 1995 Application of Transilient Turbulence Theory to a Mesoscale Meteorological Model. Proceedings of the *11th Symposium on Boundary Layers and Turbulence*. American Meteorological Society, Charlotte.

Lopes, A. 1993 Modelação Numérica e Experimental do Escoamento Turbulento em Topografia Complexa: Aplicação ao Caso de um Desfiladeiro. PhD. Thesis. Departamento de Engenharia Mecânica. Faculdade de Ciências e Tecnologia da Universidade de Coimbra. pp. 320.

Miranda, A. I. (1997) Estudo dos Efeitos na Qualidade do Ar das Emissões dos Incêndios Florestais. PhD. Thesis. Departamento de Ambiente e Ordenamento da Universidade de Aveiro. To be published.

Miranda, A. I., Coutinho, M. and Borrego, C. 1994a Forest Fire Emissions in Portugal: A Contribution to Global Warming? *Environmental Pollution*, **83**, pp. 121-123.

Miranda, A. I., Borrego, C. and Viegas, D. 1994b Forest Fire Effects in the Air Quality. In *Air Pollution II. Volume: Computer Simulation*. Ed. Baldasano, Brebbia, Power and Zannetti. Computational Machanics Publications. Southampton, Boston. pp. 191-199.

Moussiopoulos N. 1989 Mathematische Modellierung mesoskaliger Ausbreitung in der Atmosphare, Fortschr.-Ber. VDI, Reihe 15, Nr. 64, pp. 307.

Moussiopoulos, N., Sahm, P. and Kessler, Ch. 1995 Numerical Simulation of Photochemical Smog Formation in Athens, Greece a Case Study. *Atmospheric Environment*, **Vol. 29.**

Rothermel, R. 1972 A Mathematical Model for Predicting Fire Spread in Wildland Fuels. USDA F.D. Research Paper. INT-115.

Van Wagner 1973 Height of Cronw Scorch in Forest Fires. *Can. J. For. Res.* **Vol. 3**, pp.373-378.

Ward, D. and Radke, L. 1993 Emissions Measurements from Vegetation Fires: A Comparative Evaluation of Methods and Results. In *Fire in the Environment. The Ecological, Atmospheric, and Climatic Importance of Vegetation Fires*. Ed. P. Crutzen, J. Goldammer. John Wiley & Sons. Chichester, England, pp. 53-76.

Ward, D., Rothermel, R. and Bushey, C. 1993 Particulate Matter and Trace Gas Emissions from the Canyon Creek Fire of 1988. Proceedings of the *12th Fire and Forest Meteorology*. Ed. Society of American Foresters. Georgia, U.S.A. pp. 62-76.

DISCUSSION

S. ZILITINKEVICH: What is prescribed in your forest fire model, the heat flux or the near surface temperature?

C. BORREGO: The heat flux is prescribed, and the surface temperature is calculated as needed.

S. ZILITINKEVICH: Then, for the calculation of the surface temperature, I would recommend a non-local formulation for the heat transfer in the convective surface layer, e.g., the one given in my talk at the present meeting.

D. STEYN: Does your model include buoyancy due to water evaporated from burnt plant material and soil in the fire plume?

C. BORREGO: Not yet. The model is still in development and the question that you pointed would be taken into account. However, it doesn't seem that the mentioned effect in the buoyancy would be significant when compared with the buoyancy due to the difference between the plume temperature and the environment temperature, which is considerable. The contribution of evaporated water to the buoyancy could be important, in the case of some large fires, in the highest levels of the convective plume, where the energy to the buoyancy comes essentially from the condensated water.

A NEW FORMULATION OF THE PROBABILITY DENSITY FUNCTION IN RANDOM WALK MODELS FOR ATMOSPHERIC DISPERSION

Anne Katrine Vinther Jensen, Sven-Erik Gryning

Risø National Laboratory
DK-4000 Roskilde, Denmark

INTRODUCTION

Simulation of atmospheric diffusion in vertically inhomogeneous skewed turbulence, which is typical for the daytime boundary layer, is investigated. Particle trajectories are simulated by using the Langevin equation. The main objective is to introduce a formulation of the probability density function (PDF) of the vertical velocity fluctuation based on a fourth order expansion after Hermite polynomials (Gram-Charlier expansion). The statistics of the turbulence are introduced by forcing the moments of the Gram-Charlier expansion to equal the parametrized moments of atmospheric turbulence.

In several previous investigations where the PDF has been constructed this way, further use was hampered by the fact that the PDF takes negative values for a range of vertical velocities. This problem is overcome in the present formulation.

The model has been evaluated under convective conditions against the watertank experiments of Willis and Deardorff (1976, 1978, 1981) and good agreement was found.

THE LANGEVIN EQUATION

A commonly used tool for simulation of diffusion processes is the Langevin equation. The one-dimensional version for the vertical velocity fluctuation w reads

$$\begin{aligned} dw &= a(z, w)dt + b(z, w)d\mathcal{W} \\ dz &= wdt \end{aligned} \tag{1}$$

Air Pollution Modeling and Its Application XII
Edited by Sven-Erik Gryning and Nadine Chaumerliac, Plenum Press, New York, 1998

where dW are increments of a Wiener process (i.e. the increments are mutually independent and Gaussian) with mean zero and variance dt. The coefficientfunction $b(z,w)$ can be determined from eq.(1) and the Lagrangian structure function,

$$b(z,w) = \sqrt{C_0 \epsilon}$$

Here C_0 is a universal constant, the value of which is still discussed (Rotach, 1996) and ϵ is the mean rate of dissipation of turbulent energy which is a function of height and atmospheric stability.

The expression for $a(z,w)$ can be derived from the Fokker-Planck equation (the Eulerian companion of the Lagrangian Langevin equation, see e.g. Tassone (1995) for a brief overview).

$$a(z,w) = \frac{C_0 \epsilon}{2} \frac{1}{P(z,w)} \frac{\partial P(z,w)}{\partial w} - \frac{1}{P(z,w)} \frac{\partial}{\partial z} \int_{-\infty}^{w} w' P(z,w') dw'$$

where $P(z,w)$ denotes the probability density function for w.

PDF FOR THE VERTICAL VELOCITIES

For neutral boundary layers the PDF for the vertical velocitites is nearly Gaussian. When convection occurs, the PDF becomes skewed because of the presence of updrafts and downdrafts. The most probable velocity (i.e. the mode) is slightly negative.

We construct a PDF with the above mentioned properties from a Gaussian distribution

$$\begin{aligned} P(z,w) = \; & G(z,w)\{c_0 H_0(z,w) + c_1 H_1(z,w) \\ & + c_2 H_2(z,w) + c_3 H_3(z,w) + c_4 H_4(z,w)\} \end{aligned} \quad (2)$$

where $G(z,w)$ is a Gaussian distribution with mean zero and variance $\sigma(z)^2$

$$G(z,w) = \frac{1}{\sqrt{(2\pi)}\sigma(z)} \epsilon^{-\frac{w^2}{2\sigma(z)^2}} \quad (3)$$

and $H_n(z,w)$ are the Hermite polynomials:

$$H_n(z,w) = \frac{1}{G} \frac{\partial^n G(z,w)}{\partial w^n}$$

Eq. (2) is a Gram-Charlier expansion, truncated after the fourth order.

The coefficients c_n are determined by the wish that the moment equations should be satisfied, i.e.:

$$\overline{w^n} = \int_{-\infty}^{\infty} w^n P(z,w) dw$$

This gives:

$$\begin{aligned} c_0 &= \overline{w^0} = 1 \\ c_1 &= \overline{w} = 0 \\ c_2 &= \tfrac{1}{2}(\overline{w^2} - \sigma^2) \\ c_3 &= -\frac{\overline{w^3}}{6} \\ c_4 &= \tfrac{1}{24}(\overline{w^4} + 3\sigma^4 - 6\sigma^2 \overline{w^2}) \end{aligned}$$

The coefficients depend on σ which is unknown and must be determined by a closure assumption, and on the 0th- to 4th-order moments which can be parametrized.

Atmospheric Turbulence

The 2nd- to 4th-order moments are parametrized as functions of dimensionless height and atmospheric stability. The dimensionless height is given by z/z_i where z_i is the boundary layer depth and the atmospheric stability is represented by u_*/w_*. Here u_* is the friction velocity and w_* is the convective velocity scale. The stability parameter u_*/w_* cannot take values below 0.1 to 0.2 since the buoyancy-generated vertical movement creates some horizontal movement as well. This means that u_* does not tend to zero under strongly convective conditions, more likely it approaches a small fraction of w_*, e.g. $0.1w_*$ to $0.2w_*$. In the present paper $(u_*/w_*)_{\text{fully convective}} = 0.2$ is used. The stability parameter is related to another commonly used stability parameter z_i/L through $(u_*/w_*)^3 = k/(-z_i/L)$, where L is the Monin-Obukhov length and $k = 0.4$ is the von Karman constant.
Following Tassone *et al.* (1994) the following parameterizations are used:

$$\frac{\epsilon z_i}{w_*^3} = 1.5 - 1.2(\frac{z}{z_i})^{1/3} + (\frac{u_*}{w_*})^3(1.07\frac{(1-z/z_i)^2}{kz/z_i} + 2.56)$$

$$\frac{\overline{w^2}}{w_*^2} = 1.5(\frac{z}{z_i})^{2/3}exp(-2\frac{z}{z_i}) + (1.7 - \frac{z}{z_i})(\frac{u_*}{w_*})^2$$

$$\frac{\overline{w^3}}{w_*^3} = 1.25(\frac{z}{z_i})(1 - 0.8\frac{z}{z_i})^2$$

For the fourth-order moment no prevalent parameterization exists. From general statistics it is known that the the fourth-order moment can be expressed as

$$\frac{\overline{w^4}}{w_*^4} = (3 + K(\frac{u_*}{w_*}, \frac{z}{z_i}))(\frac{\overline{w^2}}{w_*^2})^2$$

where $K(\frac{u_*}{w_*}, \frac{z}{z_i})$ is kurtosis for the considered PDF. K should equal zero in the Gaussian limits $z/z_i \rightarrow 0$ (near the ground) and $u_*/w_* \rightarrow \infty$ (going towards neutral conditions). Tassone *et al.* (1994) argues that K should equal 0.5 in the convective limit. An ad-hoc expression for $K(\frac{u_*}{w_*}, \frac{z}{z_i})$ is constructed from a parabola:

$$K(\frac{u_*}{w_*}, \frac{z}{z_i}) = 2(\frac{z}{z_i})(1 - \frac{z}{z_i})\frac{1}{(\frac{u_*}{w_*})}(\frac{u_*}{w_*})_{\text{fully convective}}$$

This expression equals zero at the bottom and the top of the boundary layer, and it takes its maximum value under fully convective conditions for $z/z_i = 0.5$ i.e. $K(\frac{u_*}{w_*} = 0.2, \frac{z}{z_i} = 0.5) = 0.5$.

Closure Assumptions

It is tempting to use $\sigma^2 = \overline{w^2}$ as closure assumption, because it would make the second-order term of eq. (2) vanish. But this has the unpleasant side-effect that the PDF takes negative values at certain w's for some (convective) stabilities.

So the question is: Is it possible to choose σ^2 so that $P(z, w)$ is positive for all w under certain conditions, i.e. special parameterizations of the 2nd- to 4th-order moment and a restricted stability interval?

It seems reasonable that though the resulting skewed PDF (eq. (2)) and the Gaussian distribution it has been constructed from (eq. (3)) need not have the same variance,

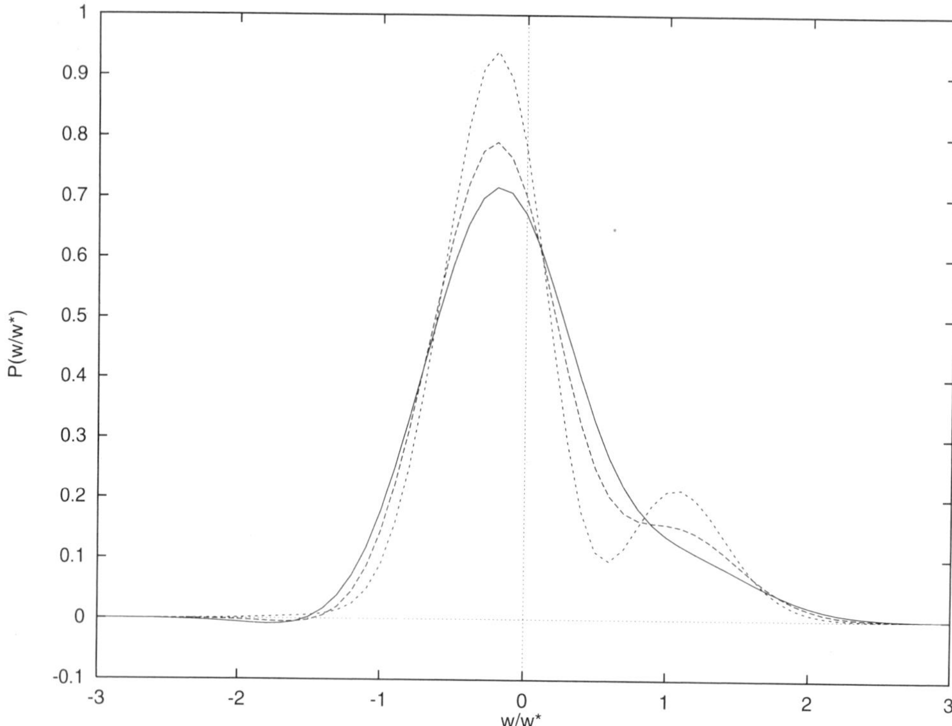

Figure 1. $P(z, w)$ for three values of f. $u_*/w_* = 0.2$, $z/z_i = 0.5$.$f = 1.0$ (solid line), $f = 0.9$ (dashed line), $f = 0.8$ (dotted line)

the variances of the two distributions must be close. Hence σ^2 is chosen to be a fraction of $\overline{w^2}$, $\sigma^2 = f^2\overline{w^2}$. This assumption does not close the system, the problem has just been transmitted into the problem of finding a proper value for f.

Numerical investigation shows that values of f less than 1 push the left tail of the PDF upwards, but it has the expense that a dive (a local minimum) on slightly positive w may occur. As f is lowered this dive becomes deeper. In fig. 1 $P(z, w)$ for three values of f can be viewed. The dive is an undesired side-effect of forcing the PDF to be positive - experimental data do not indicate a dive; at most a point of inflection. Fig. 2 shows the analytical PDF together with those obtained from the watertank experiments by Willis and Deardorff (1985).

Here it must be emphasized that the way of constructing the PDF by expanding after Hermite polynomials is an ad-hoc approach that (when comparing to measurements) seems to approximate experimental data quite well.

This is the reason why we choose to take that specific value for f, which just boundary layer and one from the top. The agreement between simulated and analytical PDF is very good in the middle of the boundary layer, while some discrepancies can be observed at the top. These discrepancies can be attributed to the perfect reflection condition; since the third-order moment does not tend to zero at the top of the boundary layer, the PDF is still skewed. As discussed by (among others) Tassone (1995) the particles reaching the upper boundary will be assigned too large negative velocities when reflected.

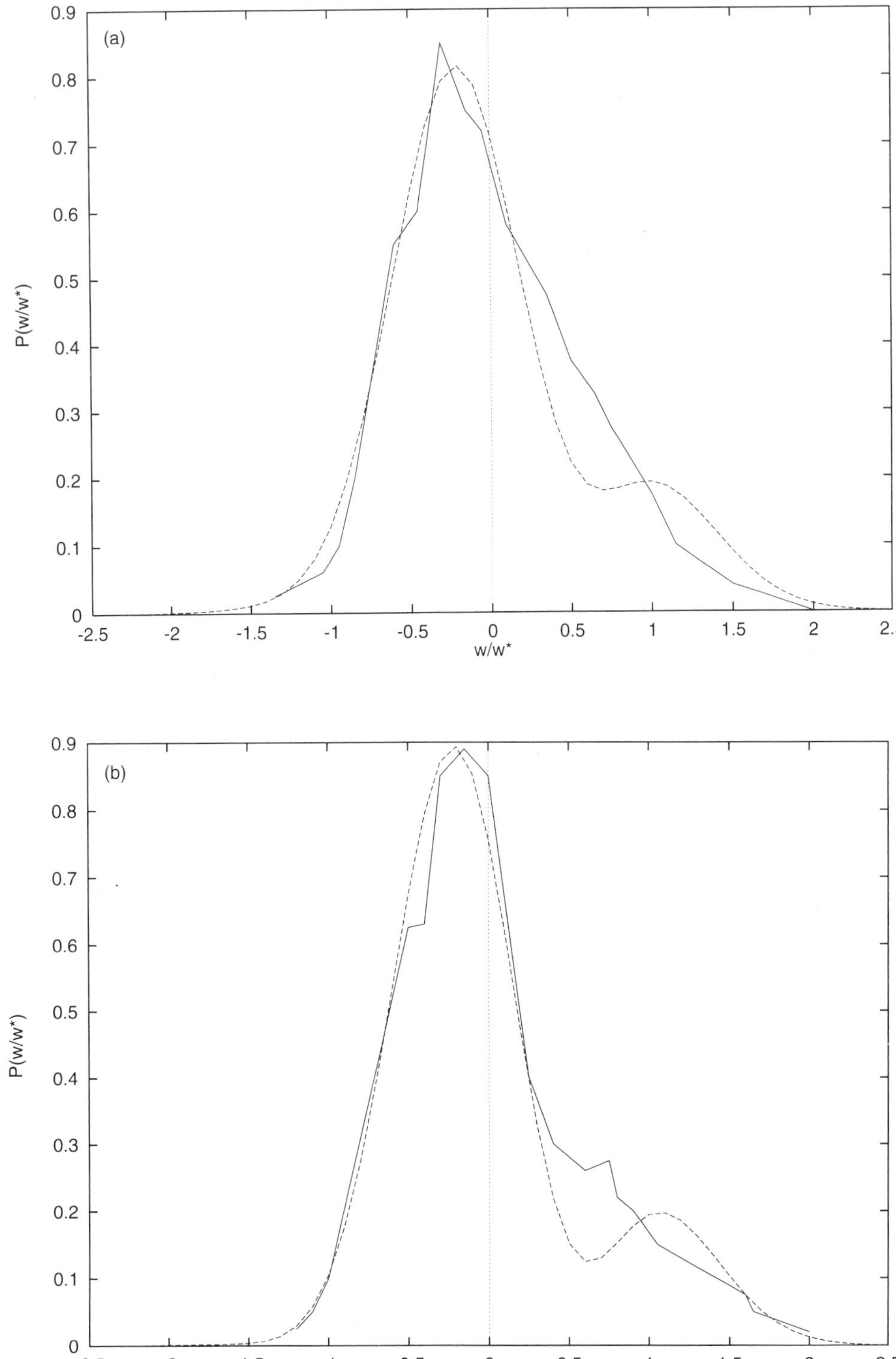

Figure 2. Comparison between analytical PDF (solid line) and experimental PDF (dashed line). $u_*/w_* = 0.2$, $f = 0.82$. (a) $z/z_i = 0.21$; (b) $z/z_i = 0.48$.

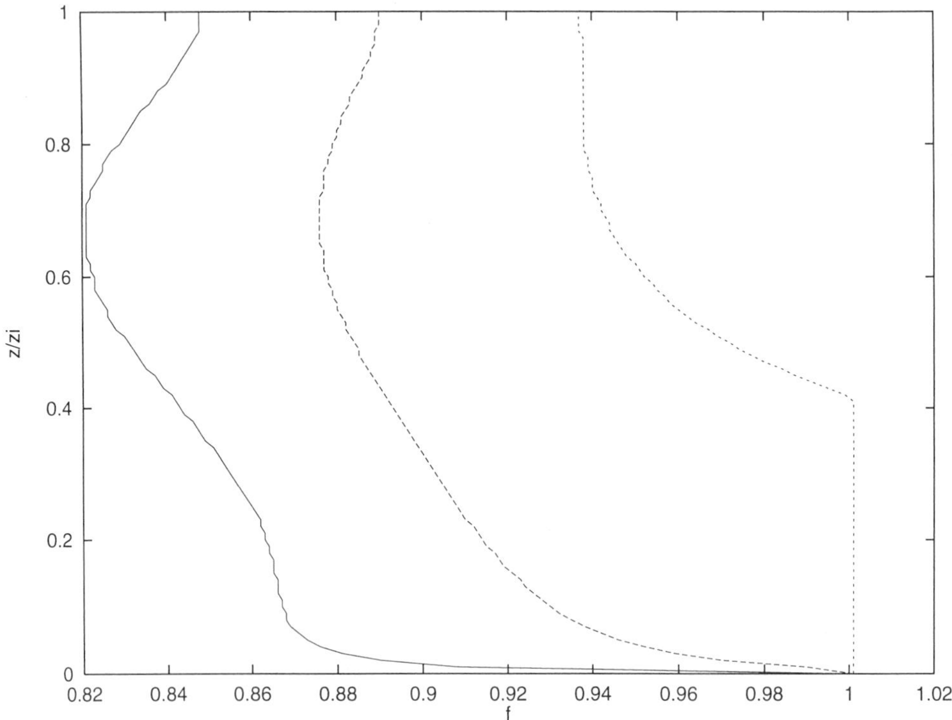

Figure 3. f as a function of height for three stabilities. Solid line: $u_*/w_* = 0.2$, dashed line: $u_*/w_* = 0.4$, dotted line: $u_*/w_* = 0.6$

RESULTS AND DISCUSSION

In order to simulate the watertank experiments (Willis and Deardorff; 1976, 1978, 1981) the model was run with the relevant source-heights: $z_s/z_i = 0.067, z_s/z_i = 0.24$ and $z_s/z_i = 0.49$. In each run 5000 particles were released at the given source-height and tracked to $T = 5$. From the simulated data the mean particle height, the standard deviation with respect to the source-height and the crosswind-integrated concentrations ensures that $P(z, w)$ becomes positive on the left side, in order not to disturb the right side too much. When it is possible, $f = 1.0$ will be chosen. Fig. 3 shows the calculated values for f as a function of height for three stabilities. Since the problem of 'negative probabilities' is larger the more skewed the PDF is, it is no surprice that f is closer to 1.0 at the top and the bottom of the boundary layer than in the middle where the PDF is most skewed.

NUMERICAL DETAILS

Before implementation in a computer program all lengths were non-dimensionalized by z_i and all velocities were scaled by w_*. These two scaling variables can be combined to form the dimensionless time, $T = t\frac{w_*}{z_i}$.
Since the aim is to reproduce the strongly convective watertank experiments of Willis and Deardorff (1976, 1978, 1981), $\frac{u_*}{w_*} = 0.2$ is used in the simulations.
We choose to use an f that is constant over the height, i.e. the value that is necessary

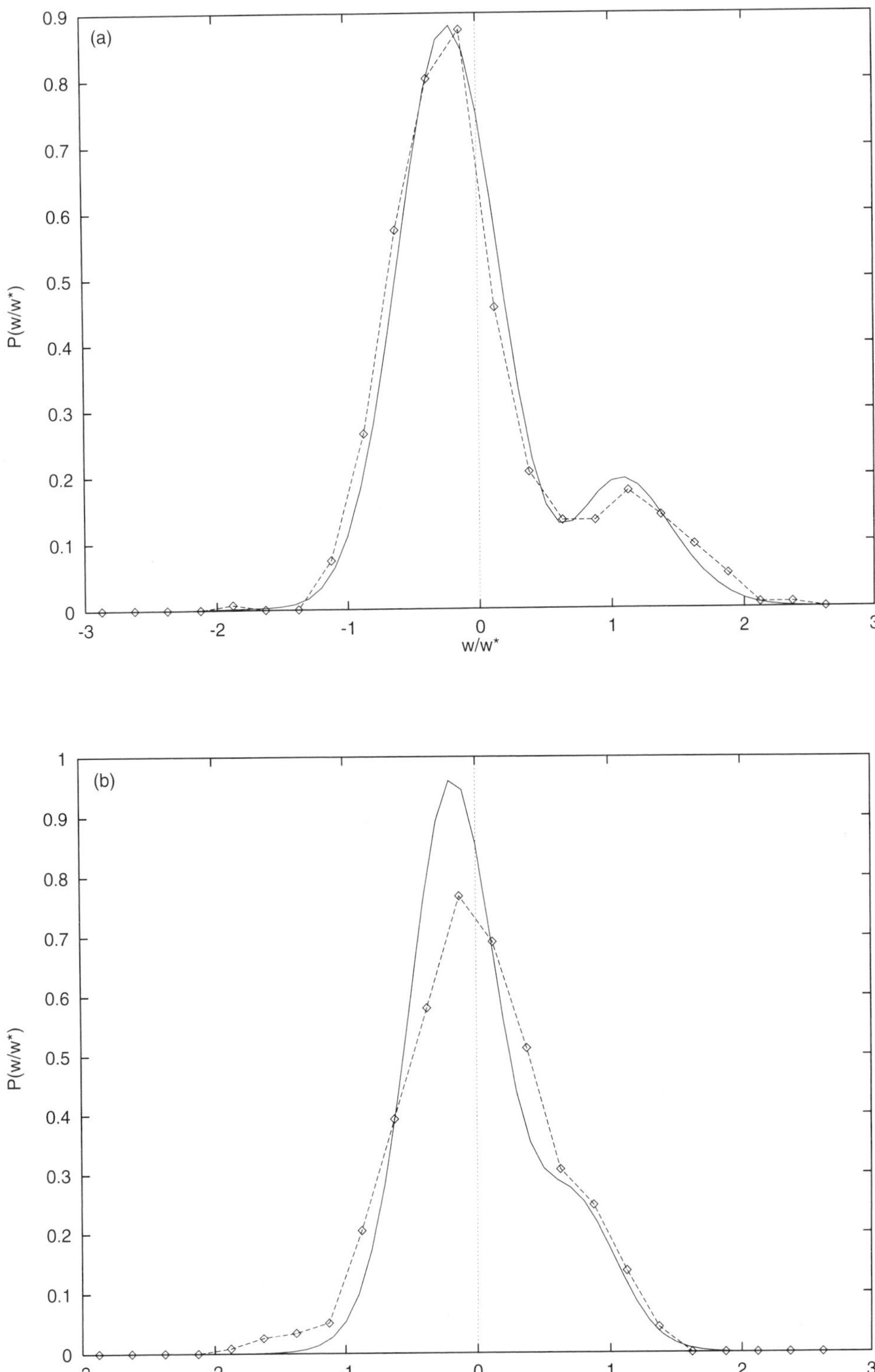

Figure 4. Simulated PDF (dashed line with diamonds) compared to the analytical input (solid line). (a) analytical: $z/z_i = 0.45$, simulated: $z/z_i \in [0.4, 0.5]$; (b) analytical: $z/z_i = 0.95$, simulated: $z/z_i \in [0.9, 1.0]$

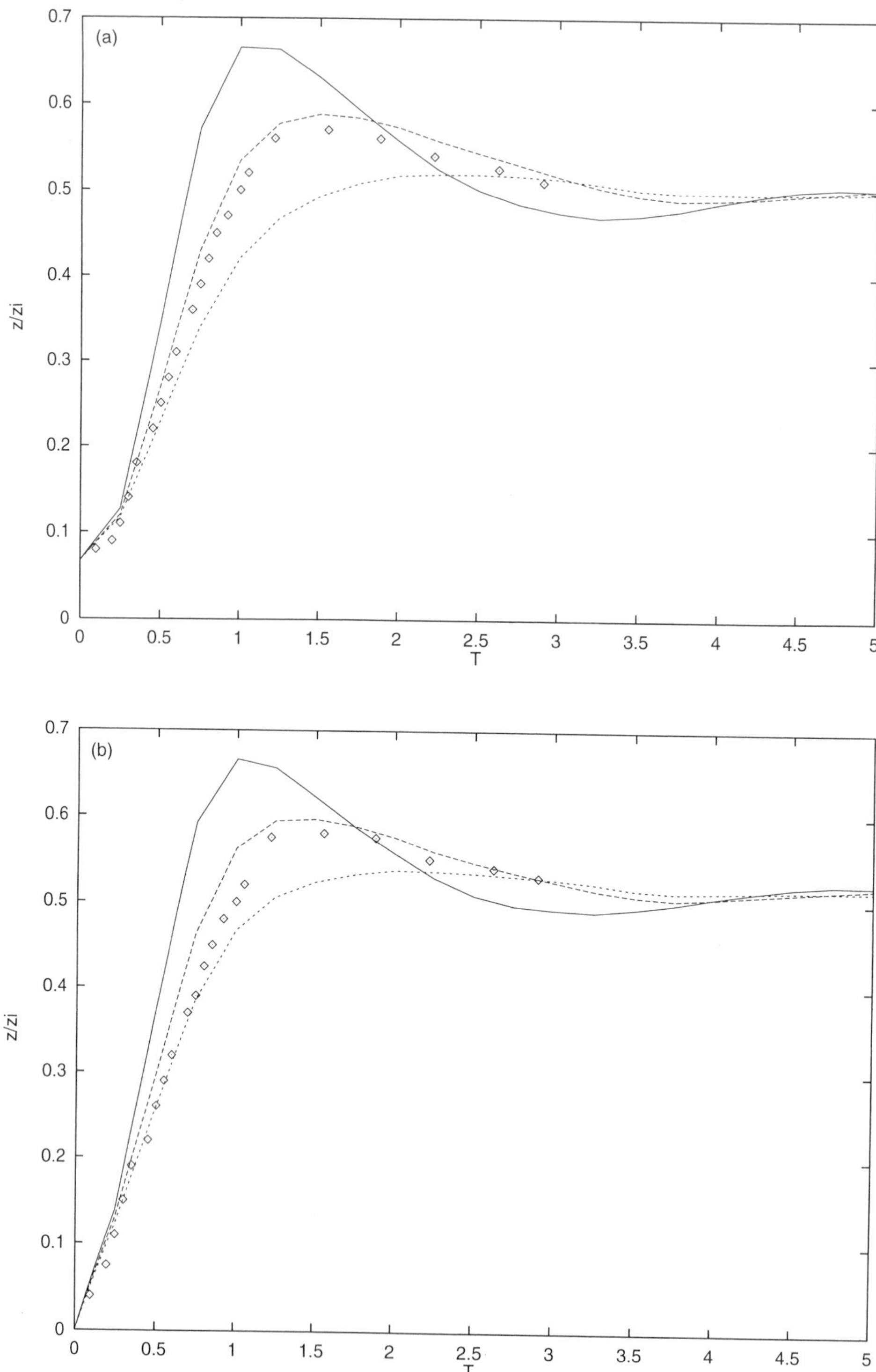

Figure 5. simulated mean particle height (a) and standard deviation (b) for a source-height $z_s/z_i = 0.067$. Solid line: $C_0 = 2$, dashed line: $C_0 = 4$, dotted line: $C_0 = 6$, diamonds: Willis and Deardorff (1976)

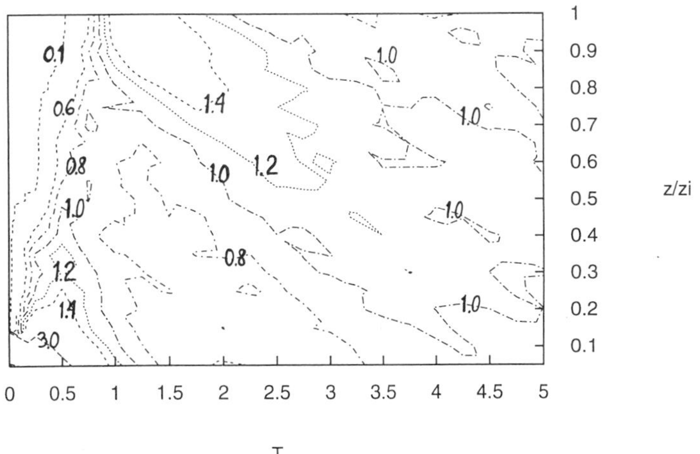

Figure 6. Simulated crosswind integrated concentrations. $z_s/z_i = 0.067$, $u_*/w_* = 0.2$, $f = 0.82$.

to keep $P(z, w)$ positive in the middle of the boundary layer is used at all heights. For the chosen stability fig. 3 gives $f = 0.82$.
Perfectly reflecting boundaries are introduced at $\frac{z}{z_i} = 0.001$ and $\frac{z}{z_i} = 0.999$.
The Langevin equation is integrated numerically by the forward Euler method, using a timestep $\Delta T = 0.001$.

It was investigated whether particles that initially are uniformly distributed remain so during the simulation. 5000 particles were distributed uniformly over the height and tracked to $T = 5$. For every $250\Delta T (= 0.25T)$ a χ^2-test was performed to check the uniformity, and the hypothesis of a uniform distribution is accepted on all levels of significance less than 5%.

The same simulation was used to check the simulated PDF: The vertical range was divided into 10 subranges of equal size. At a given time histograms of the velocities of the particles positioned in each subrange were calculated and graphically compared to the analytical input. Two examples are shown in fig. 4, one from the middle of the were calculated for every $0.25T$. Only results from the near-ground source ($z_s/z_i = 0.067$) will be shown. In fig. 5 the simulated mean particle height and standard deviation for three choices of C_0 are shown together with the observed values obtained by Willis and Deardorff (1976). When the particles become well-mixed all three of the simulated mean heights approach $z/z_i = 0.5$, and the simulated standard deviations approach an equilibrium value too. As discussed by Rotach (1996) a smaller value of C_0 increases the overshooting, while a larger value counteracts this phenomenon. Fig. 5 indicates that the model shows the best performance when $C_0 = 4$ is used. This is also the case for the two other release heights (not shown). This value of C_0 is twice as big as the value of $C_0 = 2$ that is commonly used for this type of simulation under convective conditions and it is thus closer to the theoretical value of C_0, which ranges from 5 to 10 (Rotach, 1996). When $C_0 = 4$ is used, the crosswind-integrated concentrations in fig. 6 are obtained. As can be seen, the plume rises from the ground until it reaches an elevated maximum around $T = 1.5$. After $T = 3$ the particles become well-mixed. As well the concentration at the elevated maximum as the height where it occurs are slightly overestimated compared to the measurements of Willis and Deardorff (1976); but it must be concluded that the model is able to simulate the main results, namely the rise of the plume, the overshooting of the mean value and the standard deviation, and that the particles end up being well-mixed.

REFERENCES

Deardorff,J.W. and Willis,G.E., 1985, *Further results from a laboratory model of the convective planetary boundary layer*, Boundary-Layer Meteorology, **32** pp. 205-236

Rotach,M.W., Gryning,S.-E. and Tassone,C., 1996, *A two-dimensional Lagrangian stochastic dispersion model for daytime conditions*, Q.J.R.Met.Soc., **122**, pp. 367-389

Tassone,C., Gryning,S.-E., Rotach,M.W., 1994, *A random walk model for atmospheric dispersion in the daytime boundary layer.* In Air Pollution Modelling and its Application X, Edited by Gryning, S.-E. and Millan,M.M., Plenum, New York and London, pp.204-250

Tassone,C., 1995, *A Lagrangian model for dispersion in the daytime boundary layer*, Ph.D.-thesis, Risø National Laboratory

Willis,G.E. and Deardorff,J.W., 1976, *A laboratory model of diffusion into the convective planetary boundary layer*, Q.J.R.Met.Soc., **102**, pp. 427-445

Willis,G.E. and Deardorff,J.W., 1978, *A laboratory study of dispersion from an eeavated source within a modeled convective planetary boundary layer*, Atmospheric Environment, **12**, pp. 1305-1311

Willis,G.E. and Deardorff,J.W., 1981, *A laboratory study of dispersion from a source in the middle of the convectively mixed layer*, Atmospheric Environment, **15**, pp. 109-117

DISCUSSION

P. SEIBERT: Why don't you simply clip the negative values of the negative PDF? Is it bad to clip them?

A. K. VINTHER JENSEN: From a mathematical point of view it is not correct to cut off the negative values of the PDF. Another reason for "forcing the PDF to be positive" is that this procedure increases the peakvalue of the mode. Comparison with the watertank experiments shows that this peakvalue (at least in convective cases) is better reproduced with this "modified Gram-Charlier expansion". The peakvalue is important since it corresponds to the fraction of particles having the most frequent velocity.

H. VAN DOP: Your choice for f in the closure relation is close to the 2nd example of Bill Physick discussed yesterday (his value for f=2/3). Can you comment on the differences between the bimodal Gaussian approach and your approach using the Hermite polynomials? Are there any computational advantages?

A. K. VINTHER JENSEN: The advantage of using the Hermite polynomials is that the coefficient a(z,w) in the Langevin equation can be calculated directly without solving any implicit systems of equations. This property is not ruined by adding higher order terms to the expansion (if higher than 4th order moments should be available).

D. STEYN: Can you comment on the physical meaning of a minimum (rather than inflection point as observed) in he right tail of the PDF? Does it imply a preferred scale f motion?

A. K. VINTH JENSEN: The Gram-Charlier expansion itself contains no information of the physics. The dive might be a result of neglecting higher order terms.

D. ANFOSSI: This is a comment about Dr. Steyns question. What is important in a stochastic Lagrangian model is not the "actual" form of the PDF (either a bi-Gaussian or a Gram-Charlier one or any else) but to be able to reproduce correctly the moments of the PDF (1st, 2nd, 3rd, 4th...). In other words the physical information is not contained in the specific form of the PDF, but only in the parametrized moments of the PDF. The second comment is about Dr. van Dops question: does the

Gram-Charlier PDF allow quicker simulations than those based on the bi-Gaussian PDF? My answer, based on numerical experiment we performed and that will be presented tomorrow morning in this conference is that with the Gram-Charlier PDF one needs about 20 to 30 percent CPU time less for a typical convective dispersion simulation.

NUMERICAL SIMULATION OF A TURBULENT REACTIVE PLUME USING A NEW ANALYTICAL MODEL FOR THE INFLUENCE OF MICROMIXING

L.Delamare[1], M.Gonzalez, A.Coppalle

Université de Rouen, UMR CNRS 6614/CORIA
Mont Saint Aignan, France

INTRODUCTION

The objective of this work is to derive a model, simple yet effective, for the influence of microscale turbulence on nonlinear chemical reactions in a turbulent reactive plumes, such as the photochemical reactions between nitric oxides and ozone.

Available models are either valid only in the case of very fast chemistry relative to the turbulent processes or they involve additionnal transport equations for turbulent covariances or even for joint p.d.f.s of reactive species and then demand an important computational effort(Lamb and Shu[1];Georgopoulos and Seinfeld[2]; Vilà-Guerau de Arellano and Duynkerke[3]; Galmarini *et al.*[4]; Gonzalez[5]). The analytical model we derived can be considered a compromise between these two approaches. We used the characteristics of the photochemical reactive plume (i.e the concentration of NO is assumed to be a lot larger than the background concentration of O_3) to simplify a general micromixing model for turbulent reactive flows initially used in chemical engineering, the I.E.M. (Interaction by Exchange with the Mean) model(Villermaux[6]; Borghi[7]; Fraigneau *et al.*[8]).

THE I.E.M. MICROMIXING MODEL AND ITS ALGEBRAIC REDUCTION

Influence of Micromixing on Nonlinear Chemical Reactions

If one considers the photochemical system NO, NO_2, O_3, the mean reaction rates associated to the mean concentrations write (Donaldson and Hilst[9]):

$$\begin{aligned}\overline{R_{NO}} &= k_1\overline{C_{NO_2}} - k_3\overline{C_{NO}C_{O_3}}\\ \overline{R_{NO_2}} &= -k_1\overline{C_{NO_2}} + k_3\overline{C_{NO}C_{O_3}}\\ \overline{R_{O_3}} &= k_1\overline{C_{NO_2}} - k_3\overline{C_{NO}C_{O_3}}\end{aligned} \qquad (1)$$

In a nonpremixed situation such as a reactive plume released from a point source, the covariance $\overline{C_{NO}'C_{O_3}'}$ is negative and the effect of turbulent fluctuations is to inhibit the $NO - O_3$ reaction. This influence is felt only if the fluctuations are not dissipated before chemistry takes place, that is if the chemistry is not slow relative to the molecular dissipation of the fluctuations. Inversely, if the chemistry is very fast compared to the dissipation, there is no interaction between the chemical processes and the turbulent ones and it is therefore possible to relate the segregation coefficient $I_S = \frac{\overline{C_{NO}'C_{O_3}'}}{\overline{C_{NO}}\cdot\overline{C_{O_3}}}$ to the intensity of the fluctuations of an inert species. A micromixing model is needed if the chemical and the dissipation time scales overlap. In this case, there is a complex interaction between chemistry and turbulence and it is necessary to estimate the state of mixing between the reactive species to predict the mean concentration fields.

[1]Tel: 33 (0)2 3514 6569. Fax: 33 (0)2 3570 8384 E-mail: Ludovic.Delamare@coria.fr

Air Pollution Modeling and Its Application XII
Edited by Sven-Erik Gryning and Nadine Chaumerliac, Plenum Press, New York, 1998

The I.E.M. Model

If one models molecular diffusion as a term of exchange with the mean environment at a frequency $1/\tau_I$, lagrangian transport equations for the instantaneous concentrations write :

$$\frac{dC_i}{dt} = \frac{\overline{C_i} - C_i}{\tau_I} + R_i, \quad i = NO, NO_2, O_3 \tag{2}$$

where $\overline{C_i}$ is given by solving the complete Eulerian transport equation (including advection and macromixing) and τ_I is proportional to the scalar dissipation time scale.

If we define ϕ, an inert species such as $\phi = 0$ in the ambient atmosphere and $\phi = 1$ at the source of emissions, the IEM equations in the composition space (ϕ, C_i) write :

$$(\overline{\phi} - \phi)\frac{dC_i}{d\phi} = \overline{C_i} - C_i + \tau_I R_i, \quad i = NO, NO_2, O_3 \tag{3}$$

with boundary conditions :

- $\phi = 0 \; C_i = C_{iA}$
- $\phi = 1 \; C_i = C_{iS}$

Solving this set of equations gives $C_i(\phi)$, the "trajectories" in the composition space.

It is then possible to relate the joint p.d.f.s of the reactive species to the p.d.f. of an inert species ϕ and one can write :

$$\overline{C_i} = \int_0^1 C_i(\phi)\overline{P}(\phi)d\phi, \quad i = NO, NO_2, O_3 \tag{4}$$

$$\overline{C_{NO}C_{O_3}} = \int_0^1 C_{NO}(\phi)C_{O_3}(\phi)\overline{P}(\phi)d\phi \tag{5}$$

where $\overline{P}(\phi)$ is the p.d.f. of ϕ.

In this work, we used simple presumed shapes for this p.d.f. depending on the two first moments $\overline{\phi}$ and $\overline{\phi'^2}$ (Borghi and Moreau[10];Fraigneau *et al.*[8]).

Reduction of the IEM procedure

We define two "branches" of integration : the "ambient" branch $(0 < \phi < \overline{\phi})$ and the "source" branch $(\overline{\phi} < \phi < 1)$. The point $(\phi = \overline{\phi})$ is a discontinuity point that has to be solved separately.

- The equilibrium point

  The point $(\phi = \overline{\phi})$ is the attractor point of the IEM system where the right hand sides of the IEM equations are all equal to zero. The resulting set of algebraïc equations then gives :

$$\begin{aligned} C_{NO}(\overline{\phi}) &= \alpha_1 + \beta X \\ C_{NO_2}(\overline{\phi}) &= \frac{\overline{C_{NO_2}} + k_3\tau_I X}{1 + k_1\tau_I} \\ C_{O_3}(\overline{\phi}) &= \alpha_2 + \beta X \end{aligned} \tag{6}$$

  where

$$X = \frac{1 - (\alpha_1 + \alpha_2)\beta}{2\beta^2}(1 - \sqrt{\delta}) \tag{7}$$

$$\delta = 1 - \frac{4\alpha_1\alpha_2\beta^2}{(1 - (\alpha_1 + \alpha_2)\beta)^2}$$

$$\alpha_1 = \overline{C_{NO}} + \frac{k_1\tau_I}{1 + k_1\tau_I}\overline{C_{NO_2}}$$

$$\alpha_2 = \overline{C_{O_3}} + \frac{k_1\tau_I}{1 + k_1\tau_I}\overline{C_{NO_2}}$$

$$\beta = k_3\tau_I(\frac{k_1\tau_I}{1 + k_1\tau_I} - 1)$$

- The "source" branch

  The nonlinearity of the system 3 can be measured by the Damköhler numbers $Da_{O_3} = k_3\tau_I C_{O_3}$ and $Da_{NO} = k_3\tau_I C_{NO}$, for the NO and the O_3 equations respectively. Maximum values of these Damköhler numbers on the "source" branch are $Da_{NOS} = k_3\tau_I C_{NO}(1)$ and $Da_{O_3E} = k_3\tau_I C_{O_3}(\overline{\phi})$ and in a typical plume situation , $Da_{NOS} >> 1$ and $Da_{O_3E} << 1$.

  Hence, one can make the following assumptions :

  1. The NO trajectory is only slightly affected by the nonlinearity of the reactions. It can thus be approximated by a simple linear relation connecting the two known limit points $(\phi = \overline{\phi}, C_{NO} = C_{NO}(\overline{\phi}))$ and $(\phi = 1, C_{NO} = C_{NO}(1))$. That is :
  $$C_{NO}(\phi) = C_{NO}(1) + (C_{NO}(\overline{\phi}) - C_{NO}(1))\frac{\phi - 1}{\overline{\phi} - 1} \tag{8}$$
  2. The O_3 chemical consumption is so fast relatively to the NO one that we can consider O_3 is at the equilibrium. That is :
  $$C_{O_3}(\phi) = \frac{\overline{C_{O_3}} + k_1\tau_I C_{NO_2}(\phi)}{1 + k_3\tau_I C_{NO}(\phi)} \tag{9}$$
  3. Using the fact that $NO + NO_2$ behaves like an inert species, we also have :
  $$C_{NO_2}(\phi) = (C_{NO}(1) + C_{NO_2}(1))\,\phi + (C_{NO}(0) + C_{NO_2}(0))\,(1 - \phi) - C_{NO}(\phi) \tag{10}$$

- The "ambient" branch

  For this branch, the maximum Damköhler numbers are $Da_{NOE} = k_3\tau_I C_{NO}(\overline{\phi})$ and $Da_{O_3A} = k_3\tau_I C_{O_3}(0)$. Practically, $Da_{O_3A} << 1$ and, except in the immediate vicinity of the source, $Da_{NOE} \approx 1$. Consequently, nonlinearity is not dominant for this branch and it is possible to linearize the system of IEM equations in order to solve it. Linearizing around $\phi = 0$ and expressing the "trajectories" as analytical series :
  $$C_{NO} = \sum_{k=0}^{n} A_k\phi^k;\ C_{NO_2} = \sum_{k=0}^{n} B_k\phi^k;\ C_{O_3} = \sum_{k=0}^{n} C_k\phi^k \tag{11}$$
  one can obtain the following series coefficients :
  $$\begin{aligned}
  A_0 &= C_{NO}(0);\ B_0 = C_{NO_2}(0);\ C_0 = C_{O_3}(0) \\
  A_1 &= \frac{\overline{C_{NO}} + k_1\tau_I B_0 - (1 + k_3\tau_I C_0)A_0}{\overline{\phi}} \\
  B_1 &= \frac{\overline{C_{NO_2}} - (1 + k_1\tau_I)B_0 + k_3\tau_I A_0 C_0}{\overline{\phi}} \\
  C_1 &= \frac{\overline{C_{O_3}} + k_1\tau_I B_0 - (1 + k_3\tau_I A_0)C_0}{\overline{\phi}} \\
  A_{k+1} &= \frac{(k-1)A_k + k_1\tau_I B_k - k_3\tau_I \sum_{j=0}^{k} A_j C_{k-j}}{(k+1)\overline{\phi}} \\
  B_{k+1} &= \frac{(k - 1 - k_1\tau_I)B_k + k_3\tau_I \sum_{j=0}^{k} A_j C_{k-j}}{(k+1)\overline{\phi}} \\
  C_{k+1} &= \frac{(k-1)C_k k_1\tau_I B_k - k_3\tau_I \sum_{j=0}^{k} A_j C_{k-j}}{(k+1)\overline{\phi}}
  \end{aligned} \tag{12}$$

  Relations 8,9,10 and 11 provide a complete analytical approximation of the "trajectories" in the composition space.

The segregation coefficient I_S

The simple presumed shapes we use for the p.d.f.s of an inert species can be defined by five general parameters :

- δ_0 : Dirac peak at $\phi = 0$. Generally occurs at the boundaries of the plume in the early stages of dispersion when the fluctuations intensity is high.

- δ_1 : Dirac peak at $\phi = 1$. Generally occurs on the axis of the p lume in the early stages of dispersion.
- α, β, γ : parameters defining respectively the values of ϕ limiting the rectangle and the height of this rectangle. This rectangle is in fact a crude approximation of a Gaussian p.d.f..

Application of relations 4 and 5 then gives :

$$\begin{aligned} < C_i > &= \delta_0 C_i(0) + \gamma \int_{\alpha}^{\overline{\phi}} C_i^A(\phi) d\phi \\ &+ \delta_1 C_i(1) + \gamma \int_{\overline{\phi}}^{\beta} C_i^S(\phi) d\phi, \quad i = NO, NO_2, O_3 \end{aligned} \tag{13}$$

$$\begin{aligned} < C_{NO} C_{O_3} > &= \delta_0 C_{NO}(0) C_{O_3}(0) + \gamma \int_{\alpha}^{\overline{\phi}} C_{NO}^A(\phi) C_{O_3}^A(\phi) d\phi \\ &+ \delta_1 C_{NO}(1) C_{O_3}(1) + \gamma \int_{\overline{\phi}}^{\beta} C_{NO}^S(\phi) C_{O_3}^S(\phi) d\phi \end{aligned} \tag{14}$$

where the superscript A corresponds to the analytical series 11 and the superscript S to the relations 8-10.

The segregation coefficient I_S is given by :

$$I_S = \frac{< C_{NO} C_{O_3} >}{< C_{NO} >< C_{O_3} >} - 1 \tag{15}$$

Closure of the mean reaction rates 1 is then provided writing :

$$\overline{C_{NO} C_{O_3}} = (1 + I_S) \overline{C_{NO}} \ \overline{C_{O_3}} \tag{16}$$

NUMERICAL SIMULATION OF A REACTIVE PLUME EXPERIMENT

In order to validate the above analytical closures, we perform comparisons with the experimental data obtained by Builtjes on a reactive plume in a wind tunnel (Builtjes[11]).

Parameters of this experiment are summarized in Table 1 and 2 (Georgopoulos and Seinfeld[2]).

Dispersion and dissipation parameters

Turbulent dispersion for this experiment is well approximated by a simple gaussian formula with the following dispersion parameters (Georgopoulos and Seinfeld[2]):

$$\begin{aligned} \sigma_y &= 0.072 x^{0.907} \\ \sigma_z &= 0.036 x^{0.907} \end{aligned} \tag{17}$$

Table 1. Builtjes experiment : NO source data

	exp.
NO concentration at the source :	3900 ppm
Source height :	0.14 m
Source exit velocity :	0.4 m/s
Source diameter :	3 mm

Table 2. Builtjes experiment : Ambient data

	exp.
O_3 background concentration :	0.35 ppm
Turbulence integral length scale :	0.3 m
Turbulence intensity :	0.1
Mean wind speed :	0.4 m/s
Boundary layer thickness :	0.8 m

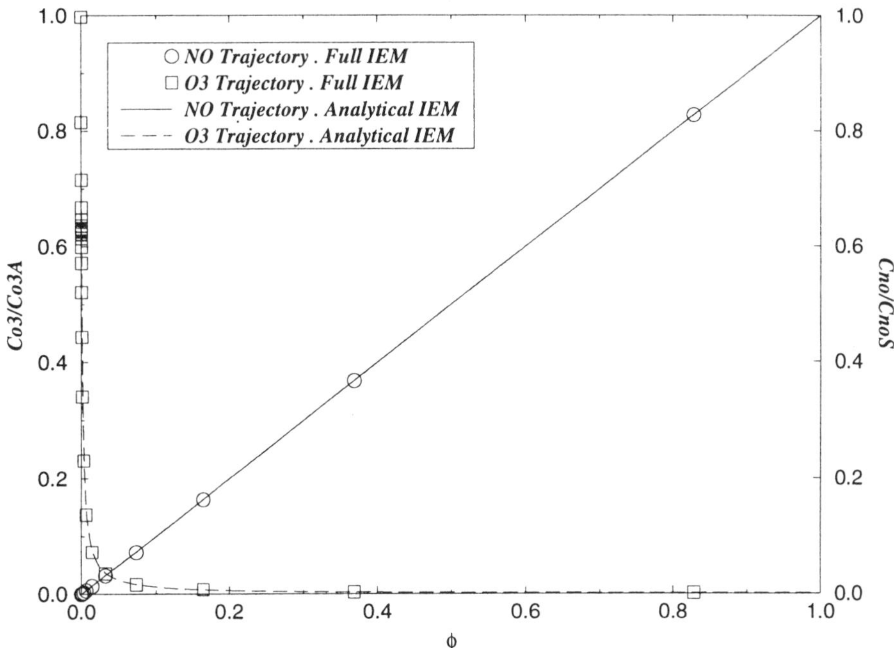

Figure 1. NO and O_3 "trajectories" at the height of the source and at a downwind distance of 0.5 m

We incorporate these relations in our simulation via equivalent turbulent diffusivities.

Another very important input of our model is the dissipation time scale τ. We use here a modified version of the model derived by Sykes,Lewellen and Parker (Sykes, Lewellen and Parker[12]). According to this model, the dissipation time scale evolves linearly with the dispersion time. That is, $\tau = \tau_0 + C_\tau t$ where τ_0 is an initial value depending on the turbulent characteristics of the flow and on the source geometry and C_τ is a constant. We obtained a good fit with the experimental data (which is very scarce concerning turbulent fluctuations) using $\tau_0 = 0$ and $C_\tau = 0.25$. The value of τ_0 is justified by the fact that in this experimental set-up, the diameter of the source is actually smaller than the Kolmogorov length scale.

We performed simulations of the reactive plume using three different models for the micromixing : the simple algebraïc covariance model of Georgopoulos and Seinfeld, the complete IEM model with full integration of the trajectories in the composition space and the analytical simplified model described above.

Analytical series 11 have been calculated up to the sixth order.

Trajectories in the composition space for NO and O_3 obtained with the full IEM model and the new analytical model are presented on Figure 1.

As expected, the NO trajectories are almost perfect straight lines and both models give the same results.

Comparisons with experimental data concerning the evolutions of the concentration of ozone and of the ratio $NO/(NO + NO_2)$ on the axis of the plume are plotted on Figures 2 and 3.

On Figure 4, we plotted the evolution of the segregation coefficient I_S

on the axis of the plume, again obtained with the three aforementioned mod els, along with recent results obtained by Gonzalez with a full p.d.f. transpor t simulation (Gonzalez[5]).

The Georgopoulos and Seinfeld model is found to underpredict the segregation coefficient in the early stages of dispersion. In fact, the assumptions we made in order to simplify the IEM equations ar e basically the sames as in the Georgopoulos-Seinfeld model, but in addition to the pure chemistry-micromixing interaction, our model, because it involves the

mean eulerian values, also takes implicitely into account the production o f covariance by the mean gradients. This effect, explicitely present in second order modelling, is expected to be important especially near the so urce and is responsible, we believe, for the slightly wrong behaviour of t he Georgopoulos and Seinfeld model.

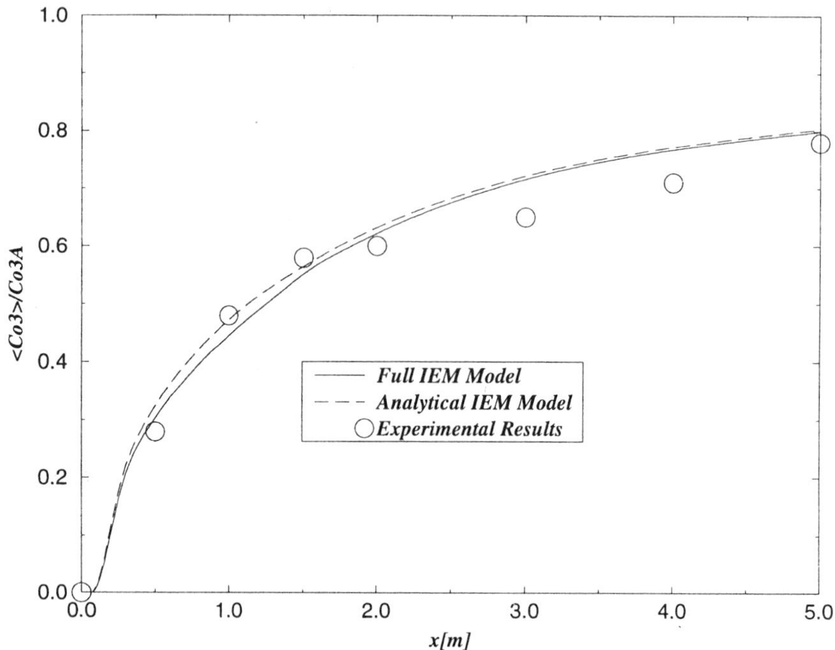

Figure 2. Ozone concentration on the axis of the plume

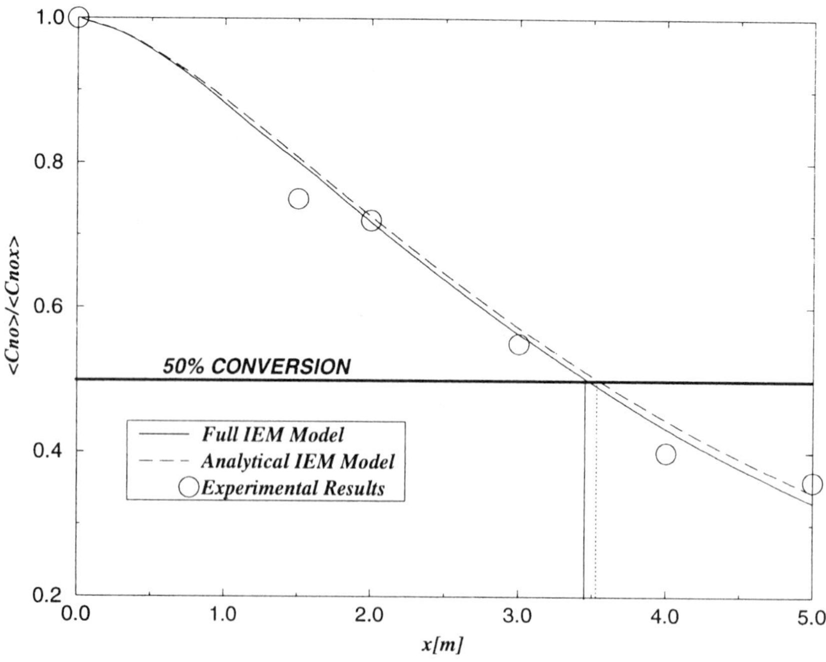

Figure 3. Ratio NO/NO_x on the axis of the plume

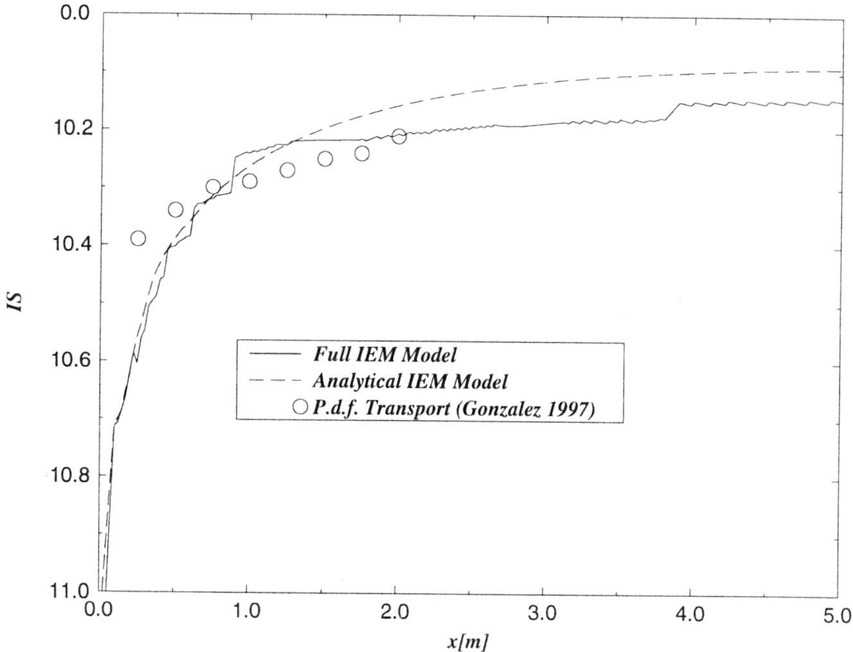

Figure 4. Segregation coefficient I_S on the axis of the plume

CONCLUSIONS

A micromixing model, formally equivalent to second order or p.d.f. transport modelling, has been degenerated in order to derive a simple analytical expression for the segregation coefficient between reactive species I_S.

Although the domain of validity of this model is necessary limited to the pure reactive plume situation, we have shown that it provides a useful and effective compromise between the simpler algebraïc models and the more complex ones involving additionnal transport equations or integrations. Computational time with this analytical model is about ten times shorter than with the full IEM model which makes it applicable to practical computations.

One can also note that the method we used to simplify the general IEM equations can be extended to other cases. This would lead to different expressions for the trajectories in the composition space but the resulting model for the segregation coefficient would still be analytical and easier to use than a more complete micromixing model.

ACKNOWLEDGMENTS

This work was funded by E.D.F (Electricté de France) through a collaboration contract #E35 L61-2L 6531AEE1793 with the C.N.R.S (Centre National de la Recherche Scientifique).

References

[1] R.G. Lamb and W.R. Shu. A model of second order chemical reactions in turbulent fluid. *Atmospheric Environment*, 12:1685–1694, 1978.

[2] P.G. Georgopoulos and J.H. Seinfeld. Mathematical modeling of turbulent reacting plumes. *Atmospheric Environment*, 20(9):1791–1807, 1986.

[3] J. Vilà-Guerau de Arellano and P.G. Duynkerke. Second-order closure study of the covariance between chemically reactive species in the surface layer. *J.Atmos.Chem.*, 16:145–155, 1993.

[4] S. Galmarini, J. Vilà-Guerau de Arellano, and P.G. Duynkerke. The effect of micro-scale turbulence on the reaction rate in a chemically reactive plume. *Atmospheric Environment*, 29:87–95, 1995.

[5] M. Gonzalez. Analysis of the effect of microscale turbulence on atmospheric chemical reactions by means of the p.d.f. approach. *Atmospheric Environment*, 31:575–586, 1997.

[6] J. Villermaux. In *Encyclopedia of fluid mechanics*, pages 375–474, West Orange, M.J., 1994. Gulf Pub. Co.

[7] R. Borghi. Turbulent combustion modelling. *Prog.Energy Combus. Sci.*, 14:245–292, 1988.

[8] Y. Fraigneau, M. Gonzalez, and A. Coppalle. The influence of turbulence upon the chemical reaction of nitric oxide released from a ground source into ambient ozone. *Atmospheric Environment*, 30:1467–1480, 1996.

[9] C. du P. Donaldson and G.R. Hilst. Effect of inhomogeneous mixing on atmospheric chemical reactions. *Envir.Sci.Technol.*, 6:812–816, 1972.

[10] R. Borghi and P. Moreau. Turbulent combustion in a premixed flow. *Acta Astronautica*, 4:321–341, 1977.

[11] P.J.H. Builtjes. Chemically reacting plume experiment for the wind tunnel. Technical Report 81-013563, Netherlands Organization for Applied Scientific Research, 1981.

[12] R.I. Sykes, W.S. Lewellen, and S.F. Parker. A turbulent transport model for concentrations fluctuations and fluxes. *J.Fluid Mech.*, 130:193–218, 1984.

DISCUSSION

D. W. BYUN: As you said, it is an interesting method. First, for clarification, what concentration units did you use for the equation (density or mixing ratio)? Second question is the generality of your method for more practical conditions, i.e. where other chemicals are presented. In your example, you have used very high NO concentrations. At this extreme, although the relation may be able to be represented analytically, the importance of the "reaction" (i.e. destruction of O_3 and NO) is dominated by the emission, so the whole solution become meaningless again. What is more important for real air quality problems is that when NO and O_3 concentrations are comparable.

L. DELAMARE: We used the very commonly used parts-per-million by volume unit (p.p.m.) for the concentrations. It should be noted that this is not really a concentration but a dimensionless volume fraction. Of course, we could have used any other unit as well. The objective of this work is to provide a model for the influence of micromixing, that is the interaction between the non linear chemical reactions and the turbulent fluctuations, not to solve all the real air quality problems. It is well known that this micromixing effect is important if the characteristic time scales of the chemical reactions and of the turbulent mixing are of the same order of magnitude. That is, if the reactions are fast enough. Hence, this effect is important in the vicinity of the source of emissions (where the NO concentrations are high) and has to be taken into account in any subgrid plume model for the treatment of strongly localized emissions (power plants, highways etc.). Of course, further downstream from the sources, numerous other phenomena become important and have to be taken into account. It is not the objective of this model and it has never been said that it was.

P. BUILTJES: Comment: The effects of micro-mixing is important in all cases when the chemical time scales are of the same order as the mixing time scales. In practice, this holds for all emissions of NO into the atmosphere (power plants, traffic) and processes of dry deposition of NO, when the NO+O_3 reaction takes place.

BUOYANT PLUME RISE CALCULATED BY LAGRANGIAN AND EULERIAN MODELLING

Stefan Heinz

Delft University of Technology
Faculty of Applied Physics, Section Heat Transfer
Lorentzweg 1, 2628 CJ Delft, The Netherlands

INTRODUCTION

Plumes emitted from stacks are usually buoyant due to temperature differences to the ambient air. Such buoyancy effects may influence strongly the turbulent mixing of the plume with the ambient air. This is essential for reactive plumes, because the mixing of compounds distributed in the ambient air and the plume determines the occurrence of reactions. Furthermore, changes in the turbulent mixing of plume and ambient flow may affect considerably the spatial patterns of the distribution of tracers and consequently their ground concentration.

These mixing processes can be described in the Eulerian approach, i.e. based on the conservation equations of mass, momentum and thermal energy, by means of assumptions on entrainment and extrainment. The idea of entrainment of air into plumes by plume-generated turbulence permits the explanation of the "two-thirds" power law for the buoyant plume rise, which is observed in a neutral stratified flow without significant turbulence.[1] For a turbulent atmosphere one finds a levelling off of the plume, that means the plume follows in the initial stage the "two-thirds" power law and at later times its mean height becomes constant. This transition and the final plume height can be explained by an extrainment proposed by Netterville, i.e. entrainment of plume material into the surrounding fluid due to the ambient turbulence.[2] This approach is able to describe the plume behaviour in complex flows, that means for an arbitrary stratification and different intensities of ambient turbulence. The applicability of this model was proved by means of field observations of plume trajectories and final rise and it was found to perform well.[2, 3]

But there are two problems related to this approach: Firstly, in particular the extrainment concept requires the knowledge of parameters which cannot be derived directly from measurements, i.e. they have to be estimated with different ad hoc assumptions which are not easy to justify.[3] Also large-eddy simulations are not instructive for the explanation of these mixing processes. Turbulence processes within the plume cannot be correctly simulated, because they occur on a subgrid scale.[4, 5] Secondly, the dispersion problem is not solved in this way. The plume is calculated as a distributed line source and a dispersion model is needed to calculate the spreading of material of this line source. This is a non-trivial problem, because of the uncertainty of this 'source strength'.

The Lagrangian approach to this problem is able to simulate both the mean plume behaviour and the dispersion of plume material as observed by van Dop.[6] The problem to explain the turbulent mixing of the plume and the ambient fluid appears here as the problem to estimate the time behaviour of the time scales which determine the dynamics of particle motion. An approach to solve this question consists in the explanation of buoyant turbulence as stochastic motion of particles and change of their temperatures.[7] Accordingly, the construction of particle models in consistency with budget equations of the turbulence permits firstly the explanation of the observed plume features, secondly, the explanation of parameters in the Eulerian approach which cannot be derived directly from measurements, and thirdly, the description of the dispersion of tracers.

These features of the Lagrangian approach are considered here in comparison with the Eulerian approach. This is done by explaining, how buoyant turbulence can be represented by a stochastic particle model. Then, the mean plume rise is calculated by these Lagrangian equations and these findings are compared with the results of the Eulerian approach.

BUOYANT TURBULENCE AS STOCHASTIC PARTICLE MOTION

One approach to explain buoyant turbulence as stochastic particle motion consists in the estimation of Lagrangian solutions of turbulence budget equations up to second-order.[7] Here, the coefficients of linear Lagrangian equations for the motion and temperatures of particles are chosen, such that budget equations of turbulence are fulfilled for the means and all the variances of the coupled Eulerian velocity-temperature fields. Consequently, the particle move and change their temperatures, where these turbulence budgets are satisfied at any time and at arbitrary positions in the flow.

Let us consider for simplicity only the vertical motion of particles in an Eulerian flow field with a mean horizontal velocity U into the x^1-direction and a mean potential temperature Θ, which depend only upon the vertical coordinate x^3. The mean vertical velocity is neglected. Lagrangian equations for the change of the particle height x_L^3 over the source, the vertical velocity U_L^3 and the potential temperature Θ_L in time t (L denotes a Lagrangian quantity) read then according to the above described approach

$$\frac{d}{dt}\langle x_L{}^3\rangle = \langle U_L{}^3\rangle, \tag{1a}$$

$$\frac{d}{dt}\langle U_L{}^3\rangle = -\frac{k_1}{4\tau}\langle U_L{}^3\rangle + \beta g\langle(\Theta_L - \Theta)\rangle, \tag{1b}$$

$$\frac{d}{dt}\beta g\langle(\Theta_L - \Theta)\rangle = -\frac{2k_3 - k_1}{4\tau}\beta g\langle(\Theta_L - \Theta)\rangle, \tag{1c}$$

where $<\cdots>$ denotes an ensemble average. The Eulerian quantity Θ is estimated at fixed heights x^3, which are replaced by the actual particle $x_L{}^3$ in the above equations. Furthermore, β is the thermal expansion coefficient, g is the acceleration due to gravity and k_1, k_3 and k_4 are parameters arising from the applied closure assumptions.[7] The values $k_1 = 8.3$, $k_3 = 6.5$ and $k_4 = 4.0$ are applied for these parameters in the calculation below. The essential quantity to be estimated is the dissipation time scale of turbulence $\tau = q^2 / (2<\varepsilon>)$, where q^2 is twice the turbulent kinetic energy (TKE) and $<\varepsilon>$ denotes the mean dissipation rate of TKE.

These equations have the same structure as those derived by van Dop,[6] but here the coefficients are estimated by the consistency with the variance budget equations. In particular the estimation of the time scales for the vertical motion of particles and their temperature changes is essential, because the time behaviour of these quantities determines the plume rise features. Usually, the particles start with a much smaller time scale than that of the ambient turbulence. This initial stage of buoyant plume rise is characterized under neutral stratification and a calm ambient turbulence by the "two-thirds" power law, that means the mean plume height grows proportional to $t^{2/3}$. At later times, τ approaches gradually to the value of the time scale of the ambient turbulence, which causes a levelling off of the mean plume height. This characteristic plume behaviour was explained by van Dop by adopting ad hoc assumptions on the time scale behaviour in these two stages. This observation enables the simulation of particle motion in accord with the similarity behaviour, but this approach raises questions on the range of applicability of these relations, the estimation of required parameters and e.g. the reflection of stratification effects.

LAGRANGIAN PLUME RISE EQUATIONS

The above described explanation of buoyant turbulence by particle motion offers the possibility to calculate τ by means of concepts of turbulence theory. In order to demonstrate the main features of this approach let us assume, that the vertical shear $\partial U / \partial x^3$ of the mean horizontal wind is constant. It is now advantageous to consider combinations of t and τ with the vertical shear, i.e. the dimensionless quantities $t' = t \cdot \partial U / \partial x^3$ and $T = \tau \cdot \partial U / \partial x^3$ are introduced. By adopting standard methods for the estimation of the mean flow frequency τ^{-1} one obtains then[8]

$$\frac{d}{dt'} T = (C_{\varepsilon 2} - 1) - 2T \cdot (C_{\varepsilon 1} - 1) \cdot \left\{ -\frac{\hat{V}^{13}}{\hat{q}^2} + \frac{\hat{V}^{34}}{\hat{q}^2} \right\}, \tag{2}$$

where $C_{\varepsilon 1} = 1.56$ and $C_{\varepsilon 2} = 1.9$ are constants. $\hat{V}^{13}$, $\hat{V}^{34}$ and $\hat{q}^2$ are dimensionless variances which satisfy the equation system

$$\frac{d}{dt'} \begin{pmatrix} \hat{V}^{13} \\ \hat{V}^{14} \\ \hat{V}^{34} \\ \hat{V}^{33} \\ \hat{V}^{44} \\ \hat{q}^2 \end{pmatrix} = \frac{1}{T} \cdot \begin{pmatrix} -k_1/2 & T & 0 & -T & 0 & 0 \\ -RiT & -k_3/2 & -T & 0 & 0 & 0 \\ 0 & 0 & -k_3/2 & -RiT & T & 0 \\ 0 & 0 & 2T & -k_1/2 & 0 & (k_1-2)/6 \\ 0 & 0 & -2RiT & 0 & -k_4 & 0 \\ -2T & 0 & 2T & 0 & 0 & -1 \end{pmatrix} \cdot \begin{pmatrix} \hat{V}^{13} \\ \hat{V}^{14} \\ \hat{V}^{34} \\ \hat{V}^{33} \\ \hat{V}^{44} \\ \hat{q}^2 \end{pmatrix}, \tag{3}$$

where the gradient Richardson number $Ri = [\beta g \, \partial\Theta / \partial x^3] / [\partial U / \partial x^3]^2$ appears, i.e. T can be calculated in this way also for unstably and stably stratified flow. The equation system (3) arises from the second-order moment equations which are proposed for the estimation of the Lagrangian equations (1a-c). The variances are normalized to twice the TKE at the initial time $q^2(t = 0)$, and the variances related to temperature fluctuations (indicated by the superscript 4) appear multiplied with a factor $\beta g (\partial U / \partial x^3)^{-1}$.

The first term on the right-hand side of (2) corresponds with the entrainment idea. T grows proportional to t', which is related to a power law for the buoyant plume rise.[6] It is worth emphasizing, that only this term appears in the original frequency equation of Kolmogorov.[8] The second term on the right-hand side of (2) depends on the state of turbulence. This term causes a decrease of T related to a levelling off of the plume (see below). These effects are just the result of the extrainment idea.

Introducing now the normalized particle height $Z = \langle x_L^3 \rangle (\partial U / \partial x^3)^2 / B_0$, the particle velocity $W = \langle U_L^3 \rangle (\partial U / \partial x^3) / B_0$ and the buoyancy $B = \beta g \langle (\Theta_L - \Theta) \rangle / B_0$, where $B_0 = \beta g \langle (\Theta_L - \Theta) \rangle (t = 0)$ is written for the initial buoyancy, the Lagrangian equations (1a-c) can be rewritten for a neutral stratification into

$$\frac{dZ}{dt'} = W, \tag{4a}$$

$$\frac{dW}{dt'} = -\frac{k_1}{4T} W + B, \tag{4b}$$

$$\frac{dB}{dt'} = -\frac{2k_3 - k_1}{4T} B. \tag{4c}$$

In order to investigate the conditions for the reproduction of the "two-thirds" power law let us consider the mean particle height Z over the source, which follows only from the first term on the right-hand side of (2). Z is then determined by

$$Z = \frac{I^{2-m_1}}{(C_{\varepsilon 2} - 1)^2 m_3} \cdot \left\{ \frac{\left(I + (C_{\varepsilon 2} - 1)t'\right)^{m_1} - I^{m_1}}{m_1} - \frac{\left(I + (C_{\varepsilon 2} - 1)t'\right)^{m_2} - I^{m_2}}{m_2} \cdot I^{m_3} \right\}, \tag{5}$$

where the abbreviations $m_1 = 2 - (2k_3 - k_1) / (4 [C_{\varepsilon 2} - 1])$, $m_2 = 1 - k_1 / (4 [C_{\varepsilon 2} - 1])$ and $m_3 = 1 - (k_3 - k_1) / (2 [C_{\varepsilon 2} - 1])$ are applied and I is written for the initial value of T. For large times one obtains

$$Z = \frac{1}{m_3 m_1} \cdot \left(\frac{I}{C_{\varepsilon 2} - 1} \right)^{2-m_1} \cdot t'^{m_1}, \tag{6}$$

when I is neglected with respect to t' and the highest power of t' is only taken into account. Hence, the "two-thirds" power law is obtained, if $m_1 = 2/3$, i.e., if

$$k_3 = \frac{k_1}{2} + \frac{8}{3} \cdot (C_{\varepsilon 2} - 1). \tag{7}$$

Provided this condition, the similarity behaviour of the buoyant plume rise appears as a consequence of the explanation of the buoyant turbulence by stochastic particle motion and the original frequency equation of Kolmogorov.

The curve (5), which follows only from the initial time scale, is shown in Figure 1 together with the "two-thirds" power law curve (6) and the curve which follows from the solution of (4a-c) combined with (2) and (3). The latter one curve is indicated by HFM to point out to the use of the homogeneous (there are no spatial transport terms) frequency model (2) and (3). The equations (4a-c) are solved for the initial conditions $Z(t = 0) = 0$, $W(t = 0) = 0$ and $B(t = 0) = 1$ and the equations (3) are solved by a Runge-Kutta procedure and with initial conditions $\hat{V}^{ij} = 1/3\, \delta_{ij}$. The initial value $I = \tau_0\, \partial U / \partial x^3$ is set to be 0.173 in accord with data of the Nanticoke plume rise measurements.[2] The curve obtained by (5) coincides in the initial stage with the HFM-curve and approaches later to the "two-thirds" power law curve (6). The HFM-curve levels off due to ambient turbulence effects. This final plume height will be discussed below after considering corresponding relations of the Eulerian approach.

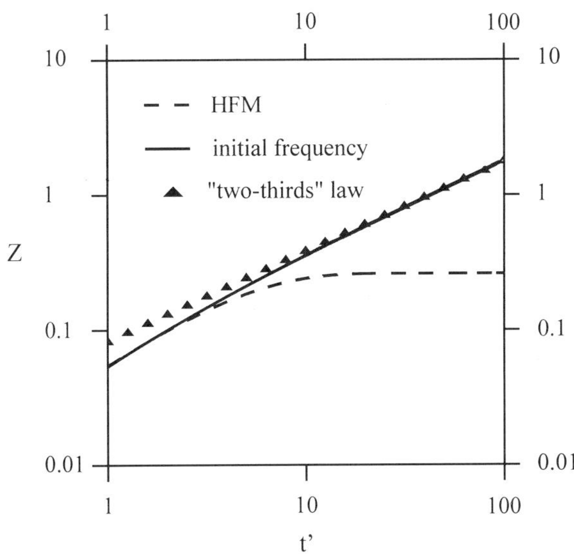

Figure 1. The dashed line gives the normalized height Z as function on t' as obtained by the equation (2). The solid curve represents the initial rise, which follows only by the initial particle frequency. The triangles represent the observed "two-thirds" power law.

EULERIAN PLUME RISE EQUATIONS

These obtained features will be compared now with the Eulerian plume rise model derived by Netterville.[2] Within this approach, the plume is considered as a superposition of an 'active plume' (the organized particle motion caused by buoyancy) around the plume centreline, and a 'passive plume' caused by turbulent dispersion of particles of the 'active plume'. The radius R of the 'active plume' is proportional to the mean plume height z over the source, $R = R_0 + \beta' z$, where β' (= 0.65) is the plume entrainment constant and R_0 is the initial plume radius in the bent-over stage. The mean plume height over the source is then given by

$$z = \left\{ \frac{3F_0}{\beta'^2\, u(f^2 + N^2)} \cdot \left[1 - \left(\cos(Nt) + f/N \cdot \sin(Nt) \right) \cdot e^{-ft} \right] + \left(\frac{R_0}{\beta} \right)^3 \right\}^{1/3} - \frac{R_0}{\beta}, \qquad (8)$$

where the contribution related to the initial momentum of the plume is neglected in comparison with the buoyancy. Here, F_0 is the initial values of the plume buoyancy, $N^2 = \beta g\, \partial\Theta / \partial x^3$ and f is a turbulence buffet frequency. The mean horizontal wind velocity u is written here small, because it is proposed to be constant within this approach.

This theory calculates the mean buoyant plume rise in dependence on the turbulence intensity and for arbitrary stratified flows. It provides the "two-thirds" power law,[9] since (8) becomes for a neutral stratification (N = 0), a laminar wind (f = 0) and a vanishing initial plume radius $R_0 = 0$,

$$z = \left\{ \frac{3F_0}{2\beta'^2\, u} \right\}^{1/3} \cdot t^{2/3}. \qquad (9)$$

The essential achievement of this theory is the explanation of the levelling off of the plume as result of extrainment, i.e. entrainment of plume material into the surrounding fluid

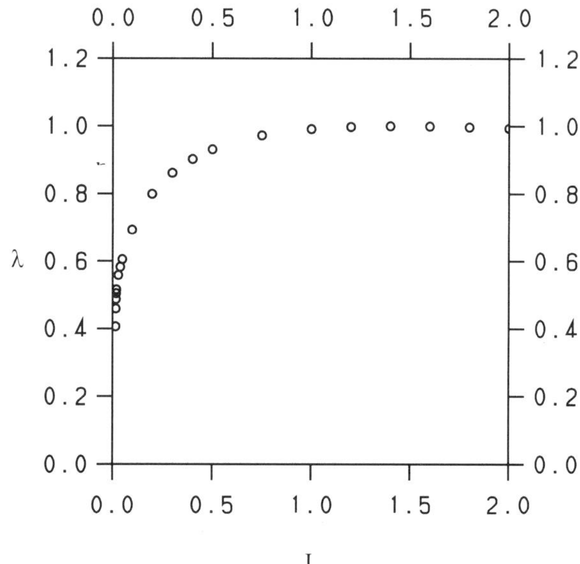

Figure 2. The factor λ in equation (13) in dependence on the normalized initial time scale.

due to the ambient turbulence. This is handled by the introduction of a turbulence buffet frequency f. This quantity is defined by $f = 2\ \beta'\ i_E\ u\ /\ \Lambda_E$, where i_E is the intensity of turbulence and Λ_E is the length scale of large-scale eddies. Neglecting again R_0, the final plume rise z is determined by (8) as

$$z = \left\{ \frac{3F_0}{\beta'^2\, u(f^2 + N^2)} \right\}^{1/3} . \tag{10}$$

Let us compare now the calculation of the plume behaviour in the Lagrangian and Eulerian approach in a non-turbulent and turbulent flow, i.e. with respect to the prediction of the "two-thirds" power law and the final plume rise, respectively. The Lagrangian approach provides the observed power law, if the relation (7) is satisfied. By adopting the definition of I it may be seen, that also the coefficients of the curves (6) and (9) are equal, when the initial particle time scale is given by

$$\tau_0 = \frac{1}{\sqrt{\beta'\beta^*}} \cdot \left(\frac{F_0}{B_0{}^3 u} \right)^{1/4} , \tag{11}$$

where

$$\frac{1}{\sqrt{\beta^*}} = \sqrt{\frac{2}{3}} \cdot \left(C_{\varepsilon 2} - 1\right) \cdot \left(1 + \frac{k_1(1 - k_3/k_1)}{2\left(C_{\varepsilon 2} - 1\right)} \right)^{3/4} \tag{12}$$

is introduced. With the above applied parameter values one finds $\beta^* = 0.65$ corresponding with the value for β applied by Netterville. For a turbulent flow, the final rise in the Lagrangian approach can be estimated numerically and is given by

$$\langle x_L{}^3 \rangle = 1.838 \cdot I \cdot \lambda(I) \cdot \frac{B_0}{\left(\partial U / \partial x^3\right)^2} , \tag{13}$$

where the curve $\lambda(I)$ is presented in Figure 8. By comparing this expression with (10) one finds, that both approaches provide the same final value, if

$$f = \beta^* \eta \cdot \frac{\partial U}{\partial x^3}, \tag{14}$$

where $\eta = 0.7\, I^{1/2}\, \lambda(I)^{-3/2}$. Consequently, by the relations (11) and (14) constraints are given for the essential parameters which determine entrainment and extrainment in the Eulerian approach, β and f. B_0 is relatively easy to derive from measurements, but this is not the case for the initial particle time scale τ_0. This problem can be avoided by applying the relation (11) as definition for τ_0. On the other hand, the relation (14) combined with (11) for the calculation of I is very helpful, because f is explained in this way by measurable quantities. Instead, this quantity is given in the Eulerian approach by $f = 2\, \beta'\, i_E\, u \,/\, \Lambda_E$, where e.g. the intensity of turbulence i_E can hardly be derived from measurements. This requires then different ad hoc assumptions in order to estimate f which are debatable.[7] Moreover, the combination of (14) with (10) represent then a simple method to calculate the final plume rise, for which a lot of different empirical formulas exist.

CONCLUDING REMARKS

The Eulerian concept provides a phenomenological theory for the buoyant plume rise in accord with measured data. It can be applied under complex conditions, i.e. for shear flows with an arbitrary stratification.[1-3] Stochastic Lagrangian models provide firstly insight into the calculation of parameters applied in the Eulerian approach as shown with respect to the estimation of the turbulence buffet frequency f, and secondly, this approach solves simultaneously the dispersion problem, that means the plume width can be estimated in contrast to the Eulerian approach, where only the radius of the 'active plume' is obtained.

By the results presented here, the extrainment idea of Netterville for a turbulent flow is proved. This permits a simple calculation of the final plume rise by means of quantities, which can be measured directly. For non-turbulent flow, the similarity behaviour of the buoyant plume rise is explained by stochastic particle motion and a frequency change according to Kolmogorov's frequency equation.

Further applications of this Lagrangian approach to plume rise simulation are in preparation. These investigations are aimed at the demonstration of the advantages for the calculation of reactive plumes by the approach to describe the turbulent mixing.

ACKNOWLEDGMENT

Many thanks to Prof. H. van Dop and Dr. D. Netterville for helpful discussions and suggestions to these topics.

REFERENCES

[1]J. C. Weil," Plume Rise". in: Lectures on Air Pollution Modeling, edited by Venkatram A. and Wyngaard J. C., American Meteorological Society, Boston, 119-166 (1988).

[2]D. D. Netterville, "Plume Rise, Entrainment and Dispersion in Turbulent Winds", Atmos. Environ. 24A, 1061-1081 (1990).

[3]G. Gangoiti, J. Sancho, G. Ibarra, L. Alonso, J. A. García, M. Navazo, N. Durana and J. L. Ilardia, "Rise of Moist Plumes from Tall Stacks in Turbulent and Stratified Atmospheres", Atmos. Environ. 31A, 253-269 (1997).

[4]F. T. M. Nieuwstadt, "A Large-Eddy Simulation of a Line Source in a Convective Atmospheric Boundary Layer - I. Dispersion Characteristics", Atmos. Environ. 26A, 485-495 (1992).

[5]F. T. M. Nieuwstadt, "A Large-Eddy Simulation of a Line Source in a Convective Atmospheric Boundary Layer - II. Dynamics of a Buoyant Line Source", Atmos. Environ. 26A, 497-503 (1992).

[6]H. van Dop, "Buoyant Plume Rise in a Lagrangian Framework", Atm. Env. 26A, 1335-1346 (1992).

[7]S. Heinz, "Nonlinear Lagrangian Equations for Turbulent Motion and Buoyancy in Inhomogeneous Flows", Phys. Fluids 9, 703-716 (1997).

[8]D. C. Wilcox, "Turbulence Modeling for CFD", DCW Industries, Inc. La Cañada, California (1993).

[9]G. A. Briggs, "Plume Rise Predictions", Lectures on Air Pollution and Environmental Impact Analyses, AMS, 59-111 (1975).

THE SEPARATION OF TRANSPORT AND CHEMISTRY IN A PHOTOCHEMICAL MODEL

Akula Venkatram[1], Shuming Du[1], Ramamurthy Hariharan[1], William Carter[1], and Robert Goldstein[2]

[1]University of California at Riverside
Riverside, CA 92521
[2]Electric Power Research Institute
Palo Alto, CA 94303

INTRODUCTION

The application of comprehensive air quality models, such as the Urban Airshed Model, is a computationally demanding exercise that has to be preceded by the compilation of extensive input data sets for emissions, meteorology, and land use for the domain of interest. These computational requirements discourage the type of numerical experimentation required for a thorough analysis of emission reduction strategies.

Even when sufficient computational resources are available, the results from comprehensive ozone models can be uncertain because of large uncertainties in both input emissions as well as the treatment of governing processes. There is a need for improving the overall approach to the modeling of photochemical air pollution. This paper suggests that a semi-empirical model can be used to supplement results from comprehensive models. The proposed semi-empirical model is based on separating transport from chemistry. In addition to increasing computational efficiency, separation of transport and chemistry facilitates the examination of the individual processes that govern ozone levels. The structure of the uncoupled model allows calibration of model results to observations by adjusting precursor emissions. This calibration can reduce some of the uncertainty in precursor emissions, and provide a rational basis for precursor emission control.

The separation of transport and chemistry allows one to use observations of ozone and its precursors at a receptor to infer aspects of the history of the air parcel associated with the observations. This information can then be used to estimate the impact of precursor emission control at a specified receptor on the ozone level at the receptor. The use of a model to interpret observations directly recognizes the effect of unavoidable uncertainties in model inputs on the results of mechanistic, emission based models. The need to minimize these uncertainties has motivated the development of alternatives to emission-based models. One such approach is the Observation-based Model (OBM) proposed by Cardelino and Chameides (1995), which uses observations to infer relationships between ozone concentrations and precursor emissions. Other approaches (Chang and Suzio, 1996; Sillman et al., 1995)) use observations of ozone and associated species to construct variables that provide insight into the likely history of the air parcel;

the results from these studies show that these indicator variables can be useful in examining the effect of region wide precursor emission control on ozone at the receptor of concern. The separation of transport and chemistry, described in this paper, is similar to these methods in that it attempts to trace the chemical history of the emissions associated with the observed ozone level. It improves upon other methods by allowing one to estimate the impact of a specified precursor emission source on the observed ozone level. Details on how this can be done are discussed in a later section.

The separation of transport and chemistry (uncoupling) results in a model is clearly an approximation to the coupled model. The usefulness of this approximation is evaluated in this paper by comparing model results from the uncoupled model with those from the coupled model. In the next section, we discuss the suggested approach to separating transport and chemistry. We then discuss results obtained from the scheme.

SEPARATING TRANSPORT AND CHEMISTRY

The underlying idea of uncoupling chemistry and transport is to replace the coupled processes of transport and chemistry by first performing the transport and then performing chemistry over a time interval corresponding to the age associated with transport. This concept is best illustrated by considering a source that emits both VOC and NO_X. To calculate the concentration of ozone at a distance x from the source, the concentrations of NO_X and VOC at the receptor are first estimated assuming that no chemistry takes place. If the wind speed is assumed to be u, the age of the molecules reaching the receptor is x/u. Then, the ozone concentration at the receptor is calculated by exposing the VOC and NO_X concentrations calculated at the receptor to sunlight over the age, x/u.

This simple idea can be generalized to more complex situations as follows. First, the NO_X and VOC concentrations are estimated in the grid model assuming that no chemistry takes place. These concentrations are calculated using the emissions and meteorology required by the comprehensive grid model. In addition to these concentrations, the ages of VOC and NO_X in each grid box are calculated. Depending on the chemical mechanism used in the model, the VOC can consist of several surrogate species. This means that each receptor is associated with a set of precursor species and associated ages. The associated ozone concentration is calculated by allowing of each of these species to react over its corresponding age: each species is injected into the reactor at time equal to the time of interest minus the age of the species. This procedure allows us to perform chemistry at any receptor of interest rather than at all grid points in the modeling domain. This can represent a several fold savings in computational time over traditional comprehensive models. It follows then that computational requirements of the chemical mechanism become much less important than in a chemistry/transport coupled model because the computations have to be performed at a limited number of specified locations. Thus, a useful method to separate transport and chemistry obviates the need for parameterized chemistry to achieve computational efficiency. The next section provides a brief description of the evaluation of the method to separate transport and chemistry.

EVALUATION OF METHOD TO SEPARATE TRANSPORT AND CHEMISTRY

The idea of separating transport and chemistry is implicit in the work of Milford et al.(1994) and Silman et al. (1995) who show that ozone levels increase with NO_y (=NO_X+oxidized nitrogen species) for NO_y levels less that about 12 ppb. Beyond these levels, ozone concentrations can decrease because of NO_X inhibition effects. The separation of transport and chemistry is consistent with this observation in that it estimates the ozone at a receptor using the NO_X transported to the receptor. Because the NO_X does not undergo further scavenging or dispersion at the receptor, the NO_y at the receptor is equal to the NO_X used in the chemistry calculation. This ensures that the predictions from SOMS are consistent with the observed relationship between ozone and NO_y at the receptor. This is illustrated in Figure 1, which plots results from several comprehensive models applied to various regions. Separation implies that the maximum ozone observed at a receptor is constrained by the concentration of the NO_y (which corresponds to the initial temperature NO_X used to create the ozone) at the receptor. If we assume that the maximum ozone is approximately

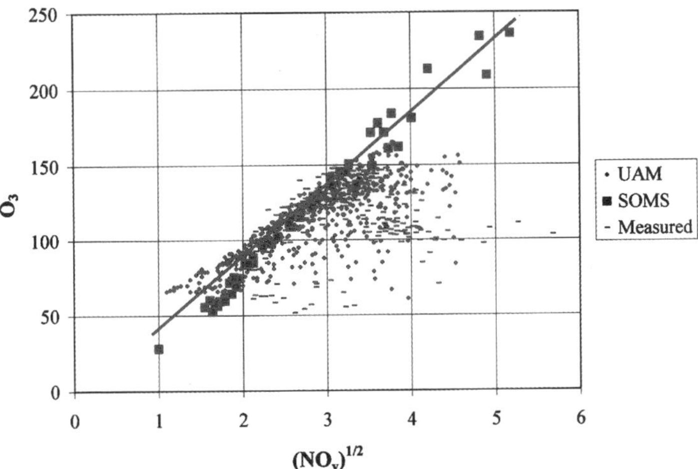

Figure 1. A comparison of the relationships between ozone and NO_y predicted by UAM for simulation on August 11, 1991 in Atlanta and SOMS. Also shown are some values measured with a helicopter.

proportional to $(NO_x)^{1/2}$ (Chang and Suzio, 1995), the line $O_3 = k(NO_y)^{1/2}$ should represent the left boundary of the set of points on the graph. This is precisely what is seen in the plot. Points close to the line represent NO_X limited conditions, while points to the far right of the line represent VOC limited conditions. We can test one aspect of the separation concept by computing maximum ozone concentrations by treating the NO_y as initial NO_x in a photochemical calculation. The results, corresponding to the CB-IV mechanism, lie close to the boundary line.

The separation of transport and chemistry requires the calculation of the "age" of a species at a location. This "age" determines the time over which the species at the location are reacted to form photochemical products. If we define

$$\phi = AC, \tag{1}$$

we can show that the "age" field can be computed from the conservation equation

$$\frac{\partial \phi}{\partial t} + \frac{\partial (F_i A)}{\partial x_i} = C. \tag{2}$$

The flux of material into the box consists of advective and diffusive components:

$$F_i = u_i C - \rho_a K^i \frac{\partial q}{\partial x_i}, \tag{3}$$

where u_i is the mean wind, K^i is the eddy diffusivity, ρ_a is the air density, and q is the mixing ratio given by

$$q = \frac{C}{\rho_a}. \tag{4}$$

When meteorological conditions can be considered to be steady-state, separation of transport and chemistry facilitates the examination of the effect of different emission scenarios on ozone. This can be seen by writing the concentration at a receptor, "r", as

$$C_r = \sum_s C_{sr}, \tag{5}$$

where C_{sr} is the contribution of the source at "s" on the receptor at "r". This contribution can be written as

$$C_{sr} = E_s T_{sr}. \tag{6}$$

where T_{sr} is the source-receptor matrix, that quantifies the effect of a source E_s at location "s" on the receptor. Every source-receptor pair is also associated with an "age" matrix, A_{sr}, which is the travel time between the source and receptor.

Because the flux operator in Equation (3) is linear, that is,

$$F_i(\sum C_{sr}) = \sum F_i(C_{sr}), \tag{7}$$

it is easy to show that the composite age, A_r, resulting from E_s, is given by

$$A_r = \frac{\sum A_{sr} C_r}{\sum C_r}. \tag{8}$$

These results show that once the steady-state source-receptor and age matrices have been calculated, we can examine the effect of a multitude of emission scenarios without recalculating transport. We can also calculate the impact of a particular source on a receptor by subtracting out the contributions of the source to the precursor concentrations and ages.

The effects of separating transport and chemistry were studied using a three-dimensional photochemical model. In the usual numerical solution of the mass conservation equation, a method called operator splitting (Yanenko, 1971) is used in which the effects of transport and chemistry are computed in successive time steps. Thus, although chemistry and transport are uncoupled in this sense, the computation of concentrations over a full time step includes the effects of both chemistry and transport. When transport and chemistry are uncoupled, ozone precursors are first transported as inert species and the ages of the pollutants following transport are obtained through the solution of the transport equation for ϕ = AC, given by Equation (2). The chemistry calculations are performed at the end of transport over a time interval corresponding to the maximum of the ages of the transported species. Each species is then injected into the chemical calculation at time equal to time of interest minus the corresponding age. We used the Carbon Bond mechanism IV, which includes 38 species, out of which 8 are transported. The differential equations governing the chemical species were solved using a semi-implicit predictor-corrector method.

For the numerical experiments, we assumed the airshed to extend 200 km along the x-direction, 200 km along the y-direction and 900 m in the vertical direction. The horizontal domain was resolved into 10 km X 10 km grid boxes, and the vertical direction was resolved using 3 layers of variable thickness. In this analysis, diffusion along the horizontal directions was neglected.

The wind field for the photochemical model was constructed using a diagnostic model, similar to the one proposed by Sherman (1978) and Kitada and Igarashi (1986). The wind field consisted of a vortex flow superimposed on a non-uniform two-dimensional flow in the x-y plane. The vortex flow was generated in the second vertical layer, and the wind field was specified such that there was shear in the vertical direction. These winds were then adjusted using the diagnostic wind model to ensure mass-consistency. The VOC and NO_X emission sources were located at ground level, and

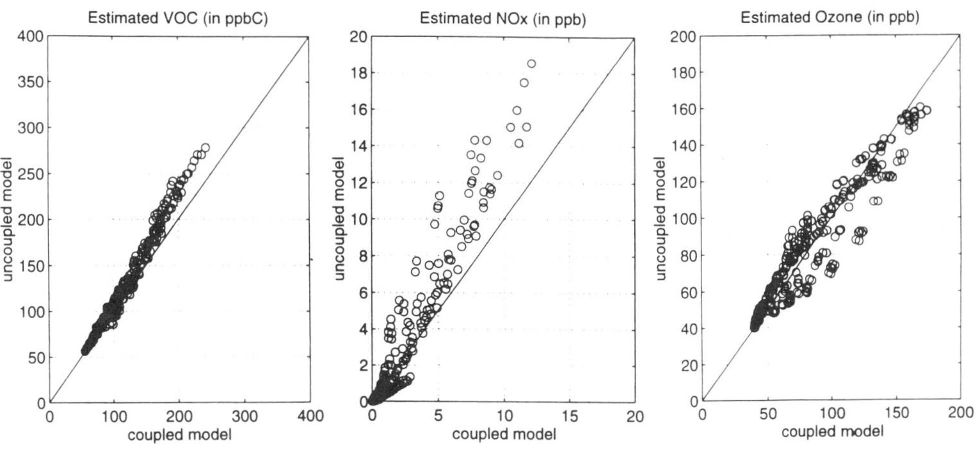

Figure 2. Comparison of ozone concentrations after 8 hours estimated by the uncoupled model with those of the coupled model for a complex mass consistent wind field.

the pollutant species were transported for approximately 8 hours. We incorporated mixing in the vertical direction by specifying a constant vertical eddy diffusivity, K_{zz}, with height.

Figure 2 plots the scatter between the coupled and uncoupled model estimates of VOC, NO_X, and O_3 at the 1200 grid cells of the computational domain at the end of the 8 hours. We find that the uncoupled model overpredicts both the VOC and NO_X concentrations compared to the coupled model. However, the r^2 between the two model estimates of ozone is close to unity ($r^2 = 0.96$).

We also compared the responses of the uncoupled and coupled models to changes in VOC and NO_X emissions. We first focused on ozone changes for base case ozone levels between 120 and 180 ppb. At 53% of the grid cells, these changes are positive for both the coupled as well as the uncoupled models; the concentration changes are close to 15 ppb for both models. In 37 % of the grid cells, both models predict decreases; the coupled model estimates an average decrease of 20% compared to 18% for the uncoupled model. In about 10% of the grid cells, the signs of the changes predicted by the coupled model disagree with those of the coupled model; these errors occur for ozone changes of less than 6 ppb.

The probability that the uncoupled model will predict the correct response to NO_X emission changes is over 90% when the ozone level is between 120 to 180 ppb. The probability for sign error is only 3% when the ozone concentration is between 80 and 120 ppb. The uncoupled model is likely to yield the wrong response 20% of the time when the ozone levels are relatively low, between 40 to 80 ppb.

The response of the uncoupled model to a 100% increase in NO_X emissions is similar to that for a 50% increase. When the ozone concentration is between 120 to 180 ppb, the uncoupled model provides the correct response 96 % of the time. The magnitudes of the responses are within a factor of 1.5 of the coupled model responses. The separation concept appears to be less useful when the magnitude of the response is less than 10 ppb or 10% of the base ozone level. When the ozone level is between 80 to 120 ppb, the response of the uncoupled model is correct about 97% of the time. The correct response percentage decreases to 80% only when the ozone level is between 40 and 80 ppb.

The results presented here suggest that separation of transport and chemistry is useful in estimating the response of ozone to VOC emission changes. Both the magnitudes and directions of the responses are correct about 98 % of the time. The incorrect responses are associated with levels of less than 3 ppb, which suggests that they are caused by numerical errors rather than those related to the underlying concept. The next section demonstrates the use of SOMS in examining the effect of precursor emission controls on ozone.

APPLICATION OF SOMS TO AN EPISODE

The Lake Michigan Oxidant Study (LMOS, Roberts et al., 1995) was a collaborative effort of several mid-western states to develop an understanding of the conditions that lead to high ozone concentrations in the Lake Michigan Air Quality Region (LMAQR), which encompasses parts of Wisconsin, Illinois, Indiana and Michigan. Portions of all four states regularly exceed the Federal standard and have been designated non-attainment areas for ozone. The LMOS program consisted of a comprehensive field measurement program during 1991, data analysis and a regional photochemical modeling program capable of simulating the formation, transport and fate of ozone and its precursors within and through the Lower Lake Michigan Air Quality Control Region under a variety of meteorological conditions.

Although a detailed meteorology dataset is available, SOMS represented meteorology in terms of a mean wind speed, wind direction, temperature and standard deviation of lateral velocity fluctuations (σ_v). Our objective here is not to simulate details of the meteorology but to estimate the values of source-receptor matrices that determine the relationships between concentrations and emission rates of ozone precursors.

Before applying SOMS to examine emission control strategies for reducing the ozone values in the LMOS region, we calibrated the model. In this exercise, we first adjusted the meteorology to achieve an overall best fit between model predicted and observed peak ozone values in the sense that the overall scatter (at all receptors) about the perfect correlation line ($C_{prediction} = C_{observation}$) is minimal. The initial choice of meteorology was guided by observations. The meteorology was then adjusted around the initial conditions until the best fit was achieved. We then adjusted the biogenic VOC concentration at each receptor to make the difference between predicted and measured ozone values as small as possible.

The results of calibrating SOMS for July 18 of the episode are shown in Table 1. We see that the adjustment of biogenic VOC concentration does allow us to explain observations. However, this calibration is not meaningful if we do not fix the values of NO_y at the receptors. This points to the need for simultaneous measurements of ozone and NO_y for realistic assessment of the effect of precursor emission control. To illustrate the usefulness of this calibration exercise, we have computed ozone concentration by reducing NO_x or VOC emissions by 25% across all anthropogenic sources.

Table 1. The calibration of SOMS by adjusting biogenic VOC concentrations and receptor classification.

					O_3 with emission reduction	
Receptor name	Observed O_3	Basecase O_3	Calibrated O_3	Biogenic VOC	25% NO_x	25% VOC
Kankakee, IL	89	122	89	48	84	86
Gary, IN	81	78	81	102	93	78
Aurora, IL	83	99	83	52	76	81
Benton Harbor East, MI	103	136	103	46	97	98
Rockford, IL	71	91	71	39	65	69
Zion 2 Mile, IL	93	169	93	28	104	73
Lake Michigan, Middle	154	126	151	200	116	152
Lake Michigan, South	112	137	112	52	107	106
Borculo, MI	170	121	135	190	112	136
Bayview, WI	95	95	95	93	119	89
Lake Michigan, North	72	127	72	33	75	66
Sheboygan, WI	99	116	99	54	93	96

CONCLUSIONS

The results from this study indicate that the separation of transport and chemistry to understand photochemical processes is a useful approximation. The direction of the estimated responses to VOC and NO_X emission changes from a model that uncouples transport and chemistry agree with results from a coupled model for more than 90% of the cases when the base ozone levels range from 80 to 180 ppb. The magnitude of these responses is within a factor of 1.5 of that from the coupled model.

The uncoupled model offers several features that can be important in understanding photochemical systems. Through improved computational efficiency, the model facilitates sensitivity studies that are crucial to the examination of the effect of inevitable input errors on ozone concentration estimates. Once the ages are computed for a set of unit emissions from all sources in the domain, we can examine different emission scenarios by simply performing box model calculations at selected receptors. This allows us to examine the effect of different chemical mechanisms independently of the transport. The results from these sensitivity studies can be used to select cases for more careful evaluation with a comprehensive model.

Separation of chemistry and transport allows the use of observations to make inferences on the impact of precursor emission controls on ozone levels at a receptor. In principle, a mechanistic photochemical model is the best tool to examine such impacts. However, the many uncertainties in both model formulation and inputs can reduce the reliability of the information provided by mechanistic model. Recognition of this has motivated the development of so called observation based models (For example, Cardelino and Chameides , 1995; Chang and Suzio, 1996; Sillman et al., 1995).

The interpretation of observations using the proposed method of separating transport and chemistry is best illustrated through an example. Consider a monitor at which the observed ozone level is 140 ppb, and the associated VOC and NO_y levels are 200 ppbC and 15 ppb respectively. If we assume that the total VOC (as moles of C per unit volume) and NO_y are affected primarily by dry deposition losses, we can use a box model to estimate the initial NO_X and VOC concentrations, and the age(time) over which these precursor concentrations need to be photolysed to yield the observed ozone concentration. The age is the only unknown variable in this calculation if we assume that we can estimate the deposition velocities of the precursors. Let us assume that the initial VOC , NO_X work out to be 250 ppbC and 20 ppb, and the corresponding age is 10 hours. This calculation requires making an assumption of the species comprising the VOC.

Now, consider a source of NO_X whose impact we want to estimate. Assume that the proposed reduction of emissions from this source results in a NO_X concentration of 5 ppb at the receptor assuming no chemistry, and the associated travel time from the source to the receptor is 2 hours. Then, reducing the source results in a NO_X concentration of (15-5)=10 ppb, and a new NO_X age of $(15 \times 10 - 5 \times 2)/10 = 14$ hours; the VOC age remains 10 hours. This information can be used to estimate the new ozone level with a photochemical box model. This simple example illustrates the approach to using the separation method to interpret observations of ozone and its precursors. In practice, if the wind field is complex, we would use the transport and age equations (Equation 2) to estimate changes in ages and the initial precursor concentrations.

This study has demonstrated the usefulness of the method to separate transport and chemistry in a photochemical model. However, we need more experience with the method to obtain understanding of its range of validity. To gain this experience, we are in the process of applying the separation method to several ozone episodes observed in the eastern United States.

Acknowledgments

This study was funded by the Electric Power Research Institute under Contract No: RP3429-02.

REFERENCES

Cardelino, C. A., and Chameides, W.L., 1995, An observation-based model for analyzing ozone precursor relationships in the urban atmosphere, *J. Air and Waste Management Assoc.* 45: 161-180.

Chang, T.Y., and Suzio, M. J., 1995, Assessing ozone-precursor relationships based on smog production model and ambient data. *J. Air and Waste Management Assoc.* 45: 20-28

Kitada, T, and Igarashi, K, 1986, Numerical analysis of air pollution in a combined field of land/sea breeze and mountain/valley wind, *J. Climate and Appl. Meteorol.* 25:767-784.

Milfod, J.B., Gao, D., Sillman, S., Bolssey, P., and Russell, G., 1994, Total reactive nitrogen (NOy) as an indicator of the sensitivity of ozone to reductions in hydrocarbon and NOx emissions, *J. Geophys. Res.* 99:3533-3542.

Roberts, P.T., Roth, P.M., Blanchard, C.L., Korc, M.E., and Lurmann, F.W., 1995, Characteristics of VOC-limited and NOx limited areas within the Lake Michigan air quality region, Technical Memorandum, STI-92322-1504-TM, Sonoma Technology, Inc., Santa Rosa, California.

Sherman, C. A., 1978, A mass-consistent model for wind fields over complex terrain, *J. Appl. Meteorol.* 17:312-319.

Sillman, S. et al., Photochemistry of ozone formation in Atlanta, GA – models and measurements, *Atmos. Environ.* 29: 3055-3066.

Yanenko, N. N., 1971, *The Method of Fractional Steps*. Springer-Verlag, New York.

DEVELOPMENT OF ADMS-URBAN AND COMPARISON WITH DATA FOR URBAN AREAS IN THE UK

D J Carruthers, H A Edmunds, C A McHugh, R J Singles

Cambridge Environmental Research Consultants Ltd,
3 King's Parade, Cambridge CB2 1SJ, UK

INTRODUCTION

ADMS-Urban is based on the Atmospheric Dispersion Modelling System (ADMS), a model developed in the UK by Cambridge Environmental Research Consultants, in collaboration with National Power and the Meteorological Office. Sponsors of the model include the UK regulatory agencies (Environment Agency and Health and Safety Executive). The original industrial source version of ADMS has already been well described elsewhere (Hunt et al., 1988 and Carruthers et al., 1992) and has been the subject of extensive validation (Carruthers et al., 1995 and Carruthers et al., 1997).

ADMS-Urban was first presented at the Fourth Workshop on the Harmonisation of Short Range Dispersion Models harmonisation in Ostende in 1996 (McHugh et al., 1996). It calculates concentrations in the atmosphere of pollutants released from the range of source types present in urban areas, namely industrial, domestic and road traffic, by representing them as point, line, area and volume sources. It is designed to allow consideration of dispersion problems ranging from the simplest (e.g. single isolated or single road) to the most complex urban problems (e.g. multiple industrial, domestic and road traffic emissions over a large urban area). It is a practical modelling tool which can assist government, planners, local authorities etc. to review and assess current air quality and potential impacts of new developments such as factories and roads. In the sections below we summarise the main characteristics of ADMS-Urban with particular emphasis given to the new features of the Urban model, we then illustrate current modelling applications.

MODEL FEATURES

The main features of ADMS-Urban are as follows:

Air Pollution Modeling and Its Application XII
Edited by Sven-Erik Gryning and Nadine Chaumerliac, Plenum Press, New York, 1998

- a non-Gaussian vertical profile of concentration in convective conditions, which allows for the skewed nature of turbulence within the atmospheric boundary layer; prediction is therefore made of the high surface concentrations near to elevated sources.
- a meteorological pre-processor which calculates the atmospheric boundary layer parameters from a variety of input data: e.g. wind speed, day, time, cloud cover or, wind speed, surface heat flux and boundary layer height. Meteorological data may be raw, hourly measurements or statistically analysed data sets.
- integration to a commercial GIS allowing an easy-to-use interactive graphical interface.
- a direct link to an Emissions Inventory database.
- versatility of applications such as comparisons with EPAQS (UK Expert Panel on Air Quality Standards), the UK National Air Quality Strategy objectives, and EU and WHO limits and guidelines; traffic planning; environmental impact assessments; "What if?" scenarios; future projections and odour predictions.
- a full range of source types - up to 1500, point, line, area, volume and road sources.
- integrated street canyon model
- the realistic calculation of flow and dispersion over complex terrain and around buildings.
- integrated chemistry model modelling reactions involving NO, NO_2 and Ozone.
- the calculation of pollutant emission rates from traffic count data, using a database of emission factors.
- the unique ability to model odour problems using short term concentration fluctuations. For a specified averaging time, the variance in concentration, and the probability that a certain concentration is exceeded, may be calculated.

Dispersion

The concentration profile from a point source is determined from analytical expressions which are Gaussian in the horizontal plane and also in the vertical plane under neutral and stable met conditions but are skewed or bi Gaussian in the vertical plane under convective conditions. Far from the source ($\sigma_z > 1.5h$), vertical gradients in concentration of the pollutant are sufficiently small that the plume can be considered to grow horizontally as a vertical wedge as if from a uniform line source of height h, rather than as a cone. Area and line sources are treated using a series of line sources aligned perpendicular to the wind, the number of line sources required being dependent on the proximity of receptor points to the source. Volume sources are treated as area sources with vertical extent.

Street Canyon

The street canyon model which is incorporated into ADMS Urban is based on OSPM (Operational Street Pollution Model) (Hertel & Berkowicz, 1989a,b,c; Hertel et al., 1990) which has been validated against data from Denmark and Norway. The model calculates the concentration by integrating the contributions from a series of line sources which represent the road and from material trapped in a recirculating region when there is a component of the wind blowing across the street. The ADMS version uses boundary layer met data generated directly by the ADMS met preprocessor and the boundary layer structure module.

The canyon model is automatically used for points which lie in roads lined with buildings with heights greater than 2m. Concentrations inside the road tend to the non-canyon results in the limits as canyon height is reduced to zero or the road width increased to over twice the canyon height. Concentrations at points outside the canyon are identical with those which would be obtained if the road were not a canyon. The model ignores end effects such as junctions. It assumes a straight length of road which has a width L, and is lined on both sides by flat-roofed buildings of height H_B.

Vehicle Induced Turbulence. The vehicle induced turbulence, σ_{w_0}, has been determined experimentally (Hertel & Berkowicz, 1989a). It is given by

$$\sigma_{w_0} = \frac{b}{(L)^{1/2}} \left(\sum_{i=h,\ell} (n_i v_i s_i) \right)^{1/2}$$

where n is the number of vehicles passing a point in the street each second, v is the vehicles' average speed, and s_2 is the horizontal area occupied by each vehicle. The subscripts h and l refer to heavy and light vehicles respectively. $b = 0.3$.

Chemistry

Vehicles and industrial sources emit a complicated mixture of chemicals including many organic compounds e.g. VOCs (Volatile Organic Compounds) and oxides of Nitrogen which are involved in reactions with Ozone. It is beyond the scope of a practical dispersion model to consider all the chemical reactions, therefore a scheme is used which models the important reactions involving Nitrogen, VOC's and Ozone. The Generic Reaction Set of equations (GRS) [11] is a semi-empirical photochemical model which reduces the complicated series of chemical reactions involving NO, NO_2, Ozone and many hydrocarbons to just seven reaction types as follows:

ROC + $h\upsilon$	$\rightarrow$	RP + ROC	(1)
RP + NO	$\rightarrow$	NO_2	(2)
NO_2 + $h\upsilon$	$\rightarrow$	NO + O_3	(3)
NO + O_3	$\rightarrow$	NO_2	(4)
RP + RP	$\rightarrow$	RP	(5)
RP + NO_2	$\rightarrow$	SGN	(6)
RP + NO_2	$\rightarrow$	SNGN	(7)

where:

ROC = Reactive Organic Compounds
RP = Radical Pool
SGN = Stable Gaseous Nitrogen products
SNGN = Stable Non-Gaseous Nitrogen products

Equations (3) and (4) represent exact chemical reactions, which are fast. The other equations represent bulk parameterisations of many different reactions.

The chemistry scheme within ADMS-Urban uses two models. Firstly a local model which is run separately for each receptor point and, secondly, a grid based 'background concentration' model which is employed principally to calculate initial concentrations or 'boundary conditions' for input into the local model. The background concentration model is a simple box model based on that described in (Singles, Sutton & Weston, 1997) and it takes account of concentrations of pollutants, either estimated or measured, upstream of the model domain, and species emitted within the model domain, but upstream of the region taken account by the local model. This local model region is receptor point specific; it extends to the edge of the model domain perpendicular to the flow, and a distance

$$x = \frac{U}{t_{ave}}$$

upstream of the receptor point, U is the wind speed at the receptor point at half the boundary layer height and t_{ave} is the minimum value of the weighted mean age (since release) of each of the pollutants calculated to be present at the receptor point (Figure 1).

User Interface and Emissions Inventory Database

The model features an intuitive graphical interface with the ArcView Graphical Information System (GIS) (ESRI, ArcView 3 technical specification). Although ADMS-Urban may be used as a stand-alone program much greater capability is achieved by using the two programs together. This allows the easy setting up of problems with the use of digital map data, and a wide range of facilities for output such as contour plots, charts and easy-to-create presentation layouts. A major benefit is that once geographical data has been entered, spatial analysis may be carried out using those data, e.g. how many homes will be affected by air pollution from a proposed city bypass?

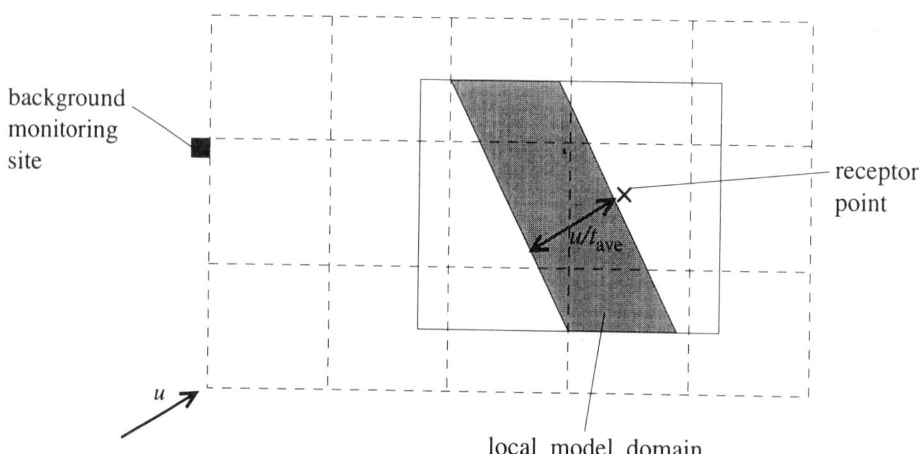

Figure 1.

The emissions inventory is a relational database developed using Microsoft Access for the storage of source and pollutant information. This database has been seamlessly integrated with ArcView for entry and geographical representation of source information. The use of standard Windows applications has allowed data gathered for other purposes to be readily available for use by ADMS-Urban. The database structure is flexible enough to allow emissions inventories such as those produced by the London Research Centre for Birmingham and London to be accessed and used by the system.

UK LOCAL AUTHORITY AIR QUALITY REVIEWS

As part of the UK National Air Quality Strategy local authorities are required to undertake a review of air quality. For authorities which have a potential air quality problem (e.g. roads with a flow of more than 25,000 per day, large industrial sources, street canyons with significant vehicle flows) an advanced assessment is required which may include monitoring and dispersion modelling. If predictions of air quality for the year 2005 suggest that air quality standards for any of seven pollutants (NO_2, SO_2, PM_{10}, Benzene, 1-3 Butadiene, lead and carbon monoxide) are likely to be breached then the local authority must declare an air quality management area for remedial action. As a first phase to implementation of the strategy, in 1996 the Department of the Environment designated 16 first phase authorities to conduct dispersion modelling. In this section briefly we describe modelling approaches being used in South Wales (Neath/Port Talbot), Cambridge and London. These illustrate the capabilities and flexibility of ADMS-Urban.

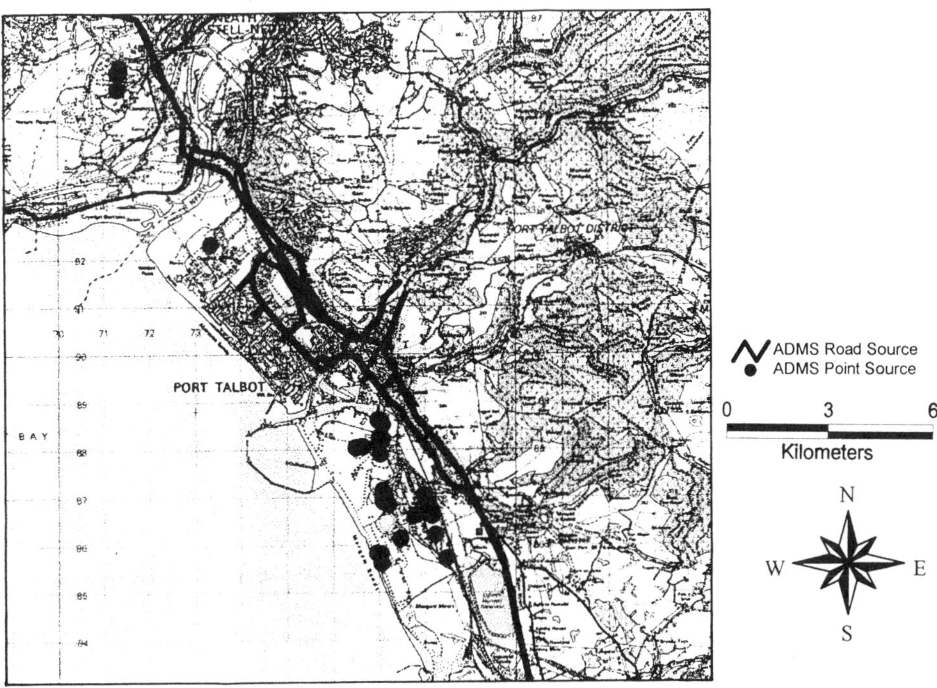

Figure 2.

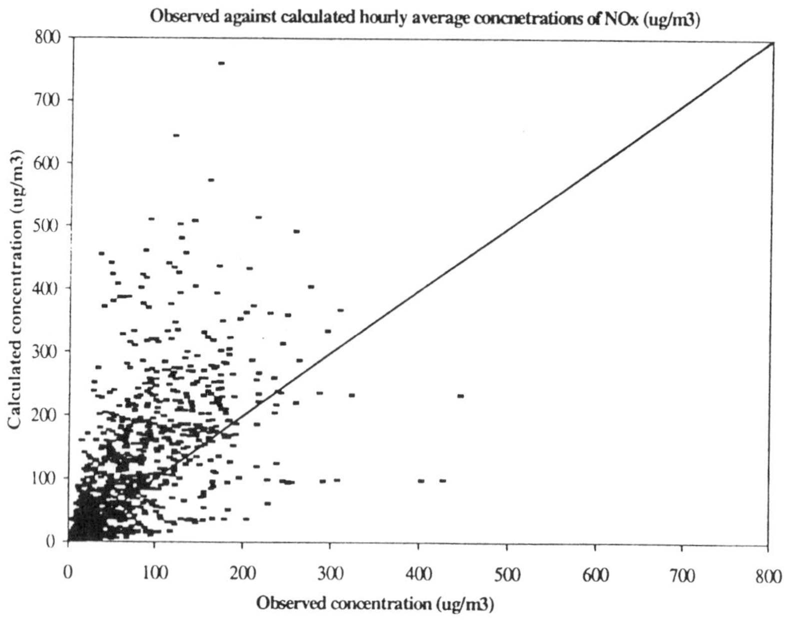
Observed against calculated hourly average concnetrations of NOx (ug/m3)
Calculated concentration (ug/m3)
Observed concentration (ug/m3)
0
100
200
300
400
500
600
700
800

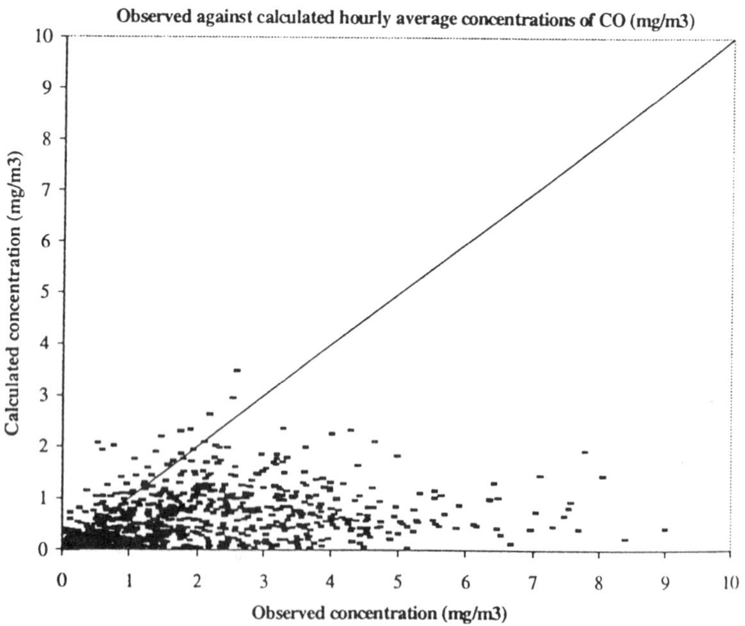
Observed against calculated hourly average concentrations of CO (mg/m3)
Calculated concentration (mg/m3)
Observed concentration (mg/m3)
0
1
2
3
4
5
6
7
8
9
10

Figure 3.

Table 1 Concentrations observed and calculated from Regent Street, Cambridge

		Maximum	99.9th Percentile	99th Percentile	98th Percentile	Mean
Concentration of NO_x ($\mu g/m^3$)	Observed	569	455	318	270	78
	Calculated	926	842	574	491	118
Concentration of CO ($\mu g/m^3$)	Observed	9.0	7.6	5.2	4.4	1.1
	Calculated	4.2	3.9	2.7	2.2	0.5

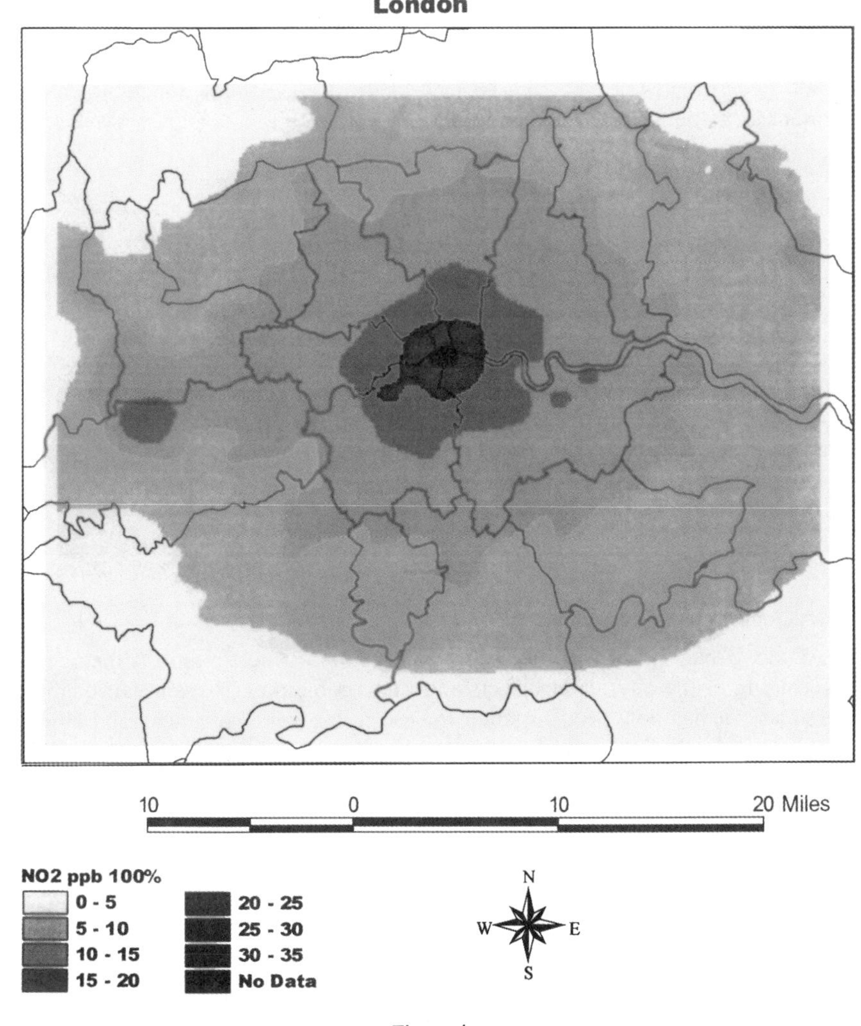

Figure 4.

Neath/Port Talbot

This is an area which includes emissions from major industrial sites (steel works, chemical works, oil refineries) from road traffic, and also significant topography (hilly terrain). The local authority has set up three continuous monitoring sites for SO_2, NO_2 and particulates and is conducting a diffusion tube survey at about 30 sites for benzene. Figure 2 shows the emission sources which are being used in the modelling. Such displays form part of the ADMS-Urban software. The study aims to compare predictions from ADMS-Urban with the measurements over a 4 month period.

Cambridge

Areas with significant concentrations of pollutants in Cambridge are confined to a few street canyons near the city centre. The local authority has conducted comparisons with monitoring data (NO_x, NO_2, CO, particulates and benzene) with the predictions of ADMS-Urban and other street canyon models. Figure 3 shows scatter plots of concentrations for NO_x and CO for a six month period; Table 1 shows high percentiles (data and model) for each of NO_x, NO_2 and CO. The underestimates of CO but overestimates of NO_x suggest incorrectly specified emissions are at least as important as model error in explaining the overall errors.

London

Local scale modelling and validation of a variety of street canyon models is being undertaken by the South East Institute of Public Health (SEIPH) for a street canyon in Central London. CERC together with ARIC (Atmospheric Research and Information Centre) are conducting modelling for two regions, namely the Central London area (20km by 20km) and the East Thames region (20km by 10km). The former is affected mainly by road traffic emissions, while the latter is subject to emissions from major industrial processes. Figure 4 shows an example of calculated average nitrogen dioxide NO_2 concentrations for the period May-July 1995. The central area has by far the highest concentrations.

CONCLUSIONS

ADMS-Urban is a practical modelling tool which applies the current understanding of the physics of dispersion to the problem of assessing air quality in urban areas. It has now been installed for use in a number of towns and cities of different sizes in the UK. Comparisons with other models and measured pollutant concentrations is underway in these areas as part of a trial scheme testing the available methods for reviewing air quality. The final results of this work should be available towards the end of 1997.

References

Carruthers, D.J., Edmunds, H.A., Bennett, M., Woods, P.T., Milton, M.J.Y., Robinson, R., Underwood B.Y. and Franklyn, C.J. (1995) Validation of the UK-ADMS Dispersion Model and Assessment of its Performance Relative to R-91 and ISC using Archived Lidar Data. Study commissioned by Her Majesty's Inspectorate of Pollution, published by the Environment Agency.

Carruthers, D.J., Holroyd, R.H., Hunt, J.C.R., Weng, W.S., Robins, A.G., Apsley, D.D. and Smith, F.B. (1992) UK-Atmospheric Dispersion Modelling System, Air Pollution Modelling and its Application IX, edited by H. van Dop and G. Kallos, Plenum Press, New York, 1993.

Carruthers, D.J., Hunt, J.C.R., Britter, R.E., Perkins, R.J., Linden, P.F. and Dalziel, S. (1991) Fast models on small computers of turbulent flows in the environment for non-expert users. Computer Modelling in the Environmental Science Edited Farmer and Rycroft, Clarendon Press.

Carruthers, D.J., McKeown, A.M., Hall, D.J. and Porter, S. (1997) Validation of ADMS against Wind Tunnel Data of Dispersion from Chemical Warehouse Fires. Submitted to Atmospheric Environment.

Environmental Science Research Institute (ESRI). ArcView 3 technical specification. http://www.esri.com/base/products/arcview/arcview.html

Hertel, O. and Berkowicz, R. (1989a) Modelling pollution from traffic in a street canyon. Evaluation of data and model development. *DMU Luft A-129.* 77p.

Hertel, O. and Berkowicz, R. (1989b) Modelling NO_2 concentrations in a street canyon. *DMU Luft A-131.* 31p.

Hertel, O. and Berkowicz, R. (1989c) Operational Street Pollution Model (OSPM). Evaluation of the model on data from St Olavs Street in Oslo. *DMU Luft A-1351.*

Hertel, O., Berkowicz, R. and Larssen, S. (1990) The Operational Street Pollution Model (OSPM). *18th International meeting of NATO/CCMS on Air Pollution Modelling and its Application.* Vancouver, Canada, 1990. 741-749.

Hunt, J.C.R., Holroyd, R.H. and Carruthers, D.J. (1988) Preparatory studies for a complex dispersion model, CERC Report HB9/88.

McHugh, C.A., Carruthers, J.J. and Edmunds, H.A. (1996) ADMS-Urban: an air quality management system for traffic, domestic and industrial pollution. Proc 4th Workshop on Harmonisation within Atmospheric Dispersion Modelling for Regulatory Purposes.

Singles, R.J., Sutton, M.A. and Weston, K.J. (1997) A multi-layer model to describe the atmospheric transport and deposition of ammonia in Great Britain. In: International Conference on Atmospheric Ammonia: Emission, Deposition and Environmental Impacts (Eds. Suton, M.A., Lee, D.S., Dollard, G.J. and Fowler, D.) *Atmospheric Environment* (in press).

Venkatram, A., Karamchandani, P., Pai, P. and Goldstein, R. (1994) The development and application of a simplified ozone modelling system. Atmospheric Environment, Vol 28, No.22, pp3665-3678.

HEMISPHERIC-SCALE MODELLING OF SULPHATE AND BLACK CARBON AND THEIR DIRECT RADIATIVE EFFECTS

Trond Iversen, Alf Kirkevåg and Øyvind Seland

Department of Geophysics, University of Oslo
P.O. Box 1022 Blindern, 0315 Oslo, Norway

INTRODUCTION

Airborne particulate matter is abundant in the atmosphere, with a typical spatial variability which reflects the relatively short atmospheric residence time on the order of a week. Natural aerosol particles originate from wind-blown dust in continental areas, from sea-spray over oceans, and from gas-to-particle conversions in the air. Man-made aerosol particles is mainly produced by gas-to-particle conversions, except for local scale dispersion of giant particles which is not included in our work.

Particles are frequently grouped into modes according to their production mechanism and typical size (Whitby, 1973; Jaenicke, 1993). The *nucleation mode* encompass particle radii between ~0.001 μm and ~0.1 μm and is produced by homogeneous nucleation; the *accumulation mode* includes particles with radii in the interval ~0.1 to ~1 μm and is produced by coagulation of smaller particles or heterogeneous condensation of gases onto pre-existing particles; and the *coarse particle mode* with radii larger than ~1 μm, is produced by mechanical processes. Coarse particles have a considerable gravitational settling velocity, whilst the small mass per particle in the nucleation mode causes efficient Brownian diffusion and coagulation. A typical "old" aerosol (several days) therefore is dominated by accumulation mode particles. Whilst the chemical composition of arbitrary nucleation or coarse mode particles reflects their origin, an accumulation mode particle typically consists of species from different sources as a consequence of coagulation. Nucleation and coarse mode particles are mostly *externally mixed*, whilst the accumulation mode particles are predominantly *internally mixed*. Aerosols influence physical processes in the atmosphere through direct interaction with (mainly solar) radiation or with the water substance. The *direct effect* of aerosols on radiative forcing is determined by the optical properties of the particles. The *indirect effect* of aerosols on radiative forcing is caused by their possible influence on the amount of activated CCNs, the number concentration and size of cloud-droplets, the optical properties of clouds (Twomey, 1977), and the cloud life-time and total cloudiness (Albrecht, 1989).

Air Pollution Modeling and Its Application XII
Edited by Sven-Erik Gryning and Nadine Chaumerliac, Plenum Press, New York, 1998

In this paper we focus on direct effects of particulate sulphate and black carbon (BC). Particulate sulphate is produced in the atmosphere by oxidation of SO_2. SO_2 is produced naturally by oxidation of biogenic sulphurous gases such as di-methyl-sulphide (DMS) from oceans (Tarrasón *et al.*, 1995), and is emitted directly from volcanic activity. Anthropogenic emissions of SO_2 are mainly concentrated to Europe, North America and South-East and East-Asia (Spiro *et al.*, 1992; Benkovitz, *et al.*, 1996). Production of BC in the atmosphere is brought about by incomplete combustion processes. Major emission sources are biomass burning and industrial processes. Two global emission surveys are presently available (Penner *et al.*, 1993; Cooke and Wilson, 1996). The numbers for BC are probably much more uncertain than those for sulphur.

Various simplified estimates of the radiative forcing of sulphate aerosols have been given with regional resolution (Charlson *et al.*, 1991; Kiehl and Briegleb, 1992), and even full climate predictions with prescribed albedo changes have been run (Mitchell *et al.*, 1995). Attempts at estimating direct radiative effects of both sulphate and BC are so far infrequent. Very preliminary indications were estimated by Iversen and Tarrasón (1995), showing a noticable contrast in forcing between the Arctic and the mid-latitudes. Later Schult *et al.* (1996) have given a global estimate with similar results. We are in the process of developing some parameterization functions linking sulphate and BC to optical properties and CCN. This paper presents preliminary results on optical properties and radiative forcing.

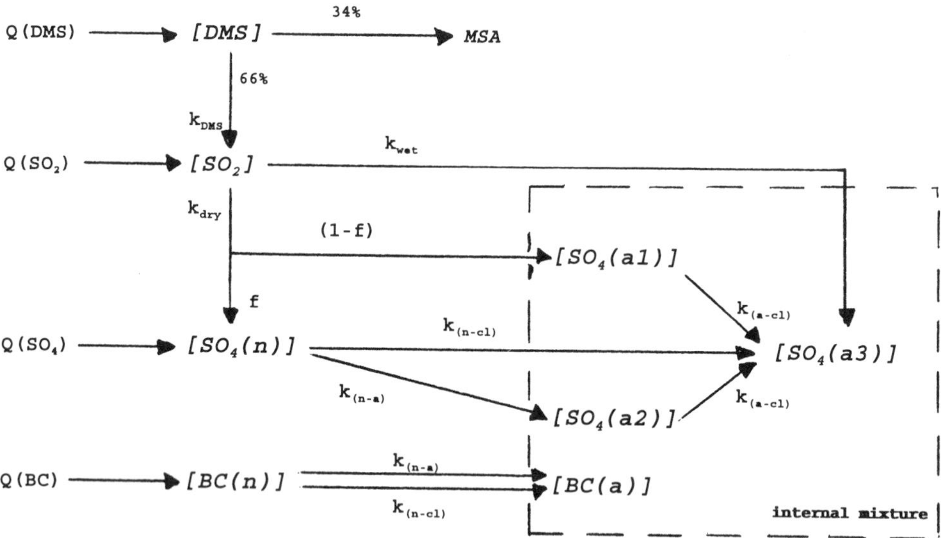

Figure 1. A flow-chart illustrating the modelled aerosol chemistry and physics. Nucleation mode (n) particles are assumed externally mixed with BC in hydrophobic mode; accumulation mode particles (a, a1, a2, or a3) are assumed internally mixed with BC in hydrophilic mode. Transformation rates (units s^{-1}) are denoted as k with subscripts DMS, dry and wet for chemical transformations, (n-a) for clear air coagulation of nucleation mode particles with accumulation mode particles, (n-cl) for in-cloud coagulation of nucleation mode particles with cloud droplets, and (a-cl) the same for accumulation mode particles. The fraction f (=0.1) is the fraction of particulate sulphate produced in clear air by nucleation. Emissions are denoted Q. Calculated components are denoted in square brackets. All components except DMS are subject to both dry and wet deposition.

HEMISPHERIC SCALE DISTRIBUTION OF SULPHATE AND BC

We have developed a hemispheric-scale model for calculation of airborne concentrations of particulate sulphate and BC. Earlier versions of the model have been used for studies of SO_2 and sulphate in the Arctic (Iversen, 1989) and of intercontinental transport (Tarrasón and Iversen, 1992; Tarrasón, 1995). The model is based on actual analyses of meteorological fields, uses a grid resolution of 150 km, and employs 10 isentropic coordinate surfaces in addition to the ground level which enters as a boundary condition. One year (1988) of meteorological data from ECMWF is used as 6-hourly input to a meteorological preprocessor which prepares the standard meteorological data for transformation to isentropic surfaces as well as calculating precipitation and all diabatic effects. Emissions of sulphur is taken from a number of sources for the anthropogenic part (NAPAP, EMEP, Kato and Akimoto (1992)), and for the natural part (Tarrasón *et al.*, 1995; Spiro *et al.*, 1992). Emissions of BC have been kindly provided by Cooke and Wilson (1996), which separates between biomass burning and fossil fuel combustion.

Figure 1 shows a sketch of the present treatment of the sulphur chemistry and BC. MSA is efficiently scavenged locally and is not carried explicitly in the model. The rate k_{DMS} accounts for oxidation of DMS to SO_2 by OH during daytime, and by the nitrate radical in darkness. The latter is assumed to occur momentarily in the lower 2 km over continents. SO_2 is oxidised to sulphate by OH in clear air (k_{dry}). In cloudy air the oxidation takes place due to O_3, H_2O_2 and catalytically by iron and manganese ions. An effective k_{wet} over a time-step of 30 minutes is estimated by a simple box calculation assuming a time-scale of replenishment of oxidants and SO_2 in clouds of one hour. Concentrations of OH, H_2O_2 and O_3 have been provided from a global photochemistry model by Berntsen (1994). Clear air coagulation rates assumes a standard accumulation-mode size and number-concentration which varies with height and area (Jaenicke, 1993). In-cloud coagulation is estimated from Ogren and Charlson (1983), assuming standard values for number-concentration and size of cloud droplets. SO_2 has a dry deposition velocity between 0.1 and 1.2 cm/s, depending on ground-surface properties. All particles have a dry deposition speed of 0.1 cm s^{-1}. The wet scavenging of SO_2 is determined by precipitation rates and wet chemistry. Wet scavenging of particles is determined in-cloud by coagulation and assumed CCN-activation, and sub-cloud by collision with precipitation elements (e.g. Hobbs, 1993).

Table 1 gives a model evaluation for stations in Europe (EMEP) for the entire 1988, and in North America (NAPAP) for the last six months. The selection of sites varies due to measurement irregularities. The model overestimates SO_2 in Europe during winter, whilst the result for sulphate shows a slight underestimation in the warmer seasons. In North America the model underestimates SO_2 in summer, with no general error trend for sulphate. In summary the seasonal variations of SO_2 is exaggerated, possibly due to overestimated seasonal variations of emissions; and the warm season concentrations of sulphate is slightly underestimated, possibly due to a too slow oxidation. Measurements of BC are not available on a regular basis. We have compared our calculations with those of Cooke and Wilson (1996) which have been compared with measurements in a climatological sense. The results are very similar, but our model yields less transport of BC, possibly due to a more efficient transformation from hydrophobic to hydrophilic mode. BC underestimations are particularly evident in the Arctic, and this is also seen by comparison with typical measurements (e.g. Hopper *et al.*, 1994; Hansen and Rosen, 1984). Vertical column burdens (Figure 2) can be used as input to simplified clear air solar radiation calculation after having estimated optical properties (extinction, single scattering albedo and asymmetry factor).

Table 1. Calculations and measurements in Europe (EMEP) and North America (NAPAP).

MONTH	EUROPEAN SITES (measurement regularity > 75% of the days)							
	SO_2 / μg(S)m^{-3}				SO_4 / μg(S)m^{-3}			
	no. of sites	calc.	meas.	corr	no. of sites	calc.	meas.	corr
Jan.	64	8.1	4.6	0.71	60	2.1	1.4	0.78
Feb.	61	5.7	4.6	0.52	62	1.0	1.5	0.54
Mar.	62	3.3	3.7	0.31	63	0.7	1.6	0.68
Apr.	62	2.7	3.7	0.35	57	1.1	1.9	0.54
May	63	1.8	2.1	0.53	57	1.1	1.7	0.54
Jun.	67	1.2	1.6	0.40	59	1.0	1.5	0.67
Jul.	68	1.0	1.1	0.59	61	1.1	1.2	0.59
Aug.	64	1.1	1.4	0.45	61	1.1	1.3	0.57
Sep.	65	1.9	1.9	0.43	60	1.3	1.4	0.64
Oct.	64	4.2	2.2	0.41	61	1.9	1.7	0.31
Nov.	64	7.2	4.7	0.49	62	1.7	1.3	0.55
Dec.	62	5.1	4.1	0.51	60	0.8	1.0	0.51
1988	69	3.7	2.9	0.68	66	1.3	1.4	0.65
	NORTH AMERICAN SITES (measurement regularity > 50% of the days)							
Jul.	64	1.3	2.8	0.48	20	1.5	2.5	0.82
Aug.	72	2.0	3.1	0.44	48	2.5	3.2	0.74
Sep.	68	3.3	3.1	0.53	48	3.0	2.0	0.70
Oct.	73	3.3	3.8	0.73	50	1.0	1.1	0.83
Nov.	69	5.7	4.5	0.66	50	0.8	1.1	0.79
Dec.	73	5.7	7.1	0.32	49	0.5	0.9	0.72
Jul.-Dec. 1988	53	3.7	4.1	0.68	18	1.4	1.7	0.56

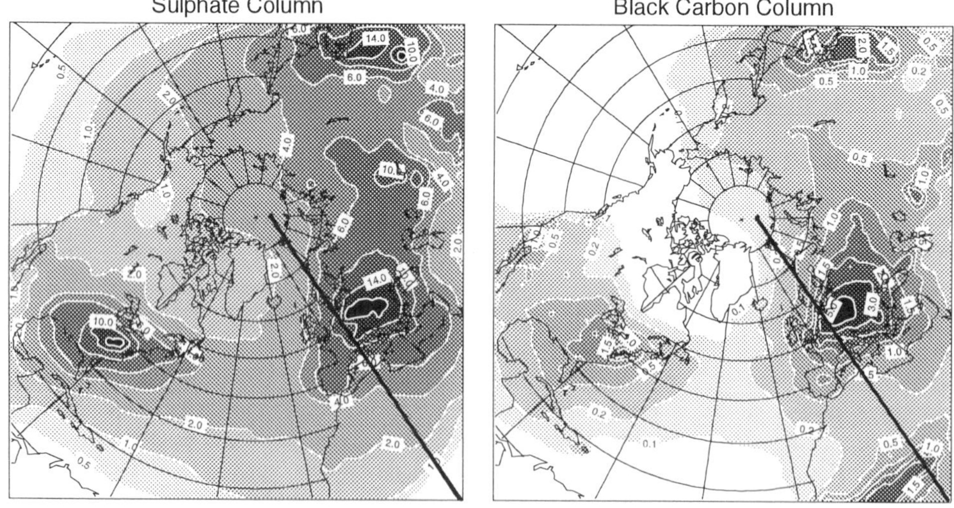

Figure 2. Calculated yearly averaged column burden of particulate sulphate (mg(SO_4)m^{-2}) and BC (mg(C)m^{-2}) for 1988.

AEROSOL OPTICAL PROPERTIES

Mie calculations (Wiscombe, 1980) have been used to estimate particle-specific optical properties for visible and near-infrared wavelengths. Input to these calculations is the complex refractive index as a function of particle size and wavelength, thus a size-distributed chemical composition of internally and externally mixed particles is needed. From d'Almeida *et al.* (1991) three typical clean-air, background particle distributions have been assumed; the clean continental aerosol over continents, the clean marine-mineral aerosol over open oceans, and a clean Arctic aerosol in the ice-covered parts of the Arctic Ocean. Aerosol mass produced by condensation on pre-existing particles (mode a1) is distributed by solving a continuity equation in the particle size-domain using a molecular diffusion coefficient for the species in question. For coagulation of nucleation mode particles Brownian diffusion coefficients are used (see Seinfeld, 1986, Ch. 8 and 10), producing mode a2 for sulphate and a for BC. The mode a3 is produced in cloud droplets which have evaporated leaving behind a larger particle than a1 and a2, and is treated as in Chuang and Penner (1995). Accumulation mode particles are assumed internally mixed, whilst nucleation mode SO_4 and BC are mixed externally. The aerosol swelling due to hygroscopicity is preliminary accounted for by assuming a radius amplification factor (r/r_0) independant of dry radius r_0, but increasing extensively with relative humidity (d'Almeida *et al.*, 1991).

The particle-specific optical parameters from the Mie-calculations are integrated over the particle size distribution arriving at spectrally resolved gross optical properties for the aerosol. Chandrasekhar averaging (Blanchet, 1982) yields optical parameters in 16 solar bands, 8 between 0.2 and 1 μm and 8 between 1 and 5 μm. Figure 3 shows optical parameters calculated for a dry aerosol along a line from the Arctic through central Europe and into Northern Africa (see Figure 2), assuming all calculated aerosols to be situated below 1000m, in the month of April 1988. The influence of the background aerosol is evident, and the low single scattering albedo in Africa is mainly caused by biomass burning. The anthropogenic and fossil fuel contribution is very clear over continental Europe with an increased optical depth, decreased single scattering albedo and an increased asymmetry factor (less sideways scattering) in visible wavelengths and decreased (more sideways scattering) in the near infrared. There are small effects in the Arctic, but the effect on the single scattering albedo is probably underestimated due to a too small BC/SO_4-ratio. Schult *et al.* (1996) estimated a ratio between 0.2 and 0.3 while our calculations yield 0.03-0.05.

Table 2 shows a more complete calculation for a smaller set of points, where swelling due to humidity is accounted for, and where a net radiative forcing on top of the atmosphere is estimated by feeding the optical parameters into a six-stream radiative transfer model (Dahlback and Stamnes, 1991). The background atmosphere assumes Rayleigh and Mie scattering and absorption by O_3 and Mie particles, as well as ground surface albedo from ECMWF. Biomass burning aerosols is estimated to heat the atmosphere considerably in the point "East Sahel", whilst the aerosols cause significant cooling in all other points. It is well demonstrated how important the effect of humidity swelling increases the forcing. The anthropogenic and fossil fuel contributions increase a factor 6 when the relative humidity increases from 0 to 90%, and as much as a factor 1.5 when increased from 80% to 90%. This large response increases the uncertainty of climate predictions based on aerosol forcing, since humidity is a major problem in all atmospheric models. Also estimated is the anthropogenic forcing when the BC/SO_4-fraction is assumed to be 0.15 instead of the model-calculated value of 0.03. This change causes the forcing to become considerable and in the direction of warming.

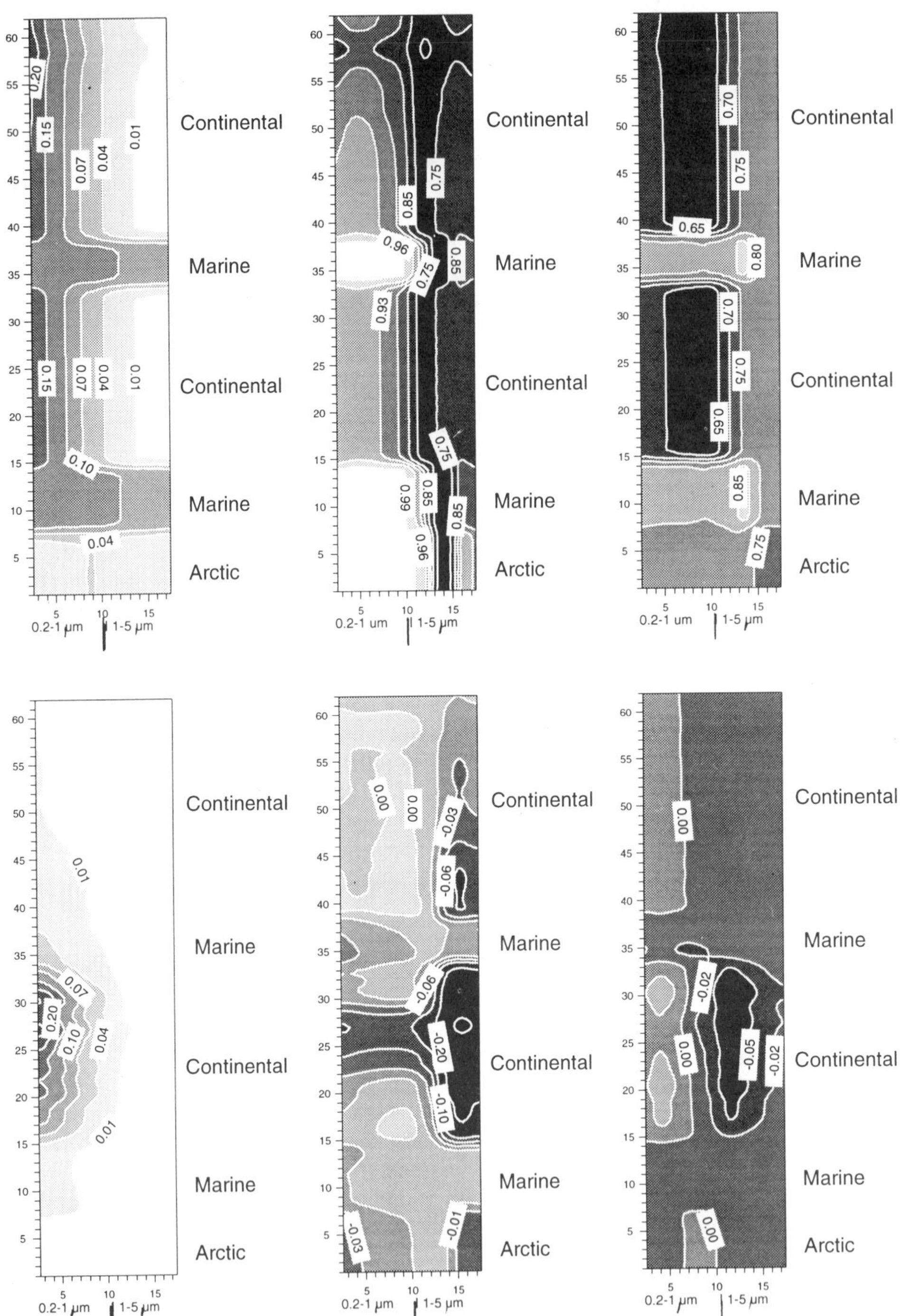

Figure 3. Optical parameters for 8 wavebands between 0.2 and 1 μm and 8 between 1 and 5 μm (x-axis) is estimates in 62 gridpoints along a line from the Arctic to North Africa (y-axis, see line in Figure 2). a), b) and c) shows the contribution from the mixture of background aerosol, natural sulphate and biomass burning BC on optical depth, single scattering ratio and asymmetry factor respectively. d), e) and f) shows the contribution from anthropogenic sulphate and fossil fuel BC on the same respective parameters.

Table 2. Top Of the Atmophere radiative forcing due to aerosols estimated in a selection of grid-points in April 1988. **0:** Forcing (W m^{-2}) due to background aerosol, **I:** Forcing (W m^{-2}) due to background aerosol + natural SO_4 + biomass BC, **II:** Forcing (W m^{-2}) due to anthropogenic SO_4 + fossil fuel BC.

POINT		Rh = 0			Rh=80%		Rh=90%	
		0	**I**	**II**	**I**	**II**	**I**	**II**
East Sahel	12°E, 15°N	0.16	2.8	0.07	-	-	-	-
East Europe	13°E, 52°N	- 9.9	- 10.2	- 4.4	- 11.9	- 9.7	-	-
North Atlantic	14°E, 60°N	-	- 3.2	- 0.3	- 9.8	- 1.3	- 13.0	- 1.8
Arctic	90°N	-	- 3.3	- 0.2	-	-	-	-
Arctic	90°N, (BC/SO_4=0.15)	-	- 3.1	1.2	-	-	-	-

CONCLUSIVE REMARKS

We have shown preliminary examples of results from a simplified aerosol model and the direct radiative effect of sulphate and BC. The aerosol model is presently being introduced in an atmospheric general circulation model, and tables are being developed to estimate optical parameters from the concentrations calculated by the model. The pronounced regional nature of the forcing may cause effects on global as well as regional atmospheric circulations, which can be more important than if the forcing was uniformly distributed. Several problems need to be addressed, such as the general slight underestimation of sulphate, the probably too short transport distance of BC, and a proper estimation of the particle swelling at high relative humidities. Also a proper spatial distribution of background aerorols is missing. Our assumptions probably exaggerate the optical contrasts between airmasses over continents, oceans and other surface types. However, this shortcoming does not seem to crucially influence the anthropogenic contribution to the optical parameters, since the contrasts are much less evident in the anthropogenic and fossil fuel increments. A good set of measurements of BC for model evaluation purposes is urgently needed.

Acknowledgments

The computer programs for Mie calculations and the radiative transfer has been provided by Dr. A. Dahlback. Calculations of oxidants have been provided by Dr. T. Berntsen. The transport calculations use meteorological data from ECMWF. Measurements have been obtained from EMEP (Europe) and NAPAP (North America). Continuous discussions with Dr. A. Dahlback, Dr. J.E. Kristjánsson and Dr. L. Tarrasón are gratefully acknowledged. The project is supported by The Research Council of Norway both through the Climate and Ozone Research Programme and the Programme for Supercomputing.

REFERENCES

Albrecht, B.A., 1989, Aerosols, cloud microphysics, and fractional cloudiness. *Science*, **245**, 1227-1230.

d'Almeida, G.A., Koepke, P., and Shettle, E.P., 1991, *Atmospheric Aerosols. Global Climatology and Radiative Characteristics.*, A. Deepak Publishing, Virginia, USA.

Benkovitz, C.M, Scholtz, T.M., Pacyna, J., Tarrasón, L., Dignon, J., Voldner, E.C., Spiro, P.A., Logan, J.A., and Graedel, T.E., 1996, Global gridded inventories of anthropogenic emissions of sulfur and nitrogen. *J. Geoph. Res.,* **101**, 29,239-29,253.

Berntsen, T.K., 1994, Two- and three-dimensional model calculations of the photchemistry of the troposphere. PhD-Thesis, University of Oslo, Oslo, Norway.

Blanchet, J.P., 1982, Application of the Chandrasekhar mean to aerosol optical parameters. *Atmos. Ocean,* **20**, 189-206.

Charlson, R.J., Langner, J., Rodhe, H., Levoy, C.B., and Warren, S.G., 1991, Perturbation of the northern hemispheric radiative balance by backscattering from anthropogenic sulphate aerosols. *Tellus*, **43AB**, 152-163.

Cooke, W.F. and Wilson, J.N., 1996, A global black carbon aerosol model. *J. Geoph. Res.,* **101**, 19,395-19,410.

Chuang, C.C. and Penner, J.E., 1995, Effects of anthropogenic sulfate on cloud drop nucleation and optical properties. *Tellus,* **47B**, 566-577.

Dahlback, A. and Stamnes, K., 1991, A new spherical model for computing the radiation field available for photolysis and heating at twilight. *Plant. Space Sci.*, **39**, 671-683.

Hansen, A.D.A. and Rosen, H., 1984, Vertical distribution of particulate carbon, sulphur, and bromine in the Arctic haze and comparison with ground-level measurements at Barrow, Alaska. *Geoph. Res. Lett.*, **11**, 381-384.

Hobbs, P.V., 1993, Aerosol-Cloud Interactions. In: *Aerosol-Cloud-Climate Interactions*, P.V. Hobbs, Ed. International Geophysics Series, **54**. Academic Press.

Hopper, J.F., Worthy, D.E.J., Barrie, L.A., and Trivett, N.B.A., 1994, Atmospheric observations of aerosol black carbon, carbon dioxide, and methane in the high Arctic. *Atmos. Environ.*, **28**, 3047-3054.

Iversen, T., 1989, Numerical modelling of the long range atmospheric transport of sulphur dioxide and particulate sulphate to the Arctic. *Atmos. Environ.*, **23**, 2571-2596.

Iversen, T. and Tarrasón, L., 1995, On climatic effects of Arctic aerosols. Institute Report Series, No. 94. Dep. of Geophysics, University of Oslo, Norway. 30 pp.

Jaenicke, R., 1993, Tropospheric Aerosols. In: *Aerosol-Cloud-Climate Interactions*, P.V. Hobbs, Ed. International Geophysics Series, **54**. Academic Press.

Kato, N. and Akimoto, H., 1992, Anthropogenic emissions of SO_2 and NO_x in Asia: Emission inventories. *Atmos. Environ.*, **26A**, 2997-3017.

Kiehl, J.T. and Briegleb, B.P., 1993, The relative roles of sulphate aerosols and greenhouse gases in climate forcing. *Science*, **260**, 311-314.

Mitchell, J.F.B., Johns, T.C., Gregory, J.M., and Tett, S.F.B., 1995, Climate response to increasing levels of greenhouse gases and sulphate aerosols. *Nature*, **376**, 501-504.

Ogren, J.A. and Charlson, R.J., 1983, Elemental carbon in the atmosphere: cycle and lifetime. *Tellus*, **35B**, 241-254.

Penner, J.E., Eddleman, H., and Novakov, T., 1993, Towards the development of a global inventory for black carbon emissions. *Atmos. Environ.*, **27A**, 1277-1295.

Schult, I., Feichter, J., and Cooke, W.F., 1996, The effect of black carbon and sulfate aerosols on the global radiation budget. Max-Planck-Institut für Meteorologie, Report No. 222. Hamburg, Germany. 23 pp.

Seinfeld, J.H., 1986, *Atmospheric Chemistry and Physics of Air Pollution.*, John Wiley & Sons, New York.

Spiro, P.A., Jacob, D.J., and Logan, J.A., 1992, Global inventory of sulfur emissions with 1°x1° resolution. *J. Geoph. Res.,* **97**, 6023-6036.

Tarrasón, L., 1995, Dispersion of sulphur in the Northern Hemisphere. PhD.-Thesis, University of Oslo, Norway.

Tarrasón, L. and Iversen,T., 1992, The influence of North-American anthropogenic sulphur emissions over western Europe. *Tellus*, **44B**, 114-132.

Tarrasón, L., Turner, S., and Fløisand, I., 1995, Estimation of seasonal dimethyl sulphide fluxes over the North Atlantic Ocean and their contribution to European pollution levels. *J. Geoph. Res.*, **100**, 11,623-11,639.

Twomey, S., 1977, The influence of pollution on the shortwave albedo of clouds. *J. Atmos. Sci.*, **34**, 1149-1152.

Whitby, K.T., 1973, In: *VIII International Conference on Nucleation, Leningrad.*

Wiscombe, W., 1980, Improved Mie scattering algorithms. *Appl. Opt.,* **19**, 1505-1509.

DISCUSSION

J. LANGNER: A major uncertainty in the global tropospheric sulphur cycle is the distribution of sulphate in the free troposphere. Did you compare your results in the free troposphere with observed data or other model calculations?

T. IVERSEN: No, we have not done much validation beyond comparing with regular surface measurements so far. I agree that this is very important since the free-tropospheric sulphate is a major component of the aerosol optical thickness away from major source areas. We have, however, seen that a major portion of the sulphate is found below 1500-2000 m over the source regions (e.g. Western Europe), whilst an increasing fraction is found above this as you move further away.

G. CARMICHAEL: Did you consider BC-emissions from aircraft? I would suggest that in the future you include this , as they are released high in the troposphere, and some estimates suggest that BC in the free troposphere may exceed sulphate levels at current growth rates in combustion of aviation fuels.

T. IVERSEN: We have used the BC-emissions of Cooke and Wilson (1996). This data-base does include a small fraction of BC from combustion of aviation gasoline, but include no information on the vertical distribution of the emissions. All emissions are assumed ground-based, and we have only used the data as they are. However, we appreciate the potential optical importance of high-level BC released above clouds of very high albedo. We will try to investigate this significance later on.

P. SEIBERT:

1) Is DMS the only marine source of S (sea-salt not considered)?
2) It appears that some features of the tropospheric model are not optimal (e.g. vertical motion from ECMWF seems not to be considered, only p-levels are used without full vertical resolution, theta-system does not allow an increased vertical resolution in the ABL, Smolarkiewicz advection scheme is somewhat outdated).

T. IVERSEN:

1) In the transport model we do not have a separate module for neither sea-salt nor crystal aerosols. When size-distributions are estimated, we use the

"climatological" estimates of background, clean-air aerosols given in the book by d'Almeida et al (1991), which separate between aerosols of different origins, including a sea-salt component.

2) It is very much a matter of taste how you define an "optimal". model. Using isentropic surfaces, the model becomes in a way optimal in the way that air-motions deviates little from motions in these surfaces (perhaps except in precipitation when many contaminants are scavenged anyway. The surfaces are tightly packed in portions of the atmosphere where the dynamics prohibits cross-isentropic mixing (e.g. stable ABL, along fronts, Arctic stratosphere). Numerical errors will be smaller in such cases as compared to quasi-horizontal coordinate surface. However, there is one important exception: there will be large concentration gradients close to emissions, and one might wish to have more coordinate surfaces there. But our comparison with measurement data in polluted areas does not reveal any serious errors, and furthermore our main interest is the more overall distributions which may give rise to significant climatic impacts.

We do employ the vertical motion from the ECMWF. Input data in 11 p-surfaces and ground surface data are used instead of model-coordinate surfaces, and I agree that model coordinate surfaces probably would have been more consistent. In particular in the ABL the p-surfaces are not really sufficiently dense. However, not all types of data are available from the ECMWF archives (e.g. the vertical distribution of precipitation) and temporal interpolation will have to be made anyway. This is why we have constructed a meteorological preprocessor with ABL-processes, precipitation and other physical processes in it.

We do not chose integration scheme based on any "popularity rate". The Smolarkiewicz (1983) scheme is possibly somewhat more diffusive than the Bott-scheme (at least that of 4th order). But the difference is not very large. We have done several tests with both schemes as well as with the pseudo-spectral technique and the semi-Lagrangian. The feature of using the advection equation with mixing-ratios rather than the mass-flux equation with mass concentrations, was the main reason to not switch to Bott. The pseudo-spectral comes out very expensive, uses all gridpoints to calculate local advection, and requires extensive techniques to account for the imposed recycling boundary conditions.

Furthermore it is not positive definite. Semi-Lagrangian could be an alternative, but we have not bothered to change, as we are quite pleased with the performance of the Smolarkicz scheme with five counter-diffusive iterations. We use the same scheme in the vertical as in the horizontal. This is difficult to do with the Bott-scheme, since the numerical "molecule" is big and the irregular distribution of coordinate surfaces makes the fourth order interpolation cumbersome. The reason why we want to use mixing ratios, is to ensure that tracers have zero material derivatives. As with the semi-Lagrangian scheme, the shortcoming is a possible lack of global mass conservation.

DEVELOPMENT AND APPLICATION OF A GLOBAL TO LOCAL MODEL HIERARCHY FOR THE DETERMINATION OF CHEMICAL PROCESSES IN THE TROPOSPHERE

Bärbel Langmann,[1] Daniela Jacob,[1] and Ralf Podzun[2]

[1]Max-Planck-Institut für Meteorologie
[2]Deutsches Klimarechenzentrum
Bundesstr. 55, 20146 Hamburg, Germany

INTRODUCTION

Computational resources usually limit the application of three-dimensional models for climate simulation as well as simulation of transport, chemical transformation and deposition of atmospheric gases and particles with respect to horizontal resolution. But many features contributing to regional climate patterns and processes involved in the transformation of atmospheric pollutants occur on spatial scales which cannot be resolved by current global or even regional models. One approach to regionalize coarse grid numerical results is the so called 'nesting' technique: large scale phenomena are simulated by coarse grid models and the results are used to provide boundary conditions for a high resolution mesoscale model simulation over the region of interest. Computational resources thus permit a higher resolution for the limited area model and, therefore, a more accurate description of topographie and turbulent processes in the planetary boundary layer. A first development and application of the nesting procedure to climate simulation is described in Dickinson et al. (1989), a first nested grid mesoscale atmospheric chemistry model has been presented by Pleim et al. (1991).

In our contribution we present the development of a global to local scale model hierarchy for the simulation of the troposphere dynamics as well as transport, transformation and deposition of photochemical and acidifying chemical species. The main focus is to couple the models with different scales by nesting of the dynamical and chemical variables and to incorporate chemical modules 'on-line' in the limited area dynamical circulation models to identify chemical-dynamical feedback mechanism. Besides the 'on-line'-branch, the 'off-line'-methodology to determine trace gas distributions in the troposphere over Europe has already been established for the regional scale model REMO. First preliminary simulations of the TRACT'92 episode focussing on the double nesting of REMO together with the chemistry-transport model will be shown.

Air Pollution Modeling and Its Application XII
Edited by Sven-Erik Gryning and Nadine Chaumerliac, Plenum Press, New York, 1998

MODEL DESCRIPTION

For the global to local scale model hierarchy we chose four major models: the global Hamburg climate model ECHAM, the regional climate model REMO, the regional chemistry-transport model CTM and the mesoscale model GESIMA.

The Global Climate Model ECHAM

The ECHAM spectral general circulation climate model (Roeckner et al., 1996) has been developed from the ECMWF numerical weather prediction model. It is available from T21 to T106 resolution. The top of the 19 layer hybrid pressure-sigma coordinate system is at 10 hPa. Vorticity, divergence, temperature, surface pressure, specific humidity and cloud water are standard prognostic variables. Chemistry modules, like the tropospheric sulfur cycle (Feichter et al., 1996), background CH_4-CO-NO_x-HO_x-photochemistry (Roelofs and Lelieveld, 1995) and a stratospheric chemistry module (Steil et al., 1997) have been implemented in ECHAM. The transport of chemical species as well as water vapor and cloud water is performed by a semi-Lagrangian scheme.

The Regional Climate Model REMO

REMO is based in the dynamical part on the weather forecast model of the German Weather Service (Majewski, 1991). Besides the physical parameterizations of the German Weather Service it includes the physical parameterization package of the global model ECHAM. Therefore, a consistent model chain of driving global climate model and regional model can be applied for climate simulations. Additionally, REMO can be forced by analyses data at its lateral boundaries, a valuable procedure for comparison studies of model results with observations. The horizontal resolution is 0.5° (1/6° if REMO is nested in REMO itself) on a spherical rotated grid. In the vertical a hybrid pressure-sigma coordinate system is used. Sensitivity studies with REMO are described in Jacob and Podzun (1997). The implementation of transport, chemical transformation and deposition modules in REMO is a main part of the ongoing model development.

The Regional Chemistry-Transport Model CTM

A modified version of the EURAD chemistry-transport model CTM (Hass, 1991) for the polluted atmosphere over Europe has been developed (Langmann and Graf, 1997). The main difference is that the meteorological driver model has been exchanged. The model HIRHAM (Christensen et al., 1996) was used - instead of MM4 - to produce information about the physical conditions of the atmosphere (horizontal wind, temperature, specific moisture, surface pressure, precipitation, cloud liquid water and cloud cover), which are passed to the CTM. To avoid interpolation errors both models, HIRHAM and CTM, employ the same vertical and horizontal coordinate system. In this model configuration the CTM was run for one winter and one summer episode. Validation and sensitivity studies of these episodes are described in Langmann and Graf (1997) and Langmann et al. (1997). As it is not intended to develop HIRHAM in the atmospheric chemistry section in Hamburg, another replacement by the model REMO as a driver model has been arranged. No technical problems occur, since HIRHAM and REMO include the ECHAM physical parameterization package and employ the same horizontal and vertical coordinate system.

The Mesoscale Model GESIMA

Tracer transport and meteorology are determined 'on-line' by the local scale three-dimensional non-hydrostatic model GESIMA (Kapitza and Eppel, 1992, Eppel et al., 1995).

A simulation of air flow and pollutant transport in the coastal region of Northern Germany is described in Eppel et al. (1992).

SIMULATION STRATEGY OF THE TRACT'92 EPISODE

The intensive field measuring campaign TRACT (Zimmermann, 1995) (September 7 - 23, 1992) which took place in the area covering the south western part of Germany and the adjacent parts of France and Switzerland with the Rhine valley in between was chosen for the development and first application of the model hierarchy described in the section before. Until now, only the regional model REMO (0.5° and 1/6° horizontal resolution) together with the CTM has been applied.

Initial and boundary conditions (every 6 h) for REMO in 0.5° horizontal resolution were derived from ECMWF reanalyses. As the orographies of the global and limited area

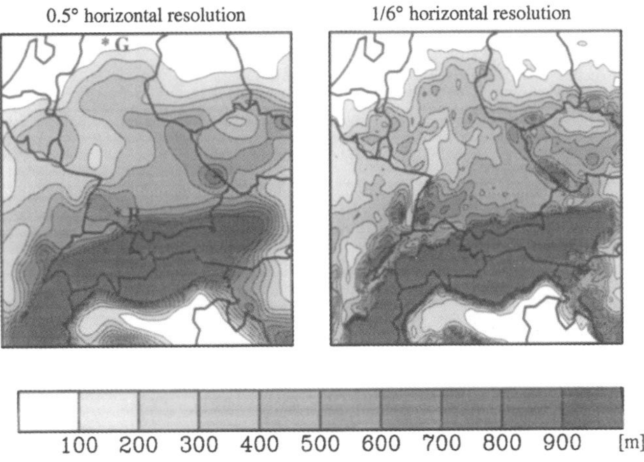

Figure 1. Orography data [m] in 0.5° and 1/6° horizontal grid spacing for the CTM-Nest domain with the TRACT area almost in the center. R marks the location of the station Rottenburg, G the station Gittrup. (The borders of the former GDR are still plotted).

model and consequently the terrain following vertical coordinates differ considerably, the global analyses data (horizontal wind, temperature, humidity and surface pressure) were interpolated horizontally and vertically to the REMO grid according to Majewski (1985). REMO was run with ECHAM physics in the 'forecast mode' (Karstens et al., 1996): Starting at 0 UTC every day a 30 h forecast was computed. The first 6 hours of the consecutive forecasts were neglected to account for a spin up time. By restarting the model every day with analyses the model state is forced to stay close to the real weather situation. This coarse grid model run provides initial and boundary conditions (every 1 h) for REMO in 1/6° horizontal grid spacing. The nesting approach is the same as described above, but the whole period (September 14 to 19,1992) was simulated without a forecast interruption ('climate mode', eg. a 6 day forecast).

Besides meteorological input data created by REMO, the CTM needs input about the emission situation, photolysis frequencies and the initial and boundary chemical composition of the atmosphere. Emission data are provided on the basis of Memmesheimer et al. (1991). Clear sky photolysis rates are created by a climatological preprocessor model (Madronich, 1987). If clouds appear in a grid box, photolysis rates are modified according to Chang et al. (1987). In the current model version with 0.5° horizontal grid spacing estimated concentration profiles of a relatively unpolluted atmosphere are used as initial and fixed boundary conditions. This approach is justified if the lateral model boundaries are located far away from source regions and if only a negligible temporal variability during the simulation episode is expected there. One day prior to the simulation period, an initialization run is started with these horizontally uniform, vertically varying clean conditions plus realistic meteorology and emission. As the CTM nesting domain with 1/6° horizontal grid spacing is much smaller than the domain with 0.5° horizontal resolution, the high variability in the chemical composition at its lateral boundaries should be taken into account. The boundary condition differ according to whether the flow is directed into or out of the nested domain. The nesting procedure of the CTM (1/6° into 0.5° horizontal resolution) is performed as described in Pleim et al. (1991). However, coarse grid model results are interpolated horizontally and vertically to the fine grid following the temperature interpolation technique of Majewski (1985).

In Figure 1 orography data as used for the model simulation with 0.5° and 1/6° horizontal grid spacing are shown for the CTM-nest domain with the TRACT area almost in the center. In the TRACT area the fine grid orography represents the Rhine valley flanked by the Black Forest and the Vosges, these features are not resolved by the coarse resolution data set.

PRELIMINARY RESULTS

Meteorology

On September 14, 1992 a frontal system crossed Germany with locally strong precipitation, especially in the southern parts. During the following days increasing high surface pressure caused large scale downward motion and sunny and warm weather, except in northern Germany where gentle maritime air masses were present, forming low clouds. In cloud free areas surface temperature decreased strongly during the night, but in the late morning hours fog dissolved again. The sunny weather stayed until the 18th of September, when a compact frontal cloud system arrived at the river Rhine. In the course of September 19 the cold front crossed the TRACT area from west and removed the warm continental air.

In Figures 2 and 3 a comparison of measured and simulated meteorological and chemical variables is shown for the two stations Rottenburg (48°27' N, 8°58' E) and Gittrup (52°03' N, 7°40' E) marked in Figure 1. Measurement data were taken from the German Federal Environmental Agency and model results from the simulation with 0.5° horizontal grid spacing. At Rottenburg on the river Neckar the maximum temperature rose to 25°C. The wind was gentle at less than 2 m/s with changing directions. At Gittrup the maximum temperature exceeded 20°C only on September 19. The wind blew with up to 5 m/s and a rapid change from a cyclonic to an anticyclonic regime took place in the night of September 16 to 17. The model closely follows the observed weather development during the episode. However, it should be kept in mind, that local measurements are compared to grid averaged model results.

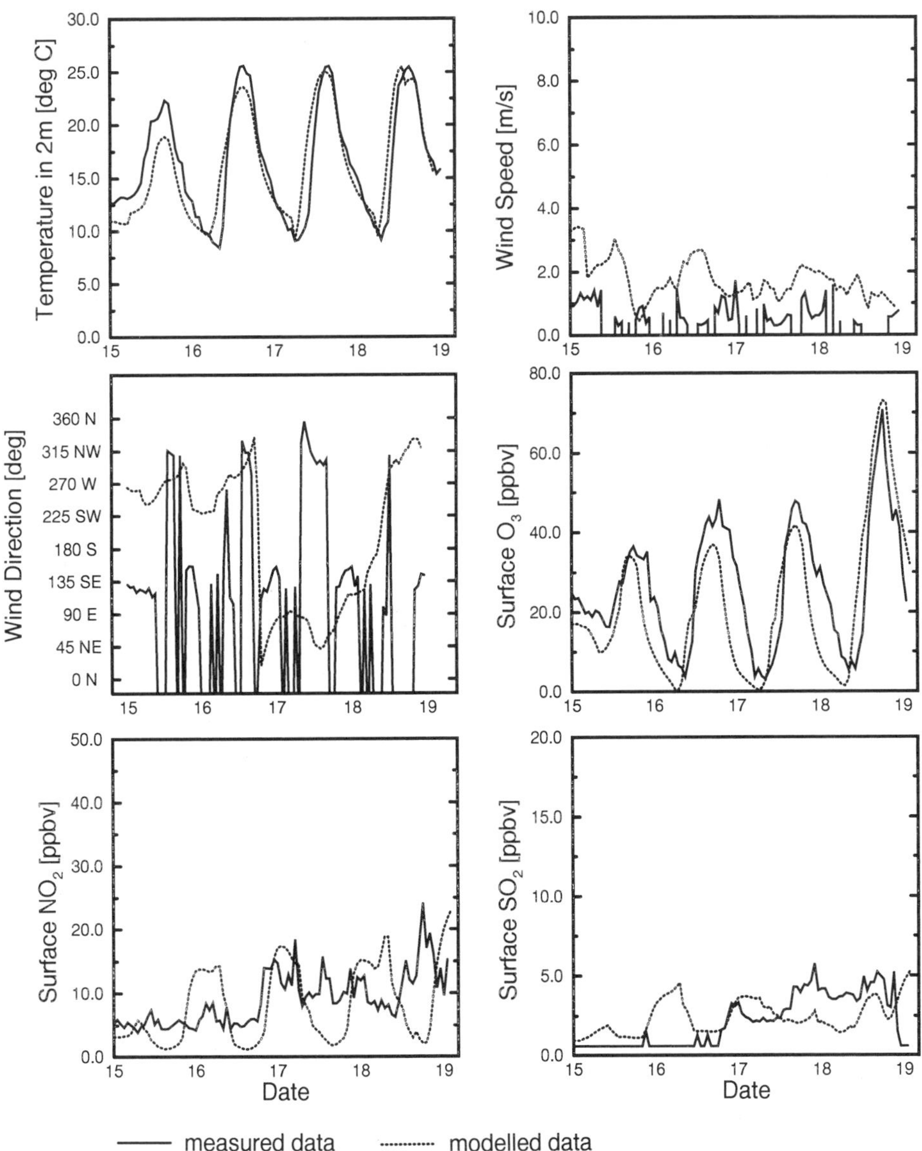

Figure 2. Measured and modelled meteorological and chemical variables at the station Rottenburg during the episode September 15 to 19, 1992.

Chemistry

The weather development during the episode is reflected in the concentrations of primary and secondary pollutants. Measured and simulated surface O_3, NO_2 and SO_2 concentrations at Rottenburg and Gittrup are shown in Figures 2 and 3. At Rottenburg modelled and measured O_3 agree very well, a maximum of 70 ppbv was reached on September 19. At Gittrup the surface O_3 level was much smaller and the model underestimates the observed

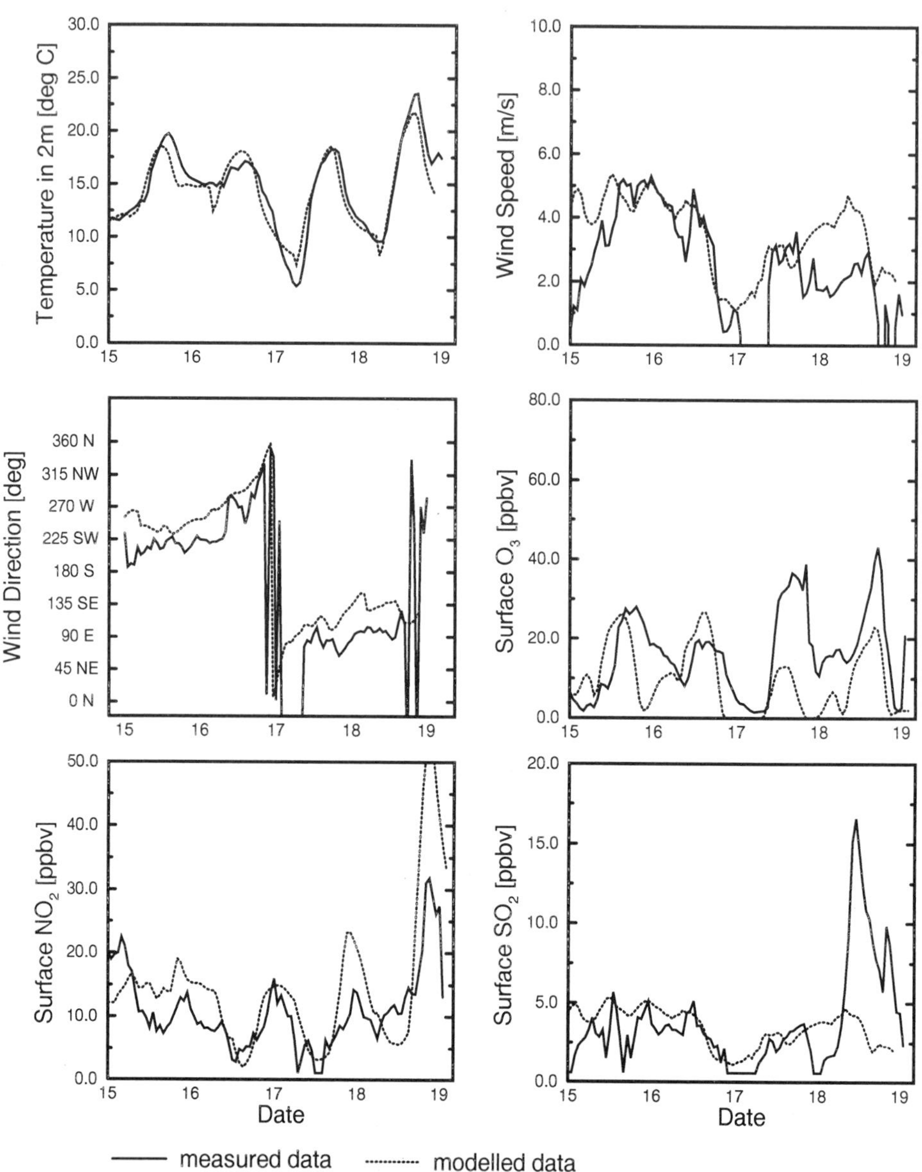

Figure 3. Measured and modelled meteorological and chemical variables at the station Gittrup during the episode September 15 to 19, 1992.

data. The model simulates a pronounced diurnal cycle of NO_2 with accumulation during the night at both stations, which is less pronounced in the measurements at Rottenburg. For surface SO_2 only small variations were observed and simulated at the two sites during the epsiode.

ACKNOWLEDGMENTS

The work is funded by the Tropospheric Research Project of the Ministry of Education and Research (BMBF) of Germany.

REFERENCES

Chang, J.S., Brost, R.A., Isaksen, I.S.A., Madronich, S., Middleton, P., Stockwell, W.R., and Walcek, C.J., 1987, A three-dimensional eulerian acid deposition model: physical concepts and formulation, *J. Geophys. Res.* 92:14681.

Christensen, J.H., Christensen, O.B., Lopez, P., van Meijgaard, E., and Botzet, M., 1996, The HIRHAM4 regional atmospheric climate model, Danish Meteorological Institute Scientific Report 96-4, Copenhagen.

Dickinson, R.E., Errico, R.M., Giorgi, F., and Bates, G.T., 1989, A regional climate model for the western United States, *Climate Change* 15:383.

Eppel, D.P., Kapitza, H., Claussen, M., Jacob, D., Koch, W., Levkov, L., Mengelkamp, H.-T., and Werrmann, N., 1995, The non-hydrostatic mesoscale model GESIMA. Part II: Parameterizations and applications, *Contr. Atmosph. Phys.* 68:15.

Eppel, D.P., Mengelkamp, H.-T., Jacob, D., Kapitza, H., and Koch, W., 1992, Nonstationary 3-D simulation of air flow and pollutant transport in the coastal region of Northern Germany and the Øresund, in: Air Pollution Modeling and its Application IX,, van Dop and Kallos, eds., Plenum Press, New York.

Feichter, J., Kjellström, E., Rodhe, H., Dentener, F., Lelieveld, J., and Roelofs, G.J., 1996, Simulation of the tropospheric sulfur cycle in a global climate model, *Atmos. Environ.* 30:1693.

Hass, H., 1991, Description of the EURAD chemistry-transport-model version 2 (CTM2), Report 83, A. Ebel, F.M. Neubauer and P. Speth, ed., Institut für Geophysik und Meteorologie Universität zu Köln.

Kapitza, H., and Eppel, D.P., 1992, The non-hydrostatic mesoscale model GESIMA. Part I: Dynamical equations and tests, *Contr. Atmosph. Phys.* 65:129.

Jacob, D., and Podzun, R., 1997, Sensitivity studies with the regional climate model REMO, *Meteorol. Atmos. Phys.*, in press.

Karstens, U., Nolte-Holube, R., and Rockel, B., 1996, Calculation of the water budget over the Baltic Sea catchment area using the regional forecast model REMO for June 1993, *Tellus* 48A:648.

Langmann, B., and Graf, H.-F., 1997, The chemistry of the polluted atmosphere over Europe: Simulations and sensitivity studies with a regional chemistry-transport-model, accepted by *Atmos. Environ.*

Langmann, B., Herzog, M., and Graf, H.-F., 1996, Radiative forcing of sulfate aerosols as determined in a regional circulation-chemistry transport model, submitted to *Atmos. Environ.*

Madronich, S., 1987, Photodissociation in the atmosphere. I: Actinic flux and the effect of ground reflections and clouds. *J. Geophys. Res.* 92:9740.

Majewki, D., 1985, Balanced initial and boundary values for a limitea area model, *Contr. Atmosph. Phys.* 58:147.

Majewski, D., 1991, The Europa Modell of the Deutscher Wetterdienst, *Seminar Proceedings ECMWF*, 2:147.

Memmesheimer, M., Tippke, J., Ebel, A., Hass, H., Jakobs, H.J., and Laube, M., 1991, On the use of EMEP emission inventories for European scale air pollution modelling with the EURAD model, in: EMEP Workshop on Photooxidant Modelling for long-range Transport in Relation to Abatment Strategies, 16. - 19. April 1991, Berlin.

Pleim, J.E., Chang, J.S., and Zhang, K., 1991, A nested grid mesoscale atmospheric chemistry model, *J. Geophys. Res.* 96:3065.

Roeckner, E., Arpe, K., Bengtsson, L., Christoph, M., Claussen, M., Duemenil, L., Esch, M., Giorgetta, M., Schlese, U., Schulzweida, U., 1996, The atmospheric general ciculation model ECHAM-4: Model description and simulation of present-day climate, MPI Report No. 218, Sep. 1996, Hamburg.

Roelofs, G.-J., and Lelieveld, J., 1995, Distribution and budget of O_3 in the troposphere calculated with a chemistry general circulation model, *J. Geophys. Res.* 100:20983.

Steil, B., Dameris, M., Brühl, C., Crutzen, P.J., Grewe, V., Ponater, M., and Sausen, R., 1997, Development of a stratospheric chemistry module for GCM's: First results of a multi-annual integration, submitted to *Ann. Geophysical.*

Zimmermann, H., 1995, Field phase report of the TRACT field measurement campaign, EUROTRAC ISS Special Publication, Garmisch-Partenkirchen.

DISCUSSION

J.A. VAN JAARSVELD: Do you expect to obtain different results from "off-line" and "on-line" modelling?

B. LANGMANN: Yes, I suppose that "on-line"-calculation of atmospheric dynamics and chemistry will give different results than "off-line"-modelling for the following reasons:

1. In "on-line"-models atmospheric chemistry processes are determined on the basis of the actual meteorological situation at every time step (e.g. every 5 minutes), whereas "off-line"-chemistry transport models receive the information about the physical conditions of the atmosphere in time intervals of 1h to 6h.
2. Inconsistencies in physical parameterizations, which exist at least in our "off-line"-model system, are avoided by "on-line"-models.

R. YAMARTINO: You have shown a triple one-way nesting of photochemical models and indicated various O_3 overpredictions. In complex meteorological episodes where mass stagnation or re-circulation of air masses is significant, the double (or triple) counting of HC and NO_x emissions on these grids can be a real problem. Two way nesting, with single counting of emissions, should be used unless it can be shown that the one-way nesting approximation is justified.

B. LANGMANN: I agree, that a one-way nesting procedure may cause problems concerning a repeated consideration of emissions, which probably could be avoided by two-way nesting techniques. If, for instance, the nested model simulation predicts inflow at the lateral boundaries whereas outflow is predicted by the coarse model, then indirectly coarse grid emissions would be included in the nested model run in addition to the nested emission inventory. In this case two-way nesting would avoid problems. Nevertheless, during the simulations presented on the meeting, the overprediction of peak surface ozone concentrations on the last day of the episode in the north-western part of the model area occurs in the coarse model simulation as well as in the higher resolution nested simulation and, therefore, has to be attributed to other model uncertainties than one-way nesting.

D. CARRUTHERS: The model calculation showed that the ozone concentration is more sensitive to model resolution than the SO_2 concentration. This is surprising in view of the fact that most SO_2 is released from point sources and that SO_2 concentrations are observed to have significant variability over small (km) scales. Can you explain what is happening? What is the resolution of the emissions inventory?

B. LANGMANN: Emission data have been provided by the Cologne EURAD group on the basis of EMEP data in 60 km and 20 km resolution in a Lambert conformal projection. By interpolating the data on the spherical rotated grid of the REMO-CTM-System with 0.5 degree and 1/6 degree resolution, a smoothed emission data set is created, especially in the high resolution case. Hopefully an improved data set or interpolation technique will be available in future. In addition, no detailed land use data set for the determination of dry deposition was included in the 1/6 degree resolution model run, coarse grid land use data was used. I suppose, that more detailed emission data and land use data could modify the nested model predictions. However, the preliminary results of the presented simulations show a minor dependence on horizontal resolution for primary pollutants like SO_2 and a higher sensitivity to the models horizontal resolution for secondary pollutants like O_3.

R. ROMANOWICZ: Where you thinking about introducing the interactions (feedback) from the small scale submodels to big scale models and the interactive running both level systems? This feedback would allow to use small scale information in estimation of global model performance and would be advantageous for both local and global predictions.

B. LANGMANN: We are also interested in two-way nesting technique and realize it, if we get enough support in future.

E. ANGELINO: Northern Italy is included in your domain of study. Did you validate your model comparing temporal series of computed O_3 concentrations with the measured ones by the existing monitoring sites in northern Italy? If you did, which was the effect of your nesting procedures on model performance in this area?

B. LANGMANN: Until now, only time series of surface ozone concentrations measured at stations of the German Federal Environmental Agency have been compared with model results. At these stations there are no significant differences in coarse and nested model results for surface ozone concentrations. Midday peak O_3 concentration can be reproduced by the models, but night-time minimum O_3 is always underpredicted. If the concentrations determined for the second model layer are compared with observations, modelled O_3 minimum concentrations are enhanced and closer to observation, whereas O_3 peak concentrations remain unchanged. Probably the models fail to determine the nocturnal boundary layer height.

ACCIDENTAL RELEASES

chairman: A. Flossmann

rapporteur: M. Rotach

AN OPERATIONAL REAL-TIME MODEL CHAIN FOR NOW- AND FORECASTING OF RADIOACTIVE ATMOSPHERIC RELEASES ON THE LOCAL SCALE

Torben Mikkelsen[1], Søren Thykier-Nielsen[1], Poul Astrup[1], Josep Moreno Santabárbara[1], Jens Havskov Sørensen[2], Alix Rasmussen[2], Sandor Deme[3], and Reinhard Martens[4]

[1] Risø National Laboratory, DK-4000 Roskilde, Denmark
[2] Danish Meteorological Institute, DK-2100 Copenhagen, Denmark
[3] Atomic Energy Research Institute, H-1525 Budapest, Hungary
[4] Gesellschaft für Reaktorsicherheit m.b.H., D-50667 Köln, Germany

INTRODUCTION

A comprehensive atmospheric dispersion modelling system, designed for real-time assessment of nuclear accidental releases from local to European scale, has been established by integrating a number of existing preprocessors, wind, turbulence, and dispersion models together with on-line available meteorology. The resulting dispersion system serves the real-time on-line decision support system for nuclear emergencies RODOS (Ehrhardt 1996; Kelly et al., 1996; Ehrhardt et al., 1997) with a system-integrated atmospheric dispersion module. This module is called MET-RODOS, Mikkelsen et al. (1997).

The present paper focusses on its local scale model chain which contains pre-processing software, local scale wind and turbulence models, and local scale dispersion models. The paper discusses the models and their integration with preprocessors and with on-line available met-data. The module generates local scale winds, turbulence levels, and dispersion and deposition patterns.

MET-RODOS

The nested meteorological model chains in MET-RODOS run on directly measured or real-time predicted weather data available via network connections to local meteorological towers and to remote collaborating operational national and international Numerical Weather Predicting (NWP) centres. MET-RODOS is designed to produce estimates of actual (real-time) and forecasted (+36 hour) ground-level air concentrations, wet and dry deposition, and ground-level gamma dose rates on all scales. It is furthermore designed to accommodate on-line available radiological monitoring data and to assist with source term determination by use of various data assimilation and back-fitting procedures.

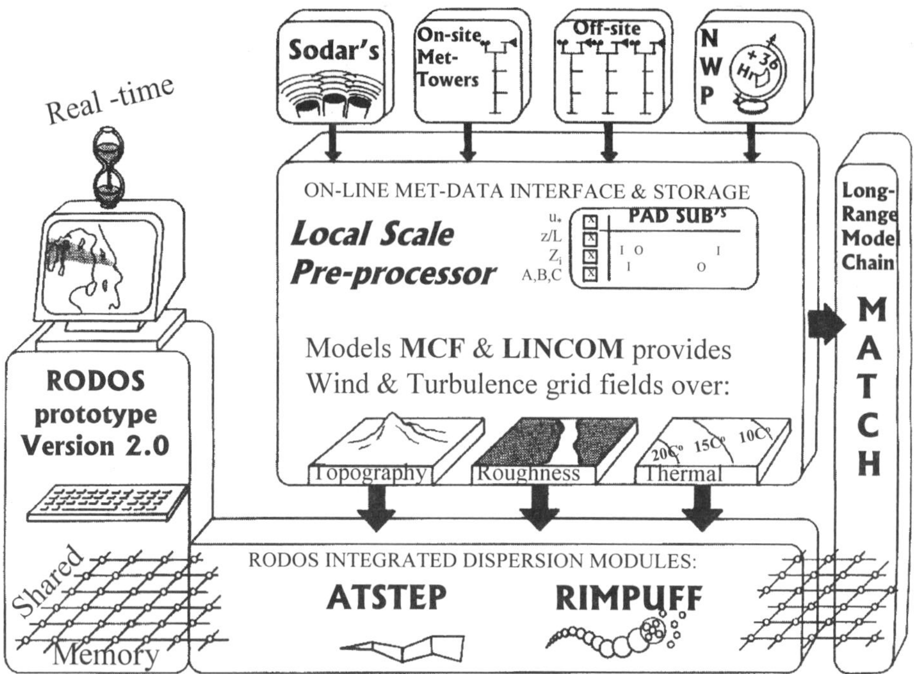

Figure 1. The MET-RODOS atmospheric dispersion module integrated with the Unix (shared-memory) based decision support system RODOS. The module has on-line interfaces to local meteorological stations and to remote Numerical Weather Prediction (NWP) centres. The Local Scale Preprocessor (LSP) interfaces the on-line available meteorological data to both the local scale dispersion models (ATSTEP and RIMPUFF) and to the long range model MATCH. The Local Scale Pre-processor contains a set of micro-meteorological preprocessing subroutines (the PAD SUB'^{s}) which are integrated with the local scale wind and turbulence models MCF and LINCOM in order to generate wind and turbulence grid files on the local scale grid.

MET-RODOS comprises a Local Scale Pre-processor LSP, a Local Scale Model Chain LCMC, and a Long Range Model Chain LRMC. It furthermore accesses two data bases: the On-Line Met-Tower Data Base OLMTDB, and the Real-Time Numerical Weather Forecast Data Base RTNWPDB. Concentration, deposition and dose rate estimates are produced simultaneously on both the local and European scales by nesting the local and the long-range atmospheric dispersion model chains and by consistent use of the real-time meteorological information available on-line to the system, see Figure 1. The MET-RODOS system is equipped with pre-processors, flow and diffusion models that have previously been selected for real-time applications within the RODOS framework (Mikkelsen and Desiato, 1993) cf. Table 1.

Table 1: Models integrated in the MET-RODOS system:

Near-range flow and dispersion models, including pre-processors:
- Meteorological preprocessor (PAD)
- Mass Consistent Flow model (MCF)
- Linearized flow model (LINCOM)
- Puff model with gamma dose module (RIMPUFF)
- Near-range segmented plume model (ATSTEP)

Mesoscale and Long-range Models:
- Eulerian K-model (MATCH) nested with puff radiation dose models (RIMPUFF)

On-line Weather Forecast data:
- Numerical Weather Prediction models (DMI-HIRLAM and SPA -TYPHOON)

The LSP maintains the RODOS real-time data base with actual and forecast local scale wind fields and corresponding micro-meteorological scaling parameters by use of pre-processor and local scale wind models. The LSMC contains a suite of local scale wind and dispersion models, from which case-specific models are selected depending on the actual topography and atmospheric stability features in question. It provides local scale deposition rates, air concentrations and cloud gamma dose rates in the local scale (about 20-30 km), and it passes on the diffusion specific parameters, such as cloud size and cloud position to the LRMC. The LRMC provides local-scale consistent trajectory and dose rate predictions on national and European scales by accessing the Numerical Weather Prediction (NWP) data stored in the real-time data base, and by integration of the LSMC outputs for its initialisation.

ON-LINE METEOROLOGICAL DATA

A real-time atmospheric dispersion based nuclear emergency system requires on-line access to both actual (real-time) measurements of the local meteorology and to numerical weather prediction data. At most nuclear sites in Europe, local-scale meteorology is on-line accessible from meteorological towers or stations located in the vicinity of the release point. Most met-towers are instrumented with wind (cup and wind vanes) and temperature sensors. As released radioactive material eventually is transported to distances beyond the local scale, also estimates of the regional (100 km) scale wind and temperature conditions are requested. Such on-line regional scale meteorology is in some European countries (Hungary, e.g.) available from a network of on-line meteorological towers.

Supplementary to measurements on the regional scale, and unique for the European scale, quantitative estimates of the actual and forecast (+36 Hr) wind and temperature conditions on the European scale are nowadays available as high-resolution limited area data distributed via computer networks all over Europe. Such data are referred to as Numerical Weather Prediction (NWP) forecasts, and are distributed on a commercial basis via national operational meteorological institutes in Europe. The MET-RODOS model chain has been designed to incorporate data from both on-line meteorological towers and high-resolution (pt. 20 km x 20 km grid) NWP data from the Danish DMI-HIRLAM model.

Atmospheric Boundary Layer (ABL) Height

The local scale model chain includes boundary layer height predictions in two different ways: By use of the PAD subroutines (Mikkelsen and Desiato, 1993) which calculates the boundary layer height based on local measurements), and 2) by analysis of the NWP data. While the former is used in MET-RODOS in diagnostic mode (now-casting), the latter will be used in forecast mode (+36 Hr).

Sørensen et al. (1997) has devised a method suited for use with DMI-HIRLAM outputs. The method is based on calculations of bulk Richardson numbers,

$$Ri_B = gz(\theta_v - \theta_s)/\theta_s(u^2+v^2) \qquad \textbf{(1)}$$

Here, z is the height, θ_s and θ_v are the virtual potential temperature at the surface and at height z, respectively, and u and v are the horizontal wind components at height z. The height of the ABL is given by the height at which the bulk Richardson number reaches a prescribed critical value.

Based on a set of about one hundred radio-soundings from Copenhagen from January to August 1994, the method was tested (Sørensen *et al.*, 1997a). The soundings represented boundary layers with a well-defined top. The observed ABL heights were obtained by inspection of profiles of temperature, humidity and wind and the optimal value for the bulk

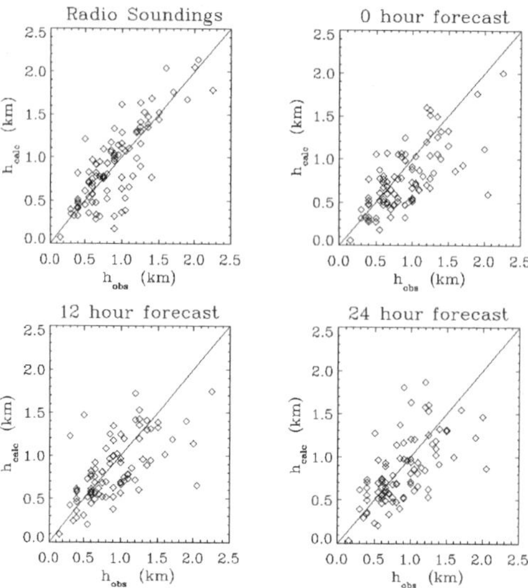

Figure 2 Scatter plots of observed and calculated ABL heights. Upper left: Observations compared with ABL height based on the model Eq.. (1) (Ri_B ~0.14). Upper right and bottom left and right: Observations compared with DMI-HIRLAM output-based ABL's (Ri_B ~0.24) at: time t = 0 (analysed field), t =+12, and at t = +24 Hrs.

Richardson number turned out to be 0.14.

The upper left part of Figure 2 is a scatter plot comparing calculated heights with observed values. The same procedure was applied to analysed DMI-HIRLAM profiles. Here, the resulting optimum critical value was 0.24. With this critical bulk Richardson number the method was applied also to forecast profiles. In the upper right part of Figure 2, a comparison is made between ABL heights calculated from analysed DMI-HIRLAM data and observed values. The lower parts of Figure 2 show similar comparisons based on forecast DMI-HIRLAM profiles.

The study also showed that the bulk Richardson number approach is not very sensitive to the critical bulk Richardson number. In fact, critical values in the range 0.1-0.4 are adequate - indicating a high robustness of the method.

This ABL height method was earlier used in the Danish Emergency Response Model of the Atmosphere (DERMA) during the real-time long-range dispersion model evaluation of the European Tracer Experiment (ETEX), as well as in the later ATMES-II phase studies (Sørensen *et al.,* 1997a; Sørensen *et al.,* 1997b; Sørensen and Rasmussen, 1995).

MODELS AND DATA STRUCTURE IN THE LOCAL SCALE MODEL CHAIN

The local scale dispersion models ATSTEP and RIMPUFF requires, in addition to on-line available primary meteorological inputs (i.e., mean wind speed and mean wind direction) also real-time determination of the dispersion controlling scaling parameters such as stability category, heat flux, momentum flux and mixing height.

Pre-processing software modules suitable for real-time applications is therefore included in the local scale model chain. On-line incoming meteorology - either from automatic meteorology stations and/or from weather forecast model grid points located near or inside the local-scale model domain - are real-time pre-processed by the LSP module to produces gridded mean (wind) and turbulence quantities (including atmospheric stability measures) at all grid points belonging to the local scale grid. A typical local-scale grid contains 41 x 41 grid points, covering an area of 40 x 40 km. Pre-processing is performed by a set of nine pre-processing routines (the PAD sub-routines) in conjunction with the diagnostic mean wind and turbulence model (LINCOM-Z_0).

Depending on the set of actually available input data for a given 10-min period, LSP automatically selects the subset of most suitable PAD-subroutines, and processes the required <atmospheric stability measures> such as: <stability category>, <Monin-Obukhov stability length scale>, <mixing heights>: (<mechanical>, <convective>), the turbulence <scaling parameters> such as: <heat-flux>, <shear-stress> and <variances>: <horizontal>, <vertical>, and also the <mean profiles> of <wind> and <temperature>.

The LSMC also runs the local scale dispersion, deposition and gamma radiation models ATSTEP and RIMPUFF and produces the "source-terms" for the long range model chain.

Linearized Wind Model: LINCOM -Hill, -Z_0, and Thermal

Detailed modelling of the wind and turbulence fields on the local scale are important for prediction of the trajectory-directions - and the time of arrival - of radioactive clouds passing across hilly terrain and over heterogeneous surfaces (e.g. land-water-land). Local-scale mean wind and turbulence fields are modelled in real-time for subsequent use during advection, diffusion and deposition of the radioactive clouds.

The integrated LINCOM model suite provides the local model chain with fast diagnostic wind fields from solving the set of linearized momentum and continuity equations with a first order spectral turbulent diffusion closure. Wind and turbulence fields are modelled in consideration of: 1) local topography (-HILL), 2) variations in the surface aero-dynamic roughness (-Z_0), and 3) non-neutral vertical thermal stratification of the atmosphere (thermal). Troen and de Baas (1986) first developed a linearized model for neutral stratified pressure-gradient driven winds over hilly terrain (LINCOM-HILL). Moreno et al. (1994) have extended the concept in order to include effects of thermally driven flows (such as valley breeze and nocturnal drainage flows (LINCOM-THERMAL). Astrup et al. (1996) furthermore extended the model concept to include the effects of changes in the surface roughness (LINCOM-Z_0). In addition to the changes in mean winds introduced by the in-homogeneity in the surface roughness, the "Z_0 version" also calculates the local turbulence levels (i.e. the surface <sheer-stress field (u_*) >) over the entire local scale grid. Figure 3 and Figure 4 shows examples of LINCOM-Z_0 generated mean and turbulence winds over Northern Zealand (Astrup et al., 1997).

LINCOM-THERMAL extends the linearized model concept to thermally driven flows.Themodel equations are the linearized Navier-Stokes equations for momentum and

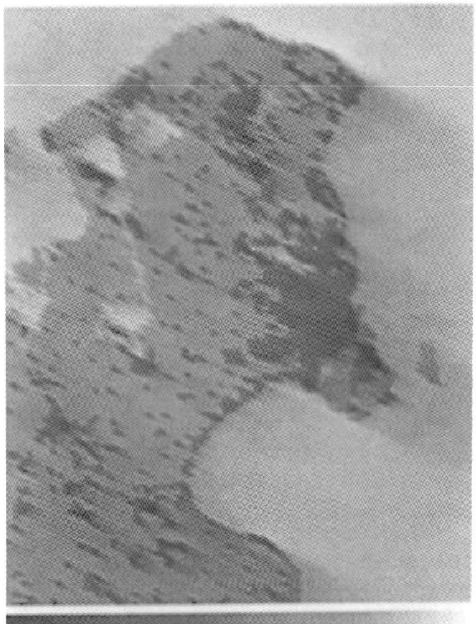

Figure 3 LINCOM-Z_0 generated 10-m winds over non-homogeneous roughness (Northern Zealand). Speeds range from 4 (dark) to 11 (bright) m/s.

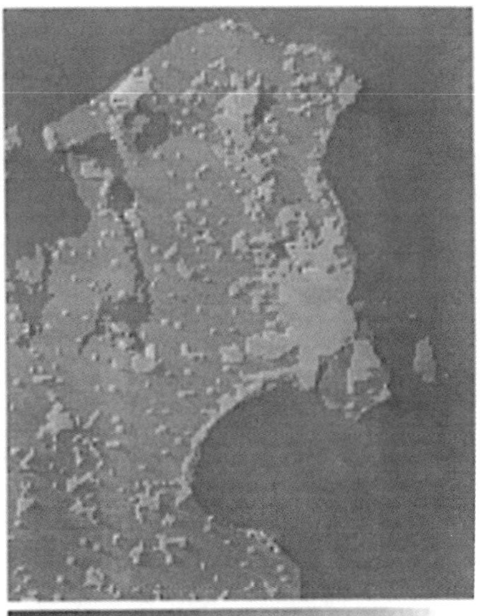

Figure 4 Surface friction distribution (u_* -field) corresponding to Figure 3. The friction velocity ranges from 0.25 (dark) to 1.0 (bright) m/s.

$$
\begin{aligned}
&(U\frac{\partial}{\partial x}+V\frac{\partial}{\partial y}-K\frac{\partial^2}{\partial z^2})\breve{u}+\frac{\partial \breve{p}}{\partial x}=0\\
&(U\frac{\partial}{\partial x}+V\frac{\partial}{\partial y}-K\frac{\partial^2}{\partial z^2})\breve{v}+\frac{\partial \breve{p}}{\partial y}=0\\
&(U\frac{\partial}{\partial x}+V\frac{\partial}{\partial y}-K\frac{\partial^2}{\partial z^2})\breve{w}+\frac{\partial \breve{p}}{\partial z}=g\frac{\breve{\theta}}{\Theta}\\
&\frac{\partial \breve{u}}{\partial x}+\frac{\partial \breve{v}}{\partial y}+\frac{\partial \breve{w}}{\partial z}=0
\end{aligned}
\qquad \textbf{(2)}
$$

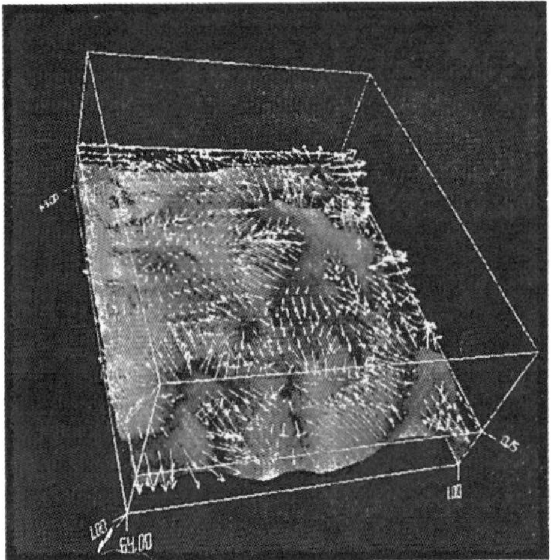

Figure 5 Drainage flow over Porton Down terrain (UK) modelled by LINCOM-THERMAL during the MADONA field trials during September and October 1992 with parameters $\breve{\theta}_0 = -1$ C° and $\lambda = 100$ m.

continuity, where U, V and Θ are the components of a horizontal background wind and potential temperature, $\breve{u}$, $\breve{v}$, $\breve{w}$, and $\breve{p}$ are the wind and pressure perturbations (*i.e.* $u = U + \breve{u}$, etc..), K is the turbulent diffusivity (modelled by a first order spectral closure scheme), $g\breve{\theta}/\Theta$ is the buoyancy term, $\breve{\theta}$ being the potential temperature perturbation. Equation (2) is solved with a (fixed) temperature field $\breve{\theta}$, which is modelled by assuming that: 1) the iso-thermals are everywhere parallel to the terrain, and 2), the vertical potential temperature profile is everywhere exponentially decaying with a depth-constant λ and a maximum temperature perturbation $\breve{\theta}_0$ at the surface. This specifies the temperature field (Santabarbara *et al.*, 1994):

$$\breve{\theta}(x,y,z) = \breve{\theta}_0(x,y)\ e^{-\frac{z-h(x,y)}{\lambda}} \qquad \textbf{(3)}$$

The integration of (2) with (3) is accomplished by Fourier transforms into a linear set of ordinary differential equations, which are then solved numerically for the mean part, and via FFT routines for the perturbation part. The result is a fast thermal perturbation code, which by estimation of only two parameters ($\breve{\theta}_0$ and λ) calculates (to first order) thermal-orography driven valley breezes and drainage flows according to the vertical temperature profile.

Figure 5 shows an example of LINCOM-THERMAL calculated drainage flow (perturbations only) during the MADONA field trials, cf. Mikkelsen et al. (1995).

Mass Consistent Wind Model MCF

Also a Mass-Consistent-Flow model (MCF) is included, which has complementary input requirements. While the dynamic-equation based LINCOM system must be driven by data from a single observation point only (eg. a met-tower or a NWP grid point) or by a weighted mean of more observations, MCF can interpolate wind measurements from a network of on-line met-towers (or NWP grid points). MCF generates mass-consistent interpolated wind fields over the entire local scale domain under the constraints of minimum flux divergence (Massmeyer et al., 1991). A user's guide in the LSP module assists the user in selecting the most suited model (LINCOM or MCF) depending on the available meteorology for a given application.

Local scale Diffusion, deposition, and gamma dose module

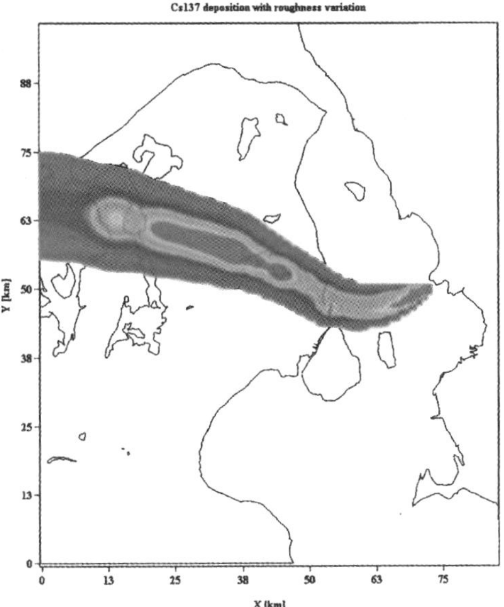

Figure 6 Deposition footprint of Cs137 over Northern Zealand calculated by RIMPUFF.

The local scale model chain integrates a puff dispersion model RIMPUFF and a segmented plume model ATSTEP. The long range model chain is established by nesting the outputs from the local scale model chain to the Eulerian long-range model MATCH , Robertson et al.,(1996).

Rimpuff (Mikkelsen et al., 1984; Thykier-Nielsen et al., 1988; Thykier-Nielsen et al. (1993a) is a fast and operational puff diffusion code, developed for real-time simulation of atmospheric dispersion during accidents. It accounts for changes in meteorological conditions (in time and space) while the accident evolves. The dispersion model is provided with a puff-splitting feature for modelling of dispersion over hilly terrain, which involves channelling, slope winds and inversion layer effects. Also a Gaussian puff-based gamma dose module is included (Thykier-Nielsen et al. 1995).

RIMPUFF is equipped with standard (Briggs) plume rise formulas and has usual inversion-height and ground-level reflection options. The diffusion parameterization in RIMPUFF is formula-based and modular. The puff advection steps and diffusion growth rates are during each time step (typically 10 seconds) determined by the puffs local wind and turbulence levels as provided by LSP. A fast set of subroutines for the calculation of the ground-level gamma dose rates from both airborne and deposited radioactive isotopes have recently been added, Thykier-Nielsen et al. (1993b), Thykier-Nielsen et al. (1995).This new feature plays an important role within RODOS for data assimilation and back-fitting procedures in conjunction with real-time radiological (gamma) monitoring data.

RIMPUFF accommodates almost any user-specified formula-based parameterization scheme for its horizontal and the vertical dispersion parameters σ_y and σ_z. It has pt. 6 optional sigma parameterization schemes included within the RODOS framework. They are, based on the co-called split horizontal and vertical σ-method, combinations of:

Deposited activity is also modelled with RIMPUFF. Dry deposition rates are treated differently for e.g. iodine vapour (elementary iodide) and iodine contaminated aerosols, and different deposition velocities can be specified depending on land use. Figure 6 shows a RIMPUFF calculated footprint of deposited radioactivity from a Cs137 plume traversing Northern Zealand. During the plume passage, the deposition rate is varied depending on the local surface characteristics (land, water, Forrest ,urban , etc).

Table 2:	Optional Sigma-parameter schemes in RIMPUFF
Mode 1-2:	Karlsruhe-Jülich height dependent σ_y and σ_z (1-hr averaged plume sigmas)
Mode 3-4:	Risø instantaneous (no averaging) true puff-diffusion sigmas σ_y and σ_z.
Mode 5:	Similarity-based plume-sigmas (σ_y and σ_z) - averaging time 10-min to 1-hr.
Mode 6:	German-French-Commission (GFC) proposed horizontal σ_y's, - for variable averaging-time between zero (instantaneous puff) and 1-hr (plume sigmas).

ACKNOWLEDGMENTS

This work has been carried out with support from the European Commission DGXII Nuclear Fission Safety Research Programme.

REFERENCES

Astrup P., N.O. Jensen and T. Mikkelsen (1997): A fast model for mean and turbulent wind characteristics over terrain with mixed surface roughness. Submitted to Radiation Protection Dosimetry, October 1996.

Ehrhardt, J. (1996): The RODOS System: Decision Support for Off-site Emergency Management in Europe. In Proceedings of the : Fourth International Workshop On Real-time Computing of the Environmental Consequences of an Accidental Release from a Nuclear Installation. Aronsborg, Sweden, October 7-11 1996. To appear in: Radiation Protection Dosimetry.

Ehrhardt, J. Brown, S. French, G.N. Kelly, T. Mikkelsen and H. Müller (1997): RODOS:Decision-making support for off-site emergency management after nuclear accidents. Submitted to "Kerntechnik" February 1997.

Kelly, G. N., J. Ehrhardt and V.M. Shershakov (1996): Decision support for off-site emergency preparedness in Europe. Radiation Protection Dosimetry, Vol. 64, No.1/2, pp129-141.

Massmeyer, K., and Martens, R. (1991): Regional Flow Fields in North Rhine Westfalia - A Case Study Comparing Flow Models of Different Complexity -.in: Air Pollution Modelling and ist Application VIII, ed. by H. van Dop and D.G. Steyn, Plenum Press, New York, pp. 301 - 309, 1991

Mikkelsen, T., S.E. Larsen and S. Thykier-Nielsen (1984). Description of the Risø Puff Diffusion Model. Nuclear Technology, Vol. 67, pp. 56-65.

Mikkelsen, T. and F. Desiato (1993). Atmospheric dispersion models and pre-processing of meteorological data for real-time application. Radiation Protection Dosimetry Vol. 50, Nos. 2-4, pp 205-218.

Mikkelsen, T., S. Thykier-Nielsen, P. Astrup, H. E. Jørgensen, J. M. Santabárbara, A. Rasmussen, J. Havskov Sørensen, L. Robertson, C. Persson, S. Deme, R. Martens, J.G.Bartzis, P. Deligiannis, N. Catsaros and Jürgen Päsler-Sauer (1997): Recent Development and Integration Status of the Comprehensive Atmospheric Transport Module MET-RODOS for Now- and Forecasting of Radioactive Airborne Spread on Local, National, and European Scales. Submitted to: Radiation Protection Dosimetry, 1997.

Mikkelsen, T. , H.E. Jørgensen, K. Nyrén and J. Streicher (1995). MADONA: Diffusion measurements of smoke plumes and of smoke puffs. In: Proceedings of the 11th Symposium on Boundary Layers and Turbulence, Charlotte, N.C., USA, March 1995, pp. 319-322.

Robertson, L., Langner, J. and M. Engardt (1996) MATCH Meso-scale Atmospheric Transport and Chemistry Modelling system. Basic model description and control experiments with 222RN. SMHI Report No. 70.

Santabarbara J.M., Sempreviva A.M., Mikkelsen T., Lai G. & Kamada R., 1994, A spectral diagnostic model for wind flow simulation: extension to thermal forcing, in *Air Pollution II*, Vol 1: Computer Simulation: Baldasano J.M., Brebbia C.A., Power H. & Zanetti P., ed., Computational Mechanichs Publications, Southampton, UK.

Sørensen, J.H., and Rasmussen, A. (1995): Calculations Performed by the Danish Meteorological Institute. In: Report of the Nordic Dispersion/Trajectory Model Comparison with the ETEX-1 Full Scale Experiments., Eds: Tveten, U. and Mikkelsen, T. Risø-R-847(EN), NKS EKO-4(95).

Sørensen, J.H., Rasmussen, A., and Svensmark, H. (1997a): Forecast of Atmospheric Boundary Layer Height Utilised for ETEX Real-time Dispersion Modelling. Accepted for publication in Physics and Chemistry of the Earth.

Sørensen, J.H., Rasmussen, A., Ellermann, T., and Lyck, E. (1997b): Mesoscale Influence on Long-range Transport; Evidence from ETEX Modelling and Observations. Submittedto: Atmos. Environ.

Thykier-Nielsen, S, Mikkelsen, T., Larsen, S.E., Troen, I., de Baas, A.F., Kamada, R., Skupniewicz, C., and Schacher, G. (1988). A Model for Accidental Releases in Complex Terrain. Proceedings of the 17th NATO/CCMS International Meeting on Air Pollution Modelling and its Application VII, Cambridge (UK), September 19-22, 1988. (Ed. H. van Dop), Plenum Publishing Corporation, 1989, 65-76.

Thykier-Nielsen, S., Mikkelsen T. and Moreno, J. (1993a): Experimental evaluation of a pc-based real-time dispersion modeling system for accidental releases in complex terrain. Proceedings from 20th International Technical Meeting on Air Pollution Modelling and its Application , Valencia, Spain, November 29 - December 3., 1993.

Thykier-Nielsen, S., Deme, S. and Láng, E. (1993b): Calculation method for gamma-dose rates from spherical puffs. Risø National Laboratory, Risø-R-692 (EN), July 1993.

Thykier-Nielsen, S., S. Deme, and E. Láng (1995). Calculation method for gamma-dose rates from Gaussian puffs. Risø-R-775(EN).

Troen, I. and de Baas, A.F. (1986):A spectral diagnostic model for wind flow simulation in complex terrain. Proceedings of the European Wind Energy Association Conference & Exhibition, pp.37-41, Rome1986.

ADVANCES IN DENSE GAS DISPERSION MODELING OF ACCIDENTAL RELEASES OVER ROUGH SURFACES DURING STABLE CONDITIONS

G. Briggs[1], R.E. Britter[2], S.R. Hanna[3], J. Havens[4], S.B. King[5], A.G. Robins[6], W.H. Snyder[7], and K.W. Steinberg[8]

[1]NOAA, U.S. DOC, on assignment to U.S. Environmental Protection Agency, MD-80, Research Triangle Park, NC 27711
[2]University of Cambridge, Trumpington Street, Cambridge, CB2 1PZ, UK
[3]EARTH TECH, 196 Baker Avenue, Concord, MA 01742, USA
[4]University of Arkansas, 700 West 20th Street, Fayetteville, AR 72701, USA
[5]Western Research Institute, 365 North 9th Street, Laramie, WY 82070-3380 USA
[6]University of Surrey, Guildford, GU2 5XH, UK
[7]NOAA, U.S. DOC, on assignment to U.S. Environmental Protection Agency, MD-81, Research Triangle Park, NC 27711. Current address: University of Surrey, Guildford, GU2 5XH, UK
[8]Exxon Research and Engineering Company, 180 Park Avenue, Florham Park, NJ 07932, USA

INTRODUCTION

A major, cooperative research project will be completed in 1997 from which an improved understanding will be gained about the dispersion of accidental, dense gas releases at industrial sites (i.e., high surface roughness) during low-wind stable meteorological conditions. The plans for this project were presented by Hanna and Steinberg (1995). Most previous research was limited to releases over smooth surfaces in nearly-neutral conditions (Hanna et al., 1993).

More specifically, the goals of this research program include the following:

- Determine the effects of a wide range of surface roughness on dense gas dispersion (DGD),
- Explore the effects of a wide range of atmospheric conditions of most concern to DGD modeling, from near neutral stability with wind speed (at a height of 1 m) of about 5 m/s, to quite stable with wind speed of about 1.5 m/s (corresponding to Pasquill "F" conditions),
- Determine the effects of wind shear and along-wind dispersion on concentration magnitude and duration downwind of short-duration releases,
- Measure the effect of plume Richardson number, Ri^* = 0 (passive plume) to about 20,

- Determine whether there are significant atmospheric stability effects on vertical DGD by repeating some of the above in a stably stratified wind tunnel flow,
- Determine more precisely the Reynolds number and Peclet number limits of good simulations of full-scale scenarios in wind tunnels.

The major elements of this research program include the following:

- Studies in several wind tunnels of vertical entrainment into dense gas clouds flowing over surfaces with roughness elements ranging in size from smaller to taller than the cloud height,
- Exploration of methods to simulate dense gas releases under a stable atmospheric boundary layer in an environmental wind tunnel by achieving appropriate parameter ranges for a flow with a Monin-Obukhov length sufficiently small and a roughness length sufficiently large,
- Development of a demonstration data-set from a series of field experiments including both short duration and continuous dense gas release during neutral to very stable atmospheric conditions for surface conditions ranging from (1) smooth, (2) uniform roughness, to (3) a combination of uniform roughness and localized very large roughness.
- Modification of scientific algorithms (e.g., vertical entrainment parameterizations) used in dense gas dispersion models and evaluations of the revised models.

An integrated philosophy was used to coordinate the field and wind tunnel elements of this project in order to enhance the usefulness of the overall data sets. As the wind tunnel experiments are now largely complete, an overview of these experiments is given. Also, some preliminary results from the completed neutral wind tunnel tests are provided, such as entrainment rate as a function of the plume Richardson number. The main field experiment known as "Kit Fox" was completed during the summer of 1995. A description of these experiments as well as a summary of the data collected are presented.

INTEGRATED EXPERIMENTAL DESIGN

We believe that the scientific conclusions of dense gas diffusion (DGD) studies would be greatly strengthened if both field and wind tunnel experiments were carried out in an integrated fashion. Wind tunnel studies are cheaper, faster, and more controllable, but for DGD the low tunnel speeds required for Richardson number (Ri) similarity impose simulation limits that are not well defined. This is especially so for DGD in stable conditions, but we saw possibilities of using advanced facilities for this purpose for the first time. Flow visualizations and plume measurements over a range of Reynolds numbers, comparisons with similar field runs, and comparison of measured dimensionless plume entrainment rates with field-validated values could both increase confidence in wind tunnel DGD studies and better define the limits of this tool.

The first step was to use neutral wind tunnels, at speeds ensuring full turbulence, to design two standard roughness arrays for use in both types of experiments. A "uniform roughness array" (URA) was developed in the wind tunnel at the U.S. Environmental Protection Agency (EPA) Fluid Modeling Facility (FMF) by measuring wind and turbulence profiles over candidate arrays (Snyder, 1995). The object was to maximize z_0/H_r, the ratio of roughness length to element height, while maintaining low element density, since we needed to cover 37,000 m^2 of field with the design. The URA is intended to represent the general effects of non-smooth land surfaces on DGD. An "equivalent roughness pattern" (ERP) was developed in the Cermak-Petersen-Peterka (CPP) wind tunnel to represent the downwind effects of a concentrated area of very large roughness, e.g. an industrial complex (Petersen and Cochran, 1995a and b). Scale models of refinery complexes were used to establish target values of downwind turbulence and passive diffusion. For maximum roughness efficiency, flat baffles facing the wind were chosen for both arrays. In both the wind tunnel and the field experiments, a line of tall spires was set up perpendicular to the wind in the approach flow in order to generate additional turbulence (the spires were 5 m tall in the field).

An earlier 1993 smooth-surface field experiment at DOE's Spills Test Facility provided experience with a dense array of collocated CO_2 (real-time) sensors and bag samplers for CO_2 and SF_6 (Egami et al., 1995). The CO_2 sensors were available with spans of 0.2 to 10% CO_2, allowing practical arc distances of about 25 to 225 m for the planned release rates. The real-time sensors allowed us to make multiple releases during favorable conditions, to measure concentration fluctuations, and to detect the passage time and peak concentrations at each arc for short duration releases (20 s). The most relevant 1995 Kit Fox field experiments had three primary goals. The first goal was to study the effect of roughness on DGD using three different surface conditions: the baseline smooth desert, the URA alone, and the URA+ERP arrays together to simulate effects of an industrial release. The second goal was to study DGD during much lower wind speeds and more stable conditions than attempted previously. The third goal was to study the effects of wind shear and along-wind dispersion on short-duration releases. The 1993 tests and continuous meteorological monitoring established that the best time to capture neutral to very stable conditions, with diminishing winds predominantly from a narrow sector, was one hour prior to and following sunset. The 1995 Kit Fox series included releases with winds at a height of 1 m of 5 m/s down to 1 m/s and stabilities from Pasquill D to F and beyond, meeting or exceeding our expectations.

Wind tunnels were used for simulations of full-scale, point source releases and for idealized DGD studies. The CPP planning studies mentioned above can be compared with actual field tests, for both continuous and 10-s releases, made with u (1m) near 5 and 2.5 m/s. However, because of the scale-down of tunnel speed required for Ri similarity, lower speed simulations were not possible because laminarization and flow instabilities would occur; this is a serious limitation in all DGD wind tunnel studies. Measurements of continuous point source DGD were also made at four wind speeds in the EPA FMF tunnel over a roughness array identical to the URA. Three series of idealized studies in three different tunnels complete our wind tunnel program. These focus on vertical entrainment because that is the most weakly supported element of present DGD modeling, especially with no previous studies over rough surfaces. To reduce the need for three dimensional measurements, which are very time consuming, and to maintain near constant plume Ri, a line source spanning the tunnel was used. The EPA FMF wind tunnel studies focused on the URA array (Snyder, 1996). Two scales were used, with elements 5 cm high and 5/6 cm high, to study entrainment for plumes both shallow and deep compared to the roughness and to better establish the minimum Reynolds number required for full-turbulence simulation. A purposely similar program for the 5 cm elements was carried out in the wind tunnel at the University of Arkansas Chemical Hazards Research Center (CHRC)(Havens et al., 1996); it used an identical physical setup but different instrumentation to check on the replicability of results. In addition, a series using roughness of a very different geometry was carried out to test the generalizability of entrainment parameterizations, e.g., ones in terms of friction velocity. Finally, a similar series of measurements over a URA type array, 2 cm high, is now in progress at EnFlo (University of Surrey) to study the effect of strong ambient stability on DGD. This is the first such attempt, and has required development of new instrumentation to measure surface heat flux.

FIELD EXPERIMENTS

The primary objective of the field experiments was to capture a matrix of finite and continuous duration releases under neutral to stable meteorological conditions over three different surface roughness configurations. To meet the objectives of the project, the field tests were designed to incorporate the results of the exploratory wind tunnel studies mentioned above, the 1993 CO_2 dispersion experiments (Egami et al., 1995), and the predictions of expected plume concentrations and geometry by dense gas models. It was necessary to measure gas concentrations, meteorological data, and source data at one second intervals prior to, during, and after release of the surrogate dense gas for three types of surface roughness: 1) the flat unobstructed desert surface, 2) the URA, and 3) the URA+ERP. When the URA and ERP roughness elements were in place, a line of spires was placed 89 m upwind of the source, in order to enhance the development of turbulence in the boundary layer over the roughness elements. These so-called Irwin spires were also used in the wind tunnel experiments. Details concerning the locations and sizes of the spires and the ERP and URA elements are given in Table 1. Because the time window available for the experiments was limited, the sequence of the surface roughness configurations used for the experiments was ordered from high (URA+ERP) to low (smooth desert). As the roughness configurations were changed, the array of meteorological towers was kept fixed, but the 95 CO_2 sensors had to be positioned lower and partially respanned for higher concentrations for anticipated changes in DGD during the smooth desert tests.

The field tests were conducted at the U.S. Department of Energy (DOE) Spill Test Facility in Nevada. Storage tanks were filled with CO_2 vaporized from a portable liquid tank to a maximum pressure of 8.85 atmospheres. A 329 m release line extended from the tank farm to a sub-surface box which had a quick-acting (<1 sec) sliding door that exposed a 1.5 m x 1.5 m opening at ground level; this provided a low-momentum source. Since concentrations are inversely related to surface roughness, flow rates were adjusted to 4 kg/s for the URA+ERP surface, 1.5 kg/s for the URA surface, and 1.0 kg/s for the smooth surface, to keep arc concentrations within the optimum ranges of the sensors.

Solid state infrared CO_2 chemical sensors were deployed at four different downwind arrays (25, 50, 100, and 225 m from the source). The first three arrays each had three towers with vertical arrays of five sensors plus additional ground level sensors, while the 225 m array consisted of only ground level sensors. Meteorological instruments, consisting of propeller/vane anemometers and temperature probes, were located on three towers, one 20 m in front of the spires, one 6 m in front of the ERP, and one 50 m downwind of the source. Towers with sonic anemometers were located 20 m upwind of the spires, 6 m upwind of the ERP, 7.5 m upwind of the source, and 50 m downwind of the source. An eight level 24 m

Table 1. Description of Locations and Sizes of Roughness Elements Used in Field Experiments.

Field Roughness Element	Farthest Upwind Location with Respect to the Source	Spatial Coverage	Number of Elements	Element Width	Element Height	Element Lateral Spacing	Element Downwind Spacing
Spires	89m	upwind edge of URA	36	0.458m bottom 0.12m top	4.87m	3.25m	not applicable
ERP	50m	39m x 85m	75	2.4m	2.4m	6.1m	8.5m
URA	89m	120m x 314m	6,600	0.8m	0.2m	2.4m	2.4m

tower with propeller/vane anemometers and temperature sensors was located approximately 100 m upwind of the source and 180 m off the centerline of the test grid.

Releases were made during 13 evenings from August 22 to September 15. At least partially successful data capture was obtained for 14 URA+ERP releases, predominately during "D" and "E" Pasquill stabilities, as determined from atmospheric Richardson numbers. For the URA surface, we count about 33 successful data captures, including about 6 each in the "E" and "F" stabilities. For the smooth surface, we count about 23 successful data captures; 7 of these were "F" stability and 3 can possibly be considered "G" stabilities. For each surface condition, about 1/3 of the releases were continuous (2 to 6 minutes) and 2/3 were short-duration (20 seconds).

The analysis of the Kit Fox field experiments is only in a preliminary stage, since the data are still being calibrated and subjected to QA/QC procedures. However, one or two runs from each stability class, roughness class, and source duration class have been analyzed and compared with the predictions of an updated version of the HEGADAS model. The results of the analyses demonstrate that 1) vertical entrainment (and hence ground-level concentrations) are strongly affected by changes in the underlying roughness, 2) ambient wind speed has the largest effect on dense gas dispersion and subsequent distribution of concentrations, 3) theoretical scaling relations developed in the wind tunnels and used in models such as HEGADAS are verified by the field observations, and 4) along-wind dispersion (for the finite duration releases) is enhanced by wind shear near the ground. The comparisons of limited observations with model predictions suggest that the updated algorithms in HEGADAS properly account for these effects.

VERTICAL ENTRAINMENT STUDIES IN NEUTRAL WIND TUNNELS

At the time of writing, the two programs of vertical diffusion measurements in neutral wind tunnels are complete, data reports are available (Havens et al., 1996: Snyder, 1996), and data analyses are in progress. The stable boundary layer program at the University of Surrey is scheduled for completion by late spring of 1997. As described in the "Design" section, the wind tunnel at the EPA FMF was used to investigate DGD over the "URA" array at two contrasting scales with element heights H_r = 5/6 and 5 cm. The small and large versions are designated "WH4-12S" and "WH4-12L" (see Fig. 1), with "4" referring to the ratio of roughness element width to H_r and "12" referring to the ratio of element spacing to H_r. Closely coordinated experiments were run in the CHRC wind tunnel, which was especially designed for dense gas studies. At CHRC, two significantly different roughness geometries were run, one identical to the WH4-12L array above and one designated as the "WH1-8" array with H_r = 3.8 cm. On the basis of earlier wind tunnel studies, the ratio z_0/H_r for the 4-12 array is about three times that of the 1-8 array; thus, the 4-12 array is far more efficient for generating increased turbulence intensity.

It is expected that when the plume depth, h, is large compared with H_r, the roughness effects may be parameterized through a single variable, z_0. However, when the elements are the same height or larger than the plume depth, entrainment will be affected by the shapes, sizes and spacings of the individual elements. Hence, the dual-scale experiments in the EPA FMF wind tunnel were designed to study plume growth both when $h >> H_r$ and when $h < H_r$ or $h \sim H_r$. The contrasting geometry studies in the CHRC wind tunnel focused on how the geometry of the elements affect entrainment rates in the latter situation. The WH4-12L array was run at both laboratories with ostensibly identical flow rates and free flow speeds. The EPA study included additional measurements to elucidate Reynolds number effects.

In both facilities, a "line" source supplied a metered rate of carbon dioxide at negligible vertical velocity through a bed of fine gravel contained within a rectangular box. This box stretched the entire width of the EPA tunnel. Because the CHRC tunnel was much wider than the EPA tunnel, interior sidewalls were used to obtain the same effective width, and the line source was identical. Its width (streamwise direction) was 10 cm. The flow structure was measured with hot-wire anemometry (EPA) and laser-doppler anemometry (CHRC). A small fraction of ethane (C_2H_6) was mixed with the carbon dioxide (CO_2) so that the downwind concentration fields could be measured with flame ionization detectors.

Prior to analyses of these wind tunnel and field data, an international group of scientific advisors agreed on certain definitions and conventions. For example, in analyzing boundary-layer profiles, we agreed to assume that the von Kármán constant $k = 0.4$. We define effective mean plume speed, $\bar{u}$, from the vertical profile of wind speed, u(z), weighted by the concentration profile: $\bar{u} = \int uCdz \div \int Cdz$; when available, crosswind integrated or summed C is used. As a practical matter, u(z) is measured outside the plume. A characteristic plume depth is defined by $h = \int Cdz \div C_s$, where C_s is the surface concentration. Vertical entrainment velocity is defined simply as $w_e = d(\bar{u}h)/dx$. Thus, when the mass flux of plume gas or tracer is conserved ($\int uCdz = Q$), then $w_e = QdC_s^{-1}/dx$, a relatively simple determination. All velocities are scaled by the friction velocity, u_*, which is measured outside the plume. The basic Richardson number definition is $Ri^* = g'h/u^{*2}$, where g' is reduced gravity ($g\Delta\rho/\rho_a$). For the two-dimensional, line source plumes used above we can derive $Ri^* = B_o/(\bar{u}\ u^{*2})$, where $B_o = g_o'Q_o/C_o$ is the line-source buoyancy flux.,

Some preliminary results, from just the EPA FMF wind tunnel, are shown in Fig. 1; we have plotted w_e/u^* versus Ri^*. One encouraging result is that, at the passive limit ($Ri^* = 0$, plotted at 0.1 in Fig. 1), our data agree well with the best available field data, Project Prairie Grass, when these data are analyzed exactly the same way: $w_e/u^* \approx 0.65$. Previously published values for this limit range from 0.4 to 1.0. Another encouraging result is the reasonably good agreement between the small and large array, provided that the tunnel speed for the small array was not dropped below 1 m/s. When it was dropped further, concentration measurements became erratic and the plume appeared to laminarize. Therefore, to maintain full-scale similarity, it appears that the minimum roughness Reynolds number for laboratory studies over sharp-edged roughness is $Re^* = u^*z_o/\nu = 1.5$. Compared to common DGD models, e.g., DEGADIS, these w_e/u^* are about 30% larger at small Ri^*. However, in a range where the models are more frequently used, $Ri^* = 0.3$ to 3, the agreement is reasonably good.

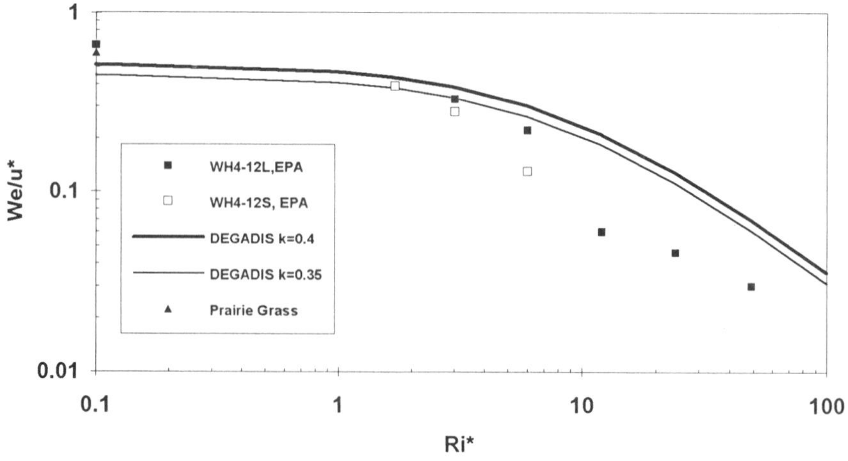

Figure 1. Entrainment velocities deduced from EPA FMF neutral wind-tunnel measurements.

PLUME STUDIES IN STABLE BOUNDARY LAYERS

Entrainment studies with dense gas plumes in stable ambient conditions are being investigated in the University of Surrey's EnFlo stratified flow wind tunnel. The objective is the same as for the neutral experiments, except that the entrainment experiments are conducted in moderately stable conditions.

Initial work concentrated on demonstrating that suitable, moderately stable boundary layers could be established, as determined by comparison of profiles of mean velocity and temperature, turbulence intensities, heat fluxes and temperature fluctuations with relationships based on measurements in the atmospheric boundary layer.

In the wind tunnel, a standard 1 m high barrier wall and a vorticity generator system (spires) were mounted at the entrance to the working section. The inlet heaters were set to provide a uniform temperature profile at the start of the cooled floor section (i.e., at x = 9 m) with the temperature difference between the free flow above the boundary layer and the cooled floor panels being held constant. For most of the work the tunnel was run at a free stream speed, u_{ref}, of 1.35 m/s, with some additional runs at speeds of 1.2 and 1.5 m/s. The floor was covered with a "WH4-12" configuration of roughness elements of height H_r = 20 mm. This configuration is similar to the one used in the EPA and CHRC wind tunnels (see above), and used in the field as the URA roughness. A combination of LDA and cold-wire instrumentation was used to measure the full set of mean flow and turbulence profiles.

A well behaved, moderately stable boundary layer was simulated with u_{ref} set at 1.35 m/s, developing quite markedly to begin with, but slowly thereafter. Between x = 16 and 18 m, log-linear profiles were found to provide a very close fit to the mean velocity and temperature data up to a height of about z = L. The boundary layer depth, δ, was about 250 mm; this is shallow but sufficient for dense gas studies. The characteristic scaling ratios, δ/L, σ_u/u^*, σ_w/u^*, and σ_T/θ^* are listed in Table 2, where L is the Obukhov (stability) length, σ_u, σ_v, and σ_T are standard deviations of longitudinal and lateral turbulent velocities and temperature fluctuations, and θ_* is the turbulent temperature scale.

Since the scaling ratios in Table 2 satisfied the boundary layer criteria set forth by the project steering group, the stable boundary layer flow was judged suitable for the dense gas entrainment studies and the tunnel was then adapted for that phase of the work. This involved extending the fetch of cooled floor for sufficient plume development, and installing a supply system for a mixture of carbon dioxide and propane. A source arrangement identical to that used in the EPA and CHRC wind tunnels was also installed. The experimental program covers both two and three dimensional plume studies, with attention focussed mainly on the former. At conclusion, the entrainment velocity versus cloud Richardson number relationship will be evaluated as a function of ambient stability conditions.

Table 2. Characteristic Properties of Stable Boundary Layer in EnFlo Wind Tunnel

x(m)	δ(mm)	δ/L	σ_u/u^*	σ_w/u^*	σ_T/θ^*
15	200	0.78	1.8 - 2.0	1.2 - 1.4	1.7 - 1.9
16	250	1.48	1.7 - 2.0	1.3 - 1.4	1.6 - 1.7
17	250	1.38	1.6 - 1.9	1.2 - 1.3	1.7 - 1.9
18	270	1.57	1.5 - 1.9	1.1 - 1.3	1.6 - 1.8

SUMMARY

A multi-component cooperative research program consisting of both field and wind tunnel experiments was designed to answer basic questions concerning dense gas dispersion over a wide range of surface roughnesses and atmospheric stabilities. The field measurements are completed and the data should be cleared through the quality assurance process by late spring 1997. Two series of experiments on vertical DGD have been completed in two neutral wind tunnels and data reports are available; partial analyses concerning Richardson and Reynolds number effects were presented here, while more finalized analyses are near completion. A similar experiment in a stably stratified wind tunnel is near or at the end of the experimental phase, with a data report and data analyses due before the end of 1997. We believe that the totality of data collected under this program will be adequate to meet the goals stated at the outset of this paper.

ACKNOWLEDGMENTS

This cooperative research is being sponsored by the Petroleum Environmental Research Forum (PERF) Project 93-16, the U.S. Environmental Protection Agency, the Western Research Institute through its Jointly Sponsored Research agreement with the U.S. Department of Energy, and the Department of Energy's support of the Chemical Hazard Research Center at the University of Arkansas.

Exxon Research and Engineering Company (ER&E) serves as contract coordinator for this PERF 93-16 Project. The other companies that are part of the Technical Advisory Committee for this PERF Project, and their technical representatives, include Allied-Signal Incorporated (Manny Vazquez), AMOCO Corporation (Doug Blewitt), Chevron Research and Technology Company (Dave Fontaine), Mobil Research and Development Company (Frank Rogers), and Shell Development Company (Dan Baker).

This paper has been reviewed in accordance with the U. S. Environmental Protection Agency's peer and administrative review policies and approved for presentation and publication.

REFERENCES

Egami, R, Bowen, J., Coulombe, W., Freeman, D., Watson, J., Sheesley, D., King, B., Nordin, J., Routh, T., Briggs, G., and Petersen, W., 1995, Controlled experiments for dense gas diffusion: experimental design and execution, model comparison. *International Conference and Workshop on Modeling and Mitigating the Consequences of Accidental Releases of Hazardous Materials*, AIChE, New York, 509-538.

Hanna, S.R., Chang, J.C., and Strimaitis, D.G., 1993, Hazardous gas model evaluation with field observations. *Atmos. Environ.*, **27A**:2265-2281.

Hanna, S.R. and K.W. Steinberg, 1995, Studies of dense gas dispersion from short-duration transient releases over rough surfaces during stable conditions. *Air Pollution Modeling and Its Application XI*, Plenum Press, New York and London, 481-490.

Havens, J., Walker, H., and Spicer, T., 1996, Data report: wind-tunnel study of air entrainment into two-dimensional dense gas plumes. CHRC, University of Arkansas, Fayetteville, AR.

Petersen, R.L., and Cochran, B.C., 1995a, Wind tunnel determination of equivalent refinery roughness patterns, CPP project 94-1152, Tasks 1-5, Ft. Collins, CO.

Petersen, R.L., and Cochran, B.C., 1995b, Wind tunnel testing of the 1995 Nevada test site field experiments, CPP project 94-1152E, Task 6, Ft. Collins, CO.

Snyder, W.H., 1995, Data report, wind-tunnel roughness array tests, EPA, Research Triangle Park, NC.

Snyder, W.H., 1996, Data report: wind-tunnel study of entrainment in two-dimensional dense-gas plumes, EPA, Research Triangle Park, NC.

MODEL ASSESSMENT AND VERIFICATION

chairman:	C. Borrego E. Batchvarova
rapporteur:	G. Cautenet M. Sofiev

NUMERICAL SIMULATION OF METEOROLOGICAL CONDITIONS FOR PEAK POLLUTION IN PARIS

Bertrand C. Carissimo

Electricité de France
Direction des Etudes et Recherches
78400 Chatou, France

INTRODUCTION

Increasingly, decisions taken for reducing urban pollution are based on some form of numerical modelling. Although the basic mechanisms for the formation of urban pollution are relatively well known, at least qualitatively, the evaluation of control strategies require accurate quantitative modelling of certain scenario, based on well documented past situations. This is the type of modelling that we will discuss here and which is quite different from the operational prediction of pollution conditions, for which the time constraint is crucial and the modelling usually simpler (and often statistical).

Air pollution in coastal cities is strongly influenced by meteorological phenomena found at coastlines (Steyn, 1996). Mountainous terrain can also play a role in the pollution over certain cities, such as Mexico (Streit and Guzman, 1996). Other cities combine both, such as Athens in Greece which has been intensely studied as Europe most polluted city (see, for example, Moussiopoulos et al., 1995, Carissimo et al., 1996). The Paris area has neither : the topography is very gentle and the city is located sufficiently far from the sea not to feel the coastal circulation. It is nevertheless subject to strong pollution episodes, in very calm meteorological conditions, both in the winter and in the summer.

After a brief description of the tools we have used, the A3UR modelling system and the MERCURE mesoscale model, we will shortly describe winter and summer atmospheric simulations of the Paris area. This will then be followed by a discussion of the difficulties we encountered in carrying out these meteorological simulations and that we will need to address in future work.

THE AIR QUALITY MODELLING SYSTEM

Various simulation systems, capable of accurate modelling of urban pollution have been developed in the US and in Europe, for example the EUMAC Zooming Model (EZM) (Moussiopoulos, 1994). One such system, A3UR, is developed in France through the

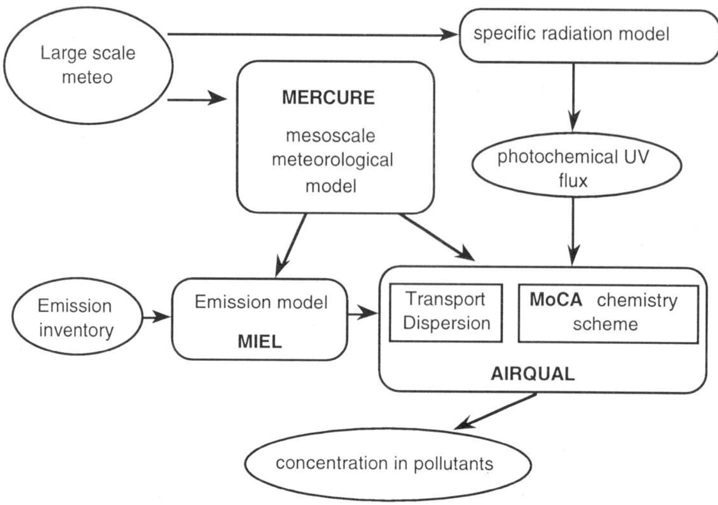

Figure 1. Schematics diagram of the A3UR modelling system

collaborative effort of three organisations (Electricité de France, Institut Français du Pétrole and Laboratoire Interuniversitaire des Systèmes Atmosphériques).

The system has a modular software structure, described in fig. 1:

- a meso-scale meteorological model (MERCURE), to compute the wind field, turbulent diffusion coefficients and other meteorological parameters from large scale meteorological fields. It can take into account different ground types and urban densities and includes a complete set of physical parameterizations.
- an emission inventory model (MIEL) describing the pollutant fluxes from automotive transportation, domestic and industrial activities (Salles et al., 1996). This model includes a mobile source inventory based on road vehicle counting together with global information on transportation fluxes extracted from statistical population data. It uses specific emission factors representative of the vehicle fleet and real driving patterns and gives hourly emissions for a typical day.
- a photochemical gas phase model (MoCA) describing the chemistry of ozone, NOx, and hydrocarbon compounds (Aumont et al., 1996). This model, with 83 species and 191 reactions, is a reduced mechanism well adapted to various air quality conditions (ranging from urban to rural conditions). For interpretative reasons, the identity of primary hydrocarbons is preserved.
- a 3D Eulerian dispersion model (AIRQUAL), describing the transport by the mean wind and turbulent diffusion of chemical species in the atmosphere, associated with a Gear type chemical equation solver.

The system has been applied to a 3-day summertime episode over the Paris area (Jaecker-Voirol et al., 1996) for which simulation results were compared to ground level concentration measurements performed by the local monitoring network.

THE MESOSCALE ATMOSPHERIC MODEL

One component of the air quality modelling system is a mesoscale meteorological model capable of realistically reproducing the conditions for peak atmospheric pollution.

MERCURE is a 3D non-hydrostatic model, based on the anelastic approximation. It solves the conservation equations for mass, momentum, scalar (potential temperature, specific humidity, and any passive pollutant concentration) and turbulent quantities on a staggered Arakawa-B grid. It takes the topography into account using a terrain following co-ordinate.

Numerics

Advection, diffusion and pressure-continuity are solved separately by a fractional time step technique:

- advection is solved with a semi-lagrangian scheme,
- diffusion uses an implicit centred finite differences scheme, solved with an alternative direction method (operator splitting in the 3 directions),
- the Poisson equation for pressure continuity is solved with a conjugate gradient method.

The spurious reflections on lateral and upper boundaries may be minimised with an absorbing layer close to the boundary .

Turbulence

Two parameterizations of turbulent diffusion are available. Both are based on the Boussinesq hypothesis, i.e. the vertical turbulent fluxes are related to the vertical gradient mean parameter values.

The first parameterization corresponds to exchange coefficients estimated with the Louis, 1979 and Louis et al 1982 formulation (one-order closure) :

$$K_{mz} = \left|\frac{\partial V}{\partial z}\right| l \Delta F_m(Ri) \qquad \text{with } l = \frac{k_a z}{1 + \frac{k_a z}{l_{inf}}} \qquad l_{inf} = 20\text{m}$$

$$K_z = \left|\frac{\partial V}{\partial z}\right| l \Delta F_h(Ri)$$

F_m and F_h are two functions of the bulk Richardson number.

The second turbulent scheme is the E-ε closure, solving the conservation equation for turbulent kinetic energy and its dissipation rate (one-and-half closure). In this case diffusion coefficients are computed as :

$$K_{mz} = K_z = C_\mu \frac{k^2}{\varepsilon} \quad (C_\mu = 0{,}09)$$

Radiative effects

The short-wave radiation scheme is based on the parameterization of Lacis and Hansen (1974). Water vapour and ozone absorption and Rayleigh diffusion are parameterized in terms of integrated transmission functions. For cloudy conditions, the multiple scattering is calculated with the two-stream approximation. The cloud absorption and scattering effects are computed with the adding method.

For long-wave radiation, the radiative transfer equation is solved with the integrated emissivity approximation. Absorptivities of Sasamori (1968) are used for water vapour, water dimers, carbon dioxide and ozone. For a cloudy sky, transmission by cloud droplets is modelled with a constant absorption coefficient. Both schemes take the cloud fraction into account.

Surface layer

The parameterization of surface turbulent fluxes of heat, humidity, and momentum, is based on the Louis (1979) and Louis et al. (1982) formulation. This formulation has the advantage of being explicit, i.e. not requiring iterative solution techniques.

Surface processes

Two methods may be used to compute temperature and humidity surface values. In the first, these variables are simply forced to values measured by the Meteo-France surface network, interpolated to model grid nodes. The second method uses a more sophisticated surface process parameterization included in MERCURE. In this case, ground surface temperature is computed according to the force-restore model (Deardorf, 1978), using an anthropogenic heat flux. Surface humidity is deduced from a simple hydrological model with two reservoirs. The local surface characteristics (roughness lengths, albedo, emissivity, etc....) are determined from a surface classification in seven categories: water, forest, bare soil, low density buildings, medium density buildings, high density buildings, and agricultural surfaces. For each type, model constants have been found in the literature. The scalar roughness length is differentiated from the aerodynamical roughness length (Garratt, 1992).

Model validation

The MERCURE model has been tested on boundary layer data (diurnal evolution during the Wangara experiment, Buty et al., 1988), and on orographic flow problems (Elkhalfi and Carissimo, 1993), by comparison with analytic solutions and with data collected during the PYRenees EXperiment (PYREX). It has also been applied to simulate the land-sea breeze cycle in the framework of the APSIS project (Carissimo et al., 1996).

WINTER AND SUMMER POLLUTION EPISODES IN PARIS

Two episodes of strong pollution in the Paris area were chosen to test the air quality modelling system :

- a winter case : 24-25 /11/93
- a summer case : 29-31/7/92

These episodes were chosen in relation with the operator of the Paris air pollution network (AIRPARIF) as representative of the different pollution conditions in the area.

For the mesoscale meteorological simulations, the objective was to have an accurate simulation with the best information available. In Table 1, we have grouped the measurements operationally available in the area. For the ground network we had 45 stations reporting temperature, 30 humidity and 24 wind for the winter episode and only 9 stations for the summer.

For the large scale meteorological forcing we had the following options :

- Meteo France operational analyses (ARPEGE model) every 3 hour,
- ECMWF operational analyses every 6 hour,
- radio-sounding from Trappes (SW suburb), every 12 hour.

We started to use the operational analyses as we did earlier (Carissimo et al., 1996) but we found them insufficiently precise, especially in the lower layers. Finally we ended up using the radio-sounding completed in the lower layers by data from the 100 m mast (Mast-1, see Table 1).

Table 1. Meteorological measurements operationally available in the Paris area

Instrument	Location	Measurements	Frequency
Radio-sounding	SW-suburb	temp.,humid., pressure, wind	12 h
100m Mast -1	SW-suburb	temp., humid., pressure, wind, rain	1 h
100m Mast -2	NW-suburb	temp., humid., wind	1 h
Eiffel Tower	Paris	temperature only	15 min
Surface network	Paris + suburb	temp., humid., solar radiation, pressure, wind	1 h

The simulation domain, a 120 km square, is shown in figure 2. The horizontal resolution is 2 km. In the vertical it varies from 10 to 300 m for a total domain height of 2 km for the winter case and 4 km for the summer. The time step is 15 s.

For the ground forcing of temperature and humidity, we have attempted to directly use the measurements available from the surface network. This has proven unsatisfactory due to poor spatial coverage, especially in the summer case and also because it is impossible to accurately estimate the strong gradients existing between 2m and the ground.

Therefore the full physics and the ground sub model were used. The only adjustment made was on the fractional cloudiness and the aerosol profile in order to obtain the correct incoming solar radiation.

With this, the diurnal cycle of temperature at 2 m (not shown) was found in good agreement with the observed values.

Figures 3 and 4 show the simulated heat island in Paris, an effect found in both the winter (4°C difference) and summer case (2°C).

For the wind (not shown), more scatter was found in the comparison with ground level measurements.

DISCUSSION

During the course of this work, we have encountered a number of difficulties that might be a little more general and that we would like to discuss here.

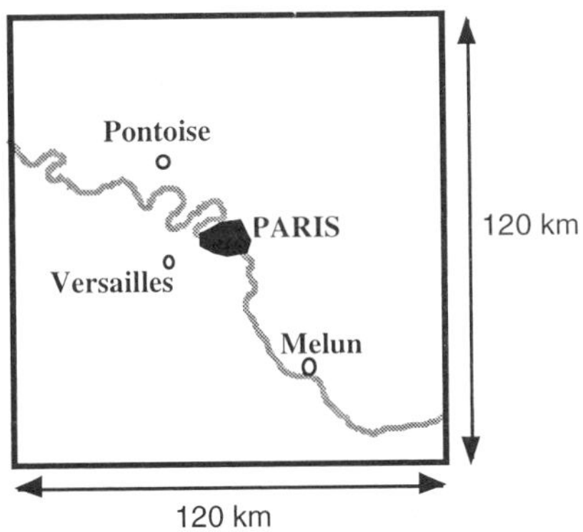

Figure 2. Numerical simulation domain taken around the Paris area.

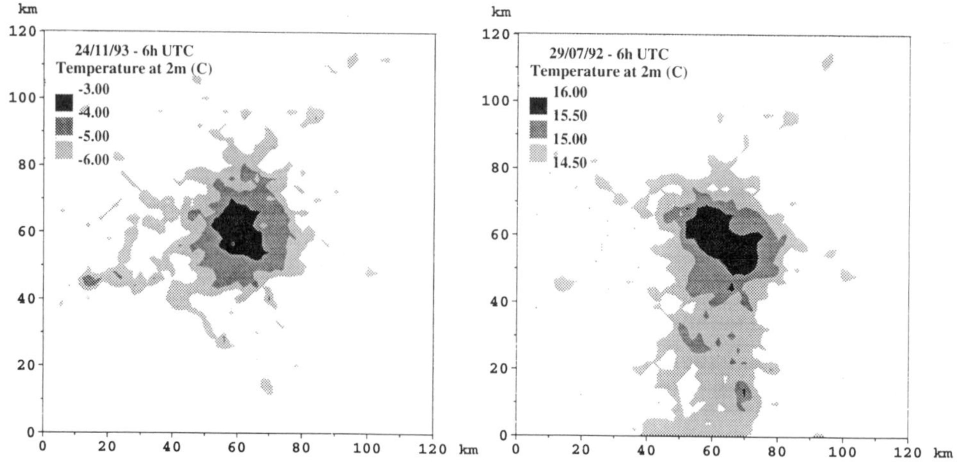

Figure 3 and 4. Simulated temperature at 2 m (°C), showing winter and summer heat islands in Paris.

Inadequate meteorological network

Current operational meteorological networks are designed with weather prediction in mind. This imply for example a focus on "hard" weather and a rather uniform geographical coverage of the country, with poor spatial and temporal resolution for altitude measurements.

Consequently, circulation on the mesoscale, such as those created in an urban environment, can be completely missed by this network.

The ECLAP campaign (Dupont et al., 1997) was designed to provide a more accurate picture of the meteorological conditions prevailing in Paris during pollution episodes. In addition to the operational network we had the following instruments :

- 2 LIDARs in Paris and the suburb,
- 2 SODARs in Paris and the suburb,
- 2 additional radio-soundings (4 per day in total),
- a 30 m mast instrumented with ultrasonic anemometers in the suburb,
- additional ground stations providing complete radiation budgets.

During this campaign, we found rather complex circulation, as illustrated in fig. 5, that could not be seen by the conventional network. In this case, the acoustic sounder sees a very rapid change in wind direction that had a dramatic impact on the boundary layer as we will see below. This illustrates the need for a refined operational network, with boundary layer profiling in and around major cities.

Formation of fog

Fog tends to form in light wind, stable conditions : these are the conditions which also lead to low dispersion of pollutants.

Urban and rural humidity differences are discussed by Oke (1987). During the day, rural humidity are higher, possibly due to greater rural evapotranspiration. At night, the city exhibits a "moisture island" similar to that of temperature, due to the rural decrease of humidity by dewfall and the urban increase by anthropogenic sources which are poorly known.

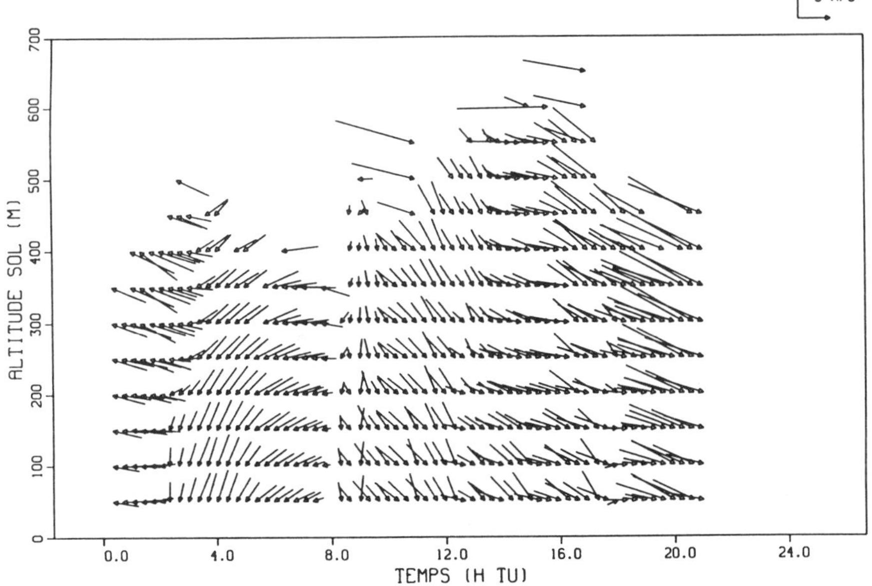

Figure 5. Time evolution of vertical profiles of horizontal winds given by the SODAR placed in the center of Paris for the 14/3/95.

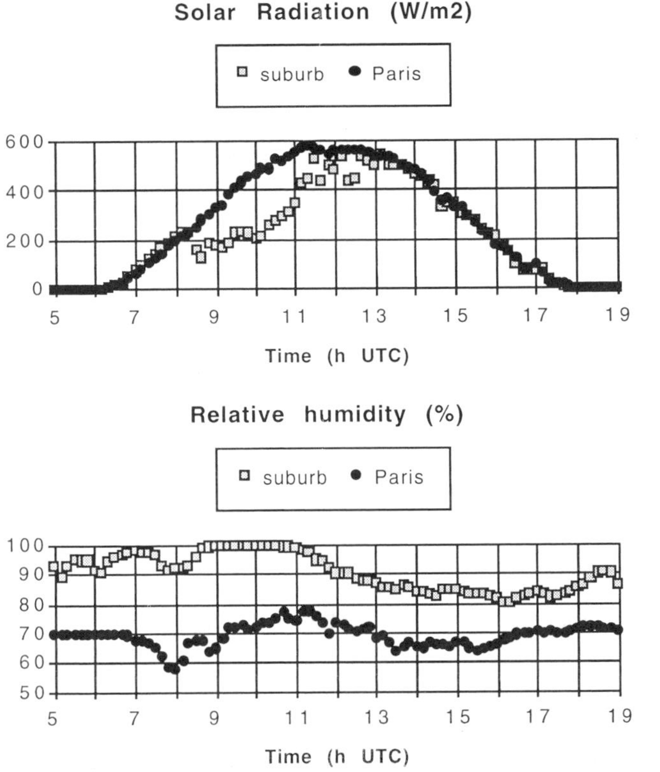

Figure 6. Time evolution of incoming solar radiation and relative humidity for the 14/3/95, corresponding to the wind shown on fig. 5

However, the city is not always more foggy. In fact, the frequency of occurence of thick fog is often less in the city than in the suburbs or the rural surroundings. This may be due both to the heat island effect and to the abundance of condensation nuclei in the city, resulting in a larger number of smaller droplets which do not produce very dense fog.

Figure 6 presents a comparison of incoming solar radiation and relative humidity for Paris and its suburb, observed on the 14/3/95, during the ECLAP campaign (Dupont et al., 1997). This evolution corresponds to that of the wind shown in fig. 5.

Between 8 and 9 h, the sudden change in wind direction with its associated drop in temperature (not shown), lead to the rapid apparition of fog in the suburb while the relative humidity in Paris remains below 80 %. Around noon, the fog gradually disappears in the suburb.

The problem is however that the simulation of the formation and dissipation of fog is still a challenge today and at the limit of capabilities of most atmospheric models (Musson-Genon, 1986, Guedalia and Bergot, 1994).

Light wind conditions

For all simulations performed, we have encountered light winds on the order of 1 m/s. These light winds are the consequence of weak large scale forcing that are easily overcome by very local forcing, unresolved by meso-scale models (with typical 1 km grid size). Plots of observed surface winds in these conditions, often show very disorganised patterns that are unlikely to be accurately simulated (at least for the wind direction).

Light winds have a particular effect on the advective terms in the equation of motion. To better see this and following Pielke (1984), we can derive the equation for the acceleration of the domain averaged wind at a given level, u_O, as a function of the average wind at the eastern and western boundaries, u_E and u_W (assuming symmetry in y to simplify the discussion) :

$$\frac{\partial u_o}{\partial t} = \frac{\overline{u}_E^{\ 2} - \overline{u}_W^{\ 2}}{2D_x} - \alpha_o \frac{p'_E - p'_W}{D_x} + R$$

where D_x is the domain length, p' is the pressure perturbation, R the remaining terms.

With $\overline{u}_E = \overline{u}_W + \Delta u$, we can compute the domain averaged acceleration (in m.s^{-1}.h^{-1}), for a given Δu as a function of D_x and u_w :

$\frac{\partial u_o}{\partial t}$ $for\ \Delta u = 1\ m.s^{-1}$ (m.s^{-1}.h^{-1})	$u_w = 10\ m.s^{-1}$	$u_w = 0\ m.s^{-1}$
$D_x = 100\ km$	0.4	0.02
$D_x = 10\ km$	4	0.2

We see here that, for a given domain size, the decrease of the average wind entering the domain (u_W) lead to an important decrease in influence on the domain averaged acceleration. This is due to the quadratic form of the advection terms.

The effect of light winds on the Coriolis term is somewhat different. This can be seen through the relative importance of the advective and Coriolis terms given by the Rossby number $\left(R_o = \frac{U^2 / L}{fU} = \frac{U}{fL} \right)$. For motions of scale on the order of L=10 km in mid-latitudes, we have : $R_o = 10$ for $U = 10\ m.s^{-1}$ but $R_o = 1$ for $U = 1\ m.s^{-1}$. This means that whereas for strong winds the Coriolis effect can be neglected, in light wind conditions, however, the advective effect become comparable to the Coriolis one.

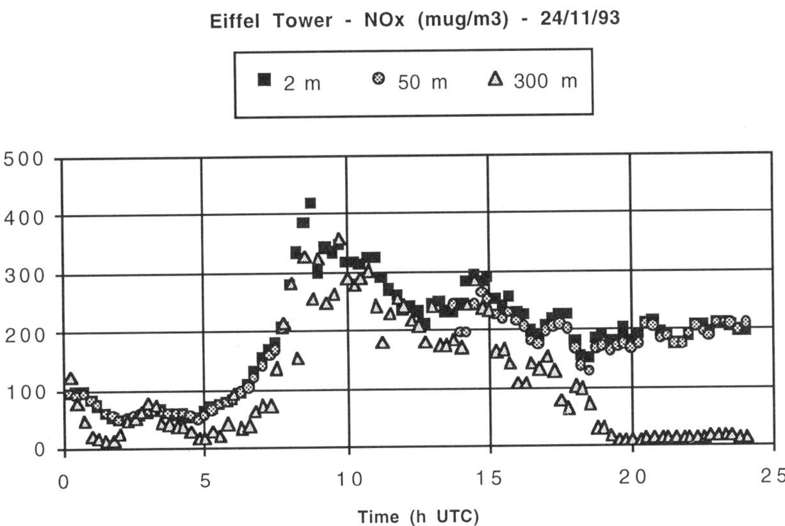

Figure 7. Diurnal evolution of NOx concentrations measured at different levels on the Eiffel Tower in Paris on the 24/11/93.

Finally, when advective and Coriolis terms are weak, small pressure errors between the eastern and western boundaries can control the wind acceleration in the domain.

Interpretation of concentration measurements

In Paris, the Eiffel Tower (300 m high) supports concentration measurements . It is therefore useful to compare the concentration time series at different levels.

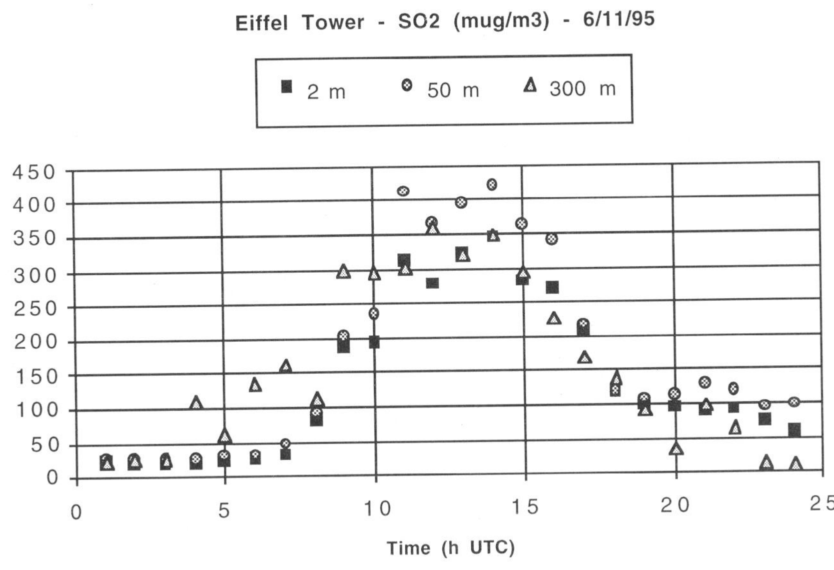

Figure 9. Diurnal evolution of SO2 concentrations measured at different levels on the Eiffel Tower in Paris on the 6/11/95.

For ground sources and if the Eiffel Tower is entirely below the height of the boundary layer (HBL), the concentration series will be comparable at all levels. If not, differences between the levels will become apparent . This can be seen on fig. 7 where diurnal evolution of NOx concentrations measurements are shown at 2, 50 and 300 m. In early morning, concentration are low and comparable at all levels. We can deduce that the HBL is above 300 m. In the evening, this is not the case anymore and the HBL is between 50 and 300 m.

In fig. 8 we have the diurnal evolution of SO2 concentrations for a different day. At first sight, the time series at different heights appear very similar and one could conclude that the HBL is above 300 m. A closer look at the early morning period indicates however that the concentrations are substantially higher at 300 m than below. It is therefore more likely that the HBL is below 300 m and that the measurements at that altitude are influenced directly by altitude sources.

In short, this illustrates the difficulty in interpreting the concentration measurements in meteorological terms when pollutants can come from different sources.

CONCLUSION

The region of Paris is sometime very polluted, although, unlike other cities, the meteorological conditions favorable to pollution are only weakly influenced by topography and thermal contrast.

Using the MERCURE mesoscale model and the A3UR modelling system, we have simulated such winter and summer episodes of peak pollution. The agreement of these simulations with the available observations is satisfactory, for example in the magnitude of the heat island effect.

During the course of this study, we have however identified a number of remaining difficulties :

- an inadequate meteorological network to properly measure the mixing height,
- the formation and dissipation of fog, which affect the suburb more than the city,
- the sensitivity of light wind conditions with weak large scale forcing to small changes in local forcing

Therefore there is an urgent need for a better meteorological network around cities especially to provide vertical profiles of the urban boundary layer accurately and frequently (for example with a Boundary Layer Profiler or a LIDAR).

In the area of numerical modelling, further progress need to be made in the direction of simulating light wind conditions and the formation and dissipation of fog.

Acknowledgments

The work presented is a team effort for which I thank all the contributors. For the Paris simulations, I have drawn on the work of Eric Dupont and benefited from discussions with him and with Luc Musson-Genon. Data from the ECLAP campaign were obtained trough a collaborative effort with the Service d'Aéronomie and the Laboratoire de Météorologie Dynamique (both from the CNRS). Concentration measurements were kindly provided by the AIRPARIF network. This work was partly financed by the PRIMEQUAL program.

REFERENCES

Aumont B., A. Jaecker-Voirol, B. Martin, G. Toupance, 1996 : Tests of Some Reduction Hypothesis Made in Photochemical Mechanisms, *Atmos. Environ.*, 30, 12, 2061-2077

Buty D., J.Y. Caneill et B. Carissimo, 1988 : Simulation de la couche limite atmosphérique en terrain complexe au moyen d'un modèle méso météorologique non hydrostatique : le code MERCURE. J. Mécanique Théorique et Appliquée, suppl. 2, 7, 35-62.

Carissimo B., 1993 : Revue des méthodes d'initialisation et de conduite aux limites possibles pour le code MERCURE. Rapport EDF/DER n° HE-33/93-024.

Carissimo B., E. Dupont et O. Marchand, 1996 : Local simulations of Land-sea Breeze cycles in Athens based on large-scale operational analyses. Atm. Env. 30, pp 2691-2704

Deardorff J.W., 1978 : Efficient Prediction of Ground Surface Temperature and Moisture with Inclusion of a Layer of Vegetation, *J. Geophys. Sc.*, 63, 1889-1903.

Dupont E., L. Menut, B.Carissimo, J. Pelon, R. Valentin and P. Flamant , 1997 : The ECLAP Experiment : Observation of the diurnal evolution of the boundary layer in Paris and its rural suburbs. Submitted to Atm. Env.

Dupont E., L. Musson-Genon, B. Carissimo, 1995 : Simulation of the Paris Heat Island during two strong pollution Events. Air Pollution 95, Porto Carras. Proceedings (Volume III :*Urban Pollution* - Edit. MM. Moussiopoulos, Power, Brebbia).

Elkhalfi A. et B. Carissimo, 1993 Numerical simulations of a mountain wave observed during the "Pyrenees Experiment" : hydrostatic / non hydrostatic comparison and time evolution. Contrib. Atm. Phys. 66,183-200.

Garratt J.R., 1992 : *The Atmospheric Boundary Layer*, Cambridge University Press.

Guedalia D. and T. Bergot, 1994 : Numerical forecasting fog. Part II : A comparison of model simulation with several observed fog events. Mon. Wea. Rev., 122, 1231-1246.

Jaecker-Voirol A., and alii, B. Carissimo, and alii, B. Aumont and alii : A 3D Regional Scale Photochemical Air Quality Model -Application to a 3 Day Summertime Episode Over Paris; *Proceedings Air pollution 1996 Toulouse*

Lacis A. and J.E.Hansen 1974 : A Parametrization for the Absorption of Solar Radiation in the Earth's Atmosphere, *J. Atmos. Sci.*, 31, 118-133.

Louis J.F. , M. Tiedtke, J.F. Geleyn, 1982 : A Short History of the PBL Parametrization at ECMWF. Proceedings, *ECMWF workshop on planetary boundary layer parametrization, Reading*, pp. 59-80.

Louis J.F., 1979 : A Parametric Model of Vertical Eddy Fluxes in The Atmosphere, *Bound. Lay. Met.*, 17, pp. 187-202.

Moussiopoulos N., 1994 : The EUMAC Zooming Model. EUROTRAC Report, Garmissch-Partenkirchen

Moussiopoulos N., P. Sahm and Ch. Kessler, 1995 : Numerical simulation of photochemical smog formation in Athens, Greece - A case study. Atm. Env. 24, 3619, 3632

Musson Genon, L., 1986 : Numerical simulation of fog with a one dimensional boundary layer model. Mon Wea. Rev., 115, 592-607

Musson Genon, L., 1995 : Comparison of different simple turbulence closures with a one-dimensional boundary layer model. Mon. Wea. Rev., 123,p 163-180

Oke T.R., 1987 : Boundary layer climates, Routledge, London and New York

Pielke R. A., 1984 : Mesoscale Meteorological modeling, Academic Press, New York

Salles J., J. Janischewski, A. Jaecker-Voirol, B. Martin, 1996 : Mobile Source Emission Inventory Model. Application to Paris Area, *Atmos. Environ.*, 30,12, 1965-1975

Sasamori T., 1968 : The Radiative Cooling Calculation for Application to General Circulation Experiments, *J. Appl. Met.*, 7, 721-729.

Stull R.B., 1988 : *An Introduction to Boundary Layer Meteorology*, Atmospheric Sciences Library, Kluwer Academic Publishers.

DISCUSSION

M. KAASIK: The night-time concentrations of O_3 are underestimated drastically. It seems that something may be wrong with the parametrization of downward transport of previously released matter?

B. CARISSIMO: This night time underestimation of O_3 has been recognized and is related to a NO_x overestimation. Several possible causes have been identified:

- the night time chemistry, which might be too simplified,
- the parameterization of deposition,
- the emission inventory,
- an underestimation of the night time mixing height.

R. SAN JOSE: We had a problem with the O_3 concentrations during the night similar to what you found in Paris. We found that it could be related to the deposition parameterization. What kind of deposition parameterization are you using in your model?

B. CARISSIMO: The parameterization we have in our model is based on Wesely (1989) and is similar to the one used in CALGRID. Due to uncertainties in the specification of the required inputs, this parameterization has been turned off for the calculations presented here. We are currently working on improvements in this area.

INVERSE MODELING OF URBAN-SCALE EMISSIONS

Gary Adamkiewicz, Peter S. Wyckoff[1], Menner A. Tatang[2], and Gregory J. McRae

Department of Chemical Engineering
Massachusetts Institute of Technology
Cambridge, Massachusetts 02139 USA
[1]Currently at Sandia National Laboratories, Livermore, CA 94551
[2]Currently at Universitas Indonesia, Depok, Indonesia 16424

ABSTRACT

Urban air pollution continues to be a problem worldwide, and there is a critical need to develop cost-effective control strategies. Current strategies are designed using air quality models that describe the formation and transport of photochemical pollutants. Unfortunately, the emissions inventories that are used in airshed modeling and control strategy design have been widely underestimated. New methods are needed to improve the quality of emissions inputs. One approach is to solve the inverse problem using existing ambient data and a photochemical urban airshed model to determine the emission field. However, the high dimensionality of spatially and temporally resolved emissions fields proves to be the primary obstacle in solving this problem.

This paper presents a new approach for the solution of the inverse problem. Emission fields used in urban airshed models can be expressed as random fields. The Karhunen-Loève (K-L) expansion can be used to approximate these random fields optimally, due to the strong correlation between points in the field. This representation greatly reduces the dimensionality of model inversion. The method is presented and applied to determining reactive organic gas (ROG) emissions in the Los Angeles basin.

AIR QUALITY PROBLEM

Control strategies that have been implemented over the past 25 years to address urban photochemical air pollution have been less successful than anticipated. As urban centers expand worldwide and experience the health and economic impacts of air pollution, the need for effective methodologies that address this issue will continue to grow.

Historically, the design of control strategies for urban and regional air pollution has been centered around the use of air quality models that predict the response of ambient pollutant concentrations to reductions in emissions. Local control requirements necessary to reduce photochemical air pollution have typically been determined by establishing the emissions control of ROG or oxides of nitrogen (NO_X) needed to reduce the maximum observed ozone concentrations to below a health-based standard, such as the National Ambient Air Quality Standard (NAAQS) in the United States. In this approach, the objective is to minimize the cost of emissions controls subject to meeting an air quality goal. There have been many obstacles to the development of control strategies, due to difficulties in quantifying precursors emissions and their transport and reaction in the atmosphere. A 1991 National Research Council (NRC) study (NRC, 1991) recommended that the methods and protocols used to develop anthropogenic and biogenic ROG emissions inventories be reassessed.

OPTIMAL FIELD DETERMINATION

Since a detailed re-assessment of all emissions sources is an expensive and difficult task, alternative methodologies should be considered. The concept of inverse modeling can be applied to this problem, using observational data to determine the actual emissions field. The goal of this approach is to vary the emissions spatially and temporally until the error between the observed and predicted ozone fields is minimized. This approach assumes that additional errors are not introduced by the model itself. Stated as a mathematical programming problem, the true emissions field can be found as follows:

$$\min_{E(\mathbf{x},t)} (\| C_m(E(\mathbf{x},t),\mathbf{x},t) - C_o(\mathbf{x},t) \|_n) \tag{1}$$

minimizing some norm of the error between the observed concentration field C_o and that predicted by the model C_m using the precursor emissions fields as the design variables.

Emissions fields used in grid based airshed models consist of thousands of data points for each species of interest and are typically highly structured, as seen in Figure 1. With each iteration requiring several hours of CPU time, an optimization over all emissions is clearly unaffordable unless the number of design variables can be reduced by several orders of magnitude. For example, the CIT Airshed model (McRae et al., 1992) describes the concentration dynamics of 35 species (16 emitted) over an 80x30 cell domain to model the Los Angeles basin. In other words, there are $24\times30\times80\times16$ possible optimization variables for the emissions estimation problem over a one-day simulation.

One method of reducing the number of variables in a field is to use an orthogonal expansion, such that the new design variables are the expansion coefficients. Such an expansion should retain as much of the field structure as possible and require the fewest number of terms possible. One efficient method is the Karhunen-Loève series expansion, which is widely known for its optimality property in approximating fields, especially when there is a strong correlation between points in the field.

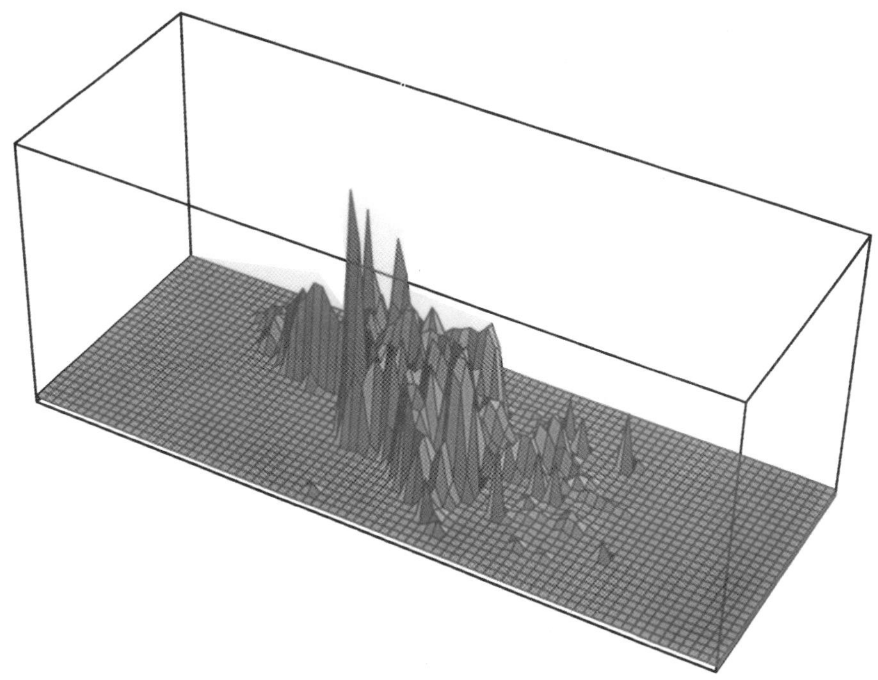

Figure 1. Typical emission field over an airshed

EMPIRICAL KARHUNEN-LOÈVE EXPANSION

Consider a field E(**x**,t) which is to be approximated by a Karhunen-Loève series expansion:

$$E(\mathbf{x},t)=\sum_{n=1}^{N}c_n\alpha_n(t)\beta_n(\mathbf{x}) \tag{2}$$

where c_n is the square root of the n-th eigenvalue, $\alpha_n(t)$ is the n-th temporal eigenfunction, and $\beta(\mathbf{x})$ is the n-th spatial eigenfunction of the correlation function of E(**x**,t). According to Karhunen-Loève series expansion properties, these eigenfunctions should satisfy the normality conditions:

$$\left\|\alpha_n(t)\right\|_t^2=\int_T\alpha_n^2(t)dt=1 \tag{3}$$

$$\left\|\beta_n(\mathbf{x})\right\|_x^2=\int_D\beta_n^2(\mathbf{x})d\mathbf{x}=1 \tag{4}$$

and also the orthogonality conditions:

$$\int_T E(\mathbf{x},t)\alpha_n(t)dt=c_n\beta_n(\mathbf{x}) \tag{5}$$

$$\int_D E(\mathbf{x},t)\beta_n(\mathbf{x})d\mathbf{x}=c_n\alpha_n(t)\,. \tag{6}$$

The last two equations are obtained by considering the eigenfunction as an element of a complete set of the orthonormal functions in the T × D space. The relationship between the eigenvalues and eigenfunctions may be derived as

$$\int_T \mathbf{C}(t,s)\alpha_n(s)ds = c_n^2\alpha_n(t) \tag{7}$$

$$\int_D \mathbf{K}(\mathbf{x},\mathbf{y})\beta_n(\mathbf{y})d\mathbf{y} = c_n^2\beta_n(\mathbf{x}), \tag{8}$$

where **C**(t,s) is the correlation matrix of the field in the temporal domain and **K**(**x**,**y**) is the correlation matrix of the field in the spatial domain. They are defined by

$$\mathbf{C}(t,s) = \int_D E(\mathbf{x},t)E(\mathbf{x},s)d\mathbf{x} \tag{9}$$

$$\mathbf{K}(\mathbf{x},\mathbf{y}) = \int_T E(\mathbf{x},t)E(\mathbf{y},t)dt\,. \tag{10}$$

When the spatial dimension is much larger than the temporal dimension, the first integral equation (9) is clearly preferable to (10). Assuming the first integral equation (9) is solved, the next step is to calculate the spatial eigenfunctions. This discussion follows a similar technique from Sirovich and Everson (1992) who use a snapshot method to calculate the spatial eigenfunctions,

$$\beta_n(\mathbf{x}) = \frac{1}{T}\int_T E(\mathbf{x},t)\alpha_n(t)dt \tag{11}$$

However, instead of using Equation 11, Equation 5 is used directly to calculate the spatial eigenfunctions. This approach seems more appropriate if the Karhunen-Loeve series expansion properties in (5) and (6) are to be ensured.

The empirical Karhunen-Loève series expansion implies the use of empirical eigenfunctions in the temporal and spatial domains, instead of analytical forms. These empirical eigenfunctions are matrices of values of the eigenfunctions at each point in T × D space, and provide a further advantage: singular value decomposition may be used to obtain the temporal eigenfunctions and eigenvalues and the first integral equation above need not be solved directly. The Karhunen-Loève series expansions can be considered as the orthogonalization of a field, such that the terms in the resulting expansion are uncorrelated, analogous to the singular value decomposition. Thus, the correlation matrix in the temporal domain is decomposed into uncorrelated terms as follows:

$$\mathbf{C}(t,s) = \sum_{n=1}^{N} c_n^2\alpha_n(t)\alpha_n(s) \tag{12}$$

The values of c_n^2 and $\alpha_n(t)$ are the diagonal and orthogonal matrices resulting from application of the singular value decomposition method to the correlation matrix:

$$\mathbf{C}(t,s) = \mathbf{U} \cdot \mathbf{W} \cdot \mathbf{V}^T \tag{13}$$

where $\mathbf{W} = \mathrm{diag}\{C_n^2\}_{n=1}^{N^*}$ and the n-th column of the orthogonal matrix U contains the values of eigenfunction $\alpha_n(t)$ at times t_i, i = 1,...,N. Thus, the eigenvalues and temporal eigenfunctions are obtained from orthogonalization of the temporal correlation matrix.

The next step is to calculate the spatial empirical eigenfunctions. Using Equation 5, we can generate the spatial empirical eigenfunctions,

$$\beta_n(\mathbf{x}) = \frac{1}{c_n} \int_T E(\mathbf{x},t)\alpha_n(t)dt \tag{14}$$

which are self-normalized.

These spatial empirical eigenfunctions along with their temporal counterparts and related eigenvalues can be used as the Karhunen-Loève series expansion of the field. In general, the empirical Karhunen-Loève series expansion can be used to approximate any field accurately. Since the complete set of eigenfunctions from that expansion spans the L_2 space (Aubry, 1991), an arbitrary regular field in L_2 space then can be theoretically approximated using the same set of eigenfunctions. In other words, in practice, chemical or physical phenomena with similar underlying structure may be represented by the same set of empirical eigenfunctions (Aubry,1993; Krischer,1993; Rowlands, 1992; Urgell,1990). Based on this fact, we can incorporate our prior knowledge of a field and use it to generate a better prediction. Retaining only the coherent structures with higher energy as our prior information in the optimization scheme suggests that we put more weight in keeping the structure of the field. Other constraints which describe the limit of physical or chemical phenomena could also be added to the problem.

CASE STUDY: ROG INVERSION

The procedure has been applied to the problem of determining the spatial and temporal structure of ROG emissions within the Los Angeles basin. As in most urban areas, the validity of the official emissions inventory for organic gases has been questioned. This example uses data collected during August 27-29, 1987 as part of the Southern California Air Quality Study (SCAQS). During this study period, a variety of special meteorological and air quality measurements were carried out to supplement the routine measurements made in the Los Angeles area.

Since most chemical mechanisms applied to urban photochemistry assign individual organic compounds to a lumped ROG category to ease computation, this estimation addresses many individual chemical compounds. The modified LCC mechanism (Lurmann, 1987) used in the CIT airshed model contains 9 lumped organic classes. Therefore, to test the procedure for the base-case organic fields, a full search over a subset of coherent structures for each lumped category field would be conducted. In order to reduce the computational burden of such a search, however, one of the ROG

Table 1. Percent of variance captured by first five eigenfunctions

	ALKE	
Eigenfunction	**Variance Captured**	**Cumulative**
First	72.3 %	72.3 %
Second	14.9 %	87.2 %
Third	4.0 %	91.2 %
Fourth	2.1 %	93.3 %
Fifth	1.7 %	95.0 %

categories, the alkene field (ALKE), was used as the test field. C_3 and higher alkenes belong to this group, including propene and trans-2-butene. The balance of lumped species was calculated using the base case ratio to ALKE emissions in each cell at each hour.

A critical step in the formulation of any norm-reduction optimization problem is the identification of an appropriate error norm. For most atmospheric modeling inverse problems, this step involves identifying a data set that is assumed to capture the "true" concentration field, as well as a spatial and temporal domain that assures the problem is well-posed. In this study, monitoring data collected during the SCAQS program were used as an estimate of the "true" ozone concentration field, assuming normally distributed measurement errors. The temporal domain of the L_2 error norm was chosen to be the 24 hourly data sets for 37 SCAQS stations on second day of the simulation, August 28, 1987. The selection of stations and the elimination of the first day's data set is intended to reduce the effect of the model's boundary and initial conditions, respectively.

The next step is to determine the number of eigenfunctions to retain in the search. This determination should balance the incremental improvement in representation of the added eigenfunction with the added computational cost of increasing the dimensionality in the search algorithm. As shown in Table 1, the first 5 eigenfunctions of the ALKE fields capture 95 percent of the ensemble's variance, or "energy." The addition of more eigenfunctions does not significantly increase the captured variance, allowing a search over the first five eigenfunctions to be acceptable. The Broyden-Fletcher-Goldfarb-Shanno (BFGS) variant of the Davidon-Fletcher-Powell (DFP) nonlinear search algorithm was used to obtain the optimal coefficients (Press et al., 1996). Many issues arise when constructing such an optimization, including: the evaluation of errors in the resulting optimized fields; the sensitivity of optimized fields to the amount of observational data used; and the "uniqueness" of optimized fields. These will be addressed in future work.

RESULTS

The search resulted in a net increase of ROG emissions of 58 percent over the entire airshed, as shown in Figure 2. A comparison between the base case predictions and the optimized results showed an improvement in predictive capability of the model for all but

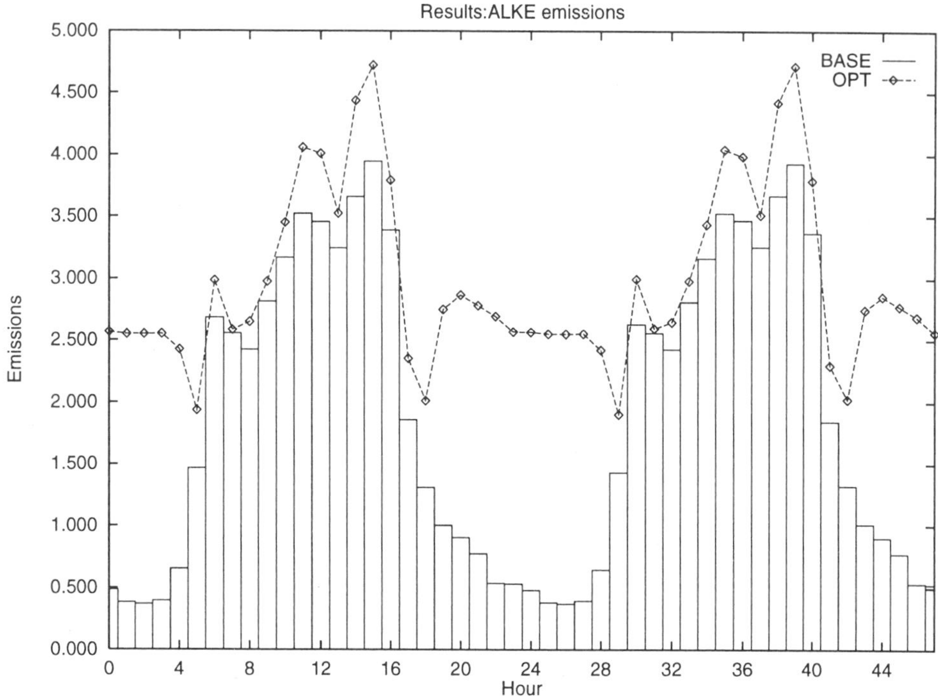

Figure 2. Time Series of Base Case and Estimated ALKE Emissions

the highest concentration ranges. The error norm was reduced by 32 percent during the hour that peak ozone levels are observed in the basin. A significant adjustment to ROG emissions during the overnight hours is evident in these results, as well as a significant spatial adjustment in central Los Angeles. These results may suggest a need to modify the underlying transportation model and associated emissions calculation.

CONCLUSIONS

Inaccuracies in base-case emissions have led to inaccuracies in the air pollution control strategies developed from them. Inverse modeling offers opportunities to provide verification of existing emissions inventories that are built from the "ground up". Optimal representation of emissions fields by Karhunen-Loève series expansions can reduce the dimensionality of the inverse problem. The results of this study are indicative of the potential errors that exist in emissions inventories worldwide. For regions where the structure of the emissions fields can be inferred from known information, the procedure described in this paper can be applied to design data collection studies using traditional sampling or remote sensing techniques.

REFERENCES

N. Aubry, R. Guyonnet, and R. Lima. Spatio-temporal analysis of complex signals: Theory and applications. *J. Stat. Phys.*, 64(3/4):683--739, 1991.

N.Aubry, W.-Y. Lian, and E.S. Titi. Preserving symmetries in the proper orthogonal decomposition. *SIAM J. Sci. Comput..*, 14(2):483--505, 1993.

K. Krischer, R.R. Martinez, I.G. Kevrekidis, H.H. Rotermund, G. Ertl, and J.L. Hudson. Model identification of a spatiotemporally varying catalytic reaction. *AIChE J.*, 39(1):89--98, 1993.

F.W. Lurmann, W.P.L. Carter, and L.A. Coyner. *A surrogate species chemical reaction mechanism for urban scale air quality simulation models*. Volumes I and II. ERT Inc., Newbury Park, CA and Statewide Air Pollution Research Center, University of California, Riverside, CA, 1987.

McRae, Gregory J., A.G. Russell and R.A. Harley, *CIT Photochemical Airshed Model: Systems Manual and Data Preparation Manual.* Carnegie Mellon University and California Institute of Technology, 1992.

National Research Council, Committee on Tropospheric Ozone Formation and Measurement. *Rethinking the Ozone Problem in Urban and Regional Air Pollution.* National Academy Press, Washington D.C., 1991.

W.H. Press, B.P. Flannery, S.A. Teukolsky, and W.T. Vetterling. *Numerical Recipes in C: The Art of Scientific Computing.* Cambridge University Press, 1992.

G. Rowlands and J.C. Sprott. Extraction of dynamical equations from chaotic data. *Physica D*, 58:251--259, 1992.

L. Sirovich and R. Everson. Management and analysis of large scientific datasets. *Int. J. Supercomp. Appl.*, 6(1):50--68, 1992.

L.T. Urgell and R.L. Kirlin. Adaptive image compression using Karhunen-Loève transform. *Signal Processing*, 21:303--313, 1990.

DISCUSSION

R. YAMARTINO: Do the negative emissions values arising form the KL expansion cause problems in your model?

G. ADAMKIEWICZ: While the individual eigenfunctions contain negative emissions, the base-case representation by a Karhunen-Loeve expansion does not. We have found that an unconstrained search procedure may encounter negative emission fields, however, the search space is limited to positive fields in the current implementation. Improvements in the search algorithm that specifically address how this constraint is incorporated are currently being explored.

M. SOFIEV: The measurements are always the source of uncertainties. Do you have any (maybe rough) estimates of the sensitivities of your approach to possible errors?

G. ADAMKIEWICZ: A full assessment of the sensitivities to measurement uncertainties is difficult, given the time required to complete each search. The error norm was designed to minimize the effect of measurement error, by incorporating a large spatial and temporal data set. The current implementation is insensitive to our current estimates of random, unbiased measurement errors; only large, systematic errors in the measurement data would produce optimized fields that are markedly different. As we make improvements in the speed of the inversion, we will attempt to assess the sensitivity of estimated fields to measurement errors.

A. HANSEN: As you know, you can sometimes obtain ostensibly "reasonable" simulation fidelity due to compensating errors. You can identify these errors only through a comprehensive diagnostic evaluation of model performance - something you have not been able to do. Therefore, your inverse modeling approach relies on taking the model as accurate simulator of the emissions-atmospheric system. Any model errors can presumably propagate backward into your corrected emissions field. Have you given any consideration as to how, through an iterative process, you might identify and correct model errors (or perhaps other input errors) as well as putative emission errors?

G. ADAMKIEWICZ: The inverse modeling algorithm we have presented assumes that the model formulation is correct and that all discrepancies between model predictions and measurements are caused by inaccuracies in the emission inventory. These assumptions are based on the premise that while errors exist in many model input parameters, the cause of the large systematic under-prediction of ozone levels is the widespread under-estimation of precursor emissions. As part of this study, we have not attempted to use the inverse algorithm to infer other sources of model errors, such as the chemical mechanism, advection scheme, etc.

E. ANGELINO: Point air pollution concentration measurements can be influenced by large scale emission sources but also by the presence of closer local emission sources. How do you take into account the contributions from the two kinds of sources to different spatially resolved inventories?

G. ADAMKIEWICZ: The method does not separate the effect of local and large-scale emissions. The method currently focuses on adjusting area emissions of ozone precursors based on ozone concentration measurements. Point source emissions are not adjusted in our method due to the relative high accuracy of point source emission rates in the Los Angeles basin. The sensitivity of local ozone levels to local area emissions on the spatial scale of the model (5x5 km) is considered to be small. Also, since the algorithm uses the principal eigenfunctions that contain the dominant structure of the base-case field, local variation should not alter the results of optimization.

MIXING HEIGHT DETERMINATION FOR DISPERSION MODELLING – A TEST OF METEOROLOGICAL PRE-PROCESSORS

Frank Beyrich[1], Sven-Erik Gryning[2], Sylvain Joffre[3], Alix Rasmussen[4], Petra Seibert[5] and Philippe Tercier[6]

[1] Brandenburg Technical University, D-03013 Cottbus, Germany
[2] Risø National Laboratory, DK-4000 Roskilde, Denmark
[3] Finnish Meteorological Institute, SF-00101 Helsinki, Finland
[4] Danish Meteorological Institute, DK-2100 Copenhagen, Denmark
[5] Institute of Meteorology and Physics, University of Agricultural Sciences, A-1180 Vienna, Austria
[6] Swiss Meteorological Institute, CH-1530 Payerne, Switzerland

INTRODUCTION

Concentrations of atmospheric trace constituents in the atmospheric boundary layer (ABL) are strongly affected by the meteorological conditions. One of the most important parameters to characterize the dispersion potential of the ABL is the mixing height (MH). In dispersion models, the MH is a key parameter needed to determine the turbulent domain in which dispersion takes place or as a scaling parameter to describe the vertical profiles of ABL-variables.

MH values derived from measurements are available, if at all, at specific sites and partly also for limited time periods only, so that parameterizations are widely used. So-called meteorological pre-processors have been developed to compute the MH as well as other ABL variables needed as input to dispersion models from routinely available data such as temperature, wind speed and solar radiation. Partly they are designed to work specifically with SYNOP and TEMP data.

The state of the art in pre-processing meteorological information for dispersion modelling has been assessed within the frame of the COST Action 710 ("Harmonization of meteorological pre-processors for atmospheric dispersion modelling") in the past two years. One of the working groups dealt with the determination of the MH. Five different pre-processors currently in use (OML, HPDM, RODOS, Servizi Territorio, FMI) have been compared with measurements using data from three different sites in Europe. Selected results of this comparison and the recommendations made are presented here. The full report of COST Action 710 is to be published by the European Commission.

ROUTINES FOR MIXING HEIGHT DETERMINATION

OML Meteorological Pre-processor

Procedures implemented in the meteorological pre-processor for the Danish dispersion model OML are described in detail in Olesen et al. (1987). Each hour, the module computes a mechanical and, during daytime, a convective MH and selects the larger one as the actual value. A minimum of 150 m is applied, however. The mechanical MH is calculated from (A1) (for all the equations, see Appendix) with c_1 = 0.25. The convective MH is obtained from the integration of (A8) with the constants A = 0.2 and B = 5. It starts with the observed temperature profile (in Europe usually from the 00 UTC radiosounding) and uses observed or parameterized values of the friction velocity u_* and the surface heat flux H_0. If a so-called convective lid is found in the noon sounding, the calculated MHs before noon are multiplied by the ratio of the base height of this lid to the calculated MH at the time of the radiosonde ascent. The integration in the afternoon is continued with the profile from the noon sounding. If a so-called sustained lid is found also in the following midnight sounding, the interpolated lid height is used as an upper bound for the MH.

HPDM Meteorological Pre-processor

The meteorological pre-processor for the U.S. Hybrid Plume Dispersion Model (HPDM) is described in detail in Hanna and Chang (1991). For night-time stable conditions it uses (A3) to calculate a mechanical MH. During daytime, two separate MH values are calculated, a purely convective MH form the integration of (A6) with the constant A = 0.2, and a second MH considering solely the convective boundary layer (CBL) growth rate due to mechanical turbulence as described by (A7). The larger of these values is taken as the daytime MH. Under neutral conditions (defined as Pasquill-Gifford stability category D), (A1) is used with c_1 =0.3. The pre-processor in its standard form does not accept observed values of the surface layer fluxes. The source code had thus to be modified in the present study in order to use measured fluxes.

Meteorological Pre-processor Library of Servizi Territorio (ST)

While the HPDM and OML met pre-processors were developed in the context of specific dispersion models, Servizi Territorio (a commercial company in Italy) offer a library of subroutines as a flexible tool which can be applied in different environments. For the computation of the MH, different equations have been considered to be used alternatively. The library offers the use of either (A1), (A2) or (A3) to calculate the neutral and stable (night-time) MH. In all these equations, the Coriolis parameter is set to 10^{-4} s^{-1}, partially absorbed in the constants. The daytime convective MH can either be calculated using (A5), the simple encroachment model, or using (A9). In both routines, a constant lapse rate above the top of the ML is assumed. This single value, which is used throughout the day, is to be taken from an early morning temperature sounding.

FMI Routine

The module used for the evaluation of the MH at the Finnish Meteorological Institute (FMI) is described in detail in Karppinen et al. (1997). The routine distinguishes between wintertime and summertime situations, since during winter the ABL in Finland is mostly stable or near neutral, even during daytime. Hence a single equation can be used throughout the complete diurnal cycle. The MH is assumed to be proportional to the friction velocity. The coefficient is derived from the MH values at 00 UTC and 12 UTC which are estimated from the respective vertical temperature profiles. During stable conditions in summertime, the growth of the nocturnal surface inversion height is calculated using the approach of Stull (1983) (A4); the MH is then estimated as the maximum of either this inversion height or a mechanical SBL height computed from (A1) with $c_1 = 0.2$. During daytime, (A10) is used to model the growth of a convective ML. After solving the daytime MH model, the MH is compared to the 12 UTC radiosounding and adjusted to the value derived from the measured temperature profile. However, no correction is applied to the hours before noon, as it is done in the OML module. The solution of the equation for the mean mixed-layer potential temperature Θ_b is continuously compared with the observed potential temperature at 2 m (Θ_2). If $\Theta_b > \Theta_2$ and Θ_s starts to decrease despite the fact that the situation should be unstable, the calculation is stopped (this may happen, e.g., in the case of strong cold air advection). The 12 UTC sounding is then used to estimate the MH for the remaining period.

RODOS Pre-processor

MET-RODOS is a comprehensive atmospheric transport and diffusion module, designed for operational use within the real-time emergency management system RODOS (Mikkelsen et al., 1996). The MH can be calculated from NWP model output utilizing a bulk Richardson number method, or from measurements of surface layer fluxes or meteorological parameters. The latter version was tested here. (A3) and (A9) are used to calculate the stable (night-time) mechanical MH, and the convective (daytime) MH, respectively. In case of an upward heat flux during night or a downward heat flux during day, a mechanical MH is computed using (A1) with $c_1 = 0.3$. The daytime mixed layer model is initialized with a starting value of 50 m for the MH, and the lapse rate is taken from the previous midnight sounding. If a strong inversion (lid) is found in the radiosonde temperature profiles at heights above 150 m a.g.l., the height of the lid is determined using a bulk Richardson number method with 0.3 as the critical Ri number. If this lid is lower than the MH calculated otherwise, it is taken as the MH. The height of the lid is not interpolated in time.

DATA

Datasets for the testing of MH routines should meet the following requirements:

- availability of radiosonde ascents (as frequent as possible);
- continuous profile information from sodar, wind profiler and/or high tower;
- availability of measured surface layer flux data (because the MH routines should be tested without superimposed effects of the parameterization of the surface fluxes);
- sufficiently uniform terrain without too much orographic influence.

It turned out that the number of available datasets fulfilling these requirements is rather limited, especially for long-term measurements. Data from three sites (Cabauw [Netherlands], Payerne [Switzerland], Melpitz [Germany]) were employed in the COST study. In the present paper, only examples from Cabauw and Melpitz are shown.

Cabauw

Cabauw (51°58'N, 4°56'E, 2 m) is a boundary layer study site operated by KNMI (Monna and van der Vliet, 1987), located in flat terrain near Utrecht, surrounded mainly by pastures and meadows. Data used in this study comprise turbulent fluxes of sensible and latent heat as derived by different methods from the energy balance measurements and a

20 m mast providing profiles of wind, temperature and humidity at several levels. A sodar without Doppler capabilities (by Aerovironment Inc.) is operated on the site since the 1970s; it is used to derive mixing heights from the backscatter profile. The data used in this study cover the period from July 1995 until January 1996. In addition, regular aerological soundings from De Bilt, which is about 25 km NE of Cabauw, have been used.

Melpitz

Melpitz (51°32' N, 12° 54' E, 87 m) is a flat-terrain site situated within relatively large agricultural fields and pastures about 40 km NE of the city of Leipzig. It was a field station of the SANA programme (study of changing air pollution situation and its impacts on highly sensitive ecosystems over Eastern Germany). The SADE-93 and SADE-94 experiments were carried out there. The analysis within the COST study focused on the three intensive observation periods in September 1993, September 1994, and October 1994. With respect to the mixing height determination, data from the Doppler-Sodar ECHO-1D and from frequent radiosoundings (64 soundings during SADE-93 and 103 soundings during SADE-1994) were used. Surface fluxes were derived from gradient measurements (8 levels) and from eddy-correlation measurements at 5.5 m and 10 m above ground.

RESULTS

Figure 1 shows examples of the MH evolution during some selected day and night periods. In addition to the pre-processor results, MH values obtained by the analysis of measured data are plotted which may serve as reference, though they can be subject of uncertainties, too. During daytime, the primary reference are MHs obtained by the parcel method from radiosoundings (level where an air parcel with near-ground temperature will become neutrally buoyant). The main reference during night-time and for shallow CBLs is a MH estimate derived from sodar backscatter intensity. In addition, MHs have been determined from radiosoundings as the level where a bulk Richardson number (across the ABL) exceeds a critical value (0.2 was used here).

Cabauw, 6/7 July 1995 (Fig. 1a)

On these days, OML produced a growing ML during the daytime, at first determined by u_*/f (growing u_*) and then by the convective model. Compared with the sodar, the early phase of this growth was overestimated by OML (because it used the u_*/f formula which is not appropriate for this situation) while it was correct in the ST and FMI models. Later on the day, OML and FMI gave MHs which co-incide with the radiosounding while the ST and RODOS model values remained lower. We cannot explain the behaviour of HPDM on the first day. The plateau of the MH given by HPDM on the second day is not shown by other models. During the second day, the FMI model produced an unexplained, unrealistic growth in the afternoon. We consider the sodar values of the night from July 6 to 7 too high, but they show nicely the growth of the ML on the morning of the July 7, co-inciding with the ST convective slab model and the morning radiosounding. However, while OML reached the correct MH at noon, ST and RODOS MHs again remained too low.

Cabauw, 12/13 November 1995 (Fig. 1b)

This period was characterized by a persistent, strong inversion at 200-300 m detected by the sodar throughout this period and also picked up by the parcel method. OML detected this inversion as a "sustained lid" and just interpolated the height of this lid linearly be-

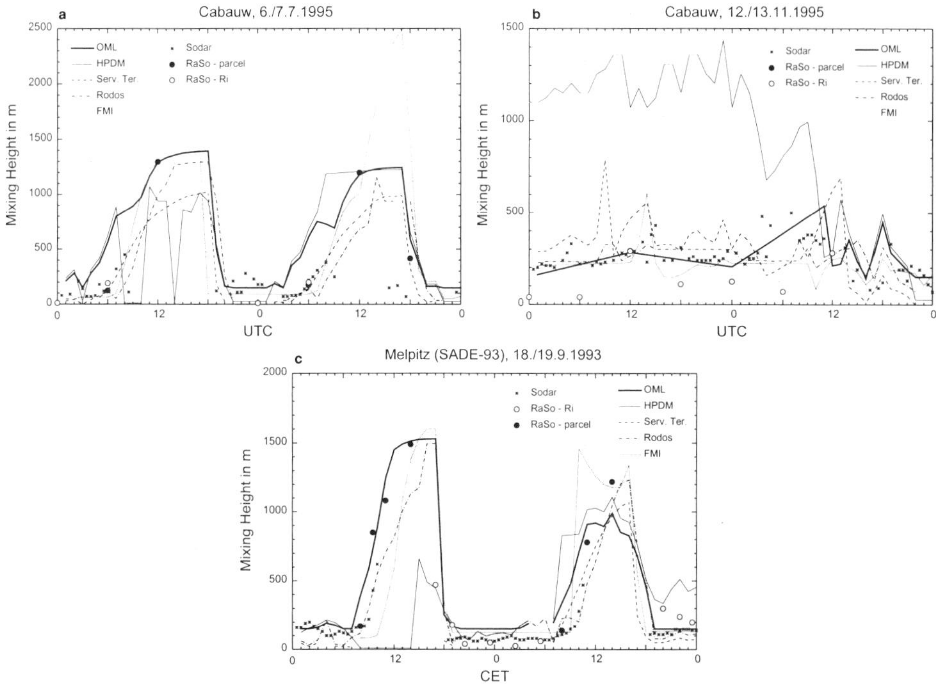

Fig. 1. Evolution of the mixing height as computed by different pre-processors (lines) and as derived with empirical methods (symbols). (a) Cabauw, 6/7 July 1995; (b) Cabauw, 12/13 November 1995; (c) Melpitz, 18/19 September 1993.

tween the radiosoundings. There was a positive heat flux only for a few hours on each of the two days, as indicated in ST slab model. In contrast to OML, strong winds with associated high u_* cause the HPDM to yield much too high MHs in the first 30 hours. The other models fluctuated around the observed values during the whole period. With the exception of 12 UTC on the second day, the Ri-number method always yielded MHs considerably lower than obtained by all other methods; the reasons for this behaviour are not clear at the moment.

Melpitz, 18/19 September 1993 (Fig. 1c)

The OML, the FMI and the RODOS models produced a strong development of the CBL up to about 1500 m, but there were differences in the onset of the growth, probably related to the initialization. The ST slab model was not activated during daytime because there was a short time with slightly positive heat flux already during the night. HPDM selected its minimum value of 10 m when the heat flux became positive in the morning until shortly after noon, for unknown reasons. On the second day, we can see that the FMI model adjusted the MH downward to the observed values of the noon sounding, but did not correct previous hours or its growth rate. HPDM displayed a very high growth rate during the second morning, which is higher than the one given by the slab model of ST and the u_*/f formula used by OML. During the second night, u_* was quite high and HPDM gave MHs exceeding those obtained by the other methods. OML found a lid in the next morning sounding that limited its MH.

CONCLUSIONS AND RECOMMENDATIONS

General

A number of meteorological pre-processors for dispersion models have been tested with real data sets, using measured surface fluxes, and have been compared with MHs derived by empirical methods. Some deficiencies were found in all the pre-processors. Partly they can be remedied by taking into account specific recommendations, but sometimes we could not explain certain unrealistic features. Developers of these pre-processors should investigate them more in depth.

For convective boundary layers, the numerical slab model (A8 or A9) is appropriate, taking into account the specific recommendations listed below.

For mechanically determined MHs, all the current pre-processors rely on similarity formulae involving the friction velocity u_* and the Coriolis parameter f, partly also the Monin-Obukhov length L_*. We regard this as not satisfactory from a physical point of view because the outer stable boundary layer (SBL) is not completely controlled by the surface fluxes and $1/f$ is certainly not the only relevant time scale for the development of the SBL. Richardson methods appear to be better suited from a physical aspect, but the necessary input for these methods is often not available. The use of numerical models (1-D or even high-resolution 3-D) may become a solution in the future. In the meantime, the similarity methods will probably still be used if continuous profile data are not available. Then, (A3) is to be preferred to the simple u_*/f approach (A1).

The performance of parameterizations and models for the MH in both convective and stable boundary layers can be improved by adjusting parameters so that the output matches observed values (e.g., by using variational methods). Corresponding techniques should be developed and implemented.

Specific Recommendations

The numerical integration of a slab model for the CBL should use the actual initial temperature profile with reasonable vertical resolution. It is not advisable to use a single or predetermined value(s) of the potential temperature gradient above the CBL. Rate equations in such models should include also the mechanical contribution to CBL growth.

Preprocessors should allow the substitution of measured data for parameterized ones at any stage. This includes that the algorithms have to be able to deal correctly with measured data, even if they do not follow idealized parameterizations (e.g., unexpected sign of H_0).

Pre-processors should be able to use all soundings available, not just those at the standard hours 00 and 12 UTC. Even in the case of standard soundings, the real launching time of the sounding should be considered, as it may vary by about 1 h around the standard time.

Pre-processors should be able to work with high-resolution radiosonde data (e.g., readings every 10 seconds) and not only with significant levels as reported in TEMP code.

Constants and parameters specific to a certain climatic region (e.g., absolute maxima or minima of the MH, or criteria to find convective lids) should be clearly documented in pre-processors, and the users should have the possibility to change such values.

Acknowledgements

The financial support of the European Commission and various national governments and institutions is gratefully acknowledged, as well as the collaboration of all the data providers whose contribution was essential for carrying out the present study.

REFERENCES

Carson, D.J., 1973, The development of a dry inversion-capped convectively unstable boundary layer. *Quart. J. Roy. Meteorol. Soc.* **99**:450.

Driedonks, A.G.M., 1981, Dynamics of the well-mixed atmospheric boundary layer. De Bilt: *KNMI Sci. Rep. WR 81-2*, 189 pp.

Gryning, S.E., and Batchvarova, E., 1990, Simple model of the daytime boundary layer height. *Proc. 9th AMS Symp. Turb. & Diff., Roskilde* : 379.

Hanna, S.R., and Chang, J.C., 1991, *Modification of the Hybrid Plume Dispersion Model (HPDM) for urban conditions and its evaluation using the Indianapolis data set.* Report No. A089-1200, Electric Power Research Institute, 3412 Hillview Av., Palo Alto, CA 94303, USA.

Karppinen, A., Joffre, S., and Vaajama, P., 1997, Boundary layer parametrization for Finnish regulatory dispersion models. *International Journal of Environment and Pollution* (in print).

Mikkelsen, T., Thykier-Nielsen, S., Astrup, P., Jørgensen, H.E., Santabárbara, J.M., Rasmussen, A., Sørensen, J. H., Robertson, L., Persson, C., Deme, S., Martens, R., Bartzis, J.G., Deligiannis, P., Catsaros, N., and Päsler-Sauer, J., 1996, The real-time on-line atmospheric dispersion module Met-Rodos. *RODOS(WG2)-TN(96)08* (can be obtained from Risø National Laboratory, DK-4000 Roskilde).

Monna, W.A.A., and van der Vliet, J.G., 1987, Facilities for research and weather observations on the 213 m tower at Cabauw and at remote locations. De Bilt: *KNMI Sci. Rep. WR 87-5*, 27pp.

Nieuwstadt, F.T.M., 1981, The steady state height and resistance laws of the nocturnal boundary layer: Theory compared with Cabauw observations. *Boundary-Layer.*

Olesen, H.R., Jensen, A.B., and Brown, N., 1987, An operational procedure for mixing height estimation. *Risø Nat. Lab. MST-Luft-A96*. Last edition 1992, 182 pp.

Stull, R.B., 1983a, A heat flux history length scale for the nocturnal boundary layer. *Tellus* **35A**:219.

Stull, R.B., 1983b, Integral scales for the nocturnal boundary layer. Part I: Empirical depth relationships. *J. Clim. Appl. Meteorol.* **22**:673.

Weil, J.C., and Brower, R.P., 1983, *Estimating convective boundary layer parameters for diffusion applications.* Report PPSP-MP-48 Prepared by Environmental Center, Martin Marietta Corporation, for Maryland Department of Natural Resources, Armapolis, MD.

Zilitinkevich, S.S., 1972, On the determination of the height of the Ekman boundary layer. *Boundary-Layer Meteorol.* **3**:141.

APPENDIX: MIXING HEIGHT EQUATIONS USED IN THE PREPROCESSORS

Mechanical Mixing Height

Similarity formula for the mixing height h used by many authors for neutral and sometimes also stable conditions:

$$h = c_1 \frac{u_*}{|f|} \tag{A1}$$

Similarity formula for stable conditions (Zilitinkevich, 1972):

$$h = c_2 \sqrt{\frac{u_* L_*}{|f|}} \tag{A2}$$

Interpolation formula between (A1) and (A2) from Nieuwstadt (1981):

$$h = 0.3 \frac{u_* / f}{1 + 1.9 h / L_*} \tag{A3a}$$

As (A3a) contains the MH on the RHS, a transformation to the following form is useful in practical applications:

$$h = \frac{L_*}{3.8}\left(-1 + \sqrt{1 + 2.28\frac{u_*}{f\,L_*}}\right) \quad \text{(A3b)}$$

In the FMI scheme (Karppinen et al., 1997), the night-time MH in summer is calculated as a function of time on the basis of formulations of Stull (1983a, 1983b):

$$h_t = h_{00}\,\frac{g_t(\Theta_M - \Theta_{00})}{g_{00}(\Theta_M - \Theta_t)}, \quad \text{with } g_t = \int_0^{t_i}(\Theta_* u_*)dt, \quad \text{(A4)}$$

where g_t is the area under the h vs. Θ curve parameterized as indicated above; the subscript 00 refers to 00 UTC. This form is based on Stull's equation: $b\,h_t\,(\Theta_M - \Theta_t) = g_t$, and the assumption that the parameter b is constant during the whole night.

Convective Mixing Height

The simple encroachment formula for the growth of a CBL ($\langle w'\Theta'\rangle_0$ denotes the kinematic heat flux at the surface, γ_Θ the potential temperature lapse rate):

$$\frac{\partial h}{\partial t} = \frac{\langle w'\Theta'\rangle_0}{\gamma_\Theta\, h}. \quad \text{(A5)}$$

A convective growth formula including the entrainment heat flux, parameterized by the surface heat flux and the constant A (Carson, 1973):

$$\frac{\partial h}{\partial t} = (1+2A)\,\frac{\langle w'\Theta'\rangle_0}{\gamma_\Theta\, h} \quad \text{(A6)}$$

The mechanical ABL growth equation according to Weil and Brower (1983) is (B=2.5):

$$h^2\Theta(h) - 2\int_0^H z\Theta\,\mathrm{d}z = 2\frac{B}{\beta}\int_0^t u_*^3(t')\,\mathrm{d}t', \quad \text{with } \beta = g/T_0 \quad \text{(A7)}$$

A formula including both convective and mechanical contribution to the growth rate (Driedonks, 1981; $\Delta\Theta$ is the potential temperature jump at the ABL top):

$$\frac{\partial h}{\partial t} = A\,\frac{\langle w'\Theta'\rangle_0}{\Delta\Theta} + B\,\frac{u_*^3}{\beta h \Delta\Theta} = \frac{A w_*^3 + B u_*^3}{\beta h \Delta\Theta} \quad \text{(A8)}$$

An alternative formulation given by Gryning and Batchvarova (1990) is ($w_* = \sqrt[3]{\beta h \langle w'\Theta'\rangle_0}$):

$$\frac{\partial h}{\partial t} = (1+2A)\,\frac{\langle w'\Theta'\rangle_0}{\gamma_\Theta\, h} + 2B\,\frac{u_*^3}{\gamma_\Theta \beta h^2} = \frac{(1+2A)w_*^3 + 2Bu_*^3}{\gamma_\Theta \beta h^2} \quad \text{(A9)}$$

The FMI summer daytime scheme (Karppinen et al., 1997) computes the MH from

$$\frac{\mathrm{d}h}{\mathrm{d}t} = C_f w_M / (C_t + Ri_M) \quad \text{(A10)}$$

where $C_f = 0.2$, $C_t = 1.5$, and $Ri_M = gh\Delta\Theta / (T\,w_M^2)$; the convective velocity scale w_M is defined by $w_M = (w_*^3 + C_\tau u_*^3)^{1/3}$ with $C_\tau = 25$.

DISCUSSION

M. KAASIK: How is the mixing height used in models, in other words, which parameters are calculated using the mixing height?

P. SEIBERT: On one hand the mixing height is needed to define the layer of the atmosphere in which turbulent diffusion is taking place. On the other hand, modern dispersion models use parameterizations based on quantities such as the convective scaling velocity w_* which is calculated from the mixing height and the surface sensible heat flux. Finally, often vertical profiles of turbulence parameters such as σ_w are assumed which of course also depend on the mixing height.

D. W. BYUN: For grid model application we have very similar problems in estimating "reasonable" boundary layer heights for air quality simulations. The methods presented in this talk are mostly based on "horizontal homogeneity" approximation. For grid models, we need to include the effects of horizontal advection as well. Yet, what is better is that meteorological dynamic models should provide PBL height as a part of a predictive quantity as was used in the simulation of atmospheric turbulence.

P.SEIBERT: No answer, was just a comment.

R. YAMARTINO: Some mixing height models make assumptions about which soundings (00 & 12 Z) represent morning and afternoon conditions (i.e. the local time). In comparing US versus European models, was a check of these interpretation differences (i.e. local solar time versus GMT-Z time) made?

P. SEIBERT: While at the conference I assumed that such a check had been made, it turned out in the meantime that unfortunately there was indeed a problem with the assignment of the 00 and 12 Z soundings. First control runs showed that some of the unreasonable behaviour in the HPDM preprocessor could be eliminated with the time correction, however, other problems remained.

THE METHODOLOGY OF THE MODEL VERIFICATION BASED ON THE COMPARISON WITH MEASUREMENTS AND WITH OTHER MODELS

M.A.Sofiev

Hydrometeorological Research Centre of RF
Stroiteley str. 4, bld. 1, app. 18, Moscow, 117311, Russia
Tel / fax+7-095-9300961, e-mail mas@glasnet.ru

INTRODUCTION

Mathematical models of atmospheric processes are a complicated matter for Quality Assurance (QA). As well as for measurements, QA procedure is oriented to numerical evaluation of the model precision, potential character of distortions and possibilities of their reduction. The procedure should not answer the question "how good this model is". Normally after obtaining an extensive numerical information about the model quality, investigator takes a decision whether current model meets the requirements of a particular task.

As it was stated in (Sofiev, 1996) QA algorithm should include at least three parts. The first part is the model analysis which consists of a set of artificial tasks (including sensitivity tests) with prescribed answers (e.g., tasks with analytically derived results). The results characterise the ability of the model to simulate simple standard problems. Sensitivity studies highlight the importance of some model sub-units, parameters, initial and boundary conditions, etc. Since the artificial tasks are completely determined by particular conditions, model peculiarities, etc., this part of the QA procedure is not considered in current paper.

The second part is the comparison of modelling results with available measurements. This way of verification is the most widely used QA procedure. But it can not be considered to be sufficient for at least three reasons:

- Measured values are often less interesting than those which can not be measured.
- For large areas it is difficult to collect a representative set of the data with appropriate spatial and temporal characteristics.
- Measurement data have limited representativeness and precision as well.

Finally, the third part of the QA algorithm is the intercomparison of the results of different models between themselves. It does not require the measurement data and is independent upon their quality and completeness. At the same time it implies the availability of comparable results from several models.

The QA algorithm presented in this paper combines the methods of both model-measurement and model-model comparison. Current approach was applied to heavy metal model intercomparison study carried out in 1994-1996 (Sofiev *et al.*, 1996).

MODEL-MEASUREMENT COMPARISON

Generally this problem is characterised by relatively small volume of the measurement data samples in comparison with model output. The second specific is different spatial features of the compared data sets - while measurements are practically point values, the model calculations are always averaged over considerable area (at least over a model grid cell). These two main difficulties require accurate choice of the statistical methods for the comparison.

Current methodology is based on minimum set of estimates of the agreement between model results and measurement data:

- Non-parametric analysis: - arithmetic means both for model and measurement sets; correlation coefficients between sets; standard deviations and/or confidence ranges for these values.

- Regression analysis - slope and bias of linear regression of model data on measurement ones; modified residual sum of squares and standard deviation of the regression slope.

It was shown (Sofiev, 1994) that regression analysis is not trivial and requires special methods. Corresponding procedures had been elaborated in (Sofiev & Galperin, 1994) and realised in task-oriented software called as Model Calibrating Environment (MCE).

The estimates of linear regression coefficients made by MCE are of robust character which is determined by the Choice Procedure of the Effective Method (CPEM) applied in MCE. Description of the methodology could be found in (Sofiev *et al,* 1995). CPEM automatically produces the standard deviation of regression slope together with other parameters.

Contrary to regression methods the non-parametric analysis involves commonly used statistics which are not robust and their direct implementation may results in mistakes. The following consideration could give the estimate of the order of magnitude of uncertainties.

Let assume that $R(x,y)$ is two-dimensional stochastic process with normal distribution function (Johnson & Leone, 1977). Let R_N be the realisation of this process with N pairs (x_i, y_i). The maximum likelihood estimate of the measure of linear relation between x and y results in well known formula for sampling correlation coefficient ρ:

(1) $$\rho = \frac{\overline{(x-\bar{x})(y-\bar{y})}}{\sigma_x \sigma_y}$$

where bar means sampling averaging, σ is sampling standard deviation.

If $N>>1$ the standard deviation of estimate (1) is:

(2) $$\sigma_\rho = \frac{1-\rho^2}{\sqrt{N}}$$

The standard deviation σ_A of sampling average estimated for the same conditions is:

(3) $$\sigma_A = \frac{\sigma_{Sample}}{\sqrt{N_{Sample}}}$$

where σ_{Sample} is the standard deviation of the sample , and N_{Sample} is the volume of the sample.

It should be stressed that formulas (1) - (2) are derived by maximum likelihood approach **only** for processes with normal distribution function. That is why application of correlation coefficient in atmospheric dispersion models assessment is rather arguable and above given accuracy estimates should be considered as lower ones (in a common case the precision of the estimates is worse).

MODEL-TO-MODEL INTERCOMPARISON

From 'classical' point of view the comparison of any mathematical objects is an establishing of some space where these objects could be considered as a points, then defining a metric in this space and finally determining the distance between these objects in this space. The situation becomes more complicated if these objects have physical meaning. In this case the space itself and the metric should be physically 'reasonable'. This requirement can be met not in all practice problems. Introduction of stochastic features to the intercomparison task results in extra requirement to above mentioned measure. It should be robust to uncertainties of mathematical representation of physical objects.

Current methodology could be considered as a step forward to the above mentioned goal. There are no systematic definitions of spaces and metrics. It can be shown that ranking of models suggested as an aggregating procedure of intercomparison results does not meet the definition of metric. But the reduction of dimensions of complicated physical objects used in presented procedure seems to be promising.

All models are considered as a sets of maps. Each map is considered as two-dimensional finite size realisation of non-observable stochastic process. The algorithm determines features of this process (model statistics) with following comparison with features of other processes (models).

Specific problem raised here is the problem of reference point for the intercomparison exercises (all participated models are of equal rights!) . It can be handled via expert estimates procedures like Delphi method (Dalkey & Helmer, 1963). It can be used for creating of artificial data set with the content soimilar to model data sets participated in the intercomparison. This new data set can be call as Statistical Average Model (SAM). In general this method involves several steps (e.g. Quade, 1970) but peculiarities of atmospheric models (namely, comparably large number of parameters and data produced) allow to reduce the procedure down to one step.

Construction of "Statistical Average Model" (SAM) - Delphi method

As other models "Statistical Average Model" (SAM) is considered as a set of maps. Following general states of experiment planning and evaluating theory (e.g. Shannon, 1975) models participated in the comparison procedure are considered as an independent non-correlated experts, and their data sets are used for creating of common model data set SAM:

$$(4)\quad SAM = \left\{ M_{SAM}(i,j)_{i,j\in G_{SAM}} \;\middle|\; M_{SAM}(i,j) = Q_K\ (M_k(i,j)),\ k=\overline{1,K} \right\}$$

where M_{SAM} is the map of the SAM set, grid G_{SAM} for SAM maps is the most precise and compatible with all others participated models, M_k, are maps of K - *th* model set, Q_K is the procedure of making of common map from individual model maps, *i, j* are indexes of the map elements. Their ranges I_G, J_G are determined by the grid dimensions of each model.

The procedure Q should meet two obvious requirements:

1. This procedure should keep the balance of pollution masses.
2. This procedure should be robust (has minimum sensitivity to large disturbances in some model data sets).

Current approach is based on median mean:

$$M_{SAM}(i,j) = \underset{k=1,K}{\text{med}} \left(M_k(i,j) \right) \tag{5}$$

The median statistics is well known as robust one, but this approach introduces an extra error which affects the balance. Order of magnitude of corresponding relative standard deviation could be (very) roughly estimated from the large figures law as the following:

$$\sigma \approx 1 / \sqrt{N_G} \tag{6}$$

where N_G is the number of cells in a grid.

This value is negligible in almost all cases, so generally median method for SAM making is more preferable than arithmetic mean.

Arithmetic mean which could be applied here is less preferable since it is not robust [Huber, (1981)]. It could be used only if there are some guaranties that all considered data are of the same origin and with the same statistic features and should not be used if there is possibility of large deviations of small number of points.

Model Statistics Calculation

One of the most important characteristics of the atmospheric dispersion models is the distance of a pollution transport. It can be calculated from the **autocorrelation function** *ACF* for all model maps separately.

Classical *ACF* is considered in the range much less than the definition one of the correlated function ϕ. In one-dimensional case it is written as follows:

$$ACF_{\Phi}(x) = corr(\Phi(\varsigma + x), \Phi(\varsigma)), \quad \varsigma \in (-\infty, \infty), \quad x \in (-\infty, \infty) \tag{7}$$

Since $ACF(x) = ACF(-x)$ the range $x \in (0, \infty)$ is enough. Extrapolation of (7) to two-dimensional case is trivial:

$$ACF_{M}(x,y) = corr(M(\varsigma + x, \upsilon + y), M(\varsigma, \upsilon)), \quad \varsigma, \upsilon, x, y \in (-\infty, \infty) \tag{8}$$

One peculiarity of the considered task is the finite range of definition of the correlated functions (maps): $\varsigma \in [0, I_G], \upsilon \in [0, J_G]$ where I_G and J_G are x- and y dimensions of the calculation grid G. In this case the correlation may be calculated only for a common range specific for each lag (x,y).

To evaluate the standard error of (8) let start from one-dimensional case (7):

$$\sigma^2(ACF(m)) = \frac{1}{N}\left(1 + 2 * \sum_{i=1}^{m-1} ACF^2(i)\right) \tag{9}$$

where N is a volume of sampling.

If $N >> 1$ then it is possible to consider (9) as a partial Riman sum of the integral

(10)
$$\sigma^2(ACF(x)) = \frac{1}{T}\left(1 + 2 * \int_0^x ACF^2(\varsigma)d\varsigma\right)$$

In (10) *ACF* is defined on real numbers set and *T* is a measure of the definition range.

Last formula could easily be extrapolated to two-dimensional case and adapted to natural number set:

(11)
$$\sigma^2(ACF(x,y)) \approx \frac{1}{N_G}\left(1 + 2 * \sum_{\upsilon}\sum_{\varsigma} ACF^2(\varsigma,\upsilon)\right)$$

where summing is taking within circle $rad(x,y) = \sqrt{x^2 + y^2}$, $N_G = I_G * J_G$.

As it is noted above *ACF* is used for evaluation of the following parameters:

- correlation radius of the emission field
- increase of the correlation radius for different substances in comparison with the emission one. This value gives the estimate of transport distance for each model.

Correlation radius is averaged from its three estimates taken at the horizontal cross-sections at levels $c = 0.67$, 0.5 and $1/e$ $(e=2.71828...)$:

(12)
$$r_c = -\sqrt{x^2 + y^2} \Big/ \ln c \qquad \Big| \; ACF(x,y) = c$$

It obviously depends upon the actual direction at each plane (e.g. it is bigger in the direction of maximum wind rose and smaller in cross one). So, the description of this radius contains two values: mean and standard deviation.

The next parameter is **histogram** H which is calculated for all output maps and for emission field:

(13)
$$H = \{H(a_l, a_{l+1})_{l=1,L-1} : H(a_l, a_{l+1}) = Num\,(M(i,j) \in [a_l, a_{l+1})\}$$

where $[a_l, a_{l+1})$ is the sub-interval of the range of values $[a_1, a_L)$ of the map $M(i,j)$ *Num* is the number of values in map $M(i,j)$ within this interval; i, j - co-ordinates of the map element.

The fourth parameter is a set of **cross-correlation functions** *CCF* between emission and each output map. *CCF* calculations are similar to (8) with minor difference.

(14)
$$CCF_{M,N}(x,y) = corr(M(\varsigma + x, \upsilon + y), N(\varsigma,\upsilon)), \quad \varsigma,\upsilon,x,y \in (-\infty,\infty)$$

As for *ACF* actual definition range is finite: $\varsigma \in [0, I_G], \upsilon \in [0, J_G]$. In this case the correlation may be calculated only for a common range (which is specific for each lag (x,y)). The estimate of the standard error of *CCF* evaluation could be calculated by (2).

Two significant parameters of the model could be taken from *CCF* consideration. The first one is the maximum value of *CCF* which describes (together with *ACF*) the influence of meteorological conditions and emission data and their relations.

The second one is the bias of CCF maximum from $(0,0)$. Non-zero lag reflects the presence of considerable non-uniformity of wind direction distribution. Simultaneously it shows the sensitivity of considered model to this effect if it is found independently (or shown by other models).

Direct Model - to - SAM Comparison

Non-parametric analysis of pairs of maps includes the evaluation of the mean values RM and standard deviations σ_{rel} of relative differences between corresponding maps of different models:

$$RM = \overline{\frac{M_1 - M_2}{M_1 + M_2}}, \quad \sigma_{rel}(M_1, M_2) = \sqrt{D\left(\frac{M_1 - M_2}{M_1 + M_2}\right)} \tag{15}$$

where σ is the standard deviation, M_1, M_2 are maps of the pair of models, $\overline{M}$ is arithmetical mean of the map M, and $D(M)$ - implies the variance calculation.

The next parameter is the cross-correlation function between corresponding maps. It is calculated in accordance with (14) with standard error (2). As for the analysis of isolated model the maximum of this function and its bias is of particular interest for the intercomparison purposes.

Regression analysis of map pairs consists of standard Method of Least Squares (MLS):

$$F_{MLS\,stand} = \sum_{i,j}\left(M(i,j) - A_{MLS\,stand} * M_{SAM}(i,j) - B_{MLS\,stand}\right)^2 \to \min_{A_{MLS\,stand},\, B_{MLS\,stand}} \tag{16}$$

where A, B are regression slope and bias, M is map, F is minimised functional.

The methods could be used not only for comparison of each model against reference (e.g. SAM) set, but also for intercomparison of two "equal-rights" models. The main difference between these two tasks is in the fact that appropriate "trust coefficient" of the reference model is higher than that of ordinary model. So, for pairs "reference model - ordinary model" the analysis should involve non-symmetric methods, but for two ordinary models symmetry of the methodology is necessary.

Symmetric MLS for consideration of the pair of ordinary models involves two steps. The first one is based on standard MLS in modified space:

$$F_{MLS\,symm} = \sum_{i,j}\left(M_2(i,j) - A^{tmp}_{MLS\,symm} * \frac{M_1(i,j) + M_2(i,j)}{2} - B^{tmp}_{MLS\,symm}\right)^2 \to \min_{A^{tmp}_{MLS\,symm},\, B^{tmp}_{MLS\,symm}} \tag{17}$$

Then temporary coefficients should be re-calculated to the initial space:

$$A_{MLS\,symm} = A^{tmp}_{MLS\,symm} \Big/ \left(2 - A^{tmp}_{MLS\,symm}\right), \quad B_{MLS\,symm} = 2 * B^{tmp}_{MLS\,symm} \Big/ \left(2 - A^{tmp}_{MLS\,symm}\right) \tag{18}$$

Residual sum of squares R is used for evaluating of precision of the regression coefficients estimates in accordance with the following equalities:

$$R = \sum_{i=1}^{N} (y_i - Ax_i - B)^2 \tag{19}$$

where A, B - are regression slope and bias correspondingly, x and y are correlated variables with N elements in the realisations.

(20) $$\sigma_A^2 = \frac{R}{N-2}\left(\frac{1}{N} + \frac{\bar{x}^2}{\sum_{j=1}^{N}(x_j - \bar{x})^2}\right)$$

σ_A here is standard deviation of the slope estimate A .

It should be noted that (20) is only an estimate of the order of deviation because it was obtained for normal distribution function which is not true for the considered problem.

MODEL RANKING PROCEDURE

Above described calculations create the large set of numbers which characterises participated models from several points of views. The final step of the considered approach is the aggregation of these data and calculation of generalised numerical estimates of the model quality.

It is suggested that each model should be characterised by two values - integrated model rank derived from model-measurement comparison; and integrated rank derived from model-to-SAM comparison. These two ranks are considered as an independent measures of the model quality.

Ranking implies the calculation of the normalised deviation *Dev* of each of the model parameters *Par* (e.g. correlation radius) from the reference point *Ref*:

(21) $$Dev_{Par} = \frac{\left|Par_{Model} - Par_{Ref}\right|}{Par_{Ref}} \times 100\%$$

Reference value of *Par* can be taken either from measurements, or from SAM. For some parameters like correlation coefficient or results of direct model-to-SAM or model-measurement comparison and some others the step (21) is not necessary since they need just normalisation. Following the same scheme the reference value for regression slope can be taken to be *1*, etc. Finally two sets of relative deviations from reference values are obtained for each model as an indication of its "distance" from measurements and from SAM. These values can be presented in a table like Table 1 (values in the table are arbitrary). Similar table can be made for model-to-SAM comparison rank.

Table 1. Example of final model quality table (for model-measurement comparison). Unit = %.

Parameter		Model 1	Model 2	SAM
Air concentr.	mean	24	59	35
	corr	18	12	12
	slope	1	59	26
Conc. in precip.	mean	8	146	23
	corr	2	27	9
	slope	35	126	42
Final rank		15	72	25

COMMENTS TO THE IMPLEMENTATION OF THE OBTAINED RESULTS

Although the analysis and implementation of the obtained data is quite specific, some important general comments could be made.

1. The precision of the figures from model-to-model intercomparison is higher than that of model-measurement ranking table. But deviations are still about 10% which should be taken into account while the interpretation of specific values.

2. SAM is an artificial data set. Its features directly follow from the procedure of its designing and from the features of participated models. It could be considered as a 'common position' of the 'experts' represented by model data sets. That is why its data are more stable than those of individual 'ordinary' model but it does not mean that this artificial model is more precise than all others.

3. Deviation from SAM means only that a specific model has different features than most of others. It does not mean that this particular parameter in this model is incorrect.

CONCLUSIONS

Statistical algorithm for the evaluation of atmospheric dispersion models is suggested and discussed. The algorithm consists of two independent parts: model-measurement and model-model comparisons. Results of both parts are aggregated by ranking procedure which produces the integrated estimates of deviations of the model characteristics from the measurement data and from other models. All considered statistics are used together with their precision estimates reflected how accurate the obtained results.

Current algorithm was developed and applied in Heavy metal model intercomparison study carried out within the scope of Co-operative Programme for Monitoring and Evaluation of the Long-Range Transmission of Air Pollutants (EMEP) in 1994-1996.

REFERENCES

Dalkey N., Helmer O. (1963), An Experimental Application of the Delphi Method to the Use of Experts, Management science, v.9, p.458

Huber P.J. (1981), Robust statistics, Wiley Series in Probability and Mathematical Statistics, John Wiley and Sons, New Your, Chichester, Brisbane, Toronto.

Johnson N.L., Leone F.C. (1977), Statistics and Experimental Design in Engineering and Physical Sciences, Vol. I, second edition, John Wiley & Sons, New York-London-Sydney-Toronto (Russian edition Джонсон Н., Лион Ф., Статистика и Планирование Эксперимента в Технике и Науке (методы обработки данных), ред. Лецкий Э.К. Мир, Москва, 1977, 610 стр.)

Quade E.S. (1970), An Extent Concept of Model, Proceedings of the 5th International O.R. Conference, J.R.Lawrence (ed.) Tavistock Publ., Ltd., London.

Shannon R.E. (1975) System Simulation Art and Science, Prentice-Hall, Inc.m Engelewood Ciffs, New Jersey, (Russian edition: Шеннон Р. Имитационное Моделирование Систем - Искусство и Наука, ред. Е.К.Масловский, Москва, Мир, 1978, 301 стр.)

Sofiev M. (1994) Statistical Properties of the Model Verification Problem and Special Methods for Comparison of Measured and Calculated Data, Proceedings of Eurotrac Symposium'94 / ed. Patricia M. Borrel *et al* - The Hague : SPB Academic Publishing. - I11, p.869.

Sofiev[1] M.A. (1996) Quality Assurance of the Model Estimates of Long-Range Air Pollution Transport and Deposition, *EUROTRAC Newsletter* 17 / 96, pp.31-33

Sofiev M., Galperin M. (1994), Robustness of Methods for Comparison of Measured and Calculated Data, Proc. of EMEP workshop on the Accuracy of Measurements, EMEP/CCC Rep.2/94

Sofiev M.A., Maslyaev A.M., Gusev A.V. (1996) Heavy metal model intercomparison. Methodology and results for Pb in 1990, *EMEP/MSC-E report 2/96*, Moscow, March 1996, pp.100.

UNCERTAINTIES IN REGIONAL OXIDANT MODELLING - EXPERIENCE WITH A COMPREHENSIVE REGIONAL MODEL

Wanmin Gong,[1] Sylvain Ménard,[2] and Xiude Lin[3]

[1]Atmospheric Environment Service, Downsview, Ontario, Canada
[2]40 Lucien Thériault, Notre-Dame-de-l'Ile Perrot, Québec, Canada
[3]Ontario Hydro, Toronto, Ontario, Canada

INTRODUCTION

In recent years comprehensive models have been used as an increasingly important tool for the understanding of "source-receptor" relationships of various atmospheric pollutants and for developing and testing emission control strategies. However one must realize that each process included in a comprehensive model can potentially introduce uncertainties in model results. Model uncertainties can arise from the formulation and numerical representation of these processes as well as from uncertainties in the required input parameters. The understanding of these model uncertainties is important in the process of model development, model evaluation and model application.

In this paper, we will present some preliminary results from a study of uncertainties in regional oxidant modelling. The oxidant model used in the study is ADOM/MC2 modelling system (Lin et al., 1996), which includes transport, emission, gas-phase chemistry, cloud micro-physics and aqueous phase chemistry, dry and wet removal. The study is based on a series of sensitivity tests designed to illustrate how some of the meteorological parameters affect model results, in particular, through processes such as transport, dry deposition and chemical transformation.

METHODOLOGY AND DESCRIPTION OF SENSITIVITY TESTS

The goal of the model uncertainty study is to assess the response of model predictions to prescribed uncertainties in model inputs and model formulations. This is usually achieved through sensitivity studies or analysis. The generalised mathematical representation of the problem is to find sensitivity coefficients of $\partial x_i/\partial k_j$, where x_i is the ith model-predicted variables and k_j is the jth model-dependant external parameter. Theoretically, formal sensitivity analysis can be pursued through either treating the sensitivity coefficients as dynamical variables and solving a separate set of differential equations, or using a Fourier

Air Pollution Modeling and Its Application XII
Edited by Sven-Erik Gryning and Nadine Chaumerliac, Plenum Press, New York, 1998

amplitude sensitivity test (FAST) method (Koda et al., 1979). However, for 3-D grid-based multiple-process comprehensive oxidant models, these approaches prove to be rather impractical. Some of the existing studies of uncertainties in oxidant modelling have been focused on a particular issue or problem.

The present study on model sensitivity/uncertainty is focused on the following aspects:

1. ***Mesoscale dynamics.*** The importance of mesoscale dynamics in regional oxidant modelling is investigated by substituting the meteorological fields in the 21 km resolution simulation, originally generated from the 21 km resolution MC2 model run, with those interpolated from the 127 km resolution MC2 run. We will refer this run as "MET-TST" in the following discussion.

2. ***Dry deposition.*** The calculations of the dry deposition employed in ADOM are based on the formulations given in Padro et al. (1991). Eight different land-use categories: water, deciduous forest, coniferous forest, swamp, cultivated land, grassland, urban and desert are considered in the calculation. Instead of simply altering the dry deposition velocities to arbitrary levels for testing the model sensitivity to the dry deposition process, the present study explores possible uncertainties in the dry deposition calculation and then attempt to assess the impact of these uncertainties in model prediction. Two tests are designed for this purpose: a). cloud and canopy wetness are turned off in the dry deposition calculation (referred as VD-TST1 hereafter); and b). grid-averaged aerodynamic resistance is calculated using boundary-layer parameters available from the dynamical model instead of treating different land-use categories individually for the aerodynamic resistance calculation (referred as VD-TST2). In VD-TST2, cloud and canopy wetness are also excluded.

3. ***Cloud process.*** Not all oxidant models include cloud micro-physics and aqueous phase chemistry. It is believed that, as far as ground level ozone is concerned, cloud and aqueous phase chemistry are not important, especially for episodic modelling with short duration and usually fair weather conditions. ADOM has two cloud modules, stratus and cumulus, to simulate cloud micro-physics and mixing. Cloud parameters (cloud cover, base and top height) from the meteorological driver are used in these calculations. Aqueous phase chemistry in ADOM (primarily sulphur chemistry) is performed whenever a cloud module is activated. It consists of 25 reactions including transfer between the gas and aqueous phases. The major oxidation reactions for SO_2 include the oxidations by H_2O_2 and O_3 and the catalytic oxidation by O_2 in the presence of Fe and Mn. Cloud parameters are also used for adjusting the actinic flux, therefore the photolysis rate, in ADOM. To assess the impact of cloud, a test run was conducted with cloud modules and aqueous-phase chemistry as well as the correction for the photolysis rate turned off. This run will be referred as "CLD-TST".

RESULTS AND DISCUSSION

All the sensitivity tests are based on an 8-day simulation from July 30 to August 6, 1988. The base-case simulation of this period is discussed in detail in Gong et al. (1997). As in the base-case, the first two days are considered as model spin-up. Therefore the results from these two days are excluded from the following discussions.

Mesoscale dynamics

To illustrate the differences in the modelled meteorology with and without the mesoscale forcing, surface wind and temperature fields for 20 Z August 1, 1988 are compared in Figure 1 between the original 21 km resolution MC2 run and those interpolated from the 127 km resolution MC2 run. The differences reflect the effect of mesoscale forcings in the Great Lakes region and the Appalachian/east coast region. Figure 2 presents the comparison of the

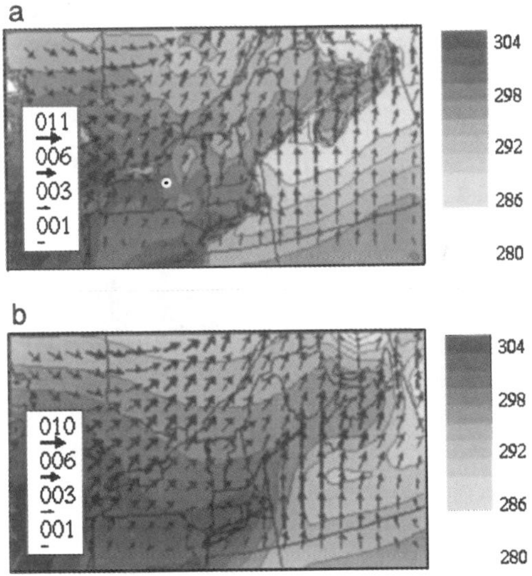

Figure 1. Surface temperature (contours) and wind (vectors) for 20 Z August 1, 1988. a). base-case (from the 21 km resolution MC2 run); b) MET-TST (interpolated from the 127 km resolution MC2 run).

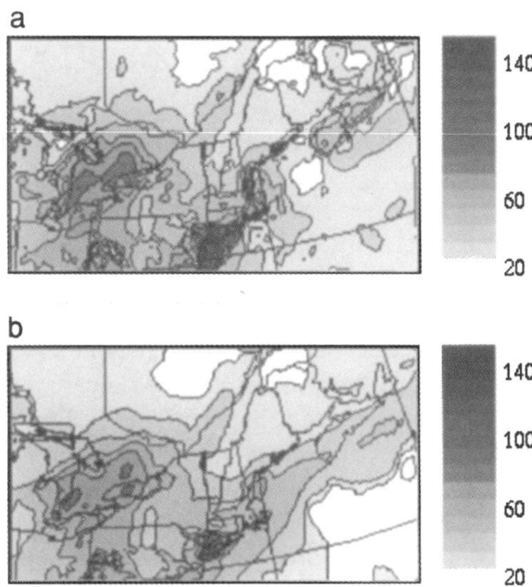

Figure 2. Model predicted ozone concentration from the first model layer at 20 Z August 1, 1988. a). base-case; b). MET-TST.

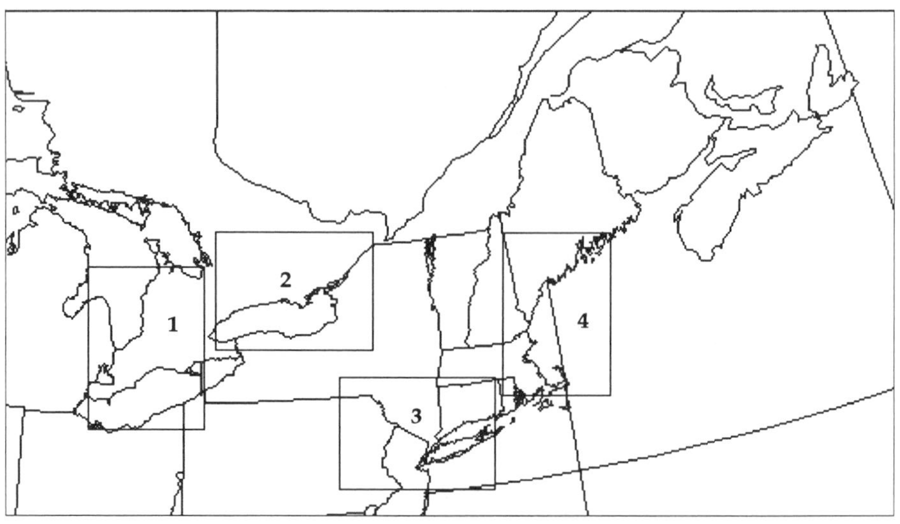

Figure 3. Definition of the 4 regions referred in the discussion.

geographical distribution of ozone in the first model layer from the base-case and the MET-TST runs (also for 20 Z August 1). Though emissions are resolved at the same resolution (21 km) in the two simulations, there are significant differences in the modelled ozone concentrations due to the difference in meteorology. Compared with the base-case, the MET-TST simulation shows less spatial variability in concentration in the areas where mesoscale dynamic and thermal forcings are pronounced. The greatest differences in modelled ozone concentration are found along the east coast where the two high ozone centres are of 133 and 94 ppb from the base-case run compared to 94 and 64 ppb from the MET-TST run.

To facilitate further discussions, we calculated time series of regional statistics (mean, maximum, minimum and standard deviations) of various fields. Four regions of interest are defined in Figure 3 for this analysis. Region 1 and 2 cover southern Ontario; region 3 and 4 cover most of the US east coast.

Figure 4 presents the time series of the regional statistics of the modelled surface level (first model layer) ozone for August 1, 1988. For region 1 and 2, the differences between the base-case and MET-TST are not too significant in terms of regional mean, maximum and minimum. The regional maximum ozone is about 10 ppb higher in the base-case for region 2. The standard deviations are comparable in the two cases. In contrast, along the east-coast, the regional maximum ozone is significantly higher in the base-case than MET-TST, over 40 percent in both region 3 and 4. This is reflected in the standard deviation as well. Figure 5 shows the regional mean, maximum and minimum of the vertical diffusivities in the lowest model layer for region 1 and 3. As seen, these regional statistical parameters are comparable for region 1 between the two cases. However, for region 3, the mean diffusivity is significantly greater for the MET-TST run than for the base-case, implying stronger vertical mixing in the former case. This is consistent with the much lower ozone level predicted by the MET-TST run in this region.

Dry deposition

Two factors affecting the dry deposition process are investigated through the test run VD-TST1, cloud and canopy wetness. Figure 6 shows the area coverage of cloud and wet

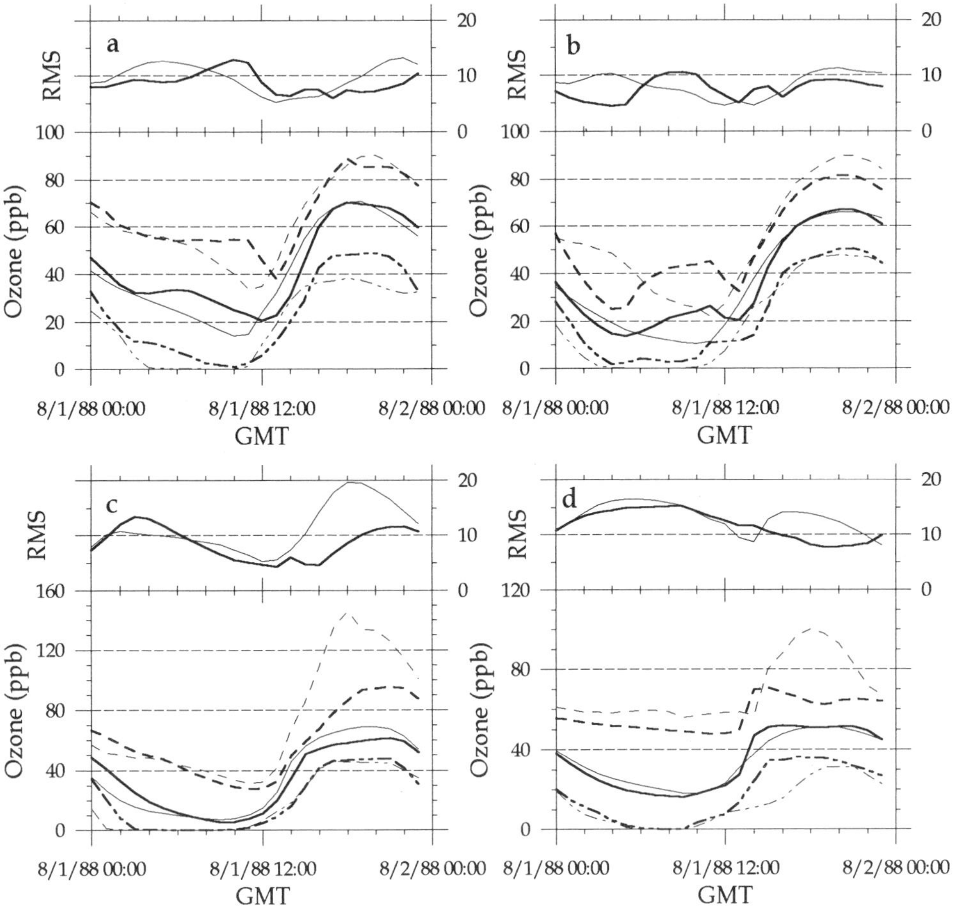

Figure 4. Regional statistics for modelled surface level ozone. a). region 1; b). region 2; c). region 3; d). region 4. Thick lines are for MET-TST run and thin lines are for the base-case. Solid lines are regional mean and standard deviation; dashed lines are regional maximum; dash-dotted lines are regional minimum.

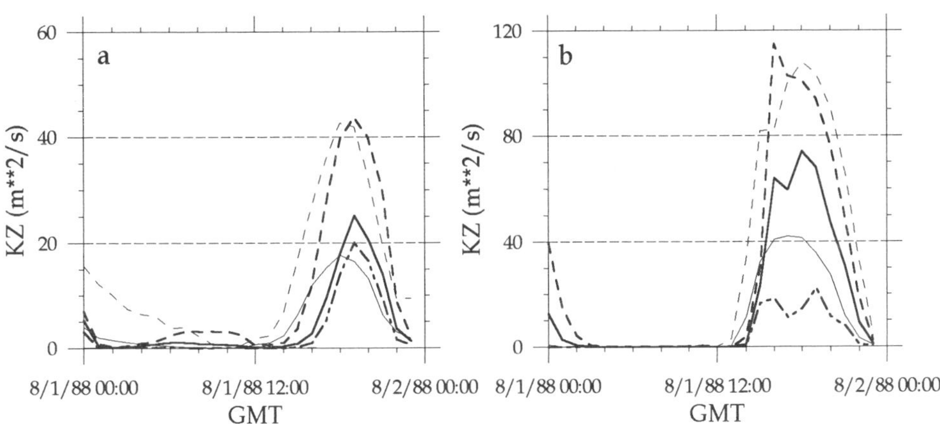

Figure 5. Vertical diffusivities for a) region 1 and b) region 3. Solid lines are for regional mean; dashed lines are for regional maximum; dash-dotted lines are for regional minimum. Thick lines are for MET-TST and thin lines are for the base-case.

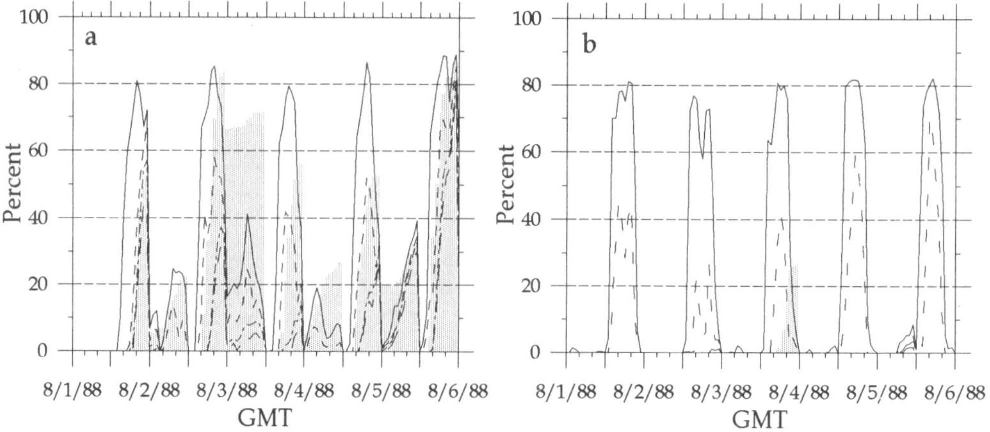

Figure 6. Regional percentage coverage of cloud (lines) and wet canopy (shaded). Solid lines: cloud fraction of at least 20% in a grid; short-dashed lines: cloud fraction of at least 40% in a grid; dash-dotted lines: cloud fraction of at least 60%; long-dashed lines: cloud fraction of at least 80% in a grid.

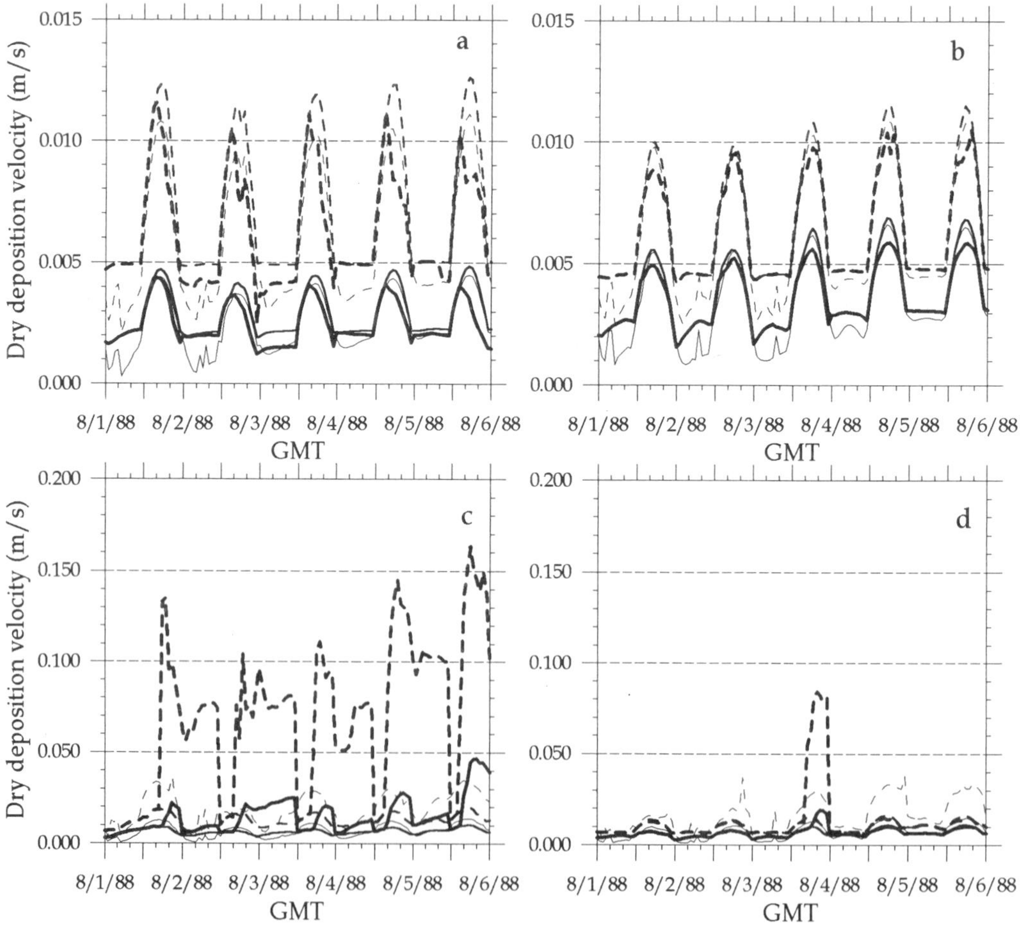

Figure 7. Regional dry deposition velocities of ozone, a) region 1 and b) region 3, and of H2O2, c) region 1 and d) region 3. Thick lines are for base-case; medium lines are for VD-TST1; thin lines are for VD-TST2. Solid lines: regional mean; dashed lines: regional maximum.

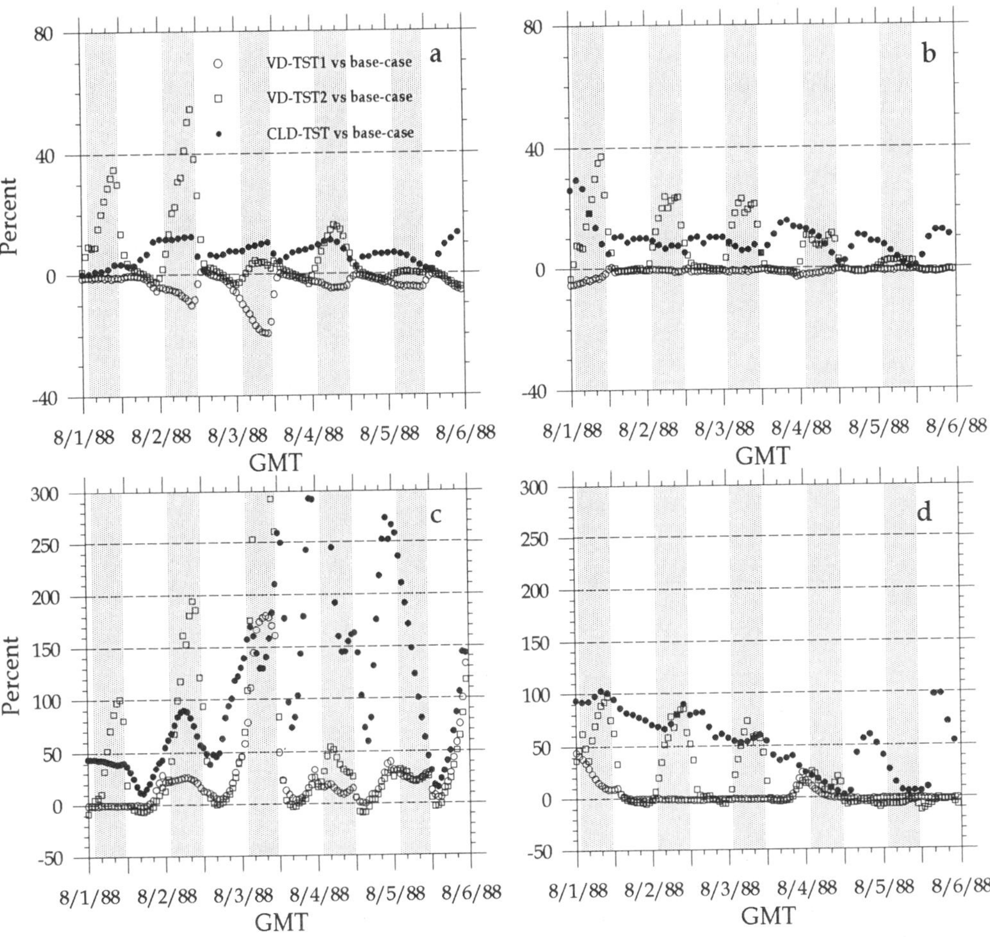

Figure 8. Relative difference between various tests and the base-case of regional mean concentration of ozone, (a). region 1 and (b). region 3, and H2O2, (c). region 1 and (d). region 4. Symbols are defined in (a).

canopy for region 1 and 3. They are ratios of number of grids with wet canopy or cloud fraction of 20, 40, 60 and 80 percent to the total number of grids in the defined region. Cloud fractions are predicted by the dynamic model MC2, while canopy wetness is linked to prognostic precipitation accumulation. The model seem to produce a fair amount of cloud. About 80 percent of grids in both region 1 and 3 have at least 20% cloud cover during daytime. Region 1 has more precipitation than region 3 during the study period. The effect of cloud on dry deposition is primarily through its effect on stomatal opening of plants. In the current dry deposition formulation, the presence of cloud reduces sunlight which will result in a reduction in stomatal opening and hence an increase in stomatal resistance. While the cloud effect on dry deposition is rather uniform to all dry deposited species, canopy wetness affects dry deposition of only the species that are more soluble, such as H_2O_2, HNO_3, HCHO etc. Canopy wetness will greatly enhance dry deposition of these species. For example, as shown in Figure 7, the dry deposition velocity of H_2O_2 can be many times greater under wet canopy conditions than those under dry canopy conditions. In contrast, the dry deposition velocity of ozone is increased (both the regional mean and maximum) when the effects of cloud and canopy wetness are removed.

Time series of the relative difference of regional mean ozone and hydrogen peroxide between VD-TST1 and the base-case for region 1 and 3 are shown in Figure 8. Generally speaking, the impact of cloud and canopy wetness on modelled ozone through dry deposition process is small in this case. As seen earlier the effect of cloud and canopy wetness is to reduce ozone dry deposition. Therefore by assuming cloud free and dry canopy in the dry deposition calculation, modelled ozone is reduced. The differences in hydrogen peroxide (a highly soluble species) between the base-case and VD-TST1, on the other hand, are very sensitive to canopy wetness. Region 3 had very little precipitation during the simulation period, and consequently the difference between the two cases are small except for late August 4 when a portion (20%) of the region is wet. In contrast, the difference between the two cases are much larger for region 1 which was relatively wet for this period.

The difference between VD-TST1 and VD-TST2 is the calculation of aerodynamic resistance. Since in most cases, canopy resistance dominates the dry deposition process during daytime, the differences in dry deposition due to aerodynamic resistance are mostly seen during nighttime. As seen in Figure 7, VD-TST2 generates considerably smaller dry deposition velocities during the first three nights. The effects of this difference in dry deposition on modelled concentrations are also seen mostly during the nighttime (Figure 8).

Cloud process

Comparisons between CLD-TST and the base-case in terms of regional mean are also included in Figure 8. Note that in this test, dry deposition is calculated as in the base-case (ie. cloud and canopy wetness are included). The impact of cloud, through the combination of aqueous phase chemistry, wet scavenging and photolysis rate correction for gas-phase chemistry, is seen to be significant. In general, the regional mean ozone concentration is increased by about 10-15% when these processes are not included, with an increase of over 30% in the modelled regional daytime maximum ozone for region 3.

SUMMARY AND CONCLUSIONS

Four tests based on the EMEFS-1 intensive period were conducted to investigate uncertainty/sensitivity in regional oxidant modelling. They mainly address the impact of meteorology. MET-TST is an overall test on the impact of mesoscale dynamics on modelled oxidants. VD-TST1 and CLD-TST are related to the impact of cloud. The study indicates that the lower predicted ozone levels over the US eastern seaboard from the MET-TST run correspond to higher vertical diffusivity from the interpolated coarse resolution meteorology over the same region. It is shown from the present study that cloud can affect the model results significantly through a number of processes (in-cloud mixing and aqueous phase chemistry, gas phase chemistry and dry deposition). The impact of cloud on ozone through dry deposition is, however, relatively insignificant in this case. This raises an interesting issue that we may have to make more effort in dealing with cloud within the content of oxidant modelling. It is shown to be important, but predicting or assimilating cloud on regional and mesoscales is a challenging problem.

REFERENCES

Gong, W., X. Lin, S. Ménard and P. Pellerin, 1997, Modelling the Canadian southern Atlantic region oxidants - a study of the EMEFS-1 period, submitted to *J. Geophys. Res.*

Koda, M., A.H. Dogru and J.H. Seifeld, 1979, Sensitivity analysis of partial differential equations with application to reaction and diffusion processes, *J. Comp. Phys.*, Vol. 30, pp 259-282.

Lin, X., W. Gong and S. Ménard, 1996, A New MC2/ADOM Modelling System and Its Application to the Study of Oxidants in the Canadian Southern Atlantic Region, in "Air Pollution Modelling and Its Applications, XI", edit. S. Gryning, and F.A. Schiermeier, pp 275-282, Plenum Publishing.

Padro, J., G. den Hartog and H.H. Neumann, 1991, An investigation of the ADOM dry deposition module using summertime O_3 measurements above a deciduous forest, *Atmos. Environ.*, Vol. 25A, No. 8, pp 1689-1704.

DEVELOPMENT AND VERIFICATION OF A MODELLING SYSTEM FOR PREDICTING URBAN NO_2 CONCENTRATIONS

Ari Karppinen,[1] Jaakko Kukkonen,[1] Mervi Konttinen,[1] Jari Härkönen,[1] Esko Valkonen,[1] Tarja Koskentalo,[2] and Timo Elolähde[2]

[1]Finnish Meteorological Institute, Air Quality Research
Sahaajankatu 20 E, 00810 Helsinki, Finland
[2]Helsinki Metropolitan Area Council
Opastinsilta 6 A, FIN-00520, Helsinki, Finland

INTRODUCTION

This paper describes the development of a modelling system for predicting the emissions, dispersion and chemical transformation of nitrogen oxides in an urban area. The system takes into account of all source categories, including stationary point and area sources, and vehicular sources. We also present a statistical comparison of predicted results and measured concentrations.

The modelling system is based on combined application of the Urban Dispersion Modelling system UDM-FMI (Karppinen et al.,1997; see also Kukkonen et al., 1996) and the road network dispersion model CAR-FMI (Härkönen et al, 1995 and 1996a) of the Finnish Meteorological Institute (FMI). The integrated modelling system includes emission models for stationary and vehicular sources, a meteorological pre-processing model, dispersion models for various source categories and a statistical analysis of the computed time series of concentrations. The programs were executed on the Cray C94 supercomputer.

The dispersion models of the system include a treatment of the chemical transformation of nitrogen oxides. A new model has been developed for evaluating the chemical interaction of pollution from a large number of individual sources. This model allows for the interdependence of urban background NO, NO_2 and O_3 concentrations and NO and NO_2 emissions from various source categories.

We have compared statistically the model predictions against the results of the air quality monitoring network of the Helsinki Metropolitan Area Council (YTV). The statistical NO_2 concentration parameters predicted by the modelling system agree well with the measured data.

THE MATHEMATICAL MODELS

The computation of the emissions

The computations included the emissions from mobile and stationary sources in the Helsinki metropolitan area. The system for evaluating the emissions of the road and street network is based on the EMME/2 transportation planning system (INRO, 1994) and new emission factors specially developed for Helsinki city traffic, combined with the emission factors of the LIISA 93 system (Mäkelä et al., 1996).

Atmospheric boundary-layer scaling

The relevant meteorological parameters for the models are evaluated using data produced by a meteorological pre-processing model (Karppinen et al., 1996). The model is based mainly on the energy budget method of van Ulden and Holtslag (1985). The model utilises meteorological synoptic and sounding observations, and its output consists of estimates of the hourly time series of the relevant atmospheric turbulence parameters (the Monin-Obukhov length scale, the friction velocity and the convective velocity scale) and the boundary layer height.

We have made use of the meteorological database of our institute, which contains routine weather and sounding observations. We used a combination of the data from the stations at Helsinki-Vantaa airport and Helsinki-Isosaari. The mixing height of the atmospheric boundary layer was evaluated based on the sounding observations at Jokioinen (90 km northwest) and the routine meteorological observations.

Urban Dispersion Modelling System (UDM-FMI)

The urban dispersion modelling system (Karppinen et al., 1996) includes a multiple source Gaussian plume model and the meteorological pre-processor. The dispersion model is an integrated urban-scale model, taking into account all source categories (point, line, area and volume sources). It includes a treatment of chemical transformation for NO_2, plume rise, downwash phenomena and the dispersion of inert particles. The model also allows for the influence of a finite mixing height.

The system computes an hourly time series of concentrations and statistical parameters, which can be directly compared to air quality guidelines. The model has been tested and validated against national urban air quality measurements. According to these comparisons the model fairly well predicts the statistical parameters defined in the national air quality guidelines. The model predictions have also been compared with the tracer experiments of Kincaid, Copenhagen and Lilleström, presented by Olesen (1995) (Karppinen et al., 1997).

Road network dispersion model (CAR-FMI)

The dispersion from a road network is evaluated with the Gaussian finite-line source model CAR-FMI (Contaminants in the Air from a Road; Härkönen et al., 1995 and 1996a). The model includes an emission model, a dispersion model and statistical analysis of the computed time series of concentrations.

The dispersion equation in the CAR-FMI model is based on the analytic solution of the Gaussian diffusion equation for a finite line source by Luhar and Patil (1989). It allows for any wind direction with respect to the road. The chemical transformation is modelled using a revised form of the discrete parcel method. The chemical model includes the basic reactions of nitrogen oxides, oxygen and ozone, but the influence of hydrocarbons and other compounds is neglected.

The predictions of the CAR-FMI model have been compared with concentration profile measurements near a major road in the city of Espoo. A discussion of these comparisons has been presented by Härkönen et al. (1996b).

Chemical interaction of pollutants originating from various sources

We have addressed the problem concerning the chemical interaction of pollution from a large number of individual sources. The modelling system was refined to allow for the chemical interdependence of the urban background concentrations and the concentrations originating from various source categories.

The regional background concentrations were evaluated based on the data from the monitoring station of Luukki, situated near the north-eastern limit of the computational region.

THE SPATIAL CONCENTRATION DISTRIBUTIONS

Figure 1 shows the computed yearly mean NO_2 concentrations at the ground level. The legend in the top left-hand corner shows the absolute values of the pollutant concentration.

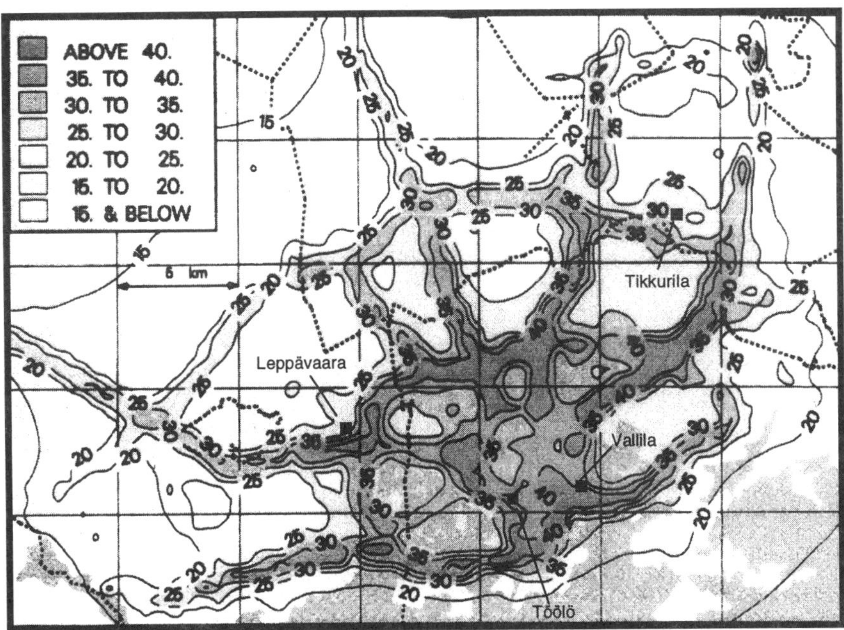

Figure 1. Predicted distribution of the yearly mean NO_2 concentration ($\mu g/m^3$) in the Helsinki metropolitan area in 1993. The location of the monitoring stations (Töölö, Vallila, Leppävaara and Tikkurila) has also been indicated. The size of the depicted area is 35 km x 25 km.

Clearly, the traffic emissions have a larger relative influence on the ground-level concentrations, compared to the stationary emissions, which are mostly released from higher altitudes. It can be shown numerically that although the contribution of traffic on the total emissions is less than a half, the ground level concentrations mostly originate from traffic sources.

The concentrations of NO_2 are strongly concentrated in the vicinity of the main roads and streets, and in the downtown area of Helsinki. The figure clearly shows the influence of the ring roads and the junctions of major roads and streets.

The computed results include also the spatial distributions of the statistical concentration parameters, which can be compared to national air quality guidelines. We also analysed an emission scenario, in which it was assumed that all gasoline driven passenger cars are equipped with a catalytic converter.

COMPARISON WITH THE DATA OF AN URBAN MONITORING NETWORK

We have compared the model predictions against the air quality measurements at YTV stations in the Helsinki metropolitan area. The monitoring network and the methods have been described by Aarnio et al. (1995). The concentrations of nitrogen oxides were measured at two urban and two suburban stations in 1993.

The urban monitoring stations, Töölö and Vallila, are located in the Helsinki downtown area. The station of Töölö is at the center of a densely trafficed junction, surrounded by several major buildings. The station of Vallila is situated in a small park, at a distance of 23 m from a busy street.

The suburban stations are located in Leppävaara, Espoo and in Tikkurila, Vantaa. The station of Leppävaara is in the city of Espoo in a shopping and residence area; the distance of the station to two major roads is approximately 200 m. The station of Tikkurila is located in a park near the centre of the city of Vantaa.

The measurement height is 4 m at the stations of Töölö, Vallila and Leppävaara, and 6 m at the station of Tikkurila.

Figure 2 presents some results of these comparisons. The predicted and measured NO_2 concentrations agree well at all the monitoring stations. The national air quality

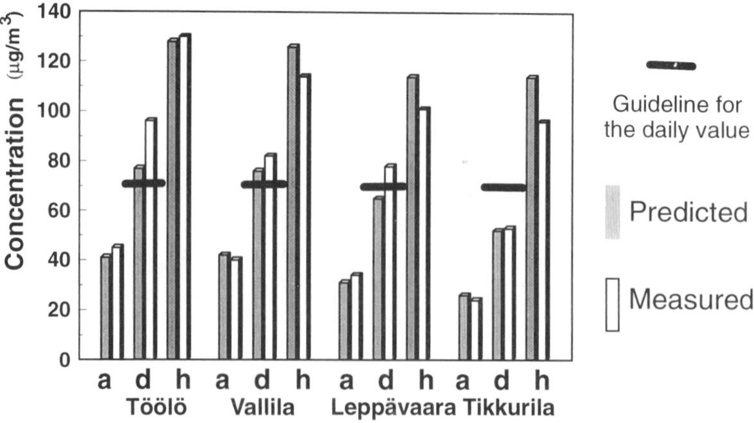

Figure 2. The comparison of predicted and measured NO_2 concentrations at four air quality measurement stations. We have compared the yearly mean concentrations (denoted by a) and the statistical concentration parameters defined in the national guidelines for the daily (d) and hourly (h) concentrations.

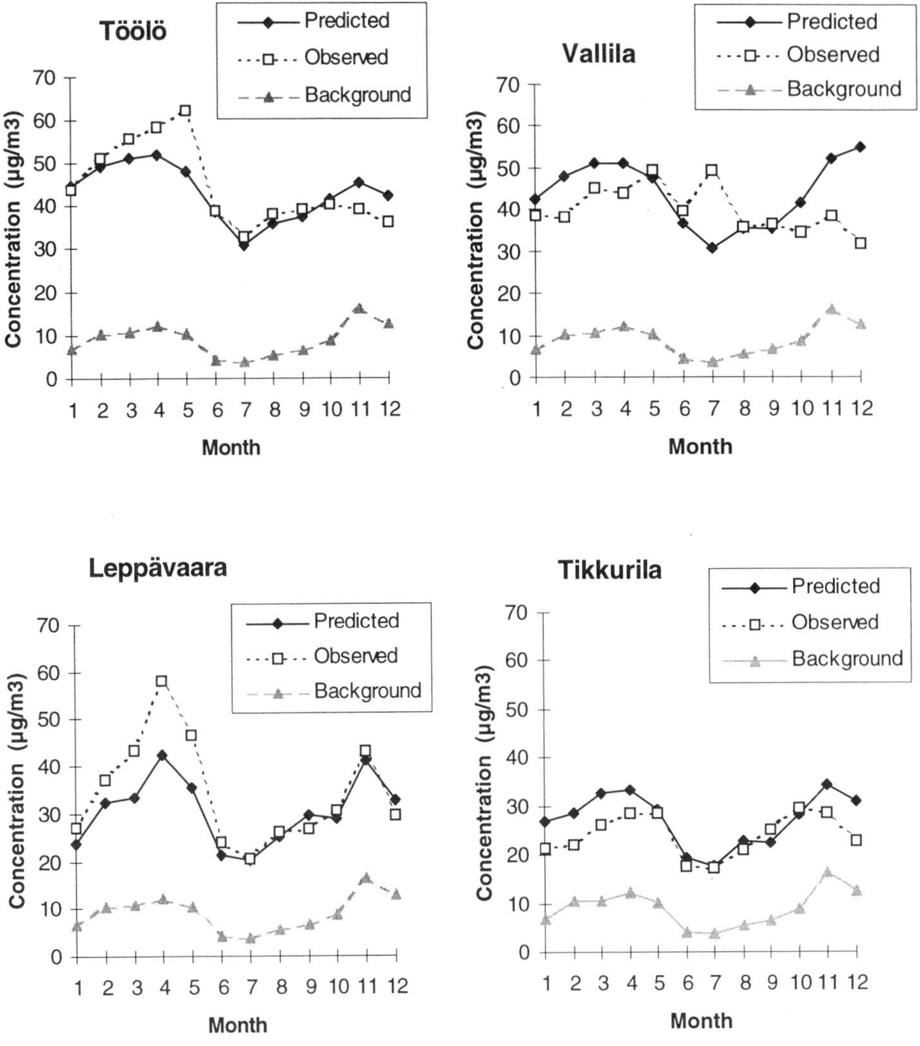

Figure 3. The monthly average predicted and measured NO_2 concentration (µg/m^3) at the four monitoring stations considered, together with the regional background concentrations.

guideline value (1996) for the daily NO_2 concentration was exceeded at three stations according to the measurements, and at two stations according to the computations.

The corresponding comparison for the NO_x concentrations showed a similar agreement, except for the monitoring station of Töölö, in which the model underpredicted the measured concentrations.

Figure 3 presents the seasonal variation of the predicted and measured NO_2 concentrations. The variation of the predicted and measured concentrations is very similar for the stations of Töölö, Leppävaara and Tikkurila. For the station of Vallila, there are some differences between the predicted and measured seasonal variation, particularly in winter.

CONCLUSIONS

We have conducted an emission inventory of the mobile and stationary sources in the Helsinki metropolitan area in 1993. Atmospheric dispersion was subsequently evaluated, resulting in hourly time series of concentrations of NO and NO_2. These time series were utilised for presenting the spatial concentration distributions of the yearly mean values and statistical concentration parameters.

The predicted results show that the annual mean concentrations of NO_2 are strongly concentrated in the vicinity of the main roads and streets, and in the downtown area of Helsinki. It can be shown numerically that although the contribution of traffic on the total emissions is less than a half, the ground level concentrations mostly originate from traffic sources. The computations indicate that the national air quality guidelines of the daily NO_2 concentrations were exceeded at some spatially limited areas, in the vicinity of the main roads and in the Helsinki downtown area.

We have compared the model predictions against the measurements of the Helsinki Metropolitan Area Council. The predicted and measured NO_2 concentrations agree well at all monitoring stations. At the monitoring stations of Töölö and Vallila, the predicted exceedences of the national air quality guidelines were also measured.

The comparison of the NO_x concentrations shows also a good agreement for the stations of Vallila, Leppävaara and Tikkurila. However, there are substantial differences of the modelled and measured values at the station of Töölö. This station is located at a densely trafficked junction, in the vicinity of several large buildings.

ACKNOWLEDGMENTS

We would like to thank our coworkers in this study, Ms. Päivi Aarnio (Helsinki metropolitan Area Council) and Mr. Juhani Laurikko (VTT Energy). This work is part of the national research programme "MOBILE - Energy and the environment in transportation". The funding from Technology Development Centre (TEKES) is gratefully acknowledged.

REFERENCES

Aarnio, P., Hämekoski, K., Koskentalo, T. and Virtanen, T., 1995. Air Quality, monitoring and air quality index in the Helsinki metropolitan area, Finland. In: Anttila, P. et al. (ed.), *Proceedings of the 10th World Clean Air Congress*, Espoo, Finland, May 28 - June 2, 1995. Vol. 2. The Finnish Air Pollution Prevention Society, Helsinki, p. 201 (4 pages).

Härkönen, J., Valkonen, E., Kukkonen, J., Rantakrans, E., Jalkanen, L. and Lahtinen, K., 1995. An operational dispersion model for predicting pollution from a road. *International Journal of Environment and Pollution*, Vol. 5, Nos. 4-6, 602 - 610.

Härkönen, J., Valkonen, E., Kukkonen, J., Rantakrans, E., Lahtinen, K., Karppinen, A. and Jalkanen, L., 1996a. A model for the dispersion of pollution from a road network. *Finnish Meteorological Institute, Publications on Air Quality* 23. Helsinki, 34 p.

Härkönen, J., Walden, J. and Kukkonen, J., 1996b. Comparison of model predictions and measurements near a major road in an urban area. In: Kretzschmar, J. and Cosemans, G. (eds.), Proceedings of the 4th Workshop on Harmonisation within Atmospheric Dispersion Modelling for Regulatory Purposes, Oostende, Belgium, 6 - 9 May 1996. Vol. 2. Vlaamse Instelling voor Technologisch Onderzoek, Mol, Belgium, pp. 453 - 460.

INRO, 1994. EMME/2 User's Manual, INRO Consultants Inc., Montréal, Canada.

Karppinen, A., Kukkonen, J., Nordlund, G., Rantakrans, E. and Valkama, I., 1997. A dispersion modelling system for urban air pollution. *Finnish Meteorological Institute, Re*port. Helsinki, 50 p.

Kukkonen, J., Härkönen, J., Valkonen, E., Karppinen, A. and Rantakrans, E., 1996. Regulatory dispersion modelling in Finland. In: Kretzschmar, J. and Cosemans, G. (eds.), Proceedings of the 4th Workshop on Harmonisation within Atmospheric Dispersion Modelling for Regulatory Purposes, Oostende,

Belgium, 6 - 9 May 1996. Vol. 2. Vlaamse Instelling voor Technologisch Onderzoek, Mol, Belgium, pp. 477 - 484.

Luhar, A, K., and Patil, R., S., 1989. A General finite line source model for vehicular pollution prediction. Atmos. Environ. 23:3, p. 555-562.

Mäkelä, K., Kanner, H. and Laurikko, J., 1996. Road traffic exhaust gas emissions in Finland - LIISA 95 calculation software. VTT Communities and Infrastructure,Transport Research, Research Notes 1772, 45 p.+ app. 51 p. Technical Research Center of Finland, Espoo (in Finnish).

Olesen, H.R., 1995. Datasets and protocol for model validation. International Journal of Environment and Pollution, Vol. 5, Nos. 4-6, pp. 693-701.

van Ulden, A. and Holtslag, A., 1985. Estimation of atmospheric boundary layer parameters for diffusion applications, J. Climate Appl. Meteor. 24, p. 1196-1207.

DISCUSSION

R. SAN JOSE: Have you made any comparison exercise to compare the results between your own emission factors and the CORINE (CORINAIR?) emission factors?

A. KARPPINEN: Yes, we have compared the results of CORINAIR and LIISA emission inventories. There are differences between the CORINAIR emission factors and our own (VTT, Finland) emission factors. We have decided to use our own factors in our studies as we believe that these factors are based on the most "up-to-date" information on our domestic vehicular emissions.

A DIAGNOSTIC COMPARISON OF EMEP AND EURAD MODEL RESULTS FOR A WET DEPOSITION EPISODE IN JULY 1990

Heinz Hass[1] and Erik Berge[2]

[1] Ford Forschungszentrum Aachen GmbH
Dennewartstr. 25
D-52068 Aachen, Germany
[2] EMEP/MSC-W, Norwegian Meteorological Institute
PB 43 Blindern
N-0313 Oslo 3, Norway

INTRODUCTION

Several models exist in Europe aiming at quantifying the transport, transformation and deposition of air pollutants. The structure and complexity of the models vary and often depend on the scope of the applications. Within the EMEP-programme a focus has been on quantifying long term loads of acidity to the European Environment and consequently the relative simple two-dimensional Acid Rain model is employed (see Eliassen and Saltbones, 1983; Barret et al., 1995). Under the EUROTRAC programme the coupled acid rain and photochemical EURAD-model (European Regional Air Pollution Dispersion model) has been developed (Hass et al., 1993, Chang et al., 1987). The purpose of the EURAD model has been to investigate the processes leading to the formation of acidifying components and ozone on the European scale on a relatively short time-scale (days to weeks).

The results from the two models have been compared for the period 2 - 10th July 1990 focusing on comparison with measurements from the EMEP network, analysis of the chemical schemes and the validity of the linearity assumption in the EMEP model. A detailed description of this comparison can be found in Hass et al. (1996) and some main results are discussed here.

SHORT MODEL DESCRIPTIONS

The version of the EMEP acid rain model employed in this study is described in Tuovinen et al., 1994). This version of the model differs from the most recent version (see Barrett et al., 1995) in that the new dry deposition module and the variable local deposition factors are not included. The EMEP acid rain model is a two-dimensional receptor oriented Lagrangian trajectory model. The model calculates dry and wet scavenging and chemical processes in air parcels with a horizontal dimension of 150km*150km which follow trajectories found from the wind speed at about 550 m height. The air parcels extend up to the top of the Planetary Boundary Layer (PBL), and the air pollutants are assumed to be well mixed in the PBL. The overall aim of the model has been to support the work under the 1979 Geneva Convention on Long Range Transboundary Air Pollution in Europe in its process to reduce the acidic loads to the European eco-systems. The design of the model reflects this aim by making the model simple enough that long term (annual) calculations can be feasible with today's computers. The model is therefore two-dimensional including only the chemical reactions that are thought to be of major importance for acid deposition. An important feature of the model is its linearity which means that export and import budgets of acidity for each country (or grid-square) in Europe can be handled in a simple manner. Such a feature is very valuable when applying the model results to assessments on the effects of emission reductions in for example one country on the depositions in other countries in Europe.

The EURAD model is an Eulerian air quality modelling system. It contains three major modules to simulate the transport, chemical transformations and the depositions of minor atmospheric constituents: the mesoscale meteorological model MM5 (Grell et al., 1994), the EURAD emission model EEM (Memmesheimer et al., 1991), the chemistry-transport model CTM2 (Chang et al., 1987; Hass et al., 1993). The choice of the modelling domain and the grid size is user specified and resolutions down to about 5 km can be used with the current version of the system. This lower limit is a restriction defined by several input data sets as e.g. the landuse data for Europe. However, applications of the model in complex terrain are limited in grid size by the hydrostatic assumption. In the vertical the sigma-coordinate allows a variable resolution too. Usually, EURAD appli-cations resolve the troposphere with 15 or more layers thereby covering the PBL with a high resolution. The EURAD model allows to nest smaller domains with high resolution into larger European scale modelling domains. This technique avoids the boundary condition problems for small scale applications and maintains the synoptic situation on all modelling domains. The meteorological model MM5 does allow a 2-way nesting technique (Grell et al., 1994), while the current version of CTM2 includes a 1-way nesting option.

EPISODE DESCRIPTION

One aim of this study was to analyse possible non-linearities in the wet deposition and thus, a period where large amounts of wet deposition occurred in North-West Europe during a summer period was selected. The episode covers the time period from 2 to 10 of July, 1990. In general, the meteorological situation for Central and Northern Europe during this summer episode can be characterised as cold and wet compared to long term statistics. During the episode 3 anticyclones crossed the domain of interest as no major high pressure system over the continent developed. The anticyclones transported cold northern Atlantic air masses to Central Europe and lead to strong precipitation. Heavy precipitation was observed at several stations within the modelling domain. Birkenes, in southern Norway, received about 5% of the total precipitation in 1990 during the considered episode. Further examples are Keldsnor (Denmark) and Deuselbach (in Southwest Germany) with about 6% and 4% of the yearly total precipitation, respectively.

At Birkenes about 6% and at Keldsnor about 5% of the total annual wet deposition of sulphur and nitrogen in 1990 were measured during these 9 days. This illustrates that such episodes are of significance to the annual budget of deposition.

COMPARISON OF MODEL RESULTS WITH EMEP NETWORK DATA

Observed data from the EMEP network for wet deposition of sulphate and nitrate together with air concentrations of SO_2, sulphate, NO_2 and nitrate were compared with results from the two models (Hass et al., 1996). In general, this comparison showed a strong under-estimation of both, sulphate and nitrate wet deposition, in the receiver area (Scandinavia) during days with heavy rainfall. In the main emitter areas an overprediction of sulphate and nitrate in air was found for both models. A bias toward too high nitrate wet deposition is seen in the EMEP model.

The behaviour of the two models at different stations is however in most cases quite well correlated. This is demonstrated in Figure 1 which displays a scatter of EURAD versus EMEP fractional differences (FD= (obs-mod)/(obs+mod)) of the episode mean depositions. The closer the stations are to the dashed diagonal in Figure 1 the more they show similar trends in both models, i.e. stations in the lower left square are strongly overpredicted and those in the upper right are underpredicted by both models.

AREA INTEGRATED EMISSIONS AND DEPOSITIONS

The depositions have been accumulated and the concentrations averaged over the episode for four different subdomains A, A1, A2 and A3. Region A brackets A1-A3 and covers a large part of north-western Europe were emissions and acid loads are high. Region A1 (southern Scandinavia) covers a major receiver area of acidity while A2 covers a major NO_x emitting area and A3 a major SO_x emitting area. The choice of the domains is motivated by one of the main aims of the study, namely to investigate eventual non-linear effects in concentrations and deposition in a typical receiver area (A1) when the transport is directed from major emitter regions (A2 and A3). Table 1 presents the results for all four domains for the EMEP and the EURAD model for the base case simulation.

Looking at the emission numbers used in the two models (Table 1) one observes that they are quite similar. This is not astonishing as both models use the same annual emission data set to start with. The largest differences are in the N-emissions in A (about 25% higher in the EMEP data) and in the S-emissions in A2 (about 14% higher in the EURAD data). The reason for the differences in N-emissions are different seasonal variations used in EURAD for low level NO_x emissions.

Larger differences can be seen in the deposition results listed in Table 1. These differences between the two models are larger for S-depositions than for N. Specially in the high sulphur emission area A3 there is a factor of 2.5 between the results. In the main area A the EURAD model deposits about 20 kt-S less than EMEP. The N-depositions in the receptor area A1 are also significantly lesser for EURAD than for EMEP. The ratios of depositions to emissions for the areas are also given in Table 1. Considering sulphur the region A1 can be clearly identified as a receptor region as about 8 times more is deposited than emitted. The sulphur source region A3 is more pronounced within EURAD. The analogous source (A2) and receptor (A1) pattern is also visible for nitrogen, however, the

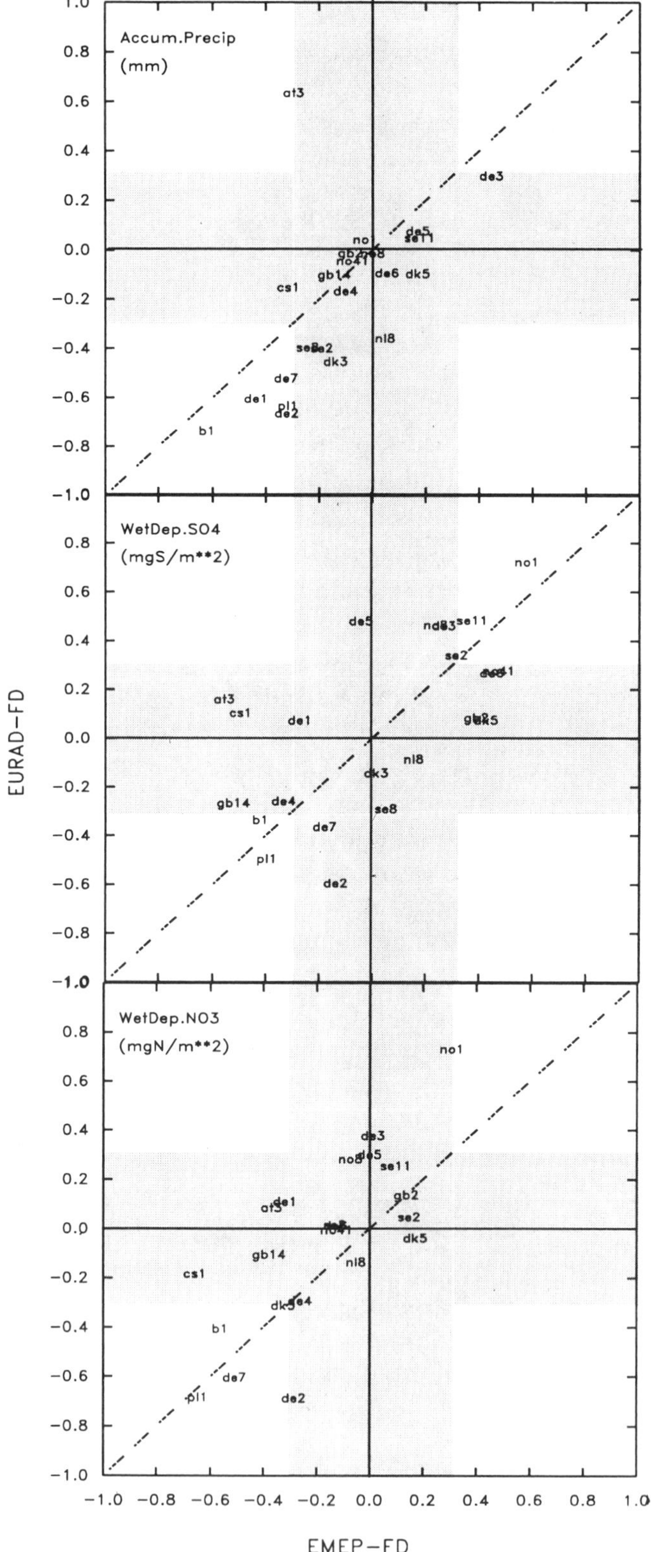

Figure 1. Fractional differences of EURAD versus EMEP for precipitation and wet depositions. Stations within shaded areas are predicted within a factor of two. The closer stations are to the dashed diagonal the more the models tend to predict a similar behaviour.

model ratios are vice versa compared to sulphur. For the large region A these numbers can only approximately be related to the fractions of the emissions which are deposited within the areas. A much more detailed budget analysis is therefore required.

The deposition efficiency relative to the EMEP model is also given in Table 1. Sulphur is deposited about 24% more effective by the EMEP model in the large area A with the highest efficiency in the sulphur source area A3. EURAD is a little more effective in sulphur deposition in the remote region A1. Although EURAD emits about 25% less N in area A the deposition is about 19% more effective compared to the EMEP model. This picture is not true for the remote area A1 where EMEP is 43% more effective. The higher effectiveness of the EURAD model for nitrogen originates from the effectiveness in the N-source area A2 and the same picture may be true for Great Britain which is also in area A but not analysed in detail. EURAD's precipitation predictions are generally higher than EMEP's analysed observations but this does not, in general, translate into a higher deposition efficiency.

NON-LINEARITY TESTS

Analysing the degree of non-linearity across the entire model domain is simply performed by changing the emissions in every grid cell by the same percentage. A reduction of 50% was chosen here as this is the range of reduction which is often discussed. No new model calculations have been performed with the EMEP-model. Instead for this particular experiment, we have made use of the linearity of the model and assumed that a 50 % reduction in the emissions in the model domain will give a 50 % reduction in concentrations and depositions as long as we assume no changes in the spatial and temporal patterns of the emissions (see Tuovinen et al., 1994). This is not strictly true because some of the concentrations and depositions will have originated from background and boundary values. Based on EURAD model results (Hass et al., 1996), the boundary conditions affect region mainly A1 while A2 and A3 are much less influenced by the boundaries.

Table 1. Area and episode integrated emissions and depositions for the base case simulations. Emissions and depositions are in Kt either S or N. Precipitation is the accumulated area averaged amount in mm. D/E: ratio of Deposition / Emission. D-Ef = (EMEP(D/E) - EURAD(D/E)) / EMEP(D/E).

	A		**A1**		**A2**		**A3**	
	EMEP	EURAD	EMEP	EURAD	EMEP	EURAD	EMEP	EURAD
Emis.-S	111.16	116.14	0.38	0.40	7.70	8.95	53.64	54.87
Emis.-N	55.27	40.74	0.86	0.84	13.71	13.31	8.42	8.36
Dep.-S	76.26	59.81	2.83	3.26	7.92	7.82	24.11	9.74
Dep.-N	32.73	28.39	2.62	1.46	3.87	4.47	5.38	5.03
D/E-S	0.67	0.51	7.45	8.13	1.02	0.87	0.45	0.18
D/E-N	0.59	0.70	3.05	1.74	0.28	0.34	0.63	0.60
D-Ef(S)	0.24		-0.09		0.15		0.60	
D-Ef(N)	-0.19		0.43		-0.21		0.05	
Precip.	32.0	44.9	59.2	74.7	29.8	57.2	27.2	31.6

Tables 2 and 3 present the area integrated sulphur and nitrogen reduction factors from scenarios with 50% reductions in SO_x and NO_x emissions, respectively, based on the calculations of the EURAD model. For the total deposition of oxidised sulphur to the four different regions A, A1, A2 and A3 one obtains a reduction of 36%, 34%, 36% and 38% respectively. The 50% NO_x reduction scenario (Table 3) yields a more close to linear relationship for the total deposition of oxidised nitrogen. The nitrogen source area A2 shows a 45% reduction response while the other areas response a little more than proportional, i.e. 51% for A and 53% for A1 and A3.

The non-linearities in the sulphur deposition are totally dominated by the wet part of the deposition (Table 2). SO_2 dry deposition in the EURAD model is close to linear except for A1 which is dominated by dry deposition of $SO_4^=$ rather than SO_2 as is the case in the other regions. However, dry deposition of $SO_4^=$ contributes little (5.4%) to the total deposition in A1. The largest non-linear response is given in the sulphate in air concentrations which are reduced by 23% in A1 to 34% in A3 given a 50% reduction in the sulphur emissions. Since the models were performed for a short period dominated by wet deposition in north-western Europe one must be careful in generalising the results to long term exposures and loads to the ecosystem. However, in areas dominated by wet deposition we anticipate that the non-linearities in the wet deposition may be of some significance with oxidant levels equal to those in the present simulations. Only small changes are seen in the sulphur components during the 50% NO_x reduction scenario. The SO_2 air concentrations are slightly lower and the sulphate concentrations are higher probably as more OH radicals are available for the gas-phase S-cycle. A slight increase in the deposition particularly close to the main sources is found.

Table 2. Area averaged concentration and integrated deposition reduction factors for the 50% SO_x reduction simulation. Boundary conditions for sulphur species were not reduced. The partition into the components are given in percentage.

	A				A1			
	EMEP		EURAD		EMEP		EURAD	
S in air	Fac.	%	Fac.	%	Fac.	%	Fac.	%
SO2	0.50	66.4	0.48	64.2	0.50	49.7	0.53	28.1
SO4	0.50	33.6	0.73	35.8	0.50	50.3	0.77	71.9
Total	0.50		0.55		0.50		0.68	
S dep.	Fac.	%	Fac.	%	Fac.	%	Fac.	%
SO2 dry	0.50	41.7	0.49	20.6	0.50	9.2	0.53	3.7
SO4 dry	0.50	1.5	0.72	6.1	0.50	1.1	0.76	6.1
SO4 wet	0.50	56.8	0.69	73.3	0.50	89.8	0.66	90.2
Total	0.50		0.64		0.50		0.66	

	A2				A3			
	EMEP		EURAD		EMEP		EURAD	
S in air	Fac.	%	Fac.	%	Fac.	%	Fac.	%
SO2	0.50	67.7	0.45	60.2	0.50	76.0	0.48	81.5
SO4	0.50	32.3	0.74	39.8	0.50	24.0	0.66	18.5
Total	0.50		0.54		0.50		0.51	
S dep.	Fac.	%	Fac.	%	Fac.	%	Fac.	%
SO2 dry	0.50	35.1	0.46	16.1	0.50	50.3	0.48	27.2
SO4 dry	0.50	1.5	0.75	5.2	0.50	1.0	0.67	6.8
SO4 wet	0.50	63.4	0.69	78.7	0.50	48.9	0.70	66.0
Total	0.50		0.64		0.50		0.62	

Table 3. Area averaged concentration and integrated deposition reduction factors for the 50% NO_x reduction simulation. The partition into the components are given in percentage.

	A				A1			
	EMEP		EURAD		EMEP		EURAD	
N in air	frac	%	frac	%	frac	%	frac	%
NO	0.50	6.7	0.35	5.7	0.50	7.3	0.50	3.2
NO2	0.50	45.3	0.45	48.9	0.50	46.6	0.51	31.7
PAN	0.50	2.3	0.96	25.2	0.50	4.0	0.83	47.4
HNO3	0.50	2.1	0.53	13.2	0.50	2.8	0.48	14.5
NO3-A	0.50	43.6	0.43	7.0	0.50	39.4	0.49	3.2
Total	0.50		0.52		0.50		0.61	
N dep.	frac	%	frac	%	frac	%	frac	%
NO2 dry	0.50	3.0	0.45	9.3	0.50	1.1	0.67	2.9
HNO3 dry	0.50	7.8	0.51	43.4	0.50	3.8	0.48	31.9
NO3 dry	0.50	4.0	0.40	1.4	0.50	1.1	0.50	0.5
NO3 wet	0.50	85.2	0.49	45.9	0.50	93.5	0.47	65.2
Total	0.50		0.49		0.50		0.47	

	A2				A3			
	EMEP		EURAD		EMEP		EURAD	
N in air	frac	%	frac	%	frac	%	frac	%
NO	0.50	8.1	0.29	7.6	0.50	6.4	0.35	4.0
NO2	0.50	51.9	0.46	59.8	0.50	44.1	0.43	43.6
PAN	0.50	2.0	1.20	13.4	0.50	2.2	0.99	31.4
HNO3	0.50	1.2	0.62	8.9	0.50	2.0	0.51	11.5
NO3-A	0.50	36.7	0.57	10.3	0.50	42.5	0.35	9.5
Total	0.50		0.50		0.50		0.52	
N dep.	frac	%	frac	%	frac	%	frac	%
NO2 dry	0.50	5.7	0.45	12.6	0.50	3.9	0.42	9.8
HNO3 dry	0.50	6.5	0.62	39.4	0.50	8.4	0.50	52.5
NO3 dry	0.50	4.7	0.56	2.0	0.50	2.8	0.33	2.1
NO3 wet	0.50	82.7	0.54	45.0	0.50	84.6	0.45	35.6
Total	0.50		0.55		0.50		0.47	

Table 3 displays the reduction response for the 50% N-reduction scenario for the individual components of the nitrogen system. Although the general reduction response is close to linear some shifts in relative importance of the different compounds are seen. The most distinct non-linear feedback is observed in the PAN and HNO_3 production and dry deposition in A2, the main NO_x emitting area. Although NO is reduced up to about 70% it is contributing only little to the nitrogen in air. Total nitrogen in air and deposition is hardly affected. Our main conclusion is that in this particular episode the non-linearities of the nitrogen chemistry and deposition is rather unimportant. Only marginal changes are seen in the nitrogen air components during the 50% SO_x reduction scenario. The most pronounced alteration is a shift from HNO_3 to NO_3^- which also results in an increased nitrate dry deposition. NH_3 and NH_4 show also a response as less NH_4 is produced in the aerosol system.

CONCLUSIONS

Results from the EMEP acid deposition and the EURAD model have been compared for an episode of in July 1990. Comparisons with measurements show a rather strong underestimation of both sulfate and nitrate wet deposition in the receiver area during heavy rainfall. A bias toward too high nitrate wet deposition is seen in the EMEP model. The behaviour of the two models at different stations is however in most cases quite well correlated. The sulfur deposition is nearly 30% larger in the EMEP model for Northwest Europe. In the main SO_2 emission area the dry and the wet deposition is 3 and 2 times more effective than in the EURAD model. Although the studied period is short, it is likely that differences in the long term sulfur deposition patterns could be expected from the two models.

The deposition of oxidized nitrogen in the EMEP model is dominated by the wet part reaching 94% in the main receiver area. In the EURAD model a much larger fraction (typically 30-40%) is dry deposited mostly through nitric acid. This reflects the fact that more aged N-compounds as nitric acid and PAN are found in the EURAD model, while the EMEP chemical scheme yields a larger production of particulate nitrate. The comparison with measurements indicates that the production of nitrate particles is too efficient in the EMEP scheme. Linearity tests were also conducted for the EURAD model by reducing first only the sulfur emissions by 50% and then secondly by reducing only the NO_x emissions by 50%. For sulfur the reduced deposition is in the range of 34% to 38% for all four sub-domains. On the other hand, the deposition of oxidized nitrogen has a close to linear response to the emission reductions.

ACKNOWLEDGMENT

EURAD is funded by the Ministry of Research and Technology (BMFT) of the Federal Republic of Germany and the Ministry of Science and Research of the Land Nordrhein-Westfalen. The EURAD project is a part of EUMAC, which is a EUROTRAC subproject. Computational support came from the Research Center Jülich (KFA), in particular from the institutes ICG2, ICG3 and the ZAM of the KFA.

REFERENCES

Barrett, K., Ø. Seland, A. Foss, S. Mylona, H. Sandnes, H. Styve and L. Tarrasón: European transboundary acidifying air pollution: Ten years calculated fields and budgets to the end of the first sulphur protocol. *EMEP/MSC-W Report* 1/95, The Norwegian Meteorological Institute, Oslo, Norway, (1995).

Chang, J.S., R.A. Brost, I.S.A. Isaksen, S. Madronich, P. Middleton, W.R. Stockwell and C.J. Walcek: A threedimensional Eulerian acid deposition model: Physical concepts and formulation. *J. Geophys. Res.*, 92, 14681-14700, (1987).

Eliassen, A. and Saltbones, J.: Modelling of long-range transport of sulfur over Europe.: A two year model run and some model experiments. *Atmos. Environ.*, 22, 1457-1473, (1983).

Grell, G.A., J. Dudhia and D.R. Stauffer: A description of the Fifth-Generation PENN State/NCAR Mesoscale Model (MM5). *NCAR Technical Note*, NCAR/TN-398+STR., 138 pp., (1994).

Hass, H., A. Ebel, H. Feldmann, H.J. Jakobs and M. Memmesheimer: Evaluation studies with a regional chemical transport model (EURAD) using air quality data from the EMEP monitoring network. *Atmos. Environm.*, 27A, 867-887, (1993).

Hass, H., E. Berge, I. Ackermann, H.J. Jakobs, M. Memmesheimer and J.-P. Tuovinen: A diagnostic comparison of EMEP and EURAD model results for a wet deposition episode in July 1990. *EMEP/MSC-W Note* 4/96, The Norwegian Meteorological Institute, Oslo, Norway, (1996).

Memmesheimer, M., J. Tippke, A. Ebel, H. Hass, H.J. Jakobs and M. Laube: On the use of EMEP emission inventories for European scale air pollution modelling with the EURAD model. In: *Proceedings of the EMEP workshop on photooxidant modelling for long--range transport in relation to abatement strategies*. Ed.: J. Pankrath, UBA, Berlin, Germany, 307-324, (1991).

Tuovinen, J.P., Barrett, K. and Styve H.: Transboundary acidifying pollution in Europe: Calculated fields and budgets 1985-1993. *EMEP/MSC-W Report* 1/94, The Norwegian Meteorological Institute, Oslo, Norway, (1994).

AN ANALYSIS OF REGIONAL DIFFERENCES IN TROPOSPHERIC OZONE OVER EUROPE

Peter J.H. Builtjes, Paul J. Esser and Michiel G.M. Roemer

TNO Institute of Environmental Sciences,
Energy Research and Process Innovation
Department of Environmental Quality
P.O. Box 342, 7300 AH Apeldoorn, The Netherlands

INTRODUCTION

It is a well known fact, based on observations and modelling studies, that there are large spatial gradients over Europe in ozone characteristics. These differences are caused by the differences in strength of the phenomena which determine the ozone concentration at a specific location: the precursor emissions and the resulting chemical production, the meteorological situation and the dry deposition. Ozone measurements show the differences in ozone characteristics, but do not reveal the reasons for the differences.

An analysis has been made making use of results of the Eulerian grid model LOTOS.

With the LOTOS-model, calculations of concentrations of ozone over Europe on grids of 0.5 lat x 1.0 long upto about 3 km have been made for the whole year of 1994 on an hourly basis. The model results are in reasonable agreement with measurements, they do show similar magnitudes and similar O_3 gradients (Builtjes 1996, Roemer e.a. 1997).

The ozone results of the LOTOS-model have been analysed for the growing season (April-September) of 1994, first by an assessment of the found regional differences in peak ozone levels, averaged ozone levels, daily amplitude and month in which the monthly averaged value has a maximum.

For selected areas budget studies have been performed, indicating the reasons for the found spatial gradients.

It should be noted that these results, next to increasing our understeanding of ozone formation processes, may also lead to the development of optimal, regionally different, ozone abatement strategies.

REGIONAL DIFFERENCES IN CALCULATED OZONE CONCENTRATIONS

Obviously, the largest differences in ozone patterns are caused by the differences due to the atmosphere being above land or sea. Because ozone is not, or hardly, soluble in water, its dry deposition over sea is very small. Over land, however, O_3 is deposited

Air Pollution Modeling and Its Application XII
Edited by Sven-Erik Gryning and Nadine Chaumerliac, Plenum Press, New York, 1998

relatively fast. Further differences will be caused by the density of precursor emissions, and meteorological conditions including the amount and strenght of UV-light. Because all these factors show regional gradients over Europe, the resulting ozone concentrations will exhibit regional differences.

In this study, an analysis is made of regional differences as they are calculated with the LOTOS-model. Four ozone characteristics are used as indicators of regional differences: First, the absolute annual maximum hourly ozone value reflecting mainly the strength of the (episodic) photochemical production (AAM), second the average over threshold value for the total ozone exposure above a certain threshold (taken here at 40 ppb, AOT40), defined as the sum of the surplus of ozone every hour with daylight (global radiation over 50 W/m^2) and over the 6 month growing season period of April to September. This indicator reflects the long term background ozone level. Third, the average daily amplitude (ADA) over the growing season, reflecting production and loss rates, and to some extent vertical exchange. The fourth indicator is the month in which the montly averaged ozone has its maximum (MMA), reflecting the longterm influence of anthropogenic levels of ozone precursors.

By analysing the LOTOS-model results and especially the calculated spatial gradients, the following threshold values have been found for the four indicators, respectively 80 ppb, 9000 ppbh, 20 ppb and > July (Builtjes and Esser, 1996).

The analysis resulted in the following classification of 3 marine, and 4 continental ozone regions.

The main resulting question is whether this ozone behaviour over the 7 regions is a consequence of a similarity within these regions in the underlying processes.

To this end budget studies have been performed over the LOTOS domain.

SET-UP OF THE BUDGET STUDIES

Budget studies are performed over selected areas over Europe, based upon the criteria described in the preceding paragraph.

The LOTOS-model calculations cover the complete year of 1994, the analysis of the budget results have been restricted to the six-month period of the growing season, April-September 1994, in which the maximum ozone production takes place. The budgets have been calculated over a certain volume, with a given surface area and a height of 2 km.

Table 1a. Characteristic ozone regions for the marine area

1)	AAM	≥	80	128 grid cells
	AOT	>≈	9000	South + East North Sea, West-East Sea, English
	MMA	>≈	July	Channel, Northern-Mediterranean
	ADA	>≈	20	
2)	AAM	<	80	48 grid cells
	AOT	≥	9000	Mid-North Sea
	MMA	<≈	July	
	ADA	<	20	
3)	AAM	<	80	691 grid cells
	AOT	<	9000	Atlantic, North + West North Sea, East + Mid East Sea,
	MMA	<≈	July	North-Botnic, soutch + West Mediterranean
	ADA	<≈	20	

Not classified: 91 grid cells (= 9%)

Table 1b. Characteristic ozone regions for the continental area

4)	AAM	≥	80	317 grid cells
	AOT	>≈	9000	South-England, North + East France, Belgium, Luxemburg
	MMA	≥	July	The Netherlands, Danmark, Germany, South + West
	ADA	≥	20	Poland, The Czec Republic, Italy
5)	AAM	<	80	254 grid cells
	AOT	>≈	9000	Nort + East Poland, Slovak Republic, Austria,
	MMA	≥	July	Switzerland, Bulgaria, Greece, East Sweden, North + West
	ADA	≥	20	Hungary, North Finland
6)	AAM	<	80	625 grid cells
	AOT	<	9000	Spain, South + West France, Ireland, North-England,
	MMA	<	July	South-Scotland, Sourth-Norway, South + Mid Sweden,
	ADA	≥	20	South-West Finland, West-Russia, South + East Hungary,
				Romania, former Yugoslavia
7)	AAM	<	80	149 grid cells
	AOT	<	9000	North-Norway, North-West Sweden, South-East Finland,
	MMA	<	July	North-Scotland
	ADA	<	20	

Not classified: 73 grid cells (= 5%)

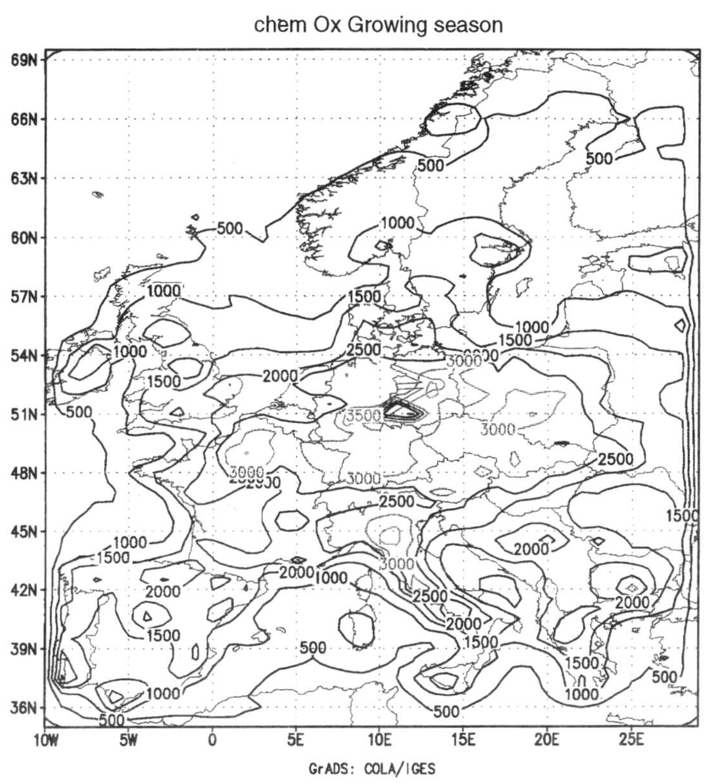

Figure 1.

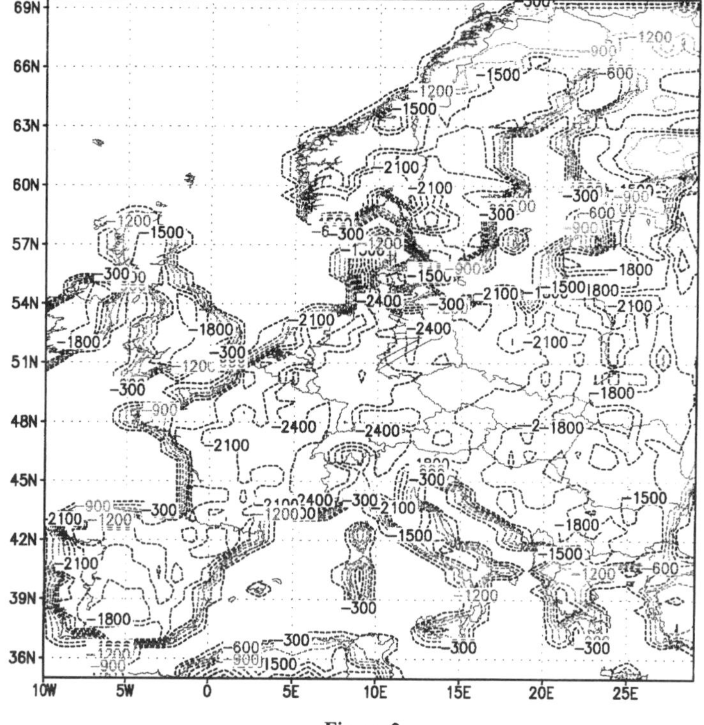

Figure 2.

In general, the highest ozone concentrations in the troposphere are found in this volume, and the processes which describe the budget of NO_x and VOC take place in the layer below 2 km (Panitz, o.a. 1996).

The budget or mass balance of O_3 over a given volume, and over a certain period in time is given by

$$\frac{\partial O_3}{\partial t} = \text{net Horizontal Advection + net Vertical Advection + net Chemical production - Dry deposition}$$

The explicit transport due to turbulence, and the influence of wet deposition are relatively small terms, and have been neglected in the calculations.

For similar approaches see Dennis (1996), Memmesheimer (1996), Memmesheimer (1997).

Taken over longer periods like a growing season or especially over a year, the change in time of O_3 will be minimal, although trends are definitely possible and the overall budget is not necessarily zero. In case a non-zero budget is found, special care should be given to the cause of this result, it might well be due to small numerical errors. In the current study, the change in time in ozone over the growing season is of the order of upto a few percent of the other terms, and will be neglected in the analysis.

THE RESULT OF THE LOTOS BUDGET STUDIES

In this paragraph, budget results for O_3 over the growing season of 1994 as determined with the LOTOS-model are presented. The major terms contributing to the

Table 2. Budget in g/h/km^2 over the growing season, marine areas

Area	Chem	Dep.	Hor. transport	Vert. transport
1.1	969	-180	-1226	443
1.2	1478	-196	-1382	116
1.3	1994	-289	-1678	-18
2.1	1045	-259	- 780	6

The next table gives the results of the continental areas.

Table 3. Budget in g/h/km^2 over the growing season, continental areas

Area	Chem	Dep.	Hor. transport	Vert. transport
4.1	2965	-2112	-158	- 613
4.2	3021	-2351	245	- 776
4.3	1916	-1871	424	- 432
5.1	1951	-1342	605	-1138
5.2	2195	-1730	340	- 705
5.3	2849	-2006	443	-1165
6.1	1731	-1929	274	4
6.2	1887	-2030	40	209

ozone budget will be given, chemical production, dry deposition, horizontal transport, vertical transport. The quantities will be presented in g/h/km^2 over the growing season, making them independent of the surface areas considered.

In Figure 1 and 2 per grid cell the terms for respectively chemical production and dry deposition are given. The spatial gradients shown in the chemical production term are large, upto nearly an order of magnitude with the highest levels in Central Europe and Northern Italy. There is no obvious increase in chemical production with lower latitute, indicating that the higher UV-light intensity and temperature at lower latitude is combined with lower emission densities than in the Northern part of Europe. Also, higher latitudes, in the growing season, experience a longer daylight period. The spatial gradients in dry deposition over land are relatively small, approaching the coast, with no deposition over water, the gradients are very sharp.

Budget studies will be presented for the following selected areas taken from the 7 categories of table 1: Mediterranean West of Italy (1,1), Mediterranean South of France (1,2), Southern Northsea (1.3), Central Northsea (2.1) and Northern Italy (4.1), Central Europe (4.2), South of England (4.3), Central Greece (5.1), Bulgaria (5.2), Austria (5.3), Spain (6.1) and South of France (6.2).

This results in 4 marine areas and 8 continental areas.

First the results of the four marine areas will be given.

A number of remarks can be made using these results. First, the four terms shown don't exactly add up to zero, at the end of the growing season there will be in general a small increase in the total budget in time, resulting in a positive $\partial O_3/\partial t$.

Secondly, chemical net production and dry deposition are strongly correlated. A high value of chemical production will lead to relatively high ozone concentrations, which will in turn lead to high deposition; a high production is combined with a high deposition.

Comparing Table 2 and Table 3 it is obvious that for marine areas, the dry deposition flux is about a factor of 10 lower than over the land. The horizontal transport results in a loss of ozone over marine areas, and to a gain of ozone, in nearly all cases over continental areas. Horizontal transport is much more important over marine areas due to the small dry deposition.

Over continental areas, there is in general an outflow of ozone to the free troposphere, over the sea after a net influx from the free troposphere is found. The large emission densities over the land surface produce a source of ozone, flowing out to the free troposphere.

To further characterize the different regions over Europe, the following remarks can be made. All marine areas look rather similar, despite the fact that they are located at different latitudes. The chemical production, the driving force for ozone formation, shows similarities between the area 1.1 and 2.1, and between the areas 1.2 and 1.3, showing a higher production for areas located more closely to the land and the emission sources. Considering the areas over land, the areas over Spain and the South of France behave clearly different than the other areas. This is caused by the different meteorological conditions which result in a partially closed box situation over these areas with little or no vertical exchange over the growing season.

The areas 4.1, 4.2 and 5.3, Northern Italy, Central Europe and Austria, all show a very high chemical production and a resulting high outflow to the free troposphere.

The other areas 4.3, 5.1 and 5.2, the South of England, Greece and Bulgaria show more moderate values.

The results show that the areas investigated in the South of Europe: Spain, Northern Italy and Central Greece do show large differences in ozone characteristics and do not belong to the same ozone structure.

CONCLUSIONS

An analysis of the regional differences in ozone patterns over Europe resulted in the following conclusions.

- The increase in photochemical production due to higher temperatures and higher UV-light intensity at lower latitudes is compensated by lower precursor emission densities at lower latitudes and lower daylights length. This leads to spatial gradients in chemical production of similar magnitude in North-South- and West-East direction over Europe.

- There is a large similarity in ozone behaviour between Northern Italy, Central Europe and Austria. Similar patterns, but with lower values for the different terms in the budget are found over the South of England, Central Greece and Bulgaria.
  Different patterns, indicating less exchange with the surroundings, are found for Spain and Southern France.

- It is remarkable that the three considered areas in Southern Europe; Spain, Northern Italy and Central Greece, show more similarities with other European areas than among themselves.

The results sofar are based only on model calculations. In the near future activities in the EU-project RIF TOZ (regional differences in tropospheric ozone) will lead to more reliable data for the summer of 1997, using also groundlevel data, vertical soundings, GOME observations, model calculations and data assimilation.

It should be stressed that the found differences and similarities over different areas in Europe will lead to optimal abatement strategies which will be different over the different areas.

ACKNOWLEDGMENT

This study is performed in the framework of the EU-DG-XII funded project RIF TOZ.

REFERENCES

Builtjes, P.J.H. *Modelling and Verification of Photo-Oxidant Formation,* Proceedings Eurotrac Symposium 1996, ed. Borrel e.c. (1996).

Builtjes, P.J.H. and P. Esser. *Regional Differences in Tropospheric Ozone,* Proceedings of the first REMAPE-workshop ed. Z. Zlatev, Copenhagen, (September 1996).

Dennis, R. *Model Evaluation Studies in the US,* Proceedings fifth US-German Workshop on the photochemical ozone problem and its control, Berlin (September 24-27, 1996).

Memmesheimer, M. e.a. *Budget Calculations for a Summer-Smog Episode as Simulated with the EURAD Model,* Proceedings Eurotrac Symposium 1996, ed., Borrell e.a. (1996).

Memmesheimer, M., A. Ebel, M. Roemer. *Budget Calculations for Ozone and its Precursors: Seasonal and Episodic Factors Based on Model Simulations,* Accepted, Journal of Atm. Chemistry (1997).

Panitz, H.J., K. Nester and F. Fiedler. *Mass Balances and Interactions of Budget Components of Chemically Reactive Air Pollutants over the Federal State of Baden-Württemberg* in: Air Pollution Modeling and its application XXI, 709, Plenum Press (1996).

Roemer, M., G. Boersen, P. Builtjes, P. Esser. *The budget of ozone and precursors over Europe calculated with the LOTOS-model,* Accepted, Atm. Env. (1997).

DISCUSSION

M. KAASIK: Why did you not compute the ozone balance for northern Europe?

P.BUILTJES: In fact, I did compute this balance, but did not show it. I focused on the south of Europe, partly because of the location of this conference.

P. SEIBERT: How sure are you that your model represents the different processes sufficiently well (e.g. vertical fluxes induced by mountains) thus how reliable are your results?

P.BUILTJES: The focus of the LOTOS model is to be an intermediate between the very complex models as EURAD, and the more simple models as EMEP-trajectory models. In my opinion, the description of the vertical fluxes, especially averaged over a longer time period as calculated with LOTOS, give at least a first order result.

R. SAN JOSE: Can you be a little more specific on the differences found over Spain and Southern France?

P.BUILTJES: The model results do show that Spain and the south of France differ in ozone behaviour from the other parts in the south of Europe. This is partly due to the specific meteorological situation, but especially also by the remarkable low CORINAIR NO_x emissions

D. FISH: Could you clarify your definition of vertical transport. If the air is transported horizontally over land, where the boundary layer is deep, over the sea, where the boundary layer is shallower, does this constitute vertical transport between the boundary layer and the free troposphere?

P.BUILTJES: No, it does not. The vertical fluxes in the model, mean vertical velocity and turbulent diffusion, are determined at a fixed height of 2 km, the chosen lid of my budget-volume.

M. BALDI: What is the definition of growing season you used in your study? Is it the same for all over Europe?

P.BUILTJES: Yes, it is. I have used in the calculation the formal ECE/EU definition of 6 month period, april-september, day-time averaged with daytime when the radiation is over 50 W/m^2

ON THE APPLICATION OF METEOROLOGICAL MODELS AND LIDAR TECHNIQUES FOR AIR QUALITY STUDIES AT A REGIONAL SCALE

Cecilia Soriano[1], José M. Baldasano[1] & William T. Buttler[2]

[1]Institut de Tecnologia i Modelització Ambiental (ITEMA)
Universitat Politècnica de Catalunya (UPC)
Apartat de Correus 508, 08220 Terrassa (Barcelona), SPAIN
[2]Los Alamos National Laboratory (LANL)
Physics Division (P-22)
MS D410, Los Alamos, NM87545, U.S.A.

INTRODUCTION

Effective air-quality studies at the regional-mesoscale range require a good understanding of the main circulatory patterns of air pollutants within the study region. When the mesoscale conditions prevail, circulatory patterns will be mainly determined by the characteristics of the terrain in the area. A region's unique orography often requires the development of unique air-quality squemes at a regional scale. That is evidenced in the frequent fact that air-quality remediation schemes developed for one region often prove useless when applied to other areas.

Two different approaches are mainly being used for this kind of studies. The primary, and historical method, utilizes field measurement of the relevant variables implied in the process, such as concentrations of pollutants emitted, or meteorological variables with an important role in their dispersion. These measurements are often made near ground level and only provide sparse, in-situ information in a study area. More recently, numerical models which can simulate meteorological conditions which cause the dispersion of air pollutants have been implemented to develop and evaluate air-quality management schemes. It is important to note that these two methodologies are not independent, and in fact, complement each other. For example, measurements are required to provide meteorological models with necessary inizialitation parameters, and also allow the posterior evaluation of the model's performance; while simulations with models provide high spatial and temporal information which is very difficult to achieve with classical measurement techniques.

Concerning this last point, recent progress in lidar techniques has led to the development of new techniques useful for regional pollution management in regions of complex orography. From the lidar information can be extracted about three-dimensional measurements of vector (winds) and scalar (elastic backscatter) fields at previously unattainable temporal and spatial densities (even with current networked surface stations and vertical soundings, such as balloons and radiosondes).

In July, 1992, ITEMA-UPC and LANL carried out the Barcelona Air Quality Initiative (BAQI) campaign. The atmosphere above Barcelona city was continuously monitored with an elastic-backscatter lidar device developed at LANL. Previous analysis of these data (Baldasano et al., 1993; Soriano et al., 1995) revealed a multilayer vertical arrangement of aerosols within the atmosphere in Barcelona. The aerosol arrangement is believed to be caused by the main processes occuring in Barcelona during the summer months because of the city's location by the sea and the surrounding complex orography. These processes are daytime convective vertical mixing, sea-breeze circulation, and vertical circulations produced by mountain thermal and mechanical effects.

In this contribution, in order to complete the description of the circulatory patterns in the area, a meteorological model is used, and its results have been compared to the three dimensional fields of winds and backscatter acquired by the lidar.

MEASUREMENT OF WIND FIELDS BY ELASTIC-BACKSCATTER LIDAR

Elastic-backscatter lidar techniques for the remote determination of wind speed and direction rely on the observation of naturally occurring aerosol density inhomogeneities. Aerosols are efficient wind tracers since they respond rapidly to changes in the air velocity. The algorithm used for the extraction of the wind information from the lidar signal is based on the maximum Cross-Correlation technique developed at LANL (Buttler and Eichinger, 1994; Buttler et al., 1994; Buttler et al., 1995; Buttler, 1996; Soriano et al., 1996). The technique uses images constructed from repetitive laser shots aimed at different points. In the case of Barcelona, the sampling pattern consisted of 4 points describing a square, what is called a four-point correlation scan.

Before the images are constructed and analyzed, the recorded signal is background subtracted and range corrected. Data are then arranged into two dimensional images of the laser return, represented as a function of range (from the lidar) and time (of the shot). Each of these 4 images (one for each of the sampling directions of the four-point correlation scan) are then low-pass filtered using a 2 dimensional (time by space) Gaussian filter in other to reduce the still significant noise of the images. The technique consists in the selection of a subimage (kernel) from within one of the images and finding its maximum correlation within another of the images. The time and range lags for which this maximum is found are used to calculate the horizontal velocity and direction of the wind responsible of the displacement of the aerosol inhomogeneity.

In the Barcelona experiment, the four points in the scan pattern were separated 2 degrees in azimuth and elevation. The lidar cycled 50 times around the four-points, taking 1.8 sec. to perform a cycle. However, given that the typical vertical velocity is much smaller than the horizontal one, the maximum cross-correlation technique cannot give results for the vertical component, and only correlations between images 1 and 2 (superirors) and between images 3 and 4 (inferiors) were performed. Therefore, only the horizontal components of the wind velocity vector have been calculated.

The velocity vectors obtained are post-processed in order to eliminate wrong results obtained from false correlations. These wrong correlations mainly occur when the kernel used to obtain the wind does not show a clear aerosol feature. This happened more often than one expected in the analysis of the data from Barcelona, since the elevation angles at which the four-angle correlation scan were acquired were often too high. Aerosol structures, whose presence is imperative for the success of this technique, are present when turbulence processes are developed. That is the reason why this maximum cross-correlation technique works better if the scans are acquired at low elevation angles, when the lidar beam travels through the

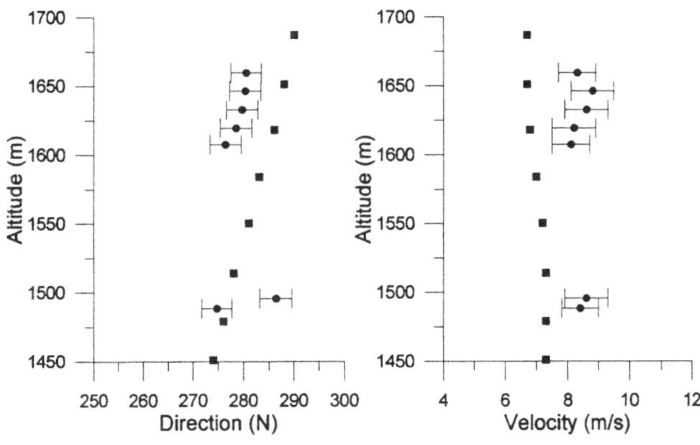

Figure 1. Wind Direction and Speed as measured by the Lidar (dots) and by the radiosonde (squares)

mixing layer, and at the central times of the day, when convection is fully developed and the lidar is able to monitor the aerosol features produced by the turbulent condition of the atmosphere. Winds acquiered at the same range from the lidar are vector-averaged and its standard deviation is calculated.

We have been able to compare the winds measured with the lidar with the ones obtained from a radiosonde launched at a near time. In fact, the lidar 4-point correlation scan was acquiered at 5:14 LST on July 28, 1992, while the radiosonde was launched at 6 LST. However, since winds at high altitudes do not change much in time, where the synoptic conditions prevail, the comparison of both profiles makes sense. Results are shown in Figure 1, where we can see the good agreement between the lidar and the radiosonde, if we take into account the standard deviation of the measurements by the lidar.

NUMERICAL SIMULATION WITH THE METEOROLOGICAL MODEL

In order to complement the information obtained by the lidar, and to be able to make a full description of the circulatory patterns of air pollutants in the Barcelona Area for a summer situation, a simulation with a mesoscale meteorological model was performed. The simulation was made with the nonhydrostatic meteorological model MEMO (Moussiopoulos, 1993). The nesting capabilities that the newer version of the model includes (a one-way scheme where the large-scale distribution of the inner grid follows from the larger simulation area) have showed to be adequate for the study being in progress. The maximum range of the lidar was approximately of 6 km from its position. That meant that the area where we were going to be able to compare the model with the lidar winds was going to be much smaller than the area needed by the model to originate the mesoscale circulations (a domain which had to include the orographical features implied in the mesoscale processes developed).

We have been able to obtain a high-density simulated wind field within the sub-region of the Barcelona Geographical Area monitored with lidar. In fact, the present simulation consists on three nested model domains (see Figure 2). A coarse grid domain, which covers and area of 80x80 km^2 with a grid resolution of 2x2 km^2. A medium grid, over a region of 40x40 km^2 with a 1x1 km^2 cell. And a fine grid, which expands 20x20 km^2 over the lidar monitored area and whose cell size is 0.5x0.5 km^2.

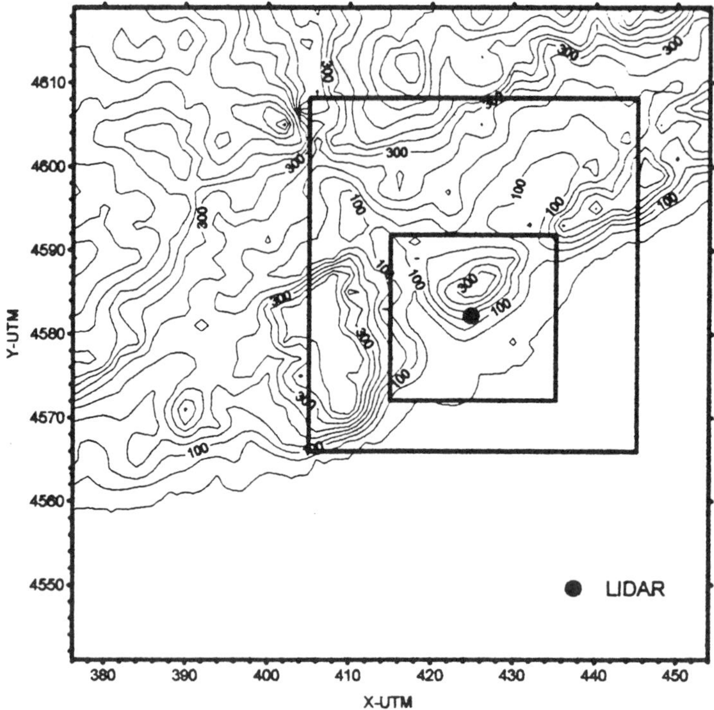

Figure 2. Nested domains (coarse, medium and fine grids) for the simulation with the meteorological model.

The simulated day chosen was July 28, 1992. The simulation started at 2100 LST (Local Standard Time) on the day before, and expanded for 27 hours. The initial state for all grids and boundary condition for middle times where derived from radiosoundings acquiered in Barcelona at 6 LST and in Zaragoza and Palma de Mallorca at 0z and 12z.

The study of the 2-dimensional fields of winds showed that the model was able to simulate the on-shore and off-shore cycles typical of the sea-breeze circulation.

On the other hand, a dispersion simulation for CO was also performed. The aim of this simulation is the comparison of the daily evolution of the distribution of this passive pollutant with the distribution of aerosols shown by the lidar. Since the emission of aerosols is associated to the emission of the typical air pollutants, specially in urban atmospheres, they can be used as tracers of the atmospheric movements and should be related to the dispersion of a low-reactive pollutant such as CO. The emission data for CO was obtained from the emission inventory EMITEMA-EIM [8]. The dispersion simulation was performed over the coarse grid (2x2 km^2 grid resolution).

JUDGING THE MODEL PERFORMANCE

The evaluation of the performance of a mesoscale meteorological model is usually carried out by comparing the results of wind velocity and direction given by the model in the first level height of the simulation domain with the measurements of these quantities from surface networked stations. Figure 3 is an example of this kind of evaluation.

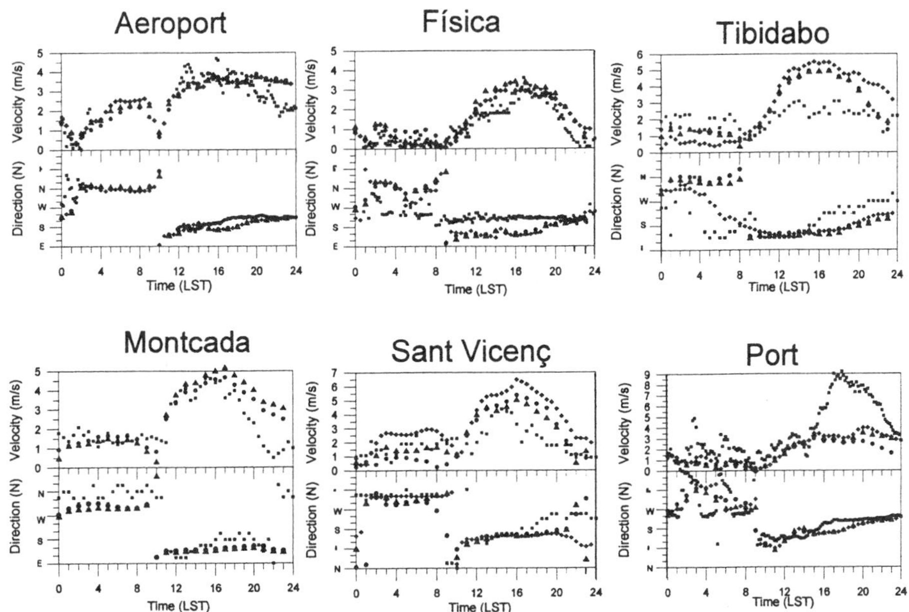

Figure 3. Wind Velocity and Direction calculated by the model and for the different grids used (coarse ●, medium ▲ and fine ◆), are compared to measurements from surface stations ■.

As we can see, simulated winds agree fairly well with measurements. The use of nested grids does not produce an important improvement in the simulated winds, which leds us to the conclusion that the resolution of the orography for the coarser grid is enough to originate the main circulatory patterns in the area.

We have been able to extend the evaluation of the model performance to high altitudes by comparing the winds calculated by the finer grid with the winds remotely-measured by the lidar. The importance of this comparison is that lidar-measured winds are free of the surface-related local features that measurements from surface stations may include. Besides, this kind of evaluation is interesting in regions like Barcelona, where phenomena with an important vertical dimensional (sea-breezes and mountain-related flows) take place. Figure 4 shows a comparison between simulated and lidar winds. Since winds from the lidar are obtained along the line of view of the laser beam, winds have been representated as a function of 'range from the lidar' in the x-axis and 'msl-altitude' in the y-axis. The modeled winds included correspond to the cells the laser beam crosses during its travel from the lidar.

A numerical study of the comparison of the modeled winds and the lidar has also been performed. Table 1 corresponds to the comparison of the lidar-winds and the model-winds at 12:30 LST. Measurements by the lidar include a deviation, which arises from the fact that the value of lidar-wind in a certain cell is the average of all the measurements the lidar acquired within the limits of the cell. Two statistics have been calculated for velocity and direction: the bias, or average of the differences between the observed and the modeled value; and the root mean squared error (rmse), calculated as the square root of the average of the squared diferences. The former informs of the model tendency to overpredict or underpredict measurements, while the latter measures the average spread of the differences. In Table 2 we can see the statistics of this comparison for different times.

At the sight of both comparison (visual and numerical), we can say that, in general, the agreement between the lidar and the model is quite good. If we first focus on the afternoon hours (12:30 to 14:30 LST), the model shows a sea-breeze circulation from the SE which is developed from the bottom of the atmosphere. The turn of the wind towards the SW (parallel

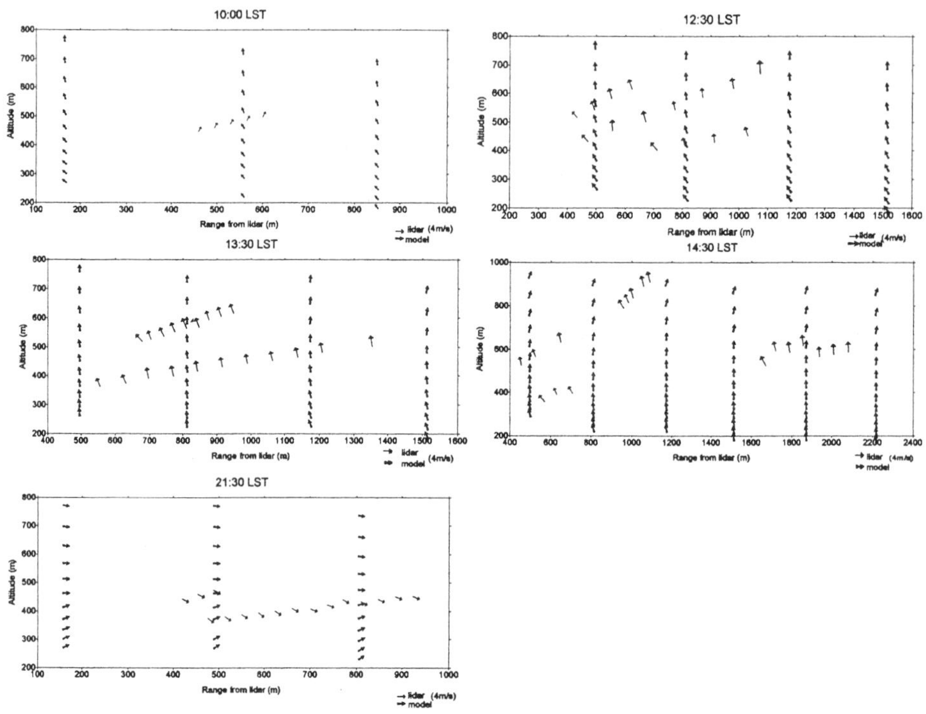

Figure 4. Horizontal wind vectors from the model finer grid are compared to the lidar winds.

Table 1. Winds measured by the lidar and calculated by the model at each of the model cells the lidar line of view crosses during its travel. Acquisition time is 12:30 LST.

model cell [x][y][z]	alt (m)	v-lidar (m/s)	d-lidar (deg)	v-model (m/s)	d-model (deg)	v-bias/rmse (m/s)	dir-bias/rmse (deg)
[19][20][11]	493.3	3.3∓0.5	164.9∓5.1	3.2	160.9	0.53/0.62	-4.8/10.8
[19][20][12]	555.0	3.7∓0.6	161.5∓6.8	3.2	166.4		
[19][20][13]	623.3	3.5∓0.6	168.6∓8.2	3.3	171.8		
[19][20][14]	698.7	4.2∓0.6	170.1∓6.9	3.4	176.3		
[19][21][12]	536.6	4.3∓0.4	178.0∓6.0	3.3	166.8		
[19][21][13]	605.1	3.6∓0.7	151.5∓10.4	3.3	172.0		
[19][21][14]	680.7	4.3∓0.8	162.3∓7.8	3.5	176.4		

Table 2. Statistics of the comparison between the lidar-winds and the model-winds

Time(LST)	vel-bias (m/s)	vel-rmse(m/s)	dir-bias(deg)	dir-rmse(deg)
10:00	-0.01	0.56	53.8	56.2
12:30	0.53	0.62	-4.8	10.8
13:30	2.15	2.20	-9.4	14.3
14:30	1.43	1.95	-12.0	12.3
21:30	-0.3	0.3	27.6	27.2

to the coast line) as the afternoon progresses is also reproduced by the model. This turn is less evident at higher levels. The statistics tells us that the model has a slight tendency to underpredict velocities, while the direction of the wind is very well simulated.

Comparison at 10:00 LST shows a circulation from sea to land in the model while the lidar reveals that the winds had still a land to sea component typical of the night-time regime. We have to take into account, however, that at this time we have the transition between night-time and day-time situations (note in the graphs that the velocity of wind is much lower). The results seem to show that the model started early in time the sea to land flow typical of the day-time regime. The model also reproduced the land-breeze with W component that is starting (again, velocities are lower) to develop at 21:00 LST.

DISPERSION SIMULATION

The simulation of the dispersion of CO has shown how the dispersion of pollutants is closely related to the flows present in the area. Another important parameter in the process is the depth of the mixing layer (ML), which tells us the width of atmosphere above the surface where pollutants are well mixed. The ML can be easily identified in the vertical cross-section of CO concentration, and also by the vertical scan from the lidar. In Figure 5 we can see vertical representations of a CO cross-sections from the model at 11:00 LST and a vertical scan from the lidar acquired at a near time. Both show the top of the ML at about 600 m above sea level.

CONCLUSIONS

This work has shown the suitability of the combination of lidar techniques and numerical simulation with meteorological models for the study of atmospheric circulations at a regional-local scale. The main contribution of this methodology to the field of atmospheric modeling comes from the ability of the lidar to remotely measure winds. This fact allows the evaluation of the model at high altitudes, where we can get rid of local effects that measurements from surface stations may include and which the model does not represent.

As far as the application to the Barcelona Area is concerned, we have shown how the model has been able to reproduce the main circulatory patterns in the area, both at a surface level and at a higher altitude. The application of nested domains has revealed that the orographic resolution given by the coarser model is enough for the model to generate the existing flows. However, the simulation at a finner domain was appropiate in our case, since we intended to compare the model and the lidar winds, which were acquired over a much reduced area.

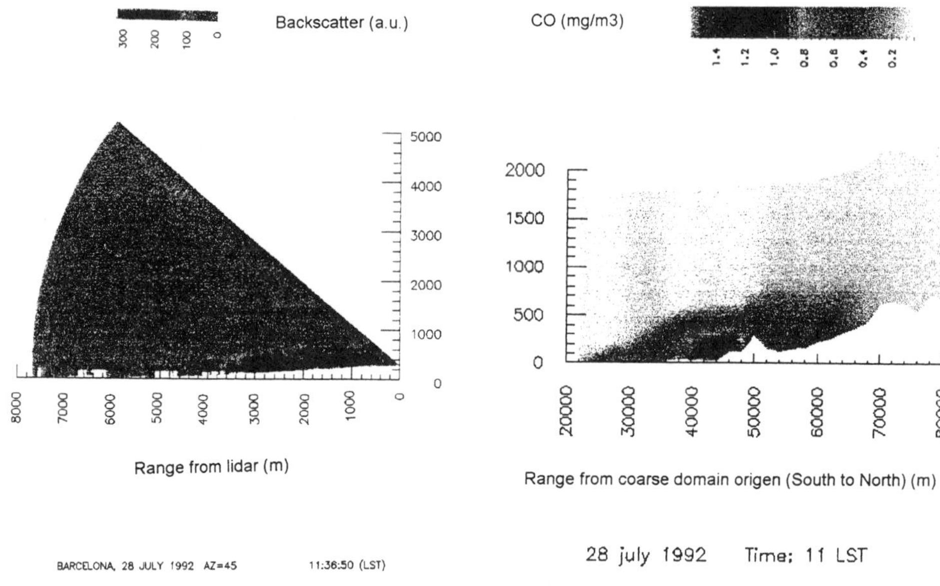

Figure 5. Lidar vertical scan at 11:36 LST and CO cross-section at 11:00 LST.

REFERENCES

Baldasano, J.M., Cremades, L. and Soriano, C., 1993, Circulation of air pollutants over de Barcelona geographical area in summertime. Angeletti, G. and Restelli, G. (editors). *Proceedings of the Sixth European Symposium on Physico-Chemical Behaviour of Atmospheric Pollutants*. Varese (Italia). Air Pollution Research Report 50. Environmental Research Programme of the European Community. pp 474-479.

Buttler W.T. and Eichinger, W.E., 1994, Wind-speed measurements with a scanning elastic-backscatter lidar, in: *Proceedings of the 21st Conference on Agriculture and Forest Meteorology and 11th Conference on Biometry and Aerology*. March 7-11, 1994. San Diego (California). American Meteorological Society .

Buttler, W.T. , Nickel, G.H., Soriano, C., Cottingame, W.B., Archuleta ,J., Smith, J. , Clark, D.A, Tellier,L.L. and Heskett,J.C.,1994, *El Paso Demostration: Wind Speed Measurements with a Scanning Elastic-Backscatter Lidar*. Los Alamos National Laboratory. LA-UR 94-2902

Buttler,W.T. , Soriano,C. , Clark, D.A.,Quick , C.R. and Oakeley,T.N,1995, *Sundland Park Border Air-Quality Study*. Los Alamos National Laboratory. LA-UR 95-86.

Buttler, W.T., 1996, *Three Dimensional Winds: a Maximum Cross-Correlation Application to Elastic Lidar Data*. Ph.D. Thesis. The University of Texas at Austin.

Costa M. and Baldasano J.M., 1995, Development of a source emission model for atmospheric pollutants of the Barcelona Area. *Atmospheric Environment*, 30A, 2: 309-318.

Moussiopoulos, N., 1994, MEMO: A non-hydrostatic mesoscale model, in: The EUMAC Zooming Model: Model Structure and Applications, N. Moussiopoulos, ed., EUROTRAC-155, Garmish-Patenkirchen, pp. 7-22.

Soriano, C., Buttler, W.T. and Baldasano, J.M.,1995, Comparison of temperature and humidity profiles with elastic-backscatter lidar data, in: *Air Pollution III (Volume 2: Air Pollution Engineering and Management*, H. Power, N. Moussiopoulos and C.A. Brebbia, eds., Computational Mechanics Publications, Southampton, UK.

Soriano, C., Baldasano, J.M. and Buttler, W.T.,1996, Lidar techniques and numerical simulation for air q uality studies: the case of Barcelona. *Eighteen International Laser Radar Conference*. Berlin, Germany, 22-26 July.

DISCUSSION

R. YAMARTINO: Do you see the sea-breeze return flow aloft with the lidar?

C. SORIANO: No, we don't. The lidar was unable to obtain any wind measurements in the return flow of the sea-breeze. It seems that the region from which the lidar was able to extract wind information from wasn't high enough to penetrate the return flow of the sea-breeze.

G. KALLOS: What kind of nesting you have performed? If it is one-way, it explains why you did not see significant improvements between the coarser and the finer grids. In one-way nesting you do not let the features to propagate and they are boundary-affected if the domain is very small.

C. SORIANO: We have performed a one-way nesting. We believe that the reason why we didn't obtain any significant improvement in the results is because the topography resolution used in the coarser domain is enough for the model to generate the mesoscale flows going on in the region and the inner topography did not introduce any significant new information with respect to the one used in the outer domain. The reason why we used nesting techniques in this work was, rather than to get better results in the modeled wind fields, to get a simulated wind field in a high density mesh. This meant that we would be able to compare the modeled winds with the winds remotely measured by the lidar for more model cells. Showing the usefulness of the lidar as a tool for model validations was one of the main goals of the work.

B. PHYSICK: I disagree with Dr Kallos' comments, as we have done similar fine-scale nesting and found differences in the winds. This is because each nest contains additional terrain and coastal features which generate additional circulations. In Cecilia's case, the dominant forcing features must have been on the 2 km grid, with no additional terrain features added by going to the smaller scale.

R. PIELKE: I agree with George Kallos with respect to the inability for nested one-way models to add much additional information if there is no additional smaller scale forcings (e.q. land use variations; terrain), and if the finer nest is so small that information advects through it and exits without time for smaller scale features to develop. The inner nest of two-way nests, of course, also has to be large enough for these small scale features to develop. Two-way nesting has the advantage, however, that some smaller scale features feed back upscale and influence features on a larger scale, as well as permits errors to propagate out of the fine nest rather that just be funneled in by one-way nesting.

R. SAN JOSE: Just a comment. We are using one-way nesting in Madrid with ANA model and we see significant differences in the wind fields.

POLLUTION AND CLOUDS OVER THE MASSIF CENTRAL

Guilène Gérémy, Wolfram Wobrock and Andrèa I. Flossmann

Laboratoire de Météorologie Physique
Université Blaise Pascal/CNRS/OPGC
24, Av. des Landais
F-63177 Aubière cedex

INTRODUCTION

Clouds play an important role in our environment. Apart from their radiative properties, this is due to the fact that they capture, transport and modify atmospheric pollutants. To understand their role in our climate is a long-standing aim. Concerning the pure liquid clouds the work has already considerably advanced, as there exist numerous models that study, e.g., the fate of sulfur, ozone or aerosol scavenged by clouds (e.g. Hall, 1980; Chaumerliac et al., 1987 a, b; Flossmann et al., 1988; Flossmann, 1991). For temperature regions from -0°C to -20°C, however, ice particles coexist with liquid water drops in the clouds and, thus, interfere in the processing of pollutants. Concerning these so-called mixed phase clouds our understanding is less advanced. There exist few model approaches which, however, lack from the fact that little is known on the fate of scavenged material in drops once the drops freeze. In order to shed light on this problem the European project CIME (Cloud Ice Mountain Experiment) has been initiated to study this problem. To support this experiment, dynamical microphysical and chemical models have been prepared which will in a later phase be used to interpret and generalize the findings.

To test the models and to study the general conditions to be expected during the experiment, the models have been applied to a dataset which has been obtained on 2 March 1994 at the same location as the CIME experiment. Unfortunately, the situation prevailed to pure liquid clouds. Furthermore, no chemical information were recorded during this experiment. However, the situation will allow to check the performance of our model hierarchy. The results will be presented below and implications for the CIME campaigns will be derived.

DESCRIPTION OF THE SITE AND THE CHOSEN DAY

The CIME experiment took place in Feb. 1997 at the summit of the Puy de Dôme, an ancient volcano of 1465m in the center of France. The Puy de Dôme which is at about 12km distance to Clermont-Ferrand belongs to a chain of ancient volcanoes oriented North/South

Air Pollution Modeling and Its Application XII
Edited by Sven-Erik Gryning and Nadine Chaumerliac, Plenum Press, New York, 1998

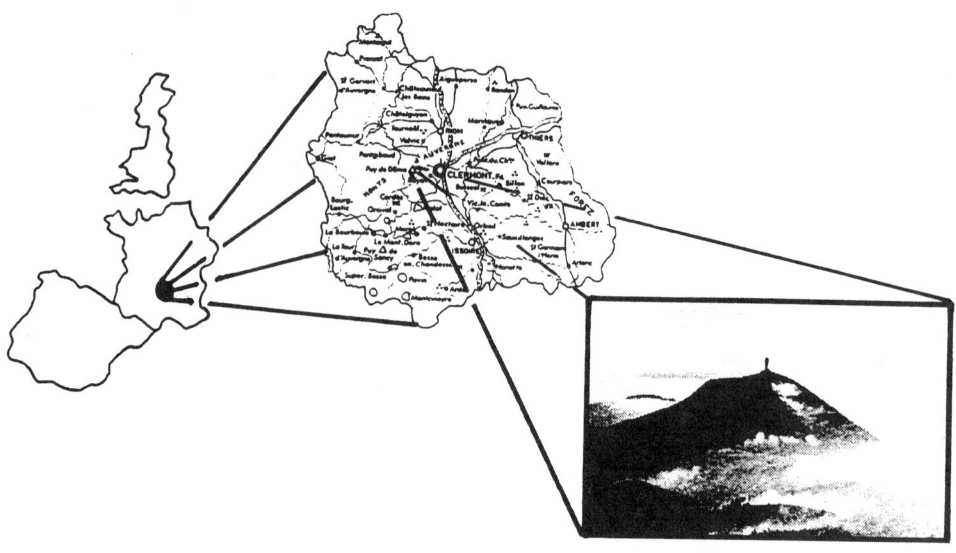

Figure 1. The geographic location of the Puy de Dôme moutain.

which are part of the «Massif Central», a large moutainous region (see Fig.1). At the summit there exists a large, fully equipped observatory also accessible in winter time. Apart from the meteorological measurements also measurement of No_x, SO_2, and O_3 are performed. Inside the observatory, there exists a wind tunnel which sucks in cloudy air. Several measuring devices inside the tunnel allow a characterisation of the cloudy air concerning, e.g., droplet spectra, ice crystal concentrations and mean droplet diameter.

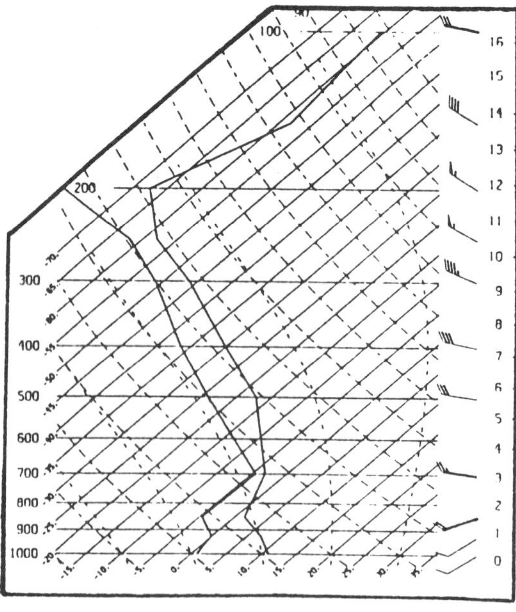

Figure 2. A skew-T-log P diagram of the temperature and humidity profile and wind for 2 March 1994 at 12 UTC, for a point situated south-west, near the Puy de Dôme, obtained from the ECMWF analysis

Furthermore, a radar ST was installed about 1.5km south and 10km east at an altitude of 394m. This radar is a Doppler radar where measurements of the vertical wind are performed through one vertical shot in a 5.6° angle. This radar allows to obtain an information on the wind field up to 10 km altitude.

For the test of our models, we choose a typical situation with westerly winds prevailing during 2 March 1994 where also microphysical measurements were performed in the wind tunnel on the Puy de Dôme. On 2 March 1994, 12 UTC France was located between two perturbations (one over England, and the other one over Italy) in a zone of weak pressure gradients (1020hPa typically at sea level). The mean flow at the surface was south-west and westerly at higher altitudes. The interpolated ECMWF sounding for a chosen point situated south-west, close to the moutain which was used for the model initialisation is presented in Fig.2. Local observations during that day seemed to indicate a stable stratification with the development of orographic clouds.

DESCRIPTION OF THE MODELS

The dynamical model

In order to realistically simulate the flow over the complex terrain upwind of the measuring site, we used the 3-D mesoscale model which was developped by Clark and coworkers (Clark, 1977; Clark and Farley, 1984; Clark and Hall, 1991). The model allows for terrain-following coordinates, nested domains and a stretched vertical cooradinate. For our purpose we used the Clark model with three domains. The largest domain which has a horizontal extension of 205*158km^2 and a grid spacing of 1.8km encompasses a large part of the Massif Central (see Fig.3). Two nested domains (coarse: 80*80km^2, grid spacing: 600m; fine: 25*20km^2, grid spacing: 200m) zoom into the area of interest and allow, e.g., the study of the formation of orographic clouds on the Puy de Dôme. The vertical resolution used increased with height (50m $\leq \Delta Z \leq$ 1000m).

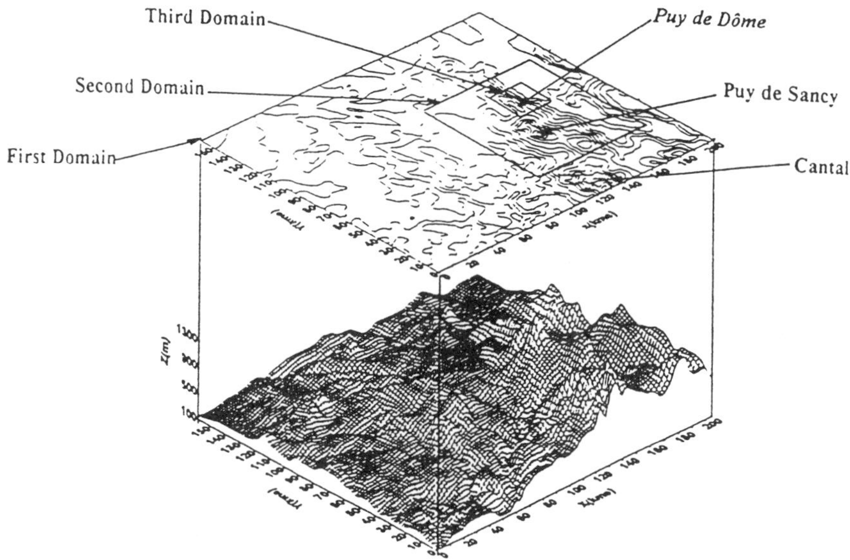

Figure 3. The location of the three domains for the model simulations.

The microphysical and chemical model

In order to study in greater detail the microphysical and chemical evolution in the clouds forming orographically at the Puy de Dôme, the DESCAM (Detailed SCAvenging and Microphysics) model (Berry and Reinhardt, 1974; Hall, 1980; Flossmann et al., 1985, 1987, 1988) was used along a backtrajectory touching the summit calculated from the 3-D wind field obtained by the Clark model. The DESCAM model allows to calculate the formation and evolution of the droplet and ice crystal spectrum in a cloud originating on a prescribed aerosol particle spectrum. The hydrometeors then can take up additional particles by impaction and scavenge gases and, thus, enable to compute the loading of cloud and rain drops with pollutant material. As the aerosol particle spectrum was not measured for the chosen day we use different aerosol particle spectra pertaining to typical air masses as proposed by Jaenicke (1988) to study the sensitivity of our results. As the meteorological conditions were not sufficiently cold during this period, no ice crystal formation was considered in the model. Also, the uptake of gases was disregarded due to the missing information on gas concentrations in the air and the drops.

RESULTS CONCERNING THE DYNAMICS

From the profiles of the vertical velocity, and the potential temperature the dimensionless Froude number for our situation can be calculated. For the lower 4km of the atmosphere, we obtain here a value of 2.75, indicating that the flow was quasi 2-dimensional in ascending the Puy de Dôme and descending at the downwind side. As to be expected from theory, our dynamic model develops a rotor at the leeward side of the mountain (see Fig.4). In addition, we see a warming of the air at the surface behind the Puy de Dôme indicating a « föhn » effect. The surface wind field, displayed in Fig.5, shows the changes in speed and direction before and behind the obstacle. Upwind of the mountain barrier wind of 1 m/s from south westerly directions prevailed. Directly behind the mountains the wind accelerated and in the « Limagne », a flat region represented at the right hand side of the figure the wind experienced a channeling towards the north.

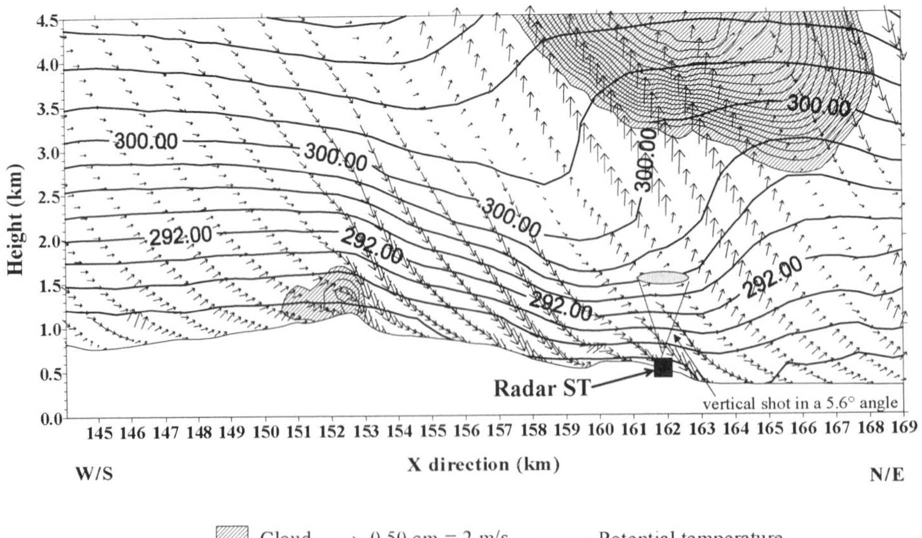

Figure 4. Cross section at 25°, following the mean flow direction. Wind, potential temperature and liquid water fields are displayed.

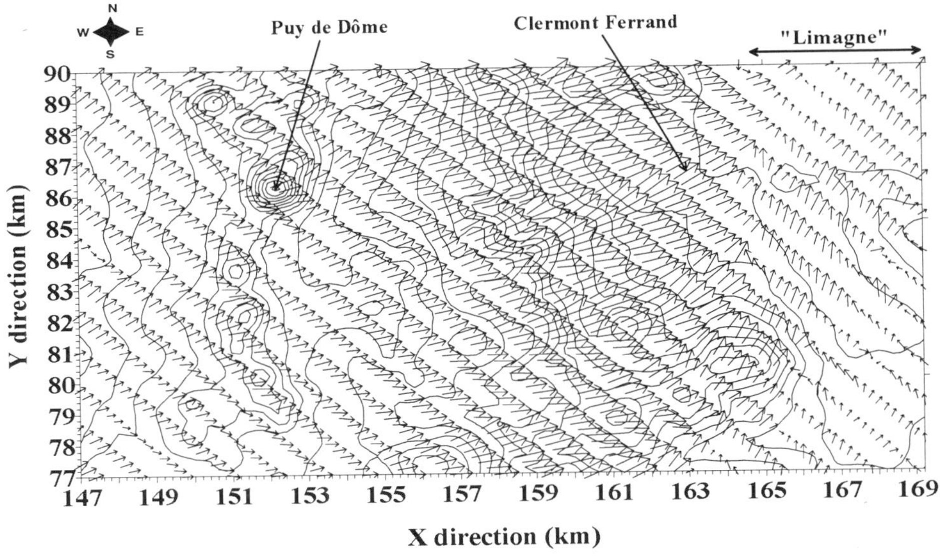

Figure 5. Surface wind field.

We tried to compare these results with the data available from the radar ST. Here, we have to consider that the radar only sees a small angle of the model domain. For the chosen day, the data of the radar encompassed a height between 1000 to 5000m. Considering the opening angle of the radar of 5.6°, and an arbitrary height of 1600m, this gives us an area of 0.019km^2. Identifying this region in Fig.4, we can state that is falls in the ascending branch of the developping rotor. This is confirmed by the radar as it also reported positive vertical velocities. The orders of magnitude of the vertical velocities are 0.52 m/s for the model results and 0.48 m/s for the radar measurements which we consider as a good agreement.

RESULTS CONCERNING THE MICROPHYSICS AND CHEMISTRY

Along a backtrajectory displayed in Fig. 6 by no. (2), an adiabatic air parcel was launched calculating the spectral microphysical and chemical evolution of the drop spectrum. As mentioned above, the spectrum of initial aerosol particle was not observed during the experiment.

However, as the wind was predominently flowing in from westerly directions and the region extending between the ocean and the Puy de Dôme is weakly populated, we can speculate that the aerosol particle spectrum represents a relatively clean air mass. Consequently, we studied the evolution of the drop spectrum as a function of a remote continental and a marine aerosol particle spectrum as proposed by Jaenicke (1988).

The resulting different droplet spectra are displayed in Fig. 7. Here, case 1, 3, and 4 use a remote continental aerosol particle spectrum with varying solubility of the particles and different number concentrations. We see here a rather dramatic influence of these parameters. Case 2 pertains to a marine aerosol particle spectrum and in case 5, we investigated the case of a different trajectory corresponding to a free convective ascent with entrainment of dry air considered. These spectra can be compared with the measurements of the PMS ASSP in the wind tunnel (curve 6). We can conclude here that the knowledge of the initial aerosol particle spectrum and the evolution of the relative humidity during the ascent is of utmost importance in determining the resulting droplet spectrum. During our investigations we noted, however, a second point: the measured droplet spectrum inside the wind tunnel is not necessarily identical to the droplet spectrum outside the wind tunnel in the

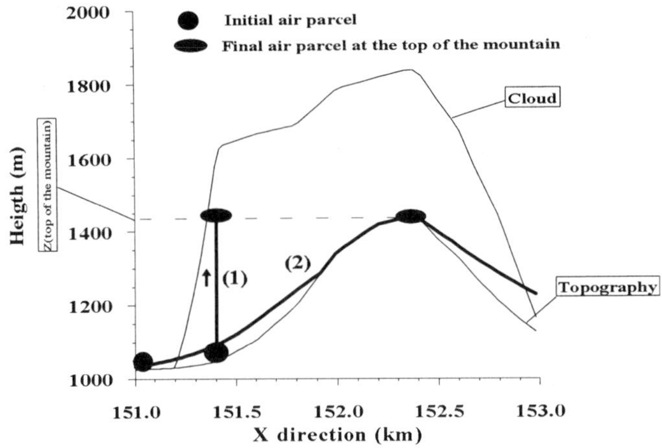

Figure 6. Trajectories following by the air parcel. (1) represents the trajectory used for calculs with entrainment, and (2), the trajectory for adiabatic calcul.

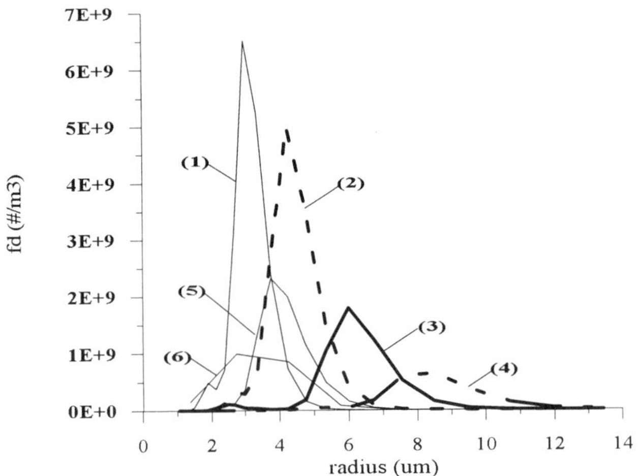

Figure 7. Cloud drop number distribution function for different types of aerosol and comparison with measurements:
(1) Remote continental, $(NH_4)_2SO_4$ (Jaenicke, 1988), solubility=50%, without entrainment;
(2) Maritime, NaCl (Jaenicke, 1988), solubility=100%, without entrainment;
(3) the same as case 1 with solubility=1%;
(4) the same as case 1 with conc/2;
(5) the same as case 1 with entrainment, along path(1) in Fig. 6;
(6) ASSP measurements

cloud. Preliminary measurements indicate a possible loss of up to 50% of the LWC. This loss can partly explain the discrepancy of the cloud drop water mass distribution displayed in Fig.8. We note, here, that the maximum of the liquid water mass is located around 4μm in the measurements as well as in the two simulations pertaining to case 1, and 5 in Fig.7 but that the total liquid water content recorded by the measurements is a factor of 3 smaller than the ones given by the model (LWC_{model}= 0.48 g/m^3 and $LWC_{PMS\ ASSP}$= 0.16 g/m^3).

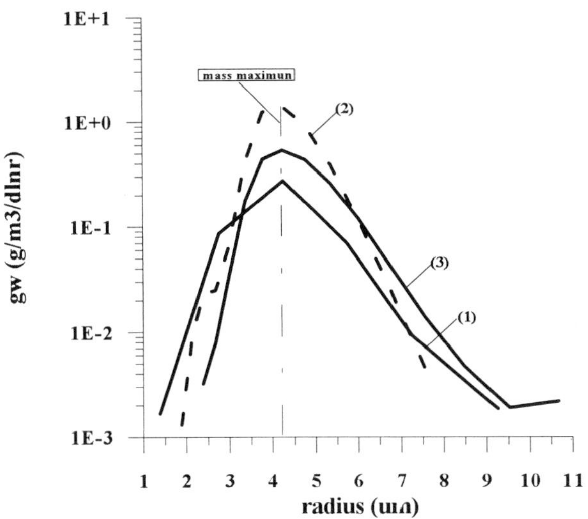

Figure 8. Cloud drop mass distribution function. Comparisons between:
(1) Remote continental, $(NH_4)_2SO_4$ (Jaenicke, 1988), solubility=50%, along path(2) in Fig. 6;
(2) the same as case 1 with entrainment, along path(1) in Fig. 6;
(3) ASSP measurements

CONCLUSIONS

We modelled the flow around the mountain Puy de Dôme, in the center of France and the clouds forming orographically and immersing its summit for the 2 March 1994. These model simulations were performed to study typical conditions of cloud formation in order to prepare the experiment CIME which was just recently terminated in Feb. 1997.

The flow round the complex terrain was simulated with the help of a 3-D mesoscale model and some aspects of the evoloving flow could be verified by measurements with radar ST. Even though more detailed comparisons with measurements would be desirable we can assume that the dynamics are reasonably well represented.

The detailed microphysical and chemical evolution of the cloud forming in the flow were simulated along a backtrajectory touching the summit. Here, we note that the knowledge of the initial aerosol particle spectrum is of utmost importance. Also, especial care needs to be taken to understand the difference of the in-cloud spectra and the ones measured inside the wind tunnel.

As the CIME campaign was just terminated we will in the future apply our models to the data obtained. Due to the many additional information available including aerosol particle and gas measurements, a more thorough comparison between model and measurements will be possible.

REFERENCES

Berry, E.X. and R.L. Reinhardt, 1974: An analysis of cloud drop growth by collection. Part I: Double distribution. *J. Atmos. Sci.*, 31, 1814-1824.

Chaumerliac, N.,E. Richard, J.-P. Pinty, and E.C. Nickerson, 1987a: Sulfur scavenging in a mesoscale model with quasi-spectral microphysics: two-dimensional results for continental and maritime clouds. *J. Geophys. Res.*, 92, 3114-4126.

Chaumerliac, N. and R. Rosset, 1987b: Pollutants scavenging in a mesoscale model with quasi-spectral microphysics. *Bound. Lay. Met.*, 41.

Clark, T.L., 1977: A small-scale dynamic model using a terrain-following coordinate transformation. *J. of Comp. Phys.*, 24, 186-215.

Clark, T.L. and R.D. Farley, 1984: Severe dowslope windstorm calculation in 2 and 3 spatial dimensions using anelastic grid nesting: A possible mechanism for gustiness. *J. Atmos. Sci.*, Vol. 41, No. 3, 329-350.

Clark, T.L. and W.D. Hall, 1991: Multi-domain simulations of the time dependent Navier-Stokes equations: Benchmark error analysis of some nesting procedure. *J. of Comp. Phys.*, 92, 456-481.

Flossmann, A.I., 1991: The scavenging of two different types of marine aerosol particles using a two-dimensional detailed cloud model. *Tellus*, 43B, 301-321.

Flossmann, A.I., W.D. Hall and H.R. Pruppacher, 1985: A theoritical study of the wet removal of atmospheric pollutants. Part I. *J. Atmos. Sci.*, 42, 582-606.

Flossmann, A.I., H.R. Pruppacher and J.H. Topalian, 1987: A theoritical study of the wet removal of atmospheric pollutants. Part II. *J. Atmos. Sci.*, 44, 2912-2923.

Flossmann, A.I. and H.R. Pruppacher, 1988: A theoritical study of the wet removal of atmospheric pollutants. Part III. *J. Atmos. Sci.*, 45, 1857-1871.

Jaenicke, R., 1988: Numerical Data and functional relationships in Science and Technology. Landolt-Börnstein New Series, V: *Geophysics and Space Research, 4: Meteorology (G. Fisher ed.) b: Physical and Chemical Properties of the air*. 391-457. Springer, Berlin.

Hall, W.D., 1980: A detailed microphysical model within a two dimensional dynamic framework: Model description and preliminary results. *J. Atmos. Sci.*, 37, 2486-2507.

THE APPLICATION OF AN INTEGRATED METEOROLOGICAL AIR QUALITY MODELLING SYSTEM TO A PHOTOCHEMICAL SMOG EVENT IN PERTH, AUSTRALIA

Martin Cope,[1] and Dale Hess[2]

[1]CSIRO Division of Atmospheric Research, Aspendale, Victoria, Australia
[2]Bureau of Meteorology Research Centre, Melbourne, Victoria, Australia.

INTRODUCTION

In this paper we present the outcomes of a meteorological and a photochemical smog simulation for the city of Perth, Australia. These simulations are designed to represent the conditions observed during the Perth Photochemical Smog Study (PPSS; PPSS 1996). Conducted during the summer of 1994, the study yielded an extensive data base suitable for the validation of both meteorological and photochemical models.

Although peak concentrations of photochemical smog were relatively low, the Perth airshed and PPSS offer a number of unique advantages. The meteorology of the region is idealised because the topography is relatively simple and the north-south coastline leads to the development of quasi- two dimensional sea breeze flow patterns. The air chemistry is less complex than that observed in many major urban centres because the majority of industrial sources are geographically separated from the urban region. Moreover, the relative isolation of the region precludes the possibility of inter-regional transport of pollutants. Lastly, the emissions inventory has recently been the subject of an extensive validation program.

METEOROLOGICAL MODELLING

The fine-scale meteorological modelling was undertaken using a quadruple-nested prognostic meteorological model. The boundary conditions were obtained from the 0.25 degree 19 level operational regional numerical weather prediction (NWP) model for the Australian Bureau of Meteorology (LAPS; Puri et al., 1997). Shown in Figure 1 is the 20-km outer grid of the fine-scale model (the 10-km, 5-km and 2.5-km grids are shown as insets), with topographic elevations displayed. The domains are centred on Perth, and are intended to provide a smooth transition between the large-scale synoptic forcing associated with the passage of a Western Australian trough and the mesoscale sea-breeze circulation. As shown in the figure, the topographic relief is moderate, with the main characteristic being an escarpment which is orientated approximately north-south and about 30 km inland. The finest grid is required to simulate the local flow fields in sufficient detail for photochemical transport modelling (which uses the horizontal domain shown in Figure 1b).

Developed from the original CSU mesoscale model (Pielke, 1974), the fine-scale meteorological model is hydrostatic, and has a height-based vertical coordinate system with 25

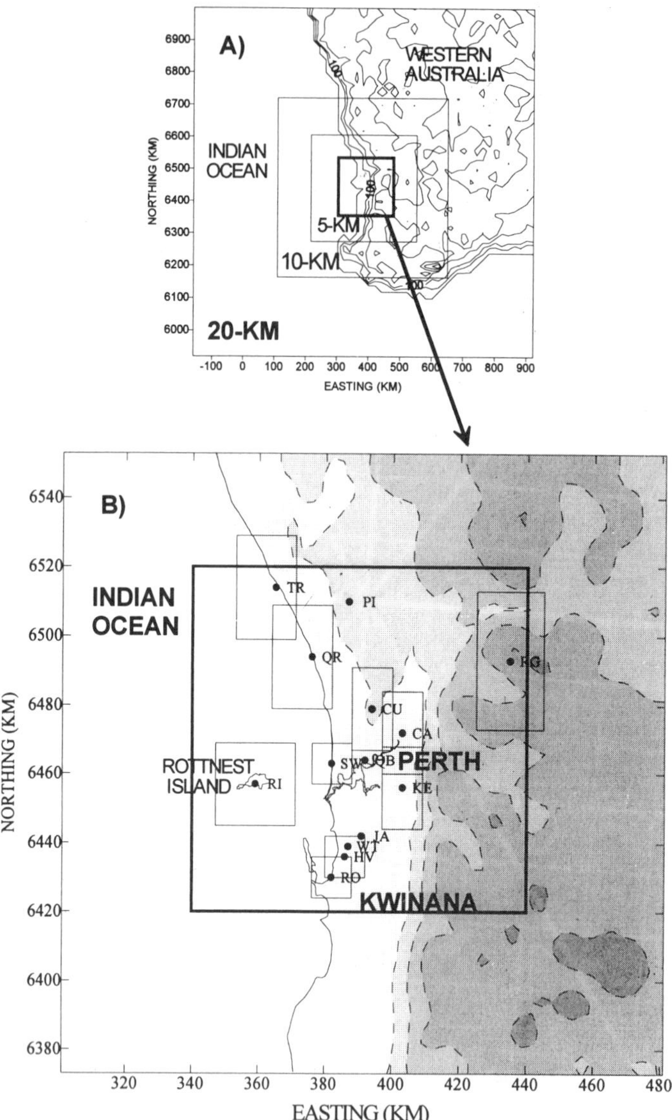

Figure 1. A) 20-km, 10-km, 5-km and 2.5-km horizontal domains used by the fine-scale mesoscale model and B) Photochemical airshed modelling domain with (inset) the Perth Photochemical Smog Study Region (PPSSR) displayed. Also shown are air quality and meteorological monitoring sites. Note the following sites: RI- Rottnest Island; SW- Swanbourne; KE- Kenwick; CA- Caversham; CU-Cullacabardee; RG- Rolling Green. The smaller rectangles denote station sub-regions, used during the air quality evaluation stage.

layers between 10 m and 5,000 m. The model physics includes a comprehensive land-surface interaction module (15 layer soil model, 5 layer canopy model with full hydrology), level 2.0 turbulence closure, and a two-stream radiation module. Advection is calculated using upwind cubic splines, a fully implicit scheme is used to solve the vertical diffusion equation, and radiation boundary conditions are applied at the lateral boundaries.

The model was driven in a hindcast mode by introducing regional scale NWP winds and temperatures and surface wind observations using four dimensional data assimilation (FDDA). In this application FDDA was used to assimilate the LAPS fields, and near-surface winds from the Western Australian EPA and Bureau of Meteorology (BoM) monitoring sites. The regional scale fields were assimilated above the boundary-layer with assimilation time constants which increased from one hour for the 20-km grid to six hours for the 2.5-km grid. This was done in order to ensure that mesoscale flow patterns in the inner grid were not excessively damped by the regional flow solutions.

The model was integrated for 72 hours to simulate the atmospheric flows during one of the major observation periods of the Perth Photochemical Smog Study, 2-4 February 1994. During the study, the existing network of surface observations was expanded and augmented by a program of special radiosonde ascents and instrumented aircraft flights. The comprehensive data set provides a useful opportunity to evaluate the meteorological models. An example is given in Figure 2 for hour 12 on 4 February (60 hours of model integration). It can be seen that the predictions of both models display good general agreement with the wind field observations. However, the fine-scale model, by virtue of its greater resolution was able to reproduce more of the smaller-scale features, such as the strength of the sea breeze and the abrupt change of wind direction above it. The vertical temperature profile has not been so well reproduced. This is likely caused by insufficient vertical resolution in the analysis which is used to initialise the NWP modelling.

A comparison of the observed and modelled near-surface winds is given in Figure 3 for hour 1 and hour 11 of 4 February 1994. The modelled winds are generated by the fine-scale model with FDDA of the LAPS fields only. It can be seen that the model has been able to successfully reproduce the general characteristics of the nocturnal and sea breeze flows.

Accuracy in modelling the sea-breeze dynamics is highly sensitive to processes associated with the land-surface interaction. The impact of alternative land-use categories, and maritime air cooling is illustrated in Figure 4, where the observed and predicted diurnal variation of temperature has been plotted for selected sites. It can be seen that the use of alternative land-use categories can strongly influence the surface temperature predictions. However, the amplitude of the diurnal temperature wave in the surface-layer is more strongly controlled by sea breeze onset time and the associated cool air advection.

In summary, the fine-scale meteorological model was able to reproduce most of the key wind field features observed on 3-4 February 1994. The observed temporal and spatial variations of the near-surface temperature field were also well simulated. However, some of the observed features in the vertical temperature profiles were not reproduced by the model, an outcome which was likely due to insufficient vertical resolution in the initial analysis.

AIR QUALITY MODELLING

The air quality was simulated using a modified version of the Carnegie Mellon - Caltech Institute of Technology (CIT) model (Harley et al., 1993). The model solves a system of coupled semi-empirical advection-diffusion equations for 35 atmospheric species and has recently been enhanced in-house through the inclusion of better numerics and physics (Cope, 1997). For the Perth simulation, the model was configured with 10 layers in the vertical, between 10 m and 2000 m, and a horizontal domain which consisted of 3600 cells of 3 km resolution (dictated by the resolution of the emissions inventory). The principal sources contained within the emissions inventory were motor vehicles, industrial and biogenic. The model was integrated for 48 hours, commencing hour 0 (local time) on 3 February. The model was initialised with clean air and the first day of integration was used to generate more realistic background concentrations, which incorporated the effect of the urban sources.

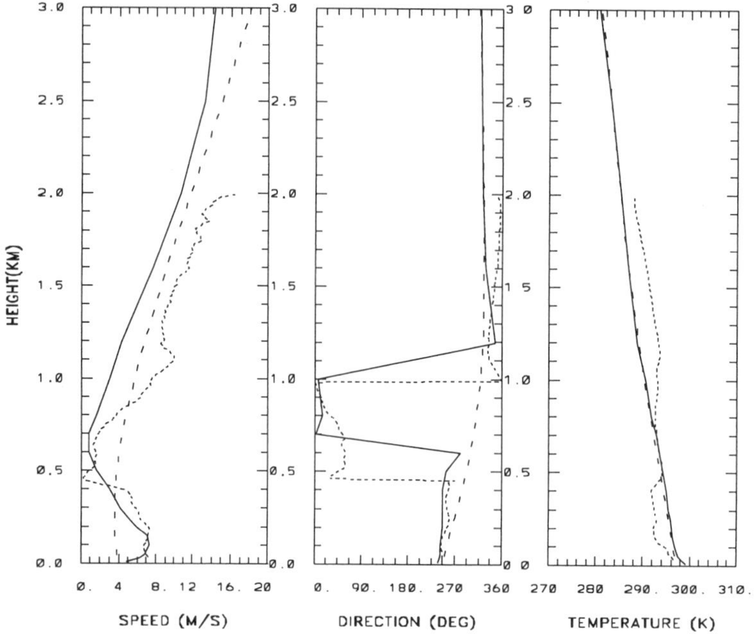

Figure 2. Comparison of the predicted and observed profiles of wind speed, wind direction and temperature for Swanbourne at hour 12 on 4 February 1994 (observation- dotted line; LAPS- dashed line; mesoscale model- solid line).

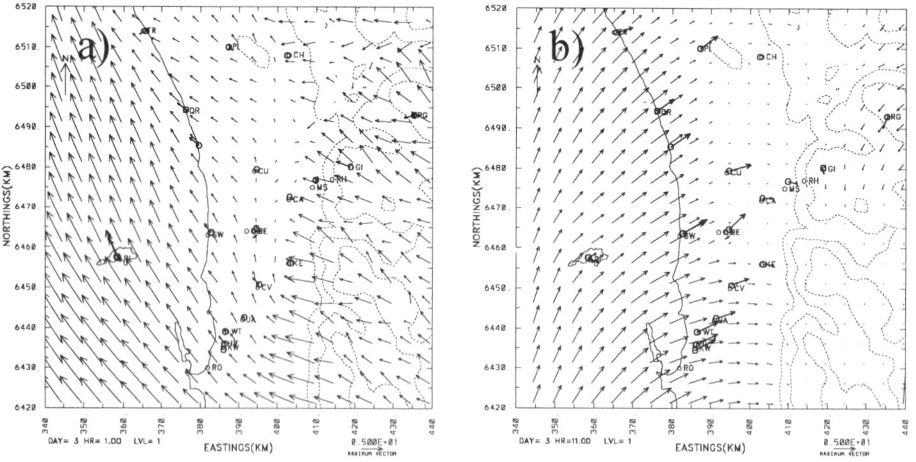

Figure 3. Observed (bold) and predicted 10m wind vectors for a) hour 1, and b) hour 11 on 4 February 1994 (2.5-km grid, every second point shown).

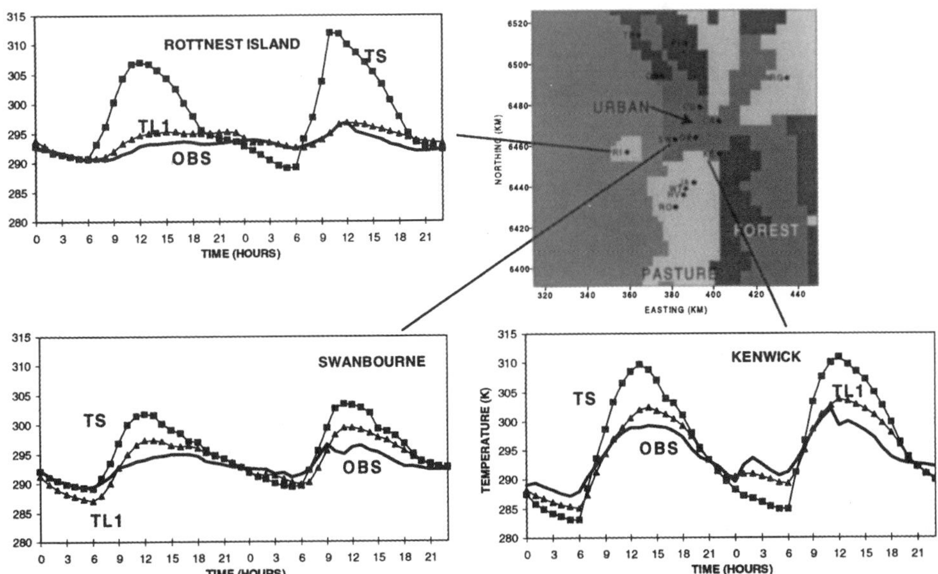

Figure 4. Predicted and observed ambient temperatures for Rottnest Island, Swanbourne and Kenwick (*TL1*- level one prediction, *TS* - surface temperature prediction; *OBS*- observed screen temperature). Also shown is the spatial distribution of land use category for the 2.5-km domain.

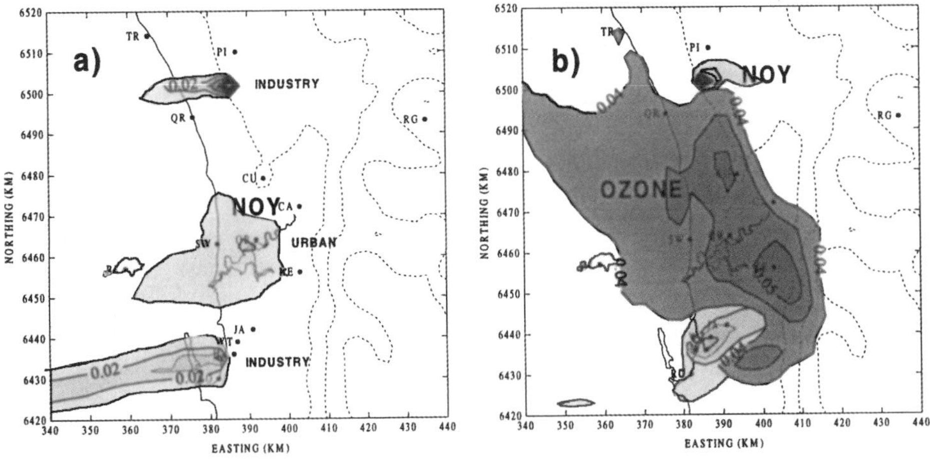

Figure 5. Estimated spatial distribution of NO_y (includes nitric oxide, nitrogen dioxide and secondary nitrate products) for (a) hour 8, and (b) NO_y (light grey) and ozone (dark grey) for hour 12 on 4 February 1994. Concentrations are given in units of ppm. Dotted lines denote 100 m interval terrain contours.

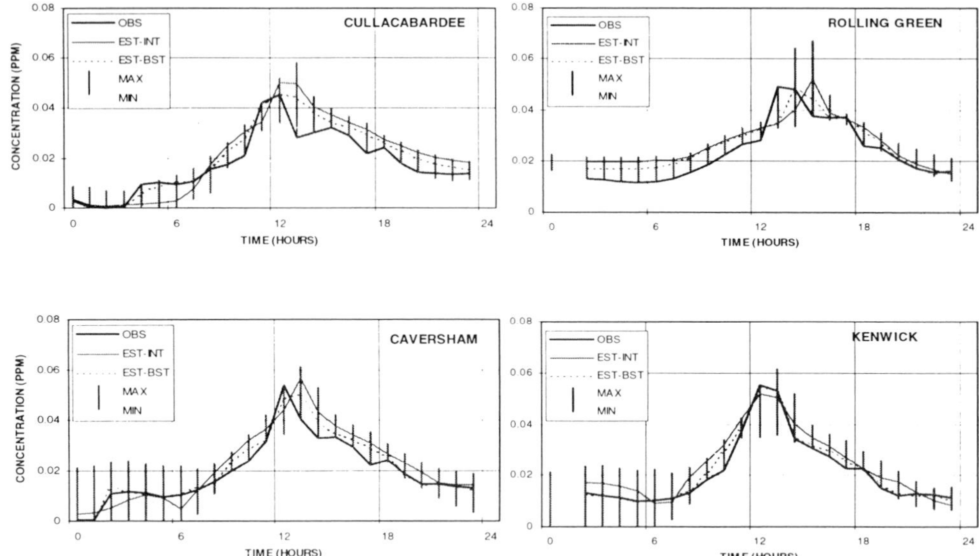

Figure 6. Observed and estimated 1-hour ozone concentrations for selected sites in the PPSS region for 4 February 1994 (OBS- observation; EST-INT estimated concentration interpolated from the four nearest model points; EST-BST best estimated concentration within 2 grid points of the observation site; MIN-MAX estimated range within a prescribed station sub-region- see Figure 1).

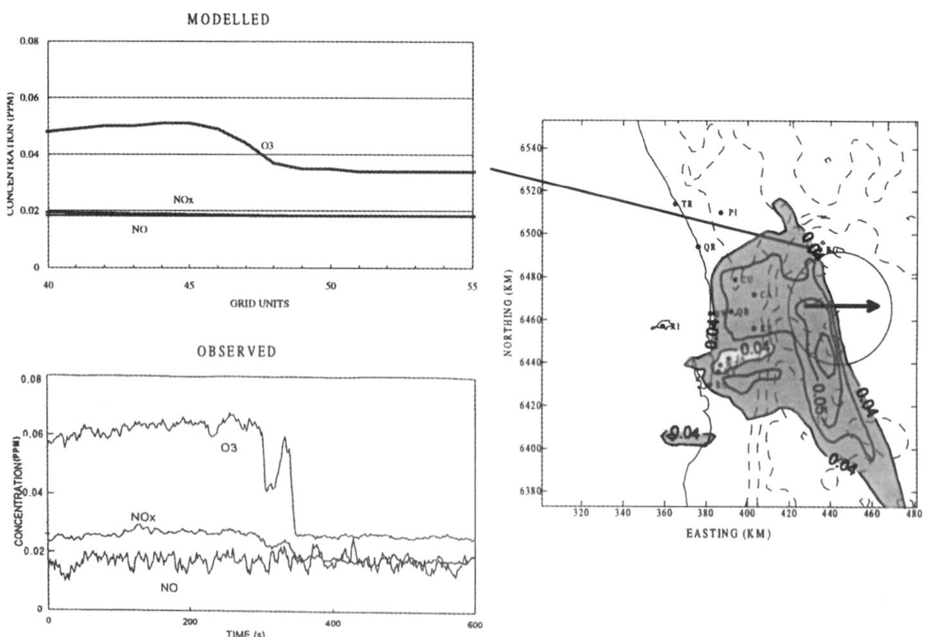

Figure 7. Observed and estimated ozone concentrations along an aircraft flight path (arrow) which traversed the sea breeze front at hour 14 on 4 February 1994.

Shown in Figure 5 is the predicted field of NO_y (nitric oxide, nitrogen dioxide and secondary nitrate products) for hour 8, and NO_y and ozone for hour 12 on 4 February. At hour 8, the urban and industrial precursor plumes are seen to be separate entities which were advected offshore by an east-northeasterly wind. By hour 12 the sea breeze had developed and penetrated about 30 km inland, transporting a pool of ozone behind the front.

The photochemical model was evaluated by comparison with near-surface (Figure 6) and aircraft (Figure 7) measurements. As shown in Figure 6, a range of site-based concentration estimates (interpolation of concentrations from the nearest four grid points, nearest grid point, best grid point and high-low range within a prescribed sub-region) were compared against observations. It was found that the timing and magnitude of the ozone peaks were reproduced with acceptable accuracy.

The model was not able to resolve the observed sharp rise in ozone concentration associated with the passage of the sea breeze front because of the inherent resolution limitations associated with an Eulerian grid-based representation of the air quality fields. This is illustrated in Figure 7, where we have compared aircraft observations and model predictions along a trajectory which intercepted the sea breeze front.

CONCLUSIONS

An air quality simulation system has been developed which uses the operational regional NWP model fields to drive a fine-scale prognostic meteorological model. The output fields from this model are then used to drive a photochemical airshed model. Because of the requirement for great precision in the meteorological fields, significant effort has been devoted to developing and refining the land-surface interaction and boundary-layer physics schemes. The photochemical airshed model was also enhanced through the inclusion of improved numerics and physics. The models were compared against a comprehensive data set obtained during the Perth Photochemical Smog Study and were able to reproduce most of the key meteorological and air quality features.

REFERENCES

Cope M.E., 1997, *Mathematical modelling of the transport and production of photochemical smog in the Port Phillip Control Region*. PhD Thesis submitted to Department of Earth Sciences, Melbourne University, Melbourne, Australia.

Harley R.A., Russell A. G., McRae G. J., Cass G.R. and Seinfeld J.H., 1993, Photochemical modeling of the Southern California Air Quality Study. *Envir. Sci. Technol. 27:378.*

Pielke, R.A., 1974, A three-dimensional numerical model of the sea breezes over south Florida. *Mon. Weather Rev.* 102:115.

PPSS, 1996,. *The Perth Photochemical Smog Study- An overview.* Western Power and Western Australia Environment Protection Authority. Final report to Department of the Environment, Perth, Western Australia.

Puri, K., Dietachmayer, G.D., Mills, G.A., Davidson, N.E., Bowen, R.A., and Logan, L.W., 1997, The new BMRC Limited Area Prediction System, LAPS. (*sub. to Aus. Met. Mag.*)

DISCUSSION

J. WALMSLEY: In Canada, we are beginning an operational air quality prediction program in the summer of 1997, using a statistical approach rather than a full photochemical modelling method that you have described. Are there any plans to try statistical models for air quality prediction in Australia?

M. COPE: At the Environment Protection Authority of Victoria, a statistically-based expert system has been in operation for over 10 years.

The system is used to forecast summer photochemical smog events and autumn-winter poor visibility events. On average, the system is able to correctly forecast an air pollution event for the following day with a 50% success rate. This relatively low success rate occurs because of the difficulty in accurately forecasting the simultaneous occurrence of the narrow set of meteorological conditions which are responsible for the development of an event.

VALUATION OF A LAGRANGIAN OPERATIONAL OZONE PREDICTION MODEL (LOOP)

J. Zimmermann, B. Fay, and R. Thehos

Deutscher Wetterdienst, FE25,
P.O.Box 100465
63004 Offenbach, Germany

INTRODUCTION

Since July 1995 German Law mandates vehicles without catalytic converters to be banned from traffic when measured ozone mixing ratios exceed 120 ppb and will be likely to exceed this value the day after. Therefore, daily ozone forecasts are needed. In order to replace currently used statistical models the DWD develops chemistry and transport models. One important restriction to be met is a time limit of 2 hours CPU-time for a three-day forecast on the DWD computers. On the Cray YMP4 available until this year this restricts the DWD to use a trajectory box model. A 3-D Eulerian model can only be used after an extensive up-grade of computer power which has just been started and will be continued during the next six years. The trajectory box model provides the additional advantage of giving insight into the couplings of emissions, transport, and chemistry due to its simple design. Modules tested in the box model could be implemented in the 3-D model later on.

Here we explain the design of the Lagrangian Operational Ozone Prediction model (LOOP), a trajectory box model coupled to the DWD weather forecast models. A comparison of model results with ozone measurements will be shown. As an example the situation on July 7, 1995, is used in comparison to measurements at 320 ground stations in Germany. We will discuss the observed discrepancies and the results of sensitivity studies with regard to possible errors of the input data.

MODEL PHILOSOPHY

The idea behind a trajectory box model is that of a homogeneous mixing layer which is in contact with the ground (emissions, deposition) (Derwent, 1990). Diurnal variation of mixing layer height creates a layer of air which is decoupled from the ground for part of the day. This residual layer can be thought to be homogeneous in a simplified approach. The two layers are represented by two boxes. Horizontal advection is accounted for in a

Air Pollution Modeling and Its Application XII
Edited by Sven-Erik Gryning and Nadine Chaumerliac, Plenum Press, New York, 1998

Lagrangian manner by movement of the boxes along trajectories. In case of LOOP there are three decoupled parts of the model: a trajectory model, an emissions module, and a two-box chemistry model.

The trajectory model

The DWD trajectory model is used to calculate backward trajectories approx. 75 mbar above the ground using forecast data of two numerical weather forecast models (EM: resolution approx. 55 km, DM: approx. 14 km). Since only boundary layer processes are modelled no vertical advection is taken into account. During episodes with high abundances of ozone trajectories with targets at the ground usually have little vertical movement (Fricke, personal communication). Meteorological parameters of the EM (or DM) are calculated along the trajectory every half hour. They include temperature, pressure, and humidity at the approximate centers of the model boxes, radiation data, and parameters characterizing turbulence in the surface layer. These data fields are passed on to the emissions module. For the results presented here only the EM was used.

The emissions module

The PHOXA data base (Veldt, 1992) provides yearly anthropogenic emissions data with area sources on a 30 km grid and point sources. Point sources are distributed over a circle with a radius of 18 km in order to account for horizontal diffusion and trajectory uncertainty. Reduction of this radius to 6 km did not alter model results for ozone by more than 5%. Diurnal, weekly, and seasonal cycles are accounted for by functions which depend on compound and emitter. SO_2, NO_2, CO, and VOC are given. Biogenic emissions are implemented using time and radiation dependencies of Geron et al. (1994) and land use data from the PHOXA data base.

Two-box model

With changing mixing layer height air is redistributed between box 1 and 2. Eddy diffusion takes place between box 2 and the free troposphere with an eddy diffusion coefficient equal to 10 m^2s^{-1}. No eddy diffusion takes place between box 1 and 2. A 30 min time step is used for transport. Clean air values are chosen as initial conditions. Initial mixing ratios for ozone are 30 ppb over the Atlantic and 60 ppb over Europe. Free tropospheric concentrations are fixed. They are 30 ppb north of 50 degree northern latitude and 60 ppb to the south.

The RADM2 chemical mechanism (Stockwell et al. 1990) is used with up-dated rate constants and supplemented by a condensed isoprene mechanism (RADM-C) (Zimmermann and Poppe, 1996). Emissions are added to the production terms and dry deposition terms enhance the loss terms of the chemical species. The differential equations are solved with the quasi-steady-state approximation (QSSA) integration scheme (Hesstvedt et al. 1978). Large time steps in the order of minutes are made possible by the use of NO_x (reactive N-compounds), O_x (odd O-compounds), and NO_y (all N-compounds) families and steady state solutions for radicals. The time step is variable with an average of three minutes and a maximum of five minutes. The integration error of ozone is less than 5%, in most cases lower than 2%.

COMPARISON WITH MEASUREMENTS

Simulations were carried out for episodes with high abundances of ozone in July 1994 and 1995. The results were compared with routine ozone measurements of the environmental agencies of the German states and of the FRG. Measurements of 320 stations were available in 1995. Only the comparison for July 7, 1995 is discussed though similar results were achieved for July 16, 1994. A scatter plot of all measurements versus simulation results shows almost no correlation (r^2=0.12) and an overprediction of ozone by 11 %. The standard deviation is 21 ppb (27 % of the average of the measurements).

However, in some regions of Germany the performance of LOOP is much better than in others. Here, Germany is divided into four regions: northern (all coastal states), southern (states of Hassia, Bavaria, Baden-Wurttemberg), western (all states at the western border that are not northern or southern states), and eastern (the other states of) Germany. The scatter plots of the comparisons between measured and calculated ozone mixing ratios for these four regions are shown in figs. 1 to 4. While considerable correlation is apparent in southern and northern Germany no correlation can be seen in western Germany. In eastern Germany there is an overprediction of ozone by 33 % which significantly exceeds the overprediction in other regions (between 1 % in western and 7 % in southern Germany).

SENSITIVITY STUDIES

Changes of the final ozone mixing ratios due to the variation of model parameters were monitored. Only very few examples are shown which represent different locations in western Germany. Bielefeld is an example of severe underprediction of ozone by LOOP (calculated 39 ppb versus measured 99 ppb). Figs. 5, 6 give an impression of the effects of changes of anthropogenic and biogenic emissions. The effect of varied initial ozone mixing ratios is shown in fig. 7. The effects of successive delay of the simulation start (shortens simulation duration) was investigated, too. In steps of 12 hours parts of the trajectory were removed. Removing all 72 hours makes the intial conditions the results of the simulation. In fig. 8 the dependence of the results on the amount of simulation time removed is shown. Other parameters are only of minor importance. Changing the deposition velocities by a factor of 2 or changing the turbulent mixing between residual layer and free troposphere by a factor of 3 varies ozone by less than 10 %.

DISCUSSION

On average the bias of results of LOOP is about 11 percent in comparison with the measurements of ozone. The tendency of slightly overpredicting ozone is not surprising

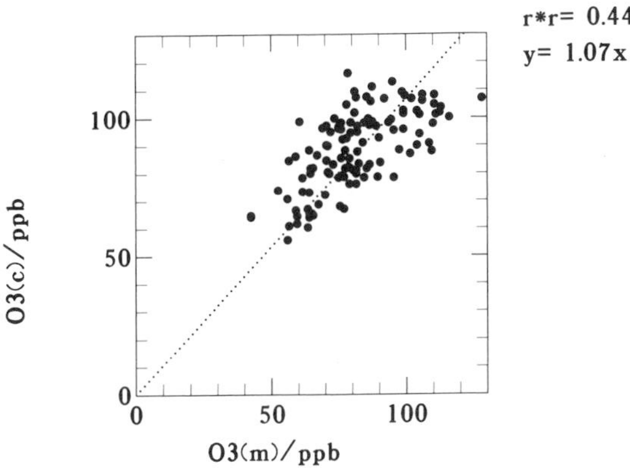

Figure 1. Scatter plot of calculated versus measured O_3 in ppb for southern Germany. r^2=0.44, standard deviation 14.8 ppb (18 %).

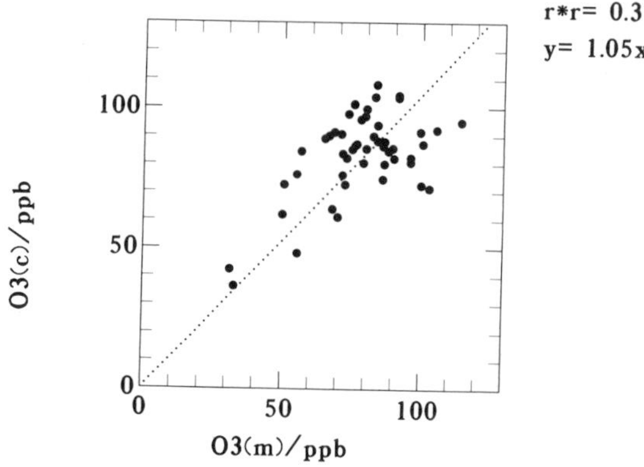

Figure 2. As fig. 1 for northern Germany. r^2=0.33, standard deviation 15.8 ppb (20 %).

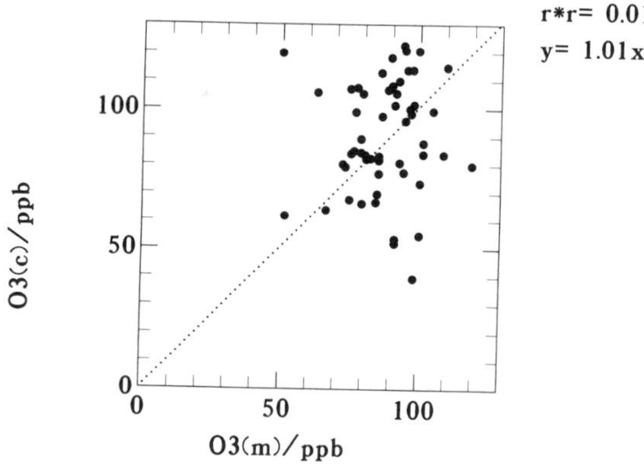

Figure 3. As fig. 1 for western Germany. r^2=0.01, standard deviation 23.0 ppb (26 %).

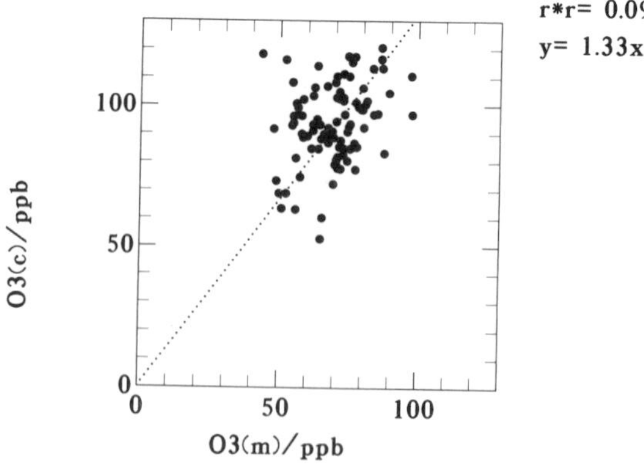

Figure 4. As fig. 1 for eastern Germany. r^2=0.09, standard deviation 28.6 ppb (41 %).

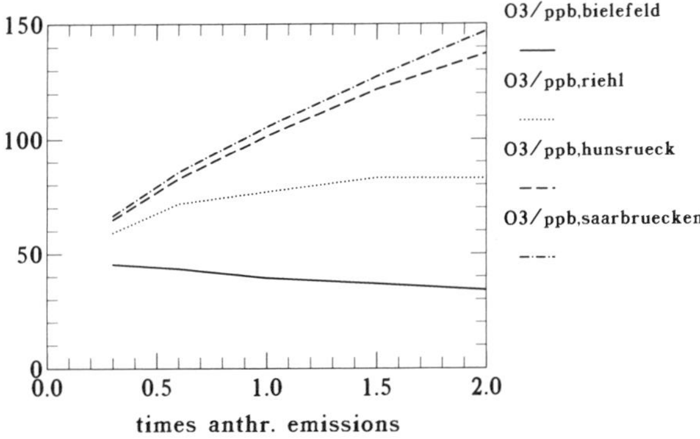

Figure 5. Calculated O_3 in ppb versus factor of change for anthropogenic emissions.

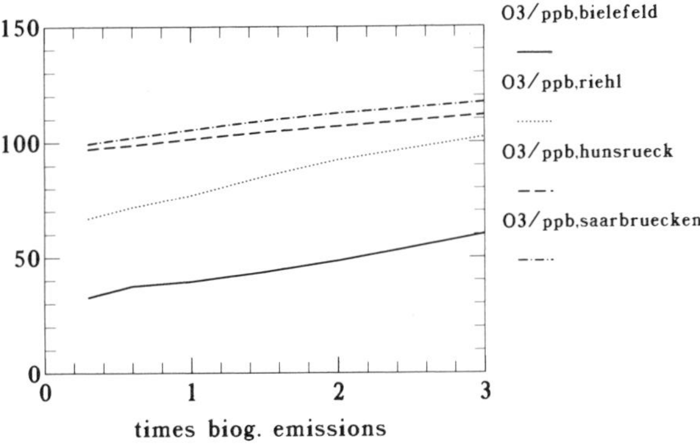

Figure 6. Calculated O_3 in ppb versus factor of change for isoprene emissions.

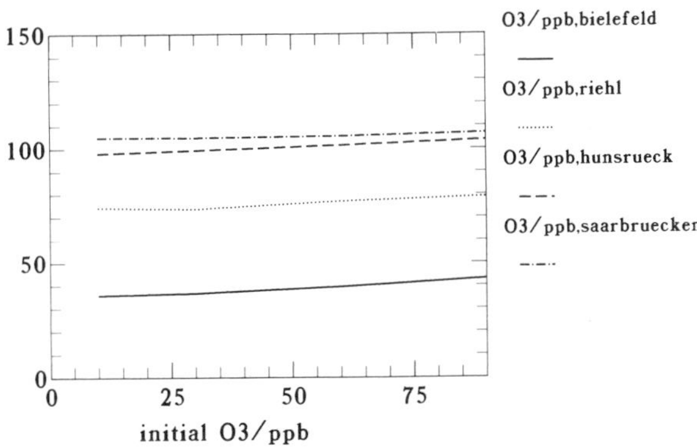

Figure 7. Calculated O_3 in ppb versus initial O_3 in ppb.

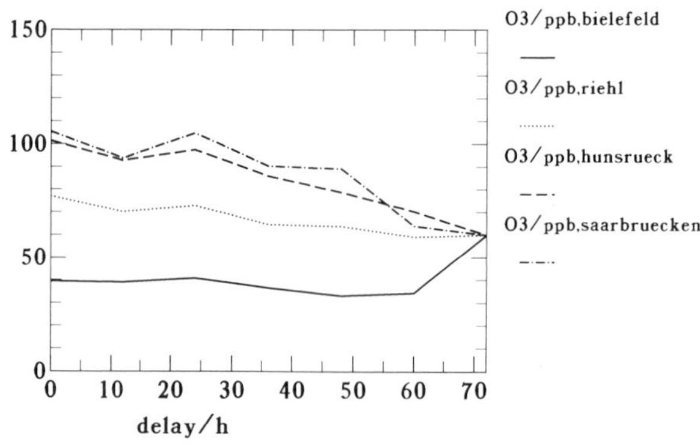

Figure 8. Calculated O_3 in ppb versus delay of start of simulation in hours (calculated in 12 hour steps).

considered that local effects (NO emissions from traffic at the measurement sites) are likely to lower measured ozone at measurement sites in the cities compared to ozone mixing ratios averaged over the grid size of the emission inventory. The measurement points for ozone are often placed near streets with heavy traffic and only 2 or 3 meters above the ground.

The correlation of measured and calculated abundances of ozone should be affected by local effects, too. However, there is only considerable correlation between measured and simulated values in southern and northern Germany. In the western part of Germany the results of LOOP and measurements sometimes differ by more than a factor of 2 and can be either too high (in some cities near the rivers Rhein and Ruhr) or too low (in rural, mountaineous regions). In the eastern part of Germany the scatter is slightly lower, but ozone is almost always overpredicted.

The sensitivity studies give evidence that the reason for the large scatter in western Germany cannot be consistently wrong emissions or initial values of ozone. If this was the case, variation of these parameters would enable simulated results to agree with measured ozone mixing ratios. For some stations (e.g. Bielefeld) this is never the case. If the first part of the simulation is discarded, the results of the simulations are only weakly altered, even if the simulation is reduced to the last day. That means, differences between the results of different trajectories are due to influences on the last day of the simulation.

One would expect that the errors of backward trajectories near their targets are low except when the wind is very calm. On the last day of the simulation average wind speeds in western Germany were 3 to 4 m/s. Therefore it is unlikely that in the last hours of the simulations severe errors were present in the trajectories. It is possible, though, that trajectory errors in the order of 10 km combined with very inhomogeneously distributed errors of the emissions could lead to large errors in simulated ozone mixing ratios. Geiß et al. (1996) demonstrated that the VOC/NO_x-ratio of the PHOXA emission inventory might be too low in urban areas. This might help to explain the observed discrepancies at some locations.

The variance of the model results unexpectedly is higher than that of the measurements. One would expect that with a low resolution of emissions model results would lack the variance that is shown by the measurements which are subject to local influences. Perhaps the model is unrealistic in that it does not take into account mixing between air traveling on different trajectories. The trajectory box model is not able to simulate the effects caused by wind shear. Vertical wind shear combined with vertical transport could lead to mixing in horizontal direction additional to advection. If this process was efficient enough to cause significant mixing within 10 hours over a distance of several 10 km, it could help to explain the lower variance in the measurements compared to the simulation results.

The sensitivity studies also revealed that other parameters like deposition velocities or turbulent exchange between residual layer and free troposphere are of minor importance for final ozone mixing ratios (5 - 10 % change in simulated ozone mixing ratios). Relative changes in the maximum mixing layer height have the same effect with different sign as changes of emissions. Errors in photolysis frequencies would have a strong, however systematic effect on the ozone mixing ratios. Since cloudiness was weak over all of Germany on July 7, 1995, changing the photolysis frequencies would not remove the scatter in the simulation results in the western part of Germany.

CONCLUSIONS

The Deutsche Wetterdienst has developed a trajectory box model (LOOP). LOOP is a time-efficient tool to predict ozone - currently with small overall bias and a standard deviation of 20 - 30%. The performance of LOOP depends on the region for which simulations are made. Discrepancies between results of LOOP and measurements are interpretated as indications that a higher resolution of emissions and, perhaps, a better representation of wind shear combined with vertical turbulent mixing is needed in order to make LOOP a reliable model for operational ozone prediction. However, LOOP promises to be a simple tool to study some processes leading to high ozone abundances.

ACKNOWLEDGMENTS

We thank the Umweltbundesamt Berlin for kindly providing ozone measurements of the Umweltbundesamt and of the Umweltämter of the Länder. This work, in part, was supported by the Bundesministerium für Bildung, Wissenschaft, Forschung und Technologie of the Federal Republic of Germany under grant 07TFS10/LT1-C.4.

REFERENCES

Derwent, R.G., 1990, Evaluation of a number of chemical mechanisms for their application in models describing the formation of photochemical ozone in Europe, *Atmos. Environ.* 24A:2615.

Geiß, H., Memmesheimer, M., Kley, D., Jakobs, H., Ebel, A., and Hass, H., 1996, Nested simulation of a summer smog episode with the EURAD model: the role of the VOC/NO_x ratio of anthropogenic emission sources, in: *Proceedings of the EUROTRAC Symposium '96,* Borrell, P.M., Borrell, P., Kelly, K., Cvitas, T., and Seiler, W., ed., Computational Mechanics Publications, Ashurst, pg. 497.

Geron, C.D., Guenther, A.B., and Pierce, T.E., 1994, An improved model for estimating emissions of volatile organic compounds from forests in the eastern United States, *J. geophys. Res.* 99:12773.

Hesstvedt, E., Hov, O., and Isaksen, I.S.A., 1978, Quasi-steady-state approximation in air quality modeling,*Int. J. Chem. Kin.*, 10:971.

Stockwell, W.R., Middleton, P., Chang, J.S., and Tang, X., 1990, The second generation regional acid deposition model chemical mechanism for regional air quality modeling, *J. geophys. Res.* 95:16343.

Veldt, C., 1992, *TNO-report* 92-118.

Zimmermann, J. and Poppe, D., 1996, A supplement for the RADM2 chemical mechanism: the photooxidation of isoprene, *Atmos. Environ.* 30:1255.

DISCUSSION

P. SEIBERT: With which wind (from which level) did you calculate the trajectories?

J. ZIMMERMANN: The trajectories were calculated using data from the model layer (constant eta-coordinate) approximately 75 mbar above ground. This layer is seen as being representative of the day time mixing layer.

J. LANGNER: It seems a bit unfair to compare the model calculations with observations often made in urban areas close to local sources of pollutants. Have you tried to make comparisons with a subset of the observations representing more rural conditions?

J. ZIMMERMANN: We are doing research to find an objective way to characterise stations according to the extent of local influences by pollution and to correct model results for this. Since we have just started research on this topic we are not able to make a corrected comparison. However, after the conference I repeated the comparison using only stations that were described by the measuring agencies to be located in rural or remote regions. 138 of 318 stations remained. Only at 4 stations disagreement between measurements and simulation were above 30%, and at 85 % of the stations the agreement was within 10%. The disagreement still was highest in the west and the east of Germany.

ADVANCED AIR QUALITY FORECASTING SYSTEM FOR CHIBA PREFECTURE

Toshimasa Ohara [1], Akihiro Fujita[2], Toshinori Kizu[2] and Shin'ichi Okamoto [3]

[1] Institute of Behavioral Sciences
Honmura-cho 2-9, Ichigaya, Shinjuku, Tokyo 162, Japan
[2] Chiba Prefecture
Ichibacho 1-1, Chiba 260-91, Japan
[3] Tokyo University of Information Sciences
Yatocho 1200-2, Chiba 265, Japan

INTRODUCTION

Chiba prefecture is located in the eastern side of Tokyo Metropolitan Area(TMA), and it is adjacent to the city of Tokyo. In this area, air pollution is a serious environmental issue, which is caused by the pollutants emitted within and around the area. In order to prevent air pollution, a new Air Pollution Monitoring and Management System was introduced by the Chiba prefectural government in April, 1996. This system consists of real-time ambient air quality and source monitoring, data processing system and an air quality forecasting system. Almost 130 air monitoring stations transmit hourly air quality and wind data to the system center. Sixty-five major industries also transmit hourly emission intensity and related source data. The advanced air quality forecasting system can predict future ambient levels of photochemical oxidant(Ox) and nitrogen dioxide(NO_2) up to 48-hours based on the predicted meteorological conditions. Photochemical air pollution warnings are issued when the hourly values of Ox exceed 120ppb and the adverse meteorological conditions are predicted to sustain these levels. When air pollution forecasting warning is issued based on the results obtained from the forecasting system, the emergency information are sent rapidly to the municipalities and to operating factories in order to mitigate the effects of high air pollution.

CHARACTERISTICS OF THE AREA

High Ox and NO_2 concentrations are frequently observed in this area in the summer and winter season, respectively. The location and topography of the TMA and Chiba prefecture are shown in Figure 1. The industrial complexes located in the Tokyo Bay area are the largest in Japan and are the major stationary sources of pollutants(NOx,VOC,SO_2, etc.), while the Tokyo Metropolis is the largest contributor of pollutants from mobile sources. In addition, the topography of the area is very complicated, which leads to complex wind patterns. Air pollution is formed as a result of complex non-linear interaction of meteorology, emission distribution and chemical reactions. Therefore it is very difficult to forecast the high concentration of the air pollution as a function of location and time.

AIR QUALITY FORECASTING SYSTEM

An advanced air quality forecasting system, based on three-dimensional mesoscale meteorological and chemical transport models, has been developed for the Chiba prefecture district so that complex and time-dependent flow field and air pollution concentration distribution can be simulated routinely.

Air quality emergent forecasting situations place many constraints on a numerical simulation model that is not required for other air quality applications. The ability to predict the distribution of air pollution concentrations in a time frame that will allow protective actions to be taken is essential in routine operation. However, much CPU time required for a numerical simulation model calculation has, until recently, precluded application for short-term forecasting. Recent advances in engineering workstations and numerical techniques have prompted the application of a numerical simulation model for air quality forecasting.

A schematic diagram of the computational environment and data flow of the advanced air quality forecasting system are shown in Figure 2. Also, input and output data used in the system are summarized in Table 1. For this system, three kinds of on-line data are used. Firstly, the regional meteorological 48-h forecasts, which is calculated by the numerical weather forecasting model of the Japan Meteorological Agency(JMA), are

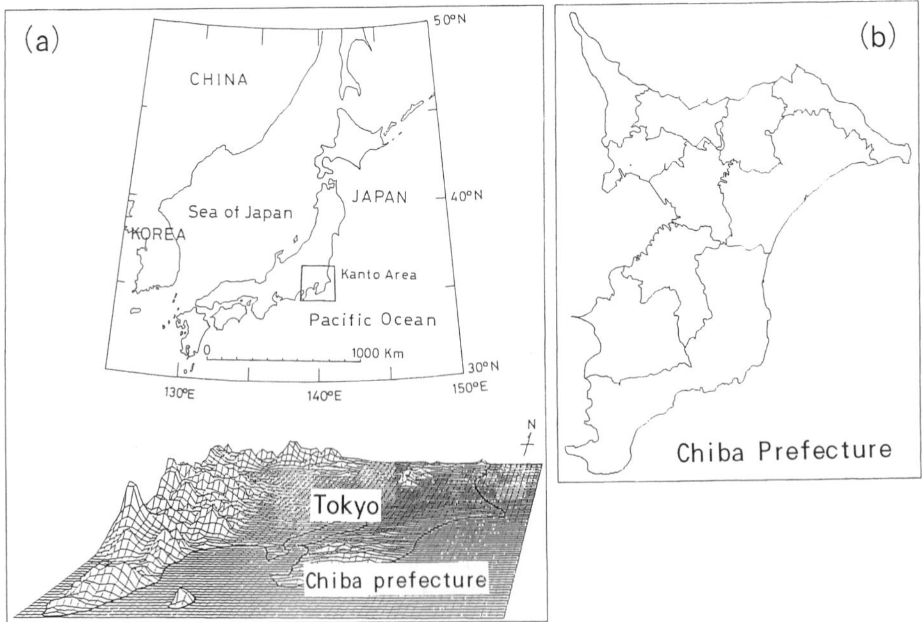

Figure 1. Location and topography of the TMA(Tokyo Metropolitan Area) and Chiba prefecture.

Table 1. Summary of input and output data.

	Content	
Input data		
Large-scale meteorology (JMA forecast model output)	WS, WD, T, RH, pressure, cloud amount, rainfall	15'x12' (surface) 30'x24' (aloft)
Emission	Emission rates(NOx and VOC) and related source data	65 major factories
Air monitoring	Air concentrations(Ox,NOx,VOC etc.) and meteorological data	almost 330 stations
Databases		
Topography	Topography, land/water ratio, land-use, govermental boader	
Air monitoring station	Site, measuring item	
Emission	Emission rate (point /area source), time variation, fingerprint	
Others	Sea surface temp., top concentrations	
Output data		
Meso-scale meteorology	U, V, W, T, RH etc. in (x,y,z,t)	
Air concentration	Ox, NO_2, NOx, VOC species in (x,y,z,t)	
Trajectory	Point, air concentration and emission rate along it	

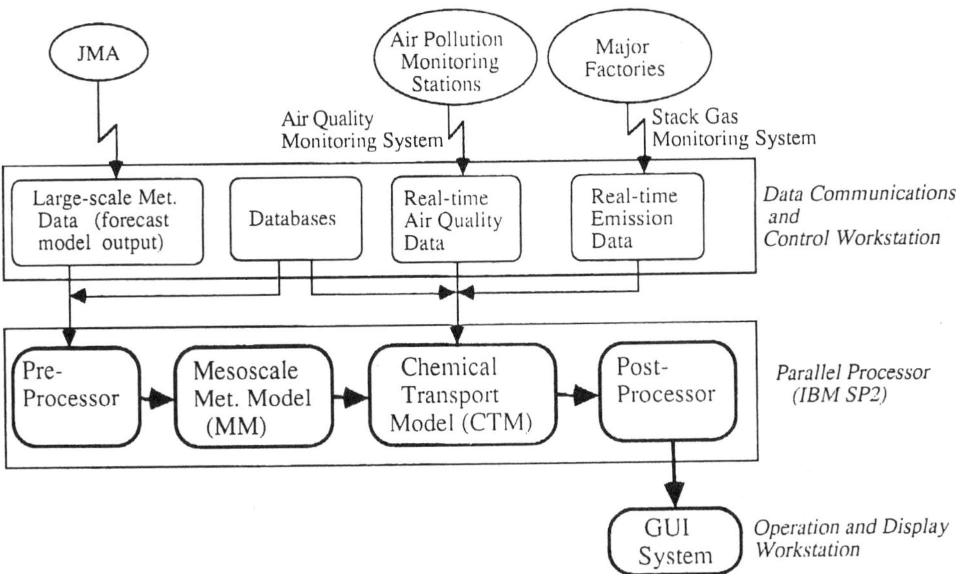

Figure 2. Schematic diagram of the data flow of the advanced air quality forecasting system.

designed to be transmitted once-daily from JMA via meteorological data provider to the communication workstation. They include hourly data for wind direction, wind velocity, temperature, humidity and cloud amount at 3-D grid base. The horizontal spacing is 0.25x0.20 degrees in surface and 0.5x0.4 degrees in aloft, and are used in the present system as the forecasting large-scale meteorological data. Secondly, hourly air quality data collected by air quality monitoring system in the Chiba prefecture and its surrounding area, are transmitted twice-daily from the data center to the workstations. The number of monitoring stations is almost 130 and 200 stations within and around the Chiba prefecture, respectively. Ox, NOx, NO_2 and VOC concentration data are used for determination of initial and lateral boundary conditions of air pollution concentration and for validation of forecasting results. Thirdly, hourly emission intensity and related source data measured at the 65 major industries in the Chiba prefecture, are transmitted once-daily from the data center to the workstations and are employed as time-dependent emission data.

In addition to these on-line data, emission data, topographic data and other data, are prepared as the data-base in the present system and are employed as the input data for forecasting calculation. The annual emission data of NOx and VOC, except the major factories connected on-line, were previously estimated for the Chiba prefecture and its surroundings at 1km grid cells. Hourly emission rates at each grid cell in the model domain are calculated in the forecasting calculation from the total annual emission using assumed patterns for the seasonal and daily variations.

MODEL DESCRIPTION

In this system, three-dimensional mesoscale model(MM) and chemical transport model(CTM) were applied. Figure 3 shows the model domain of MM and CTM. MM using four dimensional data assimilation(FDDA) is used as a meteorological driver to generate 3-D flow and diffusion fields as input to the CTM. The Colorado State University Mesoscale Model(CSUMM: Pielke,1974; Ulrickson and Mass, 1990) was used in this system. The model is hydrostatic and consists of the equations of motion, moisture and continuity within a 3-D, terrain-following coordinate system. It includes a thermodynamic equation, a diagnostic equation for pressure, and a surface heat budget. The forecasting large-scale flow data, which is calculated by the short-period numerical weather prediction model of JMA at the previous day, is used in the FDDA. 3-D pollutant

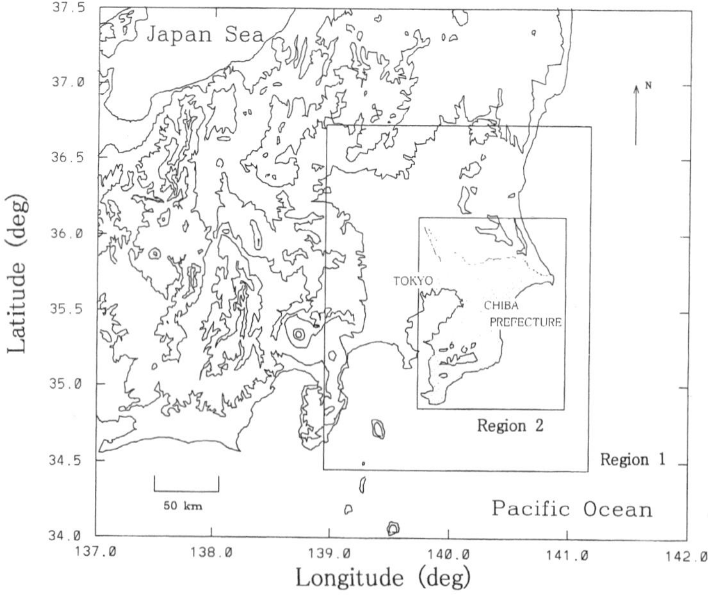

Figure 3. Map of model domain. Outer region indicates MM domain. Region 1 and 2 indicates the CTM domain for coarser and finer grids, respectively.

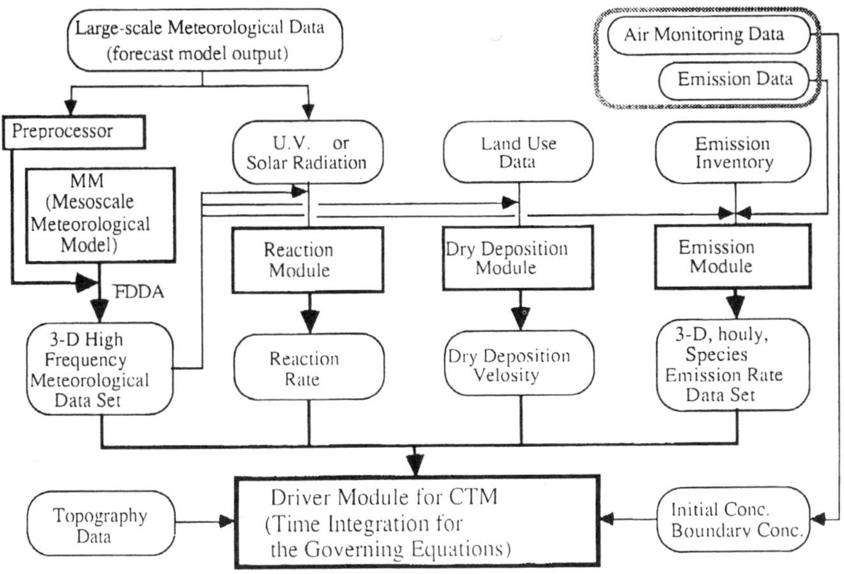

Figure 4. Data Processing flow chart for CTM.

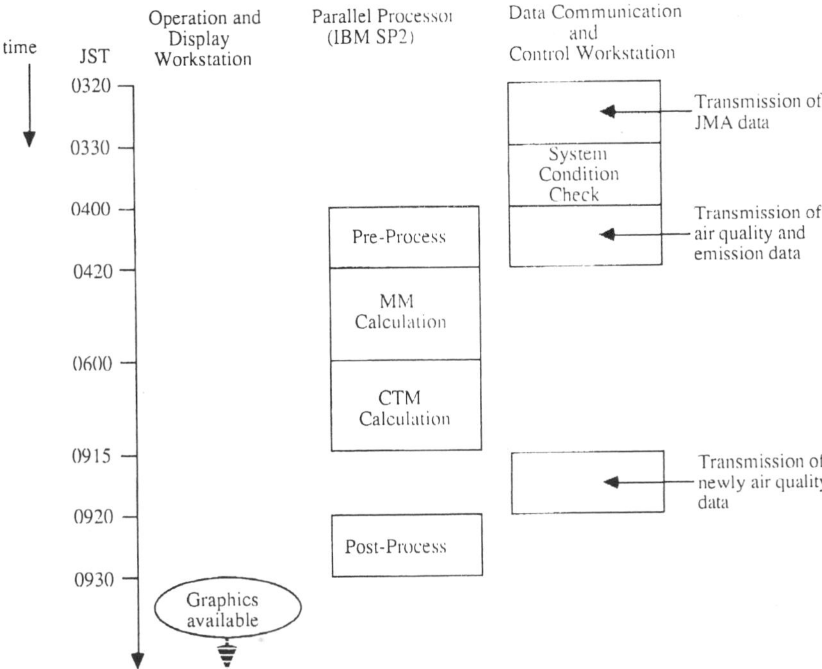

Figure 5. Schematic diagram of the operational procedure of the advancing air quality forecasting system.

distribution is predicted by the newly developed two-way nesting CTM by Ohara et al.(1997). The numerical model includes transport, diffusion, dry deposition, emission and chemical reaction processes. Figure 4 depicts the data processing flow chart for CTM. Dry deposition processes were modeled using a resistance theory of Wesely(1988) in which the surface condition depend on land use data. CBM-IV of Gery et al.(1989) is used as a chemical reaction scheme. Two-way nested grid approach is used in present system. Finer grids were embedded in coarser grids for more detailed representation of advection/diffusion, chemistry, and emissions around the Chiba prefecture.

OPERATIONAL PROCEDURE

Figure 5 shows a diagram depicting the operational procedure of this air quality forecasting system. The present system receives automatically three kinds of on-line data past 0300 Japan Standard Time(JST) every day, and then start a calculation after having checked those data. The system calculates future concentration distributions for photochemical Ox and NO_2 for each hour from 0300JST to 2400JST of the next day. Then it transforms the raw simulation data into the secondary summarized data in post-processing and transmits those to the graphic system. The graphic system using GUI(Graphic User Interface) supplies a variety of judgment data to be necessary for the administrative person to decide emergent response. In order to finish calculating until an administrative judgment at 0930JST every morning, a parallel processor with four node IBM RS/6000 590 research workstations are used for a prediction calculation.

SYSTEM RESULTS

To evaluate the performance of the advanced air quality forecasting system, a simulation was performed in thirty high pollutant periods of 1994. These periods are fourteen high Ox periods in summer season and sixteen high NO_2 periods in winter season.

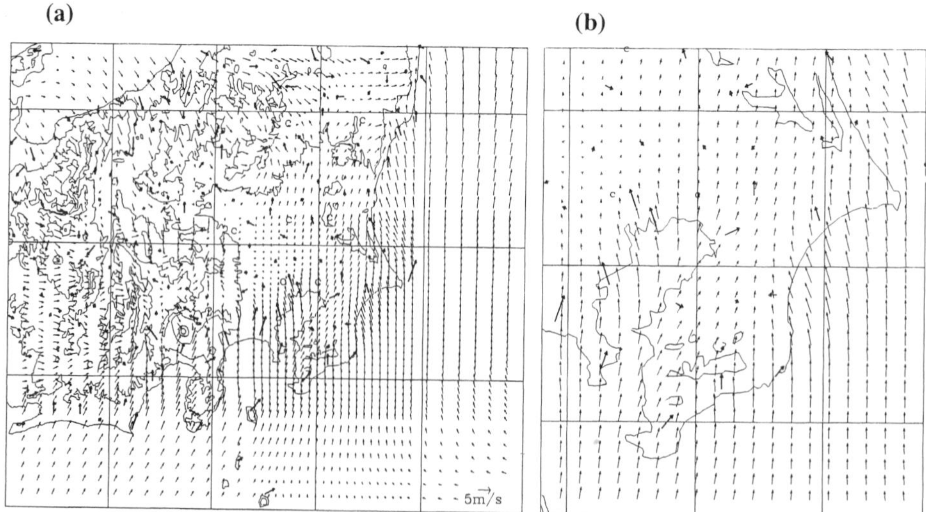

Figure 6. Surface wind vectors in the (a) Chiba prefecture area and (b) whole domain at 1200JST, 7 August 1994. Bold vector indicates the observed wind.

Figure 6 shows an example of the surface wind vector, which is predicted based on the large-scale meteorological data predicted on the previous day, over the whole domain and the Chiba prefecture area at 1200JST on 7 August 1994. The flow pattern are consistent between the observed and predicted results qualitatively. As shown in this figure, it is well possible to predict the surface flow pattern by MM, if the large scale wind results calculated by JMA models are predicted reasonably. In general, the performance of the predicted results for the mesoscale wind pattern, is dependent on the those of the large-scale predicted results.

Figure 7(a) indicates the comparison between the predicted O_3 and observed Ox concentrations for daily maximum in ten sub-regions shown in Figure 1(b) within the Chiba prefecture at summer season. And also, the similar results for the highest daily averaged NO_2 concentrations in ten sub-regions at winter season were obtained as shown in Figure 7(b). The predicted concentration are the values in the first layer(Δz =25m) calculated by the CTM based on the meteorological data predicted on the previous day. For the Ox results, the predicted values are consistent with the observed values. On the other hand, the result for NO_2 is better than Ox. The evaluation score of the prediction results over 120ppb for the daily maximum Ox and 60ppb for the highest daily averaged NO_2, is 85% and 73%, respectively. This difference between the Ox and NO_2 results are in accordance with the difference of the dependence on the meteorological data for both pollutants. Ox is the secondary substance produced by photochemical reactions and its spatial and time variations are strongly dependent on the regional air flow patterns. On the other hand, high concentration of NO_2 were observed near the emission area, because its dispersion was highly dependent on the micro-scale meteorology. Therefore, Ox prediction is more difficult than NO_2. In order to predict the Ox and NO_2 concentrations, it is extremely important to obtain good predicted meteorological conditions. Conversely, it is possible to predict the Ox and NO_2 concentrations, if large-scale meteorological conditions can be predicted with high accuracy.

The routine operation of the present system started from April, 1996. Table 2 gives the Ox results to be forecasted at the previous day in last summer season(April to October). The performance of the forecasting results under the threshold levels of the 120ppb for the daily maximum Ox, is 97%, 38% and 12% for hit, distinctive and grasped score, respectively. As a whole, the forecast concentrations tend to underestimate. On the other hand, Figure 8 indicates the comparison between the forecasted and observed results for the highest daily averaged NO_2 concentrations in ten sub-regions within the Chiba prefecture at winter season. The forecasted concentrations were calculated on the previous day. The relationship between the forecasted and observed concentrations is reasonable, nevertheless there are more and less scattering.

Table 2. Ox results forecasted at the previous day. A number of days, in which the daily maximum Ox concentration in every ten sub-regions shown in Figure 1(b) was over or under 120ppb, were accumulated in forecast and observation combination.

		Observation	
		over 120ppb	under 120ppb
Forecast	over 120ppb	6 (a)	10 (b)
	under 120ppb	44 (c)	1580 (d)

Hit score = (a+d)/(a+b+c+d)=0.97
Distinctive score= a/(a+b)=0.38
Grasped score = a/(a+c)=0.12

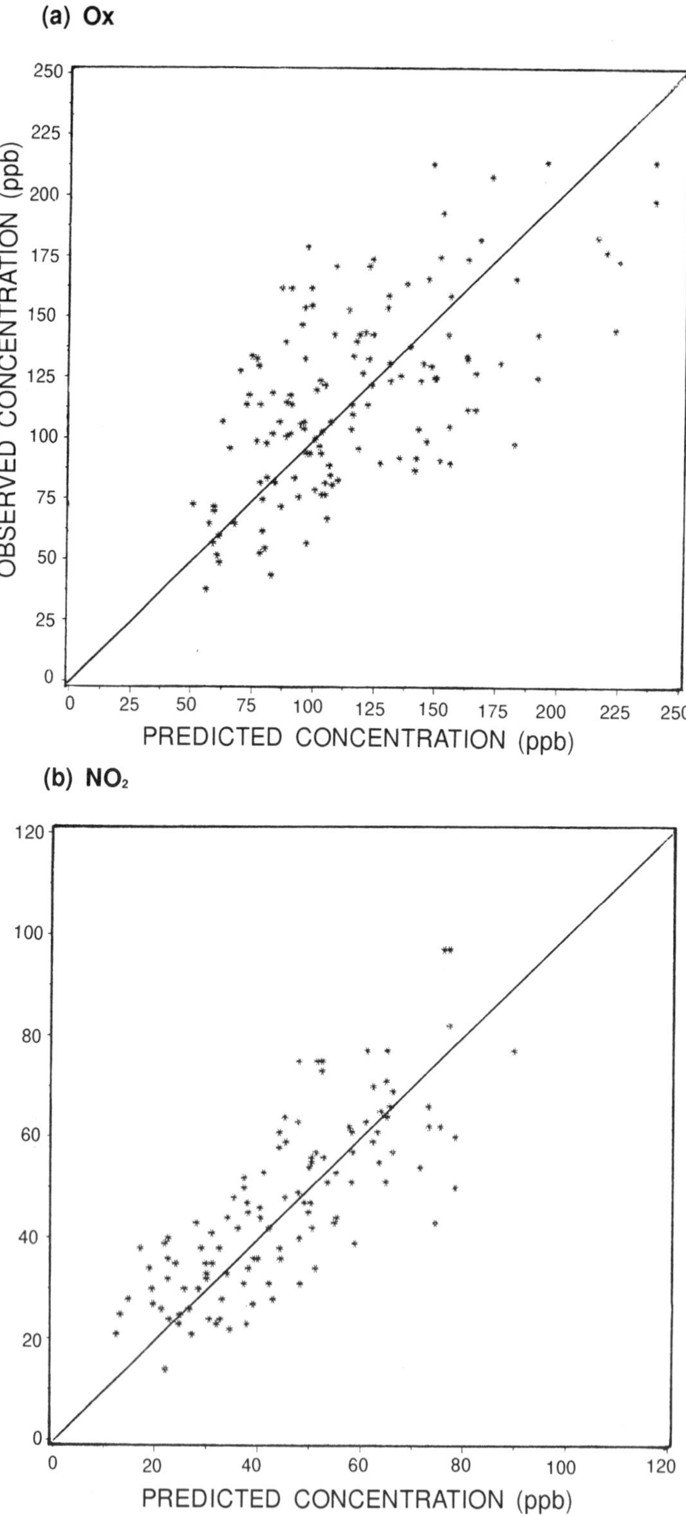

Figure 7. Comparison between the predicted and observed values for (a)the daily maximum Ox concentration at summer season and (b)the highest daily averaged NO_2 concentration at winter season, in ten sub-regions shown in Figure 1(b) within the Chiba prefecture.

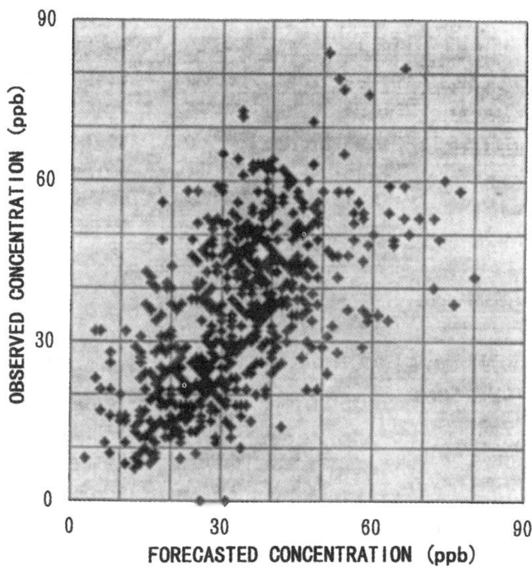

Figure 8. Comparison between the forecasted and observed results for the highest daily averaged NO_2 concentrations in ten sub-regions shown in Figure 1(b) within the Chiba prefecture at winter season.

CONCLUSION

An advanced air quality forecasting system was developed in order to predict the future ambient levels of photochemical oxidant(Ox) and nitrogen dioxide(NO_2) in Chiba prefecture. In this system, the real-time ambient air quality and major emission source data, and the forecasted large-scale meteorological data of the Japan Meteorological Agency were used together with the data-base for emission data, topography data and others. Three-dimensional mesoscale numerical model and chemical transport model were applied.

According to the episode simulation in 1994 and the results of routine operation for one year from April 1996, the system has a potential to be able to forecast the high air pollution reasonably in the case that the large-scale meteorological condition are well forecasted. Hereafter, it is very important to analyze and improve this system on the basis of the these forecasting data storage.

REFERENCES

Gery,M.W., Whitten,G.Z., Killus,J.P. and Dodge,M.C., 1989, A photochemical kinetics mechanism for urban and regional scale computer modeling, J. Geophys. Res., 94, 12925-12956.

Ohara,T., Wakamatsu,S., Uno,I., Ando,T., Izumikawa,S., Kannari,A. and Tonooka,Y., 1997, Development and validation of numerical model for photochemical oxidants., J. Jpn. Soc. Atmos. Environ., 32, 6-28. (in Japanese)

Pielke,R.A., 1974, A three dimensional numerical model of the sea breezes over South Florida, Mon. Wea. Rev., 102, 115-134.

Ulrickson,B.L. and Mass,C.F., 1990, Numerical investigation of mesoscale circulations over the Los Angles basin Part I A verification study, Mon. Wea. Rev., 118, 2138-2161.

Wesely,M.L., 1988, Improved parameterizations for surface resistance to gaseous dry deposition in regional-scale numerical models, EPA 600/3-88/025.

DISCUSSION

D. A. HANSEN: Experience with empirically based, regression type forecasts of next day ozone, using today's ozone and forecasts of tomorrow's weather, in Los Angeles and other cities in the US has shown this approach to be fairly successful. This raises two questions:

1. Have you tried this approach as well as photochemical modeling forecasts and compared the relative skill of the two approaches?

2. Is the photochemical modeling approach worth the extra money and effort in terms of its skill of prediction?

T. OHARA: 1. In this system, statistical forecasts of present day ozone are performed as well as the photochemical modeling approach, and the performance of two approaches are almost similar. However, statistical forecasts of next day ozone have never been tried. We would like to compare the relative skill of the two approaches in the near future.

2. If the photochemical modeling approach can be applied in a forecasting system, it is expected to improve a skill of prediction of next day O_x and NO_2. Furthermore, the approach has an ability to specify an emission control area and make a comparative study of quantitative effects by several kinds of emergent countermeasures, therefore it is possible to determine a reasonable and effective countermeasure supported by a scientific knowledge.

R. SAN JOSE: Can you give more details how you distribute the job into the four nodes of the SP2/IBM parallel machine?

T. OHARA: The parallel implementation are performed only for the integration of the chemistry portion of the CTM, which is the major computational load of the forecasting system. In the chemistry loop of the CTM, the whole model domain are decomposed into four collections having the same number of vertical columns, and then a collection are sent to a node for the chemistry computations.

EVALUATION OF ATMOSPHERIC TRANSPORT AND DISPERSION MODELS IN HIGHLY COMPLEX TERRAIN USING PERFLUOROCARBON TRACER DATA

Mark C. Green

Desert Research Institute
Las Vegas, NV 89119

INTRODUCTION

Project MOHAVE is a large monitoring, modeling, and data analysis study whose main goal is to assess the effects of the Mohave power plant (MPP), a large coal-fired facility in southern Nevada, upon visibility in the southwestern United States, in particular at Grand Canyon National Park (Pitchford and Green, 1997). Additional goals of Project MOHAVE include estimating the effects of other sources upon visibility in the southwestern United States, and evaluating atmospheric transport and dispersion models, and receptors models. One of the key design features of Project MOHAVE was the release of perfluorocarbon tracers (PFTs) at the Mohave power plant and other locations during a 30 day winter intensive study and a 50 day summer intensive study. Tracer and aerosol measurements were made at over 30 locations (mostly 24-hour average concentrations); upper air measurements were made with radar wind profilers, sodars, and rawinsondes; optical monitoring included transmissometers, nephelometers, and time-lapse photography.

Project MOHAVE is conducted as a partnership between the U.S. Environmental Protection Agency and Southern California Edison Company (operator of the MPP). EPA and SCE outlined a protocol for a joint EPA-SCE study; the protocol included a provision that artificial tracer data would not be made available until conceptual models have been formulated and exercised based upon other data collected. This sequestration of data provided for an opportunity for a blind experiment in which source and receptor models were developed and applied prior to their evaluation using tracer data. In this paper, the ability of the transport and dispersion models, as applied, to predict concentrations of tracer released from MPP at two receptor sites in Grand Canyon National Park, during summer 1992 is described.

DESCRIPTION OF STUDY AREA AND TRACER EXPERIMENT

The Project MOHAVE study area, with locations for tracer release and sampling, is shown in Figure 1. The Mohave power plant is located in the lower Colorado River Valley at an altitude of 213 meters MSL. The lower Colorado River Valley runs north-south and is flanked on both sides by mountains and ridges that rise approximately 1000 meters above the river. The valley is relatively flat for a few kilometers east-west, then rises first gradually, then abruptly to the ridges. The horizontal separation of the ridges on either side of the valley is typically 30-40 kilometers near the Mohave power plant, and gradually narrows to about 10 km at about 90 km north of the power plant. At 90 km north of MPP, the Colorado River is dammed (Lake Mead) and backs up to the east until the western boundary of Grand Canyon National Park. The Grand Canyon receptor site of Meadview (MEAD), 100 km NNE of MPP, is located on a small mesa at 900 meters MSL at the eastern end of Lake Mead, and is about halfway in elevation between the lake and a plateau to the east. The Grand Wash Cliffs rise 500 m abruptly a few km east of Meadview. The other Grand Canyon receptor site Hopi Point (HOPO), 280 km NE of MPP, is located on the south rim of the Grand Canyon, on the Coconino Plateau at an elevation of 2160 m MSL.

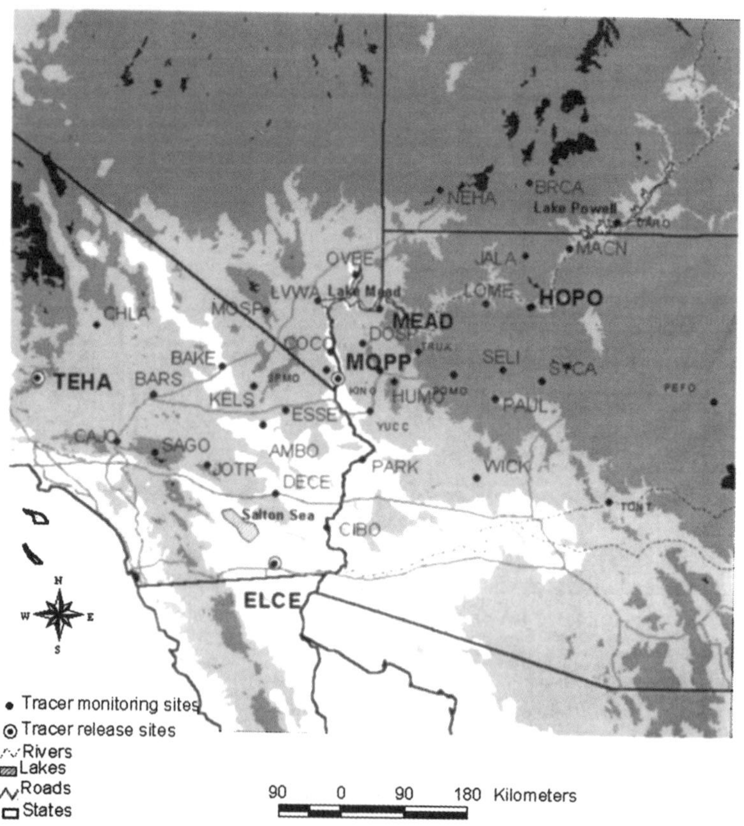

Figure 1. Project MOHAVE study area, including locations for tracer release and sampling.

The PFT ortho-perfluorodimethylcyclohexane (oPDCH) was released continuously from the stack of MPP during the 30-day winter and 50-day summer intensive study periods. Forty-five percent of the oPDCH consists of the isomer ortho-cis (oc) PDCH, which has a background of about 0.5 parts per quadrillion (10^{15}) or femtoliters per liter (fL/L). The tracer was released from the MPP stack at a rate proportional to power production. This allows the calculation of the amount of sulfur from the MPP associated with a given tracer concentration. During the summer study, tracer emission rates were nearly constant over most of the study period.

The PFT samples were over 24-hours at most sites; the Meadview and Hopi Point sites had 12-hour samples. In addition, at Meadview 15-minute samples are available for a three week period. The samples were drawn onto a tube packed with absorbent; during analysis the tube is heated, desorbing the PFTs for analysis by electron capture gas chromatography. The sampling and analysis methodology is described in detail by Dietz (1987).

The quality of the PFT data was assessed using data from collocated samples at Meadview (2 samplers) and Hopi Point (3 samplers). The correlation coefficient among the collocated data was 0.99 at Meadview and 0.88 at Hopi Point. Because of the greater distance of Hopi Point from MPP (and less frequent transport), more ocPDCH concentrations were near the detection limit, resulting in a lower correlation coefficient. The root-mean-square error (RMSE) between the collocated samples was 0.074 fL/L at Meadview and 0.051 fL/L at Hopi Point. The RMSE values are a measure of the uncertainty of the tracer concentration, which must be considered when evaluating the transport and dispersion models against the tracer data. Because the tracer concentrations during episodes were an order of magnitude greater than the uncertainty, the tracer data are useful for model evaluation.

DESCRIPTION OF TRANSPORT AND DISPERSION MODELS USED

Desert Research Institute(DRI)/Colorado State University(CSU) and DRI/semi-Gaussian

A three-dimensional second-moment closure mesoscale meteorological model (Enger et al., 1993; Enger and Koracin, 1994)) provided the meteorological fields with a telescoping horizontal grid resolution ranging from 700 m to about 7 km. The meteorological model was used to create a library of meteorological fields covering summer conditions. For every 10 degree increment in wind direction, for each of three geostrophic wind speeds (3, 7, and 12 m s^{-1}) at 700 hPa diurnally varying meteorological fields were generated. Radar wind profiler data were then used to assign each half-hour to the "library" wind field most closely matching the observations. The meteorological model assumptions include hydrostatic equilibrium, no condensation or clouds, and idealized typical thermal profiles.

The meteorological fields provided input for two dispersion models: a Lagrangian particle dispersion model (Uliasz and Pielke, 1993); and a semi-Gaussian trajectory-type dispersion model (Enger, 1990) both of which utilize turbulence quantities simulated with the meteorological model.

VISHWA

The VISHWA model (**Vis**ibility and **H**aze in the **W**estern **A**tmosphere) is an Eulerian grid based air quality model developed to predict concentrations of sulfate and other visibility affecting compounds at Grand Canyon National Park and other southwestern U.S. national parks. Its development and use was sponsored by the Electric Power Research Institute (EPRI) and a consortium of western U.S. utilities. The VISHWA model is a variation of the Acid Deposition and Oxidant Model (Venkatram, et al., 1992). VISHWA used meteorological fields obtained from Colorado State University, which used the Regional Atmospheric Modeling System (RAMS). The RAMS wind fields, which were available with 12 km grid spacing were processed by a mass consistent diagnostic wind field model to a grid with 10 km spacing.

HAZEPUFF

HAZEPUFF is a Lagrangian puff model that calculates transport, dispersion, dry deposition, oxidation of SO_2 to sulfates, light scattering, and contrast change resulting from the release of emissions from a point source and the subsequent impact at a receptor. The model approximates the release of emissions by a series of puffs which are transported by the prevailing wind patterns. HAZEPUFF used interpolation of upper air wind measurements (by rawinsondes and radar wind profilers) made in support of Project MOHAVE.

RESULTS

Meadview

Graphical comparisons of predicted and observed tracer concentration will be presented first, followed by a brief statistical comparison of predicted and observed concentrations. Time series plots of predicted and observed concentrations of ocPDCH at Meadview are shown in Figure 2.

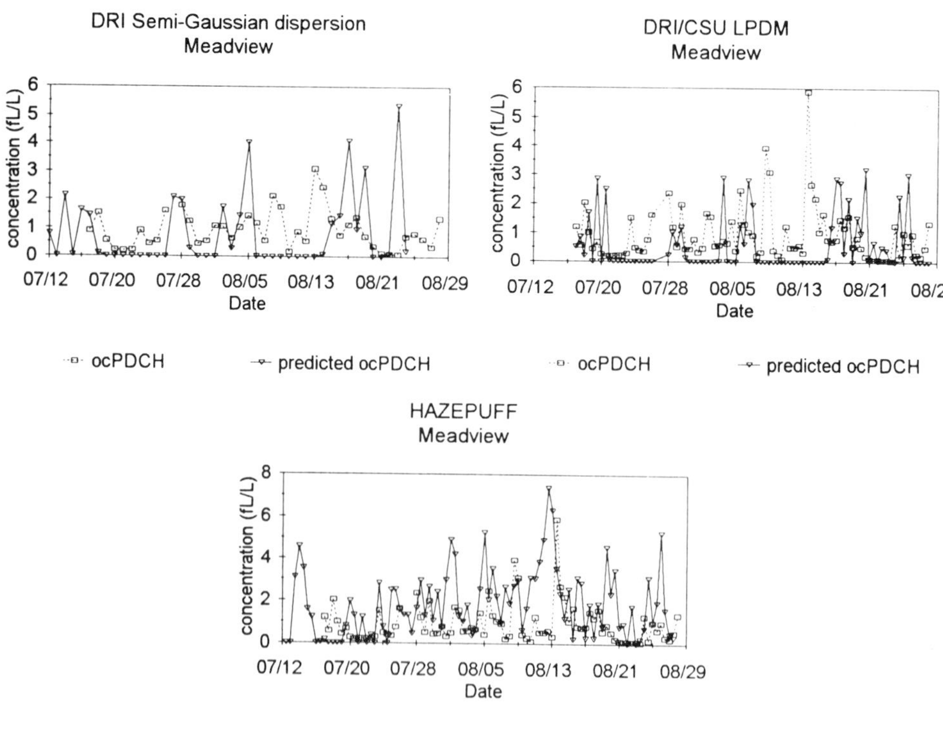

Figure 2. Time series of predicted and observed ocPDCH concentrations at Meadview.

Table 1. Summary statistics of model performance.

	mean	standard deviation	coefficient of variation	bias	r	RMSE	RMSE/ average ocPDCH
Meadview 12 hour averages							
observed ocPDCH	0.91	0.93	1.02				
DRI/CSU	0.55	0.93	1.71	0.60	-0.04	1.39	1.53
HAZEPUFF	1.88	1.60	0.85	2.07	0.14	1.99	2.18
Meadview 24 hour averages							
observed ocPDCH	0.95	0.68	0.72				
DRI/semi-Gaussian	0.77	1.40	1.81	0.81	0.00	1.56	1.65
Hopi Point 12 hour averages							
observed ocPDCH	0.20	0.17	0.83				
VISHWA	0.24	0.41	1.74	1.19	0.30	0.40	1.97
HAZEPUFF	0.57	0.68	1.21	2.83	0.43	0.73	3.62

The DRI/semi-Gaussian dispersion model (24 hour concentrations) predicted extended periods with no tracer at Meadview, which was not consistent with observations. The period 08/05 through 08/13 had zero predicted tracer, yet the observations showed elevated tracer concentration on all but one day, including the highest concentration measured during the study. High predicted tracer concentrations did not generally correspond to high measured concentrations, either. The DRI/CSU (Lagrangian particle dispersion model) model (12 hour predictions) showed many of the same periods with essentially zero predicted tracer, in contrast to the observations. Predicted peaks in ocPDCH did not match measured peaks. The HAZEPUFF results did not show as many periods with near zero tracer concentrations as the other models. The HAZEPUFF predictions were in general substantially higher than measured concentrations. As with the other models, there does not appear to be much relationship between measured and predicted concentrations. Table 1 presents some summary statistics of model performance. For observations and predictions, the mean concentration, standard deviation, and coefficient of variation (standard deviation/mean) are presented. For the model predictions the bias (average of predicted/average of observed), correlation coefficient (r), root-mean-square error (RMSE), and normalized RMSE (RMSE normalized by average observed concentration) are presented. For comparison, the RMSE of collocated observations at Meadview was 0.07 fL/L. The statistics for the observations are computed separately for the 12 and 24 hour averaged concentrations. For the 12 hour periods, only those cases with results from both models and with valid observations are included; for the 24 hour period, only cases with model results and valid observations are used.

The correlation coefficients between predicted and observed concentrations at Meadview range from -0.04 to 0.18; thus 0 to 2 percent of the variance is explained by the models. The biases show that the HAZEPUFF model had a tendency to overpredict while the DRI/CSU model, and to a lesser extent the DRI/semi-Gaussian dispersion model tended to underpredict. The DRI/CSU and DRI/ semi-Gaussian dispersion models had coefficients of variation about twice as high as the observations, demonstrating a difference in the shape of their frequency distributions from the frequency distribution of the observations (the models predicted mostly zero or high values, not enough moderate values). The HAZEPUFF model had a similar coefficient of variation as the observations.

Hopi Point

Time series plots of observed and predicted concentrations of ocPDCH at Hopi Point are shown in Figure 3. Both the HAZEPUFF and VISHWA models have higher peak predictions than were observed. The summary statistics are shown in Table 1. Again, the HAZEPUFF model tended to overpredict concentrations, with a bias of 2.83. Average concentrations predicted by VISHWA were close to the observed (bias of 1.19); however the coefficient of variation for VISHWA was about twice that observed. Thus, VISHWA had more very low concentration predictions and higher maximum concentrations than the observations. The coefficient of variation for HAZEPUFF was about 50% greater than for the observations. Correlations between predicted and observed concentration were better than at Meadview, with values for r of 0.43 for HAZEPUFF and 0.30 for VISHWA, accounting for 18 and 9 percent of the variance in observed concentrations, respectively.

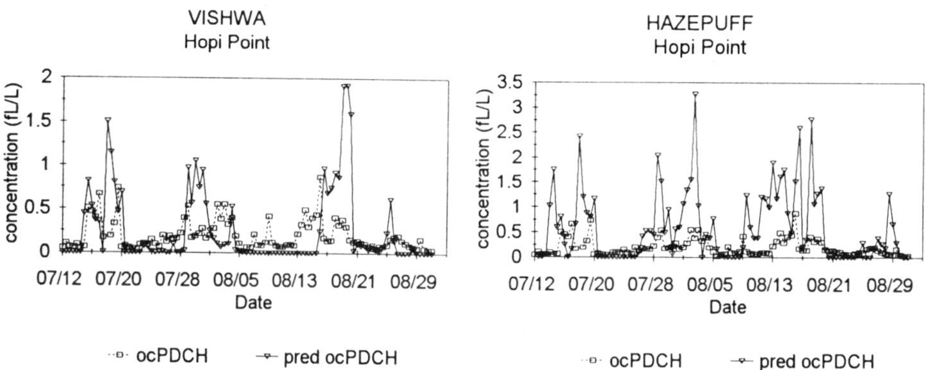

Figure 3. Time series of predicted and observed ocPDCH concentrations at Hopi Point.

DISCUSSION

In this section, some discussion into possible reasons for the poor model performance is presented. Additionally, the usefulness of the models from a clients' perspective is considered. The study area, in particular between the Mohave power plant and Meadview, is in very highly complex terrain; for models to adequately resolve the terrain features and to correctly account for localized slope flows, channeling, etc. is indeed challenging. Inspection of average predicted concentrations for the DRI/semi-Gaussian dispersion model showed strong channeling along the Colorado River with transport mainly to the north. The Meadview site has low average predicted concentrations, being east of the area with high predicted concentrations. The observations show that Meadview is frequently impacted by the MPP emissions; although the tracer observations are somewhat sparse, average concentrations appear to decrease significantly to the east of the Meadview site. The model does show some predicted concentrations higher than any observed. The model appears to be underestimating dispersion. The near zero correlation between model predictions and results suggest other problems as well. Rather than running the model for each hour of the study period, a library of simulated meteorological fields with assumed thermal profiles was used; this methodology may also be contributing to the differences between observed and predicted concentrations. Uncertainties in plume rise on a given day could lead to substantial transport errors because there is often significant vertical wind shear within the range of plume heights. Because the same meteorological inputs were used for the DRI/CSU model, similar explanations may be appropriate for its poor performance.

The VISHWA model used winds based upon the CSU/RAMS model at 12 km resolution. The CSU RAMS winds were evaluated against upper air measurements by the meteorology subcommittee of the Grand Canyon Visibility Transport Commission (GCVTC, 1995). The subcommittee found that the winds met their minimum acceptance criteria of being within 30 degrees in wind direction and within 4 m s^{-1} in wind speed, compared to observations, for less than 50% of the cases. Two conclusions were reached; 1) the model showed an excess of southwesterly winds at the expense of other directions, especially southerly; and 2) the model did not properly account for the effect of channeling of the flow by terrain. The lower than measured frequency of southerly flow, especially below mountain ridges may have been caused by inadequate resolution of the terrain features due to the coarseness of the model grid in relation to the topographic features.

The HAZEPUFF model performed simple linear interpolation of a few upper air wind measurements; thus terrain was accounted for only to extent that it is reflected in the wind data used. The model also appears to have underestimated dispersion. Twenty-four hour averaged concentration predictions compared better with observations than did 12 hour predictions, indicating a possible bias in transport time.

From a clients' perspective, the models were unsatisfactory. Even though it may seem unfair or overly strict to expect high correlations between model predictions and observations paired in space and time, this is precisely what the users of the results (clients) need. For Project MOHAVE and the Grand Canyon Visibility Transport Commission, it is desired to know on a daily basis what major sources or source areas were impacting a receptor. If a day is particularly polluted, it is not sufficient to know only the general direction of transport; to most effectively improve air quality at the location of interest, it is necessary to know which sources of pollution would be most helpful to control. If tracer data were not available for Project MOHAVE, we would have been faced with a wide range of predicted impacts from MPP, none of which were correct. Had we used only one model and naively assumed it was correct, we would have made erroneous conclusions. For the GCVTC, because the windfield evaluation showed that the windfields usually failed the acceptance criteria, seasonally averaged estimates of impacts were used rather than daily or short-term impacts. This is not the desired approach; visibility is an instantaneous event; an observer cannot average the view over a season, but the model performance limited the assessment to seasonal effects. Even then, it was not established that the seasonal averages were unbiased.

It is not practical to expect to use tracer studies every time an assessment of transport and dispersion is needed. However, for Project MOHAVE, only by using the tracer data will it be possible to realistically assess the impact of the Mohave power plant on visibility at the Grand Canyon. Other challenges are involved in the assessment, especially determining the conversion of SO2 to sulfate and accounting for deposition. The tracer data has provided the critical transport and dispersion elements, a task for which the models failed.

SUMMARY

Transport and dispersion models used to predict perfluorocarbon tracer concentrations at key receptors at Grand Canyon National Park performed poorly. Model predicted episodes showed little relation to observed episodes. The models also appeared, in general, to underpredict dispersion. From a clients' perspective, the models performed unacceptably; in contrast, the tracer data was of high quality and useful for demonstrating transport and dispersion of emissions from the Mohave power plant.

REFERENCES

Dietz, R.N., 1987, Perfluorocarbon tracer technology. In Regional and Long-range Transport of Air Pollution, Lectures of a course held at the Joint Research Center, Ispra, Italy, 15-19 September 1996, Elsevier Science Publishers B.V. Amsterdam, 215-247.

Enger, L., 1994, Simulation of dispersion in moderately complex terrain - Part C: A dispersion model for operational use. *Atmos. Environ.*, 24:2457.

Enger, L., Koracin, D., and Yang, X., 1993, A numerical study of the boundary layer dynamics in a mountain valley. Part 1: Model validation and sensitivity experiments, *Bound. Layer Meteor.*, 66:357.

Enger, L., Koracin, D., 1994, A numerical study of the boundary layer dynamics in a mountain valley. Part 2: Observed and simulated Characteristics of the atmospheric stability and local flows, *Bound. Layer Meteor.*, 69:249.

GCVTC, 1995, Evaluation of wind fields used in the Grand Canyon Visibility Transport Commission analysis. Available from the Western Governors Association, Denver, CO.

Pitchford, M., and Green, M., 1997, Analyses of sulfur aerosol size distributions for a forty-day period in summer 1992 at Meadview, Arizona, *J. Air & Waste Manage. Assoc.*, 47:136.

Uliasz, M., and Pielke, R.A., 1993, Implementation of a Lagrangian particle dispersion model for mesoscale and regional air quality studies, In *Air Pollution*, P.Zenetti, Editor, Computational Mechanics Publications, Southhampton, 157-164.

Venkatram, A., Karamchandani, P., Pai, P., and Goldstein, R., 1992, The development of the Acid Deposition and Oxidant Model (ADOM). Environ. Pollut., 75:189.

DISCUSSION

D. STEVENSON: What was the total release of perfluorocarbon tracers? Do you feel justified in using such an environmentally unfriendly tracer?

M. C. GREEN: The total release of the perfluorocarbon tracers was 1000 kg or 1 metric ton. This included releases from 2 sites in the winter for 30 days each and 3 sites in the summer for 50 days each. While PFCs are strong greenhouse gases, the amount released is insignificant. In the U.S. alone in 1997, an estimated 27,000 metric tons of PFCs and HFCs (hydrofluorocarbons) were released to the atmosphere.

S. GONG: In your tracer experiment, have you measured the tracer concentrations in both the gas and particle phases? It might provide valuable information to simulate tracer concentrations?

M. C. GREEN: At the ambient conditions, we expect that all the tracer material at the sampling sites was in the gas phase. We would not expect condensation of the tracer or reaction with particles.

J. H. SØRENSEN For the second ETEX tracer gas experiment, the models over-predicted by a factor of ten or more. This is not explained (yet) but is probably due to a cold front passing during the release. Did you experience similar meteorological conditions in your experiment? And if so, how did the models respond?

M. C. GREEN: As this study was conducted in mid-summer, no substantial cold front passages occurred.

UNCERTAINTY ESTIMATION OF AN AIR POLLUTION MODEL

Renata Romanowicz, Helen Higson, Ian Teasdale and Ian Lowles

Westlakes Research Institute
Moor Row
Cumbria, CA24 3LN
United Kingdom

INTRODUCTION

This paper addresses the problem of uncertainty analysis of the predictions of an atmospheric dispersion model. The proposed method uses Bayesian conditioning of the predictions on the available observations and enables consideration of the uncertainties which influence model predictions such as uncertainty of observations and limitations of the model. As an illustration, the proposed methodology is applied to a simple, Gaussian short range atmospheric model (Clarke, 1979). This is commonly referred to the R91 methodology, after the name of the report on which it is based, and shall be referred to as such throughout this paper. The choice of the model was dictated by its availability, simplicity and widespread use. However, the proposed approach has much wider application and can be used for different air dispersion models such as UKADMS (Carruthers et al., 1992) or any other model with comparatively short times of computations and for which observations are available. The simplifications used in dispersion modelling are very substantial and comprehensive data are very limited. This means that it is important to make best possible use of the information which is available - in the form of distributions, value ranges, specialist opinions and common sense. Also many uses in nuclear safety rely on the concept of risk and it is important that uncertainty modelling is applied at all.

The resulting uncertainties of predictions can be presented in the form of confidence limits or can have the form of predictive posterior probability densities. The aim of the work is to assess the uncertainty of the predictions following both from the measurement errors and the simplifications used in the model. The proposed methodology permits all those uncertainties in the derivation of the model predictions to be accounted for and can handle multiple, complex interactions in the model structure.

It should be stressed here, that the aim of this work is not the comparison or validation of the model performance but the assessment of the uncertainties involved in the model predictions, originating from the uncertainty of the available observations and the limitations of the model. To date the only uncertainty analysis related to dispersion models reported in the literature concerned very marginal sensitivity analysis of model responses to the changes of model input parameters made during the model inter-comparison exercises when analysing the reasons for the differences between the model predictions (Olesen, 1995).

ATMOSPHERIC DISPERSION MODEL AND DATA

R91 is a simple and widely used Gaussian plume model applicable for short to medium range dispersion up to approximately 30 km. The model assumes that the dispersion of material is described by a Gaussian distribution characterised by standard deviations in both the horizontal and vertical directions. The standard deviations, better known as diffusion parameters, are mainly dependent on the atmospheric stability. The R91 model uses Pasquill's (1961) scheme for describing atmospheric stability which gives six stability categories ranging from A, very unstable, through to F, very stable, based on measurements of wind speed, cloud cover, time of day and time of year. An estimate of the mixing layer height is assigned to each stability category and it is used by the model when predicting the effects on ground-level concentrations caused by multiple reflections of the plume at both the surface and the top of the mixing layer.

The main factors influencing the diffusion process in the atmosphere are not directly measurable. This is a turbulent phenomenon: the basic theoretical model can not be solved in a closed form. Therefore a similarity analysis is performed which relies on a number of scaling parameters (friction velocity, Monin-Obukhov length) which are not accessible to experiment directly. A simpler empirical model developed by Pasquill and modified by Smith (Pasquill and Smith, 1983) describes those processes using easily measurable atmospheric variables. Those relations have the form of 6 categories covering all meteorological conditions and are related to wind speed at 10m height. The error in the choice of the weather category is regarded as the main source of error of the predictions as it leads to different model behaviour. The determination of cross-wind spread - the standard deviation of the concentration distribution in the horizontal direction is another source of error, as the corresponding relations used in the model are derived from experiments conducted in flat terrain from elevated sources and the variables involved depend on the sampling times and height of measurements. In spite of these structure-dependent errors there are also expected measurement errors, originating from the variability of the processes in time and space, or the necessity of introducing some data prediction where there is a lack of appropriate measurements. The other essential source of errors comes from the very limited range of data on the simulated variables, thus not allowing proper calibration of the model empirical parameters. In this respect the observations should also be treated as a random process (Lee and Irwin, 1995).

The experimental dataset used for this study is the Copenhagen dataset selected from the Model Validation Kit used for recent model inter-comparison exercises (Olesen, 1995). The data consist of 23 sets of ground level concentration measurements from 1 hour releases of SF_6 tracer. The experiment was performed in neutral and unstable meteorological conditions. The tracer was released without buoyancy from a tower at height of 115 m. The measurements were taken at 10 distances from the source varying from 1.9 km to 6 km, in up to three cross-wind series of tracer sampling units. The value of roughness coefficient was estimated as 0.6 m. Meteorological measurements included vertical profiles of wind speed at 10 and 100 meters height.

BAYESIAN MONTE CARLO UNCERTAINTY ESTIMATION

The proposed methodology for uncertainty assessment of an atmospheric dispersion model is based on a Bayesian Monte Carlo approach. This method was used to evaluate the uncertainty of the predictions in hydrology as Generalised Likelihood Uncertainty Estimation (GLUE) methodology (Beven and Binley, 1992, Romanowicz et al., 1994). It reformulates the inverse problem into the estimation of posterior probabilities of model responses and uses Monte Carlo grid sampling of the parameter space. The method is based on Bayes formula (Box and Tiao, 1992):

$$f(\theta \mid \mathbf{z}, D) = \frac{f(\theta)\, f_L(\mathbf{z} \mid \theta)}{f(\mathbf{z}, D)} \qquad (1)$$

where $\mathbf{z}$ is the vector of observations, $f(\theta|\mathbf{z},D)$ is the posterior distribution (probability density) of sets of the randomly chosen parameter values θ given the output data and input variables, $f(\theta)$ is the prior probability density of the parameter sets and input variables together, $f(\mathbf{z},D)$ is a normalisation factor and $f_L(\mathbf{z}|\theta)$ represents a likelihood measure for z given the parameter set θ obtained from the forward modelling. In this formulation we assumed that all the errors involved with input observations can be modelled and incorporated into the forward modelling. Any parameters needed to model input errors are enclosed in the parameter vector θ and treated in the same way. The same applies to errors in the observation of output variables.

Equation (1) can be applied sequentially as new data become available and the existing posterior distribution, based on (N-1) calibration sets, is used as the prior for the new data in the Nth calibration set. Thus

$$f(\theta|\mathbf{z}_1,...,\mathbf{z}_N,D_1,..,D_N) \propto f(\theta|\mathbf{z}_1,...,\mathbf{z}_{N-1}\ D_1,..,D_{N-1})L(\theta|\mathbf{z}_N,D_N) \qquad (2)$$

where $L(\theta/\mathbf{z}_N,D_N)$ is the information about θ from the Nth calibration set.

Errors between the observations and simulation results together with the assumed prior distributions of parameters are used to build the posterior likelihood, reflecting the model performance (Romanowicz et al, 1996). In this way it is possible to incorporate the information from observations from different time periods and/or sites using the Bayesian updating of equation (2). Models with parameter sets that are not considered to fit the data will be given a low or zero posterior likelihood value. The predictions of the remaining (retained) models are weighted by the posterior likelihood associated with that parameter set.

An important step in this procedure is the choice of an adequate model for the prediction errors which would reflect their statistical features and allows their transformation into independent residual errors required in the Bayesian updating. For independent errors the evaluation of a posterior probability density function consists of the T multiplications of the conditional probability function for each observation of the error

ε_t given previous data, where T is the number of independent simulated events (different time moments or spatial locations).

Under these assumptions the likelihood function is defined as:

$$f(\theta,\phi|\mathbf{z}) = C \times f(\theta,\phi) \times (2\pi\sigma^2)^{-T/2} \exp[-\frac{1}{2\sigma^2}\sum_{t=1}^{T}\varepsilon(\mu,\theta)^2] \quad (3)$$

where **z** denotes the observations (concentrations and wind speed at 10 m), C is the normalising constant and $f(\theta,\phi)$ denotes the prior distribution of the parameter sets.

The cumulative distribution of the error term at any time, given a particular set of dispersion and statistical model parameters θ and ϕ, is then given by:

$$P(\delta_t < \delta|\theta,\phi) = \Phi\left(\frac{\delta-\mu}{\sigma}\right) \qquad t=1,\ldots,T \quad (4)$$

where Φ is a standard normal distribution function N(0,1).

The resulting predictive distribution of tracer concentrations $\mathbf{Z}_t$ conditioned on the calibration data **z** in the discrete (parameter set) case, with additive error ($\delta=y-y_t$, where y and y_t are realisations of simulated and observed concentrations respectively), will be then given by

$$P(Z_t < y|\mathbf{z}) = \sum_{\theta}\sum_{\phi}\Phi\left(\frac{y-\mu-y_t}{\sigma}\right)f(\theta,\phi|\mathbf{z}) \quad (5)$$

From the relation (5) one can evaluate the confidence limits for the concentrations at the observation sites. Note that in this procedure the parameters only need to be treated as sets of values. Any parameter interactions are then reflected implicitly in the calculated posterior distributions. The marginal distributions for individual parameters or groups of parameters can be calculated by an integration of the posterior distribution over the rest of the parameters if needed (Romanowicz et al., 1994).

APPLICATION TO AN AIR POLLUTION MODEL

The methodology was applied using the results of the Copenhagen experiment (Gryning, 1981; Olesen, 1995). Sensitivity analysis of the R91 model showed that the parameters which influence its performance the most are wind speed at 10 m, roughness length and effective release height. The measurements are used for the conditioning of both the results of simulation and the estimate of the wind speed at the release height. Two numerical experiments are performed. In the first one it is assumed the choice of the weather category is correct and the wind speed is deviated uniformly in the ranges corresponding to given category. Each observation is compared only with the simulations performed within the corresponding stability category. In the second experiment the wind speed values will be deviated within the wider ranges corresponding to two near neutral stability categories. The model will choose the stability condition depending only on the wind speed value. The conditioning on the observations will be done without the use of

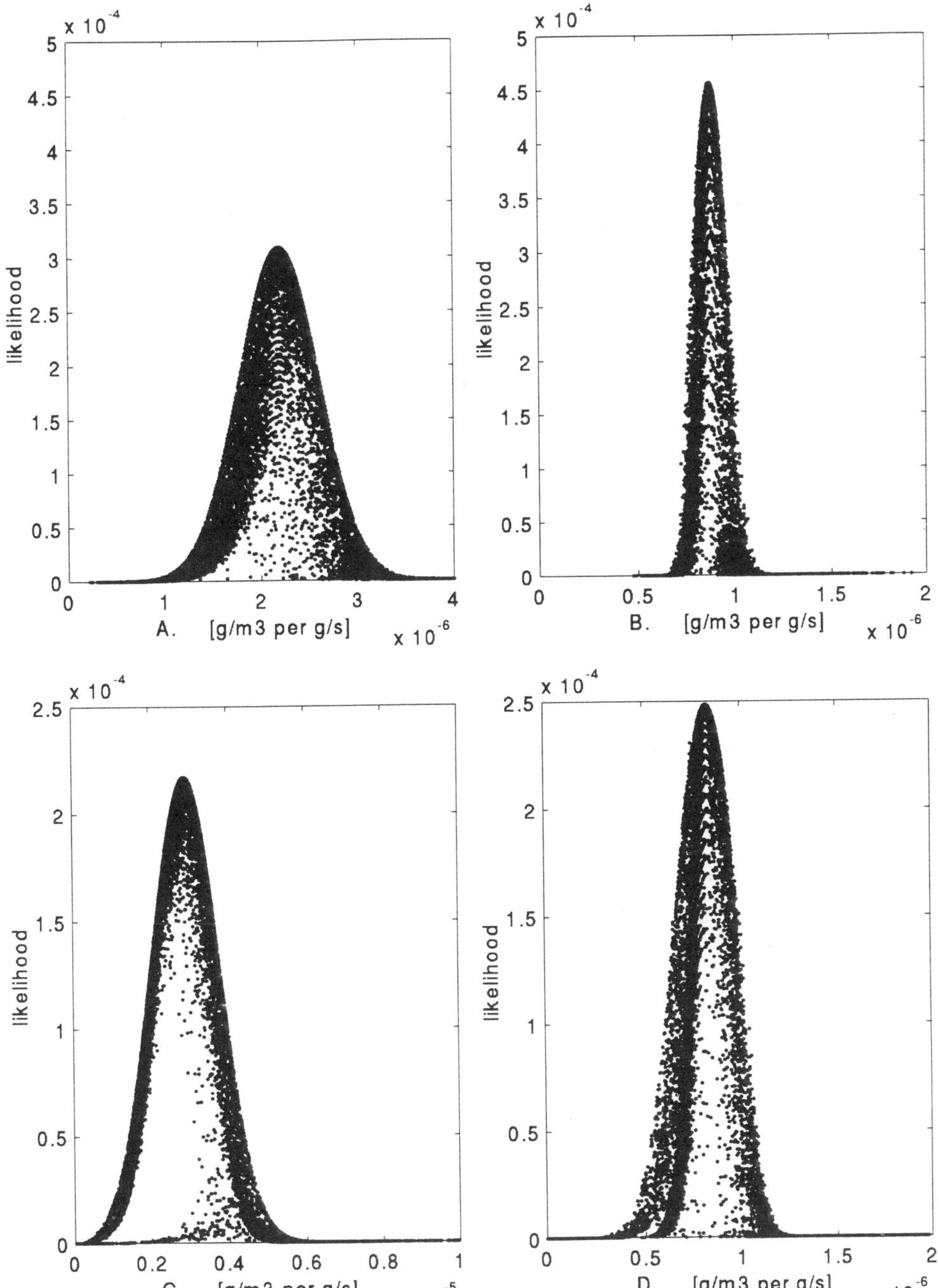

Figure 1. Scatter plot of the posterior likelihood of ground level concentration predictions, 1st numerical experiment: A - 1.9 km; B - 6 km; Scatter plot of the posterior likelihood of ground level predictions, 2nd numerical: C - 1.9 km; D. - 6 km.

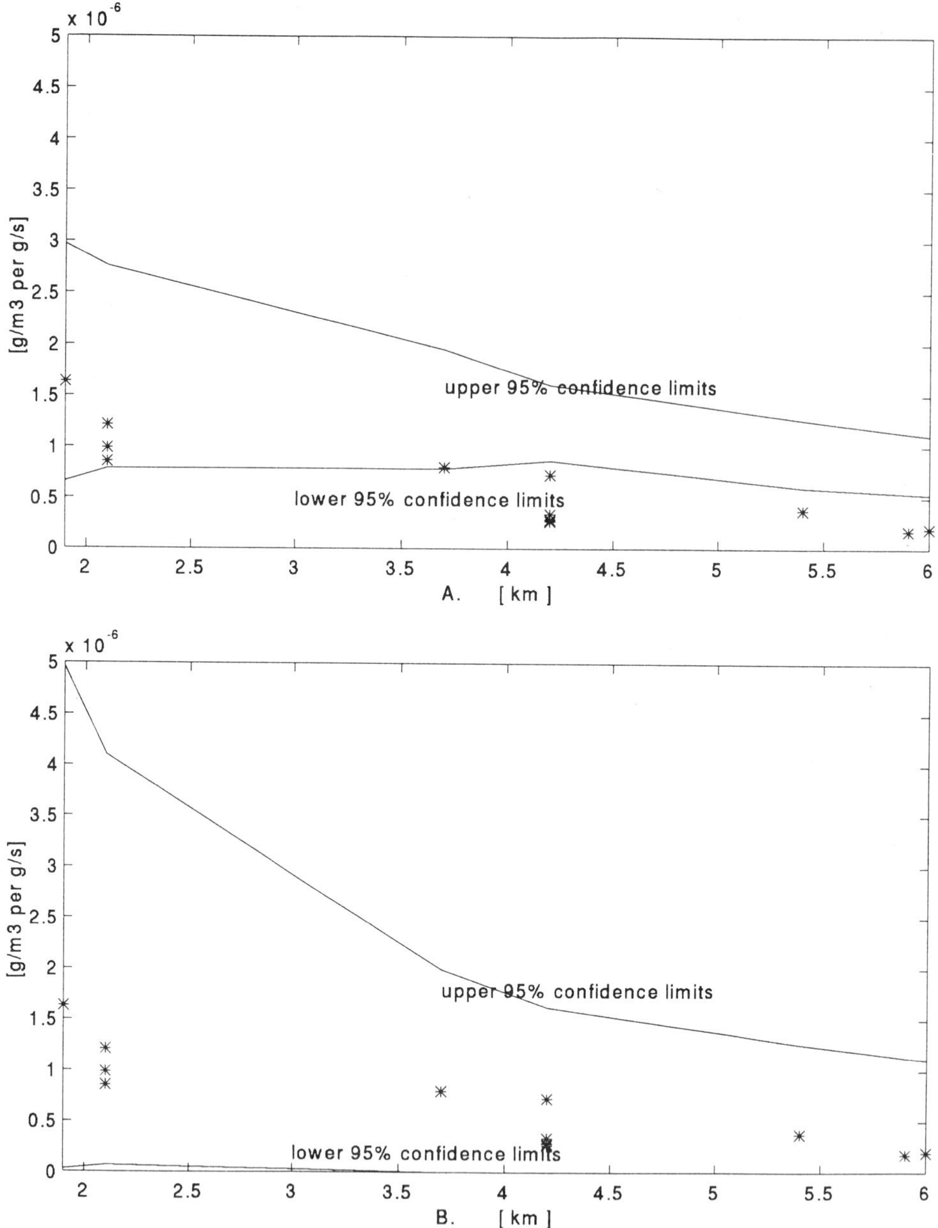

Figure 2. 95% confidence limits of the R91 concentration predictions as a function of distance from the source; A. - 1st numerical experiment; B - 2nd numerical experiment; * denotes Copenhagen observations for neutral stability.

the information on stability conditions, thus introducing certain error into the posterior probability derivation.

The parameters and their assumed prior/initial distributions are summarised in the following table:

Parameter	Distribution	Characteristic parameters
Roughness length	Uniform	0.0001-0.9 m
Wind speed at 10m	Uniform	range depends on stability class
Effective wind parameter	log-normal	mean=0.39; std=0.3

The effective wind parameter, which represents the power of the empirical relation estimating the wind speed at the release height is treated as a random input variable with log-normal distribution. Roughness length and wind speed at 10m are treated as random calibration parameters with uniform distribution over the possible range of realistic values. As the error between the observed and estimated concentrations is specified for the given distances from the source, the posterior probabilities obtained from the equation (5) for each distance are used to form the posterior of the resulting concentrations at those distances.

In the second numerical experiment the wind speed at 10m height is deviated in the wider ranges corresponding to both neutral and slightly unstable stability categories. All the rest of the model parameters remain unchanged. The error resulting from the stability condition estimation is introduced and its influence on the errors of model predictions is shown.

The resulting joint posterior probability scattered diagrams for near field (distance less than 5 km) and far field (distance equal or greater than 5km) for the concentration predictions for both numerical experiments are shown on Fig.1a,b and Fig.1c,d. Note that for the sake of clarity different scales were used. The posteriors were obtained using equation (2) for the probability updating. The confidence limits for the predictions are shown as a function of the distance from the source. Fig. 2a shows the case when the estimation of stability conditions is treated as certain and Fig. 2b shows the case with included error for the stability condition estimation. The second experiment gave wider confidence limits corresponding to greater flexibility of the model simulations. The posteriors obtained from the experiment can be used in the validation phase to predict the concentrations of the independent experiment. However, due to the random nature of the atmospheric processes it is difficult to find an experiment which would correspond to the conditions specified by the Copenhagen observations.

SUMMARY AND CONCLUSIONS

There is currently a growing understanding of the need to account for uncertainties as an intrinsic part of any environmental modelling. This particularly concerns the predictions of atmospheric dispersion models. Not only are the models simplified to a great degree and the input measurements random but also the observations of concentrations on the ground should be treated as a random process. Hence there is an obvious need for the research that would allow identification of the impact of those uncertainties in model predictions and give some indications of possible ways forward. In the paper the application of the Bayesian uncertainty estimation evaluating the uncertainty of dispersion predictions is presented. The simple Gaussian short-range model used in the calculations is rather limited and was chosen due to its availability and simplicity of computations. Model predictions were conditioned on the observation data from Copenhagen experiments. As a result, posterior probabilities and confidence limits for the predictions were obtained. The results show differences between model predictions and observations which cannot be corrected by the choice of model parameters (some observations lie outside the confidence limits) in the case when the ranges of wind speed deviations correspond to given stability conditions. However, when the range of the wind speed deviations is wider, and the model is allowed to set the stability conditions depending on the wind speed, the obtained confidence limits (i.e. also the extreme ranges of predictions) are wide enough to enclose the observed values. This work should be treated as a preliminary analysis leading towards the improvement of the modelling methodology.

Acknowledgments

Dr. Olesen is thanked for providing the Model Validation Kit data which enabled the application of the described methodology to dispersion modelling. The work was supported by British Nuclear Fuels Ltd.

REFERENCES

Beven K.J., and Binley, A., 1992, The future of distributed models: model calibration and uncertainty prediction, *Hydrol. Process.*, **6**, 279-298.

Box G.E.P., and Tiao, G.C., 1992, Bayesian Inference in *Statistical Analysis*, Wiley, Chichester.

Carruthers D.J., Holroyd, R.J., Hunt, J.C.R., Weng, W.S., Robins, A.G., Apsley, D.D., Thompson, D.J., and Smith, F.B., 1994, UK-ADMS: A new approach to modelling dispersion in the earth's atmospheric boundary layer, *J. Wind Eng. Ind. Aerodyn.* **52**, 139-153.

Clarke R.H., 1979, A model for short and medium range dispersion of radionuclides released to the atmosphere, NRPB report R91.

Gryning S.E., 1981, Elevated source SF_6-tracer dispersion experiments in the Copenhagen area, Riso report R446, Riso National Laboratory.

Lee R.F. and Irwin, J.S., 1995, Methodology for a comparative evaluation of two air-quality models, *International Journal of Environment and Pollution*, **5**, 723-733.

Olesen H.R., 1995, The model validation exercise at Mol: overview of results, *International Journal of Environment and Pollution*, **5**, 761-784.

Romanowicz, R., Beven, K., and Tawn, J., 1994, Evaluation of Predictive Uncertainty in Nonlinear Hydrological Models Using a Bayesian Approach, in: Statistics for the Environment 2, Water Related Issues, eds. V. Barnett and K. F. Turkman, 297-315.

DISCUSSION

G. COSEMANS: You used the data of the Copenhagen SF_6 dispersion experiments from Model Validation Kit (MVK) and the R91- dispersion model. Do you know the figures of 'merit' (mean, bias, nmse, cor, fa2, f_b, fs) as calculated by the Model Evaluation Protocol in the MVK for the R91-model?

R. ROMANOWICZ: The goal of the paper was not the comparison of the R91 model with other models, hence the figures of 'merit' were outside the scope of this work. The work was aimed at a preliminary investigation of the application of the Bayesian Uncertainty estimation methodology to an atmospheric dispersion model. Derived predictive posterior distribution contains all the information about the process used in the evaluation of model predictions. The main advantage of the method lies in the ability to combine uncertainty originating from different sources (input uncertainty, structural and observation errors) in estimation of the prediction of concentrations. The predictions have the form of posterior predictive probability from which all the distribution parameters can be evaluated.

REGULATORY-ORIENTED FEATURES OF THE KINEMATIC SIMULATION PARTICLE MODEL

Arno Graff[1], David Strimaitis[2], and Robert Yamartino[2]

[1]im Umweltbundesamt
Bismarckplatz 1
D-14191 Berlin, Germany

[2]EARTH TECH, Inc.
196 Baker Avenue
Concord, MA 01742 USA

INTRODUCTION

Realizing the need for a dispersion model which explicitly considers non-steady state emissions and arbitrarily complex, space-time varying meteorological conditions, and the desirability of having a model which can yield the probability distribution function (PDF) of concentrations rather than simply the ensemble mean, the German Federal Environmental Agency (UBA) funded the development and preliminary testing of an advanced atmospheric dispersion model, referred to as the Kinematic-Simulation-Particle (KSP) Model. The KSP modeling system (Yamartino et al.; 1993-6, and Strimaitis et al., 1995) includes dry and wet removal processes, many other of the processes (e.g., stack and building downwash, plume rise, partial lid penetration) included in the CALPUFF (Scire et al.; 1990, 1995) puff model, and computes the space-time average and percentile concentrations required by German and EC legislation; however, KSP also has the intrinsic capacity to predict the short-term concentration fluctuations that are critical to the assessment of odor and hazardous chemical exposure problems. A key model input involves the 3-d gridded mesoscale wind fields and accompanying 2-d fields specifying boundary layer quantities. These data may be provided by an external wind field/PBL model, or, in the case where only single-point measures of wind and stability are available, an internal flow and turbulence generator creates a vertical profile which emulates the dispersion conditions specified in the German regulatory guideline (TA-Luft). The current KSP model also accesses data produced by the CALMET diagnostic model (Scire et al., 1990; and U.S. EPA, 1995), and an interface to the GESIMA prognostic model has been developed by Reimer et al. (1994) that allows the modeled TKE budgets to be used in the computation of turbulence fields. Expansion of this interface to permit METRAS (Schluenzen et al., 1994) generated winds and turbulence to be utilized is also envisioned.

The KSP model's sub-grid-scale (SGS) flow field generator relies on explicit yet evolving, periodic, and divergence-free functions to quantify the larger scale eddies responsible for pollutant dispersion. This approach is in contrast to traditional Lagrangian particle model approaches (e.g., Janicke 1990, 1995) which use independent Langevin equations for each of a particle's three degrees of freedom, but is the key to KSP's ability to compute instantaneous concentrations and higher temporal moments (e.g., C^n). The KS formulation considered was based on the approach of Fung et al. (1990) for isotropic turbulence but was modified and extended to the more general atmosphere containing height dependent variations in wind velocity (i.e., shear) and turbulence plus turbulence anisotropies. This was accomplished by first transforming Fung's isotropic eddies into component eddies lying in the x-z, y-z, and x-y planes, respectively. The amplitudes of each of these components could then be independently varied to achieve the desired vertical and horizontal plume dispersion rates. The current KSP model uses eddies as small as 12.5m and a sum over up to five eddy sizes, with wavelengths as large as 1600m, for the most unstable dispersion conditions. The relative admixture of the various wavelength components in KSP is constrained to yield realistic turbulence spectra.

Several operational enhancements to KSP-1.6 facilitate its application to episodic single-source releases, including tracer datasets. Such applications typically involve controlled emissions of a tracer from a single source for "short" periods of time (e.g. minutes to several hours). Concentrations are typically sampled along "arcs" at several fixed distances, centered on the source, and the dataset may include crosswind-integrated concentrations along such arcs. Furthermore, meteorological measurements at the site typically provide more information than is available in a "routine" application of KSP.

KSP-1.6 now allows specialized treatments for observed wind and turbulence measurements, it can produce peak and crosswind-integrated concentration estimates in rings around a source, and it has the ability to automatically simulate multiple realizations of each scenario or trial in a dataset in order to obtain an ensemble of predictions. Similar features had previously been available only in a special 2-d analysis program that invokes the kinematic simulation subroutines of KSP, but not the plume rise, particle emission, tracking, and sampling subroutines. This analysis program generated the results previously reported for the Lillestroem and the Copenhagen tracer datasets.

Details of the KSP formulation were presented at the previous NATO/CCMS meeting (Yamartino et al., 1995) and preliminary evaluation of the model using the Lillestrøem, Norway (Haugsbakk and Tonnesen, 1989; Gronskei, 1990) and Copenhagen, Denmark (Gryning, 1981; Gryning and Lyck, 1984) tracer experiments was presented (Yamartino et al., 1996) at the recent Harmonization Workshop in Oostende. The present paper focusses on the results of applying the KSP model to simulate the Prairie Grass, surface release tracer experiments.

OVERVIEW OF THE TRACER STUDIES

Unlike the two datasets previously used to gauge the performance of the KSP model (Lillestroem and Copenhagen), the Prairie Grass dataset includes releases that were made very near the ground (0.46 m), over the full range of atmospheric stability, with far more individual trials (44). And although the dispersion from surface-level releases is somewhat easier to document and analyze than that from elevated releases, near-surface releases do test KSP formulations in a particularly challenging regime. Near the ground, wind speeds and turbulence velocities "go to zero", wind shear can be quite strong, and the length scales of turbulence are smaller than those found at greater heights.

The Prairie Grass experiments were conducted during 1956 over flat, homogeneous terrain in O'Neil, Nebraska, USA. The site was characterized by a roughness length of 0.006 m (Barad, 1958). Releases of SO_2 lasted for 10 minutes, and samplers positioned along arcs located at distances of 50, 100, 200, 400, and 800 m collected data on the mass of SO_2 that reached the impingers at 1.5 m above the surface during a period long enough for the "cloud" to pass completely by. The samplers were placed at 2° intervals at all but the 800 m arc, where the interval was lowered to 1°. After analysis, the mass was reported in the form of 10-minute average concentrations. Samplers were also positioned at 9 levels on 6 towers distributed along the 100 m arc. The levels included 0.5, 1.0, 1.5, 2.5, 4.5, 7.5, 10.5, 13.5, and 17.5 m. In all, 68 trials were conducted. The height of release was set at 0.46 m in trials 1-62, and 1.5 m in trials 63-68.

Slow-response meteorological data were obtained at 2m above the ground, at 2 locations (near the release and about 450 m downwind). Average wind speed and direction, and the standard deviation of the wind direction were reported for 10-minute and 20-minute periods, centered on the period of tracer release. At 450 m this period was shifted by the mean transport time from the source to the anemometer.

Several other sources of meteorological data were obtained during the trials. These include rawinsondes, aircraft soundings, weather observations, and micrometeorological data.

The primary source of the data used to construct the modeling dataset comes from Barad (1958). Later re-analyses of the original data were performed by van Ulden (1978) and Nieuwstadt (1980). Nieuwstadt (1980) computed values of u_* and L for the convective trials from profiles of wind velocity and temperature measured between 0.25 and 16 m. Mixing heights for these trials were estimated on the basis of temperature profiles from radiosondes and aircraft soundings. Nieuwstadt (1980) computed w_* from these derived parameters. van Ulden (1978) reports u_* and L for the non-convective trials, and these were computed using the same procedures. We believe that the roughness length used in these procedures was the 0.008 m value reported by van Ulden (1978). Briggs (1982) reviewed the Prairie Grass data, and identified trials in which the concentration field was not well documented. The 44 trials that he considered adequate comprise the dataset used to evaluate KSP. Crosswind-integrated concentrations and σ_y (second moment of the crosswind concentration distribution) for the convective periods were computed by Nieuwstadt (1980), and the crosswind-integrated concentrations for the non-convective periods were computed by van Ulden (1978).

Data from these sources were assembled by Hanna et al. (1990), and their tabulation forms the basis for the dataset used here. They report the 10-minute average 2m wind speed and direction measured near the release, the corresponding σ_θ, the temperature difference from 2 to 16 m (16-2), the mixing height, release rate, u*, w*, L, the scaled concentration (C/Q) for both the peak observed value along each arc and the crosswind-integrated value, and the second moment of the concentration along each arc. They also report a Pasquill-Gifford stability class, a power-law exponent for the wind speed profile, and an estimate of the standard deviation of the vertical elevation angle of the wind, σ_ϕ. The stability class is estimated from the site roughness length and L using the Golder (1972) nomogram, and the power-law exponent and σ_ϕ are estimated from surface similarity relations.

APPLICATION OF KSP TO THE PRAIRIE GRASS STUDY

KSP was configured as follows for this application:

Receptor boxes:	20m on a side, and 2m tall with an additional 1m tapered region on top.
Receptor rings:	20m thick, and 2m tall with an additional 1m tapered region on top. The first ring is centered at a radius of 10m, and the last is centered at a radius of 830m from the release.
Timestep:	2 seconds, with no adaptive substeps.
Particles:	5 particles are released every 2 seconds.
Averaging Time:	10 minutes
Replications:	41 simulations performed for each of the 44 trials. Only the first simulation includes the effect of "start-up", and this simulation is dropped from the analysis.

Because these releases were neutrally buoyant, no effects of plume rise are simulated.

KSP was then run in a variety of modes involving different assumptions about the wind and turbulence profiles. Unfortunately, the measured turbulence data, such as σ_ϕ, and the micrometeorological quantities extracted from these data tend to make the meteorological conditions appear more 'neutral' than they may have been.

DISCUSSION OF RESULTS

Some of the more interesting results arose from applying KSP in its 'TA-Luft mode' (i.e., using TA-Luft wind profile exponents and turbulent intensity profiles inferred

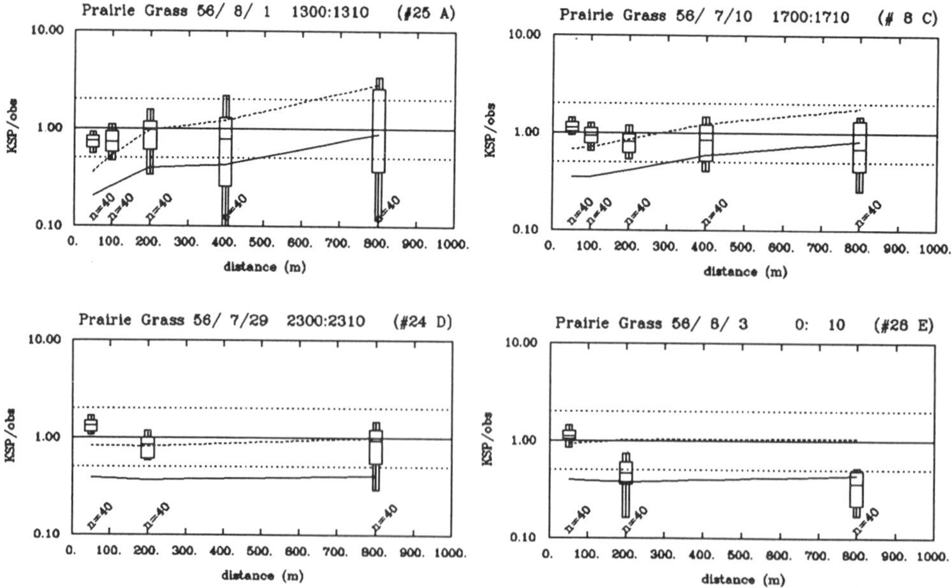

Figure 1. KSP crosswind-integrated concentrations normed by observations for four Prairie Grass trials having different atmospheric stabilities. See text for additional details.

from the AUSTAL-86 dispersion curves). The modelled crosswind-integrated (CWI) concentrations normalized by observations for several of the Prairie Grass experiments are presented in the four frames of Figure 1, along with curves indicating the scaled predictions of a Gaussian plume model employing the σ_y and σ_z dispersion curves from AUSTAL-86 (solid curve) and Pasquill-Gifford (dotted curves). These whisker plots, denoting the 10th, 25th, 50th, 75th, and 90th percentiles, show the range of CWI concentrations predicted by KSP. The experiments are ordered by increasing stability (i.e., A to E) and one notes a trend from some KSP overprediction for the unstable experiments to underprediction for the stable cases. Away from these extremes in stability one finds KSP's predictions more reasonable. One also notes that the Gaussian plume model predictions are neither superior to nor necessarily comparable to KSP results from its TA-Luft mode.

Of particular interest in these applications of the KSP model is the range of concentrations in the whisker plots. The range of concentrations likely to be observed in unstable, convective conditions is greater than that for more neutral conditions and much greater than that seen for more stable conditions. This behavior is a direct consequence of the spectral shapes assumed within the KSP model. Given by the spectral density models of Hojstrup (1982) and Moraes (1988), the larger contribution of low-frequency turbulence under unstable conditions causes greater variability in the predicted concentrations.

CONCLUSIONS

The KSP model provides the short-range (i.e., < 30 km), multi-source simulation capabilities of the TA-Luft regulatory model, AUSTAL '86, in current use, plus it provides the:

- ability to function on considerably larger mesoscale domains (e.g., to 10^3 km) encompassing wide variations in meteorology, land use and terrain;

- ability to simulate situations (e.g., weak wind, complex terrain, coastal recirculations) that are not treated in detail within a Gaussian model such as AUSTAL-86;

- ability to interface to several diagnostic/prognostic meteorological models that reflect the state-of-the-science of flow and turbulence field simulation and provide a more adequate treatment of boundary layer meteorology;

- flexibility to permit new scientific developments to be rapidly incorporated through a highly modular model structure; and

- capability of computing instantaneous concentration fields, higher concentration moments, $C^n(t)$, and concentration variability measures (e.g., variances, intermittency) as well as the more traditional time/ensemble average concentration measures.

The KSP modeling system is currently being subjected to an extensive evaluation process to determine any biases and to quantify modeling uncertainties. Such an evaluation process is also the key model acceptance criteria to emerge from the series of three European Community sponsored model 'harmonization' workshops held to date. In this paper we have considered only a portion of the Prairie Grass tracer experiments but have demonstrated the increasing variability of concentrations with degree of atmospheric instability and can link this to the correspondingly increasing, long-wavelength content of the turbulence.

ACKNOWLEDGMENT

This work was supported through a prime contract to the Meteorological Institute of the University of Hamburg by the German Federal Environmental Agency (Umweltbundesamt) as part of the Environmental Research Plan of the Federal Ministry of Environment, Nature Conservation, and Reactor Safety.

REFERENCES

AUSTAL-86, 1987: Richtlinie zur Durchführung von Ausbreitungs-rechnungen nach TA Luft mit dem Programmsystem AUSTAL86, November 1986, ©1987 Bundesanzeiger Verlagsges m.b.H., Köln, Germany.

Barad, M.L., (Ed.), 1958: Project Prairie Grass, A field program in diffusion, Vol. 1, *Geophysical Research Papers*, No. 59, AFCRC-TR-58-235(I), Air Force Cambridge Research Center, Bedford, MA.

Briggs, G.A., 1982: Similarity forms for ground source surface layer diffusion, *Bound. Lay. Meteorol.*, **23**, 489-502.

Eppel, D., H. Kapitza, M. Claussen, D. Jacob, L. Levkov, H.T. Mengelkamp, and N. Werrmann, 1994: The non-hydrostatic mesoscale model GESIMA. Part II: parameterizations and application. To be published in *Beitr. Phys. Atmosph.*

Fung, J.C.H., J.C.R. Hunt, N.A. Malik and R.J. Perkins, 1992: Kinematic simulation of homogeneous turbulent flows generated by unsteady random Fourier modes. *J. Fluid Mech.*, **236**, 281-318.

Grønskei, K.E., 1990: Variation in dispersion conditions with height over urban areas - results of dual tracer experiments, 9th AMS Symposium on Turbulence and Diffusion.

Gryning, S.E., 1981: Elevated source SF_6 tracer dispersion experiments in the Copenhagen area. Riso-R-446, Riso National Laboratory, 187 pp.

Gryning, S.E. and E. Lyck, 1984: Atmospheric dispersion from elevated sources in an urban area: Comparison between tracer experiments and model calculations. *J. Cl. Appl. Meteor.*, **23**, 651-660.

Hanna, S.R., J.C. Chang, and D.G. Strimaitis, 1990: Uncertainties in source emission rate estimates using dispersion models. *Atmospheric Environment*, **24A**, 2971-2980.

Haugsbakk, I. and D.A. Tønnesen, 1989: Atmospheric Dispersion Experiments at Lillestrøm, 1986-1987 Data Report, Lillestrøm, Norwegian Institute for Air Research (NILU OR 31/89).

Hojstrup, J., 1982: Velocity spectra in the unstable planetary boundary layer. *J. Atmos. Sci.,* **39**, 2239-2248.

Kapitza, H., and D. Eppel, 1992: The non-hydrostatic mesoscale model GESIMA. Part I: dynamic equations and tests. *Beitr. Phys. Atmosph,* **65**, 129-146.

Janicke, L., 1990: Ausbreitungsmodell LASAT-C: Handbuch Version 1.00. (In German)

Janicke, L., 1995: Ausbreitungsmodell LASAT: Referenzbuch zu Version 2.51, Primelweg 8, D-88662, Überlingen, FRG. (In German)

Moraes, O.L.L., 1988: The velocity spectra in the stable atmospheric boundary layer. *Boundary Layer Meteorol.,* **43**, 223-230.

Nieuwstadt, F.T.M., 1980: Application of mixed-layer similarity to the observed dispersion from a ground-level source, *J. Appl. Meteorol.*, **19**, 157-162.

Reimer, E., B. Scherer, W. Klug, and R. Yamartino, 1994. A Meteorological Data Base System for Next Generation Dispersion Models and a Lagrangian Particle Model Based on Kinematic Simulation Theory. *Proceedings of the Workshop on Intercomparison of Atmospheric Dispersion Modeling Systems*, Mol, Belgium.

Schluenzen, K.H., K. Bigalke, U. Niemeier and K. von Salzen, 1994. The mesoscale transport- and fluid-model 'METRAS' -- model concept, model realization - Meteorological Institute, University of Hamburg, METRAS Technical Report, 1, pp 150.

Scire, J.S., E.M. Insley and R.J. Yamartino, 1990: Model formulation and user's guide for the CALMET meteorological model. Prepared for the California Air Resources Board. Sigma Research Corporation, Concord, MA.

Scire, J.S., D.G. Strimaitis and R.J. Yamartino, 1990: Model formulation and user's guide for the CALPUFF dispersion model. California Air Resources Board, Sacramento, CA.

Scire, J.S., D.G. Strimaitis, R.J. Yamartino and X. Zhang, 1995: A User's Guide for the CALPUFF Dispersion Model. EARTH TECH Document 1321-2 prepared for the USDA Forest Service, Cadillac, MI.

Strimaitis, D., R. Yamartino, E. Insley, and J. Scire, 1995. A User's Guide for the Kinematic Simulation Particle Model. EARTH TECH Document 1274-2 prepared for the Free University of Berlin and the Umweltbundesamt, Berlin, FRG.

U.S. EPA, 1995. A User's Guide for the CALMET Meteorological Model. EPA-454/B-95-002.

van Ulden, A.P., 1978: Simple estimates for vertical diffusion from sources near the ground, *Atmospheric Environment*, **12**, 2125-2129.

Yamartino, R., D. Strimaitis, M. Spitzak, and W. Klug, 1993. Development and Preliminary Evaluation of a Dispersion Model Based on Kinematic Simulation Theory. *Proceedings of the Workshop on Intercomparison of Atmospheric Dispersion Modeling Systems*, Manno, Switzerland.

Yamartino, R., D. Strimaitis, and A. Graff, 1995. Advanced Mesoscale Dispersion Modeling Using Kinematic Simulation. *Proceedings of the 21st NATO/CCMS ITM on Air Polluting Modeling and its Application*, Baltimore, MD, Nov. 6-10.

Yamartino, R., D. Strimaitis, J. Scire, E. Insley, and M. Spitzak, 1996. Final Report on the Phase I Development of the Kinematic Simulation Particle (KSP) Atmospheric Dispersion Model. EARTH TECH Report 1274-3 prepared for the Free University of Berlin and the Umweltbundesamt, Berlin, FRG.

Yamartino, R., D. Strimaitis, and A. Graff, 1996. Evaluation of the Kinematic Simulation Particle Model Using Tracer Experiments. *Proceedings of the Fourth Workshop on Harmonization within Atmospheric Dispersion Modelling for Regulatory Purposes*, Oostende, Belgium, May 6-9.

DISCUSSION

T. MIKKELSEN: At Risø and NERI in Denmark, we have recent experimental, short-term concentration fluctuation data sets based on calibrated lidar measurements. These data sets, called BOREX '92, BOREX '94 and BOREX '95, contain measurements on relative dispersion, concentration fluctuations, and exceedance statistics that are of relevance to the evaluation of your KS approach.

R. YAMARTINO: We have heard of these experimental data and would be quite interested to use them in our ongoing evaluations of the KSP model.

H. VAN DOP: Does the modified KSP model reproduce the observed higher velocity moments, such as w^2, w^3, and w^4, for convective turbulence?

R. YAMARTINO: The new KS sub-model for convection is based on a periodic array (i.e., in x and y) of updraft cores having a specified spatial extent and peak (i.e., central) speed, surrounded by larger zones of lower velocity downdrafts. A single assumed maximum updraft speed will then yield a positively skewed distribution of w velocities. Further assumption of some distribution in these peak updraft speeds then yields a complete distribution of vertical velocities. We know that the velocity ratio, w_+/w_-, and the updraft/downdraft areal ratio are matching observations, but we have not yet looked at the various velocity moments.

H. VAN DOP: Is this model able to give a realistic description of relative dispersion in homogeneous and inhomogeneous turbulence?

R. YAMARTINO: The earlier work of Fung et al. indicated that the KS formulation does give a realistic description of relative dispersion for homogeneous turbulence. The present KS model for inhomogeneous turbulence in sheared mean flows is considerably more complex, but does appear to yield a reasonable description of total plume dispersion; however, we have not looked explicitly at relative diffusion in this case.

H. OLESEN: Please be careful when you use crosswind integrated concentrations (CWICs) from the Kincaid data set. CWICs were included in the first version of the "Model Validation Kit", but were omitted from the later versions because they had not been assigned a quality indicator.

R. YAMARTINO: We have looked in detail into the Kincaid data set, both as contained in the "Model Validation Kit" and as contained in the EPRI Data Center archive. As a result we have considered only those experiments having the highest quality indicator. While this indicator strictly applies only to the peak concentration and not the CWICs, the conditions for receiving the highest quality indicator (i.e., a plume that is neither 'spotty' or 'near the edges of the monitoring network') suggest that the CWICs should be well determined.

TOOLS FOR MODEL EVALUATION

H.R.Olesen

National Environmental Research Institute
P.O. Box 358
DK-4000 Roskilde
Denmark

INTRODUCTION

At the previous ITM in Baltimore, the author presented a paper entitled "Toward the establishment of a common framework for model evaluation" (Olesen, 1996). It described various ongoing activities aimed to improve model evaluation methods. Further, it discussed problems about the way that evaluation is normally practised, and suggested ways to improve matters.

The present paper serves as a follow-up on the previous, and supplements it with a discussion of a few important problems omitted from the earlier treatment. This paper is motivated by a desire to keep the modelling community informed about some ongoing activities, which may eventually have far reaching consequences. Specifically, there is work under way in the International Standards Organisation (ISO) which potentially can have a great impact.

The paper can stand alone, but for a thorough understanding of the issues discussed it should be read in conjunction with some of the other papers referred to in the text.

The paper includes the following elements:

- A recapitulation of difficulties in model evaluation.
- Present status of the so-called Model Validation Kit.
- The role of standards organisations with respect to model evaluation.
- A detailed discussion of one particular problem with the use of the Model Validation Kit: According to the methodology used, observed arcwise maxima are compared *directly* to modelled centerline values. Results from this type of comparison do not have a straightforward interpretation. There are some basic conceptual issues involved (How should an ensemble mean be defined? What is a "perfect model"?), and the discussion has important practical implications.
- A brief discussion: How can performance evaluation results from several data sets be combined?

DIFFICULTIES IN MODEL EVALUATION

In this paper we will not provide the reader with a complete discussion of model evaluation methodologies and the problems associated with them. However, to place the following discussion in its proper perspective we will recapitulate some essential problems which make model evaluation difficult. These difficulties and pertinent reactions to them have been analysed in previous papers by the author.

The major difficulties in model evaluation can be summarised by the following few statements

- *The appropriate evaluation method cannot be uniquely defined.*
  The pertinent evaluation method depends on the context of the application. An array of methods, useful for various purposes, should be available.
- *Input data sets are limited - they reflect only few of the possible scenarios.*
  Users want models which apply to a broad range of conditions, whereas the data sets used for design and verification of models have many limitations. One sensible response to this difficulty is to use *many data sets* for model evaluation. In parallel, one should develop an understanding of model behaviour in order to ensure that a model gives the right answer for the right reason.

- *Processing of input data for model evaluation is far from trivial.*
  Hard and careful work is required. It must be recognised, though, that modellers retain a freedom to choose between several possible variants of input data.
- *The luxury of independent data sets can rarely be afforded.*
  Only few comprehensive, high quality data sets exist. No model can be expected to preserve its virginity towards the few available data sets. Therefore, one must require models to be tested against *many* data sets.
- *There are inherent uncertainties.*
  The atmospheric dispersion processes are stochastic, but models can be expected only to predict ensemble averages, not the results of specific realisations. Some aspects of this will be discussed in detail below.

The difficulties listed are discussed, i.a., in the above mentioned paper for the previous ITM (Olesen 1996), as well as in Olesen (1994; 1995a; 1997). A number of measures intending to overcome these difficulties have been pointed to. As an overall conclusion, it would greatly enhance the productivity of the modelling community if tools for model evaluation were made generally available, so that the modelling community in practice would be able to use a common frame of reference. Such tools include *carefully prepared data sets, model evaluation software, and model evaluation protocols.* The so-called *Model Validation Kit* - which will be discussed below - can serve as one initial step in order to establish such a frame of reference.

Speaking in practical terms, the World Wide Web presents some promising new opportunities to promote the use of common tools.

STATUS: THE MODEL VALIDATION KIT

During the past few years, a series of activities have been going on in Europe within the field of short-range atmospheric dispersion modelling, under the overall title of "Harmonisation within Atmospheric Dispersion Modelling for Regulatory Purposes". The aim of this work is to modernise models for regulatory purposes, as well as to achieve increased co-operation and standardisation within modelling. Model evaluation is a key issue in this respect. Four so-called "workshops" (which have actually evolved to large international meetings) have been held in the period 1992-1996.

Within the framework of the "Harmonisation" initiative, work has been conducted in order to start establishing a toolbox of recommended methods for model evaluation. The basis for this work has been the so-called *Model Validation Kit.* This set of materials has been used at the various workshops as the basis for model evaluation exercises. Over the past few years, the kit has been distributed to more than one hundred research groups.

The Model Validation Kit is a practical tool, meant to serve as a common frame of reference for modellers, but with well recognised limitations. In its original form, it is a collection of three experimental data sets, as well as software for model evaluation, based on the work of Hanna et al. (1991). The current version of the Model Validation Kit basically dates from 1994, though some extra software tools were added in 1995. It is described by Olesen (1995b), and is available at no charge from the author.

A supplement to the original data kit is now also available. This supplement comprises data from an experimental campaign in an urban area (Indianapolis in the USA). Similarly to the three data sets in the original Model Validation Kit, this experiment deals with tracer emission from a single, elevated source.

The toolkit and its supplement are suitable for the task of evaluating operational short-range models. Their limitations can be summarised as follows:

- The type of problem primarily dealt with is relatively simple, namely a case where *a single source emits a non-reactive gas in homogeneous terrain.*
- Only *four experimental data sets* are considered.
- Further, the emphasis is on comparisons where a) *arc-wise maxima* are compared to plume centre-line concentrations, and - to some extent - b) *cross-wind integrated concentrations* are considered.

The tools can be used when comparing models, and also to diagnose the strengths and weaknesses of models; however, the limitations of the tools should not be forgotten. As pointed out in the series of papers mentioned above, there are many problems in performing model evaluations, and model evaluation results should be used with prudence.

The Model Validation Kit will presumably evolve further during the forthcoming years. In the autumn of 1997, a workshop in Cambridge will focus on experiences with the newest additions to the kit. The work is expected to be continued at the fifth International meeting on *Harmonisation within Atmospheric Dispersion*

Modelling for Regulatory Purposes in Rhodes in May 1998. Information on the current state of the Model Validation Kit is available on the World Wide Web (http://www.dmu.dk/AtmosphericEnvironment/harmoni.htm).

STATUS: STANDARDISATION ORGANISATIONS

It would seem pertinent for the existing standards organisations to take a role in the efforts to establish a common framework for model evaluation. Some initiatives have been taken in this respect.

In October 1996, the so-called ISO TC 146/SC5 had a meeting where it was decided to form an ad hoc work group, named "Model Evaluation Methods". ISO is the International Standards Organisation, TC 146 is the Technical Committee on "Air Quality", and SC5 stands for Sub Committee 5, which is on Meteorology. The ad hoc work group on model evaluation mainly consists of persons who have been involved in the model evaluation work at the "Harmonisation" workshops. The chairman of the work group is J.S. Irwin of the US EPA. The work group has chosen to adopt status as an "ad hoc" group, because it felt that it would yet be premature to establish an "official" work group. It is dangerous to designate standards for complex procedures before they have been thoroughly tested in practice, and according to the rules of ISO a proposal for a standard should be produced within 18 months from the formation of an official work group. Thus, the ad hoc work group is presently working without fixed time constraints, while tackling various problems and testing procedures which will be explained in the next sections.

In addition to the ISO, the ASTM (American Society for Testing and Materials) is considering to establish standards relating to performance evaluation of atmospheric dispersion models. The ASTM is an influential, voluntary standards development organisation which is open to members from all over the world. There is an overlap between the persons involved in the ASTM and the ISO work, so results obtained in one of these organisations are likely to propagate to the other.

Persons having a serious interest to participate in the work of the ad hoc work group of ISO TC146/SC5 are welcome to contribute.

POTENTIAL IMPROVEMENTS TO THE METHODOLOGY OF THE MODEL VALIDATION KIT

From the beginning, the Model Validation Kit was not meant as a complete evaluation toolbox. Rather, it has served as the basis for some demonstration exercises where lessons could be learned, both concerning model evaluation methods and concerning individual models.

In the present form of the kit, it does not address certain important issues. Especially two issues are of concern:

1) According to the methodology used, observed arcwise maxima are compared *directly* to modelled centerline values. Results from this type of comparison do not have a straightforward interpretation.
2) The kit does not include procedures to *combine* evaluation results from several data sets to any kind of composite performance index.

These two issues are being addressed in the ISO TC 146 ad hoc work group "Model Evaluation Methods". The first issue will be discussed here at some length, while the second will be touched upon only briefly.

Though the discussion may seem rather academic at times, it does have important practical implications., There is a discrepancy between the expectations of a model user and the capabilities of a modeller. The model user's standpoint will typically be: "I want a model which predicts concentrations correctly". On the other hand, even if a modeller has a perfect model to offer, he can state nothing but: "Here you have a perfect model. However, *it is a characteristic feature of a perfect model that it underpredicts the highest concentrations!*". This set of problems will be discussed in the following.

PROBLEMS WITH THE DIRECT USE OF OBSERVED MAXIMA

According to the methodology used in the Model Validation Kit, arcs of monitors are considered where all monitors in an arc have approximately the same distance from the source. An example of the layout of an experiment is shown as Figure 1. It displays the geographical distribution of measured ground-level tracer concentrations during one particular hour of the Kincaid experiment.

For each arc, the observed maximum is compared directly to the modelled value in the plume centerline at ground level. There are two problems in doing this. They can be designated:

- The ensemble mean problem;
- The receptor spacing effect.

These will be discussed in the next two subsections. A third subsection will discuss the concept of a "perfect model", and then a possible alternative methodology - use of near-centerline concentrations - will be introduced.

The ensemble mean problem

When validating a model against an experimental data set, a large scatter between observations and predictions is to be expected. One of the reasons for this is the stochastic nature of the problem: the atmosphere is turbulent. The collection of observations represent individual realisations of an event, whereas a "perfect model" presumably is a model which can predict ensemble means correctly.

Let us consider in some more detail what this means. An ensemble mean of arcwise maximum values at a certain arc can in principle be found in the following way:

Pick a set of meteorological scenarios, which have similar characteristics - in other words, they can be said to represent the same event.

Consider the observed concentrations along the chosen arc for each of these scenarios. We wish to form an ensemble average of the arcwise maximum concentration. It is not *a priori* completely clear how this is done.

Method (i): Ensemble average based on arcwise maxima. One obvious way to proceed (method *(i)*) is to find the arcwise maximum for each realisation, *no matter where on the arc it occurs.* Then, form the average over the ensemble of arcwise maxima.

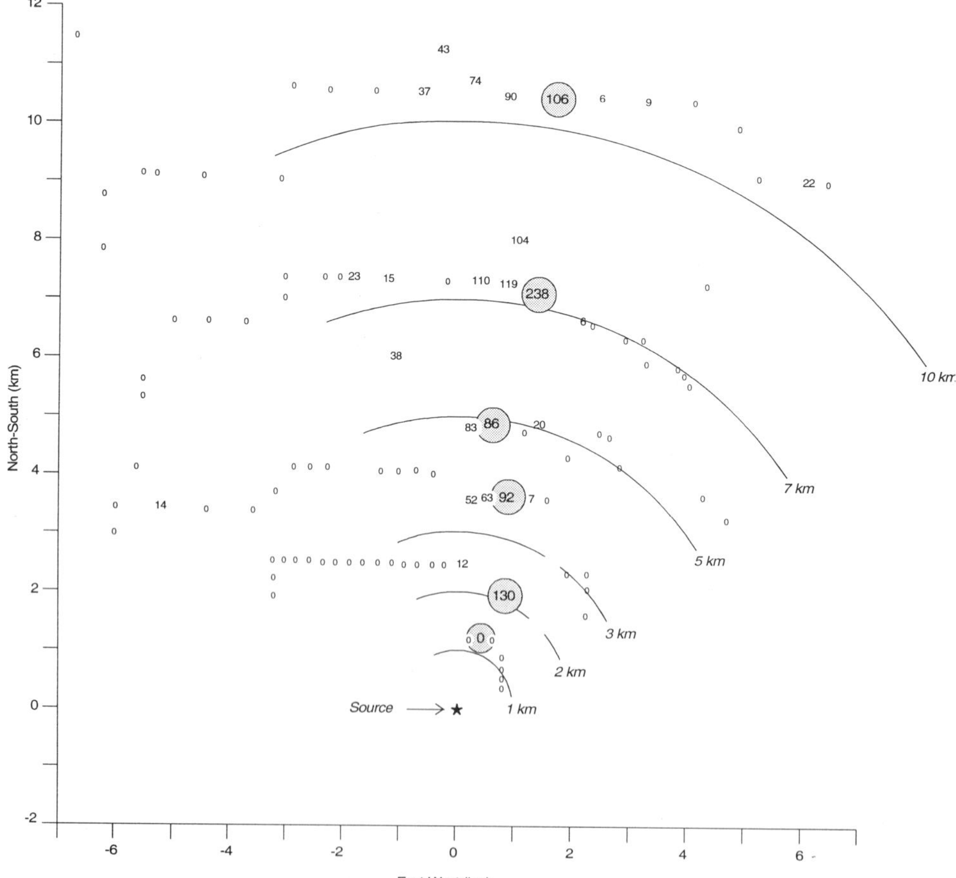

Figure 1 Geographical distribution of measured concentrations at Kincaid, 22 May 19981, 10-11 hours. Values are in ppt, and the arcwise maxima are enclosed in circles.

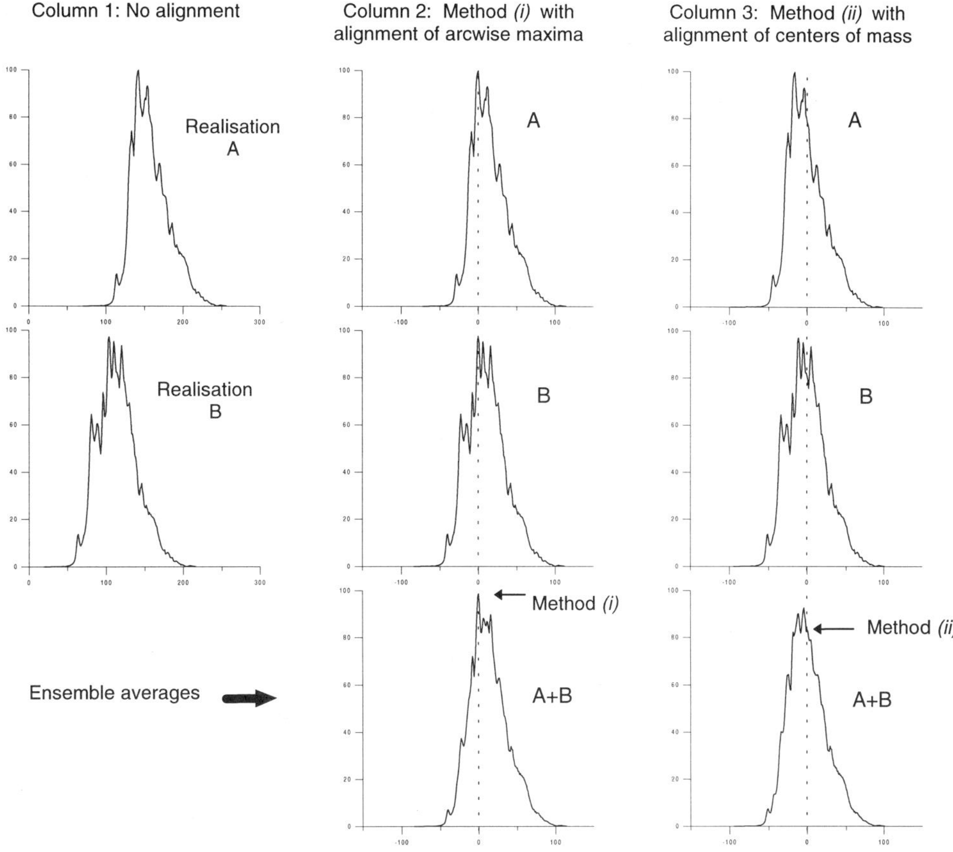

Figure 2 Formation of an ensemble mean over a collection of only two realisations. Two methods are illustrated.

The procedure is illustrated graphically in Figure 2, columns 1 and 2 (for a detailed, arbitrary data set). The concentration profiles from the individual realisations (column 1) are first aligned *with respect to the maximum* (column 2), and subsequently superimposed (bottom). In the example chosen for illustration, the "ensemble" consists of only two realisations.

Method (ii): Ensemble average based on centerline concentrations. An alternative way to form an ensemble average (method *(ii)*) would be the following: For each realisation find its center of mass. Then align the concentration profiles *with respect to the center of mass*. Subsequently, perform an averaging over the ensemble of realisations. This is illustrated in Figure 2, column 3.

This method will again allow us to identify an ensemble average concentration, but in a different sense than before. The ensemble average according to method *(ii)* will be lower than (or equal to) the ensemble average according to the first definition. The value derived by method *(ii)* can be precisely described as *an ensemble average of centerline concentrations.*

Properties of the two ensemble averages. Which of these two definitions is to be preferred? This is a question of its usefulness in practical, real-world problems. The question will be addressed in the following.

However, no matter which of the two definitions we choose, there is one important implication: if we have a perfect model (i.e., it perfectly predicts ensemble averages), and if we have a reasonable number of realisations of the same event, then there will exist observed values greater than our model prediction. Thus, our perfect model will at first glance *seem to underpredict the very highest concentrations.* The statement holds true for both definitions *(i)* and *(ii)*, but the underprediction will be most pronounced if the model is perfect according to definition *(ii)*.

Considering this effect alone, one should not be worried if the Model Validation Kit produces graphs like the one shown in Figure 3.

In the graph, the distribution of observed and modelled arcwise maxima are compared for a certain model. This results in a so-called quantile-quantile plot. The data are ordered by rank, so for instance the highest observed concentration is paired with the highest modelled concentration, no matter when or on which arc they occur.

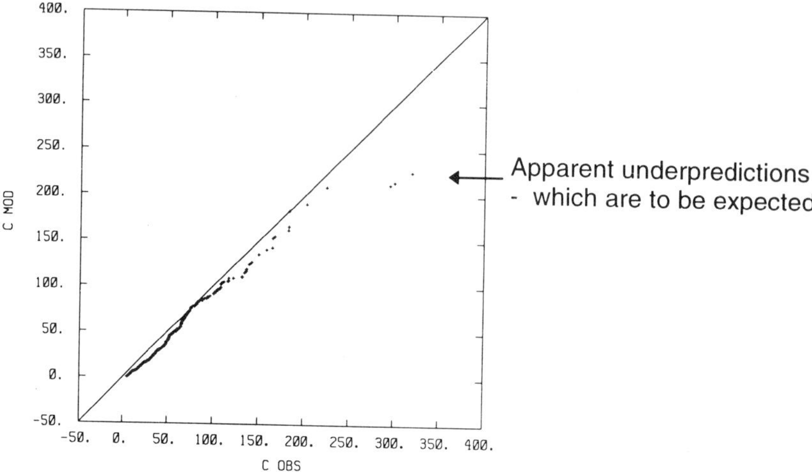

Figure 3 Sample quantile-quantile plot of observed versus modelled concentrations. Observed values are arcwise maxima, modelled values are centerline values. The arrow points to apparent underpredictions which are to be expected for a "perfect model".

The receptor spacing effect

In the section above, we considered *one* consequence of the methodology where observed maxima are compared directly to modelled. We found that underprediction of the highest concentrations are to be expected. However, there is another consequence, which tends to counteract the first.

In available data sets from tracer experiments, we have *discrete nets* of monitors. When we compare a modelled centerline maximum to an observed maximum, we will typically compare a (modelled) *centerline* value to an (observed) *off-centerline* value. The effect is that a model which is perfect for predicting *arcwise maximum values* will have a tendency to overpredict the *observed values*. This tendency applies to *any* pair (c_{mod}, c_{obs}), not just the highest concentrations.

This effect is most severe for a sparse net of monitors.

Defining a "perfect" model

The two problems discussed above - the ensemble mean problem and the receptor spacing effect - are closely connected to our way of defining a "perfect model". We have at least two choices:

(a) a perfect model is one which correctly predicts ensemble averages of *arcwise maximum* values (method *(i)*).

(b) a perfect model is one which correctly predicts ensemble averages of *arcwise centerline* values (method *(ii)*).

When using observed arcwise maxima for a direct comparison to modelled maxima as it is done in the Model Validation Kit, we implicitly use definition *(a)* - alignment of maximum values. One disadvantage of definition *(a)* is that the receptor spacing effect is much more pronounced than if we use definition *(b)*. Because of receptor spacing, we will always have difficulties in determining the true maximum value from observations, and consistently underestimate it.

Definition *(b)* of a perfect model - alignment of centerline values - has the advantage that it is not so sensitive to receptor spacing. We can do a better job in estimating the *centerline* value along an arc than in estimating the *maximum* value.

But we should be aware of the properties of our "perfect model". The perfect "*(b)*" model will predict lower concentrations than the perfect "*(a)*" model, and we will experience many more underpredictions from a perfect "*(b)*" model than from an "*(a)*" model. The properties of *(a)* and *(b)* models in respect to other quantities (crosswind integrated concentrations etc.) are also of interest, but will not be discussed here.

Using near-centerline concentrations

Irwin and Lee (1997) have discussed evaluation methods where they avoid the use of observed arcwise maxima. In the ISO work group "Model Evaluation" Irwin has provided ideas for a revised methodology for model evaluation (as yet unpublished). The essence of the methodology is that we should consider not only *one* observed value (the maximum) per arc, but consider several "near-centerline" concentrations. The proposed methodology represents a consistent way to handle the ensemble mean problem, and it will to a large extent free us from the receptor spacing problem. However, implicitly, definition *(b)* for a perfect model is assumed when adopting that methodology. It should be kept in mind that a perfect model according to this definition consistently yields lower concentrations than a perfect model according to definition *(a)*.

The procedure is as follows when we wish to form a data set for model evaluation:

- Take an experimental data set with a good coverage of samplers along the monitoring arcs.
- Consider an arc, and accept it or reject it for further processing. For instance, it will be rejected if there are unacceptably few monitors.
- Determine the center-of-mass and the lateral dispersion, σ_y, for the arc.
- Select "near-center-line" concentrations. The selection criteria could be that $|y| < 0.67\ \sigma_y$ (y is the lateral distance from the center of mass).

The above steps can be performed once and for all for a given experimental data set. When the concentration data set is to be used for evaluation of a model, the modeller is required to compute concentrations corresponding to all the observed "near-centerline" concentrations. In doing this, he must assume that the plume centerline is aligned with the center of mass of the observed concentrations. He then runs his model in order to find concentration values at the various near-centerline points.

At this step in the procedure, we have a data set ready to be used for analyses of model performance. The data set can be used to diagnose model behaviour - for instance by applying the various tools in the Model Validation Kit. The data set can also be used to deduce statistical performance indices, as the next section describes.

COMBINING PERFORMANCE EVALUATION RESULTS

There are many good reasons why a model should be evaluated versus not only one, but many data sets. That, however, leaves us with a problem: how should the results of the various analyses be combined into something manageable?

Lee and Irwin (1995) have worked on this problem. In the ISO work group, the following methodology (due to Irwin) is presently being explored:

- For each experiment divide the data set into regimes, defined by such criteria as downwind distance from the source, atmospheric stability etc.
- For each regime, construct frequency distributions of modelled and observed near-centerline values.
- Compute the fractional bias (FB) at selected percentiles of the frequency distribution. For instance, the fractional bias at the 90th percentile is defined[1] as

  $$FB_{90} = 2\,(C_{mod90} - C_{obs90})\,/\,(C_{mod90} + C_{obs90})$$

  Here, C_{mod90} is the 90th percentile among the modelled concentration values (for the regime in question). It is interesting - and of great practical significance - that the fractional bias at a low percentile, like the 50th, should be zero for a perfect model, whereas the fractional bias for *the very highest percentiles* should be expected to be negative (implying underprediction)! It is as yet an open question *how much* bias should be expected at the various percentiles.
- On the basis of such FB values, a composite performance index can be derived. In forming this index, one should pay due respect to the fact that the number of data points in each regime and data set varies. This can be done by applying a bootstrap resampling procedure. Bootstrap resampling allows us to establish variances for the FB values. These variances can in turn be used to assign a weight to each FB. We end up by combining the FB values at various percentiles (e.g. FB_{50} and FB_{90}) for each regime and data set, using a prescribed weighting scheme.

The steps described comprise the essence of the method. There remains, however, much work to be done in order to gain experiences with the method and judge about its usefulness for real-world problems. As it is evident from the previous sections, there are many details to be considered, and one should always use model evaluation results with prudence. A well-designed composite performance index may be of help, but it can not stand alone.

CONCLUSIONS

The main purpose of this paper is to stimulate discussions in the modelling community, so we can progress towards a set of commonly accepted concepts and tools in model evaluation. A considerable part of the paper has been devoted to a discussion of methods to form ensemble averages. Almost any paper in

---

[1] In literature, FB is often seen defined with a different convention than here, where the sign is reversed.

literature tacitly assumes that we all agree what we are talking about when we use the term "ensemble average". But actually, we do not!

We also have to make clear what we will require from a "perfect model". If we use maximum centerline concentrations to define a perfect model, we are left with some problems. These may to some extent be overcome by using "near-centerline concentrations" instead. However, a "perfect model" as defined by using near-centerline concentrations underpredicts in far more cases than a "perfect model" as defined by using arcwise maximum concentrations. Because of the discrepancy between the various types of "perfect models", the matter is of clear interest to the regulatory community.

ACKNOWLEDGMENTS

The author wishes to acknowledge John Irwin, who has been a valuable source of inspiration with his many contributions to the ISO work group. Thanks is also due to Steve Hanna who has contributed in numerous ways to the work with the Model Validation Kit.

REFERENCES

Hanna, S.R., Strimaitis, D.G. and J.C. Chang, 1991, Hazard Response Modeling Uncertainty (a Quantitative Method). Vol. I: User's Guide for Software for Evaluating Hazardous Gas Dispersion Models. Sigma Research Corporation, Westford, Ma.

Irwin, J.S. and Lee, R.F., 1997: Comparative evaluation of two air quality models: Within-regime evaluation statistic, 4th Workshop on Operational Short-range Atmospheric Dispersion Models for Environmental Impact Assessment in Europe, Oostende, Belgium, May 1996. To appear in *Int. J. Environment and Pollution.*

Lee, R. and Irwin, J.S, 1995, Methodology for a comparative evaluation of two air-quality models. Workshop on Operational Short-range Atmospheric Dispersion Models for Environmental Impact Assessment in Europe, Mol, Belgium, Nov. 1994, Int. J. Environment and Pollution, Vol. 5, Nos. 4-6, pp. 723-733.

Olesen, H.R., 1994, European coordinating activities concerning local-scale regulatory models, in: Air Pollution Modeling and Its Application X, pp. 481-489. Edited by S-E. Gryning and M.M. Millan, Plenum Press, New York.

Olesen, H.R., 1995a, The model validation exercise at Mol. Overview of results. Workshop on Operational Short-range Atmospheric Dispersion Models for Environmental Impact Assessment in Europe, Mol, Belgium, Nov. 1994, *Int. J. Environment and Pollution,* Vol. 5, Nos. 4-6, pp. 761-784.

Olesen, H.R., 1995b, Data sets and protocol for model validation. Workshop on Operational Short-range Atmospheric Dispersion Models for Environmental Impact Assessment in Europe, Mol, Belgium, Nov. 1994, Int. J. Environment and Pollution, Vol. 5, Nos. 4-6, pp. 693-701.

Olesen, H.R., 1996, Toward the establishment of a common framework for model evaluation, in: Air Pollution Modeling and Its Application XI, pp. 519-528. Edited by S-E. Gryning and F. Schiermeier, Plenum Press, New York.

Olesen, H.R., 1997: Pilot study: Extension of the Model Validation Kit, 4th Workshop on Operational Short-range Atmospheric Dispersion Models for Environmental Impact Assessment in Europe, Oostende, Belgium, May 1996. To appear in *Int. J. Environment and Pollution.*

SENSITIVITY ANALYSIS OF LAGRANGIAN STOCHASTIC MODELS FOR CBL WITH DIFFERENT PDF'S AND TURBULENCE PARAMETERIZATIONS

E.Ferrero[1], D.Anfossi[2]

[1]Universita' di Torino, Dip. Scienze Tecn. Avanz., Alessandria, Italy
[2]C.N.R., Istituto di Cosmogeofisica, Torino, Italy

INTRODUCTION

It is known (Thomson, 1987) that Ito's type stochastic models (LS) satisfy the well-mixed condition and hence are physically consistent. An Eulerian probability density function (PDF) of the turbulent velocities, as close as possible to the actual atmospheric PDF, must be prescribed in order to specify the model. Unfortunately these models have a unique solution in one-dimension only (Sawford and Guest, 1988). For this reason the present study will focus on one-dimensional diffusion simulation.

Four LS models have been considered. Two of them make use of the bi-Gaussian (bG) PDF, given by a mixture of two Gaussian PDFs. The parameters of the first model (named LB3) were determined according to the closure scheme suggested by Luhar and Britter (1989) that takes into account up to the third order moment of vertical velocity. The second model, proposed by Du et al. (1994), includes the fourth order moment and is named SWY. The third and fourth models, based on a Gram-Charlier series expansion truncated to the third and fourth order (named GC3 and GC4), were developed by us and are here proposed.

To assess how much the simulation results depend upon the input turbulence field, models were run with two different parameterization schemes for the vertical profiles of the second and third moment and Lagrangian decorrelation time ($\overline{w^2}$, $\overline{w^3}$ and τ).

The simulation results of the eight combinations of models and turbulence were compared to the Willis and Deardorff (1976, 1978 and 1981) convective water tank experiments (hereafter referred as WD). The comparison is based on the normalised values of ground level concentration (g.l.c.) as a function of normalised downwind distance.

OUTLINE OF THE MODELS

WD was relative to the vertical dispersion of passive particles in stationary horizontally homogeneous and vertically inhomogeneous turbulence with no mean flow. The corresponding Ito's type stochastic model is the following (Thomson, 1987):

Air Pollution Modeling and Its Application XII
Edited by Sven-Erik Gryning and Nadine Chaumerliac, Plenum Press, New York, 1998

$$dw = a(z,w)\ dt + \sqrt{2\ B_0(z)\ dt}\ d\mu \qquad (1)$$

$$dz = w\,dt \qquad (2)$$

where w is the vertical velocity, z the displacement, $d\mu$ has zero mean and unit variance. a(z,w), that depends upon the chosen PDF, P(z,w), must be determined from the Fokker-Planck equation for stationary condition; $B_0(z)$ has the following expression (Luhar and Britter, 1989; Weil, 1990):

$$B_0 = \frac{C_0\ \varepsilon}{2} = \frac{\overline{w^2}}{\tau} \qquad (3),$$

where C_0 is a universal constant whose value was not yet established (Rodean, 1994) and ε is the ensemble-average rate of dissipation of turbulent kinetic energy.

As above mentioned, two models out of the four considered in the present analysis assume the PDF to have a bi-Gaussian expression:

$$P(z,w) = A \cdot N_A(w_A,\sigma_A) + B \cdot N_B(w_B,\sigma_B) \qquad (4)$$

(where A+B=1, A >0 , B >0 and N_A, N_B are Gaussian PDFs with means w_A, w_B, and standard deviations σ_A, σ_B). The specification of the various parameters (A, B, w_A, w_B, σ_A and σ_B,) depends on the closure scheme adopted. Their expressions in the case of LB3 and SWY models are not reported here and can be found in the quoted references. These closures are based on the idea that A and B may be associated to the fractions of the area occupied by updrafts and downdrafts (or, equivalently, to the probability of occurrence of updrafts and downdrafts) and $w_A, \sigma_A, w_B, \sigma_B$ to the vertical velocity fluctuations within the updrafts and downdrafts respectively. However it may be worthwhile to recall that De Baas et al. (1986) and Anfossi et al. (1996) pointed out that the physical information is not contained in the specific form and/or in the relative weight of the two single Gaussian distributions contributing to the resultant PDF, but only in the resultant PDF moments.

Concerning the two models we developed making use of the Gram-Charlier PDF (GC), let us recall that this last, truncated to the fourth order, has the following expression (Kendall and Stuart, 1977):

$$P(x,z) = \frac{e^{-\frac{x^2}{2}}}{\sqrt{2\pi}}\left(1 + C_3 H_3 + C_4 H_4\right) \qquad (5)$$

where H_3 and H_4 are Hermite polynomials and C_3 and C_4 their coefficients, whose expressions are:

$$H_3 = x^3 - 3x \qquad (6a)$$

$$H_4 = x^4 - 6x^2 + 3 \qquad (6b)$$

$$C_3 = 1/6\,\overline{\mu^3} \qquad (6c)$$

$$C_4 = 1/24\left(\overline{\mu^4} - 3\right) \qquad (6d),$$

where $\sigma_w = \sqrt{\overline{w^2}}$, $x = w/\sigma_w$ and $\overline{\mu^3}$, $\overline{\mu^4}$ are the standardised moments of w. In case of Gaussian turbulence, equation (10) reduces to the normal distribution (C_3 and C_4 become equal to zero). Introducing eqs. (5 and 6) into the Fokker-Planck equation and solving, we found:

$$a(z,w) = \sigma_w \frac{\frac{1}{\tau}(T_1) + \frac{\partial \sigma_w}{\partial z}(T_2)}{T_3} \qquad (7)$$

where

$$T_1 = -3C_3 - x(15C_4 + 1) + 6C_3x^2 + 10C_4x^3 - C_3x^4 - C_4x^5 \qquad (8a)$$

$$T_2 = 1 - C_4 + x^2(1 + C_4) - 2C_3x^3 - 5C_4x^4 + C_3x^5 + C_4x^6 \qquad (8b)$$

$$T_3 = 1 + 3C_4 - 3C_3 x \; - 6C_4 x^2 + C_3 x^3 + C_4 x^4 \qquad (8c)$$

This model is called GC4 and the one in which C_4 is set equal to zero (i.e. truncation to the third order) is named GC3. Eq. 7 was also independently obtained by Tampieri (personal communication).

It is known that GC may produce negative probabilities in the tails of the distribution. Within these regions of negative probabilities, particles are likely to attain velocities of enormous amount (Flesch and Wilson, 1992) and, consequently, GC may have a restricted domain of applicability. Despite these reservations, GC is proposed here under the hypothesis that the small negative probabilities occur rather rarely so as to permit to discard these cases without violating, in practice, the well-mixed condition. On the other hand, previous works indicated that GC shows a good correspondence to the experimental reality (see, for instance: Frenkiel and Klebanoff,1967; Antonia and Atkinson, 1973; Nagakawa and Nezu, 1977; Durst et al., 1992) and fits PDFs observed in the atmospheric surface layer better than the bi-Gaussian PDF (Anfossi et al., 1996). It is to remember that even the bG may give some unrealistic velocities, due to numerical problems. For this reason Luhar and Britter (reference from Wilson and Flesh, 1993) imposed a threshold value to the admissible velocities by cutting the tails of the distribution. An advantage of using GC instead of bG is that no "ad hoc" closure assumption is necessary. Another advantage is that their inclusion in the Ito-type stochastic equation is very easy and it is possible to consider without any difficulty any order of input Eulerian moment. Furthermore we found that their use needs less computation time than that of bG.

TURBULENCE PARAMETERIZATION

The $\overline{w^2}$, $\overline{w^3}$ and τ vertical profiles, which are input to the models, have been prescribed according to two parameterization schemes suggested by De Baas et al. (1986) and Rodean (1994). They are named here DVN-TU and R-TU, respectively. Their expressions, derived from measurements in WD and in the real atmospheric CBL, are not reported here. GC4 needs also the profile of $\overline{w^4}$. Since neither many data nor parameterizations for $\overline{w^4(z)}$ are available in the literature, we set:

$$\overline{w^4(z)} = 3.5 \left(\overline{w^2(z)}\right)^2 \qquad (9),$$

basing ourselves on measurements performed by Lenschow et al. (1994). It is worth noting that SWY assumes $\overline{w^4(z)} = 3 \left(\overline{w^2(z)}\right)^2$, i.e. assumes the Gaussian closure.

BRIEF OUTLINE OF EXPERIMENT AND SIMULATION CONDITION

In WD experiment, stationary inhomogeneous convective turbulence without mean flow was reproduced. A non-buoyant emission was released from a line source at $z_s / z_i = 0.067$ (h1 hereafter), 0.24 (h2) and 0.49 (h3), where z_s is the source height. The diffusion of particles was measured with respect to a dimensionless time $T = w_* t / z_i$, where w_* is the convective velocity scale and t is time elapsed from the particle release. Dimensionless crosswind concentration C_y as a function of time T was then expressed as a function of dimensionless distance $X = (w_* x)/(z_i u)$, where u=x/t. To compare WD laboratory measurements and our model simulations, C_y and X have been computed as: $\chi_y = (n_p z_i)/(N_p \Delta z)$ and $X = (w_* k \; \Delta t)/(z_i)$, in which N_p is the number of emitted particles, n_p the number of particles counted in the vertical layer $\Delta z = 0.05\ z_i$ adjacent to the ground, Δt the simulation time step and k the number of time steps elapsed from the particle emission. 30,000 particles were released in each run with an initial velocity distribution that has the same moments as the input Eulerian turbulent velocity distribution at that height.

Regarding the reflection of particle velocities at the top and bottom of the computation domain, we used the method proposed by Anfossi et al. (1997).

SIMULATION RESULTS

Fig.s (1 a,c,e) show the comparison among the predicted dimensionless values of cross-wind integrated $\chi_y(X)$ trend, obtained by the four models (LB3, SWY, GC3 and GC4) using DVN-TU, for the three WD experiments (h1, h2 and h3) and those observed. The same graphs, but referring to R-TU, are shown in Fig.s (1 b,d,f).

In the h1 case (source located near the ground) all models using DVN-TU capture the overall shape of the cross-wind integrated g.l.c. trend with a satisfactorely accuracy even if all overestimate the peak value. However GC models overestimate less than bG models. The agreement with the peak value is better in the case of R-TU (particularly for LB3) but the shape of the g.l.c. is largely overestimated by all the models.

Regarding the experiment (h2) in which the source is located at about 1/4 of the CBL height, it clearly appears that R-TU yields wrong results, whereas DVN-TU performs significantly better. In particular GC3 and GC4 fit quite well the observed cross-wind integrated g.l.c.s, whereas LB3 and SWY overestimate both the peak concentration value (of about 13 and 25 per cent, respectively) and its position.

Concerning the third experiment (h3), in which tracer was emitted at about one half of the CBL height, it can be noted that R-TU again yields wrong results. In this case DVN-TU though performing better than R-TU does not fit the observations as well as in the previous case. Again GC3 and GC4 are preferable to LB3 and SWY.

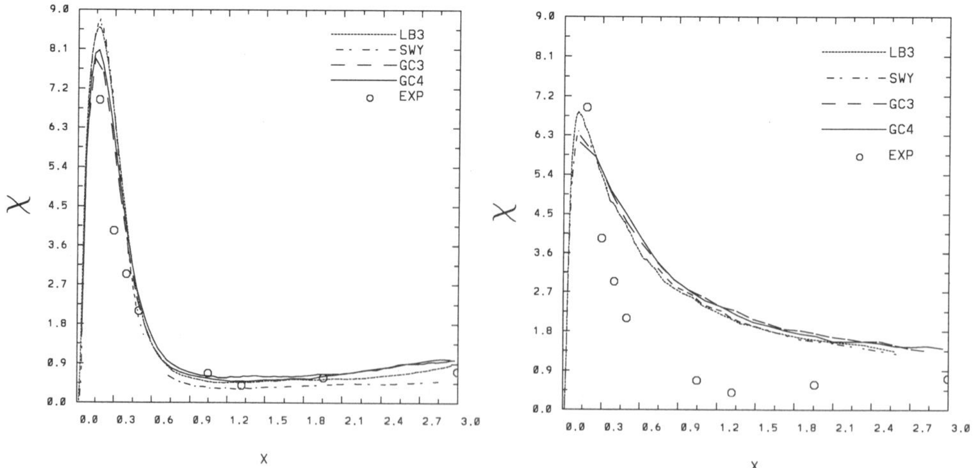

Figure 1 a - h1 and DVN-TU parameterization

Figure 1 b - h1 and R-TU parameterization

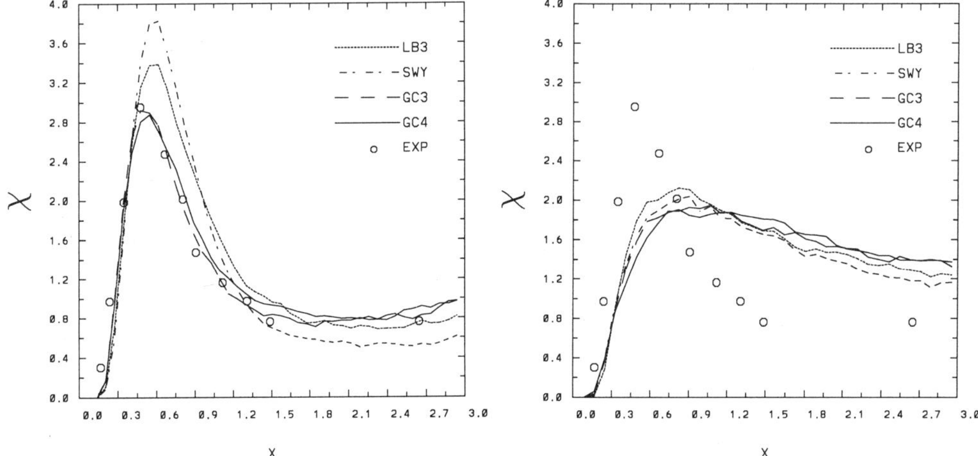

Figure 1 c - h3 and DVN-TU parameterization

Figure 1 d - h3 and R-TU parameterization

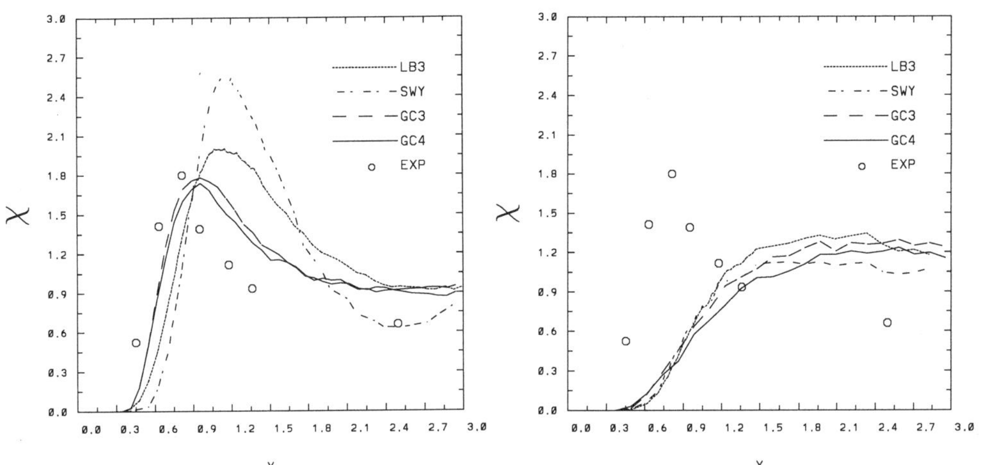

Figure 1 e - h3 and DVN-TU parameterization

Figure 1 f - h3 and R-TU parameterization

Table 1. Statistical indexes computed grouping the 3 WD experiments

Turbulence	model	FB	NMSE	CC	FA2
DVN-TU	LB3	0.11	0.20	0.83	85.2
DVN-TU	SWY	0.12	0.35	0.76	76.9
DVN-TU	GC3	0.04	0.07	0.93	92.6
DVN-TU	GC4	0.04	0.07	0.92	92.6
R-TU	LB3	- 0.05	0.29	0.34	63.0
R-TU	SWY	- 0.08	0.23	0.34	59.2
R-TU	GC3	- 0.06	0.38	0.29	63.0
R-TU	GC4	- 0.07	0.42	0.23	51.9

To get a further and more quantitative insight on the accuracy of the various simulations, a model evaluation was performed by means of the statistical package MODIA (Morselli and Brusasca, 1991). Observed and predicted $\chi_y(X)$ from the three experiments (h1, h2 and h3) were grouped together. In order to rend all the concentration data comparable, they were normalised, by dividing by the observed peak value within each experiment sub-group. Fractional bias (FB), normalised mean square error (NMSE), correlation coefficient (CC) and percentage of model predictions falling within a factor of two (FA2) were computed and presented in Table 1. A detailed inspection of Table 1 confirms the general impression obtained from Fig.s 1 a-f, namely: DVN-TU is superior to R-TU; GC models and DVN-TU give the best results; GC3 and GC4 give very similar simulation results, whereas LB3 is, in general, more accurate than SWY.

Considering all together the results depicted in Fig.s (1 a-f) and in Table 1, the following conclusion may be drawn:

i) - Among the four models considered in the present analysis, those making use of the PDFs based on Gram-Charlier series expansions, GC3 and GC4, gave results more accurate than those which use a bG PDF.

ii) - GC3 and GC4 behave in an approximately similar way; this means that GC4 was not superior to GC3. This result suggests that, including the fourth order moment in convective conditions, is not important. However it is our opinion that accounting for the fourth moment might be important in simulating dispersion in those conditions in which the skewness approaches zero but the kurtosis significantly differs from 3.0 as in neutral and stable conditions. Thus an Ito's type random walk model based on a fourth order closure might be used as a general tool for simulating airborne dispersion in any stability condition provided this moment is known. This result may also mean that more research is needed in order to better prescribe the $\overline{w^4}$ profile.

iii) - An important result of this analysis is that turbulence parameterizations have a significant impact on the model prediction of concentration field. The same result was also found in a previous analysis (Ferrero et al., 1997) in which LB3 and a similar Lagrangian particle model, developed by Weil (1990), were compared. Both these models use the bG PDF but differ in the closure scheme adopted to calculate the bG coefficients. In that paper four different turbulent parameterizations were used as input to the models and their performances were evaluated against the h2 experiment considered in the present work. Also these conclusion suggests that efforts towards further investigation, both experimental and theoretical, about the CBL profiles of $\overline{w^2}$, τ and $\overline{w^3}$ would be certainly useful.

iv) - Since turbulence parameterization has a significant influence on CBL dispersion, the correct estimate of turbulence profiles is an important point, even from a practical point of view. Therefore it is important to state which turbulence parameterization leads to results in closer agreement with measurements.

v) - Even if reservations on possible occurrences of negative values in the tails of GC PDFs, advised to handle these PDFs with care, the present study showed that neglecting the rather unlikely non physical probabilities is inconsequential in practical applications. In the present work it was found that the frequency of these discarded cases is about 10^{-5}. Such a low frequency allows the PDF not to be re-normalised. Indeed it was found that the well-mixed condition is always verified with a degree of accuracy comparable to that obtained in the simulations with the bi-Gaussian PDF. The biggest advantage in principle of the GC model over the bG is that no closure assumption is needed. It is worth adding that the model based on Gram-Charlier PDF demands less CPU time than those based on the bi-Gaussian PDF (typically 1/4 time less).

In conclusion it may be suggested that Ito's type models using a PDF derived from a Gram-Charlier expansion truncated to the third or fourth order moment may be preferred in practical applications.

ACKNOWLEDGMENTS

ENEL/CRAM (Milan, Italy) is greatly acknowledged for its financial support through the Research Contract R25TC0115/00.

REFERENCES

Anfossi D., Ferrero E. , Tinarelli G. and Alessandrini S., 1997, A simplified version of the correct boundary conditions for skewed turbulence in Lagrangian particle models, *Atmos. Environ.*, 31, 301-308

Anfossi D., Ferrero E., Sacchetti D. and Trini Castelli S., 1996, Comparison among empirical probability density functions of the vertical velocity in the surface layer based on higher order correlations, Boundary Layer Meteorology, 82, 193-218

Antonia R.A. and Atkinson J.D., 1973, High-order moments of Reynolds shear stress fluctuations in a turbulent boundary layer, *J. Fluid Mech.*, 58, part 3, 581-593

De Baas H.F., Van Dop H., and Nieuwstadt F.T.M., 1986, An application of the Langevin equation for inhomogeneous conditions to dispersion in a convective boundary layer, *Quart. J. Roy. Meteor. Soc.*, 112, 165-180

Du S., Wilson J.D. and Yee E., 1994, Probability density functions for velocity in the convective boundary layer and implied trajectory models, *Atmos. Environ.*, 28, 1211-1217

Durst F., Jovanovic J. and Johansson T.G., 1992, On the statistical properties of truncated Gram-Charlier series expansions in turbulent wall-bounded flows, *Phys. Fluids*, A 4, 118-126

Ferrero E., Anfossi D., Tinarelli G. and Trini Castelli S., An intercomparison of two turbulence closure schemes and four parametrizations for stochastic dispersion models, Nuovo Cimento 20C, 315-329

Flesch T.K. and Wilson D.J., 1992, A two-dimensional trajectory simulation model for non-Gaussian inhomogeneous turbulence within plant canopies, *Boundary Layer Meteorology*, 61, 349-374

Frenkiel F.N. and Klebanoff P.S., 1967, Higher order correlations in a turbulent field, *Phys. Fluids*, 10, 507-520

Kendall M. and Stuart A., 1977, *The advanced theory of statistics*, MacMillan, New York

Lenschow D.H., Mann J., Kristensen L., 1994, How long is long enough when measuring fluxes and other turbulence statistics, *J. Atm. Ocean. Techn.*, 661-673

Luhar A.K. and Britter R.E., 1989, A random walk model for dispersion in inhomogeneous turbulence in a convective boundary layer, *Atmos. Environ.*, **23**, 1191-1924

Morselli M.G. and Brusasca G., 1991, MODIA: Pollution dispersion model in the atmosphere, Environmental Software Guide, 211-216.

Nagakawa H. and Nezu I, 1977, Prediction of the contributions to the Reynolds stress from bursting events in open-channel flows, *J. Fluid Mech.*, 80 part 1, 99-128

Rodean H.C., 1994, Notes on the Langevin model for turbulent diffusion of "marked" particles, UCRL-ID-115869 Report of Lawrence Livermore National Laboratory

Sawford B.L. and Guest F.M., 1988, Uniqueness and universality of Lagrangian stochastic models of turbulent dispersion, *8th Symposium on Turbulence and Diffusion*, San Diego, CA, A.M.S., 96-99

Tampieri F. (personal communication)

Thomson D.J., 1987, Criteria for the selection of stochastic models of particle trajectories in turbulent flows, *J. Fluid Mech.*, 180, 529-556

Weil J.C., 1990, A diagnosis of the asymmetry in top-down and bottom-up diffusion using a Lagrangian stochastic model, *JW. Atmos. Sci.*, 47, 501-515

Willis G.E. and J. Deardorf, 1976: A laboratory model of diffusion into the convective planetary boundary layer. *Q.J.R. Meteor. Soc.*, 102, 427-445

Willis G.E. and J. Deardorf, 1978, A laboratory study of dispersion from an elevated source within a modelled convective boundary layer, *Atmos. Environ.*, 12, 1305-1311

Willis G.E. and J. Deardorf, 1981, A laboratory study of dispersion from a source in the middle of the convective mixed boundary layer', *Atmos. Environ.,* 15, 109-117

Wilson J.D. and Flesch T.K. (1993) Flow boundaries in random-flight dispersion models: enforcing the well-mixed condition. J. Appl. Meteor., **32**, 1695-1707

POSTER SESSION

MODELLING OF POLLUTANTS DISPERSION FROM A CITY ROAD TUNNEL

Miroslav Jicha[1], Jaroslav Katolicky[1], Raimund Almbauer[2]

[1] Technical University of Brno, Technicka 2, 61669 Brno, Czech Republic
[2] University of Technology of Graz, Inffeldgasse 25, 8010 Graz, Austria

INTRODUCTION

An outlet portal of a city road tunnel is one of crucial point sources of pollutants to be taken into account when modeling dispersion of pollutants within a city. Namely in case of local scale modeling a correct knowledge of velocity and concentration profiles at the outlet portals may play a non-negligible role. And these profiles are established in a dominant way by the traffic density and speed. In the first part of the paper we are going to present a new Lagrangian - Eulerian approach to moving vehicles througout a traffic tunnel and as a result to give values of flow rate entrained into and out of the tunnel and velocity and concentration profiles at the outlet tunnel portal. In the second part of the paper, dispersion of pollutants from the Plabutsch highway tunnel in Graz (Austria) is predicted for various boundarv conditions.

NUMERICAL METHOD

A set of partial differential equations for an incompressible, steady turbulent flow is solveċ using the control volume discretization method with a standard k-e model of turbulence.

The SIMPLER algorithm according Patankar, 1980 is employed for the pressure - velocity coupling, enhanced with the block - correction method to speed convergence. An extra source term in the equation results from the interaction between the continous phase of the surrrounding air and a "discrete" phase of moving vehicles. This discrete phase is treated in a special manner using Lagrangian approach according Crowe et al., 1977, see also Jicha et al., 1996. This treatment calculates the interaction source term as a result of two - way coupling between both phases and enables to account for different density of traffic and speed of cars and their geometrical shape.

RESULTS

In the Fig. 1 a typical velocity profile is presented at the outlet portal of the tunnel and also the flow rate of the air as a function of the density of traffic and the length of the tunnel.

In the Fig. 2 there is an example of concentration field calculated around the Plabutsch tunnel portal in Graz. Numerical predictions were performed for various boundary conditions including

Air Pollution Modeling and Its Application XII
Edited by Sven-Erik Gryning and Nadine Chaumerliac, Plenum Press, New York, 1998

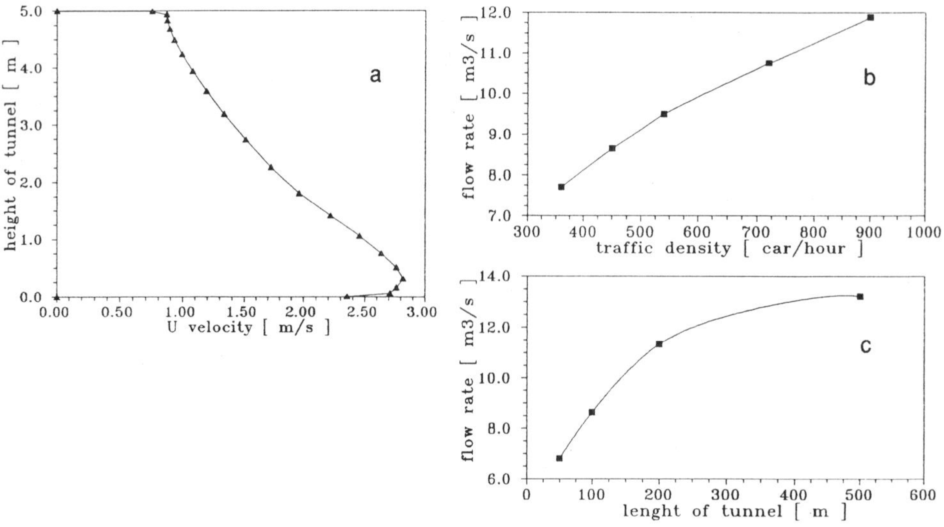

Fig.1 Velocity profile (a) - speed of cars 45 km/hour, (b) flow rate for tunnel length of 100 m, speed of cars 45km/h, (c) flow rate, speed of cars 60 km/h.

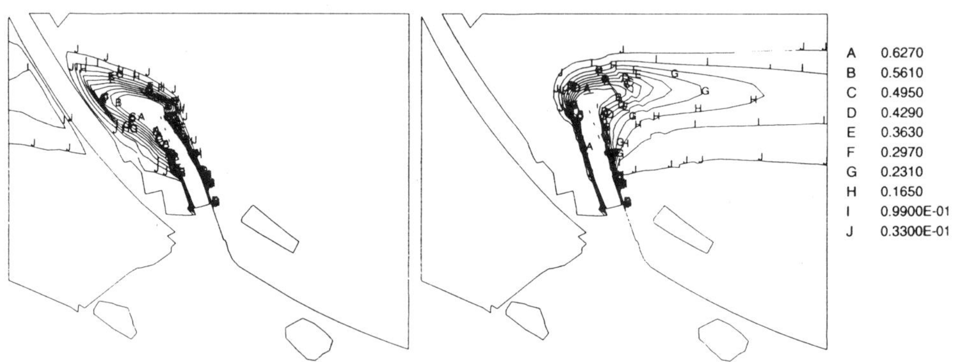

Fig. 2 Concentration distribution around Plabutsch tunnel - East wind, West wind, 2 m/s

direction of wind, boundary velocity profile, thermal stratification of the atmosphere and surface roughness. The area close to the tunnel was modeled correctly in accordance with the actual geometry of surrounding buildings.

ACKNOWLEDGMENTS

A grant COST 615.40 from the COST CITAIR program, Action 615 funded by the Czech Ministry of Education and a grant from the Austrian Ministry for Research & Development under the contract No 45.406/2-IV/3a/95 are gratefuly acknowledged.

REFERENCES

Patankar, S.V., 1980, Numerical Heat Transfer and Fluid Flow, Hemisphere Publ. Corp.,

Crowe, G.T., Sharma, M.P., Stock, D.E., 1977, The partical-source-in cell model for gas-droplet flows, J. of Fluid Engineering, pp. 325-332

Jicha M., Katolicky, J., Patankar, S.V., 1996, Modelling of pollutants distribution at city road tunnel outlet, Proc. of 4th Workshop on Harmonization within atmospheric dispersion modelling, Vol. 2, pp.353-360, Oostende, Belgium

CANCER RISK FROM TRAFFIC-EMITTED AIR POLLUTANTS

S.P. Angius[1], M.G. Santini[1], M. Tamponi[2], and R.E. Gallini[1]

[1]U.S.S.L. 18, P.M.I.P., U.O. Fisica e T.A., Brescia
[2]U.S.S.L. 7, P.M.I.P., U.O. Fisica e T.A., Oggiono (Lecco)

INTRODUCTION

Because of widespread concern about the effects of the exposure of urban populations to a large number of air pollutants, a method allowing a quantitative evaluation of the number of excess cancer cases caused by individual substances is of great interest. It makes it possible to compare the effects of different compounds, more difficult to distinguish in epidemiological studies, and it can be a useful tool to evaluate the impact of possible abatement strategies.

Since vehicular traffic is one of the main sources of air pollution in urban areas, the method described here concentrates on primary pollutants emitted by transportation sources. In particular, the expected cancer incidence due to benzene and benzo(a)pyrene has been calculated. Benzene is included by the IARC in group 1, as it induces leukaemia in humans, while exposure to BaP has been related to respiratory tract cancer (WHO 1987).

The method utilizes a modified version of the E.P.A. APRAC-3 dispersion model (Angius et al., 1995) for the calculation of ambient concentrations. The exposure is estimated on the basis of the population distribution, with a correction for indoor/outdoor permanence time. The number of expected cancer cases in the metropolitan area of the city of Brescia, in northern Italy, is then calculated, and the confidence interval estimated.

RESULTS AND DISCUSSION

The 15x12 km^2 area considered was divided into 1-square-kilometer grids, four seasonal trends of average mean 1-hour concentrations of benzene and BaP were calculated for each grid square, and a correction was introduced to account for the difference between indoor and outdoor exposure. The uncertainty on the results was estimated, based on the uncertainties associated with the various parameters entering into the calculation. The maximum error introduced with the dispersion model estimates (less than a factor of two) is significantly less than the uncertainty in the risk factor estimates, which range from $4x10^{-6}$ (WHO 1987) to $53x10^{-6}$ (SCAQMD 1988) for a lifetime exposure to 1 $\mu g/m^3$ of benzene and from $0.11x10^{-5}$ to $1.4x10^{-5}$ per year per 1 ng/m^3 of BaP (EPA 1984). For this reason, the calculation of cancer incidence was repeated using the lowest and highest available risk factor estimates.

Exposure to current concentrations of benzene is expected to cause between 7±1 and 92±13 cases of leukaemia over 70 years, while between 4.5±0.7 and 57±8 cancer cases are estimated to occur from a lifetime exposure to traffic-emitted BaP. Distribution maps of the expected yearly cancer incidence are shown in fig. 1. An outline of the urban area is also indicated.

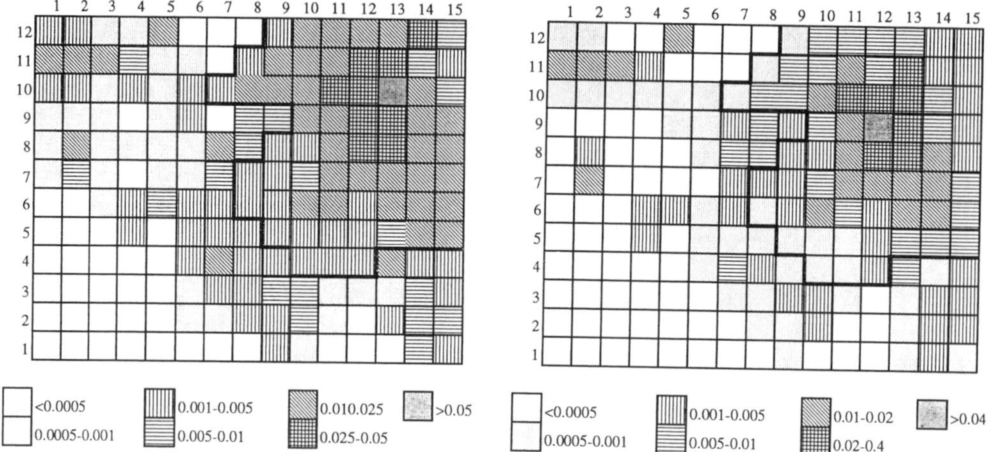

Figure 1. Distribution maps of the estimated yearly cancer incidence for benzene (left) and BaP (right).

The maximum risk occurs in the city downtown area, where both the highest population density and the highest pollutant concentrations are found.

Considering that the emission inventory assumes that 26% of vehicles have diesel engines, these results indicate that a switch from gasoline- to diesel-powered cars would not cause an overall decrease in cancer incidence.

CONCLUSIONS

The method presented here provides estimates of the level of risk associated with exposure to traffic-emitted primary pollutants in the area under study, and gives useful indications of the uncertainty to be expected in this type of calculations. It can be a useful tool in urban traffic planning and in the choice of emission reduction measures.

REFERENCES

Angius S., Angelino E., Castrofino G., Gianelle V., Tamponi M., and Tebaldi G., 1995, Evaluation of the effects of traffic and heating reduction measures on urban air quality, *Atmos. Environ.,* **29B**, 3477-3487.

E.P.A., 1984, Review and evaluation of the evidence for cancer associated with air pollution, EPA-450/5-83-006R.

S.C.A.Q.M.D.,1988, Proposed Rule 1401. New Source Review of Carcinogenic Air Contaminants. El Monte, California.

WHO, 1987, Air quality guidelines for Europe, WHO Regional Publications, European Series no.23.

MODELLING ATMOSPHERIC PARTICLE CONCENTRATIONS ON THE BASIS OF ANTROPOGENIC EMISSIONS

J.A. van Jaarsveld and E.D.G. van Zantvoort

National Institute of Public Health and the Environment
Laboratory of Air Research
P.O. Box 1, 3720 BA, Bilthoven, The Netherlands

INTRODUCTION

Health effects in relation to airborne suspended particulate matter (SPM) receive nowadays much attention in Europe and North America. The inhalable fraction of SPM consists mainly of particles with sizes below 10 μm (PM10) and is thought to be originating from antropogenic activities mainly. These activities lead partly to direct particle emissions and partly to the production of secondary particles such as ammonia-sulphate and ammonia-nitrate.

In order to explain ambient concentrations in terms of sources and source contributions, an inventory has been made of primary dust sources in the Netherlands and also (but with less detail) for Europe (Berdowski *et al.*, 1997). The OPS model is used to describe the dispersion, transport and deposition of primary SPM as a function of the particle size distribution of the various contributing sources. Concentrations of secondary aerosols are also modelled on the basis of sulphur and nitrogen emissions.

The atmospheric behaviour of particles

Transport and dispersion of small particles is similar to that of gases. If one wants to explain current levels of particle concentrations, then it is necessary to take into account also influences of sources far away from the receptor. In such a case the atmospheric residence time of particles must be taken into account which is strongly related to the physical properties of the particles, of which the particle diameter is the most important. The route to forming particles containing SO_4^{2-}, NO_3^- and NH_4^+ is through direct gas-to-particle conversion and through evaporation of cloud droplets in which previously conversion has taken place. Newly formed particles are usually smaller than 0.01 μm (Aitken particles). If the gas condenses on existing particles (e.g. heterogeneous processes), the median size of these particles will also be relatively small because small particles have the highest specific surface area. Through processes such as coagulation, small particles will grow and will finally be concentrated in a 0.1 - 1 μm range, the so-called accumulation mode.

Dry deposition of particles. Very small particles (diameter < 0.1 μm) deposit primarily through Brownian diffusion and large particles (diameter > 2 μm) through gravitational settling. Particles with diameters in the range of 0.1 - 1 μm are deposited by such processes as interception, impaction and phoretic mechanisms. In this range of diameters the characteristics of the surface become very important. Key factors are vegetation height and roughness length, leaf area index, sizes of collecting elements, such as needles, leaves and hairs, leaf stickiness and wetness in combination with turbulent velocities in the overlying atmospheric surface layer.

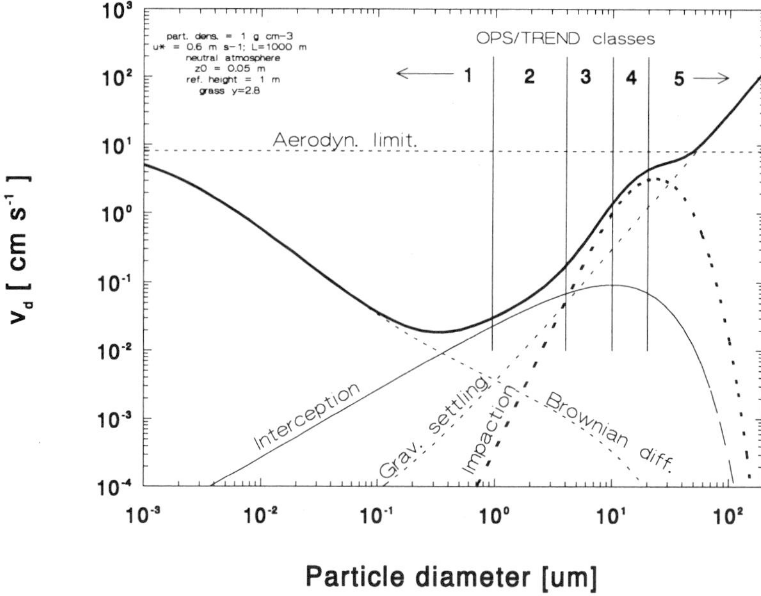

Figure 1. Predicted dry deposition velocities of particles as a function of particle size due to the different processes. Values are representative for grassland during neutral atmospheric stability. The classes used in the OPS model are also indicated.

Wet scavenging of particles. Wet scavenging of aerosols is an efficient process. Two main processes can be distinguished: the uptake of particles in cloud water with subsequent removal during precipitation (in-cloud scavenging) and uptake of particles by falling rain drops (below cloud scavenging). The below cloud process depends strongly on the size of the particles, where particles around a mass-median diameter of 0.5 μm are least efficiently scavenged. Table 1. gives an overview of the wet scavenging parameters as employed in the calculations. In Figure 1. the relation between particle size and dry deposition velocity is plotted using a model which describes Brownian diffusion, inertial impaction and gravitational acceleration as a function of roughness length, friction velocity and particle density on the basis of relations given by Slinn (1983).

The model

The model used in this study is the OPS (Operational Priority Substances) model (Van Jaarsveld, 1995). This model is intended for the simulation of time-averaged concentrations and depositions on a local to regional scale due to atmospheric emissions.

Particle modelling in the OPS model Because of the very different atmospheric residence times of particles of different sizes it is necessary to describe dynamically the dry deposition process as a function of particle size. In the OPS model, five particle size classes are used, each characterised by a (monodisperse) particle size with corresponding properties calculated by the semi-empirical model of Sehmel and Hodgson (1979) which gives similar results as the more theoretical model of Slinn (1983). Concentrations and depositions are calculated for each of these classes and weighted with the percentage of the total particle mass appointed to the individual classes. Such an approach is especially useful for the modelling of primary-emitted particles because they usually cover a broad range of particle sizes, often including a significant fraction large particles. Particle growth is not incorporated in the present model but is implicitly assumed to take place in the lowest size-class ($d < 1$ μm). In Table 1. various properties of the classes are shown.

Size-distributions. Although data from the literature indicates that different emission processes produce different particle-size distributions, it was not a straight forward case to translate the available data to characteristic distributions for the emission categories distinguished in this study. Moreover, it

Table 1. Properties of the particle size classes with respect to dry and wet deposition for land surfaces.

Class	Size range	Median aerodyn. diameter	Initial mass distribution for primary particles	Scavenging ratio	Mean scavenging rate Λ [a])	Mean dry deposition velocity v_d [c])	Mean atmos. residence time $T_{1/2}$ [b])
	μm	μm	%		s^{-1}	$m\ s^{-1}$	h
1	< 0.95	0.2	42	1.2×10^5	2.0×10^{-6}	0.00065	63
2	0.95 - 4	1.5	33	10^6	1.5×10^{-5}	0.0025	11
3	4 - 10	6	14	10^6	1.5×10^{-5}	0.0071	8
4	10 - 20	14	6	10^6	1.5×10^{-5}	0.0132	6
5	> 20	40	5	10^6	1.5×10^{-5}	0.067	2

[a]) scavenging ratio during precipitation
[b]) $T_{1/2} = \ln 2 / (v_d / z_i + \Lambda)$, where z_i is the mixing layer height
[c]) dry deposition velocity for 50 m height and a roughness length of 0.15 m

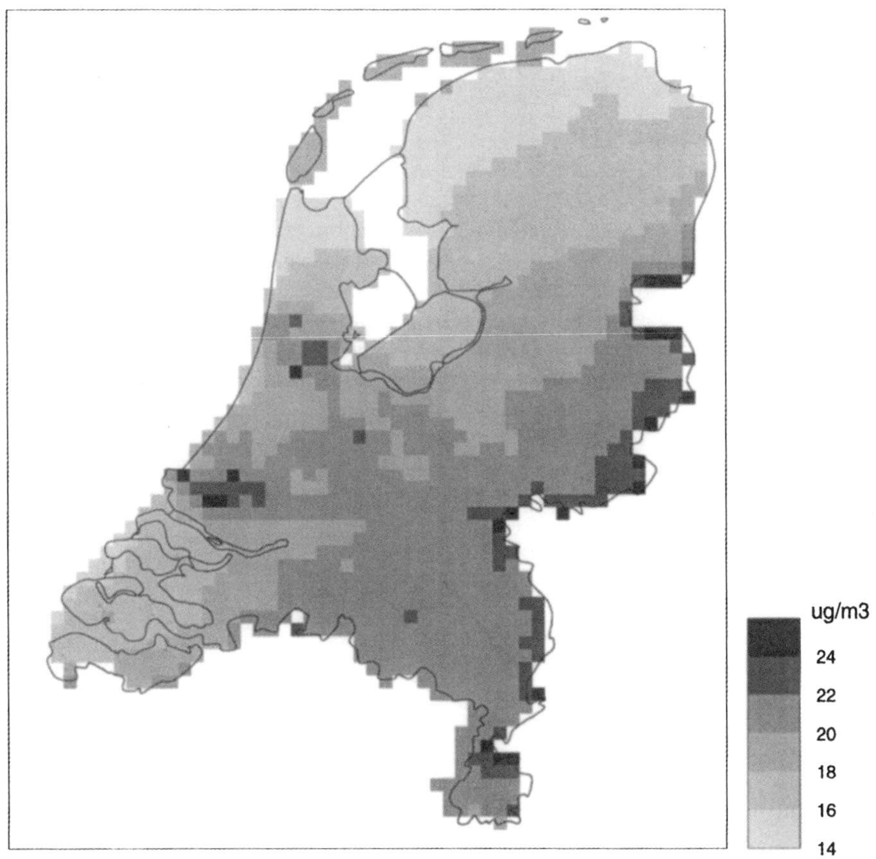

Figure 2. PM10 concentration distribution over the Netherlands modelled on the basis of 1993 emissions

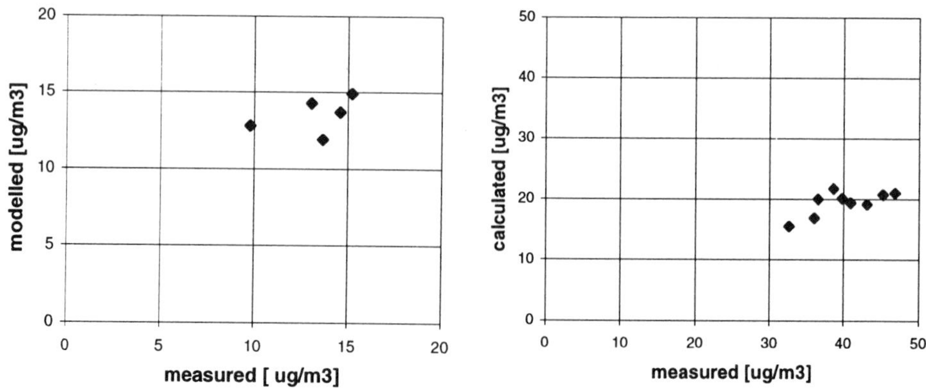

Figure 3. Comparison of calculated and measured aerosol concentrations. Left: calculated secondary aerosols versus measured secondary aerosol. Right: calculated secondary + primary aerosol versus measured non-urban PM10 levels in the Netherlands.

Table 2. Contribution of source categories to PM10 concentrations in the Netherlands ($\mu g\ m^{-3}$)

Category	Primary PM10		Secondary PM10		Total
	Dutch sources	Foreign sources	Dutch sources	Foreign sources	
Industry	0.6	1.2	0.3	5.3	7.4
Energy	1.1	1.1	0.2	0.7	3.1
Traffic	3.2	1.1	1.4	3.0	8.7
Agriculture	0.1	0.5	1.1	1.2	2.9
Other	0.8		0.1		0.9
Total	5.8	3.9	3.1	10.2	23

was not possible in the present model to use more than one distribution in a single model run. Therefore a single distribution is used for all sources in all calculations. This distribution is based on measurements on coal-fired power plants and given in Table 1. For secondary aerosols it is assumed that these are mainly surface-distributed over primary particles. Because of the higher specific surface area of smaller particles, secondary particles are mainly in class 1 and 2.

RESULTS

The calculated distribution of antropogenic aerosols is given in Figure 2. The contribution of different source categories is given in Table 2. The sum of the calculated secondary and primary aerosol plus an (unknown) contribution of non-antropogenic aerosol should be comparable with measured PM10 levels. Such a comparison is given in Figure 3. It turns out that the measured PM10 is almost a factor 2 higher than the calculations. This difference is not fully attributable to soil-generated dust and/or sea salt which was not taken into account. Since calculated concentrations of sulphate, nitrate and ammonium do agree well with their measured counterparts, it is likely that the levels of primary emitted particles are strongly underestimated by the model. Sensitivity runs with the model using different size distributions indicate that the larger part of the underestimation must be due to too low primary emissions.

REFERENCES

Berdowski J.J.M., Mulder W., Veldt C., Visschedijk A.J.H. and Zandveld P.Y.J. (1997) Particulate emissions (PM10 - PM2.5 - PM0.1) in Europe in 1990 and 1993. TNO report, TNO Delft the Netherlands. *In preparation*

Sehmel G.A. and Hodgson W.H. (1980) A model for predicting dry deposition of particles and gases to environmental surfaces. *AIChE Symposium Series* 86, 218-230.

Slinn W.G.N (1983) Predictions for particle deposition to vegetative surfaces. *Atmospheric Environment* **16**, 1785-1794.

van Jaarsveld J. A. (1995) Modelling the long-term atmospheric behaviour of pollutants on various spatial scales. Ph.D. Thesis, University of Utrecht, The Netherlands.

WORST CASE AIR POLLUTION STUDIES IN THE PLOIESTI-PRAHOVA INDUSTRIAL AREA BASED ON THE RIMPUFF DISPERSION MODEL

D. Gultureanu[1], B. Gultureanu[2], T. Mikkelsen[3], S. Thykier-Nielsen[3]

[1] Department of Physics, "Petroleum-Gas" Univ. of Ploiesti, 2000-Ploiesti, Prahova, ROMANIA
[2] Department of Computer Sciences, "Petroleum-Gas" Univ. of Ploiesti
[3] Department of Wind Energy and Atmospheric Physics, RISØ National Laboratory, DENMARK

The paper presents the results of research work started at RISØ National Laboratory, Denmark and continued at "Petroleum-Gas" University of Ploiesti, Romania regarding application of the real-time episode dispersion model RIMPUFF to the Ploiesti - Prahova industrial area with the motivation to understand and quantify severe air pollution scenarios of ground level air concentrations resulting from the different sources of petro-chemical origin located in the industrial area.

Records of one hour mean wind and temperature measurements were taken from the nearby Ploiesti meteorological station and used as input data to the combined flow and diffusion model LINCOM/RIMPUFF for a sequence of severe emissions and wind flow episodes.

The mesoscale dispersion model RIMPUFF (Thykier-Nielsen and Mikkelsen, 1984) is a fast and operational computer code suitable for real time simulation of environmentally hazardous materials and gases released to the atmosphere. RIMPUFF is a puff diffusion code designed for real-time simulation of puff and plume dispersion during time and space changing meteorology.

LINCOM, the flow model driving the dispersion model RIMPUFF, extends the theory for neutral, potential flow over a small hill to mesoscale complex terrain (Troen and de Baas, 1986). LINCOM is a fast diagnostic, non-hydrostatic dynamic flow model based on the solution of linerized versions of the three momentum equations and the continuity equation with a first order spectral turbulent diffusion closure.

Sources were considered inside the Ploiesti industrial area. For the simulations we considered a 60km x 60 km domain modelled on a 60 x 90 grid.

The LINCOM/RIMPUFF model has been run on 10 cases including examples of typical daytime convection and also including some less typical, but from an pollution point of view very interesting cases, especially because of the on-shore wind conditions.

Air Pollution Modeling and Its Application XII
Edited by Sven-Erik Gryning and Nadine Chaumerliac, Plenum Press, New York, 1998

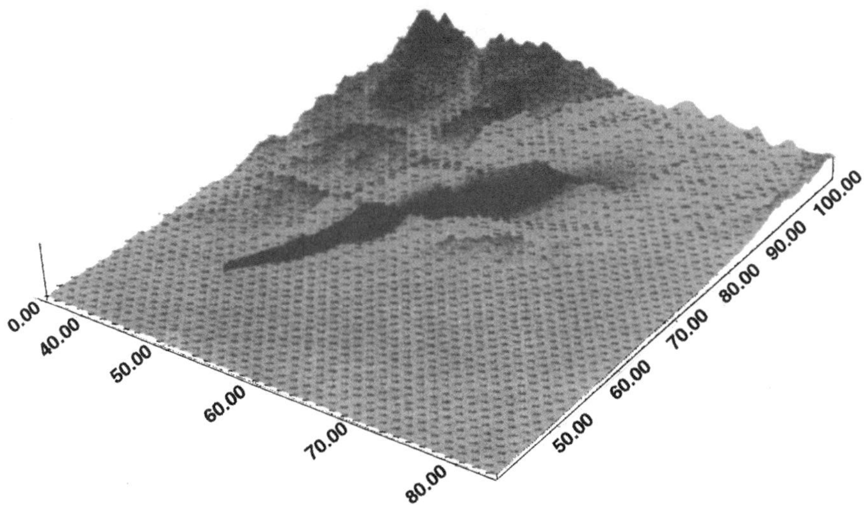

Three-dimensional map of the Ploiesti - Prahova area considered in the simulation with windfield and concentrations integrated over one hour for a typical daytime convection.

The study encounters the modelling of some possible worst case dispersion scenarios in the area. The results are in good agreement with local observations made in the past by Governmental Air Quality Agency.

We present a typical dispersion scenarios of dispersion in the Ploiesti - Prahova region:

REFERENCES

Mikkelsen, T., Larsen, S.E. and Thykier-Nielsen, S., 1984, Description of the Risoe puff diffusion model, Nuclear Technology, 67, 56-65

Mikkelsen, T., Thykier-Nielsen, S., Troen, I., et al., 1988, A hard model for complex terrain, Preprints from Eight Symposium on Turbulence and Diffusion, San Diego, CA, April 25-29, 180-185

Thykier-Nielsen, S, Mikkelsen, T., Kamada, R., and Drake, S.A., 1990, Wind flow model evaluation study for complex terrain, Proceedings of Ninth Symposium on Turbulence and Diffusion, American Meteorological Society, Boston, MA, USA

Troen, I. and de Baas, A. F., 1986, A spectral diagnostic model for wind flow simulation in complex terrain, Proc. Euro. Wind Energy Assoc. Conf., Rome, October 7-9

THE IMPORTANCE OF LAND USE CLASSIFICATION ON MESOSCALE AIR QUALITY MODELS: A SENSITIVITY STUDY OVER MADRID

Roberto San José, Juan F. Prieto, Carlos Franco and Rosa M. González[1].

Group of Environmental Software and Modelling.
Computer Science School - Technical University of Madrid
Boadilla del Monte- 28660, Madrid (Spain)

(1) Department of Meteorology.
Complutense University
Ciudad Universitaria, Madrid (Spain)

The land use classification is usually produced by hand using different detailed maps which have been produced by aerial picturing or direct processes. The use of satellite images has been incorporated widely in the recent years. We present a land use classification which has been made by using a LANDSAT-5 image (October, 1987). The ANA (San José et al., 1994) mesoscale air quality system is used to study and compare the results on ozone concentrations at surface level when the automatic land use classification by using the satellite image is used and when the handmade classification is used. We present results on four different Madrid monitoring stations for ozone for both land use classifications. Two important results can be concluded: the changes on the ozone concentrations due to different land use classifications are important and can lead to different and wrong conclusions. The satellite information when applied the automatic land use classification seems to be more precise and accurate than the handmade land use classification. A full and complete statistical study should be done in the future however computational power limitations are an important obstacle for such a experiment.

During long times, the application of environmental models has been difficult on obtaining the land use maps due to the great amount of human power which is required (Pielke (1984)). New generation of computers has allowed the development of tools to carry out automatically the land use classification. We have developed a set of software applications based on the Ormeño et al. (1994) and Ormeño (1991) and Castillo M.P. (1993) methodology under UNIX environment over a LANDSAT-5 thematic mapper image which corresponds to August, 15, 1985. The automatic procedure (Mather (1989)) has been applied over three different aspects: generation of candidates to training areas, training of the classificator and classification of the image (Curran et al., 1990). All the applications are independent and can be applied simultaneously however the results of some applications are used as input for other applications. The implementation is in C++ and the exchange of information is made by using files. A buffer system is used to optimize the accesses to hard disk.

The LANDSAT-5 image has been restricted to the ANA (Atmospheric mesoscale Numerical pollution model for regional and urban Areas) model domain which is 80 x 100 km. The pixel resolution is 30 x 30 m^2. The TM sensor stores the information from seven spectral bands so that the image is composed by seven different independent images. The image is classified in 14 different land use

Air Pollution Modeling and Its Application XII
Edited by Sven-Erik Gryning and Nadine Chaumerliac, Plenum Press, New York, 1998

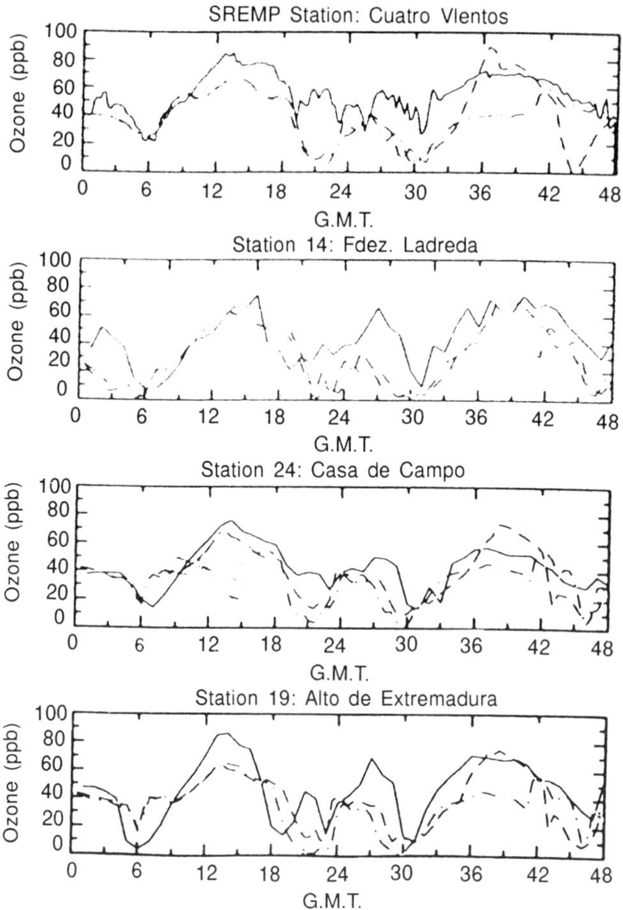

Fig. 1. Observed data ——, satellite land use — — —, handmade land use —— – ——

types: caducous, perennial, mixed, olive, garden, bush, vineyard, fruit, pasture, rice, unirrigated, water, urban and suburban. These land use types are particulary interested for air quality modelling. The biogenic emissions are considered because the mixed, caducous and perennial forest types are specifically included. The water, unirrigated, urban, etc. types are also important for determining the different surface balance fluxes which can be very different on these different types of surfaces. These differences lead to important differences on the vertical mixing and advection at different heights. These processes lead to a specific local meteorology and photochemistry which is an essential key for understanding and modelling the urban and regional air quality. The Figure 1 shows the differences between observed ozone data on 5-6/06/95 at four different stations at Madrid metropolitan area -one of them is the SREMP station which is mantained by our Laboratory and funded with the SREMP-EC-DGXII Project- and the modelled ozone concentrations with the handmade land use classification and the satellite land use classification which was produced with the software application presented in this contribution (REMO). Results shows important differences between both methods which justify the importance of a good quality land use classification and the quality of the classification produced with REMO tool.

REFERENCES

- Octubre, 1987. "LANDSAT THEMATIC MAPPER (TM) CCT Formats Standards", EPO/64-567. Revisión 2.

- M. P. Castillo, Diciembre, 1993. "Análisis y Clasificación de Imágenes Multiespectrales del Satélite Landsat (Thematic Mapper)". Trabajo Fin de Carrera.

- P. J. Curran, G. M. Foody, K. Ya. Kondratyev, V. V. Kozoderov y P. P. Fedchenko, 1990. "Remote Sensing of Soils and Vegetation in the USSR", Ed. Taylor & Francis.

- Paul M. Mather, 1989. "Computer Processing of Remotely Sensed Images; an introduction", Ed. John Wiley & Sons.

- R. San José, L. Rodriguez, J. Moreno, M. Palacios, M. A. Sanz, M. Delgado, 1994. "Eulerian and Photochemical Modelling over Madrid Area in a Mesoscale Context". Ed. Computational Mechanics Publications.

- Roger A. Pielke, 1984. "Mesoscale Meteorological Modeling", Ed. Academic Press.

- S. Ormeño, 1991. "Realce y Clasificación Automática de Imágenes de Satélite", E.U.I.T. Topográfica (UPM).

- S. Ormeño, H. Madrona, M. Castillo, J. M. Hernández, M. Delgado & R. San José, 1994. "Methodologhy for Land-use Type Classification of Multispectral Images". Ed. Computational Mechanics Publications. Envirosoft/94.

STUDY OF THE DECREASE OF AIR POLLUTION CONCENTRATION LEVELS WITH THE METEOROLOGICAL FACTORS IN İSTANBUL

Mete Tayanç[1], Mehmet Karaca[2], Arslan Saral[3] and Ferruh Ertürk[3]

[1] Marmara University, Dept. of Environmental Eng., Göztepe, İstanbul, Turkey.
[2] İstanbul Technical University, Eurasia Inst. Earth Sciences, Maslak, İstanbul, Turkey.
[3] Yıldız Technical University, Dept. of Env. Eng., Beşiktaş, İstanbul, Turkey.

INTRODUCTION

The aim of this study is twofold: First, to gain perspective for the assessment of the spatial distribution of one of the air pollutants, sulfur dioxide, in the region by the use of a statistical modelling scheme, kriging; and second, to show the decrease of sulfur dioxide concentrations over İstanbul by elaborating on the reasons of this decrease in terms of the meteorological factors.

RESULTS

Developing an accurate and detailed picture of sulfur dioxide concentration distribution in İstanbul is a crucial component of air pollution assessment and serves multiple purposes. It is a result that links the low quality fuel consumption for industrial and heating purposes of the region with the damage it causes to the environment.

Monthly sulfur dioxide concentration levels from 15 monitoring stations are used to obtain the spatial patterns of air pollution. An optimum interpolation technique, kriging, is used to obtain spatial distribution fields of sulfur dioxide over İstanbul (Tayanç et al., 1997). Figure 1 a and b depicts the spatial distribution of the monthly averages of the sulfur dioxide concentration levels over İstanbul for December 1994 and January 1995, respectively. It is intuitively clear that the sulfur dioxide concentration levels for the heating season months over the İstanbul metropolitan area are well over the long-term standard of WHO, 50 $\mu g/m^3$ (WHO, 1987). Some parts of the region were experiencing concentrations much greater than 300 $\mu g/m^3$, above which an increase in mortality can be established (Hatzakis et al., 1986; Karaca et al., 1995). Two maximum regions are observed; one in the European side and the other over the Asian side. The peak region over the European side includes Golden Horn Valley, Taksim-Şişli, and Topkapı-Eminönü areas. On the Asian side Göztepe-Bostancı area received major threaten from sulfur dioxide pollution. These areas are generally characterized by very high residential population densities. However, sulfur dioxide concentration levels in the other months of the year (april to october) remain at low levels or slightly exceeding the long-term standard. This result yields a fact that İstanbul's sulfur dioxide pollution is mainly a heating season phenomenon, owing to the consumption of low quality fuel.

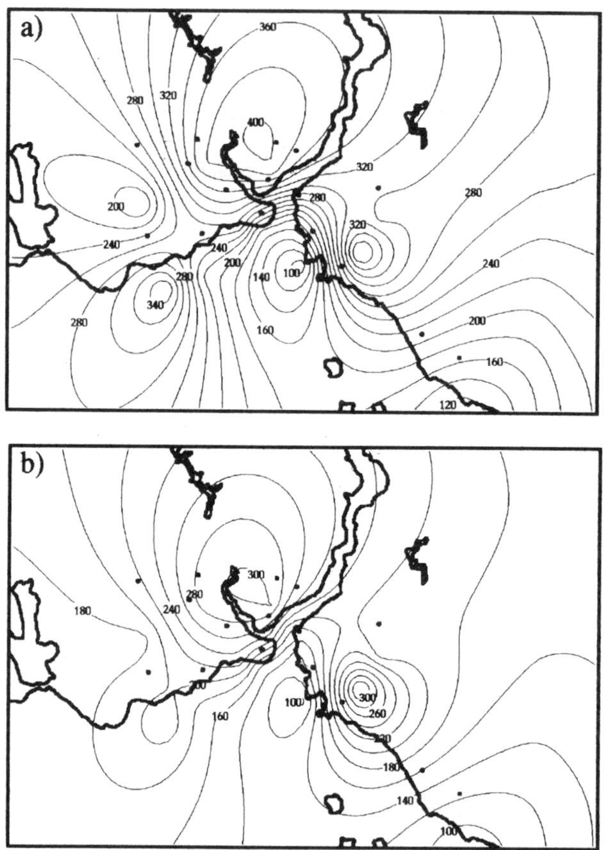

Figure 1. Sulfur dioxide concentration patterns ($\mu g/m^3$) a) December 1994, b)January 1995

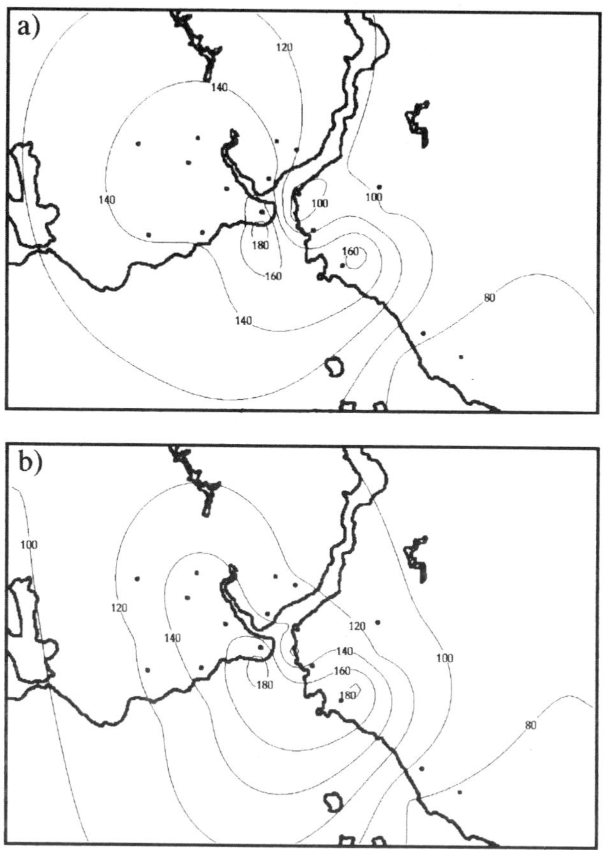

Figure 2. Sulfur dioxide concentration patterns ($\mu g/m^3$) a) December 1995, b)January 1996

However, Figure 2 a and b presents much lower levels of air pollution for the same heating season months of the following season: December 1995 and January 1996. Maximum concentration regions having sulfur dioxide levels greater than 300 $\mu g/m^3$ have disappeared in these months. This leads us to investigate the possible causes of the decrease in sulfur dioxide concentrations.

The impact of the atmospheric circulation and stability conditions on the dilution of air pollution has long been known. Here, we calculated daily ventilation indices which are the products of daily average wind speeds and mixing heights obtained from the data of upper air sounding at 00.00 UTC of Göztepe synoptic station. Results revealed that ventilation is greater in the 1995-1996 heating season. Thus, an increase in the ventilation recently explains the decrease in sulfur dioxide levels to a considerable extend. However, switching from the use of coal to natural gas for heating in many buildings, probably, is the major cause in the decrease of air pollution levels. A natural gas distribution pipeline network has been constructed to cover the regions having severe air pollution problems and it has been operated since 1995. Owing to the reasons that Turkey has very little reserves of natural gas, the majority of the clean fuel has been imported.

REFERENCES

WHO, 1987, *Air quality guidelines for Europe*, WHO Regional Publications, European Series No. 23, Copenhagen.

Karaca, M., Saral, A., Tayanç, M., and Ertürk, F., 1996, Analysis of air pollutants in İstanbul: a preliminary study, *Air Pollution Modeling and Its Application XI,* Gryning and Schiermeier, ed., Plenum Press, New York.

Hatzakis, A., Katsouyanni, K., Kaladidi, A., Day, N., and Trichopoulos, D., 1986, Short-term effects of air pollution on mortality in Athens, *Int. J. Epidemiol.*, 14:73-81.

Tayanç, M., Karaca, M., and Yenigün, O., 1997, Annual and seasonal air temperature trend patterns of climate change and urbanization effects in relation with air pollutants in Turkey,*J. Geophys. Res.,* D2, 102:1909.

EXPERIMENTAL RESULTS ON FLUCTUATIONS IN CONCENTRATION FROM PAIRS OF SOURCES

Belinda M. Davies[1], David J. Thomson[1], Chris D. Jones[2], Ian H. Griffiths[2]

[1] U.K. Met. Office (APR)
London Road, Bracknell
Berkshire RG12 2SZ
England

[2] Defence Evaluation and Research Agency, CBD
Porton Down, Salisbury
Wiltshire SP4 OJQ
England

INTRODUCTION

Estimates of fluctuations in concentration resulting from multiple sources are complicated by the non-independence between the contributions to the concentration from each of the sources. Such fluctuations are important for toxic, inflammable and odorous releases and, with new standards for common air pollutants such as SO_2 being introduced which are based on averaging times less than 1 hour, are becoming important for routine emissions too.

Here we present results from short range dual tracer experiments conducted over flat terrain. A key quantity of interest is the cross-correlation between the fluctuating concentrations as this determines the way in which the variance of the combined concentration is related to that of each concentration separately. Results are presented for this quantity as a function of downwind distance and source separation. Previous studies with multiple sources include the wind tunnel experiments by Warhaft (1984) and the atmospheric studies by Sawford et al. (1984) which were conducted with much slower response detectors than used here.

EXPERIMENTAL DETAILS

Two experimental trials were conducted, one at Cardington, England (May, 1996) and the other at Nevada in the U.S.A. (August, 1996). Two tracers consisting of ammonia and propane were released. The experimental technique allowed for simultaneous measurement of the separate instantaneous concentrations of each tracer by using two different recently developed detection systems, namely the UVIC detector which uses ultra-violet photo-ionisation (which responds to the ammonia and not the propane), and the Flame Ionisation Detector (FID) (for which the converse applies). Both de-

Air Pollution Modeling and Its Application XII
Edited by Sven-Erik Gryning and Nadine Chaumerliac, Plenum Press, New York, 1998

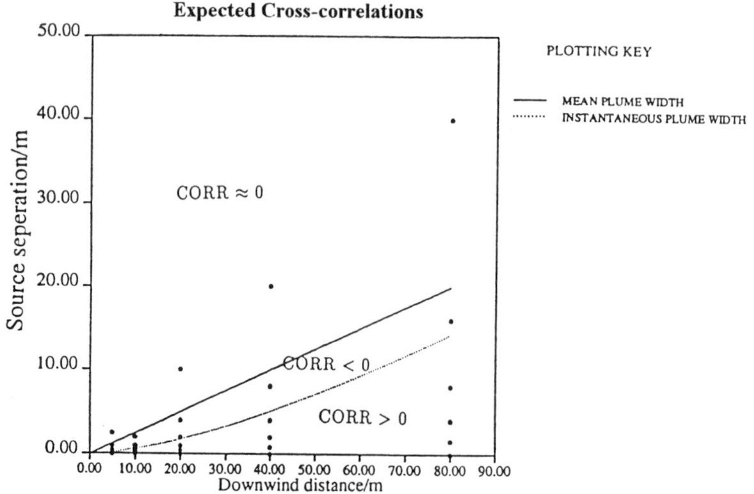

Figure 1. A schematic diagram of the expected variation of cross-correlations with source separation and downwind distance in relation to the mean and instantaneous plume widths.

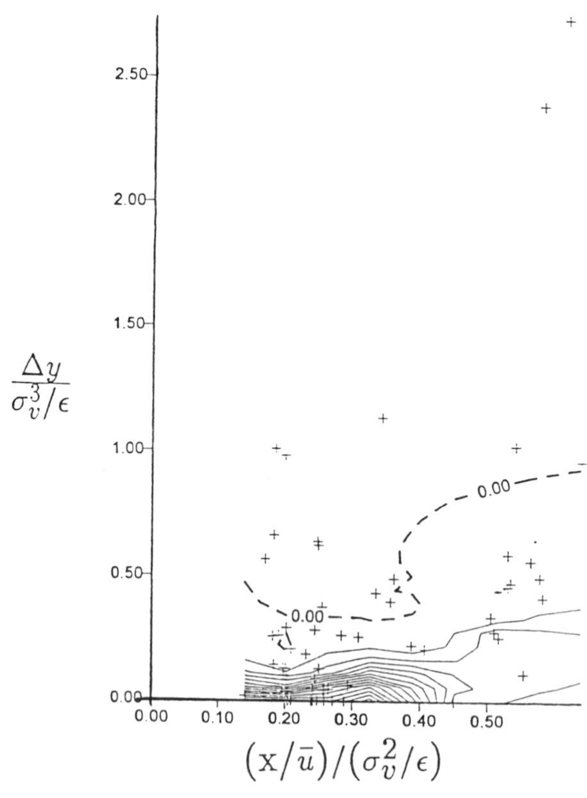

Figure 2. Cross-correlations from the Cardington trials as a function of $\Delta y/(\sigma_v^3/\epsilon)$ and $(x/\bar{u})/(\sigma_v^2/\epsilon)$ (the contour spacing is 0.05).

tectors have high sensitivities (1.0ppm) and fast response times (0.01s) and can be co-located. Experiments were conducted with various source separations and with five detector pairs positioned at varying downwind distances (between 5 and 80m). The sources and receptors were at a height of 4m. Most of the runs were conducted with lateral source separation although a few were conducted with longitudinal separation.

RESULTS

Figure 1 is a schematic representation of the expected cross-correlations in relation to the mean and instantaneous plume widths at various source separation/downwind distance configurations. Figure 2 shows how the cross-correlation from the Cardington trials varied with source separation Δy (scaled by σ_v and ϵ (the dissipation rate)) and downwind distance x (expressed as a 'travel time' $x/\bar{u}$ and then scaled by σ_v and ϵ). 17 different source separation/downwind distance configurations were used to obtain this graph with the downwind distance varying from 5 to 20m. The crosses on the graph show the different run conditions (63 in this plot) and the contours show how the cross-correlation varied with these different run conditions (132 cross-correlations were used to construct these contours).

We found generally the cross-correlation rapidly decreased from almost 1.0 when the sources were co-located to typically 0.01 for a source separation of x/20. Although we did detect negative correlations they were not very negative (typically -0.01 with a source separation of x/2) as might be expected given the highly intermittent nature of the concentration time series.

REFERENCES

Sawford B.L., Frost C.C. and Allan T.C., 1985, Atmospheric boundary-layer measurements of concentration statistics from isolated and multiple sources. *Boundary layer Meteorology*, **31**:249.

Warhaft Z., 1984, The interference of thermal fields from line sources in grid turbulence, *J. Fluid Mech.*, **144**:363.

THE EUROPEAN TRACER EXPERIMENT DESCRIPTION AND SUMMARY OF THE ETEX PROJECT

Francesco Girardi[1], Giovanni Graziani[1], Werner Klug[2], and Katrin Nodop[1]

[1] Joint Research Centre, Environment Institute, I-21020 ISPRA (Va), Italy
[2] Mittermayerweg 21, D-64289 Darmstadt, Germany

INTRODUCTION

The European Tracer Experiment (ETEX) is jointly organised by the European Commission, the International Atomic Energy Agency and the World Meteorological Organization. The project was established to evaluate the validity of long-range transport models for real time application in emergency management and to assemble a database which will allow the evaluation of long-range atmospheric dispersion models in general (Klug et al., 1993) ETEX main objectives were to conduct a European-scale tracer experiment, and test the capability of meteorological institutes responsible for emergency response to forecast in real time atmospheric dispersion (Girardi et al., 1997). Following this, the Atmospheric Transport Model Evaluation Study II (ATMES II) will simulate the tracer experiment with the same meteorological input data for all models.

EXPERIMENTS

The first tracer release in ETEX took place in October 1994, and the second release in November 1994 (Nodop et al. 1997). Perfluorocarbon tracers (Dietz, 1986) was released into the atmosphere in Monterfil, Brittany, and the air was analysed for its tracer contents at 168 stations in 17 European countries. At each site 3 hour samples were taken for a total length of 72 hours after the release. Upper air tracer measurements is available from 3 aircraft. Meteorological routine and additional observations at ground and upper air were gathered for a 7 day period around each release time. The tracer release characteristics, the tracer concentrations at ground and upper air, the meteorological observations, trajectories derived from constant level balloons and the meteorological input fields for LRT models are assembled in the ETEX database. The ETEX database is accessible via the Internet.

Both of the tracer experiments were very successful. All preparations were finished in time, the operation of samplers, shipment of samples for tracer determination at Ispra, and the GC analysis went according to plans, although it took a much longer time than anticipated. The results from the ground sampling network allowed the cloud evolution as it travelled across Europe.

Air Pollution Modeling and Its Application XII
Edited by Sven-Erik Gryning and Nadine Chaumerliac, Plenum Press, New York, 1998

REAL TIME MODEL EVALUATION

In the real-time modelling part (Graziani et al., 1997) twenty eight participants forecasted the tracer plume location as faxed concentration contours at 24, 48 and 60 h after the release. Furthermore the models predicted the concentrations every 3 h at the 168 ground stations with forecasted and analysed meteorological fields as input. For the first release, the concentrations calculated by the models, were in the range of the measurements, indicating that the main features of the flow and of the dispersion were correctly reproduced by some of the participants. Even in this case, however, there are large differences in concentration values as given by the models. Differences between model results and measured concentrations are much larger for the second release, characterised by a more complex meteorological situation, indicating the importance of an adequate description of the concentration field close to the source.

ATMES II

This study dealt only with the first release. The majority of the 34 participants were able to perform this new calculation using the meteorological field provided by the ECMWF. In general, the results of the comparison are definitely improved with respect to the real-time simulations. Even in those cases, however, the comparison does not give good results for 12 hours, when the number of sampling sites affected is small and the mesoscale effects play their role.

CONCLUSIONS

ETEX provides a unique experimental database for validating long-range atmospheric transport models. In general modellers proved their capacity to provide rapid forecasts in real time. The agreement of cloud position between model forecasts and measurements is satisfactory. None of the models, however, provided consistently accurate times of arrivals and maximum concentrations. Further statistical analysis of the great amount of information collected is still to be done.

ACKNOWLEDGMENTS

ETEX is sponsored by the EC, the IAEA and the WMO. The project is managed by the JRC, Environment Institute, and which has responsibility for the experiments and the evaluation of the models' performances. ETEX was made possible by the enthusiastic participation of the national weather services and responsible institutes inside and outside Europe. Special thanks go to the site personnel operating the samplers. The release site was made available by the University of Rennes, Radio Communications Faculty. Their contributions are gratefully acknowledged.

REFERENCES

R. N. Dietz, Perfluorocarbon Tracer Technology, in: Sandroni (ed.), Regional and Long-range Transport of Air Pollution, 215-247 Elsevier Science Pub. (1986)

F. Girardi, G. Graziani, W. Klug, K. Nodop, European Tracer Experiment, Description of the ETEX Project, EUR Report in prep., CEC Ispra 1997

G. Graziani, W. Klug, F. Girardi, K. Nodop, The European Long Range Tracer Experiment (ETEX), a Real-time Model Intercomparison, in Proc. American Nuclear Society Sixth Topical Meeting on Emergency Preparedness and Response, San Francisco (1997)

W. Klug, F. Girardi, G. Graziani,, K. Nodop, ETEX, a European Atmospheric Tracer Experiment, in Proc. Top. Meeting on Environmental Transport and Dosimetry, ANS, Charleston SC. 1993, pp. 147-149.

K. Nodop, R. Connolly, F. Girardi, European Tracer Experiment, First Release in October 1994, EUR Report in prep., CEC Ispra 1997

TRAJECTORY AND DISPERSION MODELS APPLIED TO EMERGENCY CASES AND TO OZONE RESEARCH STUDIES

Daniel A. Schneiter

Swiss Meteorological Institute (SMA),
CH 1530 Payerne, Switzerland

In case of nuclear or chemical accidents, local authorities have to be rapidly informed on pollutant immissions which could harm the population. They need to know the extent of affected area, appearance time of the pollutant, stagnation duration and levels of forecasted concentrations.

In a country with a complex topography like Switzerland, only the application of a high resolution meteorological model can offer forecasts of regional flows with sufficient accuracy. The SM (Swiss Model), which is similar to the DM (Deutschland-Modell) of the DWD (German Weather Service) is used with a grid of 14 km, allowing a good simulation of the wind channelling between the Jura mountains and the Alps. Forecasts over 48-hour period are operationally computed twice a day on the basis of initialization input data supplied every 12 hours by the EM (European model) from the DWD.

Some predicted forward and backward trajectories are automatically computed by the TRAJEK model after each SM integration, in order to test daily the operational system. In case of accidental emission, a manual procedure allows introduction of coordinates, altitude of the site and time of beginning of release. Forward trajectories are computed to obtain the forecasted displacement of pollutants.

The LPDM (Lagrangian Particle Dispersion Model) also developed by the DWD, is subsequently involved to furnish more precise data on pollutant immissions, taking into account the amount of emitted species, when they are known. Atmospheric dispersion is simulated by a stochastic method, based on the calculation of a great number of particle trajectories from the release point.

Verification on the trajectory model aims to test its ability to simulate measured trajectories observed by constant volume balloons (CVB) during projects like POLLUMET, TRACT or ETEX. Some vertical displacements, due to local convective air ascent cannot be simulated by hydrostatic models but have a strong influence on the CVB behaviour in unstable weather conditions.

Studies on backward trajectories calculated for the FREETEX project aim to better determine the origin of ozone concentration fluctuations observed at the Jungfraujoch (46.33N /7.59E, 3580 m).

INVERSE DISPERSION MODELLING BASED ON TRAJECTORY-DERIVED SOURCE-RECEPTOR RELATIONSHIPS

Petra Seibert

Institute of Meteorology and Physics, University of Bodenkultur, Türkenschanzstr. 18, A-1180 Vienna, Austria

Introduction

Inverse dispersion modelling means to derive information such as source strengths of emissions from measured concentration and/or deposition data of trace constituents, using a dispersion model. If source-receptor relationships are linear, they can be determined from a single model run, and the inversion corresponds to solving a linear system of equations. Examples for this approach are Enting and Newsam (1990), Brown (1993), and Hein et al. (1996). Non-linear source-receptor relationships require an iterative solution using an adjoint model. The calculation of large source-receptor matrices is very time consuming. Therefore, simple trajectories have been used widely to derive information on sources from pollution measurements. So-called potential source contribution functions have been calculated, e.g., by Zeng and Hopke (1989); similar techniques were employed Seibert and Jost (1994) and Stohl (1996). These are statistical methods, not based on dispersion models, and not exploiting the full information contained in the data sets. However, trajectories can be viewed as primitive Lagrangian dispersion models, and source-receptor matrices can be derived from them for use in formal inverse modelling. One has to be aware, however, that systematic shortcomings (e.g., neglection of precipitation scavenging or vertical exchange) will lead to systematic errors in the results, a problem shared with the statistical approach.

Data and Method

Results shown are based on daily measurements of the particulate sulfate concentration in air at 12 EMEP stations and a research station in Northern Finland in 1992. Three-dimensional back trajectories with 96 h length were calculated at 6 h intervals for these stations, arriving at 500 m above ground. This is a data set similar to that of Stohl (1995).

The domain was divided into a grid with 1° mesh size. The source-receptor relationship can be written as $\mathbf{c} = \boldsymbol{G}\,\mathbf{q}$, where $\mathbf{c}$ denotes the vector of observed concentrations and $\mathbf{q}$ the vector of unknown sulfur emission rates. The elements of the matrix $\boldsymbol{G}$ are

$$g_{ik} = h^{-1} \sum_j \tau_{ij} \delta_{jk} \exp(-\lambda \Delta t_{ij}) \,,$$

where h is the mixing height (assumed as 1000 m), τ_{ij} is the residence time of trajectory j in grid i, δ_{jk} describes the contribution of trajectory j to observation k (1 for trajectories

Air Pollution Modeling and Its Application XII
Edited by Sven-Erik Gryning and Nadine Chaumerliac, Plenum Press, New York, 1998

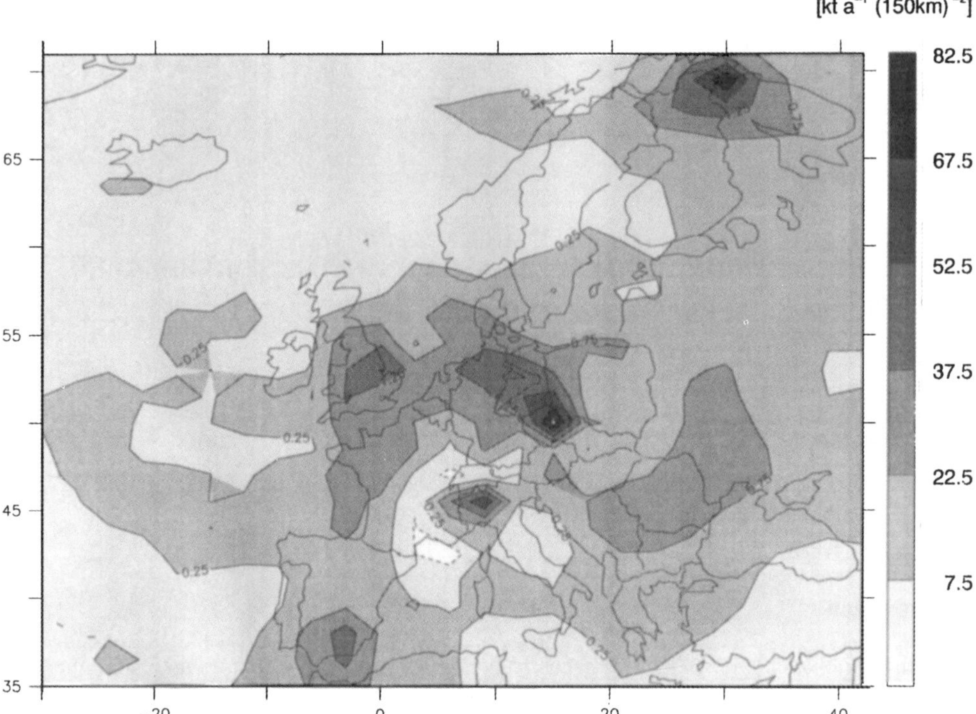

Figure 1. Geographical distribution of 1992 sulfur emission rates as obtained by inverse modelling.

arriving during the sampling period, 0 otherwise), λ is a decay constant describing the mean residence time of sulfur in the atmosphere, here set to 0.0058 h^{-1}, corresponding to 5 d half-life), and Δt_{ij} is the travel time between grid i and the receptor for trajectory j. The matrix equation cannot be solved directly, because in general the dimensions of **c** and **q** are not equal. Because some grid cells are rarely or never reached by trajectories, and because the information contained in different sample–trajectory pairs can be (approximately) linearly dependent, additional constraints have to be imposed to overcome this ill-posedness (Menke, 1984). The constraint used was to minimize the area integral of the Laplacian of **q**, weighted by the inverse of the root of the number of trajectories passing through the central grid cell. Another constraint was that the modelled concentrations have to be close to the observed ones in a least-squares sense; exact equality cannot be expected with **q** constrained to be smooth. This yields the final matrix equation to be solved:

$$(G^T G + \varepsilon W)\,\mathbf{q} = G^T\,\mathbf{c}.$$

W is a linear operator realising the smoothness constraint; ε regulates the relative weights of the two contributions to the cost function, representing the deviations from data fit and smoothness, respectively. It was determined on a try-and-error basis, so that a reasonably smooth field **q** was obtained. Objective determination of ε is an important future task.

Results

In the resulting **q** field (Fig. 1), the main source regions (Black Triangle, England, Kola Peninsula, Po Valley) are well located and have strong gradients. However, peak values are much too small as compared to the EMEP emission inventory. This is probably to be attributed to systematic errors when using trajectories instead of full dispersion models.

Acknowledgment

I want to thank my colleague Andreas Stohl for kindly providing the data set used. EMEP MSC-W made available the measurements at the EMEP stations.

References

Brown, M., 1993, *J. Geophys. Res.* **D 98:**639.

Enting, I.G. and Newsam, G.N., 1990. *J. Atmos. Chem.* **11:**69.

Hein, R., Crutzen, P.J., and Heimann, M.,1996, *An inverse modeling approach to investigate the global atmospheric methane cycle.* MPI f. Meteor. Hamburg, Rep. No. 220, and *Glob. Biogeoch. Cycl.* (in print).

Menke W. (1984): *Geophysical Data Analysis: Discrete Inverse Theory*. Orlando: Academic Press, 260 pp.

Seibert, P., and Jost, D.T., 1994, *EUROTRAC-Newsletter* **14:**14.

Stohl, A., 1996, *Atmos. Environ.* **30**:579.

Zeng, Y., and Hopke, P.K., 1989, *Atmos. Environ.* **23:**1499.

MODELLING DISPERSION PARAMETERS IN A PLANETARY BOUNDARY LAYER DOMINATED BY CONVECTION

G.A. Degrazia

Departamento de Física, Universidade Federal de Santa Maria-97.119.900 - Santa Maria, RS, Brasil

INTRODUCTION

The Gaussian plume model concept is still important for estimating ground-level concentrations due to tall stack emissions and is usually suitable for regulatory use in air quality models . Improved dispersion algorithms in updated Gaussian models calculate the dispersion parameters σ_y and σ_z in terms of distinct scaling parameters for turbulence. In this work, we use the convective similarity theory and Taylor's statistical diffusion theory to derive a general expression for the dispersion parameters in a planetary boundary layer dominated by convection. The novelty is use Taylor´s theory to construct a relation for the dispersion parameters that is described only in terms of the universal characteristics of the turbulent field in a convective boundary layer (CBL), eliminating in this manner, the necessity of dispersion parameters fitting from diffusion experiments.

THE MODEL

From Taylor´s statistical diffusion theory and spectra of turbulent kinetic energy the following dispersion parameters can be expressed by

$$\frac{\sigma_z^2}{z_i^2} = \frac{0.093}{\pi}\int_0^\infty \frac{\sin^2\left(2.96\psi_\varepsilon^{1/3} X n'\right)}{\left(1+n'\right)^{5/3} n'^2} dn' \quad ; \quad \frac{\sigma_y^2}{z_i^2} = \frac{0.21}{\pi}\int_0^\infty \frac{\sin^2\left(2.26\psi_\varepsilon^{1/3} X n'\right)}{\left(1+n'\right)^{5/3} n'^2} dn' \qquad (1)$$

where $X \equiv w_* x/U z_i$ can be thought of as a nondimensional time since it is the ratio of travel time (x/U) to the convective time scale (z_i/w_*). x is the dimensional distance downwind.

As a test for the model we are going to incorporate parameterizations (1) with $\psi_\varepsilon{}^{1/3}$ = 0.97 (Kaimal et al., 1976) in the Gaussian plume approach. The performance of model has been evaluated against experimental ground-level

Air Pollution Modeling and Its Application XII
Edited by Sven-Erik Gryning and Nadine Chaumerliac, Plenum Press, New York, 1998

Table 1. Statistical evaluation of model results

Model	NMSE	r	FB	FS
$c_y(x,0)$	0.08	0.87	0.10	0.31
c(x,0,0)	0.08	0.88	0.06	0.07

NMSE: Normalized mean square error; **r**: correlation coefficient; **FB**: Fractional bias; **FS**: Fractional standart deviation

concentrations using tracer SF6 data from dispersion experiments carried out in the northern part of Copenhagen, described in Gryning et al. (1987) and Gryning and Lyck (1984). The following dispersion experiments were here utilized: exp. no. 1; exp. no. 2; exp. no. 3; exp. no. 4; exp. no. 6; exp. no. 7; exp. no. 8; exp. no. 9. Table 1 presents some indices statistical that result of the comparison between the model and the observed ground-level concentrations, where cy(x,0) is the ground-level crosswind integrated concentration and c(x,0,0) is the centerline ground-level concentration.

As one can easily note, the statistical indices illustrate that the model performs quite well. Finally, it is important to note that the dispersion parameters as given by (1) are obtained from field experiments without observation of the dispersion parameters.

ACKNOWLEDGMENTS

We acknowledge financial support provided by Conselho Nacional de Desenvolvimento Científico e Tecnológico (CNPq) of Brazil.

REFERENCES

Gryning, S.E. and Lyck, E., 1984, Atmospheric dispersion from elevated sources in an urban area: comparison between tracer experiments and model calculations, J. Climate Appl. Meteorol. 23: 651.

Gryning, S.E., Holtslag, A.A.M., Irwin, J.S. and Siversten, B., 1987, Applied dispersion modelling based on meteorological scaling parameters. Atmos. Environ. 21:79.

Kaimal, J.C., Wyngaard, J.C. Haugen, D.A., Coté, O.R., Izumi, Y., Caughey, S.J. and Readings, C.J., 1976, Turbulence structure in the convective boundary layer, Atmos. Sci. 33:2152.

PARAMETRIZATION OF SL DIFFUSION PROCESSES ACCOUNTING FOR SURFACE SOURCE ACTION

D. Syrakov[1] and D. Yordanov[2]

[1]National Institute of Meteorology and Hydrology, 1784 Sofia, Bulgaria
[2]Geophysical Institute, 1113 Sofia, Bulgaria

Last years, a great number of dispersion models were developed. They possess different features and need different computer resources. Recently, the PC-oriented Eulerian multi-layer model EMAP (Eulerian Model for Air Pollution) was created in Bulgaria and applied to different pollution problems. The vertical diffusion block of the model uses a 2nd order implicit scheme including dry deposition as a bottom boundary condition, realised on a non-homogeneous (log-linear) staggered grid. The experiments with EMAP show that, if the concentration at the first computational level is used for calculation of the dry deposition flux, the deposited quantity changes when the height of this level is changed. It is obviously that the roughness level concentration is necessary as to calculate properly dry deposition. It is not possible to have a model level at this height, because roughness usually changes from one grid point to another. On the other hand, because of the steep gradients in the surface layer (SL), many levels must be introduced near the ground for adequate description of pollution profiles. This would increase memory and time requirements without any practical need. Usually, the first computational level is placed high enough above roughness. As a result, a good estimate for the roughness level concentration is necessary, determined on the base of the calculated concentrations. The problem becomes much more complex when surface sources of pollution have to be treated. Such are the processes of evaporation and re-emission of tracer under consideration. A proper parametrization of the diffusion processes in the surface layer can avoid these difficulties. An effective SL parametrization, based on similarity theory, was elaborated and tested by Syrakov and Yordanov (1996a,b). It allows to have the first computational level at the top of this layer. Here, an upgrade of this parametrization is presented, taking into account the presence of a continuous surface source.

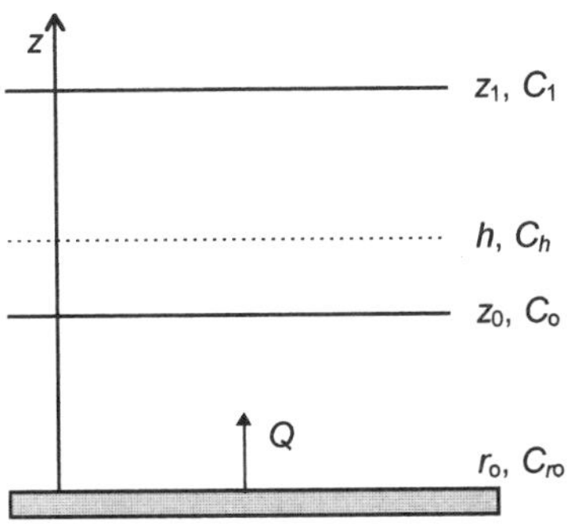

Let us have a vertical grid, hanged over the ground surface with roughness r_0, the boundary level (zero level) being at height z_0 and the first inner level - at height z_1, as shown in the sketch. For the fluxes in SL, the following can be written (Aloyan at al., 1981)

$$\kappa z\frac{d|\mathbf{u}|}{dz} = u_*\varphi_u(\varsigma)\,,\qquad z\frac{d\theta}{dz} = \theta_*\varphi_\theta(\varsigma)\,,\qquad z\frac{dC}{dz} = C_*\varphi_\theta(\varsigma)\,,\qquad \varsigma=\frac{z}{L}, \tag{1}$$

where κ=0.4 is Karman constant, $\mathbf{u}$, θ and C are the horizontal wind velocity, potential temperature and concentration, u_*, θ_* and C_* are the friction velocity, temperature and concentration scales, $\varphi_u(\varsigma)$ and $\varphi_\theta(\varsigma)$ - universal functions, L - Monin-Obukhov length. The integrals of Eqs.(1) from the ground surface r_0 to height z are given in Eq.(2) and the vertical exchange coefficients - in Eq.(3):

$$\kappa|\mathbf{u}(z)| = u_*f_u(\varsigma),\qquad \theta(z)-\theta(r_0)=\theta_*f_\theta(\varsigma),\qquad C(z)-C(r_0)=C_*f_\theta(\varsigma)\,, \tag{2}$$

$$k_u(z)=u_*\kappa z/\varphi_u(\varsigma),\qquad k_\theta(z)=k_c(z)=u_*\kappa z/\varphi_\theta(\varsigma)\,. \tag{3}$$

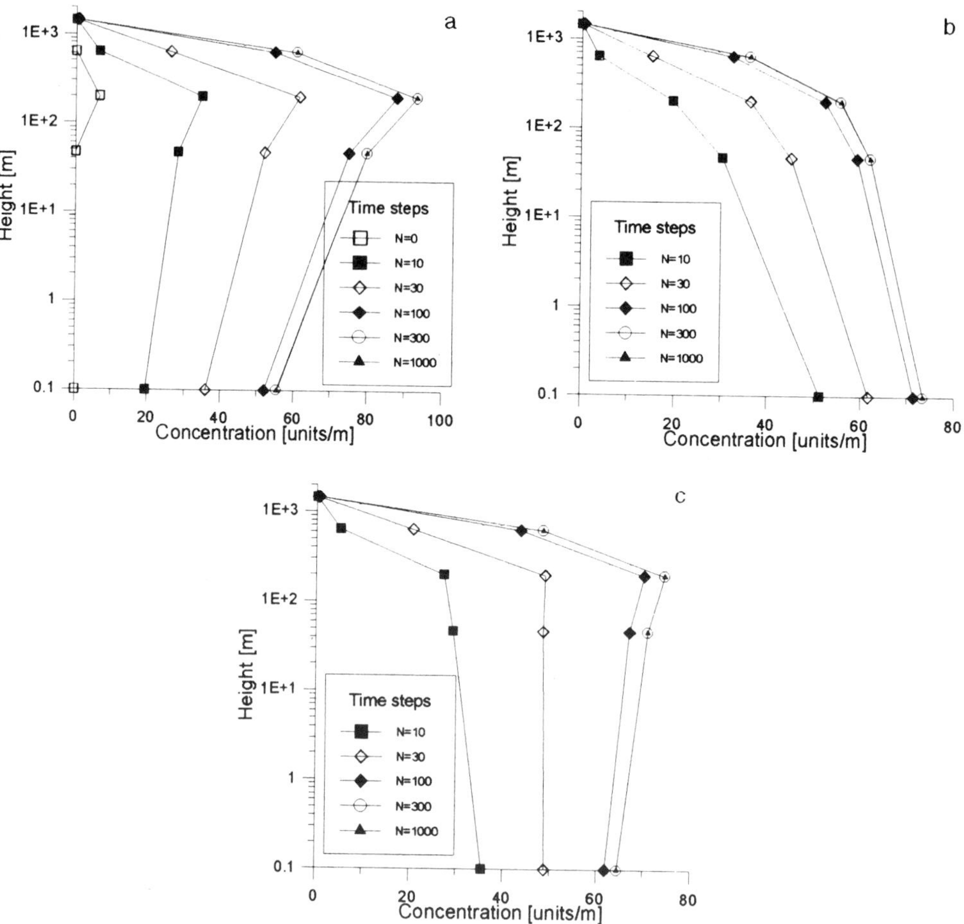

Figure 1. Concentration profiles for the cases of high source (a), surface source(b) and high+surface sources (c).

Aloyan et al. (1981) have shown on the base of Eqs.(1)-(3) that the tracer flux at height h, which is necessary as a bottom boundary condition for the diffusion equation, can be expressed by $(k_h = k_\theta(h))$

$$k_h dC / dz = \alpha(C_h - C_{ro}), \qquad a = \kappa u_* / f_\theta(\varsigma_h)\,. \tag{4}$$

According to the main SL characteristics, the constancy of fluxes with height, Eq.(3) can be equivalized to the dry deposition flux $V_d C_{ro}$, V_d being the dry deposition velocity. When a surface source with strength Q is acting, the equation is

$$V_d C_{ro} - Q = \alpha(C_h - C_{ro}),\ \text{i.e.}\ \ C_{ro} = \gamma C_h + Q/(\alpha + V_d), \qquad \gamma = \alpha/(\alpha + V_d). \tag{5}$$

The flux at height h must be discretized when used as a numerical boundary condition. After replacing C_{ro} from Eq.(5) in Eq.(4), the following relation is obtained

$$K(C_1 - C_0) = \gamma V_d C_h - \gamma Q = \gamma V_d(aC_0 + bC_1) - \gamma Q, \qquad K = k_h/(z_1 - z_0), \tag{6}$$

where C_h is presented as a linear combination from C_0 and C_1. The coefficients a and b follow from (2):

$$a = [f_\theta(\varsigma_1) - f_\theta(\varsigma_h)]/[f_\theta(\varsigma_1) - f_\theta(\varsigma_0)], \qquad b = [f_\theta(\varsigma_h) - f_\theta(\varsigma_0)]/[f_\theta(\varsigma_1) - f_\theta(\varsigma_0)]\,. \tag{7}$$

Eq.(6) is transformed to provide the bottom boundary condition for solving the vertical diffusion equation

$$C_0 = B_1 C_1 + B_0 Q, \qquad B_1 = (K - \gamma V_d b)/c, \qquad B_0 = \gamma/c, \qquad c = K + \gamma V_d a\,. \tag{8}$$

The replacement of (6) and (8) in (5) leads to en expression for C_{ro}, i.e. to the dry deposition estimate

$$C_{ro} = G_1 C_1 + C_0 Q, \qquad G_1 = \gamma(aB_1 + b), \qquad G_0 = (\alpha a B_0 + 1)/(\alpha + V_d)\,. \tag{9}$$

Eqs.(8) and (9) are basic in the proposed SL parametrization. It can be noticed that C_{ro} depends linearly on the first level concentration and on the surface source strength, coefficients depending on the turbulence and stability through u_*. and the universal temperature profile $f_\theta(\xi)$. When Q=0, a simple proportionality is obtained; the parametrization case, presented in Syrakov and Yordanov (1996a,b).

In Figure 1, examples of the calculated profiles evolution are presented. A 4-level case is shown, the lowest level placed at the height of 50 m (top of SL). Eq.(8) is used as a bottom boundary condition. The surface concentrations (roughness height of 0.1 m supposed) are calculated after Eq.(9). At the top, an "open boundary" condition is applied. The continuous source strength is 1 unit/s, when unique sources act (a,b), and 0.5 unit/s in the mixed case (c). The high source is at 200 m, the time step is 0.5 h. The dry deposition velocity is set to 0.01 m/s. The presented profiles are rather realistic unless the coarse vertical resolution. It can be noticed, that after 300 time steps, stationarity is reached in all cases.

The parametrization is tested in 1D version by introducing logarithmicaly new and new computational levels in SL. The calculated surface concentrations and dry deposited quantities do not change significantly at various conditions - different types of sources, turbulence and stability conditions, especially at large times. This shows the good quality of the proposed parametrization and it is used in the 3D Eulerian dispassion model EMAP for estimating regional loads of harmful pollutants.

REFERENCE

Aloyan A.E., D.L.Yordanov, V.V.Penenko, 1981: A numerical model of pollutant transport in the atmospheric boundary layer, *Meteorology and Hydrology*, No 8, Gidrometeoizdat,M.

Syrakov D. and D. Yordanov, 1996a: On the surface layer parameterization in an Eulerian multi-level dispersion model, in: *Bulgarian contribution to EMEP, Annual report for 1995*, NIMH, Sofia-Moscow, January 1996.

Syrakov D. and D. Yordanov, 1996b: On the surface layer parametrization in an Eulerian multi-level model, *Proceedings of the 4th Workshop on Harmonisation within Atmospheric Dispersion Modelling for regulatory purposes*, v.1, 6-9 May 1996, Oostende, Belgium.

Yordanov D., D. Syrakov, G. Djolov, 1983: A barotropic planetary boundary layer, *Boundary Layer Meteorology*, 25,. 363-373.

AEROSOL DYNAMICS MODELLING IN THE REGIONAL CHEMISTRY TRANSPORT MODEL SYSTEM EURAD

Ingmar J Ackermann[1], Heinz Hass[1] , A. Ebel[2], M.Memmesheimer[2] and H.J. Jakobs[2]

[1] Ford Forschungszentrum Aachen GmbH
Dennewartstr. 25
D-52068 Aachen, Germany
[2] EURAD-project, University of Cologne
Aachener Str. 201-209
D-50931 Cologne, Germany

INTRODUCTION

The particulate matter suspended in the atmosphere is strongly linked to numerous air pollution problems. These include the -direct and indirect- influences on the radiative budget of the atmosphere, the potential for adverse health effects, the influence of particles on the long-range transport of air pollutants and the interaction of aerosol particles with clouds. Therefore air quality models have to take into account particle formation, transport and deposition with respect to aerosol chemistry as well as aerosol dynamics. Due to the strong interactions between the gas phase and the aerosol phase, the aerosol model has to be fully coupled to a gas phase model. The Modal Aerosol Dynamics modelling technique as a time and memory efficient method provides a suitable approach to realise that purpose even in complex three-dimensional Eulerian models.

MODEL DESCRIPTION

To fulfil the task of modelling the particle size distribution and aerosol chemistry the Modal Aerosol Dynamics model for Europe (MADE) has been developed. Based on RPM, the Regional Particulate Model (Binkowski and Shankar, 1995) which has been developed further and adapted for European conditions, MADE currently describes the size distribution of the submicrometer aerosol by two overlapping modes assuming a lognormal distribution within each mode. The smaller mode (Aitken mode) represents the freshly formed particles whereas the larger one represents the aged aerosol in the accumulation mode. The size distribution is modified by emissions of particles, nucleation processes,

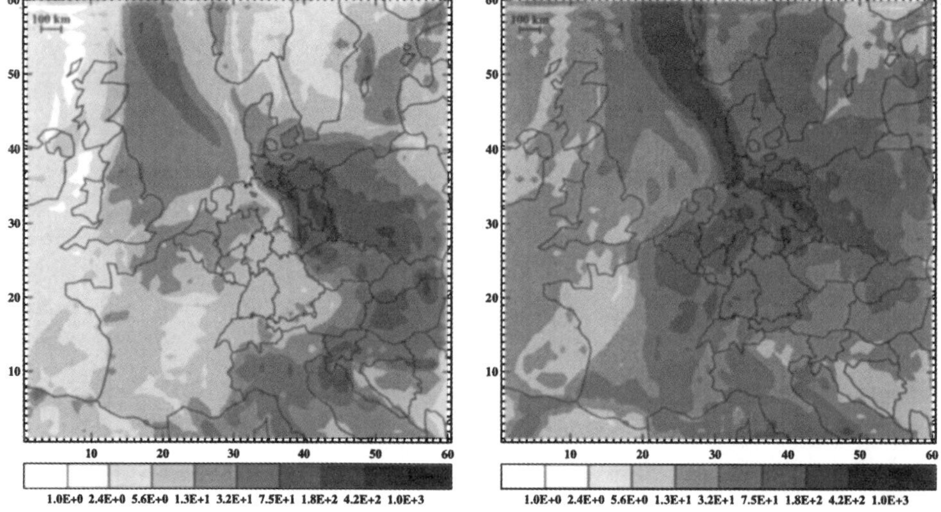

Figure 1. Aerosol surface concentration in the lowest model layer at 26.07.94 12 GMT in the Aitken mode (left) and the accumulation mode (right) in µm^2/cc.

coagulation of and condensation on existing particles and -size dependent- dry deposition. Aerosol chemistry is treated in the sulphate-nitrate-ammonia system with water in equilibrium (Ackermann et al., 1995). The aerosol calculations of MADE are performed on-line within the chemistry-transport model CTM2 (Hass et al. 1993).

THREE-DIMENSIONAL APPLICATION

Figure 1 gives an example result from the first three-dimensional simulation with MADE. Shown are the aerosol surface concentrations over the model domain after two days of simulation. Whereas the Aitken mode distribution is dominated by the source areas, the distribution in the accumulation mode shows a band of enhanced concentrations along the Scandinavian west coast, indicating long-range transport of an aged aerosol plume.

ACKNOWLEDGMENT

EURAD is funded by the Ministry of Research and Technology (BMFT) of the Federal Republic of Germany and the Ministry of Science and Research of the Land Nordrhein-Westfalen. The EURAD project is a part of EUMAC, which is a EUROTRAC subproject. Computational support came from the Research Center Jülich (KFA), in particular from the institutes ICG2, ICG3 and the ZAM of the KFA.

REFERENCES

Ackermann, I.J.: MADE: Entwicklung und Anwendung eines Aerosol-Dynamikmodells für dreidimensionale Chemie-Transportsimulationen in der Troposphäre. PhD-thesis, University of Cologne, Mitt. aus dem Inst. für Geophysik und Meteorologie, Heft 115, 1997.

Binkowski, F.S. and U. Shankar: The regional particulate model 1. Model description and preliminary results. J. Geophys. Res., 100, 26191-26209, 1995.

Hass, H., A. Ebel, H. Feldmann, H.J. Jakobs and M. Memmesheimer: Evaluation studies with a regional chemical transport model (EURAD) using air quality data from the EMEP monitoring network. *Atmos. Environm.*, 27A, 867-887, (1993).

NEW DEVELOPMENT IN MODELING OF URBAN AIR POLLUTION: RUSSIAN AND US EXPERIENCE

Eugene Genikhovich

Main Geophysical Observatory
194021 St.-Petersburg
Russia

INTRODUCTION

A new generation of the atmospheric dispersion models is under development in the USA and Europe (Berkowicz et al., 1986; Perry et al., 1994). These Gaussian models employ parametrizations of dispersion coefficients and other variables with physical parameters rather than with stability categories. Seemingly more direct approach to introduction of physical parameters into dispersion models can be based on solution on the advection-diffusion equation. Results of such an approach are presented in this paper; ways of generalization of the developed flat-terrain model for account of building-downwash effects are also discussed here.

A NEW MULTIPLE-SOURCE DISPERSION MODEL AND ITS APPLICATIONS

A newly developed dispersion model MSM is based on numerical solution of the 3D steady-state advection-diffusion equation (ADE). Coefficients in ADE are components of the wind velocity and eddy diffusivities. Taking an appropriate model for the crosswind eddy diffusivity, this equation is split in two equations corresponding, respectively, to the crosswind integrated concentration field ("line source") and crosswind distribution (the last one having the Gaussian solution). Coefficients in the first equation are parametrized inside and outside the surface layer using the Monin-Obukhov similarity theory and corresponding results for the planetary boundary layer. Convection regimes are considered separately. Using finite-difference techniques, this equation is numerically solved with given set of governing parameters, and dimensionless results are stored in form of tables. Procedures of interpolation with these tables are developed which can be used to reconstruct a solution at arbitrary values of governing parameters with errors inside those of numerical techniques in use. The initial plume rise is estimated using the Briggs' (1984) and Berlyand's (1991) formulae for the stable and unstable thermal stratifications, respectively. The multiple-source concentration field is calculated as superposition of plumes from single sources. A meteorological interface is developed to evaluate governing parameters using data of meteorological observations.

The MSM model was tested upon the Kincaid experimental data set; concentration fields from 154 sources calculated with this model were compared with those calculated with the Gaussian model. A simplified version of the MSM model for calculation of the long-term averaged concentration fields was developed and verified using data of field measurement in Raahe (Finland) and Volgograd (Russia).

An approach to modeling of the building-downwash effects by application of an operator transforming the concentration field in the absence of buildings into the concentration field in the presence of the building was developed by Berlyand et al. (1987) and Genikhovich and Snyder (1994). It was found that corresponding models perform better as soon as the plume rise is estimated as gradual rather than as final one, and that effects of the wind shear was to be taken into account (Genikhovich and Snyder, 1995; Genikhovich, 1996). Both, non-Gaussian and Gaussian, versions of the building-downwash models were validated using results of wind-tunnel tests and field experiments.

New technologies of applications of dispersion models were developed. They can be used not only for regulatory purposes but also for solution of such problems as:
- combining the results of measurements and calculations in order to map the concentration field using modeled fields as interpolators, and
- developing the strategy for siting the monitoring stations in local and urban scales.

REFERENCES

Belyand, M.E.,1991, *Prediction and Regulation of Air Pollution. Kluwer., Dordrecht*

Berlyand, M.E., Genikhovich, E.L., Gracheva, I.G., and Tsarev, A.M., 1987, Account for buildings in atmospheric diffusion models, in: *WMO Conference on Air Pollution Modelling and Its Applications, Leningrad, 1986*, v.3., Envir. Pollut. Monitor. Res. Progr. No 49, WMO/TD No 187, Geneva

Berkowicz, R., Ohlesen, J.R., and Torp, U., 1986, The Danish Gaussian air pollution model (OML): Description, test and sensitivity analysis in view of regulatory applications, in: *Air Pollution Modeling and Its Applications*, V.C. De Wispleaire, F.A. Schiermeier, and N.V. Gilliani, Eds., Plenum, New York

Briggs, G.A., 1984, Plume rise and buoyancy effects, in: *Atmospheric Science and Power Production*, D. Randerson, Ed., DOE/TIC-27601, U.S. Dept. of Energy, Washington

Genikhovich, E.L., 1996, Local-similarity description of trajectories of plumes and jets in neutrally stratified turbulent shear flow, in: *Air Pollution Modeling and Its Application XI*, S.-E. Gryning and F.A. Schiermeier, Eds., Plenum, New York

Genikhovich, E.L., and Snyder, W.H., 1994, A new mathematical model of pollutant dispersion near a building, in: *Eighth Joint Conference on Applications of Air Pollution Meteorology With A&WMA, Nashville, TN, Jan. 23 - 28*. AMS

Genikhovich, E.L., and Snyder, W.H., 1995, Trajectories of plumes and jets in turbulent shear flow under near-neutral conditions, in: *Ninth AMS Symposium on Meteorological Observations and Instrumentation, Charlotte, NC,* AMS

Perry, S.R., Cimorelly, A.J., Lee, R.F., Pain R.J. ,Venkatram, A., Weil, J.C., and Wilson, R.B., 1994, AERMOD: a dispersion model for industrial source applications. 94-TA23.04. *A&WMA 87th Annual Meeting, Cincinnati, Ohio*, A&WMA

IMPROVEMENT OF AIR POLLUTION FORECASTING MODELS USING FEATURE DETERMINATION AND PATTERN SELECTION STRATEGIES

Marija Božnar and Primož Mlakar

Jožef Stefan Institute
Jamova 39
SI-1000 Ljubljana, Slovenia

ABSTRACT

Air pollution forecasting models are a helpful tool for controlling pollution around sources such as large thermal power plants. Recently we developed a neural network - based, short-term SO_2 pollution forecasting model for measuring sites around the Šoštanj Thermal Power Plant. The most important problems that should be solved in order to improve the model performance are feature determination and pattern selection. We developed several methods to solve these two problems.

DESCRIPTION OF THE AIR POLLUTION FORECASTING MODEL

Forecasting of air pollutant concentrations is still an interesting problem especially in the surroundings of large thermal power plants that do not have efficient desulphurisation plants installed. In the past we have developed a neural network - based forecasting model. The model was applied to prediction of ambient SO_2 concentrations at the automatic air quality measuring stations around Šoštanj Thermal Power Plant that lies in complex terrain in the north-east of Slovenia.

As a platform for the model a multilayer Perceptron neural network with two hidden layers was used. The model has about 20 inputs (depending on the particular measuring station), also called input features, which are measurements of meteorological parameters and SO_2 ambient concentrations for the present and previous half hour averaging intervals. The output of the model is a forecast of the SO_2 concentration at the particular station for half hour in advance (short term prediction). Every station observed has its own model.

The process of model construction is a procedure in which neural network weights are adjusted with a backpropagation algorithm using a training set of patterns (vectors of known historical pairs of input and output feature measurements). For the training set, patterns should be selected that represent as many of the significant pollution situations as possible that appear at the particular measuring station. When the process of learning is completed, the model can be tested on independent data sets or used on - line at the power plant.

Air Pollution Modeling and Its Application XII
Edited by Sven-Erik Gryning and Nadine Chaumerliac, Plenum Press, New York, 1998

DETERMINATION OF FEATURES

The first problem that should be solved is the determination of appropriate measurements - input features for the model - because not all the available measurements are suitable. We developed several methods for feature determination. In previous years correlation analysis and expert knowledge were mainly used for feature determination.

In addition to these mostly heuristic technique we adapted three techniques known from pattern recognition theory for use in the field of air pollution modelling. This techniques are saliency metrics, sequential forward selection and sequential backward selection of features.

Determining the saliency metrics for the features is a technique based on the Perceptron neural network. The neural network is trained with patterns compounded of all possible features. After training, saliency metrics are computed for each feature from the neural network weights. More important features have higher values of saliency metrics. In this way an ordered list of features is obtained. Sequential forward selection and sequential backward selection of features are procedures that maximise the probability of successful classification of patterns compounded of the examined features. In the sequential backward selection procedure first all the features are taken into account. At each successive step, the least important feature is removed. This is the feature that reduces the probability of successful classification as little as possible. The result of the procedure is an ordered list of features. For determination of the probability density function, the Parzen estimator is used. Sequential forward selection is a similar algorithm that works in the opposite direction.

PATTERN SELECTION STRATEGIES

Basically the forecasting model is taught with patterns that represent SO_2 concentrations and the meteorological history of the observed polluted site. Two types of pattern selection strategies were developed. The models built according to these feature determination and pattern selection strategies were tested in the case of SO_2 pollution for two different automatic measuring stations around the Šoštanj Thermal Power Plant and compared with reference models. The results showed a significant improvement of the model's forecasting capabilities (relatively up to 20% better probability of successful forecasting of high SO_2 concentrations).

The basic idea of all the pattern selection strategies developed is to find clusters of similar patterns and to represent uniformly all clusters in a training set of patterns. In this way it is ensured that the model learns rare but important pollution situations as well as frequent ones. The first method is based on meteorological knowledge about air pollution mechanisms. From the available patterns, equal numbers of patterns are selected for all known meteorological mechanisms that cause pollution at a particular site. The second method (the Kohonen neural network based method) does not require expert meteorological knowledge. Dividing of patterns into clusters of similar ones is successfully performed by a self - organising Kohonen neural network.

CONCLUSION

All the methods described contribute to a significant improvement of SO_2 forecasting around the Šoštanj Thermal Power Plant. None of these methods have yet been reported in the field of air pollution modelling, so they will probably be of great interest to modellers.

REFERENCE

M. Božnar, M. Lesjak, P. Mlakar, *A Neural Network-Based Method for Short-Term Predictions of Ambient SO_2 Concentrations in Highly Polluted Industrial Areas of Complex Terrain*, Atmospheric Environment, 27B, pp.221-230, (1993).

NEURAL NETWORK SIMULATIONS IN AN INTEGRATED EXPERT SYSTEM FOR AIR POLLUTION MODELING

Viktor Pocajt, Radmila Cvijović

Faculty of Technology and Metallurgy
Karnegijeva 4, 11000 Belgrade, Yugoslavia

INTRODUCTION

A complex expert system for air pollutant dispersion modeling **ScalEx** (**Scale** up **Ex**pert) is currently being developed at Faculty of Technology and Metallurgy in Belgrade. The project goal is to develop a powerful software system as an engineering toolbox for modeling air pollutant dispersion under various conditions and terrain configurations. **ScalEx** is planned to be an integrated software system of the new generation, putting together expert system, advanced database technologies, hypermedia, conventional numeric routines and neural networks (Pocajt, 1996).

Neural network module **NeuroScale** contains set of neural networks. Different neural network architectures are being trained and tested on three well-known experimental data sets (Kincaid, Copenhagen and Lilestrem).

NEURAL NETWORKS AND SIMULATION EXERCISE

Neural networks are relatively new technique based on computer simulation of activities of human brain. Neural networks perform undeterministic modelling, i.e. modelling without defining mathematical relations between variables explicitly (Wasserman, 1989). To obtain capability of giving adequate predictions, neural network is being trained with the sets of correct input data and results. In the process of training, network adapts its internal structure according to problem under the consideration.

In process of neural network development, careful expert analysis is necessary for determining significant influence parameters on output results (e.g. wind and temperature profiles, wind fluctuations, mixing height), their implicit relations and conducting network training process, in order to obtain good generalization - predictions with input data that network have never "seen" before.

Neural network simulation exercise was performed on Copenhagen data set, provided in Model Validation Kit (Olesen, 1994). Classic and quite simple three-layer backpropagation network architecture was applied (**NeuroScale** release A.C.01). Neural network was implemented in Microsoft Windows environment using NeuroShell 2 as development tool. Network parameters were tuned for very complex and noisy problem and proprietary algorithm for training conducting was used (Ward System Group, 1993).

Maximum arcwise concentrations and cross-wind integrated concentrations (normalized with emission) were predicted by neural network. Results, shown in Table 1, are satisfactory compared to other models (Olesen, 1995). An example of scatter plot is shown in Figure 1.

Table 1. Results of neural network modelling exercise. Set of 22 experimental results was used (one results was provided to network as "dummy" parameter).

	Maximum arcwise conc.		Cross-wind integr.conc.	
	Observed	Model	Observed	Model
Mean	613,68	617,24	439,64	432,82
Sigma	461,92	396,43	246,44	207,38
Bias	0,00	3,56	0,00	7,02
Norm. Mean Square Error	0,00	0,05	0,00	0,06
Correlation	1,000	0,942	1,000	0,871
Fractional bias	0,000	0,006	0,000	0,016

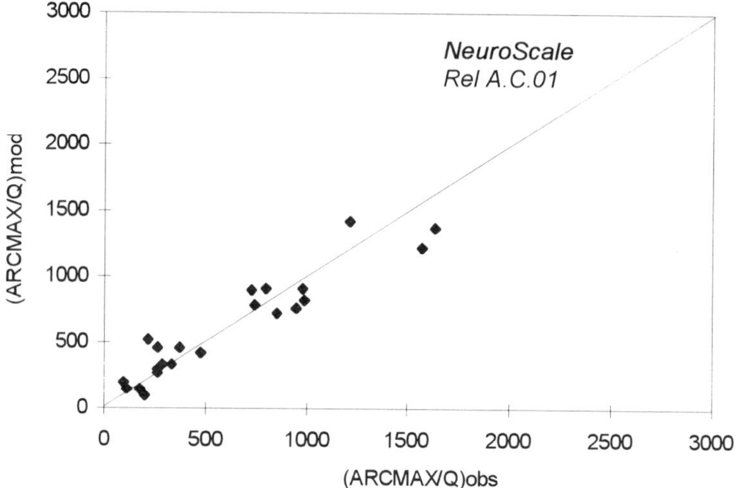

Figure 1. Scatter plot of normalized concentrations. Concentration unit is 10^{-9} s/m^3.

FURTHER DEVELOPMENTS

Neural networks can be successfully applied in air pollutant dispersion modelling, at least in some situations. However, to deploy full potential of this modelling technique, more development have to be taken.

It is necessary to implement meteorological preprocessing in some extent. More simulations have to be made to determine when to use observed instead of preprocessed parameters (e.g. friction velocity, mixing height, σ_v, σ_w) and vice versa.

Problem of generalization still remains as the most significant. It is clear that set of neural networks have to be developed for modelling different meteorological and terrain situations, since the problem of atmospheric pollutant dispersion is far to complex to be captured with one neural network. Corresponding set of guidelines and rules for "swathing" between different neural networks have to be established and coded.

REFERENCES

Olesen H.R., 1994, *Model Validation Kit*, National Environmental Research Institute, Riso.

Olesen H.R., 1995, The model validation exercise at Mol: overview of results, *Int.J. Environment and Pollution*, Vol.5, No 4-6, pp 761-784.

Pocajt V., Cvijović R., 1996, Development of an integrated system for air pollutant dispersion modelling, *Proc. 4th Workshop on Harmonization within Atm.Disp.Modelling for Regulatory Purposes*, Oostende.

Ward System Group, 1993, *NeuroShell 2 Users Guide*, Ward System Inc., Frederick, Maryland.

Wasserman Ph., 1989, *Neural Computing Theory and Practise*, Van Nostrand Reinhold, New York.

A DISPERSION MODEL FOR GROUND-LEVEL AND ELEVATED RELEASES FROM CONTINUOUS POINT SOURCES

Ana G. Ulke and Nicolás A. Mazzeo

Dept. of Atmospheric Sciences - University of Buenos Aires
Buenos Aires - Argentina

The application of a dispersion model of pollutants released from near surface and elevated continuous point sources is presented. The model is based on the bidimensional semiempirical equation, with vertical profiles of wind and eddy diffusivity for the atmospheric boundary layer. The suggested model makes use of a continuous description of the dispersion processes in the different regimes of the atmospheric boundary layer and needs meteorological input parameters that can be estimated from routine measurements[1]. The model is used to simulate the dispersion of non-buoyant, non-depositing releases from several source heights in a variety of atmospheric stability and surface roughness conditions. The observational data were obtained in tracer experiments carried out at Copenhagen (Denmark)[2] , Lillestrøm (Norway)[3], Hanford (USA)[4] and Cabauw (The Netherlands)[5]. The predicted cross-wind integrated concentrations were compared with the observed ones. All concentrations were normalized by source strength. Current quantitative measurements and techniques of model evaluation were obtained and applied[6,7].

The comparison was done for each dataset and for the full data base. A statistical evaluation of model performance was done, computing various quantitative measures. Confidence intervals on these performance measures were determined by the bootstrap resampling procedure. The obtained statistics are shown in **Table 1**.

Table 1. Statistics for normalized cross-wind integrated concentrations (s/m^2)

data		model	mean	sigma	bias	nmse	cor	fa2	fb	fs
Copenhagen	(N= 23)	obs	4.49E-04	2.39E-04	0.00E+00	0.00	1.000	1.000	0.000	0.000
		mm	4.39E-04	1.94E-04	9.75E-06	0.07	0.869	0.957	0.022	0.210
Hanford -	(N= 26)	obs	1.60E-02	1.61E-02	0.00E+00	0.00	1.000	1.000	0.000	0.000
Lillestrøm		mm	1.93E-02	1.51E-02	-3.33E-03	0.23	0.876	0.692	-0.188	0.060
Cabauw	(N= 23)	obs	5.19E-04	3.38E-04	0.00E+00	0.00	1.000	1.000	0.000	0.000
		mm	5.20E-04	3.65E-04	-1.39E-06	0.18	0.810	0.870	-0.003	-0.075
full data	(N= 72)	obs	6.08E-03	1.22E-02	0.00E+00	0.00	1.000	1.000	0.000	0.000
		mm	7.28E-03	1.28E-02	-1.20E-03	0.59	0.922	0.833	-0.179	-0.051

For the Copenhagen dataset it is observed a mean underestimation of normalized concentrations. For the Hanford-Lillestrøm data, it is observed a mean overestimation of the measured concentrations. For the Cabauw data, the measures show a mean overestimation of concentration values. The statistics for the full dataset show a mean overestimation of normalized observed concentrations.

Air Pollution Modeling and Its Application XII
Edited by Sven-Erik Gryning and Nadine Chaumerliac, Plenum Press, New York, 1998

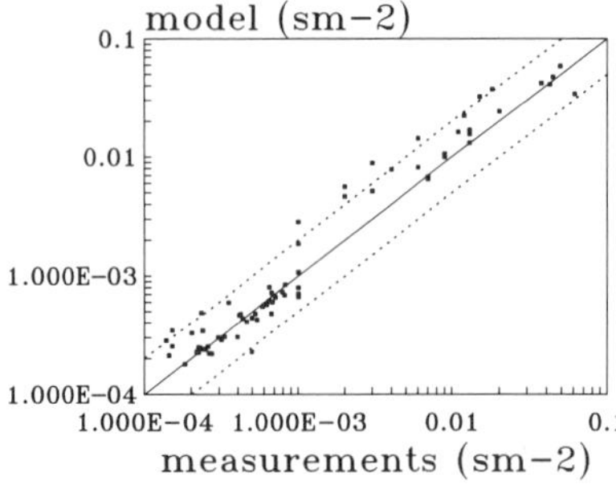

Fig.1 Scatter plot of normalized cross-wind integrated concentrations

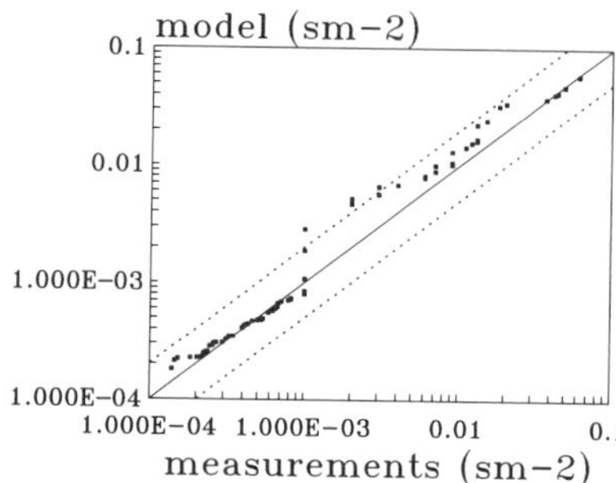

Fig.2 Quantile-quantile plot of normalized cross-wind integrated concentrations

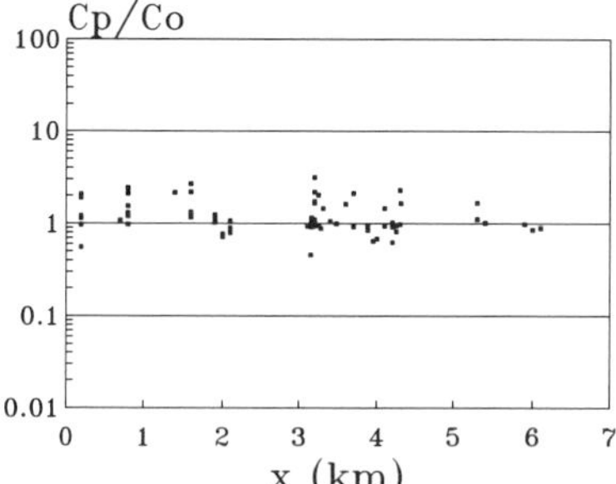

Fig.3 Residual plot of normalized cross-wind integrated concentrations analyzed in terms of distance

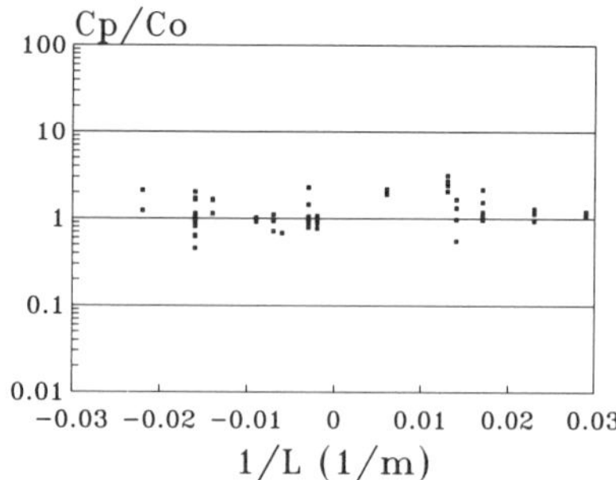

Fig.4 Residual plot of normalized cross-wind integrated concentrations analyzed in terms of stability

Evaluation results are presented using scatter plots, quantile-quantile plots and residual (predicted/observed) plots for the full dataset. The comparison between observed normalized concentrations and the predicted ones is illustrated in **Fig. 1**. There is a good correspondence between model estimates and observations. It is observed in general a different behavior for the smallest and the highest concentrations. It can be seen a slight trend to underestimate the smallest values and to overestimate the highest values. It is interesting to note that the highest values (greater than 1.2 10^{-3}) belong to the Hanford-Lillestrøm dataset (near neutral and atmospheric stable conditions), and the smallest values (less than 1.2 10^{-3}) are from the Copenhagen and the Cabauw datasets (near neutral and unstable conditions). The Q-Q plot **(Fig. 2)** shows that the distribution of model predictions compare well with the distribution of observations, with the highest observations being overpredicted by a small quantity. However, almost the entire distribution is within a factor of two for this unpaired comparison. The residual plot in **Fig. 3** shows a generally unbiased model estimates with distance. There is some scatter about a residual of 1.0, with a slight tendency toward overprediction at the closest distances. However, this overestimation does not reach very significative values. **Fig.4** shows the residual as function of atmospheric stability (L: Monin-Obukhov length). There is a slight bias toward underprediction during unstable atmospheric conditions and the opposite relation in stable conditions. However, the departures are not important.

REFERENCES

1. A.G. Ulke, *Diffusion-deposition of pollutants released in the atmospheric boundary layer*, PhD Thesis, University of Buenos Aires, Argentina(1992).

2. S.E. Gryning and E. Lyck, Atmospheric dispersion from elevated sources in an urban area: comparison between tracer experiments and model calculations. *J. Climate Appl. Met.*, 23: 651(1984).

3. B. Sivertsen and T. Bøhler, Verification of dispersion estimates using tracer data, *NILU Report TR 19/85*. The Norwegian Institute for Air Research, Lillestrøm, Norway(1985).

4. J. C. Doran and T. W. Horst, An evaluation of Gaussian plume-depletion models with dual-tracer field measurements, *Atmos. Env.*, 19:939(1985).

5. H. van Duuren and F. T. M. Nieuwstadt, Dispersion experiments from the 213m high meteorological mast at Cabauw in the Netherlands, *Atmospheric Pollution, Proc. of the 14th International Colloquium*, Paris, M. M. Benarie, ed., 77(1980).

6. S. R. Hanna, J. C. Chang and D.G. Strimaitis, Hazardous gas model evaluation with field observations, *Atmos. Env.*, 27A:2265(1993).

7. H. R. Olesen, Toward the establishment of a common framework for model evaluation,in *Air Pollution Modeling and Its Application XI*, S.E. Gryning and F.A. Schiermeier, ed., Plenum Press, New York, 519(1996).

GROUND-BASED SOLAR EXTINCTION MEASUREMENTS TO IDENTIFY SOOT- LIKE COMPONENTS OF ATMOSPHERIC AEROSOL

Sergey OSHCHEPKOV[1] and Harumi ISAKA[2]

[1]Stepanov Institute of Physics, Belarus Academy of Sciences, 68 F Scorina Avenue, Minsk, 220072 Belarus

[2]Laboratoire de Météorologie Physique URA 267/CNRS, Blaise Pascal University, 24 Avenue des Landais, Aubière Cedex, 63177 France

INTRODUCTION

Soot-like components of atmospheric aerosol is known to be one of an important factor of air pollution. This matter is especially high priority near the sources of industrial pollution and the composition of atmospheric aerosol would be the most quantity here. In contrast to usual background atmospheric aerosol, solar radiation is especially absorbed by soot-like components in passing through the atmosphere.

The paper is devoted to possibility of identification and retrieval of microphysical parameters of soot-like components with background atmospheric aerosol from ground-based solar extinction measurements. The exploration is carried out in scope of numerical tests using as a first approximation Mie theory to reproduce synthetic optical properties of atmospheric aerosol. Size spectra and spectral dependencies of real and imaginary part of refractive index for various types of aerosol at wavelength of 0.4 to 1μm have been chosen according to Ref.[1] recommended by Radiation Commission working group on a Standard Radiation Atmosphere [2].

RESULTS

Information content of spectral extinction measurements with respect to microphysical parameters is investigated. To do that we use Fisher formalism in statistical estimation theory presenting a measure of the information content thorough the quantity which is reciprocal of amplification coefficient of optical measurements error in solving an inverse problem according to the method [3]. A meaning of the amplification coefficients which we present below is rather simple: it constitutes the relative error in retrieving certain parameters per unit relative error of optical measurements. As an example, Fig.1 presents the amplification coefficient in estimating aerosol volume content as a function of ratio of soot volume concentration to volume concentration of one of another component (dust-like, water-soluble, sulfate aerosol). Solid curves characterize amplification coefficients under the condition that aerosol optical properties (specific spectral dependencies of extinction coefficient for each components) are *apriori* known ones in solving the

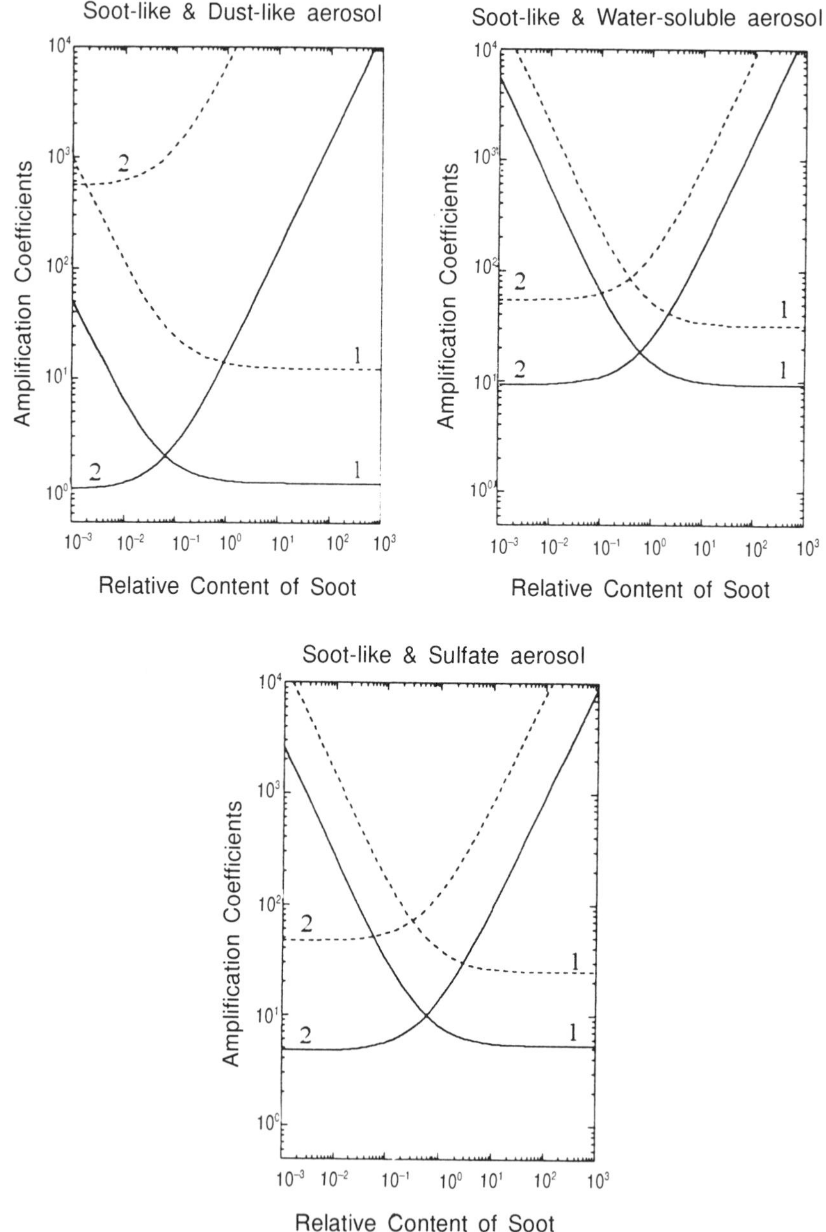

Figure 1. Amplification coefficients in retrieving of volume content of soot-like (1) and one of another kind of aerosol (2) presented in atmosphere with soot simultaneously from spectral measurements of extinction coefficient.

inverse problem and dashed curves correspond to estimate the above specific characteristics together with the volume content. As may be seen from Fig.1, in particular, at the same volume content of the components the amplification coefficients are around 10 if the specific extinction coefficients are known and increase by a factor of 10 in the second case. It holds for the additional background component such as water-soluble and sulfate aerosol to be compatible with soot-like particles in size. For much more lager particles (dust-like aerosol) retrieval error of soot content becomes significantly less as indicated in the left part of Fig.1.

In additional, the sensitivity of multi-wavelength solar extinction data in estimating the composition of atmospheric aerosol has been tested by using iterative inversion method described by Oshchepkov and Isaka[3]. It allows both particle size spectra of the above bicomponent aerosol in histogram presentation to be retrieved simultaneously.

REFERENCES

1. G.A. d'Almeida, P. Koepke, and E.P. Shettle, *Atmospheric Aerosols. Global Climatology and Radiative Characteristics*, A. Deepak Publishing, Hampton, Virginia USA (1991).
2. J. Lenoble and C. Brogniez, A comparative review of radiation aerosol models, *Beitr. Phys. Atmos.*,57: 1 (1984).
2. S.L. Oshchepkov, and O.V. Dubovik, Specific features of the method of laser diffraction spectrometry in the condition of anomalous diffraction, *J.Phys.D: Appl.Physics*, 26: 728 (1993).
3. S.L. Oshchepkov and H. Isaka, Studies of inverse scattering problem for mixed-phase and cirrus clouds, in: *Satellite Remote Sensing and Modeling of Clouds and Atmosphere*, SPIE-The International Society for Optical Engineering , 2961: 5, Taiormina (1996).

HIGH METEOROLOGICAL TOWERS: ARE THEY STILL NEEDED?

Guido Cosemans

VITO, Boeretang 200
B 2400 MOL, Belgium

SUMMARY

A meteorological pre-processor can handle input data of various kind: routine meteorological data, data from small meteorological masts of 10 or 30 meters high, heat flux measurements and so on... Does the quality of the pre-processor output depend on the set of input parameters used? This was investigated by comparing a pre-processors output for different input parameters with measurements from a high meteorological tower.

The meteorological pre-processor used

Wind speed at 69 meter height and potential temperature difference between 114 m and 8 m height are calculated using the standard wind speed profile and potential temperature profile equations:

$$u(z) = u_*/0.4\ [\ \ln (z-d)/z_0 - \psi_m(z/L)\ ]$$

$$(\theta(z)-\theta(0))/\ \theta_* = (1/0.4)\ [\ \ln(z/z_o) - \psi_h\ (z/L)\ ]$$

where u_*, L, ψ_m and ψ_h are calculated with the KNMI software library for the calculation of surface fluxes (Beljaars, Holtslag and Westrhenen, 1989, KNMI report TR-112).

Also calculated is a bulk Richardson number based stability index E_i, which ranges from 1 (stable) over 3 (neutral) to 6 (very unstable). E_i is calculated from $S=(\theta(114)-\theta(8))/u^2(69)$ (Bultynck and Malet, 1972, Tellus, Vol 24, pp. 455-472).

The data sets used

Data used are of the year 1985. Three different sets of input data have been used. A first set of input data are the three-hourly routine observations from the synoptic station Kleine Brogel (KB) in Northeast Belgium. The observations were carried out at a military airport. Surface roughness length is in the order of 0.5 meter, which is surprisingly high. The second and third set of input data were taken from the lower levels of the meteorological tower of Mol, situated about 30 km west of Kleine Brogel.

Air Pollution Modeling and Its Application XII
Edited by Sven-Erik Gryning and Nadine Chaumerliac, Plenum Press, New York, 1998

Results

Figure 1 shows results for three different sets of input parameters. Shown are 3D views of the joint frequency distribution of measured (tower) versus pre-processor (library) calculated potential temperature difference (-2.5 till 5. by 1.25 °C), wind speed at 69 m (0 m/s till 9 m/s by 1.5 m/s) and stability index (1 till 6 by 1). 'N' is the number of hours for which the pre-processor could calculate u_* and L for the given input parameters. 'sc' is the percentage of hours for which the measured and calculated value fall into the same class, which is on the diagonal. Frequencies on the diagonal are not shown, the diagonal places are indicated by a minus-sign ('-'). Frequencies less than 1% are represented by dots ('.'). The 'whiter' the 3D view, the better the agreement between measurement and pre-processor output.

Figure 1 shows an interesting feature. If wind speed and temperature are measured at two levels (input parameter set 2 and 3), the best strategy is to use the measured wind profile to determine the surface roughness length, and to give only one wind speed, together with the surface roughness length and the temperatures, to the pre-processor. This results in highest values for 'sc' and 'N'.

Input parameter set 1	**Input parameter set 2**	**Input parameter set 3**
Wind speed 10 m Kleine Brogel Surface roughness length KB Cloudiness Kleine Brogel Solar altitude	Wind speed 24 and 49 m Mol Temperature 8 m and 24 m Mol	Wind speed 24 m Mol Temperature 8 m and 24 m Mol Surface roughness length Mol

Potential temperature 114m - 8m

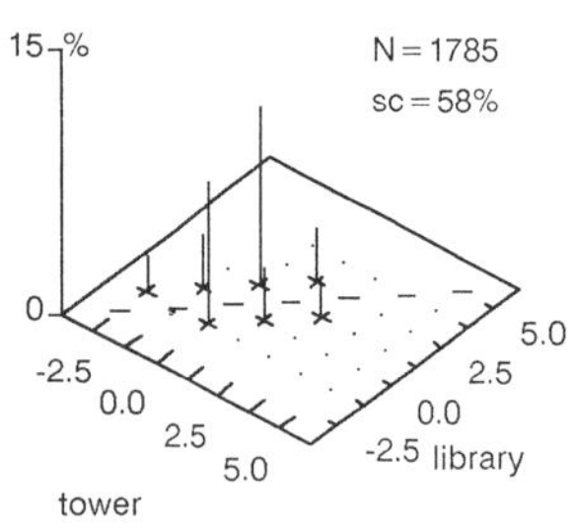

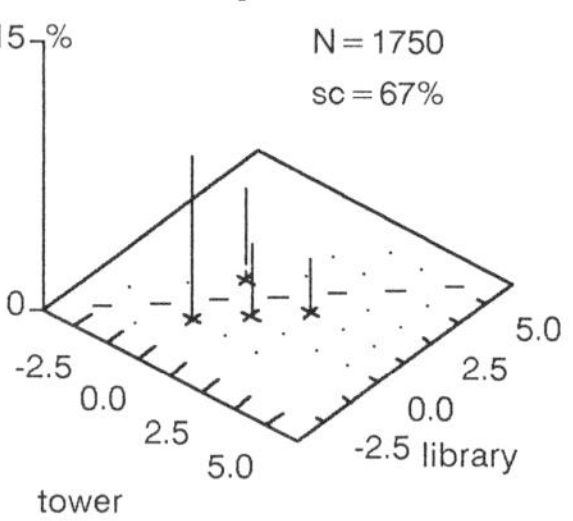

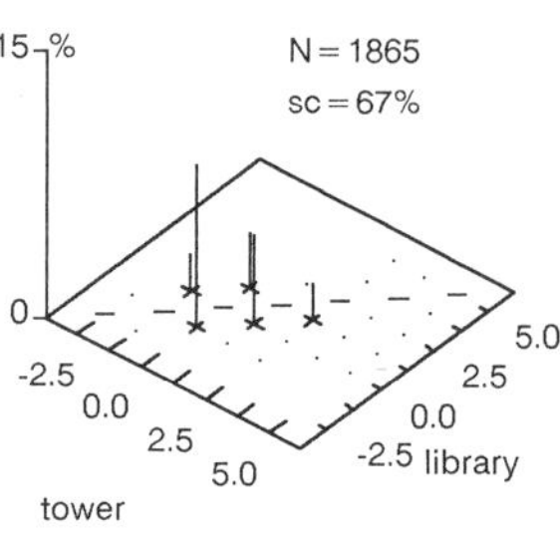

Wind speed at 69 m

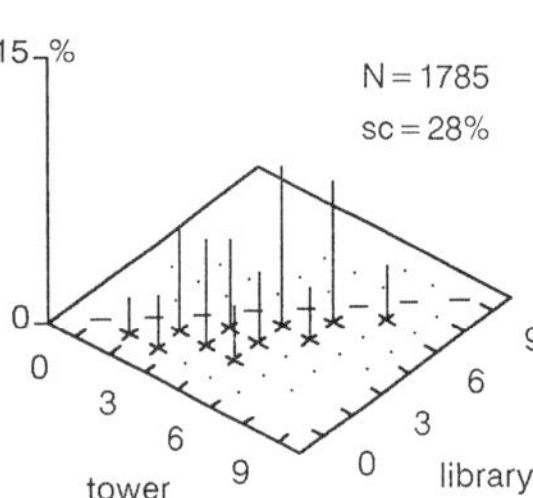

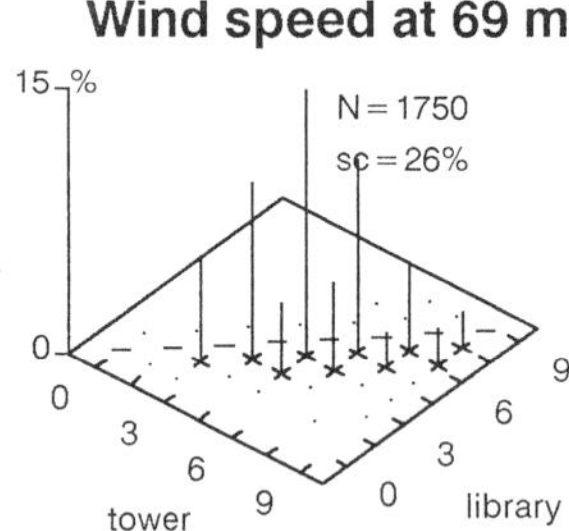

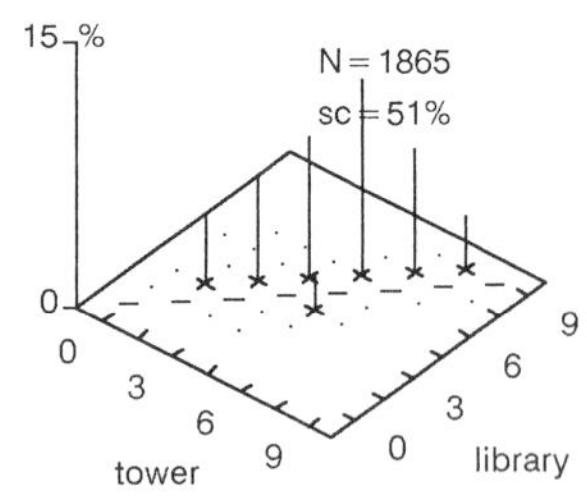

Richardson bulk number based stability class index

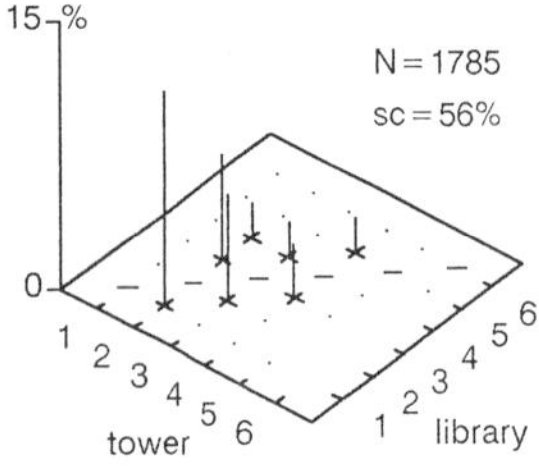

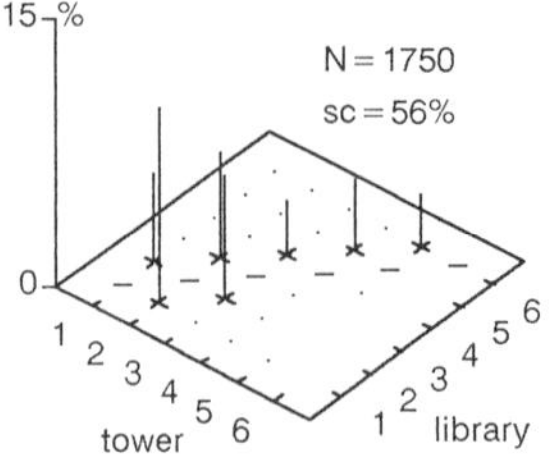

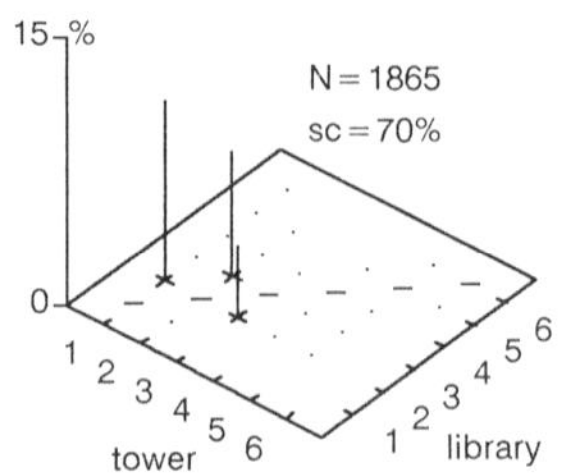

Figure 1.

VALIDATION OF LOCAL AND REGIONAL AIR POLLUTION MODELS IN NORTHERN WINTER CONDITIONS

Knut E. Grønskei,[1] Sam E. Walker,[1] Marko Kaasik,[2] Veljo Kimmel,[2] and Rein Rõõm[2]

[1]Norwegian Institute for Air Research
P.O. Box 100, N-2007 Kjeller, Norway
[2]Department of Environmental Physics, University of Tartu
EE2400 Tähe 4, Tartu, Estonia

MODELS

The Gaussian type air pollution dispersion and deposition model AEROPOL is developed at the Tartu Observatory, Estonia (Kaasik and Rõõm, 1997). The multiple reflections with partial adsorption from the underlying surface and capping inversion, as well as the initial rise of a buoyant plume with the inversion penetration, are included. The wet deposition is included considering different features of rain and snow. The concentrations of pollutants, dry and wet deposition fluxes could be computed. The computing code is written in Turbo Pascal and designed for operational use in Windows'95.

The numerical dispersion model AEROFOUR is developed at Tartu Observatory (Kaasik and Rõõm, 1997). The mixed-spectral approach (Beljaars et al., 1987) is used to solve the equation of stationary and horizontally homogenous turbulent diffusion. The spectral diffusion coefficients (Prahm et al., 1979) are introduced for better performance near the source, where the strongly correlated motions of a narrow plume prevail.

The time-dependent dispersion model EPISODE is developed at the Norwegian Institute for Air Research (Air Pollution and ..., 1991). The dispersion and wind-forced advection of Gaussian puffs are modeled on a three-dimensional grid. Dry and wet deposition are included. The computing code is written in FORTRAN.

EXPERIMENTAL METHODS

The dispersion experiment in plain suburb landscape (Lillestrøm, 20 km East from Oslo) was carried out by the Norwegian Institute for Air Research in January and February 1987. The underlying surface was snow-covered, temperature was 7 - 20 degrees of Celsius below zero. Weak winds and stable atmospheric stratification were prevailing. The sulphur hexafluoride SF_6 was released from 36 m high mast and bromtrifluormethane $CBrF_3$ from 1 m height (102 - 104 mg/s each). Most of the receptors (automatic air samplers) were situated on 2 - 3 nearly 90° arcs downwind from the source at distances 150 - 900 m and 2 m height. Wind velocity and temperature were measured at heights 10 and 36 m, fluctuations of wind components and temperature at 10 m. 8 measurement series, 15 minutes each, were performed during 4 days (Haugsbakk and Tønnesen, 1989).

The snow samples near the industrial enterprises in North-East Estonia have been collected and analyzed during winters 1984/85 (Voll et al, 1988) , 1993/94 and 1995/96 (Kaasik and Rõõm, 1997). The deposition fluxes of Ca^{2+}, Mg^{2+},SO_4^{2-} and various heavy metals, estimated from the snow samples, exceed near of pollution sources several times the background level. Nearly 50 % larger fluxes were observed in the forested landscape, compared with the open one, probably due to more intense turbulent transfer over the rough canopy. The measurement campaign by using passive monitors and other tools of measurements of air pollutants for getting model validation kit is planned to perform during next years in the North-East Estonia.

COMPARISON

The surface-level concentrations computed using the AEROFOUR model are in fair agreement with the Lillestrøm experiment with except two cases for ground-level source in the conditions of very stable atmospheric stratification and large wind shear, when the model tends to underestimate drastically. The lateral dispersion is underestimated in most of cases when the Monin-Obukhov theory for diffusivity is used (Kaasik and Rõõm, 1997). The diffusivities calculated from the thermal gradient and fluctuation measurements contradict to the Monin-Obukhov theory (10 times higher in some cases), but give better performance as a rule. Probably, the situation at the research site is affected by downslope winds and gravity waves generated over the hills situated at a few kilometers upwind from the site.

The deposition loads computed for the North-East Estonia using the AEROPOL model agree in general with the observations, but the overestimation occurs near relatively low stacks (100 m and less). The probable cause is the weakening of turbulence near the underlying surface, which cannot be considered directly in the Gaussian computation scheme. The correlation coefficients between modeled and measured loads based on the data from winter 1984/85 (80 receptor points) are +0.37 for SO_4^{2-} and +0.55 for Ca^{2+} (the later is an indicator of fly ash). The comparison of EPISODE model with these data is in progress.

REFERENCES

Air Pollution and Short-term Health Effects in an Industrialized Area in Norway. Main Report, 1991, Norwegian Institute for Air Research, Lillestrøm..

Beljaars, A.C., Walmsley, J.L., and Taylor, P.A., 1987, A mixed-spectral finite-difference model for neutrally stratified boundary-layer flow over roughness changes and topography. *Boundary-Layer Meteororol.,* 38:273.

Haugsbakk, I., and Tønnesen, D.A., 1989, *Atmospheric Dispersion Experiments at Lillestrøm 1986-87. Data-report.* Norwegian Institute for Air Research, Lillestrøm.

Kaasik, M., and Rõõm, R., 1997, *Estonian Science Foundation, Grant 186. Air Pollution Modelling and Forecast. Final Report.* Tartu Observatory, Tõravere.

Prahm, L., Berkowicz, R., and Christensen, O., 1979, Generalization of K theory for turbulent diffusion . Part II: spectral diffusivitiy model for plume dispersion. *J. Am. Meteorol. Soc.,* 18:273.

Voll, M., Trapido, M., Luiga, P., Haldna, U., Palvadre, R., and Johannes, I., 1988, Transport of atmospheric wastes from energetic devices and oil-shale processing enterprises. *The Natural State and Development of Kurtna Lake District,* M. Ilomets ed., Valgus, Tallinn (in Estonian).

A PARAMETRIZATION METHOD FOR THE ATMOSPHERIC BOUNDARY LAYER APPLIED IN EXTREMELY STABLE CONDITIONS

Ari Karppinen and Sylvain M. Joffre

Finnish Meteorological Institute,
PB 503, FIN-00101 Helsinki

INTRODUCTION

We present numerical results of the boundary layer model used for regulatory purposes in Finland, and compare these with measurements from field campaigns and routine measurements. The parameterisation schemes used in the dispersion models of the Finnish Meteorological Institute (FMI) are based on the energy flux method of van Ulden and Holtslag (1985), and the parameterisation of the boundary layer height is based on classical boundary layer models with a separate treatment for convective and stable conditions.

DESCRIPTION OF THE FMI SCHEME

In the van Ulden-Holtslag's scheme, the turbulent heat and momentum fluxes in the boundary layer are estimated from synoptic weather observations. The original method has been slightly modified, as at high latitudes the net radiation at the surface correlates better with the sunshine duration than with the cloud cover.

The FMI-method divides the net radiation into three parts: solar short-wave radiation, blackbody radiation from clouds and ground, and long-wave radiation of (isothermal) atmosphere. Short-wave radiation is approximated by a regression equation which uses observed hourly sunshine time as the explaining variable in the regression model. The radiation from clouds is modelled by another regression equation , which uses the total cloudiness and cloud height as explaining parameters.

Energy partition in the FMI-model utilises the modified Priestley-Taylor model (van Ulden and Holtslag, 1985) which divides the evaporation into 2 parts. One component is strongly correlated with the difference between net radiation and the flux to the ground, while the second part is not correlated with the equilibrium evaporation. Consequently, there is only two empirical parameters to be evaluated in the FMI-model. These parameters

Air Pollution Modeling and Its Application XII
Edited by Sven-Erik Gryning and Nadine Chaumerliac, Plenum Press, New York, 1998

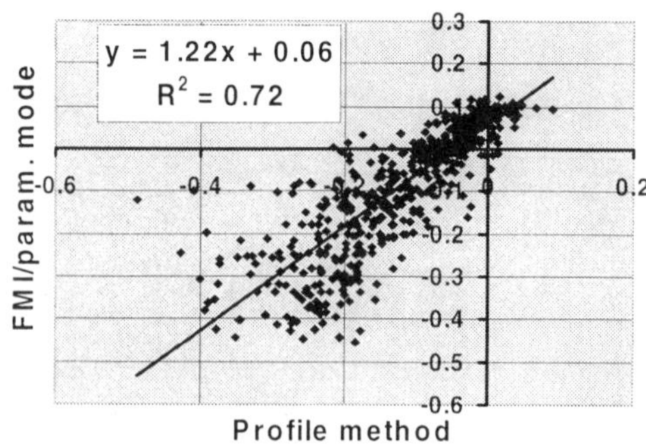

Figure 1. Comparison of the temperature scale estimates at Lillklob, Finland.

depend on surface moisture conditions which are estimated using synoptic weather codes and the amount of rain.

The parameterisation of the mixing height (MH) uses actual radiosoundings and the previously calculated surface turbulence parameters. The summer MH parameterisation is based on a slab model for daytime and modelling the integral heat flux at night. For wintertime the MH evolution is driven by mechanical turbulence.

COMPARISON WITH MEASUREMENTS

We have compared turbulence parameters such as the friction velocity u_*, the temperature scale θ_* (Figure 1) and the inverse Monin-Obukhov length L^{-1} calculated with the FMI-scheme against values obtained from local mast-measurements at Lillklob, Finland. We also discuss results from a comparison study of the FMI scheme for u_* and θ_* under wintertime conditions of northern Sweden (Dittmann et al., 1997).). Our numerical results are in general in good agreement with measurement data for the surface turbulence fluxes, although there are cases, especially under strongly stable conditions of the long winter nights, where the difference is substantial. On the other hand, it is sometimes difficult to ensure that the measurements can always accommodate for the low and intermittent turbulent intensity

Finally we present some results of the comparison of mixing height estimates performed in the COST 710-project (Beyrich et al., 1997) where the FMI scheme was compared to sodar and radiosonde measurements in, i.a., Cabauw (the Netherlands) for wintertime conditions. The FMI scheme for the MH compared very well with sodar data for an episode of 2 days in November at Cabauw and with other models run for the same period. During this period the MH remained rather low between 200 and 500 m.

REFERENCES

Beyrich, F., Seibert, P., Gryning, S-E., Joffre, S., Rasmussen, A. and Tercier, P., 1997. Mixing height determination for dispersion modelling. *COST 710 Action Working Group 2 report*, European Commission, (in print).

Dittmann, E., Johansson, P., Karppinen, A., Musson-Genon, L., Omstedt, G., Tercier, P. and Pechinger, U., 1997. Harmonisation in the preprocessing of meteorological data for atmospheric dispersion models. *COST 710 Action Working Group 1 report*, European Commission, (in print).

van Ulden, A.P. and Holtslag, A.A.M , 1985. Estimation of Atmospheric Boundary Layer Parameters for Diffusion Applications. *Journal of Climate and Applied Meteorology*. 24, p. 1196.-.1207.

INQUIRY ON METEOROLOGICAL DATA PRE-PROCESSING FOR DISPERSION MODELING: A EURO REFERENCE BENCHMARK EXERCISE

Katalin E. Fekete[1], Dezső J. Szepesi [2]

[1]Hungarian Meteorological Service
P.O.Box 38, Budapest, Hungary, H-1525
[2]Consultants on Air Resources Management
Katona J. u. 41, Budapest, Hungary, H-1137

INTRODUCTION

COST Action 710 is a voluntary program of the European Committee for the "Harmonization in the Preprocessing of Meteorological Data for Dispersion Models". The idea of carrying out an international benchmark exercise was endorsed by the Management Committee of the Action at the end of 1995 year. As responsible scientist for this exercise D. J. Szepesi of Hungary was appointed.

The goal of this benchmark exercise was to collect information from the most experienced experts active in the field of meteorological data pre-processing systems (MDPS) and hereby to help the establishment of a reference system for European countries. Such reference MDPS might serve harmonization purposes at international as well as at national levels. Basic principles of model harmonization was summarized earlier by Irwin (1992).

QUESTIONNAIRE STUDY AND ITS RESULTS

The inquiry was prepared mostly by authors of this paper together with other consultants. This benchmarking exercise concerns all types of sources, scales, pollutants and models. Major sections of the Inquiry are: Preliminary analysis, Detailed analysis, General aspects, Topographical analysis, Meteorological simulation, Source term, Flow models, Supplementary aspects and Continuation of work.

In February 1996 numerous (~300) copies of the Inquiry were distributed by the Secretariat of COST Action 710 to modellers active on this field and organisations interested around the world. Preliminary results were presented at Ostende meeting (Szepesi and Fekete, 1996).

Table 1. Some general aspects of atmospheric dispersion models

	Local and urban scale	Regional and continental scale
	Most widely used characteristics	
Number of models	48	30
Major applications	Local: Continuous point s. Urban: Urban background	Regional background
Status of models	Operational, diagnostic, short-term	
Dispersion model types	Classic Gaussian plume	Lagrangian particle
Dispersion technique	PBL scaling or Pasquill-Gifford-Turner type*	

*second most widely used

Up till now many modellers from different countries of the world responded sending altogether 61 descriptions on their models. Assessment on hundreds of model characteristics were performed. Some general aspects of atmospheric dispersion models are shown in Table 1. The appraisal showed: 1) Using boundary layer parameters for atmospheric dispersion modelling is the current trend. 2) In the coming years for new generation dispersion models harmonized meteorological data pre-processing systems ought to furnish the necessary input. 3) In the meantime, however, data needs of state-of-the-art national models and simple schemes using local measurement should not be ignored.

The questionnaire study made possible to identify major trends and most often used schemes of variables of MDPS. The research work was also supported by OTKA (No. T 014859) in Hungary. The COST Action 710 inquiry could not achieve fully its aim mostly due to (a) Lack of interest inside the Action therefore the benchmark exercise on harmonization remained unfinished. (b) Individual modellers were often counter interested to any harmonization effort.

Regarding development of MDPS for new generation dispersion models our proposals are the followings:

- Based on COST 710 background assessments - as a follow-up work - systematic harmonization should be carried out co-ordinated by an organization in charge of European harmonization efforts (e.g. EEA).
- FDDA based meteorological and boundary layer parameter fields for real time and climatological dispersion applications could be most economically provided primarily by National Meteorological Services.
- Statistics of historically processed PGT type dispersion data have to be compared nationally to new generation harmonized scaling parameters and impact on concentration data should be assessed.

REFERENCES

Irwin, J.S., 1992, Summary of 19th NATO/CCMS ITM discussion on harmonization of atmospheric dispersion models, *1st Workshop on Harmonisation within Atmospheric Dispersion Modelling for Regulatory Purposes, Riso, May 6-8 1992.*

Szepesi, D.J. and Fekete, K.E., 1996, Inquiry on meteorological data pre-processing for complex terrain dispersion modelling: A euro reference benchmark exercise, *Pre-print of 4th Workshop on Harmonisation within Atmospheric Dispersion Modelling for Regulatory Purposes, Oostende, 6-9 May 1996.*

EVALUATION OF SO_2 AND NO_2 CONCENTRATION LEVELS IN LITHUANIA USING PASSIVE DIFFUSIONAL SAMPLERS

D.Perkauskas, A.Mikelinskiene, B.Giedraitis and V.Juozefaite
Institute of physics, Savanoriu 231, 2028 Vilnius, Lithuania

INTRODUCTION

Sometimes there is a need for determination of air pollution levels at remote places far from local sources with rather good spatial distribution evaluation. It is useful to use diffusive sampler where the sorbed amount must be proportional to the ambient concentration of the gas. The gas is transported to the sorbent by molecular diffusion and the diffusion rate is made constant by placing the impregnated filter inside a tube[1].

RESULTS

The method of passive diffusional samplers (sorbents) have been used for evaluation of SO_2 and NO_2 concentration levels in Lithuania. SO_2 and NO_2 concentrations have been analysed in 50 (1994), 100 (1995) and 100 (1996) samplers for each component exposed in duration 1-2 months in rather clean, mainly forestry, sites of republic. As it seems, the rates of SO_2 flux varies inside the interval 0.0-138.0 mg/m^2month with frequent of repeatability in interval 5.0-25.0 mg/m^2month. The NO_2 flux rates evaluated with passive sorbents varied within 0.0-6.0 mg/m^2month with major repeatability within the interval of 1.0-3.0 mg/m^2month. The results for 1995 are presented in Figure1. The distribution of SO_2 and NO_2 fluxes depends mainly from regional transport of air masses and local distribution of power and industrial regions.

REFERENCES

1. A.Galvonaite, B.Giedraitis, R.Girgzdiene, A.Girgzdys, A.Juozaitis, K,Kvietkus, A.Milukaite, D.Perkauskas, J.Sakalys, D.Sopauskiene, V.Ulevicius and E.Vebra . Background Air Pollution Monitoring in Lithuania. Vilnius (1990).

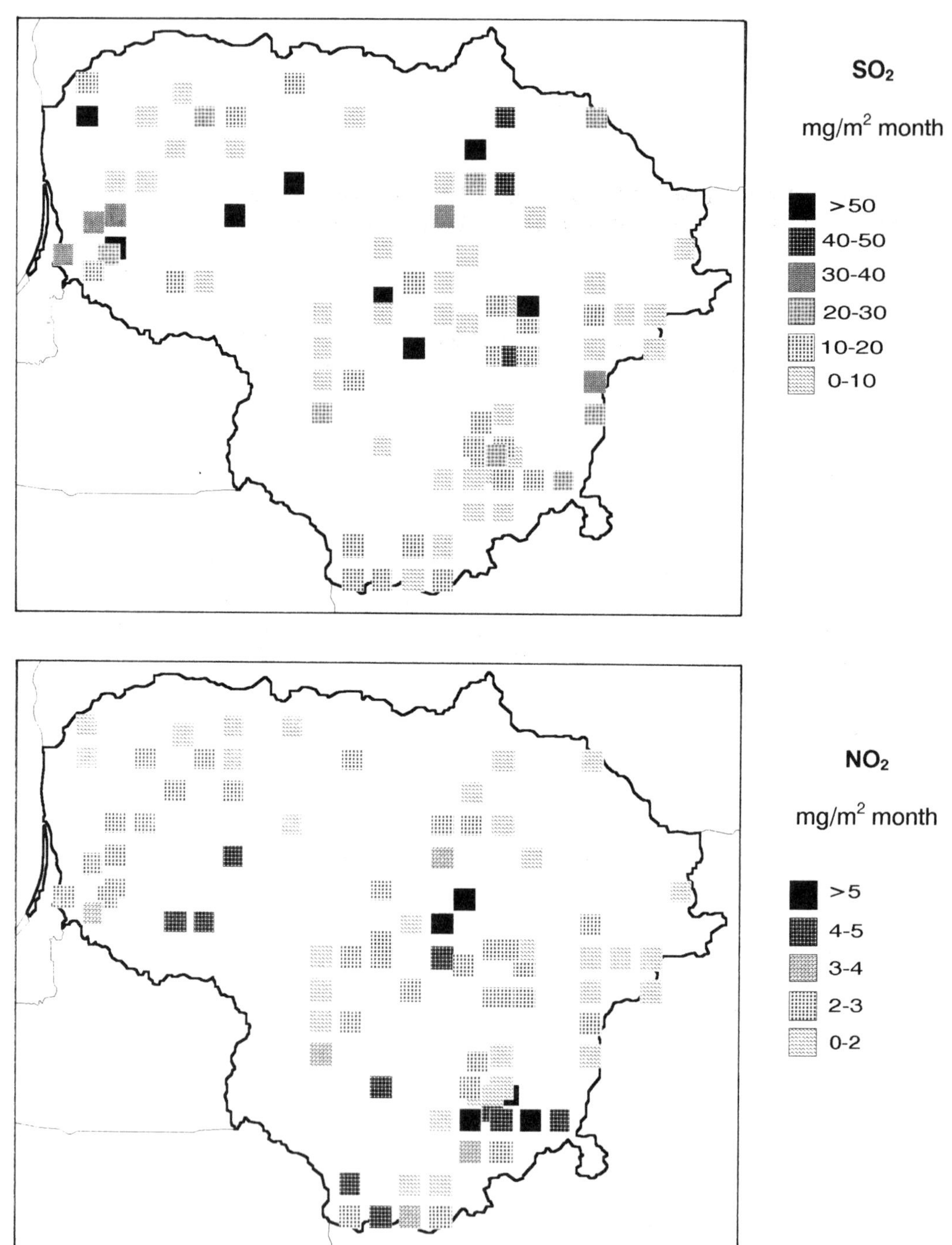

Air pollution by NO_2 and SO_2 in Lithuania (1995)

MONITORING AND CONTROL MODEL FOR THE POLLUTANTS CONCENTRATIONS EMITTED FROM INTERFERING SOURCES

Traian Pop, Octavian Olteanu, Livia-Mihaela Pop

Department of Fluid Mechanics and Pollutants Dispersion,
Research & Engineering Institute for Environment
Spl.Independentei nr.294, Bucharest 78, Romania

INTRODUCTION

Exceeding of the limit mass rate values admitted at interfering emission sources for the same type of pollutant can determine the growth of concentration level in the imssion area over the admissible limits.

The problem that must be solved, in this case, is the following one: to identify the source(s) responsible of exceeding the limit values of the outputs (concentrations) imposed at emission, by using the 24 hours average value of the concentration ($\overline{C}$) measured at a control point located in the imission area, under meteorological parameters conditions investigated during the measurement. These exceeding can occur when some facilities for air pollution removal are out of order ($\eta_S=0$). This way of addressing the problem is necessary, taking in account the reduced number of emissions measuring and control instruments that in Romania exists at present.

MATHEMATICAL MODEL

The model consists in the identification of the local source(s) that could cause pollutants concentrations exceeding at a control point, according to the following criteria:

- maximum influence source during the interested time rate;
- the same influence sources.

The influence function of the source "*s*" at "*i*" moment and at a control point can be written as:

$$H_{i,s} = \varepsilon_S(1-\eta_{i,s})\ \xi_{i,s}\ F_{i,s} \qquad (1)$$

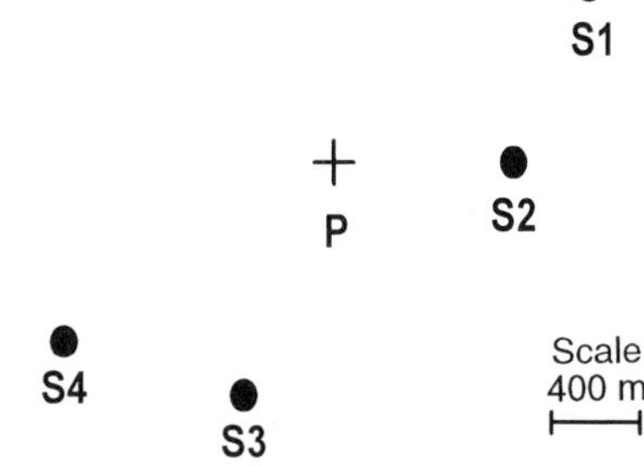

Figure 1. Control point's (P) and sources' location.

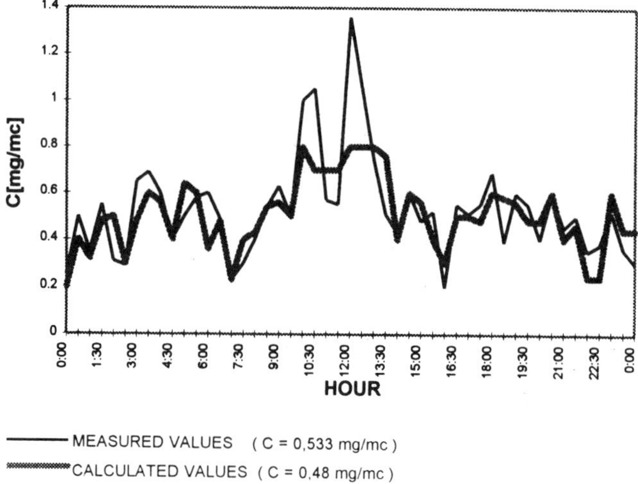

Figure 2. Comparative graph concerning measured and calculated values of concentrations at the control point (P).

Taking in account $s = 1,k$ sources, the influence function according to the interfering emissions sources at "i" moment is given by:

$$H_i = \sum_{S=1}^{k} \varepsilon_S (1 - \eta_{i,S}) \xi_{i,S} F_{i,S} \qquad (2)$$

in which: $\varepsilon_S = Q_S \ / \ Q_S^{\max}$; Q_S - emission mass flow of the pollutant; $\eta_{i,s}$ - air pollution removal facilities in service efficiency; $\xi_{i,S}$ - schedule's function of the polluting source; $F_{i,s}$ - sources partial influence function.

Average values of the influence functions in the interested time rate can be determined using the formulae (1), (2) in maximum emission conditions ($\eta_s=0$).The maximum influence sources identification in condition of $\eta_s=0$ is made by finding out: max $(\overline{H}_S^{\max})_{S=1,k}$ in which: $\overline{H}_S^{\max}$ - is the average value of the influence function of "s" source.

The sources that have almost the same influence at a control point can be identified using MANN-WHITNEY TEST to the $H_{i,s}^{\max}$ values generated in the conditions mentioned above.

RESULTS AND CONCLUDING REMARKS

This mathematical model was tested into an application concerning four emission sources located as shown in Figure 1. NOx concentrations generated by the source s=2, identified as being in abnormal functioning ($\eta=0$), and measured and recorded concentrations of NOx in 24 hours are shown in Figure 2. In the conditions of accurately $H_{i,s}$ values, this model could represent a proper instrument in local air quality monitoring and control.

PARTICIPANTS

The 22nd NATO/CCMS International Technical Meeting on Air Pollution Modelling and Its Application. Clermont-Ferrand, France, 2-6 June 1997

ARGENTINA

Ulke A. G.

Department of Atmospheric Sciences
University of Buenos Aires
Pabellon II - Piso II
Ciudad Universitaria
1428 Buenos Aires
ulke@cw.at.fcen.uba.ar

AUSTRALIA

Cope M.

CSIRO
Division of Atmospheric Research
Station st.
Aspendale, 3195 Victoria
mec@dar.csiro.au

Physick B. L.

CSIRO Atmospheric Research
Private Bag n° 1
Aspendale, 3195 Victoria
wlp@dar.csiro.au

AUSTRIA

Pechinger U.

ZAMG
Hohe Warte 38
1190 Vienna
pechinger@zamg.ac.at

Seibert P.

Institute of Meteorology and Physics
University of Agricultural Science (BOKU)
Tuerkenschanz str. 18,
1180 Vienna
seibert@mail.boku.ac.at

BELGIUM

Cosemans G.

VITO-CETAP
Boeretang 200
2400 Mol
cosemang@vito.be

Dutrieux A.

ATM-PRO SPRL
rue Saint André 5
1400 Nivelles

BULGARIA

Batchvarova E.

National Institute of Meteorology and Hydrology
66 Tzarigradsko Chaussee
1784 Sofia
ekaterina.batchvarova@meteo.bg

Syrakov D.

National Institute of Meteorology and Hydrology
66 Tzarigradsko Chaussee
1784 Sofia
dimiter.syrakov@meteo.bg

CANADA

Beattie B.

Environment Canada
1496 Bedford Highway
Bedford, NS B4A 1E5
billie.beattie@ec.gc.ca

Gong S.

Atmospheric Environment Service
ENVIRONMENT CANADA
4905 Dufferin Street
Downsview, Ontario M3H 5T4
sunling.gong@ec.gc.ca

Gong W.

Atmospheric Environment Service
4905 Dufferin Street
Downsview, Ontario M3H 5T4

Pottier J. L.

Environmental Conservation Branch
Science Division
7Tth floor, 1200 West 73rd Avenue
Vancouver, BC V6P 6H9
joanne.pottier@ec.gc.ca

Steyn D. G.

The University of British Columbia
1984 West Mall
Vancouver, B.C. V6T 1Z2
douw@geog.ubc.ca

Walmsley J. L.

Atmospheric Environment Service
4905 Dufferin Street
Downsview, Ontario M3H 5T4
john.walmsley@ec.gc.ca

CZECH REPUBLIK

Bubnik J.

Czech Meteorological Institut
Na Sabatce 17
14306 Prague
bubnik@chmi.cz

Jicha M.

Technical University of Brno
Mechanical Engineering
Technicka 2
61669 Brno
jicha@kinf.fme.vutbr.cz

Macoun J.

Czech Hydrometeorological Institute
Na Sabatce 17
14306 Prague 4

DENMARK

Gryning S.-E.

Wind Energy and Atmospheric Physics Department
Risø National Laboratory
4000 Roskilde
sven-erik.gryning@risoe.dk

Kiilsholm S.

Danish Meteorological Institute
Lyngbyvej 100
2100 Copenhagen Ø
sk@dmi.min.dk

Mikkelsen T.

Wind Energy and Atmospheric Physics Department
Risø National Laboratory
4000 Roskilde
torben.mikkelsen@risoe.dk

Olesen H. R.

National Environmental Research Institute
Frederiksborgvej 399,
P.O.Box 358
4000 Roskilde
luhro@dmu.dk

Sørensen J. H.

Danish Meteorological Institute
Lyngbyvej 100
2100 Copenhagen Ø
jhs@dmi.dk

Thykier-Nielsen S.

Wind Energy and Atmospheric Physics Department
Risø National Laboratory
4000 Roskilde
soeren.thykier@risoe.dk

Vinther Falk Anne Katrine

Wind Energy and Atmospheric Physics Department
Risø National Laboratory
4000 Roskilde
annekatrine.vinther@risoe.dk

ESTONIA

Kaasik M.

Deptartment of Environmental Physics
University of Tartu
Tahe str. 4
2400 Tartu
marko@apollo.aai.ee

FINLAND

Harkonen J.-V.

Finnish Meteorological Institute
Sahaajankatu 20E
00810 Helsinki
jari.harkonen@fmi.fi

Joffre S.

Finnish Meteorological Institute
PB 503
00101 Helsinki
sylvain.joffre@fmi.fi

Karppinen A.

Finnish Meteorological Institute
P.O. Box 503
00101 Helsinki
Ari.Karppinen@fmi.fi

Nikmo J.

Finnish Meteorological Institute
Sahaajankatu 20 E
00810 Helsinki
juha.nikmo@fmi.fi

Salonoja M.

Finnish Meteorological Institute
P.O.Box 503
00101 Helsinki
mika.solonoja@fmi.fi

FRANCE

Audiffren N.

Laboratoire de Meteorologie Physique
Universite B. Pascal - CNRS
24 avenue des Landais
63177 Aubiere Cedex

Brocheton F.

L.I.S.A.
61 av. du Général de Gaulle
94010 Creteil
brocheton@univ-paris12.fr

Carissimo B.C.

Direction des Etudes et Recherches EDF
Depart. Environnement,
Groupe Meteorologie et Climat
6, Quai Watier, B.P. 49,
78401 Chatou Cedex
bertrand.carissimo@der.edfgdf.fr

Cautenet G.

Laboratoire de Météorologie Physique
Université Blaise Pascal - CNRS/OPGC
24, avenue des Landais
63177 Aubiere Cedex
cautenet@opgc.univ-bpclermont.fr

Cautenet S.

Laboratoire de Meteorologie Physique
Université Blaise Pascal/CNRS/OPGC
24, avenue des Landais
63177 Aubiere Cedex
cautenet@opgc.univ-bpclermont.fr

Chaumerliac N.

LAMP/OPGC
24, avenue des Landais
63177 Aubiere Cedex
chaumerl@opgc.univ-bpclermont.fr

Coppalle A.

INSA de Rouen,
URA 230 Coria
76821 Mont St. Aignan

Delamare L.

Université de Rouen
URA CNRS 230/CORIA
76821 Mont Saint Aignan
Ludovic.Delamare@coria.fr

Despiau S.

ISITV
Université Toulon Var
Avenue G. Pompidou
B.P. 56
83162 La Valette du Var Cedex

Dupont E.

EDF
Direction des Etudes et Recherches
Groupe Météorologie et Climat
6, quai Watier - B.P. 49
78401 Chatou Cedex
eric.dupont@der.edfgdf.fr

Flossmann A.

Laboratoire de Meteorologie Physique
Universite Blaise Pascal/CNRS/OPGC
24, Avenue des Landais
63177 Aubiere Cedex
flossman@opgc.univ-bpclermont.fr

Gonzalez M.	UNR CNRS 6614/CORIA 76821 Mont-Saint-Aignan gonzal@celestin.coria.fr
Guelle W.	CEA - LMCE Groupe Teledetection, CE Saclay L'Orme des Merisiers (Bat 701) 91191 Gif-sur-Yvette guelle@lmce.saclay.cea.fr
Henriet A.	PSA Peugeot Citroën DRAS/DIR 2, route de Gisy 78140 Velizy
Isaka H.	LAMP/OPGC 24, avenue des Landais 63177 Aubiere Cedex isaka@opgc.univ-bpclermont.fr
Mestayer P.	Labo. des Mécaniques des Fluides Ecole Centrale de Nantes B.P. 92101 44321 Nantes Cedex 3 patrice.mestayer@ec-nantes.fr
Monote G.	TRANSOFT International Centre d'Affaires Intégral 82, rue de Paris 93804 Epinay Sur Seine transoft@pobox.oleane.com
Tedeschi G.	ISITV av. G. Pompidou BP 56 83162 La Valette du Var Cedex http://isitv.univ-tln.fr/~tedeschi
Troude F.	EDF-DER 6 quai Watier BP 49 78401 Chatou Cedex
Wendum D.	EDF-DER 6 quai Watier BP 49 78401 Chatou Cedex denis.wendum@der.edfgdf.fr

GERMANY

Ackermann I.J.	Ford Forschungszentrum Aachen GmbH Dennewartstrasse 25 52068 Aachen
Fay B.	Deutscher Wetterdienst Frankfurter Strasse 135 63067 Offenbach bfay@dwd.d400.de
Graff A.	Umweltbundesamt Bismarckplatz 1 14193 Berlin graff@uba.de
Kaminski T.	Max-Planck-Institut fur Meteorologie Bundesstrasse 55 20146 Hamburg kaminski@dkrz.de
Kastner-Klein P.	University of Karlsruhe Kaiserstr. 12 76128 Karlsruhe petra.kastner-klein@bau-verm.uni-karlsruhe.de
Klug W.	Institut für Meteorologie Technische Hochschule Darmstadt Hochschulstrasse 1 64289 Darmstadt dg5q@hrzpub.th-darmstadt.de
Langmann B.	Max-Planck Institute for Meteorology Bundesstrasse 55 20146 Hamburg langmann@dkrz.de
Martens R.	GRS mbH Schwertnergasse 1 50667 Köln mar@grs.de
Nester K.	Research Centre Karlsruhe Postfach 3640 76021 Karlsruhe nester@imkhp6.fzk.de
Petersen G.	GKKS Research Centre Institute of Hydrophysics Max-Planck-Strasse 21502 Geesthacht Gerhard.Petersen@gkss.de

Rafailidis S.

Meteorological Institute
Universitat Hamburg
Bundesstrasse 55
20146 Hamburg
rafailidis@dkrz.de

Renner E.

Institute for Tropospheric Research
Permoserstrasse 15
4303 Leipzig
renner@tropos.de

Schell B.

University of Cologne
EURAD Project
Achener Strasse 201-209
50931 Köln
bs@eurad.uni-koeln.de

Thielen H.

GRS mbH
Schwertnergasse 1
50667 Köln
thi@grs.de

Wichmann-Fiebig M.

Environment State Agency NRV
Wallneyer Str. 6
45133 Essen

Wolke R.

Institute Tropospheric Research
Permoserstrasse 15
04303 Leipzig
wolke@tropos.de

Zilitinkevich S.

Institute for Hydrophysics
GKSS Research Centre
Max Planck Strasse
21502 Geesthacht
sergej.zilitenkevich@gkss.de

Zimmermann J.

Deutscher Wetterdienst,
FE 25
P.O. box 100465
63004 Offenbach
zimmermann@f4.za-offenbach.dwd.d400.de

GREECE

Kallos G.

University of Athens
Department of Applied Physics
Laboratory of Meteorology
Panepistimioupolis, building PHYS-V
15784 Athens
kallos@etesian.dap.uoa.gr

Kirchner F. — Laboratory of Heat Transfer and Env. Engg
Aristotle University of Thessaloniki
Box 483
54006 Thessaloniki

Melas D. — Laboratory of Atmospheric Physics
Aristotle University of Tessaloniki
54006 Thessaloniki
melas@olymp.ccf.auth.gr

HUNGARY

Fekete K. E. — Hungarian Meteorological Service
P.O. Box 38
1525 Budapest
h11275fek@ella.hu

ISRAEL

Lacser A. — Head Environ. Sci. Div.
IIBR POB 19
Ness Ziona 70450
avilac@parker.inter.net.il

ITALY

Anfossi D. — CNR
Istituto di Cosmogeofisica
Corso fiume 4
I-10133 Torino
anfossi@to.infn.it

Angelino E. — Provincia de Milano
Sel. Ecologia PHIP
Cor 50 di porta Vittoria 27
Milano
cop.milano@galactica.it

Baldi M. — Inst. of Atmospheric Physics - CNR
Piazza Luigi Sturzo, 31
00144 ROMA
baldi@atmos.ifa.rm.cnr.it

Bellasio R — ENVIROWARE srl
Centro Direzionale Colleoni
Palazzo Andromeda, 1
20041 Agrate Brianza (MI)
http:www.enviroware.com

Desiato F. ANPA
Agenzia Nationale
Via Vitaliano Brancati 48
00144 Roma
desiato@aosf01.anpa.it

Ferrero E. Dipartimento di Scienze e Tecnologie Avanzate
Univ. di Torino
Corso Borsalino 54
Alessandria
ferrero@unial.it

Golinelli M. Servizio Meteorologico
c/o Regione Emilia Romagna
6 v. le Silvani
40122 Bologna
envi@meteo2.arpamet.regione.emilia-romagna.it

Graziani G. Joint Research Centre, Environment Institute
Commission of the European Communities
Via Fermi, TP 321
21020 Ispra (VA)
giovanni.graziani@jrc.it

Miffeis G. P.M.I.P. Oggiono-USSL n°7
Via Roma 22
24049 Verdello

Riccio A. Departimento di Chimica
Università di Napoli "Federicoll"
Via Mezzocannone 4
80134 Napoli

Straume A. G. Environment Institute
Commission of the European Communities
Joint Research Centre, TP 510
21020 Ispra (VA)
anne.straume@jrc.it

Thunis P. Environment Institute
Commission of the European Communities
Joint Research Centre, TP 510
21020 Ispra (VA)

JAPAN

Katatani N. Yamanashi University
4-3-11 Takeda
Kofu 400
katatani@esi.yamanashi.ac.jp

Kitada T.
Department of Ecological Engineering
Toyohashi University of Technology
Tempaku-cho
Toyohashi 441
kitada@earth.eco.tut.ac.jp

Ohara T.
Institute of Behavioral Sciences
2-9 Honmura-cho
Ichigaya,shinjuku-ku
Tokyo 162
tohara@ibs.or.jp

Okamoto S.
Tokyo University of Information Sciences
1200-2 Yatoh-cho
Wakaba-Ku, Chiba 265

Uno I.
National Institute for Environmental Studies
Japan Environment Agency
Onogawa 16-2
Tsukuba 305
iuno@sun90c.nies.go.jp

NEW ZEALAND

Fisher G.
National Institute of water
Atmosphere Tesearch (NIWA)
Box 109 695
NewMarcket
Auckland
f.fisher@niwa.cri.nz

NORWAY

Bartnicki J.
Norwegian Meteorological Institute (DNMI)
P.O. box 43 blindern
0313 Oslo Blindern
jerzy.bartnicki@dnmi.no

Iversen T.
Department of Geophysics
University of Oslo, Blindern
P.O.box 1022
0315 Oslo
trond.iversen@geofysikk.uio.no

Olendrzynski K.
The Norvegian Meteorological Institute
Air Pollution Group (EMEP)
P.O. Box 43 Blindern
0313 Oslo
krystof@dnmi.no

POLAND

Markiewicz M.

Warsaw University of Technology
Inst. of Environmental Engineering Systems
ut. Mowowiejska 20
00-653 Warsaw
maria@jowisz.iis.pw.edu.pl

PORTUGAL

Borrego C.

Department of Environment and Planning
University of Aveiro
3810 Aveiro
borrego@ci.ua.pt

ROMANIA

Gultureanu D.

Department of Physics
Petroleum-Gas University of Ploiesti
39, Bucuresti blvd,
P.O. Box 10
2000 Ploiesti Prahova
irena@csd.univ.ploiesti.ro

Pop L.

Ministry of Environment
Dept of Fluid Mechanics & Pollutants Dispersion
ICIM, Spl. Independentei nr 294
Cod 77703 Sector 6
Bucharest 78
icim@sunu.rnc.ro

Pop T.

Ministry of Environment
Dept of Fluid Mechanics & Pollutants Dispersion
ICIM, Spl. Independentei nr 294
Cod 77703 Sector 6
Bucharest 78
icim@sunu.rnc.ro

RUSSIA

Keiko A.

Siberian Energy Institute, RAS
130, Lermontov Rd
Irkutsk, 664033
root@sei.irkutsk.su

Sofiev M. A.

Hydrometeorological Centre of Russia
Stroiteley str.4-1-18
Moscow 117311
mas@glasnet.ru

SLOVENIA

Boznar M.	Jozef Stefan Institute Jamova 39 1000 Ljubljana marija.boznar@ijs.si
Mlakar P.	Jozef Stefan Institute Jamova 39 1000 Ljubljana primoz.mlakar@ijs.si

SOUTH AFRICA

Burger L. W.	Environmental Management Services CC P.O. BOX 52668 Wierda Park 0149 Centurion hawk@iaccess.za

SPAIN

Gonzales-Barras R.	University Complutense Ciudad Universit. s/n Facultad Fisica Madrid 28022
Salvador R.	CEAM Parque Tecnologico Calle 4, sector Oeste 46980 Paterna (Valencia) rsalvador@itema.upc.es
Santabarbara J. M.	ITEMA-UPC Apartat de Correus 508 08220 Terrasa (Barcelona) jmoreno@itema.upc.es
San Jose R.	Group of Environmental Software and Modelling Computer Science School Campus de Montegancedo Boadilla del Monte 28660 - Madrid roberto@fi.upm.es
Toll I.	ITEMA Aptado Correos 508 08220 Terrassa (Barcelona) toll@itema.upc.es

Soriano C.

ITEMA-UPC
Apartat de Correus 508
08220 Terrasa (Barcelona)
csoriano@itema.upc.es

SWEDEN

Bergström R.

SMHI - If
601 76 Norrkoping
rbergstr@smhi.se

Johansson P.-E.

Defence Research Establishment (FOA)
90182 Umeå
johansson@ume.foa.se

Langner J.

SMHI
60176 Norrköping
jlangner@smhi.se

SWITZERLAND

de Haan P.

Swiss Federal Institute of Technology
GIETH
Winterthurerstrasse 190
8057 Zürich
dehaan@geo.umnw.ethz.ch

Rotach M.

Swiss Federal Institute of Technology
GIETH
Winterthurerstrasse 190
8057 Zürich
rotach@geo.umnw.ethz.ch

Schneiter D.

Swiss Meteorological Institute
Station Aérologique
1530 Payerne
dsc@sap.sma.ch

THE NETHERLANDS

Builtjes P.

TNO-MEP
P.O. box 342
7300 AH Apeldoorn
p.j.h.builtjes@mep.tno.nl

Heinz S.

Delft University of Technology
Faculty of Applied Physics
Section Heat Transfer, Lorentzweg 1
2628 CJ Delft
heinz@wt.tn.tudelft.nl

van Jaarsveld J. A. | National Institute of Public Health
and Environmental Protection
P.O. Box 1
3720 BA Bilthoven
hans.van.jaarsveld@rivm.nl

van Dop H. | Institute for Marine and Atmospheric Research
Utrecht University (IMAU)
Department of Physics and Astronomy
P.O. Box 80005
3508 TA Utrecht
h.vandop@fys.ruu.nl

TURKEY

Tayanc M. | Marmara University
Department of Environmental Engineering
81040 Goztepe
Istanbul
tayanc@gaia.geol.itu.edu.tr

UNITED KINGDOM

Carruthers D. | Cambridge Environmental Research Consultants
3, King's Parade
C82 1SJ Cambridge
david.carruthers@cerc.co.uk

Cai X. | School of Geography
University of Birmingham
B15 2TT Edgbaston, Birmingham
x.cai@bham.ac.uk

Collins W.J. | Atmospheric Processes Research
Meteorological Office
London Road
RG12 2SZ Bracknell, Berkshire
wjcollins@meto.gov.uk

Davies B. | Meteorological Office
London Road
RG12 2SZ Bracknell,Berkshire
bmdavies@meto.gov.uk

Fish D. J. | Department of Meteorology
University of Reading
PO Box 243
RG6 6BB Reading
d.j.fish@reading.ac.uk

Jones C. D.	PLSD HA, CBD Porton Down, SP4 0JL Salisbury,
Malcolm A.	Rm 155 Atmos. Processes Res. The Meteorological Office London Road RG12 2SZ Bracknell,
Nelson N.	The UK Meteorological Office London Road RG12 2SZ Bracknell, Berkshire nnelson@meto.gov.uk
Romanowicz R.	Westlakes Research Institute The Princess royal building Westlakes Science and Technology Park Moore Row, CA24 3LN Cumbria renatar@westlakes.ac.uk
Stevenson D. S.	Meteorological Office London Road Bracknell, RG12 2SZ Berkshire
Tasker M.	ICI Technology Room 424 WA7 4QF Runcorm martin_tasker@ici.co.uk

U.S.A.

Adamkiewicz G.	MIT 77 Massachussetts Ave. Room 16-315, Cambridge, MA 02139 gadamkie@mit.edu
Bornstein R.	Department of Meteorology San Jose State University San Jose, CA 95192-0104 pblmodel@sjsuvm1.sjsu.edu
Byun D. W.	NOAA/EPA 4201 Bldg, MD-80 Research Triangle Park, NC 27711 bdx@hpcc.epa.gov

Carmichael G.R.
University of Iowa, Center for Global & Regional Environmental Research and Department of Chemical & Biocemical Engineering
125 A Chem bld
Iowa City, IA 52240
gcarmich@icaen.uiowa.edu

Green M.
Desert Research Institute
755 East Flamingo road
Las Vegas, NV 89119
green@snsc.dri.edu

Hansen D. A.
Electric Power Research Institute
3412 Hillview Avenue
P.O. Box 10412
Palo Alto, CA 94303
ahansen@epri.com

Irwin J. S.
Air Quality Planning and Standard
NOAA (E.P.A./OAQPS
US EPA (Mail Drop 14)
Research Triangle Park, NC, 27711
irwin.john@epamail.epa.gov

Mcnider R.
Earth System Science Laboratory
University of Alabama in Huntsville
Huntsville, Alabama 35899
dick.mcnider@atmos.uah.edu

Nowak D. J.
USDA Forest Service
Northeastern Forest Experiment Station
5 Moon Library, SUNY-CESF
Syracuse, NY 13210
djnowak@mailbox.syr.edu

Pielke R.
Colorado State University
Department of Atmospheric Sciences
Fort Collins, Colorado 80523-1371
dallas@hercules.atmos.colostate.edu

Porter P.-S.
University of Idaho
1776 Science Center Drive
Ioaho Falls 83401
porter@if.uidaho.edu

Pryor S.
Climate and Meteorology Program
Department of Geography
Indiana University,
Bloomington, IN 47405

Rao S. T.
New York Department of Environmental Conservation
Office of Science and Technology
50 Wolf Road, room 198
Albany, NY. 12233-3259
strao@dec.state.ny.us

Robe F. R.
Earth Tech
196 Baker Avenue
Concord, MA 01742-2167
frr@src.com

Schiermeier F. A.
Atmospheric Sciences Modeling Division
U.S.EPA/NOAA
Research Triangle Park, NC 27711
schiermeier.francis@epamail.epa.gov

Steinberg K. W.
Exxon Research and Engineering Company,
180 Park Avenue,
Florham Park, NJ 07932-0101

Wilkinson J.
Daniels Laboratory Georgia Inst. of Technology
200 Bobby Dodd Way
Atlanta, GA 30332-0512
jwilkins@isis.ce.gatech.edu

Yamartino R.
Earth Tech
196 Baker Avenue
Concord, MA 01742-2167
rjy@src.com

YUGOSLAVIA

Pocajt V.
Faculty of Technology and Metallurgy
Karnegijeva 4
P.O.B. 494
11000 Belgrade
ini@eunet.yu

AUTHOR INDEX

SUBJECT INDEX